Lecture Notes in Computer Science   12565

More information about this subseries at http://www.springer.com/series/7409

Giuseppe Nicosia · Varun Ojha ·
Emanuele La Malfa · Giorgio Jansen ·
Vincenzo Sciacca · Panos Pardalos ·
Giovanni Giuffrida · Renato Umeton (Eds.)

Machine Learning, Optimization, and Data Science

6th International Conference, LOD 2020
Siena, Italy, July 19–23, 2020
Revised Selected Papers, Part I

 Springer

Editors
Giuseppe Nicosia (iD)
University of Catania
Catania, Italy

Varun Ojha (iD)
University of Reading
Reading, UK

Emanuele La Malfa (iD)
University of Oxford
Oxford, UK

Giorgio Jansen
University of Cambridge
Cambridge, UK

Vincenzo Sciacca
ALMAWAVE
Rome, Italy

Panos Pardalos (iD)
University of Florida
Gainesville, FL, USA

Giovanni Giuffrida (iD)
University of Catania
Catania, Italy

Renato Umeton (iD)
Harvard University
Cambridge, MA, USA

ISSN 0302-9743   ISSN 1611-3349 (electronic)
Lecture Notes in Computer Science
ISBN 978-3-030-64582-3   ISBN 978-3-030-64583-0 (eBook)
https://doi.org/10.1007/978-3-030-64583-0

LNCS Sublibrary: SL3 – Information Systems and Applications, incl. Internet/Web, and HCI

This Springer imprint is published by the registered company Springer Nature Switzerland AG
The registered company address is: Gewerbestrasse 11, 6330 Cham, Switzerland

Preface

The 6th edition of the International Conference on Machine Learning, Optimization, and Data Science (LOD 2020), was organized during July 19–23, 2020, in Certosa di Pontignano (Siena) Italy, a stunning medieval town dominating the picturesque countryside of Tuscany. LOD 2020 was held successfully online and onsite to meet challenges posed by the worldwide outbreak of COVID-19. Since 2015, the LOD conference brings academics, researchers, and industrial researchers together in a unique multidisciplinary community to discuss the state of the art and the latest advances in the integration of machine learning, optimization, and data science to provide and support the scientific and technological foundations for interpretable, explainable, and trustworthy AI. Since 2017, LOD adopted the Asilomar AI Principles.

LOD is an annual international conference on machine learning, computational optimization, and big data that includes invited talks, tutorial talks, special sessions, industrial tracks, demonstrations, and oral and poster presentations of refereed papers.

LOD has established itself as a premier multidisciplinary conference in machine learning, computational optimization, and data science. It provides an international forum for presentation of original multidisciplinary research results, as well as exchange and dissemination of innovative and practical development experiences.

The LOD conference manifesto is the following:

"The problem of understanding intelligence is said to be the greatest problem in science today and "the" problem for this century – as deciphering the genetic code was for the second half of the last one. Arguably, the problem of learning represents a gateway to understanding intelligence in brains and machines, to discovering how the human brain works, and to making intelligent machines that learn from experience and improve their competences as children do. In engineering, learning techniques would make it possible to develop software that can be quickly customized to deal with the increasing amount of information and the flood of data around us."

The Mathematics of Learning: Dealing with Data

Tomaso Poggio (MOD 2015 & LOD 2020 Keynote Speaker) and Steve Smale

"Artificial Intelligence has already provided beneficial tools that are used every day by people around the world. Its continued development, guided by the Asilomar principles of AI, will offer amazing opportunities to help and empower people in the decades and centuries ahead."

The AI Asilomar Principles

The AI Asilomar principles have been adopted by the LOD conference since their initial formulation, January 3–5, 2017. Since then they have been an integral part of the manifesto of LOD community (LOD 2017).

LOD 2020 attracted leading experts from industry and the academic world with the aim of strengthening the connection between these institutions. The 2020 edition of LOD represented a great opportunity for professors, scientists, industry experts, and research students to learn about recent developments in their own research areas and to learn about research in contiguous research areas with the aim of creating an environment to share ideas and trigger new collaborations.

As chairs, it was an honor to organize a premiere conference in these areas and to have received a large variety of innovative and original scientific contributions.

During LOD 2020, 16 plenary talks were presented by the following leading experts from the academic world:

Yoshua Bengio, Université de Montréal, Canada (A.M. Turing Award 2018)
Tomaso Poggio, MIT, USA
Pierre Baldi, University of California, Irvine, USA
Bettina Berendt, Technische Universität Berlin, Germany
Artur d'Avila Garcez, City, University of London, UK
Luc De Raedt, KU Leuven, Belgium
Marco Gori, University of Siena, Italy
Marta Kwiatkowska, University of Oxford, UK
Michele Lombardi, University of Bologna, Italy
Angelo Lucia, University of Rhode Island, USA
Andrea Passerini, University of Trento, Italy
Jan Peters, Technische Universität Darmstadt, Max Planck Institute for Intelligent Systems, Germany
Raniero Romagnoli, Almawave, Italy
Cristina Savin, Center for Neural Science, New York University, USA
Maria Schuld, Xanadu, University of KwaZulu-Natal, South Africa
Naftali Tishby, The Hebrew University of Jerusalem, Israel
Ruth Urner, York University, Canada
Isabel Valera, Saarland University, Max Planck Institute for Intelligent Systems, Germany

LOD 2020 received 209 submissions from 63 countries in 5 continents, and each manuscript was independently reviewed by a committee formed by at least 5 members. These proceedings contain 116 research articles written by leading scientists in the fields of machine learning, artificial intelligence, reinforcement learning, computational optimization, and data science presenting a substantial array of ideas, technologies, algorithms, methods, and applications.

At LOD 2020, Springer LNCS generously sponsored the LOD Best Paper Award. This year, the paper by Cole Smith, Andrii Dobroshynskyi, and Suzanne McIntosh titled "Quantifying Local Energy Demand through Pollution Analysis" received the LOD 2020 Best Paper Award.

This conference could not have been organized without the contributions of exceptional researchers and visionary industry experts, so we thank them all for participating. A sincere thank you also goes to the 47 subreviewers and the Program Committee, formed by more than 570 scientists from academia and industry, for their valuable and essential work of selecting the scientific contributions.

Finally, we would like to express our appreciation to the keynote speakers who accepted our invitation, and to all the authors who submitted their research papers to LOD 2020.

September 2020

Giuseppe Nicosia
Varun Ojha
Emanuele La Malfa
Giorgio Jansen
Vincenzo Sciacca
Panos Pardalos
Giovanni Giuffrida
Renato Umeton

Organization

General Chairs

Giorgio Jansen	University of Cambridge, UK
Emanuele La Malfa	University of Oxford, UK
Vincenzo Sciacca	Almawave, Italy
Renato Umeton	Dana-Farber Cancer Institute, MIT, USA

Conference and Technical Program Committee Co-chairs

Giovanni Giuffrida	University of Catania, NeoData Group, Italy
Varun Ojha	University of Reading, UK
Panos Pardalos	University of Florida, USA

Tutorial Chair

Vincenzo Sciacca	Almawave, Italy

Publicity Chair

Stefano Mauceri	University College Dublin, Ireland

Industrial Session Chairs

Giovanni Giuffrida	University of Catania, NeoData Group, Italy
Vincenzo Sciacca	Almawave, Italy

Organizing Committee

Alberto Castellini	University of Verona, Italy
Piero Conca	Fujitsu, Ireland
Jole Costanza	Italian Institute of Technology, Italy
Giuditta Franco	University of Verona, Italy
Marco Gori	University of Siena, Italy
Giorgio Jansen	University of Cambridge, UK
Emanuele La Malfa	University of Oxford, UK
Gabriele La Malfa	University of Cambridge, UK
Kaisa Miettinen	University of Jyväskylä, Finland
Giuseppe Narzisi	New York University, USA
Varun Ojha	University of Reading, UK

Steering Committee

Giuseppe Nicosia	University of Cambridge, UK, and University of Catania, Italy
Panos Pardalos	University of Florida, USA

Technical Program Committee

Jason Adair	University of Stirling, UK
Agostinho Agra	Universidade de Aveiro, Portugal
Hernan Aguirre	Shinshu University, Japan
Kerem Akartunali	University of Strathclyde, UK
Richard Allmendinger	The University of Manchester, UK
Paula Amaral	NOVA University Lisbon, Portugal
Hoai An Le Thi	Université de Lorraine, France
Aris Anagnostopoulos	Sapienza Università di Roma, Italy
Fabrizio Angaroni	University of Milano-Bicocca, Italy
Davide Anguita	University of Genova, Italy
Alejandro Arbelaez	University College Cork, Ireland
Danilo Ardagna	Politecnico di Milano, Italy
Roberto Aringhieri	University of Turin, Italy
Takaya Arita	Nagoya University, Japan
Ashwin Arulselvan	University of Strathclyde, UK
Martin Atzmueller	Tilburg University, The Netherlands
Martha L. Avendano Garrido	Universidad Veracruzana, Mexico
Bar-Hen Avner	Cnam, France
Chloe-Agathe Azencott	Institut Curie Research Centre, France
Kamyar Azizzadenesheli	University of California Irvine, USA
Ozalp Babaoglu	Università di Bologna, Italy
Jaume Bacardit	Newcastle University, UK
Rodolfo Baggio	Bocconi University, Italy
James Bailey	The University of Melbourne, Australia
Marco Baioletti	Università degli Studi di Perugia, Italy
Elena Baralis	Politecnico di Torino, Italy
Xabier E. Barandiaran	University of the Basque Country, Spain
Cristobal Barba-Gonzalez	University of Malaga, Spain
Helio J. C. Barbosa	Laboratório Nacional de Computação Científica, Brazil
Thomas Bartz-Beielstein	IDEA, TH Köln, Germany
Mikhail Batsyn	Higher School of Economics, Russia
Peter Baumann	Jacobs University Bremen, Germany
Lucia Beccai	Istituto Italiano di Tecnologia, Italy
Marta Belchior Lopes	NOVA University Lisbon, Portugal
Aurelien Bellet	Inria, France
Gerardo Beni	University of California, Riverside, USA
Katie Bentley	Harvard Medical School, USA

Erik Berglund	AFRY, Sweden
Heder Bernardino	Universidade Federal de Juiz de Fora, Brazil
Daniel Berrar	Tokyo Institute of Technology, Japan
Adam Berry	CSIRO, Australia
Luc Berthouze	University of Sussex, UK
Martin Berzins	SCI Institute, The University of Utah, USA
Manuel A. Betancourt Odio	Universidad Pontificia Comillas, Spain
Hans-Georg Beyer	FH Vorarlberg University of Applied Sciences, Austria
Rajdeep Bhowmik	Binghamton University, SUNY, USA
Mauro Birattari	IRIDIA, Université Libre de Bruxelles, Belgium
Arnim Bleier	GESIS - Leibniz Institute for the Social Sciences, Germany
Leonidas Bleris	The University of Texas at Dallas, USA
Maria J. Blesa	Universitat Politècnica de Catalunya, Catalonia
Christian Blum	Spanish National Research Council, Spain
Martin Boldt	Blekinge Institute of Technology, Sweden
Fabio Bonassi	Politecnico di Milano, Italy
Flavia Bonomo	Universidad de Buenos Aires, Argentina
Gianluca Bontempi	Université Libre de Bruxelles, Belgium
Ilaria Bordino	UniCredit R&D, Italy
Anton Borg	Blekinge Institute of Technology, Sweden
Anna Bosman	University of Pretoria, South Africa
Paul Bourgine	Ecole Polytechnique Paris, France
Darko Bozhinoski	Delft University of Technology, The Netherlands
Michele Braccini	Università di Bologna, Italy
Juergen Branke	University of Warwick, UK
Ulf Brefeld	Leuphana University of Lüneburg, Germany
Will Browne	Victoria University of Wellington, New Zealand
Alexander Brownlee	University of Stirling, UK
Marcos Bueno	Radboud University, The Netherlands
Larry Bull	University of the West of England, UK
Tadeusz Burczynski	Polish Academy of Sciences, Poland
Robert Busa-Fekete	Yahoo! Research, USA
Adam A. Butchy	University of Pittsburgh, USA
Sergiy I. Butenko	Texas A&M University, USA
Sonia Cafieri	ENAC, France
Luca Cagliero	Politecnico di Torino, Italy
Stefano Cagnoni	University of Parma, Italy
Yizhi Cai	The University of Edinburgh, UK
Guido Caldarelli	IMT Lucca, Italy
Alexandre Campo	Université Libre de Bruxelles, Belgium
Mustafa Canim	IBM Thomas J. Watson Research Center, USA
Salvador Eugenio Caoili	University of the Philippines Manila, Philippines
Timoteo Carletti	University of Namur, Belgium
Jonathan Carlson	Microsoft Research, USA
Luigia Carlucci Aiello	Sapienza Università di Roma, Italy

Celso Carneiro Ribeiro	Universidade Federal Fluminense, Brazil
Alexandra M. Carvalho	Universidade de Lisboa, Portugal
Alberto Castellini	University of Verona, Italy
Michelangelo Ceci	University of Bari, Italy
Adelaide Cerveira	Universidade de Tras-os-Montes e Alto Douro, Portugal
Uday Chakraborty	University of Missouri-St. Louis, USA
Lijun Chang	The University of Sydney, Australia
Antonio Chella	University of Palermo Italy
Rachid Chelouah	Université Paris Seine, France
Haifeng Chen	NEC Labs, USA
Keke Chen	Wright State University, USA
Mulin Chen	Xidian University, China
Steven Chen	University of Pennsylvania, USA
Ying-Ping Chen	National Chiao Tung University, Taiwan, China
John W. Chinneck	Carleton University Ottawa, Canada
Gregory Chirikjian	Johns Hopkins University, USA
Miroslav Chlebik	University of Sussex, UK
Sung-Bae Cho	Yonsei University, South Korea
Stephane Chretien	ASSP/ERIC, Université Lyon 2, France
Anders Lyhne Christensen	University of Southern Denmark, Denmark
Andre Augusto Cire	University of Toronto Scarborough, Canada
Philippe Codognet	Sorbonne University, France, and The University of Tokyo, Japan
Carlos Coello Coello	IPN, Mexico
George Coghill	University of Aberdeen, UK
Pierre Comon	Université Grenoble Alpes, France
Sergio Consoli	EC Joint Research Centre, Belgium
David Cornforth	Newcastle University, UK
Luís Correia	Universidade de Lisboa, Portugal
Paulo J. da Costa Branco	Universidade de Lisboa, Portugal
Sebastian Daberdaku	Sorint.Tek, Italy
Chiara Damiani	University of Milano-Bicocca, Italy
Thomas Dandekar	University of Würzburg, Germany
Ivan Luciano Danesi	Unicredit Bank, Italy
Christian Darabos	Dartmouth College, USA
Elena Daraio	Politecnico di Torino, Italy
Raj Das	RMIT University, Australia
Vachik S. Dave	WalmartLabs, USA
Renato De Leone	Università di Camerino, Italy
Kalyanmoy Deb	Michigan State University, USA
Nicoletta Del Buono	University of Bari, Italy
Jordi Delgado	Universitat Politecnica de Catalunya, Spain
Rosario Delgado	Universitat Autonoma de Barcelona, Spain
Mauro Dell'Amico	University of Modena and Reggio Emilia, Italy
Brian Denton	University of Michigan, USA

Ralf Der	MPG, Germany
Clarisse Dhaenens	University of Lille, France
Barbara Di Camillo	University of Padova, Italy
Gianni Di Caro	IDSIA, Switzerland
Luigi Di Caro	University of Turin, Italy
Giuseppe Di Fatta	University of Reading, UK
Luca Di Gaspero	University of Udine, Italy
Mario Di Raimondo	University of Catania, Italy
Tom Diethe	Amazon Research Cambridge, UK
Matteo Diez	CNR, Institute of Marine Engineering, Italy
Ciprian Dobre	University POLITEHNICA of Bucharest, Romania
Stephan Doerfel	Micromata GmbH, Germany
Carola Doerr	CNRS, Sorbonne University, France
Rafal Drezewski	University of Science and Technology, Poland
Devdatt Dubhashi	Chalmers University, Sweden
George S. Dulikravich	Florida International University, USA
Juan J. Durillo	Leibniz Supercomputing Centre, Germany
Omer Dushek	University of Oxford, UK
Nelson F. F. Ebecken	University of Rio de Janeiro, Brazil
Marc Ebner	Ernst Moritz Arndt - Universität Greifswald, Germany
Tome Eftimov	Jožef Stefan Institute, Slovenia
Pascale Ehrenfreund	George Washington University, USA
Gusz Eiben	VU Amsterdam, The Netherlands
Aniko Ekart	Aston University, UK
Talbi El-Ghazali	University of Lille, France
Michael Elberfeld	RWTH Aachen University, Germany
Michael T. M. Emmerich	Leiden University, The Netherlands
Andries Engelbrecht	University of Pretoria, South Africa
Anton Eremeev	Sobolev Institute of Mathematics, Russia
Roberto Esposito	University of Turin, Italy
Giovanni Fasano	Università Ca' Foscari, Italy
Harold Fellermann	Newcastle University, UK
Chrisantha Fernando	DeepMind, UK
Cesar Ferri	Universidad Politecnica de Valencia, Spain
Paola Festa	University of Napoli Federico II, Italy
Jose Rui Figueira	Instituto Superior Tecnico, Portugal
Lionel Fillatre	Université Côte d'Azur, France
Steffen Finck	FH Vorarlberg University of Applied Sciences, Austria
Christoph Flamm	University of Vienna, Austria
Salvador A. Flores	Center for Mathematical Modelling, Chile
Enrico Formenti	Université Côte d'Azur, France
Giorgia Franchini	University of Modena and Reggio Emilia, Italy
Giuditta Franco	University of Verona, Italy
Emanuele Frandi	Cogent Labs, Japan
Piero Fraternali	Politecnico di Milano, Italy
Alex Freitas	University of Kent, UK

Valerio Freschi	University of Urbino, Italy
Enrique Frias Martinez	Telefonica Research, Spain
Walter Frisch	University of Vienna, Austria
Nikolaus Frohner	TU Wien, Austria
Rudolf M. Fuchslin	Zurich University of Applied Sciences, Switzerland
Antonio Fuduli	University of Calabria, Italy
Ashraf Gaffar	Arizona State University, USA
Carola Gajek	University of Augsburg, Germany
Marcus Gallagher	The University of Queensland, Australia
Claudio Gallicchio	University of Pisa, Italy
Patrick Gallinari	University of Paris 6, France
Luca Gambardella	IDSIA, Switzerland
Jean-Gabriel Ganascia	Pierre and Marie Curie University, France
Xavier Gandibleux	Universite de Nantes, France
Alfredo Garcia Hernandez-Diaz	Pablo de Olavide University, Spain
Jose Manuel Garcia Nieto	University of Malaga, Spain
Jonathan M. Garibaldi	University of Nottingham, UK
Paolo Garza	Politecnico di Torino, Italy
Romaric Gaudel	ENSAI, France
Nicholas Geard	The University of Melbourne, Australia
Martin Josef Geiger	Helmut Schmidt University, Germany
Marius Geitle	Østfold University College, Norway
Michel Gendreau	Polytechnique Montréal, Canada
Philip Gerlee	Chalmers University, Sweden
Mario Giacobini	University of Turin, Italy
Kyriakos Giannakoglou	National Technical University of Athens, Greece
Onofrio Gigliotta	University of Naples Federico II, Italy
Gail Gilboa Freedman	IDC Herzliya, Israel
David Ginsbourger	Idiap Research Institute, University of Bern, Switzerland
Giovanni Giuffrida	University of Catania, Neodata Group, Italy
Aris Gkoulalas-Divanis	IBM Watson Health, USA
Giorgio Stefano Gnecco	IMT School for Advanced Studies, Italy
Christian Gogu	Universite Toulouse III, France
Faustino Gomez	IDSIA, Switzerland
Teresa Gonçalves	University of Evora, Portugal
Eduardo Grampin Castro	Universidad de la Republica, Uruguay
Michael Granitzer	University of Passau, Germany
Alex Graudenzi	University of Milano-Bicocca, Italy
Vladimir Grishagin	Lobachevsky State University of Nizhni Novgorod, Russia
Roderich Gross	The University of Sheffield, UK
Mario Guarracino	ICAR-CNR, Italy
Francesco Gullo	UniCredit R&D, Italy
Vijay K. Gurbani	Illinois Institute of Technology, USA

Steven Gustafson	Noonum Inc., USA
Abbas Haider	Ulster University, UK
Jin-Kao Hao	University of Angers, France
Simon Harding	Machine Intelligence Ltd, Canada
Kyle Robert Harrison	University of New South Wales Canberra, Australia
William Hart	Sandia National Laboratories, USA
Inman Harvey	University of Sussex, UK
Mohammad Hasan	Indiana University Bloomington and Purdue University, USA
Geir Hasle	SINTEF Digital, Norway
Glenn Hawe	Ulster University, UK
Verena Heidrich-Meisner	Kiel University, Germany
Eligius M. T. Hendrix	Universidad de Malaga, Spain
Carlos Henggeler Antunes	University of Coimbra, Portugal
J. Michael Herrmann	The University of Edinburgh, UK
Jaakko Hollmen	Stockholm University, Sweden
Arjen Hommersom	Radboud University, The Netherlands
Vasant Honavar	Penn State University, USA
Hongxuan Huang	Tsinghua University, China
Fabrice Huet	University of Nice Sophia Antipolis, France
Hiroyuki Iizuka	Hokkaido University, Japan
Hisao Ishibuchi	Osaka Prefecture University, Japan
Peter Jacko	Lancaster University Management School, UK
Christian Jacob	University of Calgary, Canada
David Jaidan	Scalian-Eurogiciel, France
Hasan Jamil	University of Idaho, USA
Giorgio Jansen	University of Cambridge, UK
Yaochu Jin	University of Surrey, UK
Colin Johnson	University of Kent, UK
Gareth Jones	Dublin City University, Ireland
Laetitia Jourdan	University of Lille, CNRS, France
Janusz Kacprzyk	Polish Academy of Sciences, Poland
Theodore Kalamboukis	Athens University of Economics and Business, Greece
Valery Kalyagin	Higher School of Economics, Russia
George Kampis	Eötvös Loránd University, Hungary
Jaap Kamps	University of Amsterdam, The Netherlands
Dervis Karaboga	Erciyes University, Turkey
George Karakostas	McMaster University, Canada
Istvan Karsai	ETSU, USA
Branko Kavsek	University of Primorska, Slovenia
Zekarias T. Kefato	KTH Royal Institute of Technology, Sweden
Jozef Kelemen	Silesian University, Czech Republic
Marie-Eleonore Kessaci	University of Lille, France
Didier Keymeulen	NASA - Jet Propulsion Laboratory, USA
Michael Khachay	Ural Federal University, Russia
Daeeun Kim	Yonsei University, South Korea

Timoleon Kipouros	University of Cambridge, UK
Lefteris Kirousis	National and Kapodistrian University of Athens, Greece
Zeynep Kiziltan	University of Bologna, Italy
Mieczysaw Kopotek	Polish Academy of Sciences, Poland
Yury Kochetov	Novosibirsk State University, Russia
Elena Kochkina	University of Warwick, UK
Min Kong	Hefei University of Technology, China
Hennie Kruger	North-West University, South Africa
Erhun Kundakcioglu	Özyegin University, Turkey
Jacek Kustra	ASML, The Netherlands
Dmitri Kvasov	University of Calabria, Italy
Halina Kwasnicka	Wrocław University of Science and Technology, Poland
C. K. Kwong	Hong Kong Polytechnic University, Hong Kong, China
Emanuele La Malfa	University of Oxford, UK
Gabriele La Malfa	Quantitative Finance, Germany
Renaud Lambiotte	University of Namur, Belgium
Doron Lancet	Weizmann Institute of Science, Israel
Pier Luca Lanzi	Politecnico di Milano, Italy
Niklas Lavesson	Jönköping University, Sweden
Alessandro Lazaric	Facebook Artificial Intelligence Research (FAIR), France
Doheon Lee	KAIST, South Korea
Eva K. Lee	Georgia Tech, USA
Jay Lee	Center for Intelligent Maintenance Systems UC, USA
Tom Lenaerts	Université Libre de Bruxelles, Belgium
Rafael Leon	Universidad Politecnica de Madrid, Spain
Carson Leung	University of Manitoba, Canada
Peter R. Lewis	Aston University, UK
Rory Lewis	University of Colorado, USA
Kang Li	Google, USA
Lei Li	Florida International University, USA
Shuai Li	University of Cambridge, UK
Xiaodong Li	RMIT University, Australia
Weifeng Liu	China University of Petroleum, China
Joseph Lizier	The University of Sydney, Australia
Giosue' Lo Bosco	University of Palermo, Italy
Daniel Lobo	University of Maryland, USA
Fernando Lobo	University of Algarve, Portugal
Daniele Loiacono	Politecnico di Milano, Italy
Gianfranco Lombardo	University of Parma, Italy
Yang Lou	City University of Hong Kong, Hong Kong, China
Jose A. Lozano	University of the Basque Country, Spain
Paul Lu	University of Alberta, Canada
Angelo Lucia	University of Rhode Island, USA

Gabriel Luque	University of Malaga, Spain
Pasi Luukka	Lappeenranta-Lahti University of Technology, Finland
Dario Maggiorini	Università degli Studi di Milano, Italy
Gilvan Maia	Universidade Federal do Ceará, Brazil
Donato Malerba	University of Bari, Italy
Anthony Man-Cho So	The Chinese University of Hong Kong, Hong Kong, China
Jacek Madziuk	Warsaw University of Technology, Poland
Vittorio Maniezzo	University of Bologna, Italy
Luca Manzoni	University of Trieste, Italy
Marco Maratea	University of Genova, Italy
Elena Marchiori	Radboud University, The Netherlands
Tiziana Margaria	University of Limerick, Lero, Ireland
Magdalene Marinaki	Technical University of Crete, Greece
Yannis Marinakis	Technical University of Crete, Greece
Omer Markovitch	University of Groningen, The Netherlands
Carlos Martin-Vide	Rovira i Virgili University, Spain
Dominique Martinez	CNRS, France
Aldo Marzullo	University of Calabria, Italy
Joana Matos Dias	Universidade de Coimbra, Portugal
Nikolaos Matsatsinis	Technical University of Crete, Greece
Matteo Matteucci	Politecnico di Milano, Italy
Stefano Mauceri	University College Dublin, Ireland
Giancarlo Mauri	University of Milano-Bicocca, Italy
Antonio Mauttone	Universidad de la Republica, Uruguay
Mirjana Mazuran	Politecnico di Milano, Italy
James McDermott	National University of Ireland, Ireland
Suzanne McIntosh	NYU Courant, NYU Center for Data Science, USA
Gabor Melli	Sony Interactive Entertainment Inc., USA
Jose Fernando Mendes	University of Aveiro, Portugal
Lu Meng	University at Buffalo, USA
Rakesh R. Menon	University of Massachusetts Amherst, USA
David Merodio-Codinachs	ESA, France
Silja Meyer-Nieberg	Universität der Bundeswehr München, Germany
Efren Mezura-Montes	University of Veracruz, Mexico
George Michailidis	University of Florida, USA
Martin Middendorf	Leipzig University, Germany
Kaisa Miettinen	University of Jyväskylä, Finland
Orazio Miglino	University of Naples Federico II, Italy
Julian Miller	University of York, UK
Marco Mirolli	ISTC-CNR, Italy
Mustafa Misir	Istinye University, Turkey
Natasa Miskov-Zivanov	University of Pittsburgh, USA
Carmen Molina-Paris	University of Leeds, UK
Shokoufeh Monjezi Kouchak	Arizona State University, USA

Sara Montagna	Università di Bologna, Italy
Marco Montes de Oca	Clypd Inc., USA
Rafael M. Moraes	Viasat Inc., USA
Monica Mordonini	University of Parma, Italy
Nima Nabizadeh	Ruhr-Universität Bochum, Germany
Mohamed Nadif	University of Paris, France
Hidemoto Nakada	NAIST, Japan
Mirco Nanni	CNR-ISTI, Italy
Valentina Narvaez-Teran	Cinvestav Tamaulipas, Mexico
Sriraam Natarajan	Indiana University Bloomington, USA
Chrystopher L. Nehaniv	University of Hertfordshire, UK
Michael Newell	Athens Consulting, LLC, USA
Binh P. Nguyen	Victoria University of Wellington, New Zealand
Giuseppe Nicosia	University of Catania, Italy
Sotiris Nikoletseas	University of Patras, CTI, Greece
Xia Ning	IUPUI, USA
Jonas Nordhaug Myhre	UiT The Arctic University of Norway, Norway
Wieslaw Nowak	Nicolaus Copernicus University, Poland
Eirini Ntoutsi	Leibniz Universität Hannover, Germany
David Nunez	University of the Basque Country, Spain
Varun Ojha	University of Reading, UK
Michal Or-Guil	Humboldt University of Berlin, Germany
Marcin Orchel	AGH University of Science and Technology, Poland
Mathias Pacher	Goethe University Frankfurt, Germany
Ping-Feng Pai	National Chi Nan University, Taiwan, China
Pramudita Satria Palar	Bandung Institute of Technology, Indonesia
Wei Pang	University of Aberdeen, UK
George Papastefanatos	Athena RC/IMIS, Greece
Luís Paquete	University of Coimbra, Portugal
Panos Pardalos	University of Florida, USA
Rohit Parimi	Bloomberg LP, USA
Konstantinos Parsopoulos	University of Ioannina, Greece
Andrea Patane'	University of Oxford, UK
Remigijus Paulaviius	Vilnius University, Lithuania
Joshua Payne	University of Zurich, Switzerland
Clint George Pazhayidam	Indian Institute of Technology Goa, India
Jun Pei	Hefei University of Technology, China
Nikos Pelekis	University of Piraeus, Greece
David A. Pelta	Universidad de Granada, Spain
Dimitri Perrin	Queensland University of Technology, Australia
Milena Petkovi	Zuse Institute Berlin, Germany
Koumoutsakos Petros	ETH, Switzerland
Juan Peypouquet	Universidad Técnica Federico Santa María, Chile
Andrew Philippides	University of Sussex, UK
Stefan Pickl	Universität der Bundeswehr München, Germany
Fabio Pinelli	Vodafone Italia, Italy

Joao Pinto	Technical University of Lisbon, Portugal
Vincenzo Piuri	Università degli Studi di Milano, Italy
Alessio Pleb	University Messina, Italy
Nikolaos Ploskas	University of Western Macedonia, Greece
Agoritsa Polyzou	University of Minnesota, USA
George Potamias	Institute of Computer Science - FORTH, Greece
Philippe Preux	Inria, France
Mikhail Prokopenko	The University of Sydney, Australia
Paolo Provero	University of Turin, Italy
Buyue Qian	IBM T. J. Watson, USA
Chao Qian	Nanjing University, China
Michela Quadrini	University of Padova, Italy
Tomasz Radzik	King's College London, UK
Gunther Raidl	TU Wien, Austria
Helena Ramalhinho Lourenço	Pompeu Fabra University, Spain
Palaniappan Ramaswamy	University of Kent, UK
Jan Ramon	Inria, France
Vitorino Ramos	Technical University of Lisbon, Portugal
Shoba Ranganathan	Macquarie University, Australia
Zbigniew Ras	The University of North Carolina, USA
Jan Rauch	University of Economics, Czech Republic
Steffen Rebennack	Karlsruhe Institute of Technology (KIT), Germany
Wolfgang Reif	University of Augsburg, Germany
Patrik Reizinger	Budapest University of Technology and Economics, Hungary
Guillermo Rela	Universidad de la Republica, Uruguay
Cristina Requejo	Universidade de Aveiro, Portugal
Paul Reverdy	University of Arizona, USA
John Rieffel	Union College, USA
Francesco Rinaldi	University of Padova, Italy
Laura Anna Ripamonti	Università degli Studi di Milano, Italy
Franco Robledo	Universidad de la Republica, Uruguay
Humberto Rocha	University of Coimbra, Portugal
Marcelo Lisboa Rocha	Universidade Federal do Tocantins, Brazil
Eduardo Rodriguez-Tello	Cinvestav-Tamaulipas, Mexico
Andrea Roli	Università di Bologna, Italy
Massimo Roma	Sapienza Università di Roma, Italy
Vittorio Romano	University of Catania, Italy
Pablo Romero	Universidad de la Republica, Uruguay
Andre Rosendo	University of Cambridge, UK
Samuel Rota Bulo	Mapillary Research, Austria
Arnab Roy	Fujitsu Laboratories of America, USA
Alessandro Rozza	Parthenope University of Naples, Italy
Valeria Ruggiero	University of Ferrara, Italy
Kepa Ruiz-Mirazo	University of the Basque Country, Spain

Florin Rusu	University of California, Merced, USA
Conor Ryan	University of Limerick, Ireland
Jakub Rydzewski	Nicolaus Copernicus University, Poland
Nick Sahinidis	Carnegie Mellon University, USA
Lorenza Saitta	University of Piemonte Orientale, Italy
Isak Samsten	Stockholm University, Sweden
Andrea Santoro	Queen Mary University of London, UK
Giorgio Sartor	SINTEF, Norway
Claudio Sartori	University of Bologna, Italy
Frederic Saubion	Université Angers, France
Khaled Sayed	University of Pittsburgh, USA
Robert Schaefer	AGH University of Science and Technology, Poland
Andrea Schaerf	University of Udine, Italy
Alexander Schiendorfer	University of Augsburg, Germany
Rossano Schifanella	University of Turin, Italy
Christoph Schommer	University of Luxembourg, Luxembourg
Oliver Schuetze	IPN, Mexico
Martin Schulz	Technical University of Munich, Germany
Bryan Scotney	Ulster University, UK
Luís Seabra Lopes	University of Aveiro, Portugal
Natalia Selini Hadjidimitriou	University of Modena and Reggio Emilia, Italy
Giovanni Semeraro	University of Bari, Italy
Alexander Senov	Saint Petersburg State University, Russia
Andrea Serani	CNR, Institute of Marine Engineering, Italy
Roberto Serra	University of Modena and Reggio Emilia, Italy
Marc Sevaux	Université de Bretagne-Sud, France
Kaushik Das Sharma	University of Calcutta, India
Nasrullah Sheikh	IBM Research, USA
Vladimir Shenmaier	Sobolev Institute of Mathematics, Russia
Leonid Sheremetov	Mexican Petroleum Institute, Mexico
Ruey-Lin Sheu	National Cheng Kung University, Taiwan, China
Hsu-Shih Shih	Tamkang University, Taiwan, China
Kilho Shin	Gakushuin University, Japan
Zeren Shui	University of Minnesota, USA
Patrick Siarry	Université Paris-Est Creteil, France
Sergei Sidorov	Saratov State University, Russia
Alkis Simitsis	HP Labs, USA
Yun Sing Koh	The University of Auckland, New Zealand
Alina Sirbu	University of Pisa, Italy
Konstantina Skouri	University of Ioannina, Greece
Elaheh Sobhani	Université Grenoble Alpes, France
Johannes Sollner	Sodatana e.U., Austria
Giandomenico Spezzano	CNR-ICAR, Italy
Antoine Spicher	Université Paris-Est Creteil, France
Claudio Stamile	Université Claude Bernard Lyon 1, France

Pasquale Stano	University of Salento, Italy
Thomas Stibor	GSI Helmholtz Centre for Heavy Ion Research, Germany
Catalin Stoean	University of Craiova, Romania
Johan Suykens	KU Leuven, Belgium
Reiji Suzuki	Nagoya University, Japan
Domenico Talia	University of Calabria, Italy
Kay Chen Tan	National University of Singapore, Singapore
Letizia Tanca	Politecnico di Milano, Italy
Sean Tao	Carnegie Mellon University, USA
Katarzyna Tarnowska	San Jose State University, USA
Erica Tavazzi	University of Padova, Italy
Charles Taylor	UCLA, USA
Tatiana Tchemisova Cordeiro	University of Aveiro, Portugal
Maguelonne Teisseire	INRAE, UMR, TETIS, France
Fabien Teytaud	Université du Littoral Côte d'Opale, France
Tzouramanis Theodoros	University of the Aegean, Greece
Jon Timmis	University of York, UK
Gianna Toffolo	University of Padova, UK
Gabriele Tolomei	Sapienza Università di Roma, Italy
Michele Tomaiuolo	University of Parma, Italy
Joo Chuan Tong	Institute of High Performance Computing, Singapore
Gerardo Toraldo	Università degli Studi di Napoli Federico II, Italy
Jaden Travnik	University of Alberta, Canada
Nickolay Trendafilov	Open University, UK
Sophia Tsoka	King's College London, UK
Shigeyoshi Tsutsui	Hannan University, Japan
Elio Tuci	University of Namur, Belgium
Ali Emre Turgut	IRIDIA-ULB, France
Karl Tuyls	The University of Liverpool, UK
Gregor Ulm	Fraunhofer-Chalmers Research Centre for Industrial Mathematics, Sweden
Jon Umerez	University of the Basque Country, Spain
Renato Umeton	Dana-Farber Cancer Institute, MIT, USA
Ashish Umre	University of Sussex, UK
Olgierd Unold	Politechnika Wrocławska, Poland
Rishabh Upadhyay	Innopolis University, Russia
Alexandru Uta	VU Amsterdam, The Netherlands
Giorgio Valentini	Università degli Studi di Milano, Italy
Sergi Valverde	Pompeu Fabra University, Spain
Werner Van Geit	Blue Brain Project, EPFL, Switzerland
Pascal Van Hentenryck	University of Michigan, USA
Ana Lucia Varbanescu	University of Amsterdam, The Netherlands
Carlos Varela	Rensselaer Polytechnic Institute, USA
Iraklis Varlamis	Harokopio University, Greece

Eleni Vasilaki	The University of Sheffield, UK
Apostol Vassilev	NIST, USA
Richard Vaughan	Simon Fraser University, Canada
Kalyan Veeramachaneni	MIT, USA
Vassilios Verykios	Hellenic Open University, Greece
Herna L. Viktor	University of Ottawa, Canada
Mario Villalobos-Arias	Univesidad de Costa Rica, Costa Rica
Marco Villani	University of Modena and Reggio Emilia, Italy
Susana Vinga	INESC-ID, Instituto Superior Técnico, Portugal
Mirko Viroli	Università di Bologna, Italy
Katya Vladislavleva	Evolved Analytics LLC, Belgium
Robin Vogel	Telecom, France
Stefan Voss	University of Hamburg, Germany
Dean Vuini	Vrije Universiteit Brussel, Belgium
Markus Wagner	The University of Adelaide, Australia
Toby Walsh	UNSW Sydney, Australia
Harry Wang	University of Michigan, USA
Jianwu Wang	University of Maryland, USA
Lipo Wang	Nanyang Technological University, Singapore
Longshaokan Wang	North Carolina State University, USA
Rainer Wansch	Fraunhofer IIS, Germany
Syed Waziruddin	Kansas State University, USA
Janet Wiles	The University of Queensland, Australia
Man Leung Wong	Lingnan University, Hong Kong, China
Andrew Wuensche	University of Sussex, UK
Petros Xanthopoulos	University of Central Florida, USA
Ning Xiong	Mälardalen University, Sweden
Chang Xu	The University of Sydney, Australia
Dachuan Xu	Beijing University of Technology, China
Xin Xu	George Washington University, USA
Gur Yaari	Yale University, USA
Larry Yaeger	Indiana University Bloomington, USA
Shengxiang Yang	De Montfort University, USA
Xin-She Yang	Middlesex University London, UK
Li-Chia Yeh	National Tsing Hua University, Taiwan, China
Sule Yildirim-Yayilgan	Norwegian University of Science and Technology, Norway
Shiu Yin Yuen	The City University of Hong Kong, Hong Kong, China
Qi Yu	Rochester Institute of Technology, USA
Zelda Zabinsky	University of Washington, USA
Luca Zanni	University of Modena and Reggio Emilia, Italy
Ras Zbyszek	The University of North Carolina, USA
Hector Zenil	University of Oxford, UK
Guang Lan Zhang	Boston University, USA
Qingfu Zhang	City University of Hong Kong, Hong Kong, China
Rui Zhang	IBM Research, USA

Zhi-Hua Zhou   Nanjing University, China
Tom Ziemke   Linköping University, Sweden
Antanas Zilinskas   Vilnius University, Lithuania
Julius Ilinskas   Vilnius University, Lithuania

Subreviewers

Agostinho Agra   Hussain Kazmi
Alessandro Suglia   Kamer Kaya
Amlan Chakrabarti   Kaushik Das Sharma
Andrea Tangherloni   Luís Gouveia
Andreas Artemiou   Marco Polignano
Anirban Dey   Paolo Di Lorenzo
Arnaud Liefooghe   Rami Nourddine
Artem Baklanov   S. D. Riccio
Artyom Kondakov   Sai Ji
Athar Khodabakhsh   Sean Tao
Bertrand Gauthier   Shyam Chandramouli
Brian Tsan   Simone G. Riva
Christian Hubbs   Tonguc Yavuz
Constantinos Siettos   Vincenzo Bonnici
David Nizar Jaidan   Xianli Zhang
Dmitry Ivanov   Xiaoyu Li
Edhem Sakarya   Xin Sun
Farzad Avishan   Yang Li
Gaoxiang Zhou   Yasmine Ahmed
Guiying Li   Yujing Ma
Hao Yu   Zuhal Ozcan
Hongqiao Wang

Best Paper Awards

LOD 2020 Best Paper Award

"Quantifying Local Energy Demand through Pollution Analysis"
Cole Smith[1], Andrii Dobroshynskyi[1], and Suzanne McIntosh[1,2]
[1] Courant Institute of Mathematical Sciences, New York University, USA
[2] Center for Data Science, New York University, USA
Springer sponsored the LOD 2020 Best Paper Award with a cash prize of EUR 1,000.

Special Mention

"Sparsity Meets Robustness: Channel Pruning for the Feynman-Kac Formalism Principled Robust Deep Neural Nets"
Thu Dinh, Bao Wang, Andrea Bertozzi, Stanley Osher and Jack Xin
University of California Irvine, University of California, Los Angeles (UCLA).

"State Representation Learning from Demonstration"
Astrid Merckling, Alexandre Coninx, Loic Cressot, Stephane Doncieux and Nicolas Perrin
Sorbonne University, Paris, France

"Sparse Perturbations for Improved Convergence in SZO Optimization"
Mayumi Ohta, Nathaniel Berger, Artem Sokolov and Stefan Riezler
Heidelberg University, Germany

LOD 2020 Best Talks

"A fast and efficient smoothing approach to LASSO regression and an application in statistical genetics: polygenic risk scores for Chronic obstructive pulmonary disease (COPD)"
Georg Hahn, Sharon Marie Lutz, Nilanjana Laha, Christoph Lange
Department of Biostatistics, T.H. Chan School of Public Health, Harvard University, USA

"Gravitational Forecast Reconciliation"
Carla Freitas Silveira, Mohsen Bahrami, Vinicius Brei, Burcin Bozkaya, Selim Balcsoy, Alex *"Sandy"* Pentland
University of Bologna, Italy - MIT Media Laboratory, USA - Federal University of Rio Grande do Sul Brazil and MIT Media Laboratory, USA - New College of Florida, USA and Sabanci University, Turkey - Sabanci University, Turkey - Massachusetts Institute of Technology - MIT Media Laboratory, USA

"From Business Curated Products to Algorithmically Generated"
Vera Kalinichenko and Garima Garg
University of California, Los Angeles - UCLA, USA and FabFitFun, USA

LOD 2019 Best Paper Award

"Deep Neural Network Ensembles"
Sean Tao
Carnegie Mellon University, USA

LOD 2018 Best Paper Award

"Calibrating the Classifier: Siamese Neural Network Architecture for End-to-End Arousal Recognition from ECG"
Andrea Patané* and Marta Kwiatkowska*
University of Oxford, UK

MOD 2017 Best Paper Award

"Recipes for Translating Big Data Machine Reading to Executable Cellular Signaling Models"
Khaled Sayed*, Cheryl Telmer**, Adam Butchy* & Natasa Miskov-Zivanov*
*University of Pittsburgh, USA
**Carnegie Mellon University, USA

MOD 2016 Best Paper Award

"Machine Learning: Multi-site Evidence-based Best Practice Discovery"
Eva Lee, Yuanbo Wang and Matthew Hagen
Eva K. Lee, Professor Director, Center for Operations Research in Medicine and HealthCare H. Milton Stewart School of Industrial and Systems Engineering, Georgia Institute of Technology, USA

MOD 2015 Best Paper Award

"Learning with discrete least squares on multivariate polynomial spaces using evaluations at random or low-discrepancy point sets"
Giovanni Migliorati
Ecole Polytechnique Federale de Lausanne – EPFL, Switzerland

Contents – Part I

Contents – Part II

Revisiting Clustering as Matrix Factorisation on the Stiefel Manifold

Stéphane Chrétien[1,2,3]($\boxtimes$) (iD) and Benjamin Guedj[4,5]

[1] Université Lumière-Lyon-II, 69676 Bron Cedex, France
stephane.chretien@univ-lyon2.fr
[2] The Alan Turing Institute, London, UK
[3] The National Physical Laboratory, Teddington TW11 0LW, UK
stephane.chretien@npl.co.uk
[4] Inria, Lille - Nord Europe Research Centre, Lille, France
benjamin.guedj@inria.fr
[5] Department of Computer Science and Centre for Artificial Intelligence, University
College London, London, UK

Abstract. This paper studies clustering for possibly high dimensional data (*e.g.* images, time series, gene expression data, and many other settings), and rephrase it as low rank matrix estimation in the PAC-Bayesian framework. Our approach leverages the well known Burer-Monteiro factorisation strategy from large scale optimisation, in the context of low rank estimation. Moreover, our Burer-Monteiro factors are shown to lie on a Stiefel manifold. We propose a new generalized Bayesian estimator for this problem and prove novel prediction bounds for clustering. We also devise a componentwise Langevin sampler on the Stiefel manifold to compute this estimator.

Keywords: Clustering · Concentration inequalities · Non-negative matrix factorisation · Gaussian mixtures · PAC-Bayes · Optimisation on manifolds.

1 Introduction

Clustering, *i.e.*, unsupervised classification, is a central problem in machine learning and has attracted great attention since the origins of statistics, via model-based learning, but recently regained a lot of interest from theoreticians, due to its similarities with community detection Arias-Castro and Verzelen (2014, 2015). On the application side, clustering is pervasive in data science, and has become a basic tool in computer science, bio-informatics, finance, metrology, to name but a few.

1.1 Historical Background

The problem of identifying clusters in a data set can be addressed using an wide variety of tools. Two main approaches can be delineated, namely the model-based

© Springer Nature Switzerland AG 2020
G. Nicosia et al. (Eds.): LOD 2020, LNCS 12565, pp. 1–12, 2020.
https://doi.org/10.1007/978-3-030-64583-0_1

approach and the non-model based approach. Techniques such as hierarchical clustering Hastie et al. (2009), minimum spanning tree-based approaches Blum et al. (2016), K-means algorithms Hastie et al. (2009), belong to the non-model based family of methods. Model-based techniques mostly rely on mixture modelling McLachlan and Peel (2004) and often offer better interpretability whilst being easily amenable to uncertainty quantification analysis. The EM algorithm Dempster et al. (1977), McLachlan and Peel (2004) is often the algorithm of choice in the frequentist approach while many Monte Carlo Markov Chain techniques have been proposed for estimation in the Bayesian setting.

In recent years, the clustering problem has revived a surge of interest in a different setting, namely community detection in random graphs. Tools from spectral graph theory and convex optimisation, combined with recent breakthrough from random matrix theory were put to work in devising efficient clustering methods that operate in polynomial time. The celebrated example of Max-Cut, a well known NP-hard combinatorial optimisation problem strongly related to bi-clustering and with a tight Semi-Definite Programming (SDP) relaxation discovered by Goemans and Williamson (1995), is an example among the many successes of the convex optimisation approach to addressing machine learning problems. SDP is the class of optimisation problems that consist in minimising a linear function over the sets of Positive Semi-Definite matrices that satisfy a set of linear (in)equalities. Goemans and Williamson (1995) subsequently triggered a long lasting trend of research in convex relaxation with many application in data science, and recent results proposing tighter relaxations to the clustering problem can be found in Guédon and Vershynin (2016), Chrétien et al. (2016), Giraud and Verzelen (2018). Some of these methods even apply to any kind of data set endowed with a relevant affinity measure computed from the pairwise distances between the data points, and share the common feature of using low-rank matrix representations of the clustering problem. The theoretical tools behind the analysing of the performance of these convex optimisation-based methods are also quite fascinating and range from random matrix theory Bandeira (2018), Vershynin (2018), concentration inequalities for quadratic forms of random vectors Rudelson and Vershynin (2013) and optimisation theory [optimality conditions, see Royer (2017)], localisation arguments in statistical learning theory Giraud and Verzelen (2018), Grothendieck's inequality Guédon and Vershynin (2016), Montanari and Sen (2015), to name but a few.

The main drawback, however, of the current lines of approach to the performance analysis of these powerful convex *SDP* and *spectral relaxations* is that they all depend on the separation between clusters, *i.e.,* the minimum distance between two points from different clusters, a crucial parameter in the aforecited analyses. In real data sets however, sufficient inter-cluster separation rarely holds and overlaps between clusters are the common situation. This leaves open the difficult problem of finding an alternative theoretical route for controlling the estimation error. On the computational side, the sample size is also a problem for SDP relaxations for which off-the-shelf software does not scale to big data. A remedy to this problem is to use the Burer-Monteiro factorisation consisting in solving in U where $X = UU^t$ is the variable of the SDP at hand Burer and Monteiro (2003). The Burer-Monteiro factorisation results in a non-convex optimisation problem whose local minimisers are global minimisers when the number of columns of U is sufficiently large Boumal et al. (2005), Burer and Monteiro (2016). In practice however, the rank of the sought matrix is simply equal to the number of clusters, and whether such small priors on the rank of the Burer-Monteiro factorisation are compatible with the local/global equivalence of the minimisers in general remains an open question to this day. A final source of frustration in our list, is that there does not seem to exist any method for quantifying the uncertainty of the results in these convex optimisation-based approaches to clustering.

In the present paper, we propose a generalized Bayesian approach to clustering which hinges on low rank estimation of a clustering matrix. We then leverage arguments from the PAC-Bayesian theory for controlling the error which does not use any prior estimate of separation. Our approach is based on the estimation of a normalised version T^* of the adjacency matrix of the clustering, which can be factorised into $T^* = U^* U^{*^t}$, where U^* has orthonormal, non-negative columns. Leveraging this structure leads to sampling on the intersection of the Stiefel manifold Edelman et al. (1998) and the non-negative orthant, which is another surprising manifestation of the power of non-negative matrix factorisation (NMF) in clustering problems. Solving this factorised version in the PAC-Bayesian setting is the sampling counterpart of the Burer-Monteiro approach to the numerical solution of high dimensional SDP. The PAC-Bayesian approach (initiated by Catoni (2004, 2007), McAllester (1998, 1999), Shawe-Taylor and Williamson (1997); see Guedj (2019) for a recent survey) moreover makes no prior use of the separation and at the same time makes it possible to obtain state-of-the-art risk bounds.

1.2 Our Contribution

The main goal of the present paper is to study the clustering problem from a low rank Stiefel matrix, i.e. matrices with orthonormal columns, view point, and present a PAC-Bayesian analysis of the related statistical estimation problem. Our approach is in particular inspired by recent work on low rank approximation for k-means Boutsidis et al. (2009), Cohen et al. (2015), where the representation of clustering using the matrix T^* is explicitly stated (although no algorithm is provided), and PAC-Bayesian bounds for Non-Negative Matrix factorisation [as introduced by Alquier and Guedj (2017) although they do not establish the link between NMF and clustering]. To the best of our knowledge, the representation in Boutsidis et al. (2009) using the matrix T^* has never been studied from a statistical learning perspective.

We present our main result (Theorem 1, which states an inequality holding in expectation on the prediction performance) in Sect. 2. Our second main result is Theorem 2, which specifies the results of Theorem 1 in the case where we assume that the family of means is incoherent. Section 3 is devoted to our algorithm (an alternating Langevin sampler which relies on computing gradients on the Stiefel manifold).

Our choice of the Langevin sampler is motivated by its faster convergence than Metropolis type samplers due to the absence of any rejection step. This is even more true in the case where one needs to account for the manifold structure of the model as is the case in the present paper.

Additional proofs are gathered in the supplementary material.

1.3   Notation

The notation used in the present paper is fairly standard. The canonical scalar product in $\mathbb{R}^d$ will be denoted by $\langle \cdot, \cdot \rangle$, the ℓ_p norms by $\| \cdots \|_p$. For matrices in $\mathbb{R}^{d \times n}$, the operator norm will be denoted by $\| \cdot \|$ and the Frobenius norm by $\| \cdot \|_F$. The Stiefel manifold of order (n, R), i.e. the set of matrices in $\mathbb{R}^{n \times R}$ with orthonormal columns, will be denoted by $\mathbb{O}_{n,R}$, and $\mathbb{O}_{n,R,+}$ will denote the subset of the Stiefel manifold $\mathbb{O}_{n,R}$ consisting of componentwise nonnegative matrices. The matrices in $\mathbb{O}_{n,R}$ will sometimes be identified with matrices in $\mathbb{R}^{n \times n}$ where the first R columns form an orthonormal family and the remaining $n - R$ columns are set to zero. The gradient operator acting on differentiable multivariate functions will be denoted by ∇.

2   Non-negative Factorisation of the Stiefel Manifold

This section is devoted to the presentation of our framework and our main theoretical result.

2.1   Model

Let data points $x_1, \ldots, x_n$ be vectors in $\mathbb{R}^d$ and let X denote the matrix

$$X = [x_1, \ldots, x_n].$$

Let $\mu_1, \ldots, \mu_K$ be $K \in \mathbb{N} \backslash \{0\}$ vectors in $\mathbb{R}^d$. We will say that x_i belongs to cluster $k \in \{1, \ldots, K\}$ if $x_i = \mu_k + E_i$ for some centered random vector $E_i \in \mathbb{R}^d$. For each $i = 1, \ldots, n$, we will denote by k_i the label of the cluster to which x_i belongs. For each k, we will denote by I_k the index set of the points which belong to cluster k and n_k its cardinality. Now, we can decompose X as $X = M + E$ with

$$M = [\mu_{k_1}, \ldots, \mu_{k_n}],$$

$$E = [\epsilon_1, \ldots, \epsilon_n]$$

More assumptions about the noise matrix E will be introduced and their consequences on the performance of our clustering method will be studied in Theorem 2.

2.2 Ideal Solution

If we try to estimate the columns of M, one simple way is to use a convex combination of the x_i's for each of them. In other words, one might try to approximate X by XT^* where T^* is a $\mathbb{R}^{n \times n}$ matrix. One simple way to proceed is to set T as the matrix which computes the cluster means, given by

$$T_{i,j}^* = \begin{cases} \frac{1}{n_k} & \text{if } x_i \text{ and } x_j \text{ belong to the same cluster } k, i.e., k_i = k_j \\ 0 & \text{otherwise.} \end{cases}$$

Thus, each column $i = 1, \ldots, n$ of XT^* is simply the mean over cluster k_i. This type of solution is well-motivated by the fact that the mean is the least-squares solution of the approximation of μ_k by the observation points. The matrix T^* defined as above enjoys the following desirable properties: (i) its rank is exactly the number of clusters (ii) it is nonnegative (iii) the columns corresponding to different clusters are orthogonal.

One important fact to notice is that the eigenvalue decomposition of T^* is explicit and given by

$$T^* = U^* U^{*t} \tag{1}$$

with

$$U^* = \left[\frac{1}{\sqrt{n_1}} 1_{I_1}, \ldots, \frac{1}{\sqrt{n_K}} 1_{I_K} \right], \tag{2}$$

and therefore, all the eigenvalues of T^* are equal to one.

Based on this decomposition, we can now focus on estimating U^* rather than T^*, the reason being that working on estimating U^* with $\hat{U} \geq 0$ will automatically enforce positive semi-definiteness of $\hat{T}$ (the estimator of T^*) and non-negativity of its components. Moreover, enforcing the orthogonality of the columns of $\hat{U}$, combined with the non-negativity of its components, will enforce the columns of $\hat{U}$ to have disjoint supports.

Adopting a generalized Bayesian strategy [inspired by Alquier and Guedj (2017)], we will then define a prior distribution on $\hat{U}$ and study the main properties of the resulting (generalized) posterior distribution.

2.3 The Latent Variable Model

In order to perform an accurate estimation, we need to devise meaningful priors which will account for the main constraints our estimator should satisfy, namely (i) nonnegativity of the entries (ii) orthogonality of the columns (iii) the columns have unit ℓ_2 norm (iv) group sparsity of the columns.

In order to simplify this task, we will introduce a *latent (matrix) variable O* with uniform distribution on the orthogonal group, and *build priors on U* that will promote group sparsity of the columns and non-negativity (component-wise).

Prior on (U, O). Let $\mathcal{U}_R$ denote the set of matrices of the form

$$
U = \begin{bmatrix} U_{1,1} & \cdots & U_{1,R} & 0 & \cdots & 0 \\ \vdots & & \vdots & \vdots & & \vdots \\ U_{n,1} & \cdots & U_{n,R} & 0 & \cdots & 0 \end{bmatrix}.
$$

Let $\mathbb{O}_{n,R}$ denote the Stiefel manifold, *i.e.*, the manifold of all matrices with R orthonormal columns in $\mathbb{R}^n$. The prior on $(U, O) \in \mathcal{U}_R \times \mathbb{O}_{n,R}$ is given by

$$
\pi_{U,O}(U, O) = \pi_{U|O}(U)\, \pi_O(O)
$$

with

$$
\pi_{U_{i,r}|O}(U) = \frac{1}{\sqrt{2\pi}\mu} \exp\left(-\frac{(U_{i,r} - |O_{i,r}|)^2}{2\mu^2} \right),
$$

for $i = 1, \ldots, n$, and $r = 1, \ldots, R$, with R being a fixed integer and π_O being the uniform distribution on the Stiefel manifold $\mathbb{O}_{n,R}$.

2.4 Generalized Posterior and Estimator

Following the approach of Alquier and Guedj (2017), we use a loss term [instead of a likelihood, hence the term "generalized Bayes", see Guedj (2019) for a survey] given by

$$
L_\lambda(U) = \exp\left(-\frac{\lambda}{2} \|X - XUU^t\|_F^2 \right)
$$

for some fixed positive parameters λ and μ. The resulting generalized posterior (also known as a *Gibbs measure*) is defined as

$$
\rho(U, O) = \frac{1}{Z_\lambda} L_\lambda(U)\, \pi_{U|O}(U)\, \pi_O(O),
$$

where Z_λ denotes the normalisation constant $Z_\lambda = \int L_\lambda(U)\, \pi_{U|O}(U)\, \pi_O(O)\, dU$. Finally we let $\hat{U}_\lambda$ denote the posterior mean of U, i.e.

$$
\hat{U}_\lambda = \int U\, L_\lambda(U)\, \pi_{U|O}(U)\pi_O(O) dU\, dO.
$$

2.5 A PAC-Bayesian-Flavored Error Bound

Our main result is the following theorem.

Theorem 1. *Let us assume that E is fixed. Let $\nu_{\min}$ and $\nu_{\max}$ be such that*

$$
\nu_{\min} \leq \min_{\tilde{U} \in \mathbb{O}_{n,R,+},\ M\tilde{U}\tilde{U}^t = M} \|E(I - \tilde{U}\tilde{U}^t)\|_F
$$

and

$$\nu_{\max} \geq \max_{\tilde{U} \in \mathbb{O}_{n,R,+}, \ M\tilde{U}\tilde{U}^t = M} \|E(I - \tilde{U}\tilde{U}^t)\|_F.$$

Then, for all $\epsilon > 0$, and for all $c_O > 0$ and $c_U > c_O$ such that

$$c_U(2 + c_U) \leq \epsilon \frac{\nu_{\min}}{\|M\| + \|E\|}, \tag{3}$$

for any $R \in \{1, \ldots, n\}$ and for c and ρ sufficiently small universal constants, we have

$$\mathbb{E}\left[\left\|M\left(T^* - \hat{U}_\lambda \hat{U}_\lambda^t\right)\right\|_F\right] \leq (1 + \epsilon) \min_{\tilde{U} \in \mathbb{O}_{n,R,+}, \ M\tilde{U}\tilde{U}^t = M} \left\{ \|M(T^* - \tilde{U}\tilde{U}^t)\|_F \right.$$

$$+ \sqrt{\frac{1}{\exp\left(\frac{\left(\sqrt{c_U^2 - c_O^2} - \sqrt{nR}\right)^2}{2}\right) - 1}}$$

$$\left. + \sqrt{\left(nR - \frac{1}{2}(R^2 + R)\right)\log(\rho^{-1}) + \log(c^{-1}) + (2 + \epsilon)\nu_{\max}} \right\}. \tag{4}$$

This theorem gives a prediction bound on the difference between the true and the estimated cluster matrices filtered by the matrix of means. Up to our knowledge, this is the first oracle bound for clustering using a generalized Bayesian NMF. Note that the oracle bound is not sharp as the leading constant is $1 + \epsilon > 1$, however ϵ may be chosen arbitrarily close to 0.

Note also that the claim that this result is PAC-Bayesian-flavored comes from the fact that the prediction machinery is largely inspired by Alquier and Guedj (2017), and the scheme of proof builds upon the PAC-Bayesian bound from Dalalyan and Tsybakov (2008). Hence we kept that PAC-Bayesian filiation, even though the bound holds in expectation.

The following Theorem gives a more precise bound in the case where the noise E is assumed to be iid Gaussian.

Theorem 2. *Assume that the dimension is larger that the number of clusters, i.e. $d > K$. Fix $\nu_{\max} > \nu_{\min} > 0$[1]. In addition to the assumptions of Theorem 1, assume that E is iid Gaussian with minimum (resp. maximum) one-dimensional variance $\sigma_{\min}^2$ (resp. $\sigma_{\max}^2$) and assume also that the μ_k have Euclidean norm less that 1 and pairwise scalar products less than μ in absolute value. Then, as long as $\mu < 1/(K - 1)$, for all $\epsilon > 0$, and for all $c_O > 0$ and $c_U > c_O$ such that*

$$c_U(2 + c_U) \leq \epsilon \frac{\nu_{\min}}{\sqrt{\left(\max_{k=1}^K n_k\right)\mu(K - 1) + 1} + \sigma_{\max}\left(\sqrt{n} + 2\sqrt{d}\right)},$$

---

[1] As the reader will be able to check, the values of $\nu_{\max}$ and $\nu_{\min}$ will play an essential role in the expression of the success probability of the method.

with probability at least

$$1 - \exp(-d) - \exp(-nu^2/8) - K,$$

with

$$K = \left(\frac{c}{\epsilon}\right)^{nR - R(R+1)/2} \left(\frac{2}{\sqrt{\pi n(n-R)}} (t_{\min}\ e/2)^{n(d-R)/4} + \exp(-t_{\max})\right), \quad (5)$$

we have

$$\sum_{k=1}^{K} \sum_{i_k \in I_k} \left(\sum_{i'_k \in I_k} T^*_{\pi, i'_k, i_k} - \hat{U}_{\lambda, \pi, i'_k} \hat{U}^t_{\lambda, \pi, i_k}\right)^2$$

$$\leq \frac{(1+\epsilon)\ \sqrt{1 + \mu(K-1)}}{1 - \mu(K-1)} \min_{\tilde{U} \in \mathbb{O}_{n,R,+},\ M\tilde{U}\tilde{U}^t = M} \|M(T^* - \tilde{U}\tilde{U}^t)\|_F$$

$$+ \sqrt{\frac{1}{\exp\left(\frac{\left(\sqrt{c_U^2 - c_O^2} - \sqrt{dR}\right)^2}{2}\right) - 1} + \sqrt{(dR - \frac{1}{2}(R^2 + R))\log(\rho^{-1}) + \log(c^{-1})}}$$

$$+ (2 + \epsilon)\nu_{\max}.$$

with[2]

$$t_{\min} = \frac{\left(\frac{\nu_{\min}}{\sigma_{\min}} + 4\epsilon\sqrt{nd + u}\right)^2}{n(n - R)}$$

and

$$t_{\max} = \left(\frac{\nu_{\max}}{\sigma_{\max}} - 4\epsilon\sqrt{nd + u}\right)^2 - \sqrt{n(n - R)},$$

This theorem shows that a bound on the difference of the cluster matrices can be obtained when the matrix of means is sufficiently incoherent. Notice that this bound is not exactly component-wise, but considers a sum over clusters, which is perfectly relevant because the matrix M does not distinguish between points in the same cluster. As expected, the smaller the coherence μ, the better the oracle bound.

The proof of Theorem 2 is deferred to the supplementary material.

3   A Langevin Sampler

In this section, we present a Langevin sampler for our estimator $\hat{U}_\lambda, \hat{O}_\lambda$[3]. Langevin-type samplers were first proposed by Grenander (1983), Grenander

---

[2] Notice that $t_{\min}$ needs to be sufficiently smaller that $2/e$ in order for the term K to become small for n sufficiently large.

[3] Notation-wise, we will identify the Stiefel manifold with the set of matrices whose first R columns form an orthonormal family and the remaining $n - R$ columns are set to zero.

and Miller (1994), Roberts and Tweedie (1996), and have attracted a lot a attention lately in the statistical learning community Brosse et al. (2017); Dalalyan (2017); Durmus and Moulines (2018).

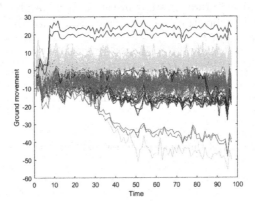

Fig. 1. Clustered satellite data representing ground movements

3.1   Computing the Gradient on the Stiefel Manifold

We start with some preliminary material about gradient computation on the Stiefel manifold from Edelman et al. (1998). The Stiefel manifold can be interpreted as the set of equivalence classes

$$[O] = \left\{ \left[O \begin{bmatrix} I_R & 0 \\ 0 & O' \end{bmatrix} \right], \quad \text{with} \quad O' \in \mathbb{O}_{d-R} \right] \right\}.$$

As can easily be deduced from this quotient representation of the Stiefel manifold, the tangent space to the Stiefel manifold at a point O is

$$T_O(\mathbb{O}_{d,R}) = \left\{ O \begin{bmatrix} A & -B^t \\ B & 0 \end{bmatrix}, \quad \text{with} \quad A \in \mathbb{R}^{R \times R} \quad \text{skew symmetric} \right\}.$$

The canonical metric at a point O is given by

$$g_c = \text{trace} \left(\Delta^t \left(I - \frac{1}{2} OO^t \right) \Delta \right).$$

For $\Delta \in T_O(\mathbb{O}_{d,R})$, the exponential map is given by $O(t) = Oe^{t\Delta} I_{d,R}$. The gradient at O of a function f defined on the Stiefel manifold $\mathbb{O}_{d,R}$ is given by[4]

$$\nabla f(O) = f_O - O f_O^t O, \tag{6}$$

where $f_O(i, i') = \frac{\partial f}{\partial O_{i,i'}}$ for any $i, i' = 1, \ldots, n$.

---

[4] This formula can be obtained using differentiation along the geodesic defined by the exponential map in the direction Δ, for all $\Delta \in T_O(\mathbb{O}_{d,R})$.

3.2   The Alternating Langevin Sampler

The alternating Langevin sampler is described as follows. It consists in alternating between perturbed gradient steps in the matrix variable O on the Stiefel manifold and perturbed gradient steps in the matrix variable U.

For clarity of the exposition, we give the formula for the gradient of $\|X - XUU^t\|_F^2$ as a function of U:

$$\nabla \left(\|X - X \cdot \cdot^t\|_F^2 \right)_U = \left((X - XUU^t)^t X + X^t(X - XUU^t) \right) U.$$

Following Brosse et al. (2018), we propose the following algorithm.

Result: A sample $\hat{U}_\lambda$ of the quasi-posterior distribution
initialise $U^{(0)} = O^{(0)}$
for $\ell = 1$ **do**

$$O^{(\ell+1)} = \exp\left(O^{(\ell)}, -h \left(\mathrm{sign}(O^{(\ell)}) \odot \left(U^{(\ell)} - |O^{(\ell)}| \right) \right) \right.$$

$$- O^{(\ell)} \left(\mathrm{sign}(O^{(\ell)}) \odot \left(U^{(\ell)} - |O^{(\ell)}| \right) \right)^t O^{(\ell)}$$

$$\left. + \sqrt{2\,h}\, Z_O^{(\ell)} \right) \right)$$

$$U^{(\ell+1)} = U^{(\ell)} - h \left(- \left((X - XU^{(\ell)}U^{(\ell)t})^t X \right. \right.$$

$$+ X^t(X - XU^{(\ell)}U^{(\ell)t}) \Big) U^{(\ell)}$$

$$\left. + \frac{1}{\mu^2} \left(U^{(\ell)} - |O^{(\ell+1)}| \right) \right) + \sqrt{2\,h}\, Z_U^{(\ell)}.$$

end

Algorithm 1. The Langevin sampler

In this algorithm the exponential function $\exp(O, H)$ at O is given by [Edelman et al. (1998) Eq. 2.45] using different notation.

4   Numerical Experiment

We provide some illustrations of the performance of our approach to clustering using a simple two component Gaussian Mixture and real satellite data. We performed an experiment with real high dimensional (satellite time series) data and one sample from the posterior after 300 iterations gave the result shown in Fig. 1.

References

Alquier, P., Guedj, B.: An oracle inequality for quasi-Bayesian nonnegative matrix factorization. Math. Methods Stat. **26**(1), 55–67 (2017). https://doi.org/10.3103/S1066530717010045

Arias-Castro, E., Verzelen, N.: Community detection in dense random networks. Annal. Stat. **42**(3), 940–969 (2014)

Bandeira, A.S.: Random Laplacian matrices and convex relaxations. Found. Comput. Math. **18**(2), 345–379 (2018)

Blum, A., Hopcroft, J., Kannan, R.: Foundations of data science. Draft book (2016)

Boumal, N., Voroninski, V., Bandeira, A.: The non-convex Burer-Monteiro approach works on smooth semidefinite programs. In: Advances in Neural Information Processing Systems, pp. 2757–2765 (2016)

Boutsidis, C., Drineas, P., Mahoney, W.: Unsupervised feature selection for the k-means clustering problem. In: Advances in Neural Information Processing Systems, pp. 153–161 (2009)

Brosse, N., Durmus, A., Moulines, E.: The promises and pitfalls of stochastic gradient Langevin dynamics. In: Advances in Neural Information Processing Systems, pp. 8278–8288 (2018)

Burer, S., Monteiro, R.D.C.: Local minima and convergence in low-rank semidefinite programming. Math. Program. **103**(3), 427–444 (2005)

Burer, S., Monteiro, R.D.C.: A nonlinear programming algorithm for solving semidefinite programs via low-rank factorization. Math. Program. **95**(2), 329–357 (2003)

Catoni, O.: In: Picard, J. (ed.): Statistical Learning Theory and Stochastic Optimization. LNM, vol. 1851. Springer, Heidelberg (2004). https://doi.org/10.1007/b99352

Catoni, O.: PAC-Bayesian Supervised Classification. Lecture Notes-Monograph Series, IMS (2007)

Chrétien, S., Dombry, S., Faivre, A.: A semi-definite programming approach to low dimensional embedding for unsupervised clustering. arXiv preprint arXiv:1606.09190 (2016)

Cohen, M.B., Elder, S., Musco, C., Musco, C., Persu, M.: Dimensionality reduction for k-means clustering and low rank approximation. In: Proceedings of the Forty-Seventh Annual ACM Symposium on Theory of Computing, pp. 163–172. ACM (2015)

Dalalyan, A.S.: Theoretical guarantees for approximate sampling from smooth and log-concave densities. J. R. Stat. Soc. Ser. B (Stat. Methodol.) **79**(3), 651–676 (2017)

Dalalyan, A.S., Tsybakov, A.B.: Aggregation by exponential weighting, sharp PAC-Bayesian bounds and sparsity. Mach. Learn. **72**(1–2), 39–61 (2008)

Dempster, A.P., Laird, N.M., Rubin, D.B.: Maximum likelihood from incomplete data via the EM algorithm. J. R. Stat. Soc. Ser. B, 1–38 (1977)

Durmus, A., Moulines, E.: Nonasymptotic convergence analysis for the unadjusted Langevin algorithm. Annal. Appl. Prob. **27**(3), 1551–1587 (2017)

Edelman, A., Arias, T.A., Smith, S.T.: The geometry of algorithms with orthogonality constraints. SIAM J. Matrix Anal. Appl. **20**(2), 303–353 (1998)

Giraud, C., Verzelen, N.: Partial recovery bounds for clustering with the relaxed k means. arXiv preprint arXiv:1807.07547 (2018)

Goemans, M.X., Williamson, D.P.: Improved approximation algorithms for maximum cut and satisfiability problems using semidefinite programming. Journal of the ACM (JACM) **42**(6), 1115–1145 (1995)

Grenander, U.: Tutorial in pattern theory. Report, Division of Applied Mathematics (1983)

Grenander, U., Miller, I.: Representations of knowledge in complex systems. J. R. Stat. Soc. Ser. B (Methodol.), 549–603 (1994)

Guedj, B.: A primer on PAC-Bayesian learning. arXiv preprint arXiv:1901.05353 (2019)

Guédon, O., Vershynin, R.: Community detection in sparse networks via Grothendieck's inequality. Probab. Theory Relat. Fields **165**(3–4), 1025–1049 (2016)

Hastie, T., Tibshirani, R., Friedman, J.: Unsupervised learning. In: The elements of statistical learning, pp. 485–585. Springer (2009)

McAllester, D.: Some PAC-Bayesian theorems. In: COLT, pp. 230–234 (1998)

McAllester, D.: PAC-Bayesian model averaging. In: COLT, pp. 164–171 (1999)

McLachlan, G., Peel, D.: Finite Mixture Models. Wiley, New York (2004)

Montanari, A., Sen, S.: Semidefinite programs on sparse random graphs and their application to community detection. arXiv preprint arXiv:1504.05910 (2015)

Roberts, G.O., Tweedie, R.L.: Exponential convergence of Langevin distributions and their discrete approximations. Bernoulli **2**(4), 341–363 (1996)

Royer, M.: Adaptive clustering through semidefinite programming. In: Advances in Neural Information Processing Systems, pp. 1795–1803 (2017)

Rudelson, M., Vershynin, R.: Hanson-wright inequality and sub-gaussian concentration. Electron. Commun. Prob. **18** (2013). 9 p

Shawe-Taylor, J., Williamson, R.C.: A PAC analysis of a Bayesian classifier. In: COLT, pp. 2–9 (1997)

Vershynin, R.: High-Dimensional Probability: An Introduction with Applications in Data Science, vol. 47. Cambridge University Press, Cambridge (2018)

Verzelen, N., Arias-Castro, E.: Community detection in sparse random networks. Annal. Appl. Prob. **25**(6), 3465–3510 (2015)

A Generalized Quadratic Loss for SVM and Deep Neural Networks

Filippo Portera[✉][iD]

Universita' degli Studi "Ca' Foscari" di Venezia, Venice, Italy
filippo.portera@unive.it

Abstract. We consider some supervised binary classification tasks and a regression task, whereas SVM and Deep Learning, at present, exhibit the best generalization performances. We extend the work [3] on a generalized quadratic loss for learning problems that examines pattern correlations in order to concentrate the learning problem into input space regions where patterns are more densely distributed. From a shallow methods point of view (e.g.: SVM), since the following mathematical derivation of problem (9) in [3] is incorrect, we restart from problem (8) in [3] and we try to solve it with one procedure that iterates over the dual variables until the primal and dual objective functions converge. In addition we propose another algorithm that tries to solve the classification problem directly from the primal problem formulation. We make also use of Multiple Kernel Learning to improve generalization performances. Moreover, we introduce for the first time a custom loss that takes in consideration pattern correlation for a shallow and a Deep Learning task. We propose some pattern selection criteria and the results on 4 UCI data-sets for the SVM method. We also report the results on a larger binary classification data-set based on Twitter, again drawn from UCI, combined with shallow Learning Neural Networks, with and without the generalized quadratic loss. At last, we test our loss with a Deep Neural Network within a larger regression task taken from UCI. We compare the results of our optimizers with the well known solver SVM[light] and with Keras Multi-Layers Neural Networks with standard losses and with a parameterized generalized quadratic loss, and we obtain comparable results (Code is available at: https://osf.io/fbzsc/wiki/home/).

Keywords: SVM · Multiple kernel learning · Deep neural networks · Binary classification and regression · Generalized quadratic loss

1 Introduction

SVM and Neural Networks methods are widely used to solve binary classification, multi-class classification, regression tasks,... In supervised binary classifica-

Supported by organization Universita' degli Studi "Ca' Foscari" di Venezia.

The original version of this chapter was revised: wrong RTS values in Table 3 and the author's e-mail address have been corrected. Additionally a link was added on the abstract. The correction to this chapter is available at https://doi.org/10.1007/978-3-030-64583-0_64

G. Nicosia et al. (Eds.): LOD 2020, LNCS 12565, pp. 13–24, 2020.
https://doi.org/10.1007/978-3-030-64583-0_2

tion learning tasks, SVM and Deep Learning methods are spread and represents the state-of-the-art in achieving best generalization performances. The work of [6] and [2] showed the potential different implementations of SVM, while there are different software to develop Deep Neural Network such as TensorFlow and PyTorch, to name a few. Our goal is to improve the generalization performances of those algorithms, considering pattern correlations in the loss function. In [3] we proposed a generalized quadratic loss for SVM, but we mathematical development was erroneous. In details, [3] presents a step from Eq. 8 and Eq. 9 that is wrong, since we said that $(\alpha + \lambda)'\mathbf{S}^{-1}(\alpha + \lambda)$ is monotonically increasing, and this is not proved. We don't try to solve this problem, we restart from problem 8 and try to find methods to solve it in its dual and primal form. Nevertheless the idea could be valid since if the matrix used to implement the loss is the identity matrix the loss reduces to the well know quadratic loss. Here we develop the loss introduced in [3] further, in the sense that we propose 2 correct optimizers for the SVM setting and, perhaps more interesting, a custom loss that can be plugged-in into a Deep Learning framework. In Sect. 2 we cover some related works about the problem we are studying. In Sect. 3 we state the mathematical problem and we present the matrix S. In Sect. 4 we report some definitions for the proposed algorithms. In Subsect. 5.1 we describe the SMOS optimization technique. In Subsect. 5.2 we characterize the RTS optimization technique. In Subsect. 5.3 we elucidate the Deep Learning framework that exploits a loss function defined with the S matrix. In Sect. 6 we present the results obtained with 2 artificially generated data-sets, 4 binary classification data-sets for SVM, and 2 larger data-sets experiments carried on with Multiple Layers Neural Nets. Finally, in Sect. 7, we draw some statements about the overall procedure and the results.

2   Related Works

We explore the use of a new loss with two different scopes: SVM and Neural Networks. The canonical SVM model was first introduced in [14], and the losses used are a class of loss functions that doesn't take into consideration pattern correlations. Several optimizers have been proposed for this model and they can be found in [2,6,15], etc...., and almost all of them are based on the linear loss version of SVM. On the other side, again, the losses used in Shallow and Deep Neural Networks are not considering pattern distribution. For the Twitter sentiment analysis task there is a work [16] that propose a Deep Convolutional Neural Network approach and [17] where they add also an attempt with LSTMs. The YearPredictionMSD data-set has been studied in [18], and in [19], to name a few.

3   The Modified Loss SVM Problem

In order to see if there is space for better generalization performances, we introduce a new loss as stated in [3]. For the rest of the paper we will use an S matrix defined as:

$$S_{i,j} = e^{-\gamma_S ||x_i - x_j||^2} \tag{1}$$

which is a symmetric, positive semi-definite, and invertible matrix (if there aren't repeated patterns).

Therefore, reconsidering Section 2 of [3], we obtain the dual problem:

$$a'1 - \frac{1}{2}\alpha'\mathbf{YKY}\alpha - \frac{1}{4C}(\alpha + \lambda)'\mathbf{S}^{-1}(\alpha + \lambda) \tag{2}$$

subject to the following constraints:

$$\alpha'y = 0 \tag{3}$$

$$i = 1, \ldots, l : \tag{4}$$

$$\alpha_i \geq 0 \tag{5}$$

$$\lambda_i \geq 0 \tag{6}$$

4   Notation

In order to have a dual problem, $\mathbf{S}$ must be invertible. If there are repeated patterns in the training set, $\mathbf{S}$ is not invertible. Thus we remove the repeated patterns from the training set.

Let:

$$i = 1, \ldots, l, \; j = 1, \ldots, l : \tag{7}$$

$$K(\boldsymbol{x}_i, \boldsymbol{x}_j) = e^{-\gamma s \|\boldsymbol{x}_i - \boldsymbol{x}_j\|^2} \tag{8}$$

$$f(\boldsymbol{x}_i) = \sum_{j=1}^{l} \alpha_j y_j K(\boldsymbol{x}_i, \boldsymbol{x}_j) + b \tag{9}$$

5   Algorithms

In the following we propose 2 different algorithms to solve the primal problem.

5.1   The SMOS Optimization Algorithm

With the same method reported in [4], we isolate the part of the dual function that depends on the updated variables:

$$\alpha_{i+1} \leftarrow \alpha_i + \varepsilon_i = \alpha_i + \nu y_i \tag{10}$$

$$\alpha_{j+1} \leftarrow \alpha_j + \varepsilon_j = \alpha_j - \nu y_j \tag{11}$$

$$\lambda_{k+1} \leftarrow \lambda_k + \mu_k \tag{12}$$

where ε is a vector of dimension l of all zeros, apart the i and j components that are, respectively, $\varepsilon_i = \nu y_i$ and $\varepsilon_j = -\nu y_j$. While μ is a vector of dimension l of all zeros, apart the k component that is equal to μ_k.

Omitting the derivation from the dual function of the optimized variables and deriving $D(\boldsymbol{\alpha} + \boldsymbol{\varepsilon}, \boldsymbol{\lambda} + \boldsymbol{\mu})$ by ν, and setting the partial derivative to 0 in order to get the maximum for a fixed μ_k, we obtain:

$$\psi = \boldsymbol{y}_i - \boldsymbol{y}_j$$

$$+ \sum_{p=1}^{l} \alpha_p \boldsymbol{y}_p K(\boldsymbol{x}_p, \boldsymbol{x}_i) - \sum_{p=1}^{l} \alpha_p \boldsymbol{y}_p K(\boldsymbol{x}_p, \boldsymbol{x}_j) +$$

$$- \frac{1}{2C} (\sum_{p=1}^{l} \alpha_p \boldsymbol{y}_i S^{-1}[p, i] - \sum_{p=1}^{l} \alpha_p \boldsymbol{y}_j S^{-1}[p, j]$$

$$+ \sum_{p=1}^{l} \lambda_p \boldsymbol{y}_i S^{-1}[p, i] - \sum_{p=1}^{l} \lambda_p \boldsymbol{y}_j S^{-1}[p, j]$$

$$+ \sum_{p=1}^{l} \mu_p \boldsymbol{y}_i S^{-1}[p, i] - \sum_{p=1}^{l} \mu_p \boldsymbol{y}_j S^{-1}[p, j])$$

and:

$$\omega = K(\boldsymbol{x}_i, \boldsymbol{x}_i) - 2K(\boldsymbol{x}_i, \boldsymbol{x}_j) + K(\boldsymbol{x}_j, \boldsymbol{x}_j)$$
$$+ \frac{S^{-1}[i, i] - 2\boldsymbol{y}_i \boldsymbol{y}_j S^{-1}[i, j] + S^{-1}[j, j]}{2C}$$

which implies:

$$\nu = \frac{\psi}{\omega} \tag{13}$$

While, fixing $\boldsymbol{a}_i, \boldsymbol{a}_j, \nu$, and deriving (14) by μ_k, we get:

$$\frac{\partial D(\boldsymbol{\alpha} + \boldsymbol{\varepsilon}, \boldsymbol{\lambda} + \boldsymbol{\mu})}{\partial \mu_k} = \tag{14}$$

$$\frac{\partial}{\partial \mu_k} - \frac{1}{4C}(\mu_k^2 S_{k,k}^{-1} + 2\mu_k \nu \boldsymbol{y}_i S_{k,i}^{-1} - 2\mu_k \nu \boldsymbol{y}_j S_{k,j}^{-1}) = 0 \tag{15}$$

then:

$$\mu_k = \frac{\nu(\boldsymbol{y}_i S_{k,i}^{-1} + \boldsymbol{y}_j S_{k,j}^{-1})}{S_{k,k}^{-1}} \tag{16}$$

These equations, (13, 16) are used for the updates described in (10, 11, and 12).

The increment variables are then clipped as follows:

If $(\boldsymbol{a}_i + \boldsymbol{y}_i \nu < 0)$ then $\nu = -\boldsymbol{y}_i(\text{previous})\boldsymbol{a}_i$
If $(\boldsymbol{a}_j - \boldsymbol{y}_j \nu < 0)$ then $\nu = \boldsymbol{y}_j(\text{previous})\boldsymbol{a}_j$
If $(\lambda_k + \mu_k < 0)$ then $\lambda_k = 0$

Let SelectPatterns(i, μ_k, k) a procedure that selects all patterns j that after an optimal ν update gives an increment above a threshold β, and dof is the acronym for "dual objective function":

$$\beta = \delta * (\text{best dof} - \text{previous dof}) + \text{previous dof} \tag{17}$$

where $0 < \delta \leq 1$, best dof is the maximum dual objective function value obtained for all $j \in [1, \ldots, l]$, previous dof is the initial value of the dual objective function without any updates on $\boldsymbol{a}_i$ or $\boldsymbol{a}_j$. We found that the optimal value of δ is 1, so only 1 pattern is selected for further real update.

While the new b is computed at each iteration with:

$$b_{\text{new}} \leftarrow \sum_{i=1|\xi_i>0}^{l} \{y_i - [f(\boldsymbol{x}_i) - b_{\text{old}}]\}/n \tag{18}$$

where n is the number of positive ξ_i's.

We introduced a solver monitor script, that eliminates the solver process whenever it employs more than 120 s to converge to a solution of the generalized quadratic loss SVM problem. This monitor is employed also to stop SVM^{light} whenever it takes too much time to converge to a solution.

5.2   The RTS Optimization Algorithm

A last attempt to solve the model problem is the Representer Theorem with S (RTS), where we consider solutions in the form:

$$f(\boldsymbol{x}) = \sum_{i=1}^{l} a_i K(\boldsymbol{x}_i, \boldsymbol{x}) + b$$

from the Representer Theorem [5], with $\boldsymbol{a}_i \in \mathbb{R}$, $i \in 1, \ldots, l$, and $b \in \mathbb{R}$. We solve the problem directly in its primal form.

For each variable $\boldsymbol{a}_i$ we take a Newton step:

$$\boldsymbol{\alpha}_{i+1} \leftarrow \boldsymbol{\alpha}_i - \frac{\frac{\partial P_S}{\partial a_i}}{\frac{\partial^2 P_S}{\partial a_i^2}} \tag{19}$$

The determination of b is the same as showed before (18). We exit the main loop whenever the problem diverges, or, for 100 consecutive steps, the Newton update is unable to lower the lowest objective function value found. One advantage of this approach is that it doesn't need the S matrix inversion. A monitor is used in order to eliminate processes that last more than 120 s.

5.3   The Deep Learning Framework

For a tutorial and survey on Deep Neural Networks you can read [7]. We use Keras 2.3.1 on TensorFlow 2.1.0 back-end in order to obtain Deep Neural Networks that are able to classify a number of input patterns of the order of $1E5$.

The initial shallow network is made of 3 layers: one input layer, one dense and regularized layer, and a sigmoid output layer. Then we add another dense layer, make algorithms comparisons, and at last, we compare our algorithms with 3 dense layers. In order to use the generalized quadratic loss we set the batch size to the entire training data-set and write a custom loss that considers pattern correlations. The S matrix used here is defined in (1). Therefore, the custom loss can be written as:

$$\text{loss} = o^{\,t}So \tag{20}$$

where o is the binary cross-entropy standard loss and t is the transposition operation.

We explore also a regression task with a neural network made of 10 dense and regularized layers. In this case the custom loss is still (20) but, the output it's the square of the error between the true value and the predicted value, and we consider only the patterns that are present in the current batch in order to build S.

6 Results

We evaluate the results on 2 artificially created data-sets, with different concentration of patterns, which could help in understanding how this loss can support in solving the proposed task. For example, a uniform distribution of patterns and the use of the generalized quadratic loss should not improve the generalization performance w.r.t. a linear loss. While a clustered distribution of patterns should highlight the benefit of the generalized quadratic loss. The uniform distribution is generated with 800 random points uniformly distributed in the 3 axes, with a random target value randomly chosen in $\{-1, +1\}$. The normal distribution is generated with 800 random points, spread with a normal distribution with 0 mean and standard deviation equals to 3 in the 3 axes, with a random target value randomly chosen in $\{-1, +1\}$. The calibration procedure is described below.

We show the two distributions on Fig. 1.

For the artificial data-sets, we don't report the results of the SMOS algorithm for a time constraint.

We also evaluate the algorithms on 5 UCI binary classification data-sets: Breast Cancer Wisconsin (Original), Haberman's Survival, Ionosphere, Connectionist Bench (Sonar, Mines vs. Rocks), Twitter Absolute Sigma 500[1], and an UCI regression data-set: YearPredictionMSD. The first 4 data-sets are used for SVM evaluation, while the Twitter and Year data-set are most suited for Multilayer Neural Network Learning. Each data-set is shuffled and divided in a training, validation, and test sets. The Year data-set is an exception, because it is not shuffled.

The training dimensions (number of patterns × number of features), and the validation and test number of patterns are reported on Table 1.

---

[1] https://archive.ics.uci.edu/ml/datasets.php.

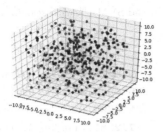 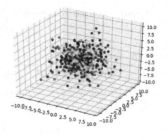

Fig. 1. A uniform distribution and a normal distribution in order to test the benefits of the generalized quadratic loss.

Table 1. Data-sets splits dimensions (the number at the right of the × symbol is the number of features d). While the number between parentheses is the number of patterns that are not repeated in the training set.

DS name	Training	Validation	Test
Uniform	550 × 3	150	100
Normal	550 × 3	150	100
Breast	400 × 9 (281)	150	149
Haberman	240 × 3 (227)	30	36
Iono	260 × 34 (260)	50	41
Sonar	130 × 60 (130)	40	38
Twit	5000 × 77	15000	120000
Year	150000 × 90	313000	51000

SMOS and RTS procedures are calibrated on three hyper-parameters: C, γ_K, and γ_S on a grid of $6 \times 6 \times 6$ with steps of powers of 10 starting from $(0.001, 0.001, 0.001)$ to $(100, 100, 100)$. For each data-set, the set of hyper-parameters that gave the best result, in term of $F1$ measure on the validation set, is saved together with the best dual variables α and best b. Then these saved variables are used to evaluate the test set, which outputs the final F1 performance on that data-set. We implement a SMO-like algorithm inspired by [4] and [2] to solve problem (2), and we name it SMOS. Besides we propose another simpler algorithm that works only in the primal problem (Sect. 5.2). At last we front the problem with an algorithm based on an objective function made only by the error term with S and no regularization term. We call this algorithm simply S. This procedure is calibrated on two hyper-parameters: γ_K and γ_S on a grid of 10×10 with steps of powers of 10 starting from $(1E - 5, 1E - 5)$ to $(10000, 10000)$.

The F1 scores on each data-set are reported in Table 2, together with the F1 score of SMOS, RTS, S, and the well known SVM^{light} software [6].

We report in Table 3 some experiments with a 10-fold cross-validation scheme, as described in [13], in order to give a greater significance level. We

Table 2. Results comparison with SMOS, RTS, S, and SVM^{light} with standard linear loss.

Algorithm	DS name	C	γ_K	γ_S	F1	Time
RTS	Uniform	100	10	10	0.589928	78 m 24 s 092 ms
S	Uniform		1E−5	0.01	**0.620690**	38 m 50 s 996 ms
SVM^{light}	Uniform	1E−05	1E−05		**0.620690**	0 m 43 s 178 ms
RTS	Normal	0.01	100	0.001	**0.722581**	51 m 29 s 542 ms
S	Normal		100	1E−05	**0.722581**	40 m 57 s 976 ms
SVM^{light}	Normal	10000	0.1		0.519206	8 m 29 s 650 ms
SMOS	Breast	0.1	0.01	1	0.961538	5 h 28 m 8 s 453 ms
RTS	Breast	0.1	100	0.001	0.938053	62 m 41 s 721 ms
S	Breast		0.001	1E−5	0.971429	6m18s446ms
SVM^{light}	Breast	1	0.0		**0.971462**	2 m 08 s 630 ms
SMOS	Haberman	0.1	0.001	10	0.885246	2 h 27 m 9 s 594 ms
RTS	Haberman	0.1	0.1	1	0.857143	75 m 51 s 646 ms
S	Haberman		0.001	0.1	0.896552	4 m 6s 346 ms
SVM^{light}	Haberman	1	0.001		0.857143	0m 18s 850 ms
SMOS	Iono	1	0.1	100	**0.958333**	6 h 22 m 10 s 739 ms
RTS	Iono	100	1	1	0.893617	47 m 38 s 634 ms
S	Iono		0.1	100	**0.958333**	6 m 59 s 161 ms
SVM^{light}	Iono	10	0.1		0.9583	6 m 15 s 790 s
SMOS	Sonar	1	1	10	0.800000	3 h 39 m 10 s 539 ms
RTS	Sonar	100	1	1	0.742857	47 m 50 s 459 ms
S	Sonar		0.1	0.1	0.820513	8 m 50 s 875 ms
SVM^{light}	Sonar	10	1		0.800000	7 m 07 s 650 ms

created 10 subsets of the training sets. Each algorithm is run 10 times, one time for each k sub training data-set. We consider the mean and the standard deviation of each resulting test F1 scores. The parameter grid explored is the same as defined before. The results with the SMOS method are not reported due to a time constraint.

Moreover, we describe some experiments in Table 4 obtained with a K and S matrices generated with multiple kernels. We call this algorithm RTS$_{\text{MKL}}$.

$$K_{MKL}(\boldsymbol{x}_i, \boldsymbol{x}_j) = \sum_{i=1}^{nK} b_i * K_i(\boldsymbol{x}_i, \boldsymbol{x}_j) \qquad (21)$$

$$S_{MKL}(\boldsymbol{x}_i, \boldsymbol{x}_j) = \sum_{i=1}^{nS} c_i * S_i(\boldsymbol{x}_i, \boldsymbol{x}_j) \qquad (22)$$

Table 3. Results comparison with RTS, S, and SVM^{light} with standard linear loss with a 10-fold cross-validation procedure.

Algorithm	DS name	F1	σ	Time
RTS	Uniform	0.607557	0.013028	45 h 06 m
S	Uniform	**0.62069**	**1.110223E−16**	4 h 45 m 08 s
SVM^{light}	Uniform	0.611796	0.811098	13 m 56 s 54 ms
RTS	Normal	0.719074	0.0156605	38 h 39 m
S	Normal	**0.717949**	**0.0**	5 h 04 m 20 s
SVM^{light}	Normal	0.534307	7.869634	1 h 01 m 23 s
RTS	Breast	**0.975627**	**0.006172**	31 h 57 m
S	Breast	0.946792	0.012488	4 h 45 m 350 ms
SVM^{light}	Breast	0.967170	0.677366	12 m 15 s 672 ms
RTS	Haberman	**0.938325**	0.033040	33 h 47 m
S	Haberman	0.865720	**0.018494**	2 h 3 m 29 s 888 ms
SVM^{light}	Haberman	0.858132	2.340863	3 m 14 s 972 ms
RTS	Iono	0.961435	0.016616	29 h 13 m
S	Iono	**0.971863**	**0.016387**	5 h 17 m 54 s
SVM^{light}	Iono	0.915835	1.784254	44 m 42 s 632 ms
RTS	Sonar	**0.934440**	**0.026761**	22 h 55 m
S	Sonar	0.800656	0.059841	5 h 15 m 18 s 655 ms
SVM^{light}	Sonar	0.843513	0.794849	1 h 10 m 2 ms 953 s

where b_i and c_i are heuristically determined as in [10], the kernels $K_i()$ and $S_i()$ are defined as RBF kernels with different γ (for the definition see (1)), and we set the maximum number of $nK = 10$ and the maximum number of $nS = 10$. We report also some experiments with an algorithm based on the Representer Theorem with a linear loss and MKL, in order to try to understand if the benefits on the generalization performance is given by the MKL technique or the use of S. We call this algorithm RT$_{\text{MKL}}$. We add that a monitor is adopted in order to eliminate SVM^{light} processes that take more than 120 s to converge, while no monitor is used for the RTS with MKL algorithm. In the last column, we report the calibration time plus the test evaluation time.

We report some experiments with different shapes of Neural Networks applied to the UCI Buzz in Social Media-Twitter data-set, in Table 5, where the algorithm is described in Sect. (5.3). Hence we name SNN the standard Shallow multi-layer Neural Network algorithm, while we name GQLSNN the Shallow Neural Network with a generalized quadratic loss. We make a grid search to find the optimal hyper-parameters. We start with one dense layer made of 5 nodes and then we increase this size until we reach 25 nodes, following steps of 5. The same sequence is applied with the case with 2 and 3 dense and regularized layers. For the GQLSNN algorithm we add another hyper-parameter, that is γ_S. The

Table 4. Results comparison with RT with MKL and linear loss, RTS with MKL and S, and SVM^{light} with standard linear loss.

Algorithm	DS name	C	γ_K	Best nKK	Best nKS	F1	Time
RT$_{\text{MKL}}$	Breast	10		1		0.953271	6 m 44,947 s
RTS$_{\text{MKL}}$	Breast	0.1		2	1	**0.972477**	61 m 6,436 s
SVM^{light}	Breast	1	0.1			0.971462	2 m 08 s 630 ms
RT$_{\text{MKL}}$	Haberman	0.001		3		0.857143	0 m 16,233 s
RTS$_{\text{MKL}}$	Haberman	0.1	0.01	3	9	**0.896552**	24 m 55,662 s
SVM^{light}	Haberman	1	0.001			0.857143	0 m 18 s 800 ms
RT$_{\text{MKL}}$	Iono	0.01		2		0.836364	0 m 23,713 s
RTS$_{\text{MKL}}$	Iono	100		2	7	0.938776	35 m 12,091 s
SVM^{light}	Iono	10	0.1			**0.9583**	0 m 50,612 s
RT$_{\text{MKL}}$	Sonar	0.01		1		**0.800000**	0 m 11,134 s
RTS$_{\text{MKL}}$	Sonar	1		2	1	0.717949	14 m 19,842 s
SVM^{light}	Sonar	10	1			**0.800000**	1 m 18,435 s

line search on this hyper-parameter spans through $[1E{-}5, \ldots, 0.1]$ with steps of powers of 10.

Each network run comprises a training procedure of 1000 epochs, with a batch size of 5000 patterns. We selected the hyper-parameters that perform at best from the whole grid, on the validation data-set. The test F1 score reported is determined by the evaluation of the best validation model on the test data-set.

For the Year data-set, described in details in [8], we trained a deeper network made of 10 dense and regularized layers, with 100 nodes for each layer. The batch size is 1000 and the number of epochs is 400. The standard Deep Neural Network setting is named DNN. For the GQLDNN algorithm the γ_S was calibrated with a line search that spans through $[1E{-}5, \ldots, 0.1]$ with steps of powers of 10. We measure the Mean Square Error, and the performance is reported in Table 6. Every Neural Network we employ uses the Adam optimizer[2].

---

[2] This software runs on a Intel(R) Core(TM) i7-6700 CPU @ 3.40 GHz with 32.084 MB of RAM, 32.084 MB of swap space, and a SSD of 512 GB.

Table 5. Results comparison with SNN, and GQLSNN exploiting S with the Twitter data-set, with the optimal hyper-parameters: the number of nodes for layers number 2 to 4, the optimal γ_S selected by the calibration procedure, the training, validation and test time.

Algorithm	nNL2	nNL3	nNL4	γ_S	Test F1	Time
SNN	15				0.88805829	1 m 5 s 274 ms
GQLSNN	25			0.1	**0.90734076**	59 m 47 s 075 ms
SNN	5	15			0.9005088	10 m 40 s 482 ms
GQLSNN	5	10		0.0001	**0.9118929**	5 h 21 m 20 s 065 ms
SNN	25	20	10		0.9080214	28 m 9 s 709 ms
GQLSNN	10	10	20	0.1	**0.91080284**	26 h 53 m 59 s

Table 6. Results comparison with DNN, and GQLDNN exploiting S, with the optimal γ_S selected by the calibration procedure.

Algorithm	γ_S	Test MSE	Time
DNN		129.3600	10 m 17 s 420 ms
GQLDNN	0.0001	**114.95802**	4 h 52 s 02 ms

7   Conclusions

We can try other S matrices, in order to furtherly generalize the results. Another improvement could be to tune the MKL coefficients with an optimization procedure as suggested by [11]. In addition, Support Vector Regression and multi-class classification with a generalized quadratic loss could be investigated.

The results obtained with the Twitter data-set, a Neural Networks with 2, 3, 4 layers, and a generalized quadratic loss, are encouraging.

For the regression setting we tried a deeper neural network on a larger data-set (UCI Year) and we found that the generalized quadratic loss performed in a similar manner than standard sum of the errors loss' square.

We still have to realize if the performances of the GQL for the Shallow and Deep Neural Networks are due to the greater number of trials needed to tune the S matrix, or to a real effect on the generalization ability of the algorithm induced by the use of the GQL. We made some preliminary experiments to establish this, but it is too premature to make an assertion.

Acknowledgments. I would like to express my gratitude to Giovanna Zamara, Fabrizio Romano, Fabio Aiolli, Alessio Micheli, Ralf Herbrich, Alex Smola, Alessandro Sperduti for their insightful suggestions.

References

1. Vapnik, V.N.: Statistical Learning Theory. Wiley, New York (1998)
2. Platt, J.: Sequential Minimal Optimization: A Fast Algorithm for Training Support Vector Machines. Advances in Kernel Methods - Support Vector Learning (1998)
3. Portera, F., Sperduti, A.: A generalized quadratic loss for Support Vector Machines. In: ECAI 2004 Proceedings of the 16th European Conference on Artificial Intelligence, pp. 628–632 (2004)
4. Aiolli, F., Sperduti, A.: An efficient SMO-like algorithm for multiclass SVM. In: Proceedings of the 12th IEEE Workshop on Neural Networks for Signal Processing (2002)
5. Schölkopf, B., Herbrich, R., Smola, A.J.: A generalized representer theorem. In: Helmbold, D., Williamson, B. (eds.) COLT 2001. LNCS (LNAI), vol. 2111, pp. 416–426. Springer, Heidelberg (2001). https://doi.org/10.1007/3-540-44581-1_27
6. Joachims, T.: Learning to Classify Text Using Support Vector Machines (2002)
7. Sze, V., Chen, Y.-H., Yang, T., Emer, J.S.: Efficient processing of deep neural networks: a tutorial and survey. Proc. IEEE **105**(12) (2017)
8. Bertin-Mahieux, T., Ellis, D.P.W., Whitman, B., Lamere, P.: The Million Song Dataset. In: Proceedings of the 12th International Society for Music Information Retrieval Conference (ISMIR 2011) (2011)
9. Lauriola, I., Gallicchio, C., Aiolli, F.: Enhancing deep neural networks via multiple kernel learning. Pattern Recogn. **101**, 107194 (2020)
10. Qiu, S., Lane, T.: A framework for multiple kernel support vector regression and its applications to siRNA efficacy prediction. IEEE/ACM Trans. Comput. Biol. Bioinf. **6**(2), 190–199 (2009)
11. Lanckriet, G.R.G., Cristianini, N., Bartlett, P., El Ghaoui, L., Jordan, M.I.: Learning the kernel matrix with semidefinite programming. In: Proceedings of the 19th International Conference on Machine Learning (2002)
12. Courcoubetis, C., Weber, R.: Lagrangian Methods for Constrained Optimization. Wiley (2003). ISBN 0-470-85130-9
13. Rodriguez, J.D., Perez, A., Lozano, J.A.: Sensitivity analysis of k-Fold cross validation in prediction error estimation. IEEE Trans. Pattern Anal. Mach. Intell. **32**(3), 569–575 (2010)
14. Cortes, C., Vapnik, V.: Support-vector networks. Mach. Learn. **20**, 273–297 (1995)
15. Fan, R.-E., Chen, P.-H., Lin, C.-J.: Working set selection using second order information for training SVM. J. Mach. Learn. Res. **6**, 1889–1918 (2005)
16. Severyn, A., Moschitti, A.: Twitter sentiment analysis with deep convolutional neural networks. In: SIGIR 2015: Proceedings of the 38th International ACM SIGIR Conference on Research and Development in Information Retrieval, pp. 959–962 (2015)
17. Cliche, M.: BB_twtr at SemEval-2017 task 4: twitter sentiment analysis with CNNs and LSTMs. In: Proceedings of the 11th International Workshop on Semantic Evaluation (SemEval 2017), pp. 573–580 (2017)
18. Hernandez-Lobato, J.M., Adams, R.P.: Probabilistic backpropagation for scalable learning of bayesian neural networks. In: ICML 2015: Proceedings of the 32nd International Conference on International Conference on Machine Learning - Volume 37, pp. 1861–1869 (2015)
19. Lakshminarayanan, B., Pritzel, A., Blundell, C.: Simple and scalable predictive uncertainty estimation using deep ensembles. In: NIPS 2017: Proceedings of the 31st International Conference on Neural Information Processing Systems, pp. 6405–6416 (2017)

Machine Learning Application to Family Business Status Classification

Giorgio Gnecco$^{(\boxtimes)}$, Stefano Amato, Alessia Patuelli, and Nicola Lattanzi

AXES (Laboratory for the Analysis of CompleX Economic Systems),
IMT - School for Advanced Studies, Piazza S. Francesco 19, Lucca, Italy
{giorgio.gnecco,stefano.amato,alessia.patuelli,
nicola.lattanzi}@imtlucca.it

Abstract. According to a recent trend of research, there is a growing interest in applications of machine learning techniques to business analytics. In this work, both supervised and unsupervised machine learning techniques are applied to the analysis of a dataset made of both family and non-family firms. This is worth investigating, because the two kinds of firms typically differ in some aspects related to performance, which can be reflected in balance sheet data. First, binary classification techniques are applied to discriminate the two kinds of firms, by combining an unlabeled dataset with the labels provided by a survey. The most important features for performing such binary classification are identified. Then, clustering is applied to highlight why supervised learning can be effective in the previous task, by showing that most of the largest clusters found are quite unequally populated by the two classes.

Keywords: Applications of machine learning · Automatic classification of family business status · Supervised learning · Clustering

1 Introduction

Recently, there was a growing interest in the applications of machine learning to business analytics. As an example, supervised learning methods were applied to the prediction of the risk of bankruptcy for firms, based on features extracted from balance sheets (see, e.g., [1, Chapter 9] and the more recent work [2]), and to financial statement fraud detection (see, e.g., [3,4]).

More generally, according to another recent trend of research, there is a growing interdisciplinary interaction between machine learning and econometrics [5], whose techniques can be also used in business analytics. For instance, various machine learning techniques such as regression trees [6] and random forests [7] were recently adapted to causal inference tasks, e.g., for the investigation of the economic effects of different policies. An extension of the causal tree framework to irregular assignment mechanisms was developed in [8]. In [9–11], the optimal trade-off, in terms of the generalization error, between sample size and precision of supervision was investigated, considering various models for which the variance of the output noise can be controlled to some extent by the researcher, by

© Springer Nature Switzerland AG 2020
G. Nicosia et al. (Eds.): LOD 2020, LNCS 12565, pp. 25–36, 2020.
https://doi.org/10.1007/978-3-030-64583-0_3

varying the acquisition cost of each supervised example. The regression models investigated therein (respectively, classical linear regression with homoskedastic measurement errors, its extension to the case of heteroskedastic measurement errors, and fixed effects panel data models) are of common use in econometrics.

In this context, in the present work, which is about the application of machine learning techniques to business analytics, some results are reported about the possibility of automatically classifying firms according to their family business status (i.e., discriminating family firms from non-family firms), using features extracted from both the results of a proprietary survey conducted on a small sample of Italian firms and corresponding balance sheet data obtained from a commercial database. As family firms and non-family firms have typically different behaviors [12], the possibility arises that family firms could be automatically identified based on a combination of features expressing accounting information. This calls for the application of machine learning techniques to this problem. As information regarding family business status is rarely publicly available, an alternative data-driven methodology could be useful to identify the firm's family business status and, as a consequence, support decision making processes by different entities, such as policy makers and financial institutes. The identification of a firm as a family firm, indeed, has relevant practical implications:

- for policy makers, to evaluate the contributions of these peculiar organizations in terms of their impact on GDP, employment, innovations, and policy effectiveness, at both the national and regional level;
- for financial institutes, to assess creditworthiness and the probability of default associated with each firm;
- for financial regulators, to monitor the stability of the financial system given that family firms represent the most common type of business worldwide.

A similar study as ours was recently performed in [13] (according to which family business status classification represents a quite novel application of supervised machine learning), but starting from a different, and much larger, dataset of Finnish firms, whose features were - however - more directly connected with the family business status, thus making the classification task easier. Another difference is that our study refers to Italy, where family firms stand out as the backbone of the national manufacturing system.

The present study shows that successful automatic classification of the family business status can be still obtained starting from a smaller dataset, and with features less directly connected with that business status. This is relevant since the cost (time) of supervision increases with the number of supervised examples, being these obtained as the result of a survey, involving firm representatives as participants (who, in general, may not return it compiled, or do it at different times). The small size of the dataset is also justified by the fact that typically, national family firm databases are either missing or not publicly available [14], due also to the presence of various non-overlapping definitions of family firm in the literature[1] [15–19]. Finally, to shed some light into the results obtained by supervised learning, an unsupervised learning technique is also applied, studying the distribution of the firm population in the feature space.

The work is structured as follows. Section 2 briefly introduces the dataset available. Section 3 reports some pre-processing steps. Section 4 describes the machine learning analysis performed, and details its results, comparing them with [13]. Finally, Sect. 5 provides some conclusions and possible directions of further research.

2 Description of the Dataset

First, family business information was collected through a proprietary survey, conducted jointly by IMT - School for Advanced Studies, Lucca and the Studies and Researches Office of Banca Intesa Sanpaolo (the largest Italian banking group), for 584 Italian firms (response rate: 27%; non-active firms/firms in liquidation were not included in the survey)[2]. Each participant in the survey was a firm representative, who answered questions related only to that specific firm (so, no analysis of inter-rater agreement was needed). The answers to two of the questions posed to the participant allowed to establish if the firm was familiar/not familiar. More precisely, taking into account the family firm definition used in [19], family firms were labeled from the survey results as those for which both the following conditions hold[3]:

- the majority share of the social capital is held - either directly, indirectly, or in mixed way - by a reference business owner, or by several persons related by family links;
- in the governance board, there are at least 2 members coming from the reference family.

---

[1] It is worth noting that the distinction between family/non-family firms depends not only on their different ownership structures, but also on other factors. Moreover, sometimes it is not immediate to understand if owners belong to the same family only based on ownership data, for two reasons: they might have different surnames, even though they belong to the same family (this is particularly common when family firms go through several generational shifts); there might be corporate groups linked by shareholding, which may make it difficult to identify individual owners. Current definitions of family firm follow one of these two approaches: the demographic approach (combining family ownership, governance and/or management) and the essence approach (the behavioral perspective on the firm's nature). Demographic approaches combining ownership and management (or governance) are the most widely used [20].

[2] The complete dataset is made of 152 companies, and is comparable in size with other datasets used in family business research [14].

[3] Here, we used a mixed demographic approach. This definition is very close to the one proposed by the European Commission (https://ec.europa.eu/growth/smes/promoting-entrepreneurship/we-work-for/family-business_en). We raised the minimum number of family members involved in the firm's governance from 1 to 2, to ensure a clearer demarcation between sole-founder and family-owned and governed firms.

Consequently, non-family firms were labeled as the remaining ones.

Then, the dataset was integrated with 45 balance sheet features (whose values were collected in the years 2013, 2014, 2015, 2016, 2017), which were retrieved from the Italian AIDA[4] database, for a subset of 83 firms belonging to the jewellery sector (representative of "Made in Italy"), in three Italian districts (Arezzo, Valenza Po, Vicenza). Since all such firms appear in both the survey and in the AIDA database, the goal of data integration was to exploit their balance sheet data extracted from the latter to classify automatically their family business status, using the supervision coming from the participants in the survey. A first selection of features was made manually, focusing on the main budget items and ratios that appear in companies' balance sheets[5]. For each among these original features, and for each firm, the following two derived features ("twin features", in the following) were generated:

- average of the feature value over the 5-years period (possibly excluding, for each firm, years for which the feature value was not available: e.g., if the feature value was available only for 4 of the 5 years, then the average was computed only over those 4 years);
- average of the absolute value of its annual change[6] over the 5-years period (again, possibly excluding, for each firm, years for which that value was not available).

The first kind of derived feature was constructed in order to remove any time dependence in the original feature values, whereas the second kind was built in order to provide a measure of their changes with respect to time. A time-series (or panel data) approach [22] was not considered, given the small size of the current dataset.

Finally, all the features were standardized, making them of mean 0 and variance 1. This was done in order to give the same a-priori importance to each feature.

3   Data Pre-processing

Due to the limited amount of data available, an initial pre-processing of the data was performed, in order to make subsequent analysis feasible, reducing the risk of incurring overfitting in its application of supervised learning techniques.

---

[4] AIDA, which stands for "Analisi Informatizzata delle Aziende Italiane" (English translation: "Computerised Analysis of Italian Firms") is a commercial database managed by Bureau van Dijk, a Moody's Analytics company, which contains a comprehensive set of financial informations on companies in Italy.

[5] This way of pre-selecting a set of features (for further successive selection of its subset) can improve the overall performance of machine learning algorithms, as shown in [21] in a different framework.

[6] Initially, for each feature, also the average of its annual change was considered, but the difference in its means over the two classes never turned out to be statistically significant.

First, the dataset was restricted to the derived features, deemed to be more representative of the behavior of each firm with respect to the original (annual) ones.

To deal with missing data, all the firms for which the value of at least one of these features was not available were removed from the dataset. This simple procedure was adopted because, in the specific case, it reduced the size of the dataset only slightly, i.e., from 83 firms in the original dataset to 80 firms in the reduced one.

Then, all the remaining features with empirical zero variance in that dataset were eliminated (being useless for binary classification), likewise those for which no statistically significant difference was found in their mean values (or in the mean values of the corresponding twin feature) between the two classes (statistical test adopted: two-sample t-test for equal means [23]; significance level adopted: 5%).

Finally, since the resulting dataset was slightly unbalanced (35 family firms versus 45 non-family firms), it was reduced, via random sub-sampling of the set of 45 non-family firms, to a balanced dataset, made of 70 firms, containing a total of 18 derived features, which are reported in Table 1. For the reader's convenience (who may be not expert in accounting), the meaning of each of the original features corresponding to the derived features reported in the table is briefly described below (see also [24,25] for more details):

- inventory: the supply of stock or goods that a firm has for sale or for production of other goods or services;
- current assets: the assets used by a firm in its ordinary work, e.g., materials, finished goods, cash and monies due, which are held only for a short period of time. They represent amounts that are cash or will be converted to cash, or that will be used up during the next year, or during the operating cycle;
- equity: the value of a firm that is owned by its shareholders, computed as the value of the firms' total assets minus the value of its liabilities. It represents contributions from owners plus earnings, less any distributions to owners;
- long-term debt: obligations to pay for services or goods already received by the firm, that are not repaid within a year;
- personnel costs: wages, salaries, and employers' costs of social security;
- intangible assets: assets that have no physical substance but still represent valuable rights, such as patents or trademarks;
- fixed assets: properties or machineries owned and used by a firm, but that are neither bought or sold as part of its regular trade;
- finance charges: total costs of borrowing. They include interests and fees;
- workforce: total number of employees in a firm.

4   Application of Machine Learning Techniques

The supervised machine learning techniques adopted for the analysis of the dataset obtained after pre-processing were:

Table 1. List of the 18 derived features (averages and average absolute values of annual changes in original features) used in the analysis. For each original feature x (e.g., inventory), $|\Delta x|$ denotes the absolute value of its change between two consecutive years.

Features			
1)	Average inventory	2)	Average $|\Delta$ inventory$|$
3)	average current assets	4)	Average $|\Delta$ current assets$|$
5)	Average equity	6)	Average $|\Delta$ equity$|$
7)	Average long-term debt	8)	Average $|\Delta$ long-term debt$|$
9)	Average personnel costs	10)	Average $|\Delta$ personnel costs$|$
11)	Average intangible assets	12)	Average $|\Delta$ intangible assets$|$
13)	Average fixed assets	14)	Average $|\Delta$ fixed assets$|$
15)	Average finance charges	16)	Average $|\Delta$ finance charges$|$
17)	Average workforce	18)	Average $|\Delta$ workforce$|$

- Linear Discriminant Analysis (LDA);
- Support Vector Machines (SVMs);
- Classification Trees (CTs).

The first one is a classical statistical technique used to perform binary classification, whereas the last two are more modern machine learning ones. Focus was given on well-known techniques (see, e.g., [26] for a description of each such technique), since they are more accessible than more recent ones to experts in accounting, who could be interested in applying the same methodology of this article to other similar studies, especially in case they were characterized by a small sample size. Among others, SVMs were chosen for their relationship with LDA (they are still linear classifiers, but in a transformed feature space), whereas CT was chosen for its high interpretability[7].

The three techniques were implemented in MATLAB, using the respective M-functions `fitcdiscr.m`, `fitcsvm.m`, `fitctree.m`, with automatic optimization of the respective hyperparameters (and default options), based on 10-fold cross-validation. In order to further reduce the risk of incurring overfitting, and also increase the level of interpretability [33] of the corresponding trained classifier (making it more trustworthy to experts in accounting), each technique was applied to a smaller-dimensional dataset made of only 3 from the original 18 features. A total number of 50 randomly selected triplets of features was considered.

The largest estimated probabilities of correct classification obtained for each of the three techniques are reported in bold, respectively, in the second, third,

---

[7] Other architectures/machine learning algorithms (e.g., enhanced k-nearest neighbors [27], multilayer perceptrons [28], random forests [29], gradient boosting algorithms like XGBoost [30], LightGBM [31], or CatBoost [32]) could be considered for further developments.

Table 2. For each supervised learning technique (2nd, 3rd, and 4th column): best triplet of features, classification performance (estimated probability of correct classification) of the best classifier (in bold) found by using that technique, and classification performances of the other two techniques applied to the corresponding triplets of features. All these classification performances are evaluated through 10-fold cross-validation.

	Best selection (LDA)	Best selection (SVM)	Best selection (CT)
Feature 1	Avg. finance charges	Avg. finance charges	Avg. inventory
Feature 2	Avg. $\lvert\Delta$ long-term debt$\rvert$	Avg. $\lvert\Delta$ personnel costs$\rvert$	Avg. $\lvert\Delta$ current assets$\rvert$
Feature 3	Avg. $\lvert\Delta$ workforce$\rvert$	Avg. $\lvert\Delta$ long-term debt$\rvert$	Avg. $\lvert\Delta$ long-term debt$\rvert$
Estimated probability of correct classification (LDA)	**0.6572**	0.6288	0.6430
Estimated probability of correct classification (SVM)	0.6497	**0.6572**	0.6160
Estimated probability of correct classification (CT)	0.6429	0.6286	**0.7429**

and fourth columns of Table 2, together with the corresponding triplets of features. For each of the entries reported in bold, corresponding results for the other classifiers, referred to the same choices of triplets of features as that entry, are also provided in the remaining entries of the associated column. Moreover, SVM was better than the other two techniques in 24 of the 50 cases, CT in 22 cases, whereas LDA in only 4 cases. However, the best performance was obtained by CT, with an estimated probability of correct classification equal to 0.7429. The most important feature for the best CT model (the one reported in Table 2) was the average inventory, being family firms in the sample typically associated with a larger value of that feature than non-family firms. Finally, given the limited amount of data at our disposal, in order to get improved estimates of the probabilities of correct classification of the LDA, SVM and CT classifiers based on the best triplets of features reported in Table 2, such models were re-trained/cross-validated 50 times, using only those triplets of features, and starting from slightly different datasets, obtained by including 35 randomly extracted examples from the most numerous class (a different subset each time), and the 35 (fixed) examples from the least numerous class. The average 10-fold cross-validation estimated probabilities of correct classification over these 50 repetitions and their empirical standard deviations are reported in Table 3. To reduce possible bias in the estimates (which is expected to be small, given

Table 3. For the best triplets of features reported in Table 2, and for each supervised learning technique: empirical average of the estimated probability of correct classification plus/minus its empirical standard deviation, both evaluated according to the procedure described in the text.

	Best selection (LDA)	Best selection (SVM)	Best selection (CT)
Feature 1	Avg. finance charges	avg. finance charges	Avg. inventory
Feature 2	Avg. $|\Delta$ long-term debt$|$	Avg. $|\Delta$ personnel costs$|$	Avg. $|\Delta$ current assets$|$
Feature 3	Avg. $|\Delta$ workforce$|$	Avg. $|\Delta$ long-term debt$|$	Avg. $|\Delta$ long-term debt$|$
Avg. estimated probability of correct classif. ($\pm$ 1 std. deviation) (LDA)	0.6238 ($\pm$0.0140)	0.6227 ($\pm$0.0128)	0.6382 ($\pm$0.0191)
Avg. estimated probability of correct classif. ($\pm$ 1 std. deviation) (SVM)	0.6148 ($\pm$0.0245)	0.6146 ($\pm$0.0226)	0.6394 ($\pm$0.0324)
Avg. estimated probability of correct classif. ($\pm$ 1 std. deviation) (CT)	0.6196 ($\pm$0.0321)	0.6184 ($\pm$0.0295)	0.7262 ($\pm$0.0295)

that only 3 features were used by the classifiers), nested cross-validation [26] will be likely considered for future analyses.

In order to shed further light on the results obtained, a cluster analysis of the original (unlabeled) dataset of 70 firms was performed, using the k-means algorithm [26], implemented in MATLAB by the M-function `kmeans.m`. A maximum number of 10 clusters was chosen, motivated by the small size of the dataset.

Table 4 reports the size of each cluster, when k-means was applied to the restricted dataset made of the best triplet of features shown in Table 4, which corresponds to the application of CT. The (a-posteriori evaluated) internal composition of each cluster in terms of the numbers of family firms/non-family firms it contains is also reported. One can see from the table that the first, second, and fourth largest clusters are quite unequally populated by the two classes, whereas the 6 smallest clusters are populated only by one class. However, the third largest cluster is nearly equally populated by the two classes. This helps understanding why the classification performance by CT was not ideal, still CT performed much better than random guessing.

Table 4. Clustering results: size of each cluster, and its internal composition in terms of numbers of family/non-family firms inside it.

	Size	No. of family firms	No. of non-family firms
Cluster 1	1	1	0
Cluster 2	1	1	0
Cluster 3	1	1	0
Cluster 4	1	1	0
Cluster 5	1	0	1
Cluster 6	2	2	0
Cluster 7	8	6	2
Cluster 8	14	7	7
Cluster 9	15	6	9
Cluster 10	26	10	16

5   Conclusions and Possible Future Developments

Results confirm that it is feasible to use machine learning techniques to identify the family business status of a firm based on a set of features derived directly from its balance sheet data. This approach can complement the traditional way of identifying family firms based on ownership, governance, and management data. As a result of the analysis, specific features were identified as significant to classify the family business status of a firm, namely: average inventory; average absolute values of the yearly variations of current assets and of long-term debt.

The results of the analysis are in accordance with the literature about the comparison between family firms and non-family firms, according to which the two kinds of firms typically differ in some aspects related to performance, such as efficiency [34] and productivity [35]. They differ also in financing costs than comparable non-family firms [36]. These aspects can be reflected in balance sheet data [12], possibly allowing a classifier to discriminate the two kinds of firms, based on a set of balance-related features.

Compared with [13], the results of the analysis show actually a worse classification performance (0.7429 versus 0.8898 as estimated probability of correct classification), but this may be attributed to the different set of features (in the present work, derived from balance sheet data), which are less directly related to the family business status of a firm. Moreover, this is compensated by a much smaller size of the dataset, in terms of both the number of firms (70 versus 7153) and the number of features (18 versus 351)[8]. As noticed in the introduction, this is relevant since the cost (time) of supervision increases with the number of supervised examples, being these obtained as the result of a survey. Moreover,

---

[8] The techniques adopted in [13] are random forests, boosting trees, and artificial neural networks. These are justified by the much larger amount of features and training data available in the dataset considered therein.

for the specific application, it is difficult to have at disposal large datasets for which the family business status is available.

Possible future developments, aimed at increasing the estimated probability of correct classification of the best obtained classifier, involve:

- increasing the size of the dataset, by including in the analysis also data with one or more missing feature values. This could be achieved either by using classifiers able to work also in the presence of missing feature values [37], or by applying suitable imputation methods to deal with such values [38, Chapter 27]. Of course, another possibility to increase the size of the dataset would be to extend the survey to a larger number of firms, taking into account that, in the specific machine learning application, the bottleneck is in the sample size of the survey, not in its combination with the AIDA database, as the latter contains information about something like 1 million Italian firms;
- applying additional machine learning techniques such as logistic LASSO regression [39] for doing feature selection in the specific case of a binary classification problem, or exploiting concepts from cooperative game theory (e.g., the Shapley value [40]) to rank features according to their importance in such classification;
- dealing directly with unbalanced datasets, e.g., giving larger weights to examples coming from the least numerous class in the dataset, or including them more than once in the training set. By doing this, there would be no need to discard a subset of examples coming from the most numerous class in the dataset;
- exploiting semi-supervised learning techniques [41], motivated by the large availability of unsupervised examples coming from the AIDA database. In particular, one could extract clusters from the latter database, then investigate whether some of the cluster so found were more populated by family-firms or by non-family firms, based on the labeled examples coming from the results of the survey.

Acknowledgements. The first author is a member of the Gruppo Nazionale per l'Analisi Matematica, la Probabilità e le loro Applicazioni (GNAMPA) of the Istituto Nazionale di Alta Matematica (INdAM), Italy. We would like to thank Giovanni Foresti and Sara Giusti of Intesa Sanpaolo Research Unit (Direzione Studi e Ricerche Intesa Sanpaolo), as part of a joint research program with the IMT - School for Advanced Studies, Lucca.

References

1. Giudici, P., Figini, S.: Applied Data Mining for Business and Industry. Wiley, Hoboken (2009)
2. Alexandropoulos, S.-A.N., Aridas, C.K., Kotsiantis, S.B., Vrahatis, M.N.: A deep dense neural network for bankruptcy prediction. In: Macintyre, J., Iliadis, L., Maglogiannis, I., Jayne, C. (eds.) EANN 2019. CCIS, vol. 1000, pp. 435–444. Springer, Cham (2019). https://doi.org/10.1007/978-3-030-20257-6_37

3. Kirkos, E., Spathis, C., Manolopoulos, Y.: Data mining techniques for the detection of fraudulent financial statements using data mining. Expert Syst. Appl. **32**(4), 995–1003 (2007)
4. Perols, J.: Financial statement fraud detection: an analysis of statistical and machine learning algorithms. Audit. J. Pract. Theory **30**(2), 19–50 (2011)
5. Varian, H.R.: Big data: new tricks for econometrics. J. Econ. Perspect. **28**, 3–28 (2014)
6. Athey, S., Imbens, G.: Recursive partitioning for heterogeneous causal effects. Proc. Natl. Acad. Sci. **113**, 7353–7360 (2016)
7. Wager, S., Athey, S.: Estimation and inference of heterogeneous treatment effects using random forests. J. Am. Stat. Assoc. **113**, 1228–1242 (2018)
8. Bargagli Stoffi, F.J., Gnecco, G.: Causal tree with instrumental variable: an extension of the causal tree framework to irregular assignment mechanisms. Int. J. Data Sci. Anal. **9**(3), 315–337 (2019). https://doi.org/10.1007/s41060-019-00187-z
9. Gnecco, G., Nutarelli, F.: On the trade-off between number of examples and precision of supervision in regression. In: Proceedings of the 4th International Conference of the International Neural Network Society on Big Data and Deep Learning (INNS BDDL 2019), Sestri Levante, Italy, pp. 1–6 (2019)
10. Gnecco, G., Nutarelli, F.: On the trade-off between number of examples and precision of supervision in machine learning problems. Optim. Lett. **3**, 1–23 (2019). https://doi.org/10.1007/s11590-019-01486-x
11. Gnecco, G., Nutarelli, F.: Optimal trade-off between sample size and precision of supervision for the fixed effects panel data model. In: Nicosia, G., Pardalos, P., Umeton, R., Giuffrida, G., Sciacca, V. (eds.) LOD 2019. LNCS, vol. 11943, pp. 531–542. Springer, Cham (2019). https://doi.org/10.1007/978-3-030-37599-7_44
12. Soler, I.P., Gemar, G., Guerrero-Murillo, R.: Family and non-family business behaviour in the wine sector: a comparative study. Eur. J. Family Bus. **7**(1), 65–73 (2017)
13. Peltonen, J.: Can supervised machine learning be used to identify family firms using a sophisticated definition? Acad. Manag. Proc. **2018**(1) (2018). 6 pages. https://doi.org/10.5465/AMBPP.2018.154
14. Beck, L., Janssens, W., Debruyne, M., Lommelen, T.: A study of the relationships between generation, market orientation, and innovation in family firms. Family Bus. Rev. **24**(3), 252–272 (2011)
15. Litz, R.A.: The family business: toward definitional clarity. Family Bus. Rev. **8**(2), 71–81 (1995)
16. Chua, J.H., Chrisman, J.J., Sharma, P.: Defining the family business by behavior. Entrepr. Theory Pract. **23**(4), 19–39 (1999)
17. Astrachan, J.H., Klein, S.B., Smyrnios, K.X.: The F-PEC scale of family influence: a proposal for solving the family business definition problem. Family Bus. Rev. **15**(1), 45–58 (2002)
18. Corbetta, G., Salvato, C.: Strategies for Longevity in Family Firms: A European Perspective. Palgrave Macmillan, London (2012)
19. Baù, M., Chirico, F., Pittino, D., Backman, M., Klaesson, J.: Roots to grow: family firms and local embeddedness in rural and urban contexts. Entrepr. Theory Pract. **43**(2), 360–385 (2018)
20. Basco, R.: The family's effect on family firm performance: a model testing the demographic and essence approaches. J. Family Bus. Strat. **4**(2), 42–66 (2013)
21. Plonsky, O., Erev, I., Hazan, T., Tennenholtz, M.: Psychological forest: predicting human behavior. In: Proceedings of the 31st AAAI Conference on Artificial Intelligence (AAAI 2017), San Francisco, USA, pp. 656–662 (2017)

22. Greene, W.H.: Econometrics Analysis. Prentice Hall, Upper Saddle River (2003)
23. Snedecor, G.W., Cochran, W.G.: Statistical Methods. Iowa State University Press, Iowa (1989)
24. Collin, S.M.H.: Dictionary of Accounting. A & C Black Publishers, London (2007)
25. Mooney, K.: The Essential Accounting Dictionary. Sphinx Publishing (2008)
26. Hastie, T., Tibshirani, R., Friedman, J.: The elements of statistical learning: data mining, inference, and prediction. Springer (2008)
27. Nguyen, B.P., Tay, W.-L., Chui, C.-K.: Robust biometric recognition from palm depth images for gloved hands. IEEE Trans. Hum. Mach. Syst. **45**(6), 799–804 (2015)
28. Haykin, S.: Neural Networks: A Comprehensive Foundation. Prentice Hall, Upper Saddle River (1998)
29. Breiman, L.: Random forests. Mach. Learn. **45**(1), 5–32 (2001)
30. Chen, T., Guestrin, C.: XGBoost: a scalable tree boosting system. In: Proceedings of the 22nd ACM SIGKDD International Conference on Knowledge Discovery and Data Mining (KDD 2016), San Francisco, USA, pp. 785–794 (2016)
31. Ke, G., et al.: LightGBM: a highly efficient gradient boosting decision tree. In: Proceedings of the 31st Conference on Neural Information Processing Systems (NIPS 2017), Long Beach, USA, pp. 3149–3157 (2017)
32. Prokhorenkova, L., Gusev, G., Vorobev, A., Dorogush, A.V., Gulin, A.: CatBoost: unbiased boosting with categorical features. In: Proceedings of the 32nd Conference on Neural Information Processing Systems (NeurIPS 2018), Montréal, Canada, pp. 6638–6648 (2018)
33. Hansen, L.K., Rieger, L.: Interpretability in intelligent systems – a new concept? In: Samek, W., Montavon, G., Vedaldi, A., Hansen, L.K., Müller, K.-R. (eds.) Explainable AI: Interpreting, Explaining and Visualizing Deep Learning. LNCS (LNAI), vol. 11700, pp. 41–49. Springer, Cham (2019). https://doi.org/10.1007/978-3-030-28954-6_3
34. McConaughy, D.L., Walker, M.C., Henderson Jr., G.V., Mishra, C.S.: Founding family controlled firms: efficiency and value. Rev. Financ. Econ. **7**(1), 1–19 (1998)
35. Martikainen, M., Nikkinen, J., Vähämaa, S.: Production functions and productivity of family firms: evidence from the S&P 500. Q. Rev. Econ. Finance **49**(2), 295–307 (2009)
36. Anderson, R.C., Mansi, S.A., Reeb, D.M.: Founding-family ownership and the agency cost of debt. J. Financ. Econ. **68**(2), 263–287 (2003)
37. Basuchoudhary, A., Bang, J.T., Sen, T.: Machine-Learning Techniques in Economics. SE. Springer, Cham (2017). https://doi.org/10.1007/978-3-319-69014-8
38. Cameron, A.C., Trivedi, P.K.: Microeconometrics: Methods and Applications. Cambridge University Press, Cambridge (2005)
39. Friedman, J., Hastie, T., Tibshirani, R.: Regularization paths for generalized linear models via coordinate descent. J. Stat. Softw. **33**(1), 1–22 (2010)
40. Choen, S., Ruppin, E., Dror, G.: Feature selection based on the Shapley value. In: Proceedings of the 19th International Joint Conference on Artificial intelligence (IJCAI 2005), Edinburgh, Scotland, pp. 665–670 (2005)
41. Chapelle, O., Schölkopf, B., Zien, A.: Semi-Supervised Learning. MIT Press, Cambridge (2006)

Using Hessians as a Regularization Technique

Adel Rahimi$^{(\boxtimes)}$, Tetiana Kodliuk, and Othman Benchekroun

Dathena Science Pte. Ltd., #07-02, 1 George St., Singapore 049145, Singapore
{adel.rahimi,tania.kodliuk,othman.benchekroun}@dathena.io
http://www.dathena.io

Abstract. In this paper we present a novel, yet simple, method to regularize the optimization of neural networks using second order derivatives. In the proposed method, we calculate the Hessians of the last n layers of a neural network, then re-initialize the top k percent using the absolute value. This method has shown an increase in our efficiency to reach a better loss function minimum. The results show that this method offers a significant improvement over the baseline and helps the optimizer converge faster.

1 Introduction

Neural Networks are heavily dependent on derivatives as a means of optimization given that they are differentiable end to end. For example, Gradient Descent [1] and its variants [2–4] minimize the network's loss function, $J(\theta)$, through its first-order partial derivatives. These derivatives, which are stored in the Jacobian matrix, represent the rates of the change in $J(\theta)$ with respect to each of the network's parameters. Still, gradient descent can have some difficulty optimizing these parameters only through the loss function's first-order derivatives as some of them can be far from their optimum.

Avoiding such issues can be done by computing the partial derivatives of the Jacobian matrix. These derivatives give us an understanding of the rate at which the gradient itself is changing, which is really useful when dealing with multiple parameters, especially when updating them is expensive. This gradient of gradients, i.e. $J(\theta)$'s second-order derivatives, is stored in the Hessian matrix as the gradient of the gradients. The Hessian is formulated in Eq. 1.

$$\mathbf{H}(f(x)) = \mathbf{J}(\nabla f(x))^T \tag{1}$$

As explained above, the Hessian matrix is based on the Jacobian's partial derivatives with respect to each network parameter. This calculation is shown in Eq. 2.

$$\mathbf{H}(f(x)) = \frac{\partial \mathbf{J}}{\partial x_1}, \cdots, \frac{\partial \mathbf{J}}{\partial x_n} \tag{2}$$

© Springer Nature Switzerland AG 2020
G. Nicosia et al. (Eds.): LOD 2020, LNCS 12565, pp. 37–40, 2020.
https://doi.org/10.1007/978-3-030-64583-0_4

As can be seen in Eq. 2, the dimensions of the Hessian matrix can be quite large, requiring $\theta(n^2)$ memory to store it. As most of the modern neural network architectures have millions of parameters, storing all of this data can be tremendously hard, thus limiting the use of second-order derivatives when training neural networks.

2 Proposed Method

To solve the limitations introduced in the first part of this paper, we propose a simple method to find and re-initialize weights that are moving too fast to their optimal point. While some alternative methods only compute estimates as a way of overcoming the computation and storage challenges posed by Hessians, they are not accurate as they are entirely based on value approximations.

Instead, we propose calculating the Hessians solely for the last n layers in the neural network as these last layers have an immediate effect on the task at hand. Large hessian values for the n layers are the indicators of an extreme slope, and therefore of high fluctuations in the gradient. The parameters corresponding to these large values are far from their optimal weight, and subsequently will not have any impact in the network.

The training process with the proposed method is as follows:

```
for Epoch in Training:
    Forward pass
    Calculate loss
    Update weights using optimizer
    for n last layers:
        Calculate Hessian matrix
        Get absolute value of each element
        Sum/Average alongside dimensions
        Get the top Hessian magnitudes (absolute values)
        Re-initialize weights
```

Weight re-initialization can be defined by users, for instance by setting all weights to 0, by using the random uniform distribution, or by using the Xavier initialization [5]. Moreover, we can either re-initialize a whole unit—corresponding to a column of weights—or a single weight.

The experiments ran on the test data allows us to conclude that summing and averaging the hessian values does not make any difference as use absolute values; therefore, either aggregation method can be used depending on the network's architecture and the task at hand.

3   Experiment

To evaluate the proposed method, we construct a classifier on the MNIST dataset [6] with 4 hidden layers, each with a ReLU activation function. The number of hidden units per layer are respectively: 1000, 1000, 500, and 200. We then follow the algorithm presented above to get the Hessian matrix (which was averaged on the additional dimensions) and re-initialize the top k percent of the weights. To analyze the effects of k on the results, we tried different values on our data— namely $k = 10, 20, 30$.

4   Results

Experiments have shown that the proposed method significantly improves the convergence time and overall accuracy in comparison to normal training. Using a small k, $k = 10$ in our case, will bring less divergence between the normal training and the proposed method. Figure 1 shows the accuracy of our method in comparison to the baseline model, which was trained without any regularization. Our model reached a higher accuracy and has a bigger area below the curve, meaning our model reaches a higher accuracy in less epochs. Setting a larger k—with $k = 20$ for example—will divert the network as shown in Fig. 2 (left plot). The difference between the plots is smaller than the previous results, i.e. for $k = 10$. Finally, using a very high k ($k = 30$) will make the training process unstable, and the network will likely perform even worse, as shown in the Fig. 2 (right plot).

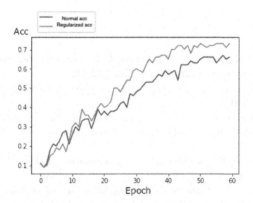

Fig. 1. Comparison of accuracy between a normal training and our proposed method with k value of 10.

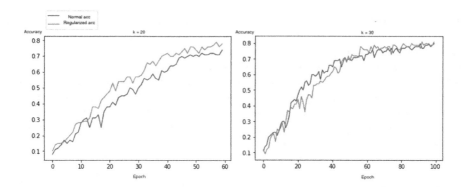

Fig. 2. Comparison of accuracy between a normal training and our proposed method with different k values (for $k = 20$ and $k = 30$).

5   Conclusion

Using the proposed method, calculating the Hessians of the last n layers of the network to re-initialize the top k percent, we can speed up the convergence of our training and tune the network's parameters that are hard to optimize. This method is more trustworthy than the alternative methods presented as it uses exact Hessians instead of simple estimations. Moreover, using only the last n layers in the network makes it less costly than conventional methods that use Hessians for the whole network. In the end, this method can be used as a simplified robust method for Neural Network regularization.

References

1. Rumelhart, D.E., Hinton, G.E., Williams, R.J.: Learning representations by back-propagating errors. Nature **323**(6088), 533–536 (1986)
2. Kingma, D.P., Ba, J.: Adam: a method for stochastic optimization. arXiv preprint arXiv:1412.6980 (2014)
3. Tieleman, T., Hinton, G.: Lecture 6.5-rmsprop: Divide the gradient by a running average of its recent magnitude. COURSERA Neural Netw. Mach. Learn. **4**(2), 26–31 (2012)
4. Robbins, H., Monro, S.: A stochastic approximation method. Annal. Math. Stat. **22**, 400–407 (1951)
5. Glorot, X., Bengio, Y.: Understanding the difficulty of training deep feedforward neural networks. In: Proceedings of the Thirteenth International Conference on Artificial Intelligence and Statistics, pp. 249–256 (2010)
6. LeCun, Y., Cortes, C.: MNIST handwritten digit database (2010)

Scaling Up Quasi-newton Algorithms: Communication Efficient Distributed SR1

Majid Jahani, Mohammadreza Nazari, Sergey Rusakov, Albert S. Berahas[✉],
and Martin Takáč

Lehigh University, Bethlehem, PA 18015, USA
albertberahas@gmail.com

Abstract. In this paper, we present a scalable distributed implementation of the Sampled Limited-memory Symmetric Rank-1 (S-LSR1) algorithm. First, we show that a naive distributed implementation of S-LSR1 requires multiple rounds of expensive communications at every iteration and thus is inefficient. We then propose DS-LSR1, a communication-efficient variant that: (*i*) drastically reduces the amount of data communicated at every iteration, (*ii*) has favorable work-load balancing across nodes, and (*iii*) is matrix-free and inverse-free. The proposed method scales well in terms of both the dimension of the problem and the number of data points. Finally, we illustrate the empirical performance of DS-LSR1 on a standard neural network training task.

Keywords: SR1 · Distributed optimization · Deep learning

1 Introduction

In the last decades, significant efforts have been devoted to the development of optimization algorithms for machine learning. Currently, due to its fast learning properties, low per-iteration cost, and ease of implementation, the stochastic gradient (SG) method [8,32], and its adaptive [19,25], variance-reduced [18,22,34] and distributed [17,31,39,44] variants are the preferred optimization methods for large-scale machine learning applications. Nevertheless, these methods have several drawbacks; they are highly sensitive to the choice of hyper-parameters and are cumbersome to tune, and they suffer from ill-conditioning [2,9,42]. More importantly, these methods offer a limited amount of benefit in distributed computing environments since they are usually implemented with small mini-batches, and thus spend more time communicating instead of performing "actual" computations. This shortcoming can be remedied to some extent by increasing the batch sizes, however, there is a point after which the increase in computation is not offset by the faster convergence [38].

Recently, there has been an increased interest in (stochastic) second-order and quasi-Newton methods by the machine learning community; see e.g., [4–6,10,11,16,21,23,29,33,35,41]. These methods judiciously incorporate curvature

© Springer Nature Switzerland AG 2020
G. Nicosia et al. (Eds.): LOD 2020, LNCS 12565, pp. 41–54, 2020.
https://doi.org/10.1007/978-3-030-64583-0_5

information, and thus mitigate some of the issues that plague first-order methods. Another benefit of these methods is that they are usually implemented with larger batches, and thus better balance the communication and computation costs. Of course, this does not come for free; (stochastic) second-order and quasi-Newton methods are more memory intensive and more expensive (per iteration) than first-order methods. This naturally calls for distributed implementations.

In this paper, we propose an efficient distributed variant of the Sampled Limited-memory Symmetric Rank-1 (S-LSR1) method [3]—DS-LSR1—that operates in the master-worker framework (Fig. 1). Each worker node has a portion of the dataset, and performs local computations using solely that information and information received from the master node. The proposed method is matrix-free (Hessian approximation never explicitly constructed) and inverse-free (no matrix inversion). To this end, we leverage the compact form of the SR1 Hessian approximations [12], and utilize sketching techniques [40] to approximate

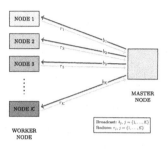

Fig. 1. Distributed computing schematic.

several required quantities. We show that, contrary to a naive distributed implementation of S-LSR1, the method is communication-efficient and has favorable work-load balancing across nodes. Specifically, the naive implementation requires $\mathcal{O}(md)$ communication, whereas our approach only requires $\mathcal{O}(m^2)$ communication, where d is the dimension of the problem, m is the LSR1 memory and $m \ll d^1$. Furthermore, in our approach the heavy computations are done by the worker nodes and the master node performs only simple aggregations, whereas in the naive approach the computationally intensive operations, e.g., Hessian-vector products, are computed locally by the master node. Finally, we show empirically that DS-LSR1 has good strong and weak scaling properties, and illustrate the performance of the method on a standard neural network training task.

Problem Formulation and Notation. We focus on machine learning empirical risk minimization problems that can be expressed as:

$$\min_{w \in \mathbb{R}^d} F(w) := \frac{1}{n} \sum_{i=1}^{n} f(w; x^i, y^i) = \frac{1}{n} \sum_{i=1}^{n} f_i(w), \qquad (1.1)$$

where $f : \mathbb{R}^d \to \mathbb{R}$ is the composition of a prediction function (parametrized by w) and a loss function, and $(x^i, y^i)_{i=1}^{n}$ denote the training examples (samples). Specifically, we focus on deep neural network training tasks where the function F is nonconvex, and the dimension d and number of samples n are large.

The paper is organized as follows. We conclude this section with a discussion of related work. We describe the classical (L)SR1 and sampled LSR1 (S-LSR1)

---

[1] Note, these costs are on top of the communications that are common to both approaches.

methods in Sect. 2. In Sect. 3, we present DS-LSR1, our proposed distributed variant of the sampled LSR1 method. We illustrate the scaling properties of DS-LSR1 and the empirical performance of the method on deep learning tasks in Sect. 4. Finally, in Sect. 5 we provide some final remarks.

Related Work. The Symmetric Rank-1 (SR1) method [15,24] and its limited-memory variant (LSR1) [28] are quasi-Newton methods that have gained significant attention by the machine learning community in recent years [3,20]. These methods incorporate curvature (second-order) information using only gradient (first-order) information. Contrary to arguably the most popular quasi-Newton method, (L)BFGS [27,30], the (L)SR1 method does not enforce that the Hessian approximations are positive definite, and is usually implemented with a trust-region [30]. This has several benefits: (1) the method is able to exploit negative curvature, and (2) the method is able to efficiently escape saddle points.

There has been a significant volume of research on distributed algorithms for machine learning; specifically, distributed gradient methods [7,14,31,39,44], distributed Newton methods [21,36,43] and distributed quasi-Newton methods [1,13,17]. Possibly the closest work to ours is VF-BFGS [13], in which the authors propose a vector-free implementation of the classical LBFGS method. We leverage several of the techniques proposed in [13], however, what differentiates our work is that we focus on the S-LSR1 method. Developing an efficient distributed implementation of the S-LSR1 method is not as straight-forward as LBFGS for several reasons: (1) the construction and acceptance of the curvature pairs, (2) the trust-region subproblem, and (3) the step acceptance procedure.

2   Sampled Limited-Memory SR1 (S-LSR1)

In this section, we review the sampled LSR1 (S-LSR1) method [3], and discuss the components that can be distributed. We begin by describing the classical (L)SR1 method as this will set the stage for the presentation of the S-LSR1 method. At the kth iteration, the SR1 method computes a new iterate via

$$w_{k+1} = w_k + p_k,$$

where p_k is the minimizer of the following subproblem

$$\min_{\|p\| \le \Delta_k} m_k(p) = F(w_k) + \nabla F(w_k)^T p + \tfrac{1}{2} p^T B_k p, \qquad (2.1)$$

Δ_k is the trust region radius, B_k is the SR1 Hessian approximation

$$B_{k+1} = B_k + \frac{(y_k - B_k s_k)(y_k - B_k s_k)^T}{(y_k - B_k s_k)^T s_k}, \qquad (2.2)$$

and $(s_k, y_k) = (w_k - w_{k-1}, \nabla F(w_k) - \nabla F(w_{k-1}))$ are the curvature pairs. In the limited memory version, the matrix B_k is defined as the result of applying m SR1 updates to a multiple of the identity matrix using the set of m most recent curvature pairs $\{s_i, y_i\}_{i=k-m}^{k-1}$ kept in storage.

The main idea of the S-LSR1 method is to use the SR1 updating formula, but to construct the Hessian approximations using sampled curvature pairs instead of pairs that are constructed as the optimization progresses. At every iteration, m curvature pairs are constructed via random sampling around the current iterate; see Algorithm 2. The S-LSR1 method is outlined in Algorithm 1. The components of the algorithms that can be distributed are highlighted in *magenta*.

Several components of the above algorithms can be distributed. Before we present the distributed implementations of the S-LSR1 method, we discuss several key elements of the method: (1) Hessian-vector products; (2) curvature pair construction; (3) curvature pair acceptance; (4) search direction computation; (5) step acceptance procedure; and (6) initial Hessian approximations.

For the remainder of the paper, let $S_k = [s_{k,1}, s_{k,2}, \ldots, s_{k,m}] \in \mathbb{R}^{d \times m}$ and $Y_k = [y_{k,1}, y_{k,2}, \ldots, y_{k,m}] \in \mathbb{R}^{d \times m}$ denote the curvature pairs constructed at the kth iteration, $S_k^i \in \mathbb{R}^{d \times m}$ and $Y_k^i \in \mathbb{R}^{d \times m}$ denote the curvature pairs constructed at the kth iteration by the ith node, and $B_k^{(0)} = \gamma_k I \in \mathbb{R}^{d \times d}$, $\gamma_k \geq 0$, denote the initial Hessian approximation at the kth iteration.

Algorithm 1 . Sampled LSR1 (S-LSR1)

Input: w_0 (initial iterate), Δ_0 (initial trust region radius).
1: **for** $k = 0, 1, 2, \ldots$ **do**
2:     Compute $F(w_k)$ and $\nabla F(w_k)$
3:     Compute (S_k, Y_k) (Algorithm 2)
4:     Compute p_k (solve subproblem (2.1))
5:     Compute $\rho_k = \frac{F(w_k) - F(w_k + p_k)}{m_k(0) - m_k(p_k)}$
6:     **if** $\rho_k \geq \eta_1$ **then** Set $w_{k+1} = w_k + p_k$
7:     **else** Set $w_{k+1} = w_k$
8:     $\Delta_{k+1} = \texttt{adjustTR}(\Delta_k, \rho_k)$ [3, Appendix B.3]
9: **end for**

Algorithm 2. Construct new (S_k, Y_k) curvature pairs

Input: w_k (current iterate), m (memory), r (sampling radius), $S_k = [\ ]$, $Y_k = [\ ]$ (curvature pair containers).
1: **for** $i = 1, 2, \ldots, m$ **do**
2:     Sample a random direction σ_i
3:     Sample point $\bar{w}_i = w_k + r\sigma_i$
4:     Set $s_i = w_k - \bar{w}_i$ and
        $y_i = \nabla^2 F(w_k) s_i$
5:     Set $S_k = [S_k \ s_i]$ and $Y_k = [Y_k \ y_i]$
6: **end for**
Output: S, Y

Hessian-Vector Products. Several components of the algorithms above require the calculation of Hessian vector products of the form $B_k v$. In the large-scale setting, it is not memory-efficient, or even possible for some applications, to explicitly compute and store the $d \times d$ Hessian approximation matrix B_k. Instead, one can exploit the compact representation of the SR1 matrices [12] and compute:

$$B_{k+1} v = B_k^{(0)} v + (Y_k - B_k^{(0)} S_k) \underbrace{(D_k + L_k + L_k^T - S_k^T B_k^{(0)} S_k)^{-1}}_{M_k} (Y_k - B_k^{(0)} S_k)^T v,$$

$$D_k = diag[s_{k,1}^T y_{k,1}, \ldots, s_{k,m}^T y_{k,m}], \quad (L_k)_{j,l} = \begin{cases} s_{k,j-1}^T y_{k,l-1} & \text{if } j > l, \\ 0 & \text{otherwise.} \end{cases} \quad (2.3)$$

Computing $B_{k+1} v$ via (2.3) is both memory and computationally efficient; the complexity of computing $B_{k+1} v$ is $\mathcal{O}(m^2 d)$ [12].

Curvature Pair Construction. For ease of exposition, we presented the curvature pair construction routine (Algorithm 2) as a sequential process. However, this need not be the case; all pairs can be constructed simultaneously. First, generate a random matrix $S_k \in \mathbb{R}^{d \times m}$, and then compute $Y_k = \nabla^2 F(w_k) S_k \in \mathbb{R}^{d \times m}$. We discuss a distributed implementation of this routine in the following sections.

Curvature Pair Acceptance. In order for the S-LSR1 Hessian update (2.2) to be well defined, and for numerical stability, we require certain conditions on the curvature pairs employed; see [30, Chapter 6]. Namely, for a given $\eta > 0$, we impose that the Hessian approximation B_{k+1} is only updated using the curvature pairs that satisfy the following condition:

$$|s_{k,j}^T (y_{k,i} - B_k^{(j-1)} s_{k,j})| \geq \eta \|s_{k,j}\| \|y_{k,i} - B_k^{(j-1)} s_{k,j}\|, \qquad (2.4)$$

for $j = 1, \ldots, m$, where $B_k^{(0)}$ is the initial Hessian approximation and $B_k^{(j-1)}$, for $j = 2, \ldots, m$, is the Hessian approximation constructed using only curvature pairs $\{s_l, y_l\}$, for $l < j$, that satisfy (2.4). Note, $B_{k+1} = B_k^{(m)}$. Thus, potentially, not all curvature pairs returned by Algorithm 2 are used to update the S-LSR1 Hessian approximation. Checking this condition is not trivial and requires m Hessian vector products. In [3, Appendix B.5], the authors propose a recursive memory-efficient mechanism to check and retain only the pairs that satisfy (2.4).

Search Direction Computation. The search direction p_k is computed by solving subproblem (2.1) using CG-Steihaug; see [30, Chapter 7]. This procedure requires the computation of Hessian vectors products of the form (2.3).

Step Acceptance Procedure. In order to determine if a step is successful (Line 6, Algorithm 1) one has to compute the function value at the trial iterate and the predicted model reduction. This entails a function evaluation and a Hessian vector product. The acceptance ratio ρ_k determines if a step is successful, after which the trust region radius has to be adjusted accordingly. For brevity we omit the details from the paper and refer the interested reader to [3, Appendix B.3].

Initial Hessian Approximations $B_k^{(0)}$. In practice, it is not clear how to choose the initial Hessian approximation. We argue, that in the context of S-LSR1, a good choice is $B_k^{(0)} = 0$. In Fig. 2 we show the eigenvalues of the true Hessian and the eigenvalues of the S-LSR1 matrices for different values of γ_k ($B_k^{(0)} = \gamma_k I$) for a toy problem. As is clear, the eigenvalues of the S-LSR1 matrices with $\gamma_k = 0$ better match the eigenvalues of the true Hessian. Moreover, by setting $\gamma_k = 0$, the rank of the approximation is at most m and thus the CG algorithm (used to compute the search direction) terminates in at most m iterations, whereas the CG algorithm may require as many as $d \gg m$ iterations when $\gamma_k \neq 0$. Finally, $B_k^{(0)} = 0$ removes a hyper-parameter. Henceforth, we assume that $B_k^{(0)} = 0$, however, we note that our method can be extended to $B_k^{(0)} \neq 0$.

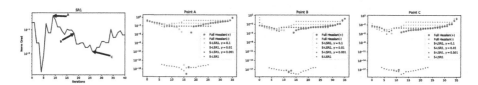

Fig. 2. Comparison of the eigenvalues of S-LSR1 for different γ (@ A, B, C) for a toy classification problem.

2.1 Naive Distributed Implementation of S-LSR1

In this section, we describe a naive distributed implementation of the S-LSR1 method, where the data is stored across $\mathcal{K}$ machines. At each iteration k, we broadcast the current iterate w_k to every worker node. The worker nodes calculate the local gradient, and construct local curvature pairs S_k^i and Y_k^i. The local information is then reduced to the master node to form $\nabla F(w_k)$, S_k and Y_k. Next, the SR1 curvature pair condition (2.4) is recursively checked on the master node. Given a set of accepted curvature pairs, the master node computes the search direction p_k. We should note that the last two steps could potentially be done in a distributed manner at the cost of $m + 1$ extra expensive rounds of communication. Finally, given a search direction the trial iterate is broadcast to the worker nodes where the local objective function is computed and reduced to the master node, and a step is taken.

As is clear, in this distributed implementation of the S-LSR1 method, the amount of information communicated is large, and the amount of computation performed on the master node is significantly larger than that on the worker nodes. Note, all the Hessian vector products, as well as the computations of the M_k^{-1} are performed on the master node. The precise communication and computation details are summarized in Tables 2 and 3.

3   Efficient Distributed S-LSR1 (DS-LSR1)

The naive distributed implementation of S-LSR1 has several significant deficiencies. We propose a distributed variant of the S-LSR1 method that alleviates these issues, is communication-efficient, has favorable work-load balancing across nodes and is inverse-free and matrix-free. To do this, we leverage the form of the compact representation of the S-LSR1 updating formula ($B_k^{(0)} = 0$)

$$B_{k+1}v = Y_k M_k^{-1} Y_k^T v, \tag{3.1}$$

and the form of the SR1 condition (2.4). We observe the following: one need not communicate the full S_k and Y_k matrices, rather one can communicate $S_k^T Y_k$, $S_k^T S_k$ and $Y_k^T Y_k$. We now discuss the means by which we: (1) reduce the amount of information communicated and (2) balance the computation across nodes.

3.1 Reducing the Amount of Information Communicated

As mentioned above, communicating curvature pairs is not necessary; instead one can just communicate inner products of the pairs, reducing the amount of communication from $2md$ to $3m^2$. In this section, we show how this can be achieved, and in fact show that this can be further reduced to m^2.

Construction of $S_k^T S_k$ and $S_k^T Y_k$. Since the curvature pairs are scale invariant [3], S_k can be any random matrix. Therefore, each worker node can construct this matrix by simply sharing random seeds. In fact, the matrix $S_k^T S_k$ need not be communicated to the master node as the master node can construct and store this matrix. With regards to the $S_k^T Y_k$, each worker node can construct local versions of the Y_k curvature pair, Y_k^i, and send $S_k^T Y_k^i$ to the master node for aggregation, i.e., $S_k^T Y_k = 1/\kappa \sum_{i=1}^{\mathcal{K}} S_k^T Y_k^i$. Thus, the amount of information communicated to the master node is m^2.

Construction of $Y_k^T Y_k$. Constructing the matrix $Y_k^T Y_k$ in distributed fashion, without communicating local Y_k^i matrices, is not that simple. In our communication-efficient method, we propose that the matrix is approximated via sketching [40], using quantities that are already computed, i.e., $Y_k^T Y_k \approx Y_k^T S_k S_k^T Y_k$. In order for the sketc.h to be well defined, $S_k \sim \mathcal{N}(0, I/m)$, thus satisfying the conditions of sketching matrices [40]. By using this technique, we construct an approximation to $Y_k^T Y_k$ with no additional communication. Note, the sketch size in our setting is equal to the memory size m. We should also note that this approximation is only used in checking the SR1 condition (2.4), which is not sensitive to approximation errors, and not in the Hessian vector products.

3.2 Balancing the Computation Across the Nodes

Balancing the computation across the nodes does not come for free. We propose the use of a few more rounds of communication. The key idea is to exploit the compact representation of the SR1 matrices and perform as much computation as possible on the worker nodes.

Computing Hessian Vector Products $B_{k+1}v$. The Hessian vector products (3.1), require products between the matrices Y_k, M_k^{-1} and a vector v. Suppose that the we have M_k^{-1} on the master node, and that the master node broadcasts this information as well as the vector v to the worker nodes. The worker nodes then locally compute $M_k^{-1}(Y_k^i)^T v$, and send this information back to the master node. The master node then reduces this to form $M_k^{-1}(Y_k)^T v$, and broadcasts this vector back to the worker nodes. This time the worker nodes compute $Y_k^i M_k^{-1}(Y_k)^T v$ locally, and then this quantity is reduced by the master node; the cost of this communication is d. Namely, in order to compute Hessian vector products, the master node performs two aggregations, the bulk of the computation is done on the worker nodes and the communication cost is $m^2 + 2m + 2d$.

Checking the SR1 Condition 2.4. As proposed in [3], at every iteration condition (2.4) is checked recursively by the master node. For each pair in memory, checking this condition amounts to a Hessian vector product as well as the use of inner products of the curvature pairs. Moreover, it requires the computation of $(M_k^{(j)})^{-1} \in \mathbb{R}^{j \times j}$, for $j = 1, \ldots, m$, where $M_k^{-1} = (M_k^{(m)})^{-1}$.

Inverse-Free Computation of M_k^{-1}. The matrix M_k^{-1} is non-singular [12], depends solely on inner products of the curvature pairs, and is used in the the computation of Hessian vector products (3.1). This matrix is constructed recursively (its dimension grows with the memory) by the master node as condition (2.4) is checked. We propose an inverse-free approach for constructing this matrix. Suppose we have the matrix $(M_k^{(j)})^{-1}$, for some $j = 1, \ldots, m-1$, and that the new curvature pair $(s_{k,j+1}, y_{k,j+1})$ satisfies (2.4). One can show that

$$(M_k^{(j+1)})^{-1} = \begin{bmatrix} (M_k^{(j)})^{-1} + \zeta(M_k^{(j)})^{-1} u v^T (M_k^{(j)})^{-1} & -\zeta(M_k^{(j)})^{-1} u \\ -\zeta v^T (M_k^{(j)})^{-1} & \zeta \end{bmatrix}$$

where $\zeta = 1/c - v^T(M_k^{(j)})^{-1}u$, $v^T = s_{k,j+1}^T Y_{k,1:l}$ and $Y_{k,1:l} = [y_{k,1}, \ldots, y_{k,l}]$ for $l \leq j$, $u = v$, and $c = s_{k,j+1}^T y_{k,j+1}$. We should note that the matrix $(M_k^{(1)})^{-1}$ is a singleton. Consequently, constructing $(M_k^{(j)})^{-1}$ in an inverse-free manner allows us to compute Hessian vector products and check condition (2.4) efficiently.

3.3   The Distributed S-LSR1 (DS-LSR1) Algorithm

Pseudo-code for our proposed distributed variant of the S-LSR1 method and the curvature pair sampling procedure are given in Algorithms 3 and 4, respectively. Right arrows denote broadcast steps and left arrows denote reduce steps. For brevity we omit the details of the distributed CG-Steihaug algorithm (Line 5, Algorithm 3), but note that it is a straightforward adaptation of [30, Algorithm 7.2] using quantities described above computed in distributed fashion.

Algorithm 3. Distributed Sampled LSR1 (DS-LSR1)

Input: w_0 (initial iterate), Δ_0 (initial trust region radius), m (memory).
Master Node: **Worker Nodes** $(i = 1, 2, \ldots, \mathcal{K})$:

1: **for** $k = 0, 1, 2, \ldots$ **do**
2:    ***Broadcast:*** w_k $\longrightarrow$ Compute $F_i(w_k)$, $\nabla F_i(w_k)$
3:    ***Reduce:*** $F_i(w_k)$, $\nabla F_i(w_k)$ to $F(w_k)$, $\nabla F(w_k)$ $\longleftarrow$
4:    Compute new (M_k^{-1}, Y_k, S_k) pairs via Algorithm 4
5:    Compute p_k via CG-Steihaug [30, Algorithm 7.2]
6:    ***Broadcast:*** p_k, M_k^{-1} $\longrightarrow$ Compute $M_k^{-1}(Y_k^i)^T p_k$, $\nabla F_i(w_k)^T p_k$, $F_i(w_k + p_k)$
7:    ***Reduce:*** $M_k^{-1}(Y_k^i)^T p_k$, $\nabla F_i(w_k)^T p_k$, $F_i(w_k + p_k)$ to $M_k^{-1}Y_k^T p_k$, $\nabla F(w_k)^T p_k$,
      $F(w_k + p_k)$ $\longleftarrow$
8:    ***Broadcast:*** $M_k^{-1}Y_k^T p_k$ $\longrightarrow$ Compute $(Y_k^i)^T M_k^{-1} Y_k^T p_k$
9:    ***Reduce:*** $(Y_k^i)^T M_k^{-1} Y_k^i p_k$ to $B_k p_k = (Y_k)^T M_k^{-1} Y_k p_k$ $\longleftarrow$
10:   Compute $\rho_k = \frac{F(w_k) - F(w_k + p_k)}{m_k(0) - m_k(p_k)}$
11:   **if** $\rho_k \geq \eta_1$ **then** Set $w_{k+1} = w_k + p_k$ **else** Set $w_{k+1} = w_k$
12:   $\Delta_{k+1} = \texttt{adjustTR}(\Delta_k, \rho_k)$ [3, Appendix B.3]
13: **end for**

Algorithm 4. Construct new (S_k, Y_k) curvature pairs

Input: w_k (iterate), m (memory), $S_k = [\,]$, $Y_k = [\,]$ (curvature pair containers).

Master Node:	**Worker Nodes** $(i = 1, 2, \ldots, \mathcal{K})$:
1: **Broadcast:** $\bar{S}_k$ and w_k $\longrightarrow$	Compute $\bar{Y}_{k,i} = \nabla^2 F_i(w_k)\bar{S}_k$
2: **Reduce:** $\bar{S}_k^T \bar{Y}_{k,i}$ to $\bar{S}_k^T \bar{Y}_k$ and $\bar{Y}_k^T \bar{S}_k \bar{S}_k^T \bar{Y}_k$ $\longleftarrow$	Compute $\bar{S}_k^T \bar{S}_k$ and $\bar{S}_k^T \bar{Y}_{k,i}$

3: Check the SR1 condition (2.4) and construct M_k^{-1} recursively using
$\bar{S}_k^T \bar{S}_k$, $\bar{S}_k^T \bar{Y}_k$ and $\bar{Y}_k^T \bar{Y}_k$ and construct list of accepted pairs S_k and Y_k

4: **Broadcast:** the list of accepted curvature pairs

Output: M^{-1}, Y_k, S_k

3.4 Complexity Analysis - Comparison of Methods

We compare the complexity of a naive distributed implementation of S-LSR1 and DS-LSR1. Specifically, we discuss the amount of information communicated at every iteration and the amount of computation performed by the nodes. Tables 2 and 3 summarize the communication and computation costs, respectively, and Table 1 summarizes the details of the quantities presented in the tables.

Table 1. Details of quantities communicated and computed.

Variable	Dimension
$w_k, \nabla F(w_k), \nabla F_i(w_k)$	
$p_k, Y_{k,i} M_k^{-1} Y_k^T p_k, B_k d$	$d \times 1$
$F(w_k), F_i(w_k)$	1
$S_k, S_{k,i}, Y_k, Y_{k,i}$	$d \times m$
$S_k^T Y_{k,i}, S_{k,i}^T Y_{k,i}, M_k^{-1}$	$m \times m$
$M_k^{-1} Y_{k,i}^T p_k$	$m \times 1$

As is clear from Tables 2 and 3 the amount of information communicated in the naive implementation $(2md + d + 1)$ is significantly larger than that in the DS-LSR1 method $(m^2 + 2d + 2m + 1)$. Note, $m \ll d$. This can also be seen in Fig. 3 where we show for different dimension d and memory m the number of floats communicated at every iteration. To put this into perspective, consider a training problem where $d = 9.2M$ (e.g., VGG11 network [37]) and $m = 256$, DS-LSR1 and naive DS-LSR1 need to communicate 0.0688 GB and 8.8081 GB, respectively, per iteration. In terms of computation, it is clear that in the naive approach the amount of computation is not balanced between the master and worker nodes, whereas for DS-LSR1 the quantities are balanced.

Table 2. Communication details.

	Naive DS-LSR1	DS-LSR1
Broadcast:	w_k	w_k, p_k, M^{-1}
Reduce:	$\nabla F_i(w_k), F_i(w_k),$ $S_{k,i}, Y_{k,i}$	$\nabla F_i(w_k), F_i(w_k), S_k^T Y_{k,i},$ $Y_{k,i} M_k^{-1} Y_{k,i} p_k, M_k^{-1} Y_{k,i}^T p_k$

Table 3. Computation details.

	Naive DS-LSR1	DS-LSR1
Worker:	$\nabla F_i(w_k), F_i(w_k), Y_{k,i}$	$\nabla F_i(w_k), F_i(w_k), Y_{k,i}, S_{k,i}^T Y_{k,i},$ $M_k^{-1} Y_{k,i}^T p_k, Y_{k,i} M_k^{-1} Y_k^T p_k$, CG
Master:	$M_k^{-1}, w_{k+1}, B_k d$, CG	M_k^{-1}, w_{k+1}

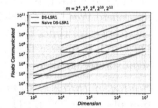

Fig. 3. Number of floats communicated per iteration for different dimension d and memory size m.

4   Numerical Experiments

The goals of this section are threefold: (1) To illustrate the scaling properties of the method and compare it to the naive implementation (Figs. 4 and 5); (2) To deconstruct the main computational elements of the method and show how they scale in terms of memory (Fig. 6); and (3) To illustrate the performance of DS-LSR1 on a neural network training task (Fig. 7). We should note upfront that the goal of this section is not to achieve state-of-the-art performance and compare against algorithms that can achieve this, rather to show that the method is communication efficient and scalable.[2]

4.1   Scaling

Weak Scaling. We considered two different types of networks: (1) `Shallow` (one hidden layer), and (2) `Deep` (7 hidden layers), and for each varied the number of neurons in the layers (MNIST dataset [26], memory $m = 64$). Figure 4 shows the time per iteration for DS-LSR1 for different number of variables and batch sizes.

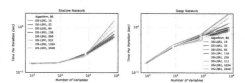

Fig. 4. Weak Scaling: Time/iteration (sec) vs # of variables; `Shallow` (left), `Deep` (right).

Strong Scaling. We fix the problem size (LeNet, CIFAR10, $d = 62006$ [26]), vary the number of nodes and measure the speed-up achieved. Figure 5 illustrates the relative speedup (normalized speedup of each method with respect to the performance of that method on a single node) of the DS-LSR1 method and the naive variant for $m = 256$. The DS-LSR1 method achieves near linear speedup as the number of nodes increases, and the speedup is better than that of the naive

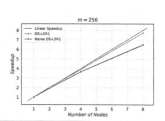

Fig. 5. Strong Scaling: Relative speedup vs # of nodes.

approach. We should note that the times of our proposed method are lower than the respective times for the naive implementation. The reasons for this are: (1) DS-LSR1 is inverse free, and (2) the amount of information communicated is significantly smaller.

Scaling of Different Components of DS-LSR1. We deconstruct the main components of the DS-LSR1 method and illustrate the scaling (per iteration) with respect to memory size. Figure 6 shows the scaling for: (1) reduce time; (2) total time; (3) CG time; (4) time to sample S, Y pairs. For all these plots, we ran 10

---

iterations, averaged the time and also show the variability. As is clear, our proposed method has lower times for all components of the algorithm. We attribute this to the aforementioned reasons.

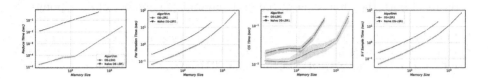

Fig. 6. Time (sec) for different components of DS-LSR1 with respect to memory.

4.2   Performance of DS-LSR1

In this section, we show the performance of DS-LSR1 on a neural network training task; LeNet [26], CIFAR10, $n = 50000$, $d = 62006$, $m = 256$. Figure 7 illustrates the training accuracy in terms of wall clock time and amount of data (GB) communication (left and center plots), for different number of nodes. As expected, when using larger number of compute nodes training is faster. Similar results were obtained for testing accuracy. We also plot the performance of the naive implementation (dashed lines) in order to show that: (1) the accuracy achieved is comparable, and (2) one can train faster using our proposed method.

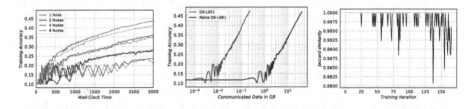

Fig. 7. Performance of DS-LSR1 on CIFAR10 dataset with different number of nodes.

Finally, we show that the curvature pairs chosen by our approach are almost identical to those chosen by the naive approach even though we use an approximation (via sketching) when checking the SR1 condition. Figure 7 (right plot), shows the Jaccard similarity for the sets of curvature pairs selected by the methods; the pairs are almost identical, with differences on a few iterations.

5   Final Remarks

This paper describes a scalable distributed implementation of the sampled LSR1 method which is communication-efficient, has favorable work-load balancing across nodes and that is matrix-free and inverse-free. The method leverages the compact representation of SR1 matrices and uses sketching techniques to drastically reduce the amount of data communicated at every iteration as compared to a naive distributed implementation. The DS-LSR1 method scales well in terms of both the dimension of the problem and the number of data points.

Acknowledgements. This work was partially supported by the U.S. National Science Foundation, under award numbers NSF:CCF:1618717, NSF:CMMI:1663256 and NSF:CCF:1740796, and XSEDE Startup grant IRI180020.

References

1. Agarwal, A., Chapelle, O., Dudík, M., Langford, J.: A reliable effective terascale linear learning system. J. Mach. Learn. Res. **15**, 1111–1133 (2014)
2. Berahas, A.S., Bollapragada, R., Nocedal, J.: An investigation of newton-sketch and subsampled newton methods. arXiv preprint arXiv:1705.06211 (2017)
3. Berahas, A.S., Jahani, M., Takác, M.: Quasi-newton methods for deep learning: forget the past, just sample. arXiv preprint arXiv: 1901.09997 (2019)
4. Berahas, A.S., Nocedal, J., Takác, M.: A multi-batch L-BFGs method for machine learning. In: NeurIPS, pp. 1055–1063 (2016)
5. Berahas, A.S., Takáč, M.: A robust multi-batch L-BFGs method for machine learning. Optim. Methods Softw. **35**(1), 191–219 (2020)
6. Bollapragada, R., Byrd, R.H., Nocedal, J.: Exact and inexact subsampled Newton methods for optimization. IMA J. Num. Anal. **39**(2), 545–578 (2019)
7. Bottou, L.: Large-scale machine learning with stochastic gradient descent. In: Lechevallier, Y., Saporta, G. (eds.) Proceedings of COMPSTAT'2010Proceedings of COMPSTAT 2010, pp. 177–186. Springer, Cham (2010). https://doi.org/10.1007/978-3-7908-2604-3_16
8. Bottou, L., Cun, Y.L.: Large scale online learning. In: NeurIPS, pp. 217–224 (2004)
9. Bottou, L., Curtis, F.E., Nocedal, J.: Optimization methods for large-scale machine learning. SIAM Rev. **60**(2), 223–311 (2018)
10. Byrd, R.H., Chin, G.M., Neveitt, W., Nocedal, J.: On the use of stochastic Hessian information in optimization methods for machine learning. SIAM J. Optim. **21**(3), 977–995 (2011)
11. Byrd, R.H., Hansen, S.L., Nocedal, J., Singer, Y.: A stochastic Quasi-Newton method for large-scale optimization. SIAM J. Optim. **26**(2), 1008–1031 (2016)
12. Byrd, R.H., Nocedal, J., Schnabel, R.B.: Representations of Quasi-Newton matrices and their use in limited memory methods. Math. Program. **63**, 129–156 (1994)
13. Chen, W., Wang, Z., Zhou, J.: Large-scale L-BFGs using MapReduce. In: NeurIPS, pp. 1332–1340 (2014)
14. Chu, C.T., et al.: Map-reduce for machine learning on multicore. In: NeurIPS, pp. 281–288 (2007)
15. Conn, A.R., Gould, N.I., Toint, P.L.: Convergence of Quasi-Newton matrices generated by the symmetric rank one update. Math. Program. **50**(1–3), 177–195 (1991)
16. Curtis, F.E.: A self-correcting variable-metric algorithm for stochastic optimization. In: ICML, pp. 632–641 (2016)
17. Dean, J., et al.: Large scale distributed deep networks. In: NeurIPS, pp. 1223–1231 (2012)

18. Defazio, A., Bach, F., Lacoste-Julien, S.: Saga: a fast incremental gradient method with support for non-strongly convex composite objectives. In: NeurIPS, pp. 1646–1654 (2014)
19. Duchi, J., Hazan, E., Singer, Y.: Adaptive subgradient methods for online learning and stochastic optimization. J. Mach. Learn. Res. 12, 2121–2159 (2011)
20. Erway, J.B., Griffin, J., Marcia, R.F., Omheni, R.: Trust-region algorithms for training responses: machine learning methods using indefinite hessian approximations. Optim. Methods Softw. 35(3), 460–487 (2020)
21. Jahani, M., et al.: Efficient distributed hessian free algorithm for large-scale empirical risk minimization via accumulating sample strategy. arXiv preprint arXiv:1810.11507 (2018)
22. Johnson, R., Zhang, T.: Accelerating stochastic gradient descent using predictive variance reduction. In: NeurIPS, pp. 315–323 (2013)
23. Keskar, N.S., Berahas, A.S.: adaQN: an adaptive Quasi-Newton algorithm for training RNNs. In: Frasconi, P., Landwehr, N., Manco, G., Vreeken, J. (eds.) ECML PKDD 2016. LNCS (LNAI), vol. 9851, pp. 1–16. Springer, Cham (2016). https://doi.org/10.1007/978-3-319-46128-1_1
24. Khalfan, H.F., Byrd, R.H., Schnabel, R.B.: A theoretical and experimental study of the symmetric rank-one update. SIAM J. Optim. 3(1), 1–24 (1993)
25. Kingma, D.P., Ba, J.: Adam: a method for stochastic optimization. arXiv preprint arXiv:1412.6980 (2014)
26. LeCun, Y., Bottou, L., Bengio, Y., Haffner, P.: Gradient-based learning applied to document recognition. Proc. IEEE 86(11), 2278–2324 (1998)
27. Liu, D.C., Nocedal, J.: On the limited memory BFGs method for large scale optimization. Math. Program. 45(1–3), 503–528 (1989)
28. Lu, X.: A study of the limited memory SR1 method in practice. University of Colorado at Boulder (1996)
29. Martens, J.: Deep learning via hessian-free optimization. ICML 27, 735–742 (2010)
30. Nocedal, J., Wright, S.J.: Numerical Optimization, 2 edn. Springer Series in Operations Research, Springer (2006)
31. Recht, B., Re, C., Wright, S., Niu, F.: Hogwild: a lock-free approach to parallelizing stochastic gradient descent. In: NeurIPS, pp. 693–701 (2011)
32. Robbins, H., Monro, S.: A stochastic approximation method. Annal. Math. Stat. 22, 400–407 (1951)
33. Roosta-Khorasani, F., Mahoney, M.W.: Sub-sampled newton methods. Math. Program. 174(1–2), 293–326 (2019)
34. Schmidt, M., Le Roux, N., Bach, F.: Minimizing finite sums with the stochastic average gradient. Math. Program. 162(1–2), 83–112 (2017)
35. Schraudolph, N.N., Yu, J., Günter, S.: A stochastic quasi-newton method for online convex optimization. In: Artificial Intelligence and Statistics, pp. 436–443 (2007)
36. Shamir, O., Srebro, N., Zhang, T.: Communication-efficient distributed optimization using an approximate Newton-type method. In: ICML, pp. 1000–1008 (2014)
37. Simonyan, K., Zisserman, A.: Very deep convolutional networks for large-scale image recognition. arXiv preprint arXiv:1409.1556 (2014)
38. Takác, M., Bijral, A.S., Richtárik, P., Srebro, N.: Mini-batch primal and dual methods for SVMs. In: ICML (3), pp. 1022–1030 (2013)
39. Tsitsiklis, J., Bertsekas, D., Athans, M.: Distributed asynchronous deterministic and stochastic gradient optimization algorithms. IEEE Trans. Autom. Control 31(9), 803–812 (1986)
40. Woodruff, D.P.: Sketching as a tool for numerical linear algebra. Found. Trends® Theoret. Comput. Sci.10(1–2), 1–157 (2014)

41. Xu, P., Roosta, F., Mahoney, M.W.: Newton-type methods for non-convex optimization under inexact Hessian information. Math. Program. **184**, 35–70 (2017). https://doi.org/10.1007/s10107-019-01405-z
42. Xu, P., Roosta-Khorasan, F., Mahoney, M.W.: Second-order optimization for non-convex machine learning: an empirical study. arXiv:1708.07827 (2017)
43. Zhang, Y., Xiao, L.: Communication-efficient distributed optimization of self-concordant empirical loss. In: Giselsson, P., Rantzer, A. (eds.) Large-Scale and Distributed Optimization. LNM, vol. 2227, pp. 289–341. Springer, Cham (2018). https://doi.org/10.1007/978-3-319-97478-1_11
44. Zinkevich, M., Weimer, M., Li, L., Smola, A.J.: Parallelized stochastic gradient descent. In: NeurIPS, pp. 2595–2603 (2010)

Should Simplicity Be Always Preferred to Complexity in Supervised Machine Learning?

Falco Bargagli-Stoffi[1,2], Gustavo Cevolani[1], and Giorgio Gnecco[1(✉)]

[1] IMT School for Advanced Studies, Lucca, Italy
{falco.bargaglistoffi,gustavo.cevolani,giorgio.gnecco}@imtlucca.it
[2] Harvard University, Cambridge, MA, USA

Abstract. In this short paper, a theoretical analysis of Occam's razor formulation through statistical learning theory is presented, showing that pathological situations exist for which regularization may slow down supervised learning instead of making it faster.

Keywords: Simplicity versus complexity · Occam's razor · Statistical Learning Theory · Vapnik-Chervonenkis bound · Sample size

1 Introduction

Statistical Learning Theory (SLT) [1] provides a solid mathematical formalization of philosophical principles such as Occam's razor, which suggests to prefer simple explanations of observed phenomena over complex ones [2, Section 9.2.5]. In the case of supervised learning, this is taken into account by formulating the training of a machine learning model as a suitable optimization problem, in which the optimization variables are its parameters. Ideally, the objective function of that problem should be the expected risk of the model, which quantifies its expected performance on a test set made of examples not used for its training, and whose labels are concealed to the learning machine in the testing phase. However, the probability distribution associated with such examples is typically unknown. In such case, the expected risk has to be replaced by its suitable estimate, which is commonly the sum of two terms: one - the empirical risk - that quantifies how well the model fits a set of input/output data used for training (whose labels are, in this case, provided to the learning machine); the other - the regularization term - that penalizes complex models over simple ones, and whose value decreases to zero as the size of the training set increases. The rationales for the introduction of this regularization term are manifold: (i) simpler models are less prone to overfitting; (ii) they are more easily explainable to human beings; (iii) they can be implemented in a more efficient way; (iv) they are more easily falsifiable in the sense of Popper [3].

G. Cevolani acknowledges support from MIUR PRIN grant no. 201743F9YE.

G. Nicosia et al. (Eds.): LOD 2020, LNCS 12565, pp. 55–59, 2020.
https://doi.org/10.1007/978-3-030-64583-0_6

Still, the question arises: should simplicity be "always" preferred to complexity in supervised machine learning? Using SLT arguments, we show that the answer is no! Specifically, we compare, theoretically, the following two situations:

a) the outputs in the training and test sets are generated by a "simple" data generating process (but the machine does not know that), to which some noise is added. Then, two families of models (which are, respectively, "simple", and "complex") are fitted to the training data, using SLT principles. Finally, the "best" of those models (according to SLT) is selected automatically;
b) as above, but in this case the outputs in the training and test sets are generated by a "complex" data generating process, to which some noise is added.

Ideally, a model corresponding to the true nature of the data generating process should be selected in both cases (i.e., a "simple" model in the first case, and a "complex" one in the second case). However, the use of a regularization term tends to make a simpler model be selected, when the size of the training set is sufficiently small. So, the following results are obtained (see Sect. 2):

– in both cases a) and b), both regularization and its absence guarantee that, with probability above any threshold smaller than 1, a model in the "correct" family is selected, provided that the size of the training set is sufficiently large;
– in the case a), regularization is beneficial for learning, in the sense that it reduces the minimal number of training examples (i.e., the minimal amount of "information") able to guarantee that a model in the correct family (the "simple" one) is selected with probability above any threshold smaller than 1, with respect to the case in which no regularization is performed (i.e., when the sole empirical risk is minimized);
– in the case b), where the correct family is the "complex" one, the opposite occurs (i.e., no regularization has a better performance guarantee). In this (pathological) case, however, one could still prefer simpler models for reasons different from better generalization ability (e.g., better interpretability, and/or easier implementation).

The results are in line with the "no free lunch theorems" [4] of machine learning (another important formalization of philosophical principles [5]), according to which all training algorithms have the same expected performance, when a suitable average over all possible supervised machine learning problems is taken.

2   Theoretical Analysis

The analysis is grounded on the well-known Vapnik-Chervonenkis (VC) two-sided upper bound on the difference between expected risk and empirical risk (the latter based on a training set of size L). Two families of models, S ("simple") and C ("complex") are considered, each parametrized by a vector ($\mathbf{w}_S$ for S, $\mathbf{w}_C$ for C, which vary respectively in two sets W_S and W_C), with possibly different dimension for each family. The expected and empirical risks of the various models are denoted, respectively, by $R_S^{exp}(\mathbf{w}_S)$ and $R_S^{emp,L}(\mathbf{w}_S)$ for the models in S, and by $R_C^{exp}(\mathbf{w}_C)$ and $R_C^{emp,L}(\mathbf{w}_C)$ for the models in C. All these risks are computed using, respectively, bounded loss functions $V_S(\cdot, \mathbf{w}_S)$ and $V_C(\cdot, \mathbf{w}_C)$ with codomain an interval $[A, B]$, which are parametrized by the two vectors.

Let h_S and h_C be the Vapnik-Chervonenkis dimensions, respectively, of the sets of loss functions $\{V_S(\cdot, \mathbf{w}_S), \mathbf{w}_S \in W_S\}$ and $\{V_C(\cdot, \mathbf{w}_C), \mathbf{w}_C \in W_C\}$, with $h_S < h_C$ (being, indeed, the models in S simpler than those in C). Finally, let $\delta \in (0,1)$ be given, and let the size of the training set be $L > h_C > h_S$. Then, according to the VC theory, the two following bounds hold with probabilities $p_S \geq 1 - \frac{\delta}{2}$ and $p_C \geq 1 - \frac{\delta}{2}$ with respect to the generation of the training set (whose examples are drawn independently, from the same probability distribution as the test examples) [1, Section 3.7]:

$$\sup_{\mathbf{w}_S \in W_S} |R_S^{exp}(\mathbf{w}_S) - R_S^{emp,L}(\mathbf{w}_S)| \leq (B - A)\sqrt{\frac{h_S \ln \frac{2eL}{h_S} - \ln \frac{\delta}{8}}{L}}, \quad (1)$$

$$\sup_{\mathbf{w}_C \in W_C} |R_C^{exp}(\mathbf{w}_C) - R_C^{emp,L}(\mathbf{w}_C)| \leq (B - A)\sqrt{\frac{h_C \ln \frac{2eL}{h_C} - \ln \frac{\delta}{8}}{L}}. \quad (2)$$

Hence, by applying the union bound technique, both bounds (1) and (2) hold simultaneously with probability $p \geq 1 - \delta$.

According to the Empirical Risk Minimization (ERM) principle [1, Section 1.5], one selects, for each family, the model that minimizes the empirical risk on that family (i.e., the one associated, respectively, with the parameter choice $\hat{\mathbf{w}}_S := \operatorname{argmin}_{\mathbf{w}_S \in W_S} R_S^{emp,L}(\mathbf{w}_S)$ and $\hat{\mathbf{w}}_C := \operatorname{argmin}_{\mathbf{w}_C \in W_C} R_C^{emp,L}(\mathbf{w}_C)$). Without significant loss of generality, we assume that these minimizers exist and are unique. Finally, of these two parameters $\hat{\mathbf{w}}_S$ and $\hat{\mathbf{w}}_C$, the one achieving the smallest between $R_S^{emp,L}(\hat{\mathbf{w}}_S)$ and $R_C^{emp,L}(\hat{\mathbf{w}}_C)$ is chosen. If the Structural Risk Minimization (SRM) principle [1, Section 4.1] is chosen, instead, then, of the two parameters, the one associated with the smallest between $R_S^{emp,L}(\hat{\mathbf{w}}_S) + (B-A)\sqrt{\frac{h_S \ln \frac{2eL}{h_S} - \ln \frac{\delta}{8}}{L}}$ and $R_C^{emp,L}(\hat{\mathbf{w}}_C) + (B-A)\sqrt{\frac{h_C \ln \frac{2eL}{h_C} - \ln \frac{\delta}{8}}{L}}$ is chosen. For simplicity, the case of ties is not considered in the following.

Let $\Delta := \min_{\mathbf{w}_C \in W_C} R_C^{exp}(\mathbf{w}_C) - \min_{\mathbf{w}_S \in W_S} R_S^{exp}(\mathbf{w}_S)$ (supposing again, without significant loss of generality, that such minima exist and are uniquely achieved). Of course, Δ is in general unknown, but in the following analysis we suppose that its modulus $|\Delta|$ (but not its sign), or at least an approximation of that modulus, is provided to the learning machine, e.g., by an oracle. We also assume that the two families of models are non-nested, and that $\Delta \neq 0$ holds.

Let both the training/test output data be generated by a particular model in S or C, with the output perturbed by 0-mean independent additive noise[1] with small variance σ^2. In this case, one has, respectively, $\Delta > 0$ or $\Delta < 0$, meaning

---

[1] Without such noise, one would have always a 0 minimum empirical risk in the correct family of models, which would make its detection easier.

in the first case that the best model in S is better than the best model in C, and in the second case that the best model in C is better than the best model in S.

Using (1) and (2), one concludes that, if the ERM principle is applied, then a model in the "correct" family (i.e., one coming from the same family from which the training/test output data are generated, even though it may not coincide with the best such model in terms of the expected risk) is selected with probability $p \geq 1 - \delta$ if

$$(B - A) \left(\sqrt{\frac{h_S \ln \frac{2eL}{h_S} - \ln \frac{\delta}{8}}{L}} + \sqrt{\frac{h_C \ln \frac{2eL}{h_C} - \ln \frac{\delta}{8}}{L}} \right) < |\Delta|. \tag{3}$$

a) Let $\Delta > 0$. In this case, if the SRM principle is applied, then a model in the correct family (which is S) is selected with probability $p \geq 1 - \delta$ if

$$2(B - A) \sqrt{\frac{h_S \ln \frac{2eL}{h_S} - \ln \frac{\delta}{8}}{L}} < |\Delta|. \tag{4}$$

b) Let $\Delta < 0$. In this case, if the SRM principle is applied, then a model in the correct family (which is C) is selected with probability $p \geq 1 - \delta$ if

$$2(B - A) \sqrt{\frac{h_C \ln \frac{2eL}{h_C} - \ln \frac{\delta}{8}}{L}} < |\Delta| \tag{5}$$

It is worth noting that, being $\Delta \neq 0$, all the bounds (3), (4), and (5) are guaranteed to hold if L is large enough. Since, for $L > h_C > h_S$, one has

$$2 \sqrt{\frac{h_S \ln \frac{2eL}{h_S} - \ln \frac{\delta}{8}}{L}} < \left(\sqrt{\frac{h_S \ln \frac{2eL}{h_S} - \ln \frac{\delta}{8}}{L}} + \sqrt{\frac{h_C \ln \frac{2eL}{h_C} - \ln \frac{\delta}{8}}{L}} \right)$$

$$< 2 \sqrt{\frac{h_C \ln \frac{2eL}{h_C} - \ln \frac{\delta}{8}}{L}}, \tag{6}$$

one can also conclude that:

- in the case a), the regularization term $(B - A) \sqrt{\frac{h \ln \frac{2eL}{h} - \ln \frac{\delta}{8}}{L}}$ (for $h = h_S, h_C$) is beneficial for learning, because the minimal size of the training set for which condition (4) - which is associated to the selection by the SRM principle of a model in the correct "simple" family - holds is smaller than or equal to the minimal size of the training set for which condition (3) - which is associated to the selection by the ERM principle of a model in same correct family - holds;
- in the case b), no regularization has a better performance guarantee, in the sense that the minimal size of the training set for which condition (5) - which is associated to the selection by the SRM principle of a model in the correct "complex" family - holds is larger than or equal to the minimal size of the training set for which condition (3) holds.

References

1. Vapnik, V.N.: The Nature of Statistical Learning Theory. Springer, New York (2000). https://doi.org/10.1007/978-1-4757-3264-1
2. Duda, R.O., Hart, P.E., Stork, D.G.: Pattern Classification. Wiley, Hoboken (2000)
3. Corfield, D., Schölkopf, B., Vapnik, V.N.: Falsificationism and statistical learning theory: comparing the Popper and Vapnik-Chervonenkis dimensions. J. Gen. Philos. Sci. 4(1), 51–58 (2009)
4. Wolpert, D.H.: The lack of a priori distinctions between learning algorithms. Neural Comput. 8, 1341–1390 (1996)
5. Lauc, D.: Machine learning and the philosophical problems of induction. In: Skansi, S. (ed.) Guide to Deep Learning Basics, pp. 93–106. Springer, Cham (2020). https://doi.org/10.1007/978-3-030-37591-1_9

An Application of Machine Learning to Study Utilities Expenses in the Brazilian Navy

Stefan Silva[1,2(✉)] and José Crispim[1]

[1] NIPE, University of Minho, Campus of Gualtar, Braga 4710-057, Portugal
id8399@alunos.uminho.pt, crispim@eeg.uminho.pt
[2] Brazilian Navy, DGOM, Rio de Janeiro 20091-000, Brazil

Abstract. The extensive Brazilian territory endows its Navy with more than 350 facilities with several distinct activities that transcend military operations. Understanding the variation of all the essential and common costs of those facilities proved to be a challenging and relevant task. This paper presents a machine learning approach to support the decision-making process based on data that represents several facilities attributes, where models were trained, and those with the best performance were further analyzed. Besides data limitations, our results show that predictions and explanations derived from the models can be applied to support decision-making within the organization and contribute with insights to improve management over its resources.

Keywords: Military expenditures · Machine learning · Decision-making · Data-driven organization

1 Introduction

Governments all around the world face a never-ending challenge to provide an increasing number of public services with the same (or less) spending at each fiscal year [1]. In order to keep reducing costs while improving the quality of public services, some researchers argue that innovations like process standardization, open data, and social media plays an essential role, allowing governmental organizations to better deal with societal changes such as population growth or economic health [2, 3]. Information and Communication Technologies are the foundation behind many of these new innovative ideas, and the evolution of its utilization in the public sector have even started the electronic government reform movement [3]. However, despite public sector innovation literature pointing out that little attention is being paid to technological process innovations [4], the benefits of big data, Artificial Intelligence (AI), and Machine Learning (ML) have begun to influence the transition towards a data-driven public sector [5]. The opportunities arising from this change are studied in [6], resulting in higher transparency and efficiency in terms of resources used.

The Brazilian Navy, like any other member of the government, is committed to the challenge of saving resources wherever possible. The technologies mentioned above can improve the usage of the data it produces to support the decisions that relate to

© Springer Nature Switzerland AG 2020
G. Nicosia et al. (Eds.): LOD 2020, LNCS 12565, pp. 60–71, 2020.
https://doi.org/10.1007/978-3-030-64583-0_7

the allocation of these resources, especially when considering the size and diversity of activities performed by the organization. Brazil is the country with the most extensive river basin in the world and owner of a 7400 km coastline where a significant amount of natural resources is located. Such territorial dimension endows its Navy with more than 70.000 personnel, distributed in more than 350 facilities throughout the territory [7, 8], which are responsible for a wide range of activities that go beyond military operations, including schools; hospitals; warehouses; maintenance, communications, and R&D centers; military facilities (e.g., headquarters, military barracks), among others.

All these facilities spend resources, at different rates, to achieve their purposes. Among these expenses are those considered common amid the facilities and essential for maintaining a proper level of operation (e.g., energy, water, gas, cleaning). In this paper, these expenses will be referred to as "utilities". According to [9], in Brazil (2018), 75% of the defense budget is allocated to personnel and social security expenditures, 12% in operations and maintenance (where utilities are included), and nearly 13% in investments. Given the rigidity of spending on personnel, maintenance assumes a critical position in each facility, where budget constraints may influence the efficiency of military forces by affecting combat readiness obtained by training [10].

This study aims to develop ML models able to support the decision-making process by understanding how these utilities expenses behave across the navy facilities. From an ML point of view, regression algorithms will be used to train models on data obtained from different sources, representing the expenses made and a limited set of facilities features. Furthermore, models with the best predictive performance will be analyzed using Interpretable Machine Learning (IML) techniques.

The paper is structured as follows. In Sect. 2, we present related work found in the literature regarding data-driven decision making, ML, and military expenditures in order to support our objective and elucidate the gap this paper is proposed to fill. In Sect. 3, the dataset creation process is described, and the methodology for training the models is introduced. Section 4 describes and discusses the results of the experiments, presenting potential uses of the models in order to support decision-making. Finally, Sect. 5 concludes the paper.

2   Related Work

Data-driven decision making is especially useful and vital to organizations that own large datasets that are interconnected and reflects past, current, and subsequent performance [11]. Previously studies found significantly higher levels of productivity of its adopters [12]. Nevertheless, many organizations do not have the capability to perform data-driven decision making [13] and so, research is needed to explore how to help them achieve this goal [14]. Despite the massive data availability, conducting an organizational change towards data-driven decision making is not an easy task. It depends on many factors like commitment and leadership, well-governed data, know-how, and specialists. Such may prove challenging to implement, mainly when managers are used to acting on experience or intuition [15].

The literature contains ML-based approaches for data-driven decision making in the most diverse areas of study, like R&D budget allocation [16], inventory classification

[17], profit margins estimation [18], energy consumption [19], demand response [20], the compressive strength of materials in engineering [21], disease prediction in medicine [22], and autonomous driving [23].

In terms of military expenditures, there are several articles and databases about spending on military organizations (e.g., SIPRI, Worldbank database) concerning: country economy [24], effects on economic growth [25], country productivity [26], and expenditure determinants [27]. However, very few studies analyze the defense budget by expenditure category. The vast majority of Defense-related studies focus on its core activity, e.g., military technologies [28], combat enhancements [29].

To the best of our knowledge, no previous studies were found regarding the specific problem exposed in this paper: a single organization, made up of hundreds of facilities with quite distinct activities from each other, trying to understand the behavior of the common and essential expenses being made, which may be similar to that of other organizations and may become somewhat characteristic of an environment with the same peculiarities.

3 Methodology

A brief graphical overview of the methodology used is given in Fig. 1, covering all the steps regarding the two main objectives of this section: the dataset creation and the methods used to train the ML models.

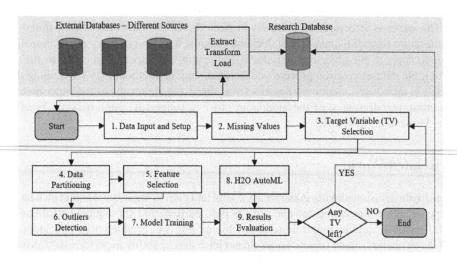

Fig. 1. Modelling framework.

3.1 Data

The data used in this work have been originally collected for non-research purposes, consisting of attributes of the facilities (i.e., features), including geographical, social, and climate data. Dealing with secondary data requires demonstrating its usefulness in this new context [30], which will be addressed by the ML models through the evaluation of the importance of the features.

The Extract, Transform, and Load process in Fig. 1 represents the three steps conducted to manipulate the data for the learning process, involving: collecting data from distinct sources, inside and outside the organization, using different methods; merging

Table 1. Data description

Feature name	Description	Format	Source
FCOS	Identification of which one of the main bodies of the navy organizational structure (OS) [7] the facility is located	Categorical (11 classes)	Org Analytics Platform
FCHI	Facility Hierarchy Level. Representation of how deep in the hierarchy proposed by the OS [7] the facility is	Categorical (07 classes)	
FCTP	Facility Type. The main activity performed	Categorical (09 classes)	
FCSP	Service Provider Indicator. Indicates a special kind of facility that does not receive an initial budget, and instead provides services for other facilities, charging them to sustain itself	Categorical (02 classes)	
FCPP	Facility Personnel	Numeric	Org Human Resources Database
STRG	Country Region where the facility is located	Categorical (05 classes)	Brazilian Institute of Geography and Statistics [31]
STPD	Population Density of the state where the facility is located	Numeric	
STHDI	An adaptation of the Human Development Index (HDI) for the Brazilian States	Numeric	
STMII	Monthly Income per Inhabitant of the state where the facility is located	Numeric	
CTPD	Population Density of the city where the facility is	Numeric	
CLPR CLMAX CLMIN CLHMD	Average monthly rainfall, maximum and minimum temperatures, and humidity measurement of the closest location relative to the facility	Numeric	Brazilian National Meteorological Institute [32]
FNPP	The facility property. Consists of the financial evaluation of all the types of equipment used	Numeric	Brazilian Integrated Financial Management System
FNSF	A square meter value for the active facility buildings	Numeric	
FNEC	Expense Category. A classification based on the nature of each type of expense	Categorical (37 classes)	
FNCC	Facility Cost. A fraction of the expense that represents what has been effectively consumed	Numeric	

all the different formats in a single structure, while removing incomplete or irrelevant parts; and inserting the data in a research database created for this study. A description of the data is shown in Table 1, with the sources, feature name, description, and format.

At the end of the load step, the resulting dataset contained 4605 rows of data, where each row represents the total expense in one category made by one facility. Unfortunately, some limitations were found, and due to compatibility issues, the FNPP and FNSF features could not be used in this dataset. In their sources, these features were stored related to groups of facilities (i.e., not individualized), making the merging process without losing information impossible. However, in order to evaluate the relative importance of the discarded features, a reduced version of the dataset was created containing only those facilities whose measurement of FNPP and FNSF could be individualized, resulting in 2232 rows.

Additionally, an analysis of FNCC and FCPP distributions shows a high degree of positive skewness due to the existence of outliers in the data, representing some few massive-sized facilities within the organization. The Box-Cox transformation [33] of the "caret" package was applied to address the skewness, resulting in a log-transformation.

3.2 Methods

The methods used to train the models were implemented through an algorithm using R [34]. Table 2 shows a brief description of each numbered step in Fig. 1, mentioning any additional packages used. In general, the steps follow a straightforward logic where it receives the data, apply transformations, train ML models, and evaluate the results. However, two aspects are worth explaining: (1) if the data received by the algorithm

Table 2. Methodology steps description

Step name	Description
1. Data input and setup	Loads data using a previously made SQL script, which indicates the set of TV contained in it
2. Missing values	A multiple imputation model in numeric features was implemented through the "mice" package [35], in order to avoid the deletion of the rows with missing values, while preserving existing relations and uncertainty
3. TV selection	A subset of the original data is created containing all the features and only one TV. Steps 3 to 9 are repeated for each TV in the data
4. Data partitioning	The models are trained using 5-repeated 10-fold cross-validation
5. Feature selection	The "Boruta" package [36] was used to remove any features statistically proved to be less relevant than random to explain the variation on the TV
6. Outliers detection	The "solitude" package was used to detect outliers. No further treatment is made once an outlier is identified
7. Model training	The "caret" package was used to train different ML models and apply additional procedures like centering, scaling, resampling, and parameter tuning. The following models were used: Multiple Linear Regression (MLR), Gaussian Kernel Support Vector Regressor (SVR), Random Forests (RF), Gradient Boosting Machines (GBM), and XGBoost (XGB)
8. H2O autoML	For this study, a time limit of 10 min has been set. The available models are RF, GBM, XGB, MLR, Deep Learning, and two Stacked Ensembles
9. Results evaluation	This step stores the R^2 score, and the normalized versions of root mean square error (nRMSE) and mean absolute error (nMAE). The best model is saved for later use

contains more than one Target Variable (TV), it will loop using one TV at a time, storing the results on how well the predictors were able to explain that specific TV for later analysis, and (2) besides the standard training of ML models, an Automatic Machine Learning platform (H2O AutoML) was also used in parallel for comparison purposes.

4 Experimental Study

4.1 Experiments for the Model Selection

The data was reorganized into two perspectives, P1 and P2, allowing to provide different structures of TV for the learning process. The variable FNEC, categorical with 37 classes, responsible for classifying the expense based on its nature, was the basis for the perspectives creation, in addition to the organization's understanding of what makes the distinction between essential and common utilities. Figure 2 illustrates how the decomposition of FNEC is related to P1 and P2.

The goal in P1 is to train two final models, one for all the essential utilities and another for all the common utilities. In P2 the algorithm will train a total of 9 models: Energy (ENE), Gas (GAS), Cleaning (CLN), Water/Sewage (WTS) and Communications (COM) for the essential utilities and Assets Maintenance (AST), Office & IT Supplies (OIS), Leases & Rents (LAR), and Copies & Reproduction (CAR) for the common utilities. The creation of P1 and P2 aims to verify if the expenses categories need to be individualized to generate better results during the learning process.

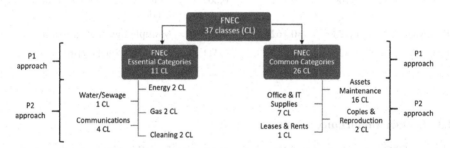

Fig. 2. FNEC decomposition for perspective creation

4.2 Performance Results

Table 3 summarizes the results by showing the metrics for the best models found on each TV. The primary metric used is nMAE, followed by nRMSE and R^2 in case of a performance tie. AutoML was able to generate models that outperform the manual training process in most of the cases using Stacked Ensembles. Surprisingly, the models trained during the P1 approach achieved lower error levels in nMAE and nRMSE than the ones in P2. One possible explanation is that in P2, the dataset for each TV is considerably smaller, thus making it difficult for the learning process to be efficient.

The findings demonstrate that there is still a moderate amount of unexplained variation in the models and that it might not be accurate enough for an effective decision support system. It suggests that additional features (e.g., the discarded FNSF and FNPP) or new measurements for the target variables (e.g., energy and water consumption instead of expense) could be used in an attempt to improve the results. The implementation of this methodology as a framework within the organization can allow it to make a more in-depth analysis, with the benefit of handling more data with little human intervention, while evaluating whether the error levels are acceptable for each situation.

Table 3. Performance results

Perspective	nMAE	nRMSE	r^2	Best model
P1 Essential	**0,071**	**0,096**	**0,559**	**AutoML - Stacked Ensemble**
P2 GAS	0,105	0,143	0,322	AutoML - Stacked Ensemble
P2 CLN	0,107	0,136	0,337	Random Forest
P2 ENE	0,075	0,108	0,634	AutoML - Deep Learning
P2 WTS	0,100	0,133	0,525	AutoML - Stacked Ensemble
P2 COM	0,090	0,120	0,438	AutoML - Stacked Ensemble
P1 Common	**0,074**	**0,099**	**0,465**	**AutoML - Stacked Ensemble**
P2 AST	0,101	0,144	0,217	AutoML - Stacked Ensemble
P2 OIS	0,100	0,136	0,267	AutoML - Generalized Linear Model
P2 LAR	0,123	0,162	0,123	Multiple Linear Regression
P2 CAR	0,125	0,164	0,411	AutoML - Stacked Ensemble

4.3 Feature Importance

The two H2O AutoML Stacked Ensembles trained in P1 will be the base for the feature importance measurement, which consists in applying a model-agnostic permutation-based technique that evaluates, regardless of the model, how its global performance degrades when the values within a feature are randomized. Figure 3 shows the ten most important features in the two models proposed, where the top 5 features in each one are the same, and they are all facilities' attributes (expense category, personnel, type, hierarchy, and OS). The results also suggest that the social and geographic related features are substantially less relevant to the outcome.

4.4 Reduced Dataset Analysis

In order to evaluate the relative importance of the discarded FNPP and FNSF features, the H2O AutoML platform was used to perform the same approach on the reduced

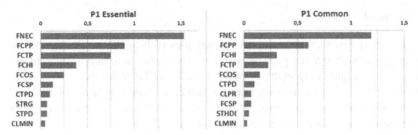

Fig. 3. Feature importance in P1

dataset, described in Sect. 3.1. Table 4 presents the results in comparison with those obtained from the complete dataset and an alternative version of the reduced dataset without FNPP and FNSF.

Table 4. Performance results on reduced dataset

Perspective	nMAE	nRMSE	r^2	Data description
P1 Essential	0,071	0,096	0,559	Complete dataset – without FNSF/FNPP
P1 Essential	0,067	0,090	0,564	Reduced dataset – without FNSF/FNPP
P1 Essential	**0,045**	**0,074**	**0,692**	Reduced dataset – with FNSF/FNPP
P1 Common	0,074	0,099	0,465	Complete dataset – without FNSF/FNPP
P1 Common	0,072	0,096	0,472	Reduced dataset – without FNSF/FNPP
P1 Common	**0,052**	**0,080**	**0,613**	Reduced dataset – with FNSF/FNPP

A significant performance improvement can be noticed in the models trained on the reduced dataset, which can be attributed to the capacity of FNPP and FNSF to explain an additional fraction of the TV. Moreover, this capacity is also supported by the feature importance technique that was applied, with results shown in Fig. 4.

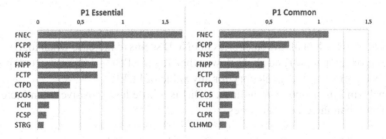

Fig. 4. Feature importance on P1 - reduced dataset

4.5   Potential Uses

The purpose of this section is to present examples of how the results can be used to support the decision-making process within the organization. Examples regarding both predictive power and interpretability of the models will be presented.

Predictions can assist in determining whether a change in a set of the facilities attributes will affect the expenses in a way beyond or below the desired limits. They can also be useful in case of activation/deactivation of facilities.

- A hospital is being reinforced with personal and equipment to deal with a recent disease outbreak, but the associated increase in utilities must stay below 25%.
- A repair center is being shut down, and its personnel and equipment relocated. The amount saved by the deactivation must be enough to cope with the associated increase in all the other facilities.

Additionally, explanations derived from the models can help decision-makers to obtain insights through the comprehension of what was taken into account to make a prediction. Besides knowing which features are relevant, another IML procedure called Partial Dependence (PD), described in [37], is useful to understand the marginal effect of a subset of features on the outcome. Figure 5 illustrates a PD plot between FCPP, FNCD, and the TV in P1 Essential. The following explanations can be extracted:

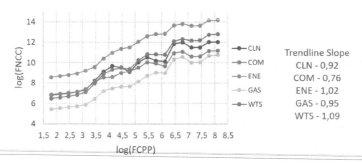

Fig. 5. Partial dependence plot in p1 essential

- The rate of growth of all the categories in FNCD seems to increase when log(FCPP) > 3,5 (approx. 30 people) and slows down when log (FCPP) > 7 (approx. 1090 people).
- ENE is the most relevant category for any value of FCPP.
- By analyzing the slope of trendlines, COM is a little less sensitive to an increase in personnel than the other categories.

In short, PD can be applied to analyze a combination of features in order to support decision-making through model interpretability. Additionally, these IML procedures can guide future data collection efforts towards an improvement in model performance and even help the organization evaluate the error levels of the models.

5 Conclusions

This study is a first attempt to holistically understand utility expenses within the Brazilian Navy, across a large set of facilities that individually and independently manage their expenses. In order to achieve this purpose, several ML models were trained on a dataset representing multiple attributes of the organization's facilities. Afterwards, an IML technique was used on the best models to identify the most relevant features based on how strongly they contribute to the model's predictive power. Finally, examples of how the predictive power and interpretability of the models can be applied to support decision-making within the organization were presented.

Admittedly, some limitations can be pointed out, and the first one is the data itself. Dealing with secondary data from distinct sources hindered the merging process, as it was necessary to deal with different formats and unnecessary parts. As a result, two features were initially discarded from the learning process due to incompatibility issues. They were included in a reduced alternative dataset that was able to elucidate that these features are relevant and could have significantly improved the model performance.

Another limitation is the lack of prior research with the same characteristics, where a single organization is composed of hundreds of facilities that demand resources at different rates to cope with utilities expenses. This fact made it preferable to take a rather basic approach focusing on identifying the relevance of ML to the solution of the problem instead of searching for a specific method that could maximize the predictive performance of the results.

Finally, some opportunities for further research could also be identified. The basic approach adopted in this study increases the interest in implementing techniques such as Clustering and Principal Component Analysis, as well as training a broader list of models that could make the algorithm deal with a more extensive set of problems. For the decision-maker, introducing facilities comparison capabilities with methods like Data Envelopment Analysis could be of great help. It would also be of great interest to evaluate the performance of the methodology as an organizational framework towards data-driven decision making. Once implemented, it can be reused with more features and little human intervention, supporting decision-makers and guiding future data collection efforts.

Acknowledgement. This paper is financed by National Funds of the FCT – Portuguese Foundation for Science and Technology within the project No. UIDB/03182/2020.

References

1. Curristine, T., Lonti, Z., Joumard, I.: Improving public sector efficiency: challenges and opportunities. OECD J. Budg. **7**, 161 (2007)
2. Damanpour, F., Schneider, M.: Characteristics of innovation and innovation adoption in public organizations: assessing the role of managers. J. Pub. Adm. Res. Theory **19**, 495–522 (2009)
3. Janssen, M., Estevez, E.: Lean government and platform-based governance-doing more with less. Gov. Inf. Q. **30**, S1–S8 (2013). https://doi.org/10.1016/j.giq.2012.11.003

4. de Vries, H., Bekkers, V., Tummers, L.: Innovation in the public sector: a systematic review and future research agenda. Pub. Adm. **94**, 146–166 (2016). https://doi.org/10.1111/padm. 12209

5. Agbozo, E., Asamoah, B.K.: Data-driven e-government: exploring the socio-economic ramifications. eJournal eDemocracy Open Gov. **11**, 81–90 (2019). https://doi.org/10.29379/jedem. v11i1.510

6. Christodoulou, P., et al.: Data Makes the Public Sector Go Round BT - Electronic Government. Presented at the (2018)

7. Marinha do Brasil: Estrutura Organizacional. https://www.marinha.mil.br/estrutura-organi zacional. Accessed 12 Jan 2020

8. Ministério da Defesa: Marinha do Brasil. https://www.defesa.gov.br/forcas-armadas/marinha-do-brasil. Accessed 12 Dec 2020

9. De Rezende, L.B., Blackwell, P.: The Brazilian national defence strategy: defence expenditure choices and military power. Def. Peace Econ. 1–16 (2019). https://doi.org/10.1080/10242694. 2019.1588030

10. Svendsen, N.G., Kalita, P.K., Gebhart, D.L.: Environmental risk reduction and combat readiness enhancement of military training lands through range design and maintenance. In: 2005 ASAE Annual International Meeting (2005)

11. Morrel-Samuels, P., Francis, E., Shucard, S.: Merged datasets: an analytic tool for evidence-based management. Calif. Manage. Rev. **52**, 120–139 (2009). https://doi.org/10.1525/cmr. 2009.52.1.120

12. Brynjolfsson, E., McElheran, K.: The rapid adoption of data-driven decision-making. Am. Econ. Rev. **106**, 133–139 (2016). https://doi.org/10.1257/aer.p20161016

13. Kumar, V., et al.: Data-driven services marketing in a connected world. J. Serv. Manag. **24**, 330–352 (2013). https://doi.org/10.1108/09564231311327021

14. Lerzan, A.: How do you measure what you can't define? The current state of loyalty measurement and management. J. Serv. Manag. **24**, 356–381 (2013). https://doi.org/10.1108/JOSM-01-2013-0018

15. Kiron, D.: Lessons from Becoming a Data-Driven Organization. MIT Sloan Manag. Rev. **58** (2017)

16. Jang, H.: A decision support framework for robust R&D budget allocation using machine learning and optimization. Decis. Support Syst. **121**, 1–12 (2019). https://doi.org/10.1016/j. dss.2019.03.010

17. Kartal, H., Oztekin, A., Gunasekaran, A., Cebi, F.: An integrated decision analytic framework of machine learning with multi-criteria decision making for multi-attribute inventory classification. Comput. Ind. Eng. **101**, 599–613 (2016). https://doi.org/10.1016/j.cie.2016. 06.004

18. Bilal, M., Oyedele, L.O.: Guidelines for applied machine learning in construction industry—a case of profit margins estimation. Adv. Eng. Informatics. **43**, 101013 (2020). https://doi.org/ 10.1016/j.aei.2019.101013

19. Robinson, C., et al.: Machine learning approaches for estimating commercial building energy consumption. Appl. Energy **208**, 889–904 (2017). https://doi.org/10.1016/j.apenergy.2017. 09.060

20. Pallonetto, F., De Rosa, M., Milano, F., Finn, D.P.: Demand response algorithms for smart-grid ready residential buildings using machine learning models. Appl. Energy **239**, 1265–1282 (2019). https://doi.org/10.1016/j.apenergy.2019.02.020

21. Yaseen, Z.M., et al.: Predicting compressive strength of lightweight foamed concrete using extreme learning machine model. Adv. Eng. Softw. **115**, 112–125 (2018). https://doi.org/10. 1016/j.advengsoft.2017.09.004

22. Xiao, Y., Wu, J., Lin, Z., Zhao, X.: A deep learning-based multi-model ensemble method for cancer prediction. Comput. Methods Programs Biomed. **153**, 1–9 (2018). https://doi.org/10.1016/j.cmpb.2017.09.005

23. Ye, H., Liang, L., Li, G.Y., Kim, J., Lu, L., Wu, M.: Machine learning for vehicular networks: recent advances and application examples. IEEE Veh. Technol. Mag. **13**, 94–101 (2018). https://doi.org/10.1109/MVT.2018.2811185

24. SIPRI - The World Bank: Military expenditure (% of GDP) | Data. https://data.worldbank.org/indicator/MS.MIL.XPND.GD.ZS. Accessed 28 Jan 2020

25. Alptekin, A., Levine, P.: Military expenditure and economic growth: a meta-analysis. Eur. J. Polit. Econ. **28**, 636–650 (2012). https://doi.org/10.1016/j.ejpoleco.2012.07.002

26. Caruso, R., Francesco, A.: Country survey: military expenditure and its impact on productivity in Italy, 1988-2008. Def. Peace Econ. **23**, 471–484 (2012). https://doi.org/10.1080/10242694.2011.608964

27. Hou, D.: The determinants of military expenditure in Asia and Oceania, 1992–2016: a dynamic panel analysis (2018). https://doi.org/10.1515/peps-2018-0004

28. Rath, M., Pattanayak, B.K., Pati, B.: Energy efficient MANET protocol using cross layer design for military applications. Def. Sci. J. **66**, 146–150 (2016). https://doi.org/10.14429/dsj.66.9705

29. Sudhakar, I., Madhusudhan Reddy, G., Srinivasa Rao, K.: Ballistic behavior of boron carbide reinforced AA7075 aluminium alloy using friction stir processing – an experimental study and analytical approach. Def. Technol. **12**, 25–31 (2016). https://doi.org/10.1016/j.dt.2015.04.005

30. Stablein, R.: Data in Organization Studies. Sage Publications, London (1999)

31. Instituto Brasileiro de Geografia e Estatística: Cidades e Estados: Rondônia. https://www.ibge.gov.br/pt/cidades-e-estados.html. Accessed 22 Jan 2020

32. Insituto Nacional de Meteorologia - INMET: BDMEP - Banco de Dados Meteorológicos para Ensino e Pesquisa. http://www.inmet.gov.br/portal/index.php?r=bdmep/bdmep. Accessed 19 July 2019

33. Osborne, J.: Improving your data transformations: Applying the Box-Cox transformation. Pract. Assess. Res. Eval. **15**, 12 (2010)

34. R Core Team: R: A Language and Environment for Statistical Computing (2017). https://www.r-project.org/

35. van Buuren, S., Groothuis-Oudshoorn, K.: {mice}: multivariate imputation by chained equations in R. J. Stat. Softw. **45**, 1–67 (2011)

36. Kursa, M.B., Rudnicki, W.R.: Feature Selection with the Boruta Package. J. Stat. Softw. **36**, 1–13 (2010)

37. Friedman, J.H.: Greedy function approximation: a gradient boosting machine. Ann. Stat. **29**, 1189–1232 (2001)

The Role of Animal Spirit in Monitoring Location Shifts with SVM: Novelties Versus Outliers

Iulia Igescu$^{(\boxtimes)}$ (iD)

National Bank of Romania, 030031 Bucharest, Romania
iulia.igescu@bnro.ro

Abstract. In 2018 a process change in the industry sector initiated at the "center" (West Europe) induced internal spillover effects at the end of the value chain (Romania) and propagated further affecting the "center" with a lag. Classical econometrics deals poorly with circular change. Support Vector Machines (SVM) can deal with that. As it can quantify small movements in "animal spirit" (coming from survey data) as novelties, that way it can monitor equilibrium changes (location shifts) ahead of hard indicators (industrial production index). This is a positive side of "animal spirit." Confronted with a Rare Event, an unexpected change like that of the Virus Crisis in 2020, the same survey data produce outliers. "Animal spirit" shows its negative side as irrational behavior.

Keywords: Location shifts · Equilibrium change · Monitoring · Support vector machines · Animal spirit · Novelties · Outliers

1 Motivation

Large shifts in economic variables, known as location shifts due to Hendry [1], occur mostly at irregular times and are a main source of concern for policy makers and scholars.

Scholars find forecasting difficult during such shifts. Hendry [1] proposes co-breaking to deal with location shifts econometrically: a linear combination of variables whose result does not depend on their dynamical properties.

Policy makers are in turn interested in monitoring location shifts, as they imply a change in equilibrium. For example policy makers would like to identify turning points that could possibly be shifting the variable to an even lower or higher location. According to Castle, Clements, and Hendry [2] one can exploit that mean change alters the growth and variance of a variable. Often such shifts start with peculiar one-period spike-like movements. However, sometimes location shifts produce no observable step, only a response that takes time to show

With support from National Bank of Romania; the usual disclaimer applies. The author would like to thank reviewers and participants at the 6th International Conference on Machine Learning, Optimization, and Data Science LOD 2020 for useful comments.

© Springer Nature Switzerland AG 2020
G. Nicosia et al. (Eds.): LOD 2020, LNCS 12565, pp. 72–82, 2020.
https://doi.org/10.1007/978-3-030-64583-0_8

up in the data. According to the above mentioned authors, this is due to inter-temporal co-breaking, increasing economic uncertainty.

With increased uncertainty, financial sector turns often to soft indicators, especially industry survey data, for forecasting and investment decision making. Survey data are faster than hard indicators. They are easily available from public sources for all countries and give information along many dimensions of a sector. Even though they are not really time series per-se, just a sum of opinions, the financial sector takes them as leading indicators. Often only newest survey counts in their investment decisions. This paper shows that while survey data are useful when change is small and planned, they fail with unexpected change.

Clements and Hendry [3] point out that leading indicators are often not causally related to the variables they lead. If they are subject to breaks, their relationship is unlikely to co-break. Scatter plots show that in fact survey indicators have different regimes, i.e. locations. One should first identify the current regime of a soft indicator. Regimes are not clearly separated, with values from one location tainting the other locations. Moreover, hard and soft indicators are not simply cointegrated, their relationship is in fact "circular". One would also suspect, as per Keynes, that survey data are subject to "animal spirit." An unusual shock should induce an unjustified confidence loss in those surveyed. Financial sector could take it as perfect foresight, generating panic in financial markets, in a case of what Farmer and Woodford [4] call "self-fulfilling prophecies". The authors have long advocated treating "animal spirit" as a fundamental, like capital and labor. Virus Crisis 2020 is an example of such a shock. It is a Rare Event, adding unexpectedly to an ongoing location shift in industry. It induced a confidence loss greater than Great Recession 2008.

As industry is organized in value chains that produce across country borders, change in one country induces change in countries down the value chain, often with a lag. Change then propagates back to the originating country. Induced by the "center" (West Europe), Romanian industry sector has been going through a slow, yet protracted process of change since 2018. There was no major warning signal (no data spike, for example). This country is preferred because it is at the end of the value chain. This industry location shift makes the object of this paper to better understand the role of "animal spirit" in the economy. Ideally one would like to identify a turning point, a way out of a bad equilibrium. In case of an additional shock - such that of a Rare Event - a policy maker must decide to discard or not this shock in future monitoring or to change the set of survey indicators and classification method previously used.

This paper caters to a policy maker interested in monitoring such a shift. It uses Support Vector Machines (SVM) introduced by Schoelkopf et al. [5] with survey data (soft data) to indirectly monitor the evolution of a hard indicator, in this case industry production index. Two soft indicators are selected. One is proxy for domestic factors and one for external factors, both inducing the location shift of the hard indicator. SVM finds a combination of samples from the current distribution of the two soft indicators to build a hyper-plane to maximize the distance between the observations from the old and the new location. It separates a new observation from the old observations, when classified as "unusual". This is called a "novelty." Smola and Schoelkopf [6] offer a good introduction into this topic.

One major weakness of SVM is that it is sensitive to outliers. SVM could not handle Virus Crisis 2020 and its massive drop in survey data levels, a sign that these values could in fact be outliers. Therefore one would expect for the hard indicator to return fast to previous levels once this Rare Even is over. The Virus Crisis would be an example of "animal spirit" tainting soft data, not a crisis with structural changes like Great Recession. Indeed, hard indicator values returned fast to original levels. In this case one could change the classification method for the duration of the crisis. Local Outlier Factor (LOF) algorithm managed to classify outliers of March-June 2020 during the Virus Crisis. However, LOF needs a bigger data sample and results were inferior to SVM during the location shift of 2018–2019. One could also look for a new set of soft indicators during this period.

2   The Basic Idea

Suppose we have n observations from the same distribution described by p-features. Take $p = 2$, as there are two soft indicators. Let us add one more observation. Detect if the newest observation measured across these two dimensions is in fact unusual, i.e. it comes from a different distribution. This is called a novelty. Novelties tend to form a cluster if they are in a low density region of the old data used for training.

Mathematically speaking, a density exists if the underlying probability measure possesses an absolutely continuous distribution function. SVM introduced by Schoelkopf et al. [5] is applicable to cases where the density of the data distribution is not even well-defined, e.g. if there are singular components. In situations where the goal is to detect novelties, it is not always necessary to estimate a full density model. Classical regression looks at how many training points fall into the region of interest. This algorithm starts by training points that fall into a region, and then estimates a region with the desired property. As there are many such regions, the algorithm imposes for the region to be small.

A frontier for the region requires a kernel and a scalar parameter, ν. This parameter corresponds to the probability of finding a new, yet regular, observation outside the frontier (this is the error parameter). There is no exact algorithm to set the band-width parameter ν. The measure of smallness depends on the kernel used. The authors use RBF (Radial Basis Function) kernel with two parameters: C and γ. The parameter C trades off accuracy against simplicity of the decision surface. A high C aims at classifying all training examples correctly, making the decision surface look rough. The γ parameter is the inverse of the radius of influence of samples selected by the model as support vectors. The lower γ, the more influence has a training sample that is far away. Even though calibration of C and γ can affect results, this paper keeps the calibration recommended by Schoelkopf et al. [5], as it fits the low size of the training sample.

2.1 Novelty Versus Outlier Detection

Support Vector Machines (SVM) and Outliers. Novelties can form a dense cluster, as long as they have a lower density than that of the training data. This is what helps separate data when distributions change. In SVM the question is how isolated the sample is, making SVM sensitive to outliers. SVM does great in small sample sizes.

Local Outlier Factor (LOF) and Outliers. In LOF the question is how isolated the sample is with respect to the surrounding neighborhood, making it able to detect outliers. LOF can also detect novelties. Technically, LOF uses RBF kernel with calibration as above. Additionally, results depend also on the number k of neighbors considered, normally around 20. It does poorly in small sample sizes, as in location shifts. In this paper k is size of the training sample.

3   Assessing Survey Data

Romanian industry is a good choice to study location shifts because it experienced a complex process of transformation: massive technological improvement, deep institutional change, and two Rare Events in only 30 years.

Soft indicators selected in this paper look at two dimensions of change, internal and external factors. *Production trend observed in recent months* (Trends Soft) is proxy for internal factors, while *Assessment of export order-book levels* (Exports Soft) for external ones. The target indicator is *Industry production index* (Industry Hard). Data are from European Commission for soft and from Eurostat for hard indicators, monthly.

Data from 2007 to June 2020 indicate some peculiarities of soft indicators. A scatterplot of a hard indicator versus a soft indicator (see below Fig. 1 with domestic trends as soft indicator) shows the big difference between a hard indicator, a time series whose values evolve in time, and a soft indicator, a survey whose values tend to move around determined bounds.

There are four regimes corresponding to the four structural breaks and two Rare Events: the Great Recession (big dots in Fig. 1) of 2008 and the Virus Crisis of 2020 (see triangles). The evolution during a Rare Event is peculiar to each event. During 2008 Great Recession both soft and hard indicators posted protracted drops, with soft indicators leading hard indicators. They formed a cluster, suggesting that industry output had moved to a lower location with lower levels of output - a bad equilibrium. During the Virus Crisis, values seem in fact outliers, as massive drops in the values of soft indicators look rather random.

The switch between regimes (location shifts) in normal times is often marked by an apparent "squeeze" in the values of the soft variable (see for example 2013 values marking a switch from the cross to the square regime in Fig. 1). Often during this switch values from the old location are invading the new location. That has become even more pronounced in recent years, when variation in the

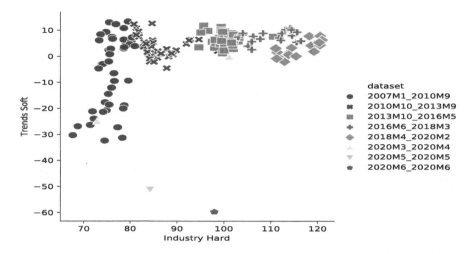

Fig. 1. Data on soft indicators (domestic factors) versus hard indicator (industrial production index) for 2007–2020 indicate four dynamical regimes and two Rare Events. Rather "squeezed" soft indicator values between regimes indicate structural breaks. In the Rare Event of 2008 there was a clear cluster (see big dots) of lower values. Industrial production experienced a location shift with a lower level of output. "Animal spirit" affects values of March-June 2020 (see triangles), indicating in fact outliers. One would expect in this case a fast recovery of the hard indicator.

soft indicators became minimal. A scatterplot of the soft indicators (see Fig. 2) illustrates that.

The recently lower volatility in soft indicators requires formal structural break tests. This is done in the next section.

3.1   Structural Breaks

Table 1 below shows the succession of breaks in the soft and hard indicators for 2018–2019 breaks.

Table 1. Breaks in hard and soft indicators.

	Exports soft	Industry hard	Trends soft	Exports soft	Trends soft
Break date	2018M3	2018M10	2019M4	2019M5	2019M7
Intercept	Negative	-	Negative	Negative	Negative
Trend	-	Negative	-	-	-
Break type	Fast	Fast	Fast	Slow	Slow

The steady and double digit wage increase in Romania since 2016 forced super-star firms to introduce labor saving technologies. The "center" of the value

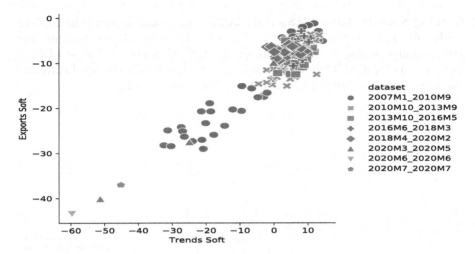

Fig. 2. Scatter plot of soft indicators in 2007–2020. In normal times values of different locations overlap. The 2020 Rare Event generated massive value drops. Only in July 2020 they started moving upwards at fast speed (see diamond). This is in contrast to industry output that had already posted double digit growth rates in May 2020. They are therefore outliers, an expression of the bad side of "animal spirit".

chain (West Europe industry) initiates change in early 2018, quantified as a fast negative break in Exports Soft and a negative break in industrial production index by the end of 2018 at the end of the value chain (Romania). That manifested itself in a negative break in Trends Soft in early 2019, a worsening in the "mood" of domestic agents. As output change became protracted, it slowly induced two further breaks in both soft indicators in the middle of 2019, further worsening the overall "mood" in the economy for both domestic and external perspectives. Classical econometric methods have a hard time exploiting small and circular data changes.

4   Detecting Novelties with SVM

SVM exploits this "squeeze" between different regimes to separate them. The first step is to define an initial interval as training and re-scale values around mean zero and standard deviation 1. The same rescaling parameters are then used for the testing sample. That is why the training interval should not be tainted with values from the new location. Second, SVM classifies new observations with small departures from the training sample as "novelties." Third, SVM orders them according to their distance from the training sample.

4.1   Choosing a Training Sample

As SVM is semi-supervised learning, one should help the system with the training sample, as it should be untainted with values from the new location.

Training Sample April 2018 - July 2019. This training sample includes all breaks after Exports Soft break of March 2018. There are 16 observations. The testing sample is August 2019-February 2020. SVM classified August 2019 as the worst observation. Only three out of five novelties were classified correctly. The new location is tainted by the old location, as seen in Fig. 3.

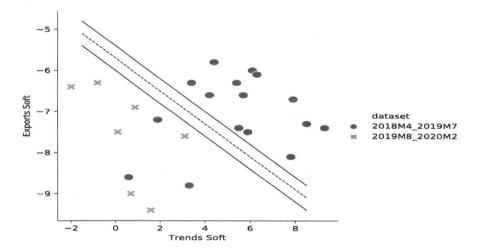

Fig. 3. Scatter plot of soft indicators regimes, as separated by SVM. Support vectors are full lines and separating hyperplane is the dashed line. With sample values from April 2018 till July 2019 as old location, too many old values invade the new location (see big dots in the territory of crosses). SVM generates too many errors with this sample choice. It seems that June and July 2019 observations are in fact from the new location.

Training Sample April 2018 - May 2019. This sample removes the last two breaks in Exports and Trends Soft from the training sample above. They are slow breaks, most likely a second-round effect of the Industry Hard indicator break in October 2018. There are now 13 observations left. The test sample this time is from June 2019 to February 2020.

4.2 Novelties Since June 2019

Schoelkopf et al. [5] recommend as calibration the following: ν is 0.1, γ is 0.1 and C is 0.5. With so few observations, one can hardly experiment with calibration. The current error parameter allows only for two errors in a training sample of 13, and one error in a testing sample of 9. The system learns the frontier and then sorts observations. Two observations out of 13 were misclassified, as they were close to the upper part of the frontier.

As diagrams below point out, domestic factors as measured by Trends Soft (a result of higher domestic wages) dominate this process of change, with most negative values. Each month, starting June 2019, SVM classifies the latest observation. If it is a novelty, it will be outside the learned frontier. The further away from the frontier, the closer to the location shift trough. The worst values so far were in October 2019 and January 2020, classified at the edge of the outermost area of the density. The observation for December bounced back, close to the training frontier. The outlook was leaning again towards an upward movement in February (see the star in Fig. 4). Bouncing away from the location shift trough implies most likely that this trough is not an attractor and the system will move back to higher output. Indeed industry output had positive growth in both January and February 2020.

All diagrams below use as source scikit-learn [7] in Python, have Exports Soft on the horizontal axis, and Trends Soft on the vertical. Training observations are white circles. Learned frontier is the dotted gray line. The newest observation is marked as a black star. Previous novelties are black circles, white stars, or white triangles.

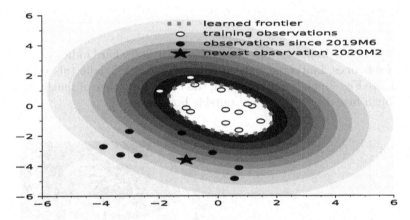

Fig. 4. After a worst October value, December moved up to 30% range. Even though January value was again in 90% distribution range, SVM in February 2020 was in the 60% range (see star). This bouncing up could signal that the system does not settle at a lower output.

5   Rare Event in March-April-May 2020

In March-April-May 2020 an exogenous shock of Rare Event type (a Virus Crisis) affected the World economy. As during Great Recession of 2008, soft indicators dropped dramatically. In March, even though most of the survey had already taken place and results were only partly affected, SVM produced a value already outside the range, hinting at an outlier.

Local Outlier Factor (LOF) classified only 8 out of 9 observations as novelties in the above selected sample (it had one error). It is inferior to SVM (with zero error) for the purpose of the studied location shift. However, LOF managed to position March value in a range close to the previous ones (see Fig. 5).

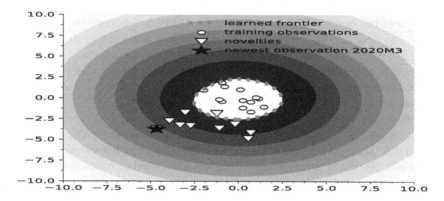

Fig. 5. LOF classifies new value of March 2020 still close to previous ones.

In April values dropped dramatically. Unlike in 2008, hard indicators dropped ahead of soft ones, hinting at outliers. SVM could not handle this value. LOF diagram (see Fig. 6), placed this value in the 70% outer range of the distribution. Even though big, the drop was not outside boundaries.

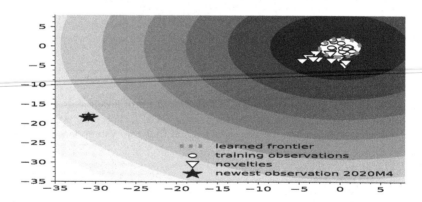

Fig. 6. New value in April 2020, even though far, not outside boundaries with LOF.

This assumption that "animal spirit" taints soft indicators gains support if one takes into account the evolution of another soft indicator, Expected Production. Even though Expected Production has not performed well as predictor for 2018–2019 location shift, it becomes a valuable source of information during this Rare Event. After a drop in March–April, values recover fast (see Fig. 7), and in June reach March levels again. The original soft indicators continued to worsen in May and June, even though by June industrial production index itself had reached March levels again, in line with the evolution of Expected Production. The explanation for this behavior could be that the first location shift was planned well in advance by the center and domestic agents had therefore information ahead of time. The Rare Event was unexpected and previous indicators, which have also a strong backward looking component, failed to handle it. One could take it as a case of "animal spirit" tainting soft indicators.

The strategy during this Rare Event is to switch to a new set of indicators. At the same time one should continue monitoring the old indicators with LOF rather than SVM, to see how fast "animal spirit" regains trust, so that one can resume monitoring the old location shift.

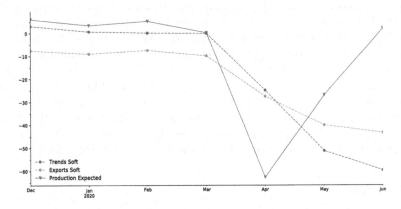

Fig. 7. By June 2020 other soft indicators such as Expected Production recovered to March levels (full line). Values of original soft indicators (Trends Soft and Exports Soft - dashed lines) continued to drop in May and June. By June 2020 industrial output had indeed reached its March level. One could take this outcome as proof of "animal spirit" tainting soft data. In a Rare Event, it is advised to change the set of soft indicators or to keep the old ones yet switch to LOF.

6   Concluding Remarks

In this paper Support Vector Machines (SVM) use a combination of soft indicators to describe the evolution of industry production index during its 2018–2019 location shift, a result of both internal factors (increased wages in Romania) and external factors (technological changes along the value chain coming from Germany and France). Even though it manifests itself as an output loss in a first stage, Romanian industry is in fact moving to a new equilibrium with higher productivity, higher wages, and increased export competitiveness. This endeavor allows a glimpse into the role of "animal spirit," as measured by survey indicators, during an equilibrium formation process.

One first conclusion is that "animal spirit" is not always ahead of change. Hard core changes reflect back into the "mood" of business participants. Hard and soft changes are circular. However, SVM can successfully exploits small "mood" changes to monitor the evolution of the hard indicator, as survey values are two months faster than the hard one. This is done indirectly, through a diagram with two soft indicators (one for domestic and one for external factors) to quantify how far they are from their original location. This is the positive side of "animal spirit".

As a Virus Crisis hit the World economy in March-April-May 2020, survey values dropped dramatically. SVM could not handle such a big drop, as its scope is to magnify small changes. Local Outlier Factor, that handles both outliers and novelties, manages to sort these values. It turns out that these values are in fact outliers. Therefore one would expect merely a heavy "psychological load" rather than quantitative consequences on current equilibrium. The industry sector should rapidly return to its previous growth path. Indeed, already in June 2020, the hard indicator moved to values similar to those of March 2020. The soft indicators that served so well in monitoring the location shift continued to drop in June 2020, giving insight into the negative side of "animal spirit": fear and irrational behavior taint soft indicators. In this case one would try to find a new set of soft indicators. Alternatively one could use LOF with the current indicators to monitor how long it takes to restore previous trust levels.

One second conclusion is that soft indicators therefore serve well only for a certain purpose and a limited amount of time. One has to carefully understand the regime (the current location) of both the hard and the soft indicator.

References

1. Hendry, D.F.: Dynamic Econometrics. Oxford University Press, Oxford (1995)
2. Castle, J., Clements, M., Hendry, D.: Overview of forecasting facing breaks. J. Bus. Cycle **12**, 3–23 (2016)
3. Clements, M., Hendry, D.: Forecasting Non-stationary Economic Time Series. MIT Press (2001)
4. Farmer, R., Woodford, M.: Self-fulfilling Prophecies and the Business Cycle. Macroeconomic Dynamics I. pp. 740–769 (1997)
5. Schoelkopf, B., Williamson, R., Smola, A., Shawe-Taylor, J., Platt, J.: Support vector method for novelty detection. In: Advances in Neural Information Processing Systems: Proceedings of the 1999 Conference http://papers.nips.cc/paper/1723-support-vector-method-for-novelty-detection.pdf (2000)
6. Smola, A., Schoelkopf, B.: A tutorial on support vector regression. Stat. Comput. **14**, 199–222 (2004)
7. Code source adapted from scikit-learn. https://scikit-learn.org/. Accessed 13 May 2020

Long-Term Prediction of Physical Interactions: A Challenge for Deep Generative Models

Alberto Cenzato[1], Alberto Testolin[1,2(✉)], and Marco Zorzi[1,3,4]

[1] Department of General Psychology, University of Padova, 35131 Padua, Italy
`alberto.cenzato@outlook.it`, {`alberto.testolin,marco.zorzi`}`@unipd.it`
[2] Department of Information Engineering, University of Padova, 35131 Padua, Italy
[3] Padova Neuroscience Center, University of Padova, 35131 Padua, Italy
[4] IRCCS San Camillo Hospital, 30126 Venice-Lido, Italy

Abstract. Anticipating salient events by actively predicting the sensory stream is a core skill of intelligent agents. In recent years, a variety of unsupervised deep learning approaches have been proposed for modeling our ability to predict the physical dynamics of visual scenes: In this paper we conduct a systematic evaluation of state-of-the-art models, considered capable of learning the spatio-temporal structure of synthetic videos of bouncing objects. We show that, though most of the models obtain high accuracy on the standard benchmark of predicting the next frame of a given sequence, they all fall short when probed with the generation of multiple future frames. Our simulations thus show that the ability to perform one-step-ahead prediction does not imply that the model has captured the underlying dynamics of the environment, suggesting that the gap between deep generative models and human observers has yet to be filled.

Keywords: Unsupervised deep learning · Recurrent networks · Generative models · Video prediction · Visual perception · Intuitive physics

1 Introduction

The predictive processing framework conceives the brain as a powerful inference engine, constantly engaged in attempting to match sensory inputs with top-down expectations (Clark 2013; Friston 2010). According to this view, cortical circuits exploit feedback loops to learn a generative model of the environment, which can be used to actively produce a "virtual version" of the incoming sensory stream. Both behavioral (Fiser and Aslin 2001) and neurobiological evidence (Berkes et al. 2011; Jia et al. 2020) suggest that the process of building such internal models is mostly unsupervised (Hinton 2007; Testolin and Zorzi 2016). However, despite the recent ground-breaking achievements of recurrent neural networks (Graves et al. 2013; Sutskever et al. 2014), modeling complex high-dimensional sequences such as video streams has been usually framed as a

A. Cenzato and A. Testolin—Equal contribution.
This work was supported by the Cariparo Excellence Grant 2018 "Numsense" to M.Z., and by the Stars Grant "Deepmath" from the University of Padova to A.T.

ⓒ Springer Nature Switzerland AG 2020
G. Nicosia et al. (Eds.): LOD 2020, LNCS 12565, pp. 83–94, 2020.
https://doi.org/10.1007/978-3-030-64583-0_9

supervised learning problem, where the goal is to produce a series of labels or natural language descriptions of the video content (e.g., Donahue et al. (2015)).

In this paper we carry out a systematic evaluation of state-of-the-art unsupervised generative models which have been proposed to model sequences of frames in high-dimensional videos. We consider a prototypical Lab setting, which involves synthetic videos whose content and dynamics can be more easily controlled. In particular, we employ the *bouncing balls* dataset, which consists of synthetic videos of 3 white balls bouncing within a constrained box of black background, without friction nor gravity. The first attempt in modeling this kind of high-order structure using unsupervised learning was based on a probabilistic generative model, extending the restricted Boltzmann machine to the sequential domain (Sutskever et al. 2009). A step forward was proposed in Gan et al. (2015), where a dynamic generative model was built using a sigmoid belief network that used a directed graph to generate sequences efficiently. Another approach was explored by Srivastava et al. (2015), which exploited a Long-Short Term Memory (LSTM) encoder-decoder network to learn both synthetic and realistic video sequences. The advantage of LSTMs over more traditional recurrent architectures lies on their ability to selectively retain temporal information depending on the input structure. Finally, inspired by modern theories of cortical computation, recent studies have tried to capture the complexity of video streams using predictive coding mechanisms (Lotter et al. 2015, 2016; Oord et al. 2018). In particular, Lotter et al. (2015) reported state-of-the-art performance on the bouncing balls dataset in the one-step-ahead prediction task.

Despite these methodological advances and the increasing model complexity, the present study reveals that none of the considered models captures the underlying dynamics of these synthetic videos, as measured by a long-range prediction task. This finding shows that one-step-ahead prediction cannot be considered a reliable indicator of model quality, thus calling for the adoption of stricter evaluation metrics based on long-range autoregressive tasks.

For the sake of completeness, we should mention the existence of alternative modeling approaches placing stronger emphasis on explicit representations in learning and processing of structured information (Goodman et al. 2016). Though such framework has also been used for learning physical dynamics from synthetic videos (Chang et al. 2017; Zheng et al. 2018), it assumes that the model is inherently endowed with some generic knowledge of objects and their interactions, and the main purpose of learning is "only" to efficiently adapt to the specific properties observed in different environments. Our focus here is instead on models where no *a priori* knowledge is built into the agent.

2   Temporal Deep Generative Models

In this section we provide a brief formal characterization of the models considered, in order to highlight their main features; for an exhaustive description the reader is referred to the original articles.

LSTM. All models are based on LSTM networks (Hochreiter & Schmidhuber, 1997) so we use LSTMs both as a baseline model and for establishing a common notation that will be used throughout the paper. The LSTM used here has *input* (i_t), *forget* (f_t), and *output* (o_t) gates with no peephole connections; its architecture is summarized in the following equations:

$$i_t = \sigma(W_{ii}x_t + b_{ii} + W_{hi}h_{(t-1)} + b_{hi})$$
$$f_t = \sigma(W_{if}x_t + b_{if} + W_{hf}h_{(t-1)} + b_{hf})$$
$$o_t = \sigma(W_{io}x_t + b_{io} + W_{ho}h_{(t-1)} + b_{ho})$$
$$g_t = \tanh(W_{ig}x_t + b_{ig} + W_{hg}h_{(t-1)} + b_{hg})$$
$$c_t = x_t \odot c_{(t-1)} + i_t \odot g_t$$
$$h_t = o_t \odot \tanh(c_t)$$

where σ is the logistic sigmoid activation function, Wx and Wh are matrix-vector multiplications and $\odot$ is vector element-wise multiplication. The LSTM retains a memory of its past inputs in the state vectors h_t and c_t, which are respectively the hidden and cell states of the network. The first one is the LSTM cell output which, in the case of stacked networks, is propagated as input to the next LSTM. c_t is the internal memory storage of the cell; the gates regulate how much of the current input to add to this state and what to remove from it. Finally g_t is a non-linear transformation of the input, whose addition to the cell state is regulated by the input gate i_t. The $60 \times 60 \times 1$ grayscale input frames X_0, $X_1, \ldots, X_t$ are flattened into a sequence of vectors of size 3600: $x_0, x_1, \ldots, x_t$. The input sequence is fed into a stack of LSTM layers, which outputs the predicted next frame $\hat{x}_{t+1}$. To predict k frames ahead, each output $\hat{x}_{t+i}$ is fed back as input to the LSTM to obtain $\hat{x}_{t+i+1}$ until $\hat{x}_{t+k}$ is reached.

ConvLSTM. First proposed by Xingjian et al. (2015), ConvLSTMs are designed to improve vanilla LSTM's ability in dealing with spatio-temporal data. In ConvLSTMs the matrix multiplications of the LSTM cell, which account for a fully connected gate activation, are replaced with a convolution over multiple channels. This means that hidden (h_t) and cell (c_t) states become tensors of shape $60 \times 60 \times$ number of channels. Prediction works as described for LSTM, but instead of using x_i as input vectors we use the X_i frames. The use of convolutional rather than fully connected gates gives a spatial prior that encourages the network to learn moving features detectors using less parameters. The LSTM equations reported above remain the same, except for the replacement of matrix-vector multiplications with convolutional operations (e.g.., $W_{ii}x_t$ becomes $W_{ii} \star x_t$).

Seq2seq ConvLSTM. Predicting future video frames can be cast as a sequence-to-sequence mapping problem: given the frames $X_0, X_1, \ldots, X_t$ as input sequence, we want to predict $X_{t+1}, X_{t+2}, \ldots, X_{t+k}$ as output sequence. The two sequences do not necessarily have the same length: in a long-term generation setting, we might want the output sequence to be as long as possible, while providing a short input sequence (i.e., the context). Sutskever et al. (2014) proposed an effective sequence-to-sequence model that maps an input sequence into a fixed-size vector, and then uses this vector to produce the output sequence. This model was first introduced in the context of natural language processing, but can be adapted to image sequences. As in Srivastava et al. (2015), we implemented it as a stack of LSTMs which map the flattened video frames $x_0, x_1, \ldots,$

x_t into the vectors (h_t^l, c_t^l), which are respectively the hidden and cell states of the lth LSTM layer. The next frame x_{t+1} can be predicted starting from the state (h_t^l, c_t^l) and giving x_t as input to another LSTM stack; for multi-steps predictions, $\hat{x}_{t+i}$ is then fed back until $\hat{x}_{t+k}$ is reached. A diagram of the model architecture is shown in Fig. 1. To make the model computationally more efficient, following Xingjian et al. (2015) fully-connected LSTMs are replaced by ConvLSTM layers and an additional 1×1 convolutional layer that receives the concatenation of all decoder's channels and outputs the predicted image.

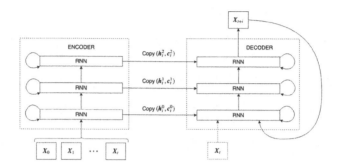

Fig. 1. Architecture of an autoregressive sequence-to-sequence model with generic RNN layers; in this work they are either LSTM or ConvLSTM layers, so we copy the hidden states (h_t^l, c_t^l), but in principle one could use any RNN.

Multi-decoder seq2seq ConvLSTM. Building on top of the seq2seq LSTM model, Srivastava et al. (2015) attached multiple decoders on the same encoder allowing multiple output sequences for the same input sequence. Carefully choosing decoders' tasks to be different, but not orthogonal, forces the encoder to learn a common representation with features that can be used by all decoders. As Srivastava et al. (2015) we used two decoders, one to reconstruct the reversed input sequence and the other to predict future frames; both tasks can be accomplished exploiting the same information: position, speed and direction of the balls at the end of the input sequence. Training the encoder together with two decoders should encourage the encoder to learn a representation that does not simply memorize the sequence, but still contains the relevant temporal context. We tried to improve the model of Srivastava et al. (2015) by using ConvLSTM layers instead of LSTM layers; this should maintain the advantages of seq2seq and multi-decoder models, while adding the spatio-temporal capabilities of ConvLSTM. Multi-decoders aside, the implementation is the same of seq2seq ConvLSTM.

RTRBM, DTSBN and Predictive Coding Models. As benchmarks, we also compare our results with the one-step prediction accuracy previously reported for the Recurrent Temporal Restricted Boltzmann Machine (Sutskever et al. 2009), the Deep Temporal Sigmoid Belief Network (Gan et al. 2015) and the predictive coding model of Lotter et al. (2015).

3 Training and Testing Methods

The best hyperparameters for each model were established using a grid-like search procedure, varying one parameter at a time and keeping the others fixed; for each architecture we varied the number of layers, hidden units and/or channels, learning rate, number of epochs and batch size. All models were trained on a training dataset with 6000 sequences of 40 frames each; hyperparameters optimization was performed on a validation dataset composed by 1200 videos, while final evaluation was carried out on a separate test set of 1200 videos. All models were implemented in PyTorch; the complete source code is available for download[1] along with a script to generate the bouncing balls dataset, adapted from Sutskever et al. (2009).

The best configuration resulting for each model was then trained using two strategies. In the simplest one, the model receives in input the frame X_i and predicts $\hat{X}_{i+1}$ at each time step i; this procedure is called *teacher forcing* because the new input is always *given*, and does not depend upon previous model predictions. One main issue with teacher forcing is that the model only observes sample frames from the training dataset; during long-term generation, though, it must autoregressively predict k frames ahead and therefore will receive its own predictions $\hat{X}_{t+i}$ as input from frame $t+1$ to frame $t+k$. Such input is not drawn from the training dataset, but from the model's frame-generating distribution; the model does not expect possible errors in $\hat{X}_{t+i}$, which quickly sum up producing increasingly worse predictions at each frame.

One way to deal with the mismatch between the training and generated distributions is to train the model on its own predictions. During training, after t ground-truth context frames, the model is given its own subsequent predictions $\hat{X}_{t+i}$ as input; the error is then computed, as for teacher forcing, in terms of the difference between the predicted and ground-truth sequences:

$$\frac{1}{k} \sum_{i=1}^{k} L(\hat{X}_{t+i}, X_{t+i})$$

where $L(\cdot)$ is the loss function. However, since at early learning stages it might be hard for the model to learn in this "blind" prediction mode[2] we also explored a *curriculum training* routine inspired by Zaremba and Sutskever (2014). Our curriculum strategy progressively blends teacher forcing into blind training, by increasing the task difficulty as the model becomes more and more accurate during learning. Training starts in a simple teacher forcing setting, where X_{t+i} is used as input to predict frames X_{t+i+1}; after 10 epochs the task difficulty is raised and to predict one of the X_{t+i+1} frames the previously predicted frame $\hat{X}_{t+i}$ is used. So 1 out of k frames is generated using the model's own prediction as input. Task difficulty is increased every 10 epochs, each time using one additional predicted frame as input. We hypothesized that curriculum training would facilitate long-term predictions, being the model gradually accustomed to perform predictions on its own generated frames.

Model performance is evaluated using the Mean Squared Error (MSE) between predicted and ground-truth frames. Moreover, we introduce an additional evaluation

---

[1] https://github.com/AlbertoCenzato/dnn_bouncing_balls.
[2] We call it "blind" because the model, when predicting frames $\hat{X}_{t+i}$ does not have any information about the ground-truth frames X_{t+i}.

metric based on the distance between the centroid of the jth predicted ball $(\hat{c}_j)$ and the centroid of its nearest neighbor ball in the ground-truth frame (c_i):

$$d(\hat{F}_{t+i}, F_{t+i}) = \sum_{j=0}^{\hat{n}-1} ||\hat{c}_j - c_j||_2 + |\hat{n}_{t+i} - n_{t+i}| \cdot 60\sqrt{2}$$

where $\hat{n}_{t+i}$ is the number of balls in $\hat{F}_{t+i}$ and n_{t+i} is the number of balls in F_{t+i}. The last term of the equation adds a penalty to $d(\cdot)$ when there is a mismatch between $\hat{n}_{t+i}$ and n_{t+i}; $60\sqrt{2}$ was chosen because it is the diagonal length of the 60×60 squared frame which is more than the maximum distance between two balls. Simple computer vision algorithms were used to determine the coordinate of balls in each frame[3]. Both MSE and centroid distance are measured for each frame, and then averaged over all sequences of the testing dataset.

4   Results

Learning the bouncing balls dataset proved to be a particularly hard optimization problem. For many hyperparameter configurations the optimizer got stuck in plateaus or local minima; we show an example of the resulting training loss (MSE) in the left panel of Fig. 4. Most of the time these plateaus were associated with minimum activation on all output units (i.e. empty, black frames).

As shown in Fig. 2, in the one-step prediction task all models (apart for the vanilla LSTM) are able to accurately predict the next frame, given 10 frames as context. Table 1 reports both average MSE and centroid distances: even the simple ConvLSTM achieves excellent performance, and more complex models do not exhibit significant improvements. As expected, the centroid distance metric is well aligned with MSE. Vanilla LSTM is the only model unable to accurately perform one-step predictions, probably due to the high number of parameters and lack of explicit spatial biases: indeed, after 30 epochs learning did not yet converge and was stopped. Crucially, in the one-step-ahead prediction task all convolutional models are well aligned with the performance reported in previous studies: in fact, the ConvLSTM accuracy is even higher than the current state-of-the-art reported by Lotter et al. (2015).

However, accurate one-step-ahead prediction does not entail that the model has captured the underlying spatio-temporal dynamics of the videos. This phenomenon is clearly visible in Fig. 3, which shows samples of generated sequences in a more challenging setting, where the next 20 frames have to be predicted given a context of 10 frames[4]. The striking result is that most of the models cannot maintain the visual dynamics for more than just a few frames. A common misprediction pattern consists in expanding the balls into larger blobs: the balls are increasingly stretched and lose their shape. From a qualitative point of view, it seems that ConvLSTM distorts the objects but

---

[3] We compute the centroids of the connected components in each frame of the sequence, by first eroding each image and removing pixels under a fixed threshold. Erosion and threshold are applied to avoid that two colliding balls are detected as one single object.

[4] This interval allows to unroll the prediction for a sufficient amount of steps and it ensures that several collisions will occur (between the balls, or with the bounding box).

is able to maintain an overall sharp appearance for the distribution of active pixels. This might explain why it achieves the lowest error in the one-step-ahead prediction task: even if the trajectory dynamics is not captured at all, the spatial consistency between early consecutive frames is almost perfect. The seq2seq ConvLSTM is clearly the model that generates video sequences with the best visual appearance, though object dynamics seems to diverge from ground-truth after about 15 predicted frames. Additional examples of generated sequences for each model are included in the online source code folder, along with longer 200-step-ahead generations.

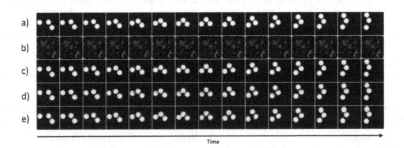

Fig. 2. Samples of one-step-ahead predictions from all the models considered: (a) ground-truth, (b) LSTM, (c) ConvLSTM, (d) seq2seq ConvLSTM, (e) seq2seq ConvLSTM multi-decoder.

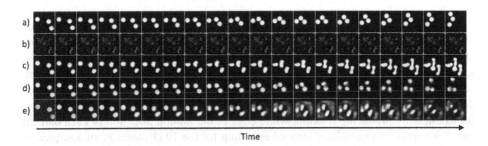

Fig. 3. Samples of 20-steps-ahead predictions from all the models considered: (a) ground-truth, (b) LSTM, (c) ConvLSTM, (d) seq2seq ConvLSTM, (e) seq2seq ConvLSTM multi-decoder.

Average MSE and centroid distances are reported in Table 2: ConvLSTM are indeed much less accurate in the long-range prediction task, and the best performance is achieved by seq2seq architectures. The right panel in Fig. 4 shows how errors rapidly accumulate when predicting 20 frames (the LSTM error has been included as a semi-random baseline). The seq2seq ConvLSTM obtains the best results: its prediction accuracy degrades gently as time passes, maintaining the best trajectory estimation.

Table 1. One-step-ahead prediction errors (mean and standard error) in terms of MSE and centroid distance in pixels. The latter is omitted for LSTM because the predicted frames are mostly filled with noise. As baseline we report the error of a null model that always predicts empty frames.

Model	MSE	Centroid distance
LSTM	111.09 ± 0.68	n.a
ConvLSTM	$\mathbf{0.58 \pm 0.16}$	$\mathbf{1.84 \pm 0.32}$
seq2seq ConvLSTM	1.34 ± 0.19	7.88 ± 0.76
seq2seq ConvLSTM multi-decoder	4.55 ± 0.40	6.80 ± 0.64
Sutskever et al. (2009)	3.88 ± 0.33	-
Gan et al. (2015)	2.79 ± 0.39	-
Lotter et al. (2015)	0.65 ± 0.11	-
Baseline (empty frame)	93.07	254.56

Table 2. Prediction errors for all models in the more challenging 20-step-ahead prediction task.

Model	MSE	Centroid distance
LSTM	111.11 ± 0.67	n.a
ConvLSTM	95.67 ± 3.31	61.76 ± 1.51
seq2seq ConvLSTM	27.13 ± 1.91	11.73 ± 0.60
seq2seq ConvLSTM multi-decoder	60.37 ± 1.79	67.88 ± 1.31

Interestingly, a qualitative analysis on the internal encoding of seq2seq architectures revealed how these models might encode the input sequence into a static state vector by creating a spatial representation of object trajectories. We tried to investigate how temporal information was encoded in this type of architectures by visualizing the channel activations corresponding to a specific input sequence. This allows to observe how the model internally uses a form of spatial encoding to capture the temporal structure, for example by producing a fading white edge in the motion direction of each ball. In Fig. 5 we show an example of internal encoding for the 10 channels h_t^3 of a sequence-to-sequence ConvLSTM multi-decoder model: channels 5 and 7 seems to exhibit a preference for the motion direction. A more striking form of encoding, though, can be obtained with a modified version of this model, where the encoder is forced to compress the whole input sequence in a tensor with only 2 channels. As shown in Fig. 6, in this case the channels encode the complete trajectory of the balls in the input sequence.

We finally investigated whether the curriculum teaching regimen, where ground-truth frames are provided during learning as prediction unfolds over time, would improve the generation performance. However, we did not observe a significant improvement in terms of metrics and/or visual appearance of the generated sequences.

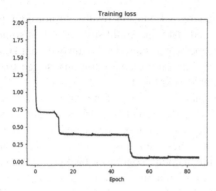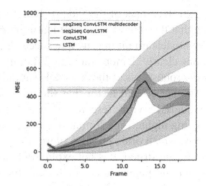

Fig. 4. Left: MSE loss during training, for one representative model. Right: curves showing how the prediction error for all models increases as frame generations are made deeper into the future. The shaded area around the lines corresponds to standard deviation of the mean error.

Fig. 5. Hidden state representation of the 3rd encoding layer for a sequence-to-sequence ConvL-STM multi-decoder model.

Fig. 6. Hidden state representation of the last encoding layer for a multi-decoder model with reduced number of channels, smaller kernels (3×3) and only this hidden state as bottleneck between encoder and decoder.

5   Conclusion

We presented an extensive empirical evaluation of several unsupervised deep learning architectures, considered among the state-of-the-art for modeling complex time series involving high-dimensional data. We focused our investigation on a popular dataset containing thousands of synthetic videos of bouncing balls, which has traditionally been used as a test bed in this domain and which can be considered as a good proxy of experiments carried out in psychophysics Labs.

Our simulations demonstrate that different approaches achieve good performance on a standard one-step-ahead prediction task. However, long-term prediction proved much harder than expected: all models considered, including those reporting state-of-the-art in the standard prediction task, fall short when probed on the task of generating multiple successive frames (for a similar setting in lexical generation, see Testolin et al. (2016)), suggesting that none of the models successfully captured the underlying physical dynamics of the videos. It should be stressed that our results do not demonstrate that these models *cannot* learn the required task: one could still argue that some of the models might in principle succeed, if properly parameterized. However, our simulations clearly demonstrate the difficulty in finding proper configurations, suggesting that the robustness of these models should be better explored.

Another crucial factor that should be considered is the loss function used to train the models. Previous research (Ranzato et al. 2014; Mathieu et al. 2015; Srivastava et al. 2015) highlighted the limitations of pixel-level loss functions such as MSE or cross-entropy, which often produce blurry images since they are focused on per-pixel appearance: image structure is not explicitly taken into account, so a black background filled with salt-and-pepper noise might have the same loss of a nice-looking frame containing objects with slightly distorted shapes or small coordinates offset. To ensure compatibility with previous results in our work we used MSE loss, but alternatives based on adversarial loss or mutual information are under investigation (Hjelm et al. 2018; Oord et al. 2018). Furthermore, although our simulations with the bouncing balls data set allow to clearly demonstrate the limitations of popular models using synthetic sequences, a similar evaluation should also be carried out with more realistic data sets (for a recent review, see Zhou et al. (2020)), where large-scale recurrent models achieve promising results (Villegas et al. 2019; Luc et al. 2020). An interesting venue for future research would be to also test predictive models that do not try to generate new frames from scratch, but rather transform pixels from previous frames by explicitly modeling motion (Finn et al. 2016) or try to first predict the high-level structure of the future sequence, and only subsequently generate the corresponding frames (Villegas et al. 2017; Xu et al. 2019).

Overall, our findings highlight the necessity to establish proper and common methodological practices to more carefully evaluate predictive generative models: achieving accurate performance in one-step prediction tasks does not necessarily imply that the model has captured the spatio-temporal structure in the data. Once a satisfactory model will be found, one critical direction for future research would then be to validate its performance against that of human observers, who are capable of tracking moving objects for a long time interval (Tresilian 1995), but also suffer from perceptual limitations when multiple objects are involved (Cavanagh and Alvarez 2005).

References

Berkes, P., Orbán, G., Lengyel, M., Fiser, J.: Spontaneous cortical activity reveals hallmarks of an optimal internal model of the environment. Science **331**(6013), 83–87 (2011)

Cavanagh, P., Alvarez, G.A.: Tracking multiple targets with multifocal attention. Trends Cogn. Sci. **9**(7), 349–354 (2005)

Chang, M.B., Ullman, T., Torralba, A., Tenenbaum, J.B.: A compositional object-based approach to learning physical dynamics. International Conference on Learning Representations (2017)

Clark, A.: Whatever next? predictive brains, situated agents, and the future of cognitive science. Behav. Brain Sci. **36**(3), 181–204 (2013)

Donahue, J., et al.: Long-term recurrent convolutional networks for visual recognition and description. In: Proceedings of the IEEE Conference on Computer Vision And Pattern Recognition. pp. 2625–2634. (2015)

Finn, C., Goodfellow, I., Levine, S.: Unsupervised learning for physical interaction through video prediction. In: Advances in Neural Information Processing Systems. pp. 64–72 (2016)

Fiser, J., Aslin, R.N.: Unsupervised statistical learning of higher-order spatial structures from visual scenes. Psychol. Sci. **12**(6), 499–504 (2001)

Friston, K.: The free-energy principle: a unified brain theory? Nat. Rev. Neurosci. **11**(2), 127 (2010)

Gan, Z., Li, C., Henao, R., Carlson, D. E., Carin, L.: Deep temporal sigmoid belief networks for sequence modeling. In: Advances in Neural Information Processing Systems. pp. 2467–2475 (2015)

Goodman, N. D., Tenenbaum, J. B., Contributors, T.P.: Probabilistic Models of Cognition (Second ed.). http://probmods.org/v2. Accessed 13 May 2019

Graves, A., Mohamed, A.-R., Hinton, G.: Speech recognition with deep recurrent neural networks. In: 2013 IEEE International Conference on Acoustics, Speech and Signal Processing. pp. 6645–6649 (2013)

Hinton, G.E.: Learning multiple layers of representation. Trends Cogn. Sci. **11**(10), 428–434 (2007)

Hjelm, R.D., et al.: Learning deep representations by mutual information estimation and maximization. arXiv preprint arXiv:1808.06670 (2018)

Hochreiter, S., Schmidhuber, J.: Long short-term memory. Neural Comput. **9**(8), 1735–1780 (1997)

Jia, X., Hong, H., DiCarlo, J.J.: Unsupervised changes in core object recognition behavioral performance are accurately predicted by unsupervised neural plasticity in inferior temporal cortex. bioRxiv (2020)

Lotter, W., Kreiman, G., Cox, D.: Unsupervised learning of visual structure using predictive generative networks. arXiv preprint arXiv:1511.06380 (2015)

Lotter, W., Kreiman, G., Cox, D.: Deep predictive coding networks for video prediction and unsupervised learning. arXiv preprint arXiv:1605.08104 (2016)

Luc, P., et al.: Transformation-based adversarial video prediction on large-scale data. arXiv preprint arXiv:2003.04035 (2020)

Mathieu, M., Couprie, C., LeCun, Y.: Deep multi-scale video prediction beyond mean square error. arXiv preprint arXiv:1511.05440 (2015)

Oord, A.V.D., Li, Y., Vinyals, O.: Representation learning with contrastive predictive coding. arXiv preprint arXiv:1807.03748 (2018)

Ranzato, M., Szlam, A., Bruna, J., Mathieu, M., Collobert, R., Chopra, S.: Video (language) modeling: a baseline for generative models of natural videos. arXiv preprint arXiv:1412.6604 (2014)

Srivastava, N., Mansimov, E., Salakhudinov, R.: Unsupervised learning of video representations using lstms. In: International Conference on Machine Learning. pp. 843–852 (2015)

Sutskever, I., Hinton, G.E., Taylor, G.W.: The recurrent temporal restricted boltzmann machine. In: Advances in Neural Information Processing Systems. pp. 1601–1608 (2009)

Sutskever, I., Vinyals, O., Le, Q.V.: Sequence to sequence learning with neural networks. In: Advances in Neural Information Processing Systems. pp. 3104–3112 (2014)

Testolin, A., Stoianov, I., Sperduti, A., Zorzi, M.: Learning orthographic structure with sequential generative neural networks. Cogn. Sci. **40**(3), 579–606 (2016)

Testolin, A., Zorzi, M.: Probabilistic models and generative neural networks: towards an unified framework for modeling normal and impaired neurocognitive functions. Front. Computat. Neurosci. **10**, 73 (2016)

Tresilian, J.: Perceptual and cognitive processes in time-to-contact estimation: analysis of prediction-motion and relative judgment tasks. Perception Psychophys. **57**(2), 231–245 (1995)

Villegas, R., Pathak, A., Kannan, H., Erhan, D., Le, Q. V., Lee, H.: High fidelity video prediction with large stochastic recurrent neural networks. In: Advances in Neural Information Processing Systems. pp. 81–91 (2019)

Villegas, R., Yang, J., Zou, Y., Sohn, S., Lin, X., Lee, H.: Learning to generate long-term future via hierarchical prediction. In: Proceedings of the 34th International Conference on Machine Learning. vol. 70, pp. 3560–3569 (2017)

Xingjian, S., et al.: Convolutional lstm network: a machine learning approach for precipitation nowcasting. In: Advances in Neural Information Processing Systems. pp. 802–810 (2015)

Xu, Z., et al.: Unsupervised discovery of parts, structure, and dynamics. In: International Conference on Learning Representations (2019)

Zaremba, W., Sutskever, I.: Learning to execute. arXiv preprint arXiv:1410.4615 (2014)

Zheng, D., Luo, V., Wu, J., D Tenenbaum, J. B.: Unsupervised learning of latent physical properties using perception-prediction networks. arXiv preprint arXiv:1807.09244 (2018)

Zhou, Y., Dong, H., El Saddik, A.: Deep learning in next-frame prediction: a benchmark review. IEEE Access **8**, 69273–69283 (2020)

Semantic Segmentation of Neuronal Bodies in Fluorescence Microscopy Using a 2D+3D CNN Training Strategy with Sparsely Annotated Data

Filippo M. Castelli[1](✉), Matteo Roffilli[3], Giacomo Mazzamuto[1,2],
Irene Costantini[1,2], Ludovico Silvestri[1], and Francesco S. Pavone[1,2,4]

[1] European Laboratory for Non-Linear Spectroscopy (LENS), University of Florence,
Via Nello Carrara 1, 50019 Sesto Fiorentino (FI), Italy
castelli@lens.unifi.it
[2] National Institute of Optics, National Research Council (INO-CNR), Via Nello
Carrara 1, 50019 Sesto Fiorentino (FI), Italy
[3] Bioretics SrL, Corte Zavattini, 11, 47522 Cesena (FC), Italy
[4] Department of Physics and Astronomy, University of Florence, Via G. Sansone, 1,
50019 Sesto Fiorentino, Italy

Abstract. Semantic segmentation of neuronal structures in 3D high-resolution fluorescence microscopy imaging of the human brain cortex can take advantage of bidimensional CNNs, which yield good results in neuron localization but lead to inaccurate surface reconstruction. 3D CNNs on the other hand would require manually annotated volumetric data on a large scale, and hence considerable human effort. Semi-supervised alternative strategies which make use only of sparse annotations suffer from longer training times and achieved models tend to have increased capacity compared to 2D CNNs, needing more ground truth data to attain similar results. To overcome these issues we propose a two-phase strategy for training native 3D CNN models on sparse 2D annotations where missing labels are inferred by a 2D CNN model and combined with manual annotations in a weighted manner during loss calculation.

Keywords: Semantic segmentation · Neuronal segmentation · Pseudo-labeling.

1 Introduction and Related Work

Quantitative studies on the human brain require the ability to reliably resolve neuronal structures at a cellular level. Extensive three-dimensional imaging of the cortical tissue is nowadays possible using high-resolution fluorescence microscopy techniques which are characterized by high volumetric throughput and produce Terabyte-sized datasets in the form of 3D stacks of images.

G. Nicosia et al. (Eds.): LOD 2020, LNCS 12565, pp. 95–99, 2020.
https://doi.org/10.1007/978-3-030-64583-0_10

Semantic segmentation of neuronal bodies in such datasets can be tackled with two-dimensional CNNs: this yields good results in neuron localization, but leads to inaccurate surface reconstruction when the 2D predictions are combined, in a z-stack approach, to estimate a 3D probabilistic map. The issue could be solved by relying on 3D models such as 3D U-Net by Cicek et al. [1] which in turn requires large-scale annotation of volumetric ground truth and consequently an unfeasible human efforts. Cicek's own semi-supervised solution introduces a voxelwise binary loss weightmap that excludes the contributions of non-labeled voxels. However, the restriction of loss contributions to a small fraction of the training data is highly inefficient and leads to long training times. To overcome these issues, namely the overwhelming effort for manual annotations and the optimization inefficiency due to very sparse input data, we propose strategy for training 3D CNN models by following *pseudo-labeling* framework introduced by Lee et al. [3]. In our setup available partial annotations (ground truth) are complemented on unlabeled areas bypseudo-labels generated by an independent 2D CNN model trained on available annotations. We are exploring the application of this training scheme in semantic segmentation of high-resolution Two Photon Fluorescence Microscopy imaging of the human cortex [2,4].

2   Methodology

Three dimensional imaging of four different mm^3-sized human neocortical tissue samples was acquired using a custom-made Two Photon Fluorescence Microscope, as described in [2]. The images were acquired in the form of multiple, partially overlapping $512 \times 512px$ stacks with a voxel size of $0.88 \times 0.88 \times 2\mu m^3$ which are fused using a custom stitching tool. Four volumes sized $100 \times 100 \times 100\mu m^3$ ($114 \times 114 \times 50px$) were manually annotated by an expert operator in a 2D slice-by-slice fashion, three samples were chosen for model training and the fourth was kept for testing. The voxel content of our volumetric dataset can be split into a *labeled* and an *unlabeled* subset as $X = X_L + X_U$: the goal of our semi-supervised learning task is to jointly use the set of all labeled datapoints $\{X_L, Y_L\}$ and the set of unlabeled ones X_U to train a segmentation model for inductive inference on previously unseen data. Our strategy makes use of a 2D CNN segmentation model trained on the available ground truth, this model is used to transductively infer pseudo-labels $\hat{Y}_U$ on the unlabeled dataset partition. Another 3D CNN model f_θ is then trained on the dense combination $\{X_L, Y_L, X_U, \hat{Y}_U\}$ of available ground truth and inferred pseudo-labels using a weighted loss function

$$L(X_L, Y_L, X_U, \hat{Y}_U; \theta) = L_s(X_L, Y_L; \theta) + \alpha L_p(X_U, \hat{Y}_U; \theta) \tag{1}$$

where loss terms L_s and L_p are respectively a *supervised* term that applies only to the labeled partition of the dataset and a *pseudo-label* term that only applies to the *unlabeled* one, while α is a tradeoff parameter. We chose to use for both terms a *binary cross-entropy* loss function: this allows for straightforward implementation using a voxel-wise loss weightmap with values 1 for labeled voxels and α for unlabeled ones.

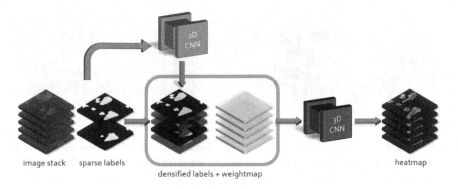

Fig. 1. Proposed training scheme: a 2D CNN model is used to infer pseudo-labels in the unlabeled parts of the training dataset, the main 3D CNN model is trained on a dense weighted combination of ground truth labels and pseudo-labels.

We test our pseudo-labeling scheme on a 3D CNN model based on 3D UNET, trained on $64 \times 64 \times 32$ patches. The auxiliary pseudo-labeling CNN is a lightweight 2D CNN with three convolutional layers, followed by two fully connected layers which we have already used in segmentation of large Two Photon Microscopy imaging datasets in Costantini et al. [2].

3   Evaluation and Results

The effectiveness of our semi-supervised approach was tested by simulating several conditions of label sparsity by using increasing numbers of 2D slices as ground truth, while treating the remaining voxels as unlabeled. We compare

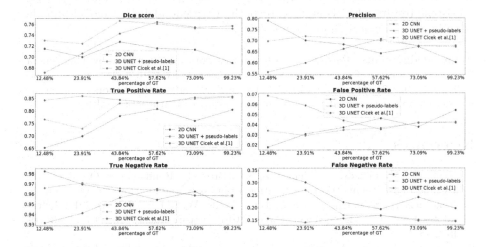

Fig. 2. Performance on an independent test set at various label sparsity conditions

Fig. 3. Visual comparison between training schemes at 43.84% of annotated voxels, from left to right: GT annotations + 3D isosurface reconstruction, 2D CNN model + 3D isosurface reconstruction, 3D CNN sparse training + 3D isosurface reconstruction, 3D CNN sparse training with pseudo-labels + 3D isosurface reconstruction.

the performances of the pseudo-labeling model, the 3D model trained with the described pseudo-labeling scheme and the same network, sparsely trained without pseudo-labels on the available annotations. By choosing a binary classification threshold of 0.5, in Fig. 2 we report on a voxel basis the statistical assessment of the semantic segmentation quality as Dice score, as well as the values of precision and the confusion matrix evaluated on the test dataset.

4   Conclusions

The results suggest that in heavily sparse labeling conditions, the model trained with our semi-supervised scheme could achieve better segmentation performances than both the 2D CNN model and the same 3D CNN model trained without pseudo-labels.

Visually, as depicted in Fig. 3, the 3D CNN model trained with our scheme achieves better reconstruction of the neurons' marginal sections and yield more consistent results when compared to both the pseudo-labeling model and the baseline 3D training without pseudo-labels.

Acknowledgements. Research reported in this work was supported by The General Hospital Corporation Center of the National Institutes of Health under award number 1U01MH117023-01. The content is solely the responsibility of the authors and does not necessarily represent the official views of the National Institutes of Health.

References

1. Çiçek, Ö., Abdulkadir, A., Lienkamp, S.S., Brox, T., Ronneberger, O.: 3D U-Net: learning dense volumetric segmentation from sparse annotation. In: Ourselin, S., Joskowicz, L., Sabuncu, M.R., Unal, G., Wells, W. (eds.) MICCAI 2016. LNCS, vol. 9901, pp. 424–432. Springer, Cham (2016). https://doi.org/10.1007/978-3-319-46723-8_49
2. Costantini, I., et al.: A combined pipeline for quantitative analysis of human brain cytoarchitecture. bioRxiv preprint 2020.08.06.219444 (2020)

3. Lee, D.H.: Pseudo-label: the simple and efficient semi-supervised learning method for deep neural networks. In: ICML 2013 Workshop: Challenges in Representation Learning (WREPL) (2013)
4. Mazzamuto, G., et al.: Automatic segmentation of neurons in 3D samples of human brain cortex. In: Applications of Evolutionary Computation. pp. 78–85 (2018)

Methods for Hyperparameters Optimization in Learning Approaches: An Overview

Nicoletta Del Buono, Flavia Esposito, and Laura Selicato[✉]

Department of Mathematics, University of Bari Aldo Moro,
via E. Orabona 4, Bari, Italy
{nicoletta.delbuono,flavia.esposito,laura.selicato}@uniba.it

Abstract. Automatic learning research focuses on the development of methods capable of extracting useful information from a given dataset. A large variety of learning methods exists, ranging from biologically inspired neural networks to statistical methods. A common trait in these methods is that they are parameterized by a set of hyperparameters, which must be set appropriately by the user to maximize the usefulness of the learning approach. In this paper we review hyperparameter tuning and discuss its main challenges from an optimization point of view. We provide an overview on the most important approaches for hyperparameter optimization problem, comparing them in terms of advantages and disadvantages, focusing on Gradient-based Optimization.

Keywords: Hyperparameter optimization · Gradient-based optimization

1 Introduction

Any automatic learning model requires that different hyperparameters have to be configured. Hyperparameters (HPs) are all the variables that govern the learning model which are set before the training process and influence the performance of the model itself. Principal HPs can be summarized according to their role and domain in the learning process (see Table 1). Deep neural networks, for instance, need to have fixed their network's architecture (e.g. size of the space, number of layers etc.), the cost function optimization algorithm (e.g. stochastic gradient descent, Adam, RmsProp, etc.), some regularization factors and so on. HP tuning requires a lot of efforts and it is usually done by hand and only accelerated by domain knowledge. Automated hyperparameter optimization (HPO) would clearly bring a solution to these problems also providing more reproducible results and facilitating fairly comparisons between different learning models [4].

N. D. Buono, F. Esposito, and L. Selicato—-Members of INDAM-GNCS research group.

G. Nicosia et al. (Eds.): LOD 2020, LNCS 12565, pp. 100–112, 2020.
https://doi.org/10.1007/978-3-030-64583-0_11

HPO problem dates back to the last decade of 1900 when several researches highlighted that a learning model with different HP configurations does not have the same performance on the same given dataset [8]. Even if HPO is a hard problem in practice, the need for automatic HPO is definitely increasing as learning applications become pervasive nowadays. The goal of this paper is twofold: firstly we propose a formalization of the problem of hyperparameter search and then we review the current state-of-the-art on HPO in terms of algorithms also proposing a taxonomy of the different techniques. Particular focus is also given to gradient-based optimization and related open problems. We will finally sketch future research direction and existing relationships between HP tuning and multi-objective optimization.

Table 1. Table of principal HPs grouped by role and domain

	Categories	Examples
Role	Architectural	Number of layers, number of hidden neurons per layer, number of trees in random forest
	Training	Max epochs, learning rate, batch size, momentum
Domain	Discrete	Number of layers, number of trees in random forest, max epochs, batch size
	Continuous	Learning rate
	Categorical	Type of regularization(L_1 or L_2)

2 Hyperparameter Optimization

An automatic learning task aims to train a model on a given data $D = D_{(train)} \cup D_{(test)} \subset \mathbb{D}$ minimizing a predefined loss function $\ell : \mathbb{D} \to \mathbb{R}$ on some test and validation sets $D_{(test)}$. Let $\mathcal{A}$ be a learning model with m HPs, $\lambda \in \Lambda^1$ and $\mathcal{A}_\lambda$ denote $\mathcal{A}$ with its HPs instantiated to λ, the HPO can be formalized as the following bi-level optimization problem:

$$\lambda^* = \underset{\lambda \in \Lambda}{\arg\min}\, F(\mathcal{A}_\lambda, D_{(test)}) \quad \text{s.t.} \quad \mathcal{A}_\lambda \in \underset{\mathcal{A}_\lambda}{\min}\, \ell(\lambda, D_{(train)}) \tag{1}$$

where $F(\mathcal{A}_\lambda, D_{(test)})$ measures the goodness of $\mathcal{A}$ trained with λ on $D_{(train)}$ and evaluated on $D_{(test)}$. It should be observed that when $\omega \in \Omega$ are the model parameters, $\mathcal{A}_\lambda$ is commonly referred to as the *inner optimization* problem and ℓ is expressed as a function of ω. For example, when we simply consider a cost function with a square ℓ_2-norm regularization term, the hyperparameter λ is a scalar and controls the penalty term, the problem (1) becomes:

$$\underset{\lambda \in \Lambda}{\arg\min}\, F(\omega(\lambda), D_{(test)}) \quad \text{such that} \quad \omega(\lambda) = \underset{\omega \in \Omega}{\arg\min}\, \ell(\omega, D_{(train)}) + \lambda ||\omega||^2 \tag{2}$$

---

[1] $\Lambda = \Lambda_1 \times \cdots \times \Lambda_m$ is the HP domain, where each set Λ_i can be real-valued (e.g., learning rate), integer-valued (e.g., number of layers), binary (e.g., whether to use early stopping or not), categorical (e.g., choice of optimizer).

In the following overview, we consider the more general formulation:

$$\operatorname*{argmin}_{\lambda \in \Lambda}\{f(\lambda) \triangleq g(\omega(\lambda), \lambda) \mid \omega(\lambda) = \operatorname*{argmin}_{\omega \in \Omega} \ell(\lambda, \omega, D_{train})\} \qquad (3)$$

where f represents the external cost function measuring the goodness of model on the test set and the internal loss function ℓ provides the better solution of the inner optimization problem in terms of training data and model parameters.

3   Overview on Existing Methods

HPO is a difficult problem to face in practice: for example, function evaluations can be extremely expensive for large models (e.g. in deep learning). It is not always clear which one among the algorithm HPs need to be optimized, and in which ranges. In addition we usually don't have access to a gradient of the loss function with respect to the HPs. The quality of a predictive model critically depends on its hyperparameter configuration, but it is unclear how these HPs interact with each other to affect the resulting model. In Black-Box type methods, no gradient is calculated, therefore the HPO algorithm cannot rely on it to reduce the validation error. Instead, it must blindly try a new configuration in the search space. Vice versa in Gradient-based methods, the model gradient is obtained with respect to HPs or an approximation thereof. Existing approaches for the HP search are sketched in the following.

Grid Search. Since the1990s Grid Search has been a way of performing HPO. It is simply a searching through a manually specified subset of the HP space of a learning algorithm [19]. This represents one of the disadvantages of this technique, because it requires the intervention of the domain expert in order to choose this subset, making the process difficult to reproduce.

Grid search algorithm must be guided by some performance metric, typically measured by cross-validation on the training set or evaluation on a held-out validation set. The parameter space of a learner may include real-valued or unbounded value spaces for certain parameters, so manually set bounds and discretization may be necessary before applying grid search. Important knowledge of the domain is required. The search space is discretized as the Cartesian products of the search space of each HPs, however increasing the resolution of discretization substantially increases the required number of function evaluations. Hence, the algorithm launches a learning for each of the HP configurations, and selects the best at the end. It is an embarrassingly parallel problem but suffers from the curse of dimensionality.

Random Search. It is a random sampling of the search space, instead of discretizing it with a Cartesian grid. The number of search iterations is set based on time or resources. This can be simply applied to the discrete setting described previously, but also generalizes to continuous and mixed spaces.

While it's possible that Random Search will not find as accurate of a result as Grid Search, it surprisingly picks the best result more often and in a fraction of the time it takes Grid Search would have taken. Given the same resources, Random Search can even outperform Grid Search, in particular when some HPs are much more important than others [3]. The method has no end, a time budget has to be specified. This algorithm suffers likewise from the curse of dimensionality to reach a preset fixed sampling density. Additional benefits over grid searching include easier parallelization and flexible resource allocation. If the execution of the model takes a long time using a lot of computing resources, Random Search as well as Grid Search is ineffective as it does not use the information acquired from all previous attempts.

A little variation of Random Search involves the use of Quasi-random number generation. This techniques have been developed for Monte Carlo algorithms and have better convergence properties than Random Search, ensuring a more uniform exploration of the space.

Model-Based Optimization. The Sequential Model-Based Optimization methods (SMBO) builds a surrogate model of the validation loss as a function of the HPs can be fit to the previous tries. SMBO is a formalism of Bayesian optimisation (BO). In the case of HP optimization BO finds the value that minimizes an objective function by building a surrogate function, probabilistic model of the function mapping from HP values to the objective assessed on a validation set, based on past evaluation results of the objective. The objective function is the validation error of a learning model with respect to the HPs. By iteratively evaluating a promising HP configuration based on the current model, and then updating it, BO aims to gather observations revealing as much information as possible about this function and, in particular, the location of the optimum. With reference to the problem (1), for each iteration i, a probabilistic model $p(F|X)$ of the objective function F based on the observed data points $X = \{(\lambda_k, y_k)\}_{k=0,\dots,i-1}$ is considered. BO uses an acquisition function $a : \Lambda \to \mathbb{R}$ based on the current model $p(F|X)$ that trades off exploration. Due to the intrinsic randomness of most learning algorithms, we assume that we cannot observe $F(\lambda)$ directly but rather only trough noisy observations $y(\lambda) = F(\lambda) + \varepsilon$, with $\varepsilon \sim \mathcal{N}(0, \sigma^2_{noise})$. BO iterates the following three steps:

1. select the point that maximizes the acquisition function $\lambda_{new} = \underset{\lambda \in \Lambda}{\operatorname{argmax}}\, a(\lambda)$
2. evaluate the objective function $y_{new} = F(\lambda_{new}) + \varepsilon$
3. augment the data $X \leftarrow X \cup (\lambda_{new}, y_{new})$ and refit the model.

The acquisition function is commonly the expected improvement (EI) over the currently best observed value $\alpha = \min y_0, \dots, y_n$:

$$a(\lambda) = \int \max(0, \alpha - F(\lambda))dp(F|X)$$

BO is efficient in tuning few HPs but its performance degrades a lot when the search dimension increases too much, up to a point where it is on par with random search [16]. One major drawback is not an embarrassingly parallel problem, in contrast to random or grid searches. Introducing parallelization is done through the use of multi-points acquisition function, which enables to find several prospective configurations at once and learn models using them simultaneously [12]. BO can only work on continuous HPs, and not categorical ones. For integer parameters, a method proposed in [11] changes the distance metric in the kernel function so as to collapse all continuous values in their respective integer. The model is built using only function evaluations, and for this reason SMBO is often considered as a Black-Box optimization method.

Hyperband. This is a bandit strategy that dynamically allocates resources to a set of random configurations and uses successive halving (SH) [13] to stop poorly performing configurations. Compared to Bayesian optimization methods that do not use multiple fidelities, Hyperband showed strong anytime performance, as well as flexibility and scalability to higher-dimensional spaces. However, it only samples configurations randomly and does not learn from previously sampled configurations. This can lead to a worse final performance than model-based approaches. Hyperband is a multi-armed bandit strategy for HP optimization that takes advantage of these different budgets by repeatedly calling SH of resources with respect to some budget limit. At the beginning, there is little/no knowledge about how possible allocations will perform, the allocation can be changed based on previous observations. Hyperband makes two simple and reasonable assumptions:

1. Budget: The amount of resources (e.g. time, computing power) is fixed/capped which is called "budget".
2. Delta: It is likely that sets of HPs that perform differently at the beginning will preserve this difference by the time training completes.

Of course, the usage of the resource budget should be maximized. Starting from the assumptions stated above, it makes sense to use a SH algorithm: start with a large population of arms and progressively cut those poorly performing arms of HPs earlier on while carrying on only with the most promising ones. By regularly discarding configurations that do not look promising, SH releases the resources that they used to take and reallocates them to the more promising arms. This way SH agrees with the formulated assumptions: it maintains a fixed budget of resources, it maximizes their usage and by SH we make use of the delta assumption [16]. Hyperband is a very interesting method, because it is an optimizer per se but can also be combined with other optimizers to improve their resource allocation. BOHB [8] combine Bayesian optimization with Hyperband.

Reinforcement Learning (RL). This is a direct approach to learn from trial and error [20] from interactions with an environment in order to achieve a defined goal. An agent in an environment takes an informed action (based on the rewards it received for the previous actions), responding to the agent's feedback. The environment returns rewards in the form of a numerical scalar value. The agent seeks to maximize rewards over time. Finally, the agent updates its history of actions and rewards and the cycle continues [24]. These methods, as well as Bayesian ones, introduce a surrogate model that predicts the validation loss for a specific HP setting, then sequentially select the next HP to test, based on a heuristic function of the expected value and the uncertainty of the surrogate model. However Bayesian methods aim at finding the lowest error in as few steps as possible, whereas in reinforcement learning there is no such constraint. The trade-off is about the waiting time. At the moment, RL-based optimizer is a remarkable class of optimizers since it gives the state-of-the art performance in terms of automatic HP tuning. They are very often combined with deep learning. The main disadvantage of RL is that it is very expensive in terms of computing resources which make these methods inaccessible for many users. Also, because the underlying models are neural networks they tend to be parametric as well.

Population-Based Methods. These are optimization algorithms that maintain a population, i.e. a set of configurations and improve this population by applying local perturbations (mutations) and combinations of different members (crossover) to obtain a new generation of better configurations. These methods can handle different data types, and are embarrassingly parallel [17] however parametric. They are based on the Darwinian concept of evolution. In order to happen, evolution needs *selection*, the best performing individuals are more likely to have children. A fitness function is used to evaluate each guess/individual and the higher the value of the function the more suitable the guess was. In order to create a new generation, we must define and follow a rule that will make the fittest individuals parents of the new generation; *heredity*, a process known as crossover is used to create the offspring of any parents. The exact way crossover works depends on the encoding of the genes; *variation*, after crossover, the offspring is made only of genes from its parents. However, to ensure variation in the next generation, some genes, parts of the encoding, can be randomly modified by mutation.

Among the best-known population-based methods are Evolutionary Algorithms (EAs) also referred to as Genetic Algorithms (GAs). These algorithms are stochastic optimization methods inspired by the biological phenomena of natural evolution. An GA maintains a population of possible solutions to the problem and simulates an evolutionary process on it, repeatedly applying a certain number of stochastic operators known by the names of mutation, recombination, reproduction and selection. Basically, GAs find the optimal solution when checking every possibility (backtracking) is not tractable. Specifically for HP tuning GAs can be simply regarded as an informed version of random search. The steps are:

1. Generate some random sets of HPs.
2. Run the learning model for each set of HPs and wait for the errors on the validation set.
3. Create new sets of HPs by mixing HPs from sets that just performed best.
4. Repeat steps 2, 3, 4 until a certain time stop criteria or a given accuracy is achieved.

GAs operate indirectly on a suitable code of the set of parameters, rather than directly on the parameters. The search for the optimal is carried out in parallel, no regularity condition is required on the objective function, they are defined by probabilistic rather than deterministic transition rules. They are performing very well on a number of datasets, on some datasets GAs can beat Bayesian methods. GAs, however, require many error evaluations $n \times m$, where n is the size of population and m is the number of generations; both terms can be of the order of hundreds or, sometimes, even thousands. In addition, GAs are sensitive to population size and also are parametric in turn, they need some HPs that need to be optimized.

4   Gradient-Based Methods

The gradient-based HPO methods estimate the optimal HPs using smooth optimization techniques such as gradient descent. These methods use local information about the cost function f in order to compute the Hypergradient (i.e. the gradient of the cost function with respect to HPs) or an approximation.

Gradient-based algorithms are used to optimize the performance on a validation set with respect to the HPs [1]. In this setting, the validation error should be evaluated at a minimizer of the training objective. However, in many current learning systems, the minimizer is only approximate. The capabilities of statistical learning methods is limited by the computing time rather than the sample size. Indeed optimization algorithms such as stochastic gradient descent show amazing performance for large-scale problems. In particular, the second order stochastic gradient and the average stochastic gradient are asymptotically efficient after a single pass on the training set [6].

Stochastic Gradient Descent (SGD). This is a very efficient method for learning linear classifiers in convex loss functions (SVM and logistic regression) [23]. Despite its presence in the field of machine learning for a long time, SGD has received considerable attention only recently in the context of large-scale learning, e.g. in text classification and natural language processing. It is an efficient and easily implemented approach, however it requires a number of HPs such as the regularization parameter and the number of iterations.

From the HPs optimization point of view, the approach can be formalized as follows. With reference to the problem (3), given $f(\lambda)$ a loss function, the SGD algorithm updates the HP λ on the basis of the hypergradient: $\lambda_{t+1} = \lambda_t - \eta_t \nabla_\lambda f(\lambda_t)$ where the learning rate η_t controls the step-size in the parameter space.

Convergence results usually require decreasing gains satisfying the conditions $\sum_t \eta_t^2 < \infty$ and $\sum_t \eta_t = \infty$.

The Robbins-Siegmund theorem establishes almost sure convergence under mild conditions [5], including cases where the loss function is not everywhere differentiable. The convergence speed of stochastic gradient descent is in fact limited by the noisy approximation of the true gradient. When the gains decrease too quickly, the expectation of the parameter estimate λ_t takes a very long time to approach the optimum. Under sufficient regularity conditions, the best convergence speed is achieved using gains $\eta_t \sim t^{-1}$. The expectation of the residual error then decreases with similar speed. The second order stochastic gradient descent multiplies the gradients by a positive definite matrix Γ_t that approximates the inverse of the Hessian: $\lambda_{t+1} = \lambda_t - \eta_t \Gamma_t \nabla_\lambda f(\lambda_t)$ Nonetheless, this modification does not reduce the stochastic noise, although constants are improved, the expectation of the residual error still decreases like t^{-1}. Future research should instead focus on the integration of gradient-based algorithm with emerging RL hyperparameter optimization approaches [25]. It may be useful to study methods that have the potential to overcome SGD, such as those based on noise reduction and second-order techniques. These methods offer the ability to attain improved convergence rates, overcome the adverse effects of high nonlinearity and ill-conditioning, and exploit parallelism.

Compute Hypergradient. A fundamental role in SGD is represented by hypergradient so we will describe some approaches to compute it.

In [18] Maclaurin et al. considered reverse-mode differentiation of the response function and proposed an approximation capable of addressing the associated loss of information due to finite precision arithmetic, reversing parameter updates. Pedregosa in [21] proposed the use of inexact gradients, allowing HPs to be updated before reaching the minimizer of the training objective. Franceschi et al. in [9] illustrate two alternative approaches to compute the hypergradient. The first one is based on a Lagrangian formulation associated with the parameter optimization dynamics. The second one is to computing the hypergradient in forward-mode and it is efficient when the number of HPs is much smaller than the number of parameters.

In general there is no closed form expression ω, so it is not possible to directly optimize the outer objective function. Depending on how the gradient is calculated with respect to HPs, there are two main types of approaches, the *Implicit differentiation*, applying the implicit function theorem, and *Iterative differentiation* approach, replacing the inner problem with a dynamical system.

Implicit Differentiation. The Implicit differentiation consists in deriving an implicit equation for the gradient using the optimality conditions of the inner optimization problem. If ℓ is a convex function, then the values $\omega(\lambda)$ are characterized by the implicit equation $\nabla_1 \ell(\omega(\lambda), \lambda) = 0$. Deriving this implicit equation with respect to λ we have $\nabla^2_{1,2}\ell + \nabla^2_1 \ell \cdot D\omega = 0$. Assuming $\nabla^2_1 \ell$ invertible, we can characterizes the derivative of ω. Combining the chain rule with this equation, we can write the following formula for the gradient of f:

$$\nabla f = \nabla_2 g + (D\omega)^\top \nabla_1 g = \nabla_2 g - (\nabla^2_{1,2}\ell)^\top (\nabla^2_1 \ell)^{-1}\nabla_1 g \qquad (4)$$

Iterative Differentiation. As proposed in [9] we see the training procedure by stochastic gradient descent as a dynamical system with a state $s_t \in \mathbb{R}^d$ that collects weights and possibly accessory variables. The dynamics are defined by the system of equations $s_t = \Phi_t(s_{t-1}, \lambda)$ $t = 1, ..., T$ where T is the number of iterations, s_0 contains initial weights and initial accessory variables, and, for every $t \in \{1, ..., T\}$, $\Phi_t : (\mathbb{R}^d \times \mathbb{R}^m) \to \mathbb{R}^d$ is a smooth mapping that represents the operation performed by the t-th step of the optimization algorithm (i.e. on mini-batch t). Note that the iterates $s_1, ..., s_T$ implicitly depend on the vector of HPs λ. Our goal is to optimize the HPs according to a certain error function E evaluated at the last iterate s_T. We wish to solve the problem (3) where $f : \mathbb{R}^m \to \mathbb{R}$ is defined at $\lambda \in \mathbb{R}^m$ as $f(\lambda) = E(s_T(\lambda))$, called response function.

The hypergradient is computed by differentiating each step of the inner optimization algorithm and then using the chain rule to aggregate the results. Since the gradient is computed after a finite number of steps of the inner optimization routine, the estimated gradient is an approximation to the true gradient. This method was first proposed in [7] where an iterative algorithm, like gradient descent, was run for a given number of steps, subsequently computing the gradient of the validation error by a back-optimization algorithm.

Reverse Mode. For the reverse-mode computation of the hypergradient under a Lagrangian perspective, we reformulate problem (3) as the constrained optimization problem

$$\min_{\lambda, s_1, \dots, s_T} E(s_T) \quad \text{subject to} \quad s_t = \Phi_t(s_{t-1}, \lambda), \quad t \in \{1, \dots, T\} \tag{5}$$

The Lagrangian of (5) is $\mathcal{L}(s, \lambda, \alpha) = E(s_T) + \sum_{t=1}^{T} \alpha_t(\Phi_t(s_{t-1}, \lambda) - s_t)$ where, for each $t \in \{1, \dots, T\}$, $\alpha_t \in \mathbb{R}^d$ is a row vector of Lagrange multipliers. The partial derivatives of the Lagrangian are given by

$$\frac{\partial \mathcal{L}}{\partial \alpha_t} = \Phi_t(s_{t-1}, \lambda) - s_t \quad \frac{\partial \mathcal{L}}{\partial s_t} = \alpha_{t+1} A_{t+1} - \alpha_t$$
$$\frac{\partial \mathcal{L}}{\partial s_T} = \nabla E(s_T) - \alpha_T \quad \frac{\partial \mathcal{L}}{\partial \lambda} = \sum_{t=1}^{T} \alpha_t B_t$$

being $A_t = \frac{\partial \Phi_t(s_{t-1}, \lambda)}{\partial s_{t-1}}$ and $B_t = \frac{\partial \Phi_t(s_{t-1}, \lambda)}{\partial \lambda}$, for $t \in \{1, \dots, T\}$. Therefore, the optimality conditions are obtained: $\alpha_t = \begin{cases} \nabla E(s_T) & \text{if } t = T, \\ \nabla E(s_T) A_T \cdots A_{t+1} & \text{if } t \in \{1, \dots, T-1\} \end{cases}$
and

$$\frac{\partial \mathcal{L}}{\partial \lambda} = \nabla E(s_T) \sum_{t=1}^{T} \left(\prod_{s=t+1}^{T} A_s \right) B_t. \tag{6}$$

Forward-Mode. The forward approach to compute hypergradient appeals to the chain rule for the derivative of composite functions, to obtain that the gradient of f at λ satisfies

$$\nabla f(\lambda) = \nabla E(s_T) \frac{ds_t}{d\lambda} \tag{7}$$

where $\frac{ds_t}{d\lambda}$ is the $d \times m$ matrix formed by the total derivative of the components of s_T (regarded as rows) with respect to the components of λ (regarded ad columns). The operators Φ_t depend on the HP λ both directly by its expression and indirectly through the state s_{t-1}. Using again the chain rule we have, for every $t \in \{1, \dots, T\}$, that

$$\frac{ds_t}{d\lambda} = \frac{\partial \Phi_t(s_{t-1}, \lambda)}{\partial s_{s-1}} \frac{ds_{t-1}}{d\lambda} + \frac{\partial \Phi_t(s_{t-1}, \lambda)}{\partial \lambda} \tag{8}$$

Defining $Z_t = \frac{ds_t}{d\lambda}$ for every $t \in \{1, \dots, T\}$ using A_t and B_t definited as above, we can rewrite (8) as the recursion

$$Z_t = A_t Z_{t-1} + B_t \qquad t \in \{1, \dots, T\} \tag{9}$$

Using (9) in (7) we obtain that $\nabla f(\lambda) = \nabla E(s_T) \sum_{t=1}^{T} (\prod_{s=t+1}^{T} A_s) B_t.$

This equation coincides with (6). From the above derivation it is apparent that $\nabla f(\lambda)$ can be computed by an iterative algorithm which runs in parallel to the training algorithm.

5   Discussion

It should be highlighted that the iterative differentiation approach should be generally preferred to the implicit one; it can, in fact, learn HPs that control the inner level optimization (e.g. step size). On the contrary, the implicit differentiation approach is more sensitive to the optimality of $\omega(\lambda)$. However it has been successfully applied to problems characterized by simply cost function whose hypergradient has a simple expression (e.g. image reconstruction problem).

Our research is currently focusing on iterative methods, in particular the development of a bilevel optimization algorithm that alternatively optimizes the Loss and Response function at each iteration.

Depending on the complexity and the dimensionality of a learning problem, some hits on the choice of the most appropriate method among those previously reviewed can be provided. In Fig. 1 the various methods are compared based on their scalability, parallelism, computational cost required by function evaluations, the presence of inner parameters, the adaptability to particular classes of variables and the use of past knowledge. With SMBO methods we can avoid the problem of parameters, which adds complexity, but it does lose effectiveness on scalability. In the context of big data, effectiveness on scalability is not a trivial aspect. Methods applicable to large-scale problems are therefore preferred. Using Gradient-based methods we can obtain great results for large-scale problems, using only local information and at least one parameter (learning rate). However, sometimes there is no access to the hyper-gradient; moreover it is not known whether the loss functions are convex and this represents a great limitations. In Black-Box methods, such as Grid Search and Random Search, instead, no gradient is calculated. Therefore the HPO algorithm must blindly try a new configuration in the search space. Through the Monte-Carlo simulation a Black-Box method can be improved, considering only the output, since it does not need any piece of information on the functional form of the Loss function. Even GAs allow us to work on a large-scale problem and have the advantage of being very parallelizable. However they are parametric, as such as are the RL algorithms, that currently constitute the state-of-the-art for HPO.

We want to point out to the reader that most conventional learning algorithms can only deal with a scalar cost function. However ML is a multi-objective task [14], but traditionally multiple objectives are aggregated to a scalar cost function. Efforts on solving ML problems using the Pareto-based multi-objective optimization methodology have gained increasing impetus, especially in conjunction with the success of GAs and population-based stochastic search methods. As mentioned above, SMBO is one state-of-the-art method to address the HPO.

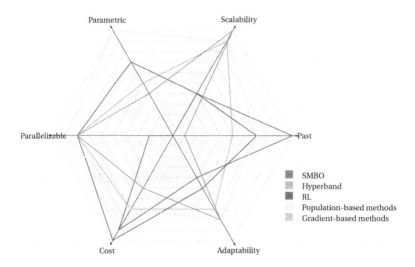

Fig. 1. Spiderweb diagram

However, resulting models often lack interpretability, as models usually contain many features with non-linear effects and higher-order interactions. The advantage of multi-objective approach lies in a more complete view of the learning problem gained by analyzing the Pareto front (PF). For any multi-objective problem, there are generally multiple Pareto-optimal solutions. The goal is to devise algorithms that will efficiently find solutions on the PF with which the practitioner can then decide manually which model to deploy.

In the multiobjective case, HPO is the problem to find the set of λ which form the PF of $f(\lambda)$, where $f : \Lambda \rightarrow \mathbb{R}^p$, $[f_1(\lambda), f_2(\lambda), \ldots, f_p(\lambda)]$. Given a particular λ_k, the Black-Box instantiates a neural network model according to the HP setting, runs training to convergence, and reports the metrics (accuracy, model size, inference speed) computed on a held-out set. The algorithm then decides what HP λ_{k+1} to search next, using information from past λ_k and $f(\lambda_k)$. This is difficult because there is no gradient information, though several classes of algorithms have been developed, including BO [15], GAs [22], and RL [25]. An interesting direction of our future work is to focus on a multi-objective approach because, at the moment, there are not very robust algorithms able to compete with the best known previously mentioned. But these methods have all the potential to overcome them.

References

1. Bengio, Y.: Gradient-based optimization of HPs. Neural Comput. **12**(8), 1889–1900 (2000)

2. Bergstra, J., Bardenet, R., Bengio, Y., Kégl, B.: Algorithms for hyper-parameter optimization. In: Shawe-Taylor, J., Zemel, R., Bartlett, P., Pereira, F., Weinberger, K. (eds.) Proceedings of the 25th International Conference on Advances in Neural Information Processing Systems (NeurIPS 2011). pp. 2546–2554 (2011)
3. Bergstra, J., Bengio, Y.: Random search for hyper-parameter optimization. J. Mach. Learn. Res. **13**, 281–305 (2012)
4. Bergstra, J., Yamins, D., Cox, D.: Making a science of model search: Hyperparameter optimization in hundreds of dimensions for vision architectures. Dasgupta and McAllester. pp. 115–123 (2013)
5. Bottou, L.: Online Algorithms and Stochastic Approximations. Cambridge University Press (1998)
6. Bottou, L.: Large-scale machine learning with stochastic gradient descent. In: Lechevallier Y., Saporta G. (eds) Proceedings of COMPSTAT 2010. Physica-Verlag HD (2010)
7. Domke, J.: Generic methods for optimization-based modeling. AISTATS **22**, 318–326 (2012)
8. Falkner, S., Klein, A., Hutter, F.: BOHB: robust and efficient hyperparameter optimization at scale. In: Dy and Krause (eds) Proceedings of the 35th International Conference on Machine Learning (ICML 2018), pp. 1437–1446 (2018)
9. Franceschi, L., Donini, M., Frasconi, P., Pontil, M.: Forward and reverse gradient-based hyperparameter optimization. In: Proceedings of the 34th International Conference on Machine Learning (ICML 2017), pp. 1165–1173 (2017)
10. Gardner, S., et al.: Constrained Multi-Objective Optimization for Automated Machine Learning (2019)
11. Garrido-Merchán, E., Hernandez-Lobato, D.: Dealing with Integer-valued Variables in Bayesian Optimization with Gaussian Processes. arXiv:1706.03673 (2017)
12. Ginsbourger, D., Le Riche, R., Carraro, L.: Kriging is well-suited to parallelize optimization. In: Tenne, Y., Goh, C.-K. (eds.) Computational Intelligence in Expensive Optimization Problems. ALO, vol. 2, pp. 131–162. Springer, Heidelberg (2010). https://doi.org/10.1007/978-3-642-10701-6_6
13. Jamieson, K., Talwalkar, A.: Non-stochastic best arm identification and hyperparameter optimization. In: Proceedings of the Seventeenth International Conference on Artificial Intelligence and Statistics (AISTATS) (2016)
14. Jin, Y., Sendhoff, B.: Pareto-based multiobjective machine learning: an overview and case studies. IEEE Trans. Syst. Man, Cybernetics-part C: Appl. Rev. **38**(3), 397–415 (2008)
15. Kotthoff, L., Thornton, C., Hoos, H., Hutter, F., Leyton-Brown, K.: Auto-weka 2.0: Automatic model selection and hyperparameter optimization in weka. J. Mach. Learn. Res **18**(25), 1–5 (2017)
16. Li, L., Jamieson, K., DeSalvo, G., Rostamizadeh, A., Talwalkar, A.: Hyperband: A novel bandit-based approach to hyperparameter optimization. J. Mach. Learn. Res. **18**(185), 1–52 (2018)
17. Loshchilov, I., Hutter, F.: CMA-ES for hyperparameter optimization of deep neural networks. In: International Conference on Learning Representations Workshop Track (2016)
18. Maclaurin, D., Duvenaud, D., Adams, R.: Gradient-based hyperparameter optimization through reversible learning. In: ICML, pp. 2113–2122 (2015)
19. Michie, D., Spiegelhalter, D., Taylor, C., Campbell, J. (eds.): Machine Learning, Neural and Statistical Classification. Ellis Horwood (1994)
20. Murgulet, D.: Independent Study Option: Automated Hyper-parameter Tuning for Machine Learning Models. Imperial College, London (2018)

21. Pedregosa, F.: Hyperparameter optimization with approximate gradient. In: ICML, pp. 737–746 (2016)
22. Real, E., et al.: Large-scale evolution of image classifiers. In: Proceedings of the International Conference on Learning Representations (ICLR) (2017)
23. Robbins, H., Monro, S.: A stochastic approximation method. Ann. Math. Stat. **22**(3), 400–407 (1951)
24. Sutton, R.S., Barto, A.G.: Reinforcement Learning: An Introduction, 2nd edn. The MIT Press Cambridge, Massachusetts (2017)
25. Zoph, B., Le, Q.: Neural architecture search with reinforcement learning. In: Proceedings of the International Conference on Representation Learning (ICLR) (2017)

Novel Reconstruction Errors for Saliency Detection in Hyperspectral Images

Antonella Falini[1]([✉])(iD), Cristiano Tamborrino[2], Graziano Castellano[1],
Francesca Mazzia[1](iD), Rosa Maria Mininni[2](iD), Annalisa Appice[1](iD),
and Donato Malerba[1](iD)

[1] Department of Computer Science, University of Bari, Bari, Italy
{antonella.falini,francesca.mazzia,annalisa.appice,
donato.malerba}@uniba.it, g.castellano10@studenti.uniba.it
[2] Department of Mathematics, University of Bari, Bari, Italy
{cristiano.tamborrino,rosamaria.mininni}@uniba.it

Abstract. When hyperspectral images are analyzed, a big amount of data, representing the reflectance at hundreds of wavelengths, needs to be processed. Hence, dimensionality reduction techniques are used to discard unnecessary information. In order to detect the so called "saliency", i.e., the relevant pixels, we propose a bottom-up approach based on three main ingredients: sparse non negative matrix factorization (SNMF), spatial and spectral functions to measure the reconstruction error between the input image and the reconstructed one and a final clustering technique. We introduce novel error functions and show some useful mathematical properties. The method is validated on hyperspectral images and compared with state-of-the-art different approaches.

Keywords: Error measures · Hyperspectral images · Saliency detection

1 Introduction

HyperSpectral Imaging (HSI) is a technology that combines the imaging properties of a digital camera with the spectroscopic properties of a spectrometer able to detect the spectral attributes of each pixel in an image. Hyperspectral sensors measure radiance, i.e. the radiant energy emitted, reflected, transmitted

The research of Antonella Falini is founded by PON Project AIM 1852414 CUP H95G18000120006 ATT1. The research of Cristiano Tamborrino is funded by PON Project "Change Detection in Remote Sensing" CUP H94F18000270006. The research of the other authors is funded by PON Ricerca e Innovazione 2014–2020, the application of the proposed method to the saliency detection task is developed in partial fulfillment of the research objective of project RPASInAir "Sistemi Aeromobili a Pilotaggio Remoto nello spazio aereo non segregato per servizi" (ARS01 00820). While the application to the change detection task is developed in partial fulfillment of the research objective of the project "CLOSE to the Earth" (ARS01 00141).

© Springer Nature Switzerland AG 2020
G. Nicosia et al. (Eds.): LOD 2020, LNCS 12565, pp. 113–124, 2020.
https://doi.org/10.1007/978-3-030-64583-0_12

or received by a surface, by a large number of regularly spaced narrow bands covering a wide spectral range. All the information about reflectance across the whole spectral range of the sensor is contained by a single pixel, producing the so called *spectral signature*. Every spectral image can be thought as a cuboid: on the *x*- and *y*- directions, the spatial information is stored, while on the *z*-direction the spectral signature is smeared. Due to the size of the spectral signature, it is crucial to develop techniques able to reduce the dimensionality of the problem while keeping the important information. Characterizing what can be considered *salient* and what can be neglected becomes another challenging task. The concept of *saliency* refers to identifying parts, regions, objects or features that first draw visual attention and hence can be considered notable and important. Many attention models derive from the "Feature Integration Theory" [15], where it is stated which visual features can be considered important to drive human attention. In [7], Koch and Ullman propose a feed-forward model to combine these features and introduce the concept of a *saliency map* as a topographic map representing *conspicuousness* of scene locations. Saliency detection methods can be divided into top-down (see e.g., [4]), bottom-up (see e.g., [5]) approaches or a combination of both (see e.g., [18]). Besides being used as a technique to identify "what draws attention first", saliency detection can be employed also for more advanced tasks, as the so called "change detection". In this case it is relevant to highlight those pixels which result different when the hyperspectral image of the same area has been acquired at different time intervals, see for example [11] and references therein for a general overview.

To compress the dimension of the problem being able to extract relevant features, usually saliency detection methods are combined with matrix factorization techniques, like Principal Component Analysis (PCA) or Independent Component Analysis (ICA). In the present work, we use *Non Negative Matrix Factorization* (NMF) and the Gaussian Mixture Model clustering (GMM) to develop a saliency detection algorithm following a bottom-up approach. The main peculiarity of NMFs lies in factorizing a given matrix into the product of two matrices with non-negative values. Thanks to this property, NMFs are especially suitable for dealing with data which are only positive or null, like for instance intensities colors, and are preferable to other factorizations like PCA or ICA. NMFs were recently introduced in the context of saliency detection in [13], to discover the latent structure of image data-set, in [14] to extract features information from each superpixel, and in [9] to learn the basis functions from images patches. Once the factorization is performed, the reconstruction error is computed by using several functions. In particular, new error measures and their mathematical properties are analyzed. The introduction of such new functions provides a valuable contribution to output the sought saliency map. Then the reconstruction error vector, together with specific configurations of the available information is processed by the clustering algorithm in order to produce the final saliency map.

The remainder of the paper is organized as follows. In Sect. 2 the definitions of the employed error functions are provided. In Sect. 3 the structure of the proposed algorithm is presented. In Sect. 4 the experimental results are shown. In Sect. 5 final conclusions and possible directions for future work are discussed.

2   Reconstruction Error Measures

Every image can be represented as a tensor $\mathcal{A}$ of size $m \times n \times k$, where m and n identify the location of every pixel, while k represents the expression of the spectrum in the adopted spectral space. The input image $\mathcal{A} \in \mathbb{R}^{m \times n \times k}$ can be rearranged as a matrix A of size $p \times k$, with $p = m \times n$. In the resulting A, the spectral information of each pixel (x, y) is stored in a suitable row. The chosen factorization algorithm produces as output matrices $W \in \mathbb{R}_+^{p \times r}$ and $H \in \mathbb{R}_+^{r \times k}$ such that $A \approx Z := WH$. The number r is called *compression index*. The r columns of the basis matrix W can be thought as the latent factors embedded into the dataset. The elements of the coefficient matrix H can be considered as the weights associated to each factor.

We call *reconstruction error* the vector defined as $RE(A, Z)$ that collects the spectral differences between every pixel (x, y) in A and the pixels in the approximation Z. The RE can be thought as a dissimilarity index between the reconstructed image and the input one. In particular, since the background is the more extensive part of an image, we expect the NMF to reconstruct it quite accurately, discarding any other detail. Measuring RE should provide an image containing the salient part of the input A, therefore, finding a good error measure it is of fundamental importance. To this end we compute the following functions:

- *Euclidean distance* (ED), corresponds to the usual 2-norm for vectors and it is computed for every row of the matrix $A - Z$;
- *Spectral Angle Mapper (SAM)* [8], for every pixel (x, y) let ℓ denote its corresponding index in A and let $A_{\ell,:}$ be the ℓ-th row of the matrix A. SAM computes the angle between the spectral vectors $A_{\ell,:}$ and $Z_{\ell,:}$ in a space of dimension equal to the number of bands k.
- *Spatial Spectral Cross Correlation (SSCC)* [17], a measure of the similarity in the spatial-spectral domain of two images within a local window $V(\ell)$ centered at the specific pixel (x, y). The window $V(\ell)$ collects the indices $\ell_{i,j}$ corresponding to the pixels $(x + I, y + J)$ with $I, J = \pm d_V$ and $d_V \geq 1$.
- *The modified Z-score Index (ZID)* from [12],
- *SAMZID* [12], a combination of a modified Z-score Index and the SAM function,
- *SAM_g_ZID* we combine the SAM function with a new ZID index where a geometric mean has been used rather than the usual arithmetic mean:

$$SAM\_g\_ZID := [scale(\sin SAM)] \times [scale(ZID\_g)].$$

ZID_g is computed as the ZID indicator from [12] with the arithmetic mean replaced by μ_{g_ℓ}, the geometric mean for the vector $A_{\ell,:}$,

$$\mu_{g_\ell} := \left(\prod_{i=1}^{k} A_{\ell,i} \right)^{1/k}$$

The geometric mean describes the central tendency. The same mean μ_{g_ℓ} is used to compute the standard deviation for the vector $A_{\ell,:}$ required by the definition of the ZID.

– *Spatial Mean SAM (SMSAM)*, we consider the angles between the spectral vectors $A_{\ell,:}$ and $Z_{\ell,:}$ for the label ℓ varying in V and then we compute their arithmetic mean:

$$SMSAM(\ell) := \frac{1}{|V(\ell)|} \sum_{\ell_{i,j} \in V(\ell)} \arccos\left(\frac{A_{\ell_{i,j},:} Z_{\ell_{i,j},:}^\top}{\|A_{\ell_{i,j},:}\|_2 \|Z_{\ell_{i,j},:}\|_2}\right).$$

The symbol $|V(\ell)|$ denotes the number of elements of $V(\ell)$.

– *Spatial Mean Spectral Angle Deviation Mapper (SMSADM)*, we compute the arithmetic mean within a local window V of the deviation angles from their spectral mean inside V,

$$SMSADM(\ell) := \frac{1}{|V(\ell)|} \sum_{\ell_{i,j} \in V(\ell)} \arccos\left(\frac{E(A)E(Z)^\top}{\|E(A)\|_2 \|E(Z)\|_2}\right).$$

For a generic matrix S, the operator $E(S)$ is defined as:

$$E(S) := (S_{\ell_{i,j},:} - \mu(S)_{V(\ell)}).$$

The quantity $\mu(S)_{V(\ell)}$ denotes the spectral mean within the window $V(\ell)$ for the matrix S.

To improve the efficiency of processing every image, a superpixel segmentation can be applied to the original image A. We use the "simple linear iterative clustering" (`SLIC`) [16] algorithm which groups pixels in separate regions according to color. With this setting we compute three new error measures, *Spx-SSCC, Spx-SMSAM, Spx-SMSADM*, which respectively are the functions SSCC, SMSAM and SMSADM, where V is given as a tile produced by the superpixel algorithm.

2.1   Mathematical Properties

In this section we show some useful properties for the proposed reconstruction errors.

It is straightforward to notice that $RE(A, Z) = RE(Z, A)$ for SAM, SSCC, SMSAM, SMSADM and their respective superpixel variants, i.e., the listed functions are *symmetric*.

Moreover, we can also prove that those error measures result *scale invariant*, i.e., $RE(\alpha A, \alpha Z) = RE(A, Z)$ for any $\alpha \in \mathbb{R}$. For the *SAM* measure we have,

$$RE(\alpha A, \alpha Z) = \arccos\left(\frac{\alpha A_{\ell,:} \alpha Z_{\ell,:}^\top}{\|\alpha A_{\ell,:}\|_2 \|\alpha Z_{\ell,:}\|_2}\right) = RE(A, Z).$$

The same steps can be applied to *SMSADM, Spx-SMSADM* and to *ZID*. Regarding SSCC, SMSADM and their superpixel version we can observe that, given a generic matrix S and $\alpha \in \mathbb{R}$,

$$E(\alpha S) = \left(\alpha S_{\ell_{i,j},:} - \mu(\alpha S)_{V(\ell)}\right) = \left(\alpha S_{\ell_{i,j},:} - \alpha\mu(S)_{V(\ell)}\right) = \alpha E(S).$$

Then, for instance, if we use SSCC, for any $\alpha \in \mathbb{R}$ we have,

$$RE(\alpha A, \alpha Z) = \frac{\displaystyle\sum_{\ell_{i,j} \in V(\ell)} E(\alpha A)E(\alpha Z)^{\top}}{\sqrt{\displaystyle\sum_{\ell_{i,j} \in V(\ell)} \|E(\alpha A)\|^2} \sqrt{\displaystyle\sum_{\ell_{i,j} \in V(\ell)} \|E(\alpha Z)\|^2}} = RE(A, Z)$$

All the proposed error measures are non negative besides *SAMZID* and *SAM_g_ZID*. Finally, we observe that any rotation or reflection applied in the spatial domain, would not change the relative location of the spectral vectors. Therefore, the RE value is the same up to the applied transformation.

3   Structure of the Algorithm

The output of our algorithm will be a binary map where pixels belonging to the salient region have value 1 and 0 otherwise. The algorithm can be summarized by the following stages:

1- The input image is factorized by using one NMF algorithm provided by the open source Python library Nimfa [19], with compression index $r = 2$. NMF algorithms return as output a *basis* matrix W and a *mixture* matrix H such that $A \approx WH$.

2- The reconstruction error RE is computed by using several functions (see Sect. 2 for the details). Whenever a window V is considered, we set $d_V = \pm 2$.

3- A clustering algorithm is employed to classify the pixels between salient and not salient ones.

4- A post-processing step is performed to partly remove the noise in the final output. More precisely, for each pixel (x, y) a square window $V(\ell)$ with $d_V = \pm 4$ is considered. The same label as the 65% of the pixels within $V(\ell)$ is assigned to pixel (x, y) under consideration.

The set-up for parameters r and d_V has been empirically done by selecting the values which gave the best output.

In the following subsections, we recall some details and mathematical background.

3.1   NMF

Images datasets usually contain non negative elements. Therefore, when stored in matrices, in order to reduce the dimensionality of the problem, it is useful to apply first some non-negative factorization algorithm (NMF). In this study we use a variant of a standard NMF which is able to enforce sparseness constraints (SNMF) on the coefficient matrix H. Indeed, since we are trying to "compress" the spectral dimension, it is meaningful to set a null weight to certain coefficients.

We solve the following minimization problem

$$\min_{W \geq 0, H \geq 0} \frac{1}{2} \left\{ \|A - WH\|_F^2 + \eta\|W\|_F^2 + \beta \sum_{j=1}^{k} \|H(:,j)\|_1^2 \right\} \tag{1}$$

where $\| \ \|_F$ denotes the Frobenius norm, $\eta > 0$ is a parameter to minimize $\|W\|_F^2$ and $\beta > 0$ is a regularization parameter to balance the trade-off between the accuracy of the approximation and the sparseness of H. In particular, the initialization of the matrices W and H is done by averaging $m/5$ random columns and $n/5$ random rows of the matrix A. Then the descent gradient method is applied in order to compute the proper W and H that minimize the cost function (1). The problem (1) is a non convex optimization problem, therefore there is no guarantee to converge to a global minimum. Hence, we run the factorization 3 times and then, the matrices W and H with the lowest objective function value are chosen (parameter n_run = 3). We set $\eta = 1$, $\beta = 10^{-4}$ and, as a stopping criterion, the maximal number of iterations is given equal to 15.

3.2   Clustering Stage

In order to classify the pixels belonging to the salient region we use the *Gaussian Mixture Model* (GMM) algorithm provided by the library scikit-learn, [10]. GMM is a "model-based" method as it relies on a probabilistic model for the data generation process. In particular, GMM assumes the points to be generated by a mixture of a finite number of Gaussian distributions. The number of components of our model is a-priori given, in particular, one cluster should gather salient pixels and another one should group the not-salient pixels, for a total of 2 clusters. At the beginning, each sample is assigned membership to the Gaussian it most probably belongs to, by computing posterior probability of each component. Then the *expectation maximization* (EM) strategy is applied to maximize the likelihood of the data, given those memberships.

As input to the GMM algorithm, for every error function, we provide the following configurations:

(1) D_W: the $RE(A, Z)$ and the basis matrix W.
(2) D_WM: the $RE(A, Z)$ and W_M where the index M denotes the column of W which outputs the maximal $\|RE(A, Z) - RE(A, W_{:,i}H_{i,:})\|_F$.
(3) D_Wm: the $RE(A, Z)$ and W_m where the index m denotes the column of W which outputs the minimal $\|RE(A, Z) - RE(A, W_{:,i}H_{i,:})\|_F$,
(4) D_D-WM: the $RE(A, Z)$ and $RE(A, W_M H_M)$.
(5) D_D-Wm: the $RE(A, Z)$ and $RE(A, W_m H_m)$.
(6) D_D-W: the $RE(A, Z)$, $RE(A, W_{:,1}H_{1,:})$ and $RE(A, W_{:,2}H_{2,:})$.

The GMM has been run with setting: the general covariance matrix is used for every component, the convergence threshold is set to 10^{-4}, k-means algorithm is used for three initializations. Moreover, in order to produce more robust results, we produce three different binary maps by running the GMM algorithm three times, where each new run starts with the last computed information. Supposing that the salient region covers a small part of the whole image, the best classification output is chosen to be the one which by far has two clusters of the same size.

4 Experimental Results

We test the proposed algorithm on the HS-SOD dataset[1] introduced in [6] and on the Hermiston dataset[2] which is a common benchmark for change detection task.

(a) image id= 26   (b) Ground Truth   (c) SAMZID,   D_WM, AUC:92.55%

Fig. 1. Output on image id = 26 from HS-SOD dataset.

(a) image id= 34   (b) Ground Truth   (c) SMSAM,   D_D-WM, AUC:95.48%

Fig. 2. Result on image id = 34 from HS-SOD dataset.

(a) image id= 2   (b) Ground Truth   (c) SMSADM,   D_D-W,
                                                   AUC:92.80%

Fig. 3. Output of the algorithm on image id = 2 from HS-SOD dataset.

(a) image id= 38   (b) Ground Truth   (c) SAM_g_ZID, D_D-W,
                                                    AUC:60%

Fig. 4. The algorithm cannot fully identify the correct saliency map on image id = 38 from HS-SOD dataset.

4.1 HS-SOD Dataset

The HS-SOD dataset is a hyperspectral salient object detection dataset consisting of 60 hyperspectral images and their relative ground-truth binary images. There are 81 spectral bands, collected in the visible spectrum (380–780 nm). Each image has 1024 × 768 pixels resolution. The performance of the algorithm is evaluated by comparing the output saliency map with the provided ground-truth by using the Borji variant of the *Area Under the Roc Curve* (AUC) [2]. In the figures we show the output of the proposed method on four images, with respect to the best cluster configuration. For each image, in (a) we display the RGB rendering, in (b) the corresponding ground-truth and in (c) the best AUC obtained with the selected error measure and cluster configuration, see Fig. 1, 2, 3, and 4.

In Fig. 4 we also decided to show an example where our method partially fails to fully detect the correct salient region. This highlights one limitation of the proposed approach which is mainly driven by identifying spectral similarities using several error measures. When the spectral features of the salient region are alike to the ones of the background, our algorithm is not able to extract the correct information and the background is not distinguished at all from the salient object.

Table 1 reports the mean and standard deviation for the AUC on the whole dataset, obtained varying the error function. The label "Best" refers to the mean and standard deviation computed by taking into account only the best configuration for every image, regardless from the used error function.

In Table 2 we show the percentage of the images classified with AUC greater than 80% (class f_1), AUC between 65%–80% (class f_2) and AUC less than 65% (class f_3) by varying the adopted error function and configuration. We observe that SMSAM has the highest percentage on f_1 for three different cluster configurations among all the error measures, while the Euclidean distance (ED) has the highest percentage on f_3 for all the cluster configurations. It is noteworthy that the obtained results are sensitive to the initialization of both the SNMF and the GMM algorithms. From the experiments we observed that by using the "Best" settings, 77% of the images were classified on class f_1, 22% on f_2 and only 2% on f_3. This result suggests the importance of defining an automatic selection algorithm for the suitable error measure.

Finally, in Table 3 we can see that the proposed algorithm, Alg_1, outperforms several techniques tested on HS-SOD. More in details, the competitors include: a preliminary version of the current approach, Alg_0, introduced by the same authors in [3]; a variant, named AISA, where an autoencoder architecture is constructed to elaborate HSI data [1], other methods analyzed in [6].

Table 1. Mean ± standard deviation for AUC-Borji computed on the HS-SOD dataset. In bold the maximum value achieved by every error measure.

RE	AUC (mean ± std)					
	D_W	D_WMax	D_Wm	D_D–WM	D_D–Wm	D_D–W
ED	**0.71** ± 0.11	0.70 ± 0.11	0.68 ± 0.12	0.68 ± 0.11	0.70 ± 0.11	0.69 ± 0.12
SAM	0.72 ± 0.12	0.74 ± 0.12	0.67 ± 0.12	**0.76** ± 0.11	0.74 ± 0.11	0.72 ± 0.13
SSCC	0.72 ± 0.13	0.72 ± 0.13	0.69 ± 0.13	0.73 ± 0.12	**0.74** ± 0.12	**0.74** ± 0.10
ZID	0.71 ± 0.12	0.73 ± 0.11	0.70 ± 0.12	**0.74** ± 0.12	0.73 ± 0.11	**0.74** ± 0.11
SAMZID	0.71 ± 0.12	0.71 ± 0.14	0.71 ± 0.13	0.73 ± 0.12	0.73 ± 0.12	**0.74** ± 0.11
SAM_g_ZID	0.72 ± 0.12	0.70 ± 0.14	0.70 ± 0.13	0.74 ± 0.12	0.74 ± 0.13	**0.75** ± 0.11
SMSAM	0.73 ± 0.12	0.75 ± 0.12	0.67 ± 0.12	**0.76** ± 0.12	**0.76** ± 0.12	0.71 ± 0.13
SMSADM	0.73 ± 0.12	0.73 ± 0.13	0.68 ± 0.13	0.73 ± 0.11	**0.75** ± 0.11	0.74 ± 0.11
Spx-SSCC	**0.73** ± 0.13	0.71 ± 0.13	0.68 ± 0.13	**0.73** ± 0.12	**0.73** ± 0.13	0.72 ± 0.12
Spx-SMSAM	0.72 ± 0.11	**0.74** ± 0.12	0.69 ± 0.13	0.73 ± 0.13	0.75 ± 0.12	0.73 ± 0.13
Spx-SMSADM	0.72 ± 0.11	0.70 ± 0.12	0.69 ± 0.13	**0.73** ± 0.13	0.72 ± 0.12	**0.73** ± 0.12
Best	0.85 ± 0.08					

4.2 Hermiston Dataset

The second dataset consists of two hyperspectral images of Hermiston City (Oregon) acquired by HYPERION sensor taken on year 2004 and 2009, respectively and a ground truth binary image. The hyperspectral images have 390×200 pixels resolution and 242 spectral bands. The proposed algorithm has been tested to detect the changes in the two pictures I_1, I_2 by determining the saliency map of

Table 2. Percentage for the classes f_1, f_2 and f_3 for each error function and configuration. In bold the maximum value achieved according to the used cluster configuration.

RE	D_W			D_WM			D_Wm			D_D-WM			D_D-Wm			D_D-W		
	f_1	f_2	f_3	f_1	f_2	f_3	f_1	f_2	f_3	f_1	f_2	f_3	f_1	f_2	f_3	f_1	f_2	f_3
ED	23	38	38	22	35	43	18	32	50	18	32	50	23	33	43	20	35	45
SAM	28	37	35	38	32	30	20	28	52	37	43	20	30	43	27	32	32	37
SSCC	**33**	25	42	33	33	33	27	35	**38**	33	37	30	37	40	23	33	47	**20**
ZID	24	35	42	28	48	**23**	23	35	42	35	40	25	25	53	**22**	35	43	22
SAMZID	28	33	38	30	25	45	**35**	25	40	33	38	28	20	40	30	30	45	25
SAM_g_ZID	28	28	43	32	22	47	30	28	42	33	38	28	35	33	32	33	47	**20**
SMSAM	32	35	33	**43**	32	25	20	25	55	**43**	35	**22**	**42**	37	**22**	30	25	45
SMSADM	30	35	35	38	30	32	22	32	47	33	40	27	35	43	**22**	33	42	25
Spx-SSCC	32	35	33	32	37	32	18	35	47	37	33	30	40	28	32	28	42	30
Spx-SMSAM	25	43	**32**	42	30	28	22	35	43	35	33	32	38	35	27	30	35	35
Spx-SMSADM	25	37	38	30	27	43	22	33	45	38	28	33	35	32	33	**38**	32	30

Table 3. Comparison of the AUC-Borji with competitors from [1,3,6]. In particular, these last ones are the following: Itti's method; a saliency detection method based on the computation of the euclidean distance (SED) and spectral angle distance (SAD); spectral group method (GS); orientation-based salient features method (OCM) in combination with: SED and GS (SOG), and, SED and SAD (SOS); the method using the spectral gradient contrast (SGC) computed by using superpixels with both, spatial and spectral gradients.

Method	Alg_1	Alg_0	AISA	Itti	SED	SAD	GS	SOG	SOS	SGC
AUC-Borji	0.8509	0.7971	0.7727	0.7694	0.6415	0.7521	0.7597	0.7863	0.8008	0.8205

Ground Truth	SAMSADM, D_D-WM, AUC:91.00%	ED, D_D-W, AUC:90.66%

Fig. 5. Results for the change detection task on Hermiston dataset.

the following input: $|I_1 - I_2| \, (|1 - I_{1s}/I_{2s}| + |1 - I_{2s}/I_{1s}|)/2$, where $I_{js} = I_j + s$, s is chosen such that the minimum value between I_1 and I_2 is 1 and the difference to 1 on the ratios has been applied in order to enhance the changes. From the results displayed on Fig. 5 we can see that the produced output and the expected ground truth are very much alike.

5   Conclusions

In the present work we propose an algorithm to detect saliency regions in hyperspectral images by employing SNMF and several reconstruction errors based on spectral and spatial measures which add a notable contribution to better characterizing what should be understood as salient. This approach has been tested on HS-SOD and Hermiston datasets. The obtained results are promising, leave room to further investigations and to broader applications in data mining tasks such as semantic discovery patterns, collaborative filtering and recommender systems design. The output of our approach is strongly dependent on the initialization steps for the SNMF and for the clustering algorithm, and on the choice of the error measure as well. Hence a possible future direction could be to automatically identify the best settings. In particular, it would be interesting to study the relations between the adopted error function and the obtained output.

Acknowledgments. Antonella Falini, Cristiano Tamborrino and Francesca Mazzia are members of the INdAM Research group GNCS. Rosa Maria Mininni is member of the INdAM Research group GNAMPA. Annalisa Appice and Donato Malerba are members of the CINI Big Data Laboratory.

This work was also developed within the project "Modelli e metodi per l'analisi di dati complessi e voluminosi" of the University of Bari.

References

1. Appice, A., Lomuscio, F., Falini, A., Tamborrino, C., Mazzia, F., Malerba, D.: Saliency detection in hyperspectral images using autoencoder-based data reconstruction. In: Helic, D., Leitner, G., Stettinger, M., Felfernig, A., Raś, Z.W. (eds.) ISMIS 2020. LNCS (LNAI), vol. 12117, pp. 161–170. Springer, Cham (2020). https://doi.org/10.1007/978-3-030-59491-6_15
2. Borji, A., Cheng, M.M., Jiang, H., Li, J.: Salient object detection: a benchmark. IEEE Trans. Image Process. **24**(12), 5706–5722 (2015)
3. Falini, A., et al.: Saliency detection for hyperspectral images via sparse-non negative-matrix-factorization and novel distance measures. In: 2020 IEEE Conference on Evolving and Adaptive Intelligent Systems (EAIS). pp. 1–8. IEEE (2020)
4. Gao, D., Vasconcelos, N.: Discriminant saliency for visual recognition from cluttered scenes. In: Advances in Neural Information Processing Systems. pp. 481–488 (2005)
5. Han, J., Zhang, D., Hu, X., Guo, L., Ren, J., Wu, F.: Background prior-based salient object detection via deep reconstruction residual. IEEE Trans. Circuits Syst. Video Technol. **25**(8), 1309–1321 (2014)
6. Imamoglu, N., et al.: Hyperspectral image dataset for benchmarking on salient object detection. In: 0th International Conference on Quality of Multimedia Experience (QoMEX) (2018)
7. Koch, C., Ullman, S.: Shifts in selective visual attention: towards the underlying neural circuitry. In: Matters of intelligence, pp. 115–141. Springer (1987)
8. Kruse, F.A., et al.: The spectral image processing system (SIPS)—interactive visualization and analysis of imaging spectrometer data. Remote Sens. Environ. **44**(2–3), 145–163 (1993)

9. Liu, J., Liu, Y.: A model for saliency detection using NMFsc algorithm. In: Jiang, X., Petkov, N. (eds.) CAIP 2009. LNCS, vol. 5702, pp. 301–308. Springer, Heidelberg (2009). https://doi.org/10.1007/978-3-642-03767-2_37

10. Pedregosa, F., et al.: Scikit-learn: machine learning in python. J. Mach. Learn. Res. **12**, 2825–2830 (2011)

11. Radke, R.J., Andra, S., Al-Kofahi, O., Roysam, B.: Image change detection algorithms: a systematic survey. IEEE Trans. Image Process. **14**(3), 294–307 (2005)

12. Seydi, S.T., Hasanlou, M.: A new land-cover match-based change detection for hyperspectral imagery. Euro. J. Remote Sens. **50**(1), 517–533 (2017)

13. Tang, J., Lewis, P.H.: Non-negative matrix factorisation for object class discovery and image auto-annotation. In: Proceedings of the 2008 international conference on Content-based image and video retrieval. pp. 105–112. ACM (2008)

14. Tao, D., Cheng, J., Song, M., Lin, X.: Manifold ranking-based matrix factorization for saliency detection. IEEE Trans. Neural Networks Learn. Syst. **27**(6), 1122–1134 (2015)

15. Treisman, A.M., Gelade, G.: A feature-integration theory of attention. Cogn. Psychol. **12**(1), 97–136 (1980)

16. van der Walt, S., et al.: The scikit-image contributors: scikit-image: image processing in Python. Peer J. **2**, e453 (2014)

17. Yang, Z., Mueller, R.: Spatial-spectral cross-correlation for change detection–a case study for citrus coverage change detection. In: ASPRS 2007 Annual Conference (2007)

18. Zheng, Q., Yu, S., You, X.: Coarse-to-fine salient object detection with low-rank matrix recovery. Neurocomputing (2019)

19. Zupan, B., et al.: Nimfa: A python library for nonnegative matrix factorization. J. Mach. Learn. Res. **13**, 849–853 (2012)

Sparse Consensus Classification for Discovering Novel Biomarkers in Rheumatoid Arthritis

Cláudia Constantino[1]⊙, Alexandra M. Carvalho[2]⊙, and Susana Vinga[1(✉)]⊙

[1] INESC-ID, Instituto Superior Técnico, ULisboa, Lisbon, Portugal
`susanavinga@tecnico.ulisboa.pt`
[2] Instituto de Telecomunicações, Instituto Superior Técnico, ULisboa, Lisbon, Portugal
`alexandra.carvalho@tecnico.ulisboa.pt`

Abstract. Rheumatoid arthritis (RA) is a long-term autoimmune disease that severely affects physical function and quality of life. Patients diagnosed with RA are usually treated with anti-tumor necrosis factor (anti-TNF), which in certain cases do not contribute to reach remission. Consequently, there is a need to develop models that can predict therapy response, thus preventing disability, maintain life quality, and decrease cost treatment. Transcriptomic data are emerging as valuable information to predict RA pathogenesis and therapy outcome. The aim of this study is to find gene signatures in RA patients that help to predict the response to anti-TNF treatment. RNA-sequencing of whole blood samples dataset from RA patients at baseline and following 3 months of therapy were used. A methodology based on sparse logistic regression was employed to obtain predictive models which allowed to find 20 genes consensually associated with therapy response, some known to be related with RA. Gene expression levels at 3 months of therapy showed no added value in the prediction of response to therapy when compared with the baseline. The analysis using Bayesian network learning unveiled significant protein-protein interactions in both good and non-responders, further confirmed using the STRING database. Structured sparse regression coupled with Bayesian learning can support the identification of disease biomarkers and generate hypotheses to be further analysed by clinicians.

Keywords: Regularized optimization · Bayesian networks · Protein-protein interaction networks.

1 Introduction

As high-dimensional data becomes increasingly available in all the fields of research, effective analytic methods are fundamental to extract the maximum

Partially funded by FCT (PTDC/CCI-CIF/29877/2017, UIDB/50021/2020, UIDB/50008/2020).

ⓒ Springer Nature Switzerland AG 2020
G. Nicosia et al. (Eds.): LOD 2020, LNCS 12565, pp. 125–136, 2020.
https://doi.org/10.1007/978-3-030-64583-0_13

scientific understanding from the data. For instance, in genetic studies, in which the number of variables (genes) is particularly large compared with the number of samples (subjects), methods to extract knowledge from the data are essential.

Rheumatoid arthritis (RA) is a common systemic autoimmune disease that severely damages physical function and quality of life. This chronic disease affects about 1% of adult citizens [1]. RA cause remains unknown, although genetic factors are responsible for a part of disease predisposition.

Nowadays, RA therapy is processed by the administration of disease-modifying anti-rheumatic drugs (DMARDs), that have shown to slow down the disease progression. However, DMARDs may have no effect on patients, and therapy with anti-tumor necrosis factor (TNF) is then recommended [2,3]. If the therapy with TNF inhibitors fails in a particular patient, an alternative agent is again chosen. It is usually very difficult to find an agent, or a combination of agents, that induce remission. So, RA patients may experience therapy with successive changes in the administration of these agents until disease remission, which can significantly worsen patient disability and can increase the cost of treatment.

The prediction of the patient's response to anti-TNF therapy is then of paramount importance, and it has been the object of study in several studies that take into account demographic, clinical, and genetic data [4,7]. Interestingly, the studied clinical baseline biomarkers do not seem to add value in the prediction of treatment response [5]. Notwithstanding, gene expression profiling may provide insight into disease pathogenesis and already showed significant changes before and after anti-TNF treatment [6], which illustrates the potential of using molecular information for prognosis. At the baseline of anti-TNF treatment, innate/adaptive immune cell-type-specific genes revealed associations with the response to treatment within 3 months of therapy [7]. Recent studies have studied and demonstrated the role of molecular data to conduct robust predictions in different human diseases [8,9]. Therefore, additional evaluation of the gene expression can be helpful to define biomarkers of outcome and response to therapy in RA.

The aim of this study is to find gene signatures in patients with RA before and after the beginning of the therapy, which may help to predict the response to anti-TNF treatment. We propose a predictive model, based on dimensionality reduction techniques and on a consensus approach, which uncover the most relevant genes to predict the response to therapy. We also use a Bayesian network methodology to discover relevant protein-protein interactions.

The paper is organised as follows. In Sect. 2 we present the RA data under study and the sparse logistic regression and Bayesian networks. Next, in Sect. 3 we present the experimental results and discuss them. Finally, we draw some conclusions and discuss future works.

2   Methods

2.1   Rheumatologic Transcriptomic Data

The dataset under study is constituted by RNA-sequencing of whole blood samples from biologic naíve RA patients. All the files are publicly available from the CORRONA CERTAIN registry [10] and at the NCBI-GEO database (GSE129705); these data were previously analysed by Farutin et al. [7]. These patients had no previous biologic agent treatment, and they are initiating therapy with anti-TNF. The transcriptomic data are composed of a set of 25,370 variables (gene expressions) measured from 63 patients at baseline (BL), and from 65 patients at 3 months (M3) after the beginning of anti-TNF treatment.

According to EULAR criteria for clinical response to therapy at 3 months [11], each patient is classified as *good responder* or *non-responder* (GR and NR, respectively). Under this classification, 36 patients were categorized as GR and the remaining 27 as NR to therapy with anti-TNF.

2.2   Sparse Logistic Regression

Binary logistic regression defines the relationship between n independent observations $\{\mathbf{X}_i\}_{i=1}^n$ and a binary outcome $\{Y_i\}_{i=1}^n$, where each observation is measured over p variables $\mathbf{X}_i = (X_{i1}, \ldots, X_{ip})^T$.

Specifically in this work, $n = 63$, $Y_i = 1$ corresponds to a GR patient and $Y_i = 0$ to a NR patient. Then, the logistic regression is given by

$$P(Y_i = 1|\mathbf{X}_i) = \frac{\exp(\mathbf{X}_i^T \beta)}{1 + \exp(\mathbf{X}_i^T \beta)}, \tag{1}$$

where β represents the regression coefficients related with the p variables and $P(Y_i = 1|\mathbf{X}_i)$ is the probability of observation i being a good responder (GR).

Logistic regression is a classical method that has shown to be competitive, with equally performing results, compared with alternative machine learning techniques in clinical and biological research [12–14].

To deal with datasets with a number of variables much higher than the number of observations, $(p \gg n)$, an initial dimensionality reduction step is fundamental. Getting an adequate generalized model can be extremely difficult in a high dimensional dataset due to the number of variables to be considered in the final model and the few observations to support the model's hypothesis. To overcome this problem, additional constraints in the cost function can be applied. Regularization methods like Least Absolute Shrinkage and Selection Operator (Lasso), Ridge regression, elastic net, and other sparsity methods, provides a sparse estimate of the unknown regression coefficients and have become a classical approach to deal with the possible non-identifiability of the regression models.

For example, Lasso regression [15] enables shrinkage and variable selection in the solutions by penalizing the sum of the absolute values of the coefficients (L1-norm), defined as:

$$\Psi(\beta) = \sum_{i=1}^{p} |\beta_i|. \tag{2}$$

Ridge regression [16] considers the L2-norm (sum of the squared error of the coefficients) penalty instead, having:

$$\Psi(\beta) = \sum_{i=1}^{p} |\beta_i^2|. \tag{3}$$

The elastic net regularization combines L1 and L2 norms with the objective of limit solution space [17]. So, the elastic net is a combination of Lasso and Ridge regression:

$$\lambda\Psi(\beta) = \lambda(\alpha||\beta||_1 + (1-\alpha)||\beta||_2^2), \tag{4}$$

where α is a controller between L1 and L2 penalties, given a fixed λ. Penalization control of the weights is given by λ.

If $\alpha = 0$, Ridge regression is applied. On the other hand, if $\alpha = 1$, we are dealing with Lasso regression. Elastic net allows the balance of sparsity with the correlation between variables, which gives to this method high flexibility for different types of datasets.

Maximum likelihood are used to estimate β coefficients. In the case of elastic net regression, the penalized log-likelihood function, with L1 and L2 weights, is the following:

$$l(\beta) = \sum_{i=1}^{n} \left\{ y_i \log P(Y_i = 1|\mathbf{X}_i) + (1-y_i)\log[1 - P(Y_i = 1|\mathbf{X}_i)] \right\} + \lambda\Psi(\beta)$$

$$= \sum_{i=1}^{n} \left(y_i\mathbf{X}_i^T\beta - \log(1 + \exp(\mathbf{X}_i^T\beta)) \right) + \lambda\Psi(\beta), \tag{5}$$

where the binary variable y_i indicates if the ith patient is a good responder (GR) ($y_i = 1$) or a non-responder (NR) ($y_i = 0$).

To fit the best predictive model to a dataset, cross-validation (CV) is an important step. It allows estimating parameters for the envisaged model to improve predictions. When regularization is being performed, the addition of a variable in the model may increase model performance. However, its inclusion may also have a high cost, and in that case, the variable should be disregarded of the final model. CV allows tuning the model parameter to perform the best feature selection and prevent overfitting. The penalty λ parameter is estimated using this CV strategy, i.e., we use the value that achieves the minimum mean cross-validation error.

Leave-one-out cross-validation (LOOCV) is another technique generally used that we apply in the present study. It is based on estimating a model by considering all observations except one, that is left out from the training set. That observation is then used to validate the predictive power of the estimated model. This procedure is run the same number of times as the number of the existing observations. In this case study, the objective is to predict whether a RA patient responds or not to the treatment with anti-TNF, i.e., a binary outcome. To evaluate the estimated model, the classifier's specificity and sensitivity trade-off in the validation set can be visualized through Receiver operating characteristic (ROC) curves. The area under the ROC curve (AUC) is then calculated as a quantitative measure of the classifier performance.

2.3   Bayesian Networks

Bayesian networks (BNs) are a rich framework to model domains with complex connections between thousands or millions of random variables. BNs are graphical models that represent a family of probability distributions defined in terms of a directed graph. The nodes of the graph contain the random variables, and the product of local connections defines a unique joint probability distribution. These local connections represent dependencies between a random variable and its parents in the graph.

Rigorously, let $\mathbf{Z} = (Z_1, \ldots, Z_p)$ be a p-dimensional random vector. A Bayesian network (BN) is a pair (G, θ) where $G = (V, E)$ is a directed acyclic graph (DAG), with nodes in V, coinciding with the random variables in $\mathbf{Z}$, and edges in E. The parameters describe how each variable relates probabilistically with its parents. Using the chain rule, we can then obtain a joint probability distribution, in a factored way, according to the DAG structure, defined as:

$$P(Z_1, \ldots, Z_p) = \prod_{j=1}^{p} P(Z_j | \mathrm{pa}(Z_j), \theta_j), \tag{6}$$

where $\mathrm{pa}(Z_j) = \{Z_i : Z_i \to Z_j \in E\}$ is the parent set of Z_j and θ_j encodes the parameters that define the conditional probability distribution (CPD) for Z_j. In the case of continuos data, Gaussian CPDs are considered.

When learning a BN, the challenge is in structure learning. With the structure fixed, parameters are quite easy to learn. Structure learning is accomplished through score-based learning, where a score is used to understand the network that best fits the data. A possible scoring criterion is the maximum likelihood. When overfitting occurs, penalisation factors are used to avoid it.

Aragam et al. (2017) developed an R package, called sparsebn [18], especially devoted to high dimensional data. For that, a sparse BN is outputted using a score-based approach that relies on regularised maximum likelihood estimation. The scoring criterion is given by:

$$\mathrm{LL}(B; \boldsymbol{X}) + \rho_\lambda(B), \tag{7}$$

where LL denotes the negative log-likelihood, B the BN, and ρ_λ is some regulariser. For continuous data, a Gaussian likelihood with L1 or minimax concave penalty is proposed.

Considering gene expression, from rheumatologic transcriptomic data, as observations of random variables, a sparse BN is a rich model to describe the underlying data. The nodes represent the genes, and connections between them correspond to gene interactions/edges. The interactions found are those that best explain the data.

3   Results and Discussion

To unravel the most relevant genes in RA patients for the response to anti-TNF therapy, two datasets of gene expression levels, from the CORRONA CERTAIN registry, were used: (i) at baseline (BL), and (ii) three months after starting the treatment with anti-TNF (M3).

Both BL and M3 datasets contain a set of 25370 variables/genes. The data were preprocessed as follows. In both datasets, the variables with zero standard deviation were firstly excluded. This resulted in a reduction to 21911 and to 22142 variables, respectively, for BL and M3 datasets. Then, the variables were log-transformed and normalised to unit variance. A vector with binary responses, with '1' for GR patients and '0' for NR patients, were further used in logistic regression (a vector for BL dataset and another for M3 dataset).

To ensure full reproducibility of our results, all the R code and data are available at https://github.com/sysbiomed/RA-CORRONA.

3.1   Identification of Response Biomarkers

Dimensionality reduction was achieved by applying sparse logistic regression using the glmnet R package [19]. For model validation, 70% of the dataset were randomly split for training the model, and the remaining 30% for the test. This procedure was repeated 100 times. The model is estimated in the training set with logistic regression, defining the λ and α parameter. CV is used to choose the λ parameter that better fit the model. The α parameters used varied between 0 and 1 with intervals of 0.1. Then, the fitted model was used to predict the response of the treatment of the test set. For each model, the ROC curve was accessed, and the AUC calculated. The ROC curve is obtained by using different values for the classification probability threshold. The median and interquartile range of the models estimated using each α in the BL data and over the 100 runs are presented in Table 1.

The AUC values obtained with sparse logistic regression at month 3 (M3) were not significantly different from those at baseline (BL), presented in Table 1 (t-test, p-values ranging from 0.07 to 0.49). Therefore, only the dataset from BL was further used.

In Farutin et al. [7], the top 100 genes over-expressed with the most significant contributions to the negative correlation with the effect of treatment are presented. We applied the same methodology as before considering those specific genes, to achieve a predictive model. In this case, only $\alpha = 0$ was used for the logistic regression, to ensure that all the genes emerge in the model. The median

Table 1. Area Under the Curve (AUC) results from the sparse logistic regression at baseline data for different α parameters. For each α, the statistics of the AUC over all 100 runs are reported, namely the median values, the interquartile range (IQR), the maximum (Max) and the minimum (Min). The best median AUCs results are highlighted in bold.

AUC	α 0	0.1	0.2	0.3	0.4	0.5	0.6	0.7	0.8	0.9	1.00
Median	0.62	**0.64**	0.60	0.60	**0.63**	0.59	0.58	0.57	0.58	0.58	0.52
IQR	0.12	0.13	0.13	0.16	0.11	0.14	0.12	0.13	0.14	0.12	0.09
Max	0.83	0.82	0.89	0.90	0.85	0.78	0.84	0.86	0.81	0.86	0.76
Min	0.37	0.45	0.42	0.43	0.44	0.42	0.39	0.38	0.41	0.37	0.36

and interquartile range (IQR) of the AUCs was 0.57 and 0.09, respectively. These results show that although these specified genes are over-expressed, they do not have better predictive power than those identified with a sparse logistic regression using all the available genes and $\alpha \in [0, 0.9]$ (Table 1).

The better predictive models were achieved with an α of 0.1 and 0.4 applied to the BL dataset, with median AUC of 0.64 and 0.63 for $\alpha = 0.1$ and $\alpha = 0.4$, respectively. These results are satisfactory, taking into account the few observations (63) in the dataset. Hereupon, these were the parameters for logistic regression further used.

To detect genes strongly associated with the response, LOOCV was applied. The intersection of genes appearing in all the predictive models calculated with LOOCV correspond to those that may have a better predictive response at BL. Amongst the 21911 genes in the dataset, 20 were repeatedly selected for model prediction, both for $\alpha = 0.1$ and $\alpha = 0.4$. The 20 identified genes are the following:

– *ALOX12B, CAPNS2, CTSG, EPHX4, EVPLL, FAM133CP, FOXD4L3, HIST1H3J, IGF2BP1, LOC339975, LRGUK, MPO, NUAK1, ODF3L2, PRKG1, PRSS30P, RAD21L1, SLC6A19, SYT1*, and *TGFB2*.

Even though none of these discovered genes were present in the top 100 genes over-expressed presented by Farutin et al. [7], some are already known to be related to the RA disease.

CTSG was previously found to participate in the pathogenesis of some autoimmune diseases, as, for instance, RA [20]. When compared with healthy controls, *CTSG* activity and concentration are augmented in the synovial fluids of RA patients. Therefore, identification of *CTSG* in our model may be associated with the response to anti-TNF therapy in RA patients.

Karouzakis et al. [21] found that *EPHX4* gene was one of the top-ranked genes differentially expressed in human lymph node stromal cells (LNSCs) during the earliest phases of RA. LNSCs are decisive in shaping the immune response in lymphoid tissue (where is initiated the adaptive immunity). This suggests that *EPHX4* could be related to the immune response to treatment.

The *MPO* gene encodes myeloperoxidase. Myeloperoxidase serum levels are encountered in inflammatory diseases, like RA. Fernandes et al. [22] observed significantly higher MPO plasma levels in RA patients. Yet, no correlation between disease activity measured by EULAR criteria and MPO expression was found. Contrarily, our analysis led to the identification of this gene as helpful for the prediction of the response to therapy.

Also, *RAD21L1* is part of the 21 upregulated by tumor necrosis factor'like ligand 1A (TL1A) [23]. TL1A is a tumor necrosis factor that influences positively the pathogenesis of autoimmune diseases, including RA.

All the hypotheses above should be confirmed in further studies with the contribution of a rheumatologist. The relevance of the remaining genes in RA was not explored in previous studies. So, we propose a subsequent investigation of the 20 genes achieved in our analysis with the response to therapy with anti-TNF agents. Moreover, the presented methods can be further added to a future benchmarking study that comprehensively assesses the performance of different classifiers.

3.2   Identification of Protein-Protein Interactions

The disclosure of gene networks regulating the response to anti-TNF treatment was performed through Bayesian network (BN) learning. The 239 variables (genes), resulting from the sparse logistic regression with $\alpha = 0.1$, were used. This parameter α was selected, taking into account the trade-off between the AUC medians (Table 1) and the identification of a reasonable number of genes to be further analysed. The baseline dataset with the 239 variables was split into two independent sets: responders at BL (R-BL; 36 observations) and non-responders at BL (NR-BL; 27 observations), each one described by two distinct Bayesian networks.

This methodology was applied using `sparsebn` R package [18]. The method `estimate.dag` was used to learn the two distinct Bayesian networks. Therein, we used default parameter settings; the `edge.threshold` parameter was specified to force the number of edges in the solution to be less or equal than the double of the nodes (239×2). The output is a set of different networks, each one with a distinct number of edges. From this set, the solution giving the number of edges equal to the number of nodes was chosen, both for R-BL and NR-BL datasets. The same was done in case the number of edges being twice the number of nodes. Figure 1 illustrates the obtained networks.

To verify if the obtained edges (pairwise gene connections) were previously identified, the STRING information was used. STRING is a database of known and predicted protein-protein interactions [24]. In our study, only protein-protein interactions with high combined score (*combined_score* > 0.7) in the STRING database were considered. The highly connected genes from STRING were then compared with the edges given by BNs.

From the 239 edges, 4 in R-BL and 2 in NR-BL are reported in STRING. There was no improvement in the number of common edges found in STRING when increasing the number of edges from 239 to 478 edges.

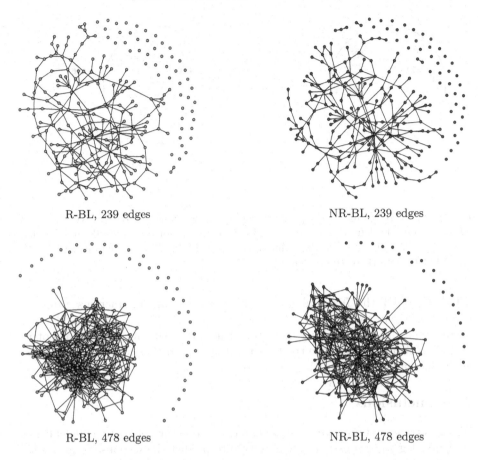

R-BL, 239 edges   NR-BL, 239 edges

R-BL, 478 edges   NR-BL, 478 edges

Fig. 1. Bayesian networks learnt from R-BL data (green nodes) and NR-BL data (red nodes). (a) BN for R-BL with 239 edges. (b) BN for NR-BL with 239 edges. (c) BN for R-BL with 478 edges. (d) BN for NR-BL with 478 edges.

The obtained Venn diagrams illustrate the overlap between the identified protein-protein interactions in R-BL, NR-BL, and STRING (Fig. 2), and are as follows:

- **R-BL ∩ STRING:** *DEFA4—CTSG*; *AZU1—MPO*; *CTSG—SERPINB10*; *DEFA4—AZU1*
- **NR-BL ∩ STRING:** *CTSG—AZU1*; *AZU1—MPO*

One overlap interaction between responders and non-responders was found, which suggests that protein-protein interaction (*AZU1—MPO*) may be relevant for both responders and non-responders. According to STRING, this interaction with a total combined score of 0.985, is associated with: 1) the co-expression of these two genes, 2) Database knowledge from the Reactome (in particular the Neutrophil degranulation pathway, in the Innate Immune System – R-HSA-

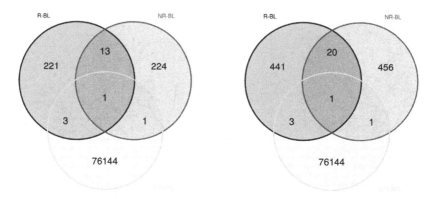

Fig. 2. Venn diagrams with protein-protein interactions in R-BL, NR-BL, and STRING. On the left, the overlap of STRING interactions with R-BL and NR-BL networks with 239 edges. On the right, the overlap of STRING interactions with R-BL and NR-BL networks with 478 edges.

6798751), and 3) the Textmining category. The remaining interactions stand out independently in the responders and non-responders. These known highly connected genes may be a strong indicator of how RA patients respond or not to anti-TNF therapy before the treatment initiation and therefore represent interesting biomarkers to be further explored.

4   Conclusions

Through transcriptomic data, we presented a satisfactory predictive model based on sparse logistic regression that may help to predict the response to anti-TNF therapy in RA patients prior to treatment initiation. A predictive model at BL showed an identical prediction performance compared with that at M3. Also, our methodology was able to unveil genes consistently associated with therapy response, which may be valuable in the expression profiling of RA patients. Some of these genes are already known to be related to RA disease. Moreover, the application of BN learning uncovered highly connected genes in responders and non-responders. The next challenge is to study promising gene signatures individually to validate biomarkers to be used in clinical practice.

References

1. Neumann, E., Frommer, K., Diller, M., Müller-Ladner, U.: Zeitschrift für Rheumatologie **77**(9), 769–775 (2018). https://doi.org/10.1007/s00393-018-0500-z
2. Radner, H., Aletaha, D.: Wiener Medizinische Wochenschrift (9), 3–9 (2015). https://doi.org/10.1007/s10354-015-0344-y
3. Smolen, J.S., Landewé, R., Breedveld, F.C., et al.: EULAR recommendations for the management of rheumatoid arthritis with synthetic and biological disease-modifying antirheumatic drugs. Ann. Rheum. Dis. **69**(6), 964–975 (2010). https://doi.org/10.1136/ard.2009.126532

4. Wijbrandts, C.A., Tak, P.P.: Prediction of response to targeted treatment in rheumatoid arthritis. Mayo Clin. Proc. **92**(7), 1129–1143 (2017). https://doi.org/10.1016/j.mayocp.2017.05.009

5. Cuppen, B.V., et al.: Personalized biological treatment for rheumatoid arthritis: a systematic review with a focus on clinical applicability. Rheumatology **55**(5), 826–839 (2016). https://doi.org/10.1093/rheumatology/kev421

6. Oswald, M., et al.: Modular analysis of peripheral blood gene expression in rheumatoid arthritis captures reproducible gene expression changes in tumor necrosis factor responders. Arthritis Rheumatol. **67**(2), 344–351 (2015). https://doi.org/10.1002/art.38947

7. Farutin, V., et al.: Molecular profiling of rheumatoid arthritis patients reveals an association between innate and adaptive cell populations and response to anti-tumor necrosis factor. Arthritis Res Ther. **21**(1), 216 (2019). https://doi.org/10.1186/s13075-019-1999-3

8. Barracchia, E.P., Pio, G., D'Elia, D., Ceci, M.: Prediction of new associations between ncRNAs and diseases exploiting multi-type hierarchical clustering. BMC Bioinf. **21**(1), 1–24 (2020). https://doi.org/10.1186/s12859-020-3392-2

9. Pio, G., Ceci, M., Prisciandaro, F., Malerba, D.: Exploiting causality in gene network reconstruction based on graph embedding. Machine Learning **109**(6), 1231–1279 (2019). https://doi.org/10.1007/s10994-019-05861-8

10. Pappas, D.A., Kremer, J.M., Reed, G., Greenberg, J.D., Curtis, J.R.: Design characteristics of the corrona certain study: a comparative effectiveness study of biologic agents for rheumatoid arthritis patients. BMC Musculoskelet Disord. **15**(1), 113 (2014). https://doi.org/10.1186/1471-2474-15-113

11. Fransen, J., van Riel, P.L.: The Disease activity score and the eular response criteria. Clin. Exp. Rheumatol. **23**(5 Suppl 39), S93–S99 (2005)

12. Pua, Y.-H., Kang, H., Thumboo, J., Clark, R.A., Chew, E.S.-X., Poon, C.L.-L., Chong, H.-C., Yeo, S.-J.: Machine learning methods are comparable to logistic regression techniques in predicting severe walking limitation following total knee arthroplasty. Knee Surgery, Sports Traumatology, Arthroscopy **28**(10), 3207–3216 (2019). https://doi.org/10.1007/s00167-019-05822-7

13. Faisal, M., Scally, A., Howes, R., Beatson, K., Richardson, D., Mohammed, M.A.: A comparison of logistic regression models with alternative machine learning methods to predict the risk of in-hospital mortality in emergency medical admissions via external validation. Health Inf. J. **26**(1), 34–44 (2020). https://doi.org/10.1177/1460458218813600

14. Kuhle, S., et al.: Comparison of logistic regression with machine learning methods for the prediction of fetal growth abnormalities: a retrospective cohort study. BMC Pregnancy and Childbirth **18**(1), 333 (2018). https://doi.org/10.1186/s12884-018-1971-2

15. Tibshirani, R.: Regression shrinkage and selection via the Lasso. J. Royal Stat. Soc.: Series B 58(1), 267–288 (1996)

16. Hoerl, A.E., Kennard, R.W.: Ridge regression: biased estimation for nonorthogonal problems. Technometrics **12**(1), 55–67 (1970)

17. Zou, H., Hastie, T.: Regularization and variable selection via the elastic net. J. Royal Stat. Soc. Series B **67**(2), 301–320 (2005)

18. Aragam, B., Gu, J., Zhou, Q.: Learning large-scale bayesian networks with the sparsebn package. J. Stat. Software **91**(11), 01–38 (2019). https://doi.org/10.18637/jss.v091.i11

19. Friedman, J., Hastie, T., Tibshirani, R.: Regularization paths for generalized linear models via coordinate descent. J. Stat. Software **33**(1), 1–22 (2010)

20. Gao, S., Zhu, H., Zuo, X., Luo, H.: Cathepsin g and its role in inflammation and autoimmune diseases. Arch Rheumatol. **33**(4), 498–504 (2018). https://doi.org/10.5606/ArchRheumatol.2018.6595

21. Karouzakis, E., et al.: Molecular characterization of human lymph node stromal cells during the earliest phases of rheumatoid arthritis. Front. Immunol. **10**, 1863 (2016). https://doi.org/10.3389/fimmu.2019.01863

22. Fernandes, R.M., da Silva, N.P., Sato, E.I.: Increased myeloperoxidase plasma levels in rheumatoid arthritis. Rheumatol. Int. **32**(6), 1605–1609 (2012). https://doi.org/10.1007/s00296-011-1810-5

23. Fukuda, K., Miura, Y., Maeda, T., Hayashi, S., Kuroda, R.: Expression profiling of genes in rheumatoid fibroblast-like synoviocytes regulated by tumor necrosis factor-like ligand 1A using cDNA microarray analysis. Biomed. Rep. **1**(1), 1–5 (2019). https://doi.org/10.3892/br.2019.1216

24. Szklarczyk, D., et al.: STRING v10: protein-protein interaction networks, integrated over the tree of life. Nucleic Acids Res. **43**, 447–452 (2015). https://doi.org/10.1093/nar/gku1003

Learning More Expressive Joint Distributions in Multimodal Variational Methods

Sasho Nedelkoski[1(✉)], Mihail Bogojeski[2], and Odej Kao[1]

[1] Distributed Systems, TU Berlin, Berlin, Germany
{nedelkoski,odej.kao}@tu-berlin.de
[2] Machine Learning, TU Berlin, Berlin, Germany
mihail.bogojeski@campus.tu-berlin.de

Abstract. Data often are formed of multiple modalities, which jointly describe the observed phenomena. Modeling the joint distribution of multimodal data requires larger expressive power to capture high-level concepts and provide better data representations. However, multimodal generative models based on variational inference are limited due to the lack of flexibility of the approximate posterior, which is obtained by searching within a known parametric family of distributions. We introduce a method that improves the representational capacity of multimodal variational methods using normalizing flows. It approximates the joint posterior with a simple parametric distribution and subsequently transforms into a more complex one. Through several experiments, we demonstrate that the model improves on state-of-the-art multimodal methods based on variational inference on various computer vision tasks such as colorization, edge and mask detection, and weakly supervised learning. We also show that learning more powerful approximate joint distributions improves the quality of the generated samples. The code of our model is publicly available at https://github.com/SashoNedelkoski/BPFDMVM.

1 Introduction

Information frequently originates from multiple data sources that produce distinct modalities. For example, human perception is mostly formed by signals that go through visual, auditory, and motor paths, and video is accompanied by text captions and audio signals. These modalities in separate describe individual properties of the observation but are also correlated with each other.

The learning of a joint density model over the space of multimodal inputs is likely to yield a better generalization in various applications. Deep multimodal learning for the fusion of speech or text with visual modalities provided a significant error reduction [16,20]. The best results for visual question answering are achieved by using a joint representation from pairs of text and image data [10]. Distinct from the fully supervised learning where the mapping between modalities is learned, multimodal learning with generative models aims to capture the

© Springer Nature Switzerland AG 2020
G. Nicosia et al. (Eds.): LOD 2020, LNCS 12565, pp. 137–149, 2020.
https://doi.org/10.1007/978-3-030-64583-0_14

approximation of the joint distribution. This enables data generation from the joint as well as conditional distributions to produce missing modalities.

The complex nature of the multimodal data requires learning highly expressive joint representations in generative models. One of the most utilized approaches for multimodal generative models is variational inference. However, these models have limitations as the search space for the best posterior approximation is always within a parametric family of distributions, such as the Gaussian [21,23,25]. Even when the complexity of the data increases, as multiple sources of information exist, the search space of the approximate posterior remains unchanged from the unimodal variant [11]. This type of parametric posterior imposes an inherent restriction since the true posterior can be recovered only if is contained in that search space, which is not observed for most data [2].

Contributions. We propose an approach to learn more expressive joint distributions in the family of multimodal generative models, which utilizes normalizing flows to transform and expand the parametric distribution and learn a better approximation to the true posterior.

First, we utilize the product-of-experts to prevent an exponential number of inference networks [7], leading to a reduced number of inference networks as one per modality. The output of the product-of-experts inference network is the parameters of the initial joint posterior distribution of the modalities, which is often Gaussian. Subsequently, we add a module that transforms this parametric distribution using continuous normalizing flows. This process produces a new and more complex joint distribution, enabling a better approximation to the true posterior in multimodal variational methods. The learned transformation allows the model to better approximate the true joint posterior. Samples from this distribution are subsequently used by the decoder networks to generate samples for each modality.

We evaluate the quality of the method to the state of the art and show improvements on multiple different datasets. We demonstrate that the quality of the generated missing modalities is higher than that of the state of the art and that our model can learn complex image transformations. Furthermore, the experiments also show that the model can be used for weakly supervised learning with a small amount of labeled data to reach decent values of the loss function.

2   Related Work

Recently, deep neural networks have been extensively investigated in multimodal learning [13]. A group of methods, such as Deep Boltzmann Machines [20] and topic models [1,8], learn the probability density over the space of multimodal inputs (i.e., sentences and images). Although the modalities, architectures, and optimization techniques might differ, the concept of fusing information in a joint hidden layer of a neural network is common.

Kingma et al. [11] introduced a stochastic variational inference learning algorithm, which scales to large datasets. They showed that a reparameterization of the variational lower bound yields a lower-bound estimator, which can be directly optimized using standard stochastic gradient methods. This paved the way for a new class of generative models based on variational autoencoders (VAE) [5].

The training of bimodal generative models that can generate one modality by using the other modality, provided decent results by utilizing variants of VAE referred to as conditional multimodal autoencoders and conditional VAEs [14,19]. Pu et al. [17] proposed a novel VAE to model images as well as associated labels or captions. They used a Deep Generative Deconvolutional Network (DGDN) as a decoder of the latent image features and deep Convolutional Neural Network (CNN) as an image encoder. The CNN is used to approximate distribution for the latent DGDN features.

However, these models cannot generate data when they are conditioned interchangeably on both modalities. Vedantam et al. [24] reported the modification of the VAE to enable a bimodal conditional generation. The method uses a novel training objective and product-of-experts inference network, which can handle partially specified (abstract) concepts in a principled and efficient manner. Perarnau et al. [15] introduced the joint multimodal VAE (JMVAE), which learns the distribution by using a joint inference network. To handle missing data at test time, the JMVAE trains $q(z|x_1, x_2)$ with two other inference networks $q(z|x_1)$ and $q(z|x_2)$.

Kurle et al. [12] formulated a variational-autoencoder-based framework for multi-source learning, in which each encoder is conditioned on a different information source. This enabled to relate the sources through the shared latent variables by computing divergence measures between individual posterior approximations. Wu et al. [25] introduced a multimodal VAE (MVAE), which uses a product-of-experts inference network and sub-sampled training paradigm to solve the multimodal inference problem. The model shares parameters to efficiently learn under any combination of missing modalities.

Common for most multimodal methods based on variational inference is the lack of having an expressive family of distributions. This limits the generative power of the models, which we address in our work.

3   Variational Inference

The variational inference method approximates intractable probability densities through optimization. Consider a probabilistic model with observations $\mathbf{x} = x_{1:n}$, continuous latent variables $\mathbf{z} = z_{1:m}$, and model parameters θ. In variational inference, the task is to compute the posterior distribution:

$$p(\mathbf{z}|\mathbf{x}, \theta) = \frac{p(\mathbf{z}, \mathbf{x}|\theta)}{\int_z p(\mathbf{z}, \mathbf{x}|\theta)} \tag{1}$$

The computation requires marginalization over the latent variables $\mathbf{z}$, which often is intractable. In variational methods, a distribution family is chosen over the latent variables with its variational parameters $q(z_{1:m}|\phi)$. The parameters that lead to q as close as possible to the posterior of interest are estimated through optimization [9]. However, the true posterior often is not in the search space of the distribution family and thus the variational inference provides only an approximation. The similarity between the two distributions is measured by the Kullback–Leibler (KL) divergence.

Direct and exact minimization of the KL divergence is not possible. Instead, as proposed in [9], a lower bound on the log-marginal likelihood is constructed:

$$\log p_\theta(\mathbf{x}) \geq \log p_\theta(\mathbf{x}) - \mathrm{KL}\left[q_\phi(\mathbf{z}|\mathbf{x}), p_\theta(\mathbf{z}|\mathbf{x})\right] \tag{2}$$
$$= \mathbb{E}_{q_\phi(\mathbf{z}|\mathbf{x})}\left[\lambda \log p_\theta(\mathbf{x}|\mathbf{z})\right] - \beta\mathrm{KL}\left[q_\phi(\mathbf{z}|\mathbf{x}), p(\mathbf{z})\right] \tag{3}$$
$$= \mathrm{ELBO}(\mathbf{x})$$

where λ and β are weights in the Evidence Lower Bound (ELBO). In practice, $\lambda = 1$ and β is slowly annealed to 1 to form a valid lower bound on the evidence [3]. In Eq. 3, the first term represents the reconstruction error, while the second corresponds to the regularization. The ELBO function in VAEs is optimized by gradient descent by using the reparametrization trick, while the parameters of the distributions p and q are obtained by neural networks.

4   Learning Flexible and Complex Distributions in Multimodal Variational Methods

Consider a set $\mathbf{X}$ of N modalities, $\mathbf{x}_1, \mathbf{x}_2, \ldots, \mathbf{x}_N$. We assume that these modalities are conditionally independent considering the common latent variable $\mathbf{z}$, which means that the model $p(\mathbf{x}_1, \mathbf{x}_2, \ldots, \mathbf{x}_N, \mathbf{z})$ can be factorized. Such factorization enables the model to consider missing modalities, as they are not required for the evaluation of the marginal likelihood [25]. This can be used to combine the distributions of the N individual modalities into an approximate joint posterior. Therefore, all required 2^N multimodal inference networks can be computed efficiently in terms of the N unimodal components. In such a case, the ELBO for multimodal data becomes:

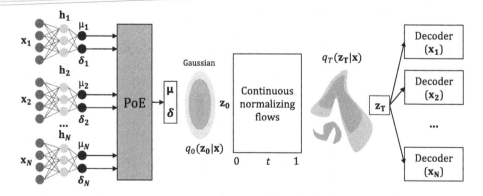

Fig. 1. Model architecture of the multimodal VAE with normalizing flows. The model can be used with any type of normalizing flows, transforming z_0 to z_T.

$$\text{ELBO}(\mathbf{X}) = E_{q_\phi(\mathbf{z}|\mathbf{X})} \left[\sum_{\mathbf{x}_i \in \mathbf{X}} \lambda_i \log p_\theta(\mathbf{x}_i|\mathbf{z}) \right]$$
$$- \beta \text{KL}\left[q_\phi(\mathbf{z}|\mathbf{X}), p(\mathbf{z}) \right] \tag{4}$$

In Fig. 1 we present the architecture of the proposed model. Here, the modalities are transformed by encoding inference networks and mapped to a predefined parametric family of distributions, such as the Gaussian. These parameters are combined in the product-of-experts network along with the prior expert [7]. The output of the product-of-experts is again a parametric distribution, which has low expressive power and is not sufficient for good approximations to the true posterior.

A better variational approximation to the posterior would provide better values for the ELBO. As the ELBO is only a lower bound on the marginal log-likelihood, they do not share the same local maxima. Therefore, if the lower bound is too far off from the true log-likelihood, the maxima of the functions can differ even more, as illustrated in Fig. 2. Due to its simplicity, one of the most commonly used variational distributions for the approximate posterior $q_\phi(\mathbf{z}|\mathbf{x})$ is the diagonal-covariance Gaussian. However, with such a simple variational distribution, the approximate posterior will most likely not match the complexity of the true posterior. According to Eq. 2, this can lead to a larger KL divergence term and thus a larger difference between the ELBO and true log-likelihood, which often leads to a low-performance generative model.

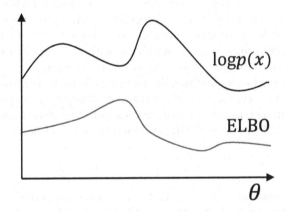

Fig. 2. Difference between the ELBO and true log likelihood. With the increase in their difference, their local maxima tend to differ more, which leads to a bias in the optimization of the model parameters.

For the optimal variational distribution that provides $KL = 0$, q matches the true posterior distribution. This can be achieved only if the true posterior is a part of the distribution family of the approximate posterior; therefore, the ideal family of variational distributions $q_\phi(\mathbf{z}|\mathbf{x})$ is a highly flexible one.

An approach to achieve such flexible and complex distributions is to utilize normalizing flows [22]. Consider $\mathbf{z} \in \mathbb{R}$ as a random variable and $f : \mathbb{R}^d \to \mathbb{R}^d$ as an invertible mapping. The function f can transform $\mathbf{z} \sim q_\phi(\mathbf{z})$ and yield a random variable $y = f(\mathbf{z})$, which has the following probability distribution:

$$q_y(\mathbf{y}) = q_\phi(\mathbf{z}) \left| \det \frac{\partial f^{-1}}{\partial \mathbf{z}} \right| \tag{5}$$

This transformation is also known as the change-of-variable formula. Composing a sequence of such invertible transformations to a variable is known as normalizing flows:

$$\mathbf{z}_{t+1} = \mathbf{z}_t + f(\mathbf{z}_t, \theta_t) \tag{6}$$

where $t \in \{0, \ldots, T\}$ and $\mathbf{z}_t \in \mathbb{R}^D$.

Generally, the main bottleneck of the use of the change-of-variables formula is the computation of the determinant of the Jacobian, which has a cubic cost in either the dimension of $\mathbf{z}$ or number of hidden units.

Rezende et al. introduced a normalizing flow, referred to as planar flow, for which the Jacobian determinant could be computed efficiently in $O(D)$ time [18]. Berg et al. [2] introduced Sylvester normalizing flows, which can be regarded as a generalization of planar flows. Sylvester normalizing flows remove the well-known single-unit bottleneck from planar flows, making the single transformation considerably more flexible.

It is simple to sample from these models and they can be trained by maximum likelihood by using the change-of-variables formula. However, this requires placing awkward restrictions on their architectures, such as partitioning of dimensions or use of rank-one weight matrices, to avoid a $O(D^3)$ cost determinant computation. These limitations are overcome with the introduction of continuous normalizing flows [6], which we refer to in the following.

The sequence of transformations shown in Eq. 6 can be regarded as an Euler discretization of a continuous transformation [4]. If more steps are added, in the limit, we can parametrize the transformations of the latent state by using an ordinary differential equation (ODE) specified by a neural network:

$$\frac{\partial \mathbf{z}(t)}{\partial t} = f(\mathbf{z}(t), t, \theta) \tag{7}$$

where t is now continuous, i.e. $t \in \mathbb{R}$. Chen et al. [4] reported that the transfer from a discrete set of layers to a continuous transformation simplifies the computation of the change in normalizing constant. In the generative process, sampling from a base distribution $\mathbf{z}_0 \sim p_{\mathbf{z}_0}(\mathbf{z}_0)$ is carried out. Considering the ODE defined by the parametric function $f(\mathbf{z}(t), t; \theta)$, the initial value problem:

$$\mathbf{z}(t_0) = \mathbf{z}_0, \quad \frac{\partial \mathbf{z}(t)}{\partial t} = f(\mathbf{z}(t), t; \theta) \tag{8}$$

is solved to obtain $\mathbf{z}(T) = \mathbf{z}_T$, which constitutes the transformed latent state. The change in log-density under the continuous normalizing flow (CNF) model follows a second differential equation, referred to as formula of instantaneous change of variables:

$$\frac{\partial \log q_t(\mathbf{z}(t))}{\partial t} = -\mathrm{Tr}\left(\frac{\partial f}{\partial \mathbf{z}(t)}\right) \tag{9}$$

We can compute the total change in log-density by integrating over time:

$$\log q_T(\mathbf{z}(T)) = \log q_0(\mathbf{z}(t_0)) - \int_{t_0}^{T} \mathrm{Tr}\left(\frac{\partial f}{\partial \mathbf{z}(t)} dt\right) \tag{10}$$

Extending Chen et al. [4], Grathwohl et al. introduced the Hutchinson's trace unbiased stochastic estimator of the likelihood, which has $O(D)$ time cost, which enables completely unrestricted architectures [6].

Using such continuous normalizing flows, we transform the joint distribution $\mathbf{z}_0$ to $\mathbf{z}_K$ to obtain a more complex distribution, as shown in Fig. 1. Lastly, $\mathbf{z}_K$ is used to reconstruct the original inputs $\mathbf{x}_1, \ldots, \mathbf{x}_N$ through the decoder networks.

Utilizing this, we derive new KL divergence in the multimodal setting. With this modification, the new ELBO (Eq. 4) has new KL divergence, which is computed by:

$$\mathrm{KL} \sim \log q_T(\mathbf{z}_T|\mathbf{X}) - \log p(\mathbf{z}_T)$$
$$= \log q_0(\mathbf{z}_0|\mathbf{X}) - \log p(\mathbf{z}_T) - \int_0^T \mathrm{Tr}\left(\frac{\partial f}{\partial \mathbf{z}_t} dt\right) \tag{11}$$

We obtain the new objective of the multimodal VAE by replacing the KL divergence term in Eq. 4 with that in Eq. 11. Training the model with the new objective gives larger generative power to model, which can be utilized to improve the learning of joint distributions in complex multimodal data.

5   Evaluation

In this section, we present multiple experiments carried out to evaluate our model performance which we name multimodal variational autoencoder with continuous normalizing flows (MVAE-CNF) on several datasets and learning tasks including MNIST, FashionMNIST, KMNIST, EMNIST, CelebA, and Flickr.

The MNIST-like unimodal datasets are transformed into multimodal datasets by treating the labels as a second modality, which has been also carried out in previous related work. We refer to the images as the image modality and labels as the text modality. For each of them, the comparison is carried out against the state-of-the-art MVAE. In [25], MVAE is compared to previous multimodal variational approaches.

For a proper evaluation, the same model architectures for the encoders and decoders, learning rates, batch sizes, number of latent variables, and other parameters are used.

Our model, MVAE-CNF, has additional parameters for the CNFs. We used the version from [6] as the CNF without amortization. As the solver for the ODE, we used the Euler method to achieve computational efficiency and demonstrate that even with the utilization of the simplest solver, the method is improved compared to the current state of the art. For the CNF settings, we used only one block of CNFs, 256 squash units for the layers of ODENet, and softplus non-linearity as the activation function for the ODENet.

For the four MNIST datasets, we used linear layers having 512 units for the encoders and decoders for both modalities. Each of the encoders is followed by μ and σ hidden layers having 128 units with Swish non-linearities. To optimize the lower bound in these data, we used annealing where the factor of the KL divergence is linearly increased from 0 to 1 during 20 out of the total of 60 epochs [3].

For the image transformation tasks, we used the CelebA dataset. For the encoders and decoders, we used simple convolutional networks composed of 4 [convolution-batch normalization] layers, whose last layer is flattened and decomposed into μ and σ with 196 units for all models. We slowly anneal the factor of the KL from 0 to 1 during 5 epochs and train for 20 epochs owing to the limitations of the available computational resources. The above CNF parameters are unchanged.

Lastly, the FLICKR dataset is composed of images and textual descriptions. We used the image and text descriptors as in [20]. All parameters of the decoders and encoders, batch size, epochs, learning rates, etc. are similar to those of the MNIST-like datasets.

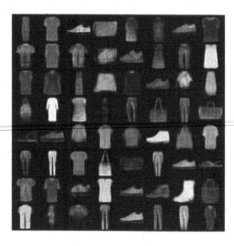

Fig. 3. Generated samples from MNIST, sampled from the conditional distribution, when $\mathbf{X}_2 = 3$.

Fig. 4. Generated samples from the Fashion MNIST dataset, sampled from the unconditioned joint distribution.

5.1 Evaluation of the Model and Generated Sample Quality

For the MNIST-like datasets, we denote the images as $\mathbf{X}_1$ and labels as $\mathbf{X}_2$ and we set $\lambda_1 = 1$ and $\lambda_2 = 50$ (Eq. 4). The up-weighting of the reconstruction error for the low-dimensional modalities is important for the learning of good joint distribution.

Table 1 shows the joint ELBO loss, computed according to 4 and 11, ELBO losses for both modalities separately, and binary cross-entropy (BCE) values for $\mathbf{X}_1$ and $\mathbf{X}_2$. Lower values for all columns imply better performance. We present the results for the four MNIST datasets. For both modalities, jointly and separately, the MVAE-CNF outperforms the MVAE in ELBO loss. As presented in the table, we evaluated the BCEs for both modalities, by ignoring the KL divergence, which provides a direct comparison of the models in terms of learning of the reconstructions. The MVAE-CNF outperforms the MVAE in all datasets. This is due to the increased capacity of the joint distribution because of the utilization of continuous normalizing flows.

Table 1. Binary cross-entropy and ELBO losses for four MNIST-like datasets

	$\mathbf{X}_1$ BCE loss		$\mathbf{X}_2$ BCE loss		ELBO joint		ELBO $\mathbf{X}_1$		ELBO $\mathbf{X}_2$	
	MVAE	MVAE-CNF	MVAE	MVAE-CNF	MVAE	MVAE-CNF	MVAE	MVAE-CNF	MVAE	MVAE-CNF
MNIST	80.9758	78.2835	0.0142	0.0078	100.9352	98.0086	99.5205	96.9436	5.0624	3.8717
FashionMNIST	225.6746	224.4749	0.034	0.0102	240.9705	239.4592	238.1776	237.4313	4.1957	3.9981
KMNIST	188.2816	182.3637	0.0072	0.070	214.2033	209.6676	210.6992	206.7195	3.8445	3.7986
EMNIST(letters)	112.2935	111.8147	0.0064	0.0062	143.9629	141.6544	138.7585	137.6635	5.4532	5.5672

Furthermore, we evaluated the qualities of the generated samples for MNIST, FashionMNIST, and KMNIST. First, we trained a simple neural network for classification. The classifier was not tuned for the best accuracy; it served only for the relative comparison of the generative models. Subsequently, from each of the MVAE and MVAE-CNF, we generated 1000 image samples conditioned on the labeled modality and utilized the classifier to predict the appropriate class. We summarize the accuracies of the predictions by using the generated data in Table 2. Notably, the MVAE-CNF can generate samples with approximately 10–50% higher qualities.

In Fig. 3 and 4, we illustrate the generated samples from the MNIST and FashionMNIST datasets. We find the samples to be good quality, and find conditional samples to be largely correctly matched to the target label. Figure 3 shows images sampled from the conditional distribution, while Fig. 4 shows generated samples from the joint distribution of the FashionMNIST dataset.

Table 2. Quality of the generated samples evaluated by a supervised classifier

Dataset	MVAE	MVAE-CNF
MNIST	0.67	0.73
FashionMNIST	0.24	0.44
KMNIST	0.11	0.26

Table 3. Learning with few examples

Dataset	Image loss	Text loss	Joint loss
Flickr 15%	1241.7388	23.5962	2571.3684
Flickr 30%	1229.2150	23.8598	2544.5045
Flickr 100%	1214.1170	23.2236	2535.6338

Table 4. Image transformations on celebA

Cross-entropy	MVAE	MVAE-CNF
IG: Image	6389.5317	6322.8523
IG: Gray	2239.9395	2197.7927
IG: Image-Gray	8548.3848	8478.5559
IE: Image	6446.3184	6315.6718
IE: Edge	1005.5892	992.2274
IE: Image-Edge	7423.9004	7289.6343
IM: Image	6396.6943	6329.7452
IM: Mask	245.4223	239.4394
IM: Image-Mask	6632.7891	6556.8282

5.2   Image Transformation Tasks

In this section, we compare the performances of the multimodal generative models on an image transformation task. In edge detection, one modality is the original image, while the other is the image with the edges. In colorization, the two modalities are the same image in color and grayscale. Similarly to the approach in [25], we apply transformations to images from the CelebA dataset. For colorization, we transform the RGB images to grayscale, for edge detection we use the Canny detector, and for landscape masks, we use dlib and OpenCV.

We used the validation subset of CelebA as training data and test subset for the computation of the scores. Both MVAE and MVAE-CNF use the same parameters, as explained above, with $\lambda_i = 1$.

In Table 4, we show the results and comparison of the different image transformation tasks. The MVAE-CNF model outperforms MVAE in all tasks. We show the results for the cross-entropy loss for the joint image-gray, image-edge, and image-mask tasks and each of the modalities separately. Furthermore, in Fig. 5, we show the unconditional sampling from the joint distribution. The samples preserve the main task. However, they are characterized as samples from other variational models with smooth and blur effects. On CelebA, the result suggest that the product-of-experts approach generalizes to a larger number of modalities (19 in the case of CelebA), and that jointly training shares statistical strength.

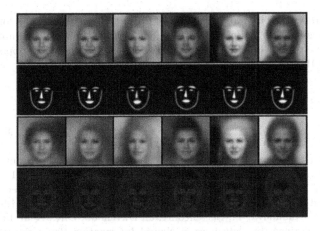

Fig. 5. Image transformations generated by using the joint distribution. From top to bottom: colorization, masks, grayscale, and edge extraction.

In Table 3, on the Flickr dataset, we show that even with 5% of matched data samples, the obtained results almost reach the same performance like that when all data are supervised. This implies that the model can learn the data with only a few supervised samples and utilize the unmatched unlabeled modalities.

6   Conclusion and Future Work

Modeling the joint distribution of multimodal data requires larger expressive power to capture high-level concepts and provide better data representations. Previous multimodal generative models based on variational inference are limited due to the lack of flexibility of the approximate posterior, which is obtained by searching within a known parametric family of distributions. We presented a new approach, which aims to improve the family of multimodal variational models. We utilized the product-of-experts as the inference network to approximate the joint data distribution, which is followed by continuous normalizing flows, which transform the parametric distribution into a more complex distribution. This provides a larger expressive power of the model. In general, these flows can be used for multimodal generative models of any architecture.

Through several experiments, we showed that our approach outperforms the state of the art on variety of tasks including the learning of joint data distributions, image transformations, weakly supervised learning, and generation of missing data modalities. Utilizing a trained classifier on the generated samples, we also show that learning more powerful approximate joint distributions improves the quality of the generated samples by more than 0.15% compared to the previous multimodal VAE methods. As future work, in support of our results, we believe that the focus should lie on efficiently obtaining better and more complex joint distributions that can learn the underlying data.

Acknowledgments. This work was funded by the German Ministry for Education and Research as BIFOLD - Berlin Institute for the Foundations of Learning and Data (ref. 01IS18025A and ref 01IS18037A). We also acknowledge financial support by BASF under Project ID 10044628.

References

1. Barnard, K., Duygulu, P., Forsyth, D., Freitas, N.D., Blei, D.M., Jordan, M.L.: Matching words and pictures. J. Mach. Learn. Res. **3**(2), 1107–1135 (2003)
2. Berg, R.V.D., Hasenclever, L., Tomczak, J.M., Welling, M.: Sylvester normalizing flows for variational inference. arXiv preprint arXiv:1803.05649 (2018)
3. Bowman, S.R., Vilnis, L., Vinyals, O., Dai, A.M., Jozefowicz, R., Bengio, S.: Generating sentences from a continuous space. arXiv preprint arXiv:1511.06349 (2015)
4. Chen, T.Q., Rubanova, Y., Bettencourt, J., Duvenaud, D.K.: Neural ordinary differential equations. In: Advances in Neural Information Processing Systems. pp. 6571–6583 (2018)
5. Doersch, C.: Tutorial on variational autoencoders. arXiv preprint arXiv:1606.05908 (2016)
6. Grathwohl, W., Chen, R.T.Q., Bettencourt, J., Sutskever, I., Duvenaud, D.: Ffjord: Free-form continuous dynamics for scalable reversible generative models. In: International Conference on Learning Representations (2019)
7. Hinton, G.E.: Training products of experts by minimizing contrastive divergence. Neural Comput. **14**(8), 1771–1800 (2002)
8. Jia, Y., Salzmann, M., Darrell, T.: Learning cross-modality similarity for multinomial data. In: 2011 International Conference on Computer Vision. pp. 2407–2414. IEEE (2011)
9. Jordan, M.I., Ghahramani, Z., Jaakkola, T.S., Saul, L.K.: An introduction to variational methods for graphical models. Mach. Learn. **37**(2), 183–233 (1999)
10. Kim, J.H., et al.: Multimodal residual learning for visual QA. In: Advances in Neural Information Processing Systems. pp. 361–369 (2016)
11. Kingma, D.P., Welling, M.: Auto-encoding variational bayes. arXiv preprint arXiv:1312.6114 (2013)
12. Kurle, R., Guennemann, S., van der Smagt, P.: Multi-source neural variational inference. arXiv preprint arXiv:1811.04451 (2018)
13. Ngiam, J., Khosla, A., Kim, M., Nam, J., Lee, H., Ng, A.Y.: Multimodal deep learning. In: Proceedings of the 28th International Conference on Machine Learning (ICML 2011). pp. 689–696 (2011)
14. Pandey, G., Dukkipati, A.: Variational methods for conditional multimodal deep learning. In: 2017 International Joint Conference on Neural Networks (IJCNN). pp. 308–315. IEEE (2017)
15. Perarnau, G., Van De Weijer, J., Raducanu, B., Álvarez, J.M.: Invertible conditional gans for image editing. arXiv preprint arXiv:1611.06355 (2016)
16. Potamianos, G., et al.: Audio and visual modality combination in speech processing applications. In: The Handbook of Multimodal-Multisensor Interfaces. pp. 489–543. Association for Computing Machinery and Morgan & Claypool (2017)
17. Pu, Y., et al.: Variational autoencoder for deep learning of images, labels and captions. In: Lee, D.D., Sugiyama, M., Luxburg, U.V., Guyon, I., Garnett, R. (eds.) Advances in Neural Information Processing Systems 29, pp. 2352–2360. Curran Associates, Inc. (2016), http://papers.nips.cc/paper/6528-variational-autoencoder-for-deep-learning-of-images-labels-and-captions.pdf

18. Rezende, D.J., Mohamed, S.: Variational inference with normalizing flows. arXiv preprint arXiv:1505.05770 (2015)
19. Sohn, K., Lee, H., Yan, X.: Learning structured output representation using deep conditional generative models. In: Advances in Neural Information Processing Systems. pp. 3483–3491 (2015)
20. Srivastava, N., Salakhutdinov, R.R.: Multimodal learning with deep boltzmann machines. In: Advances in Neural Information Processing Systems. pp. 2222–2230 (2012)
21. Suzuki, M., Nakayama, K., Matsuo, Y.: Joint multimodal learning with deep generative models. arXiv preprint arXiv:1611.01891 (2016)
22. Tabak, E., Turner, C.V.: A family of nonparametric density estimation algorithms. Commun. Pure Appl. Math. **66**(2), 145–164 (2013)
23. Vedantam, R., Fischer, I., Huang, J., Murphy, K.: Generative models of visually grounded imagination. arXiv preprint arXiv:1705.10762 (2017)
24. Vedantam, R., Fischer, I., Huang, J., Murphy, K.: Generative models of visually grounded imagination. CoRR abs/1705.10762 (2017), http://arxiv.org/abs/1705.10762
25. Wu, M., Goodman, N.: Multimodal generative models for scalable weakly-supervised learning. In: Advances in Neural Information Processing Systems. pp. 5575–5585 (2018)

Estimating the F_1 Score for Learning from Positive and Unlabeled Examples

Seyed Amin Tabatabaei[1,2]([✉]), Jan Klein[3], and Mark Hoogendoorn[1]

[1] Department of Computer Science, Vrije Universiteit Amsterdam,
Amsterdam, Netherlands
{s.tabatabaei,m.hoogendoorn}@vu.nl
[2] Elsevier B.V., Amsterdam, Netherlands
[3] Centrum Wiskunde & Informatica, Amsterdam, Netherlands
j.g.klein@cwi.nl

Abstract. Semi-supervised learning can be applied to datasets that contain both labeled and unlabeled instances and can result in more accurate predictions compared to fully supervised or unsupervised learning in case limited labeled data is available. A subclass of problems, called Positive-Unlabeled (PU) learning, focuses on cases in which the labeled instances contain only positive examples. Given the lack of negatively labeled data, estimating the general performance is difficult. In this paper, we propose a new approach to approximate the F_1 score for PU learning. It requires an estimate of what fraction of the total number of positive instances is available in the labeled set. We derive theoretical properties of the approach and apply it to several datasets to study its empirical behavior and to compare it to the most well-known score in the field, LL score. Results show that even when the estimate is quite off compared to the real fraction of positive labels the approximation of the F_1 score is significantly better compared with the LL score.

1 Introduction

There has been a keen interest in algorithms that can learn a good classifier by using both labeled and unlabeled data. The field addressing such data is called semi-supervised learning (cf. [2]). Semi-supervised learning algorithms exploit the labeled data just like supervised learning algorithms do, but in addition take the structure seen in the unlabeled data into account to improve learning. Based on this combination, the algorithms are able to surpass the performance of fully supervised and unsupervised algorithms on partially labeled data (see e.g. [7]).

One category of problems in semi-supervised learning focuses on learning from datasets that only have positively labeled and unlabeled data, referred to as Positive-Unlabeled (PU) learning. PU learning is seen in multiple application domains (see e.g. [2,10,12]). The F_1 score is a prominent metric in classification problems in general, because taking both the precision and recall into account is desirable. This allows one to select the best model. However, in PU learning, since there are no negatively labeled examples available it is impossible to directly

© Springer Nature Switzerland AG 2020
G. Nicosia et al. (Eds.): LOD 2020, LNCS 12565, pp. 150–161, 2020.
https://doi.org/10.1007/978-3-030-64583-0_15

compute the F_1 score. Attempts to mitigate this problem have been proposed. For example, the LL score (cf. [5]) shows approximately the same behavior as the F_1 score without the need to have negatively labeled examples. However, in absolute terms, it can be quite off from the real F_1 score.

In this paper, we present a novel approach to estimate the F_1 score for a PU learning scenario. This estimator assumes an additional piece of information on top of the performance on the positively labeled data, namely an estimation of what fraction of labeled cases is available compared to the entire number of positive samples in the dataset. This assumption is in many cases not unrealistic and we show that even when the estimation is somewhat off, the proposed estimator still performs better than the popular LL score. We mathematically specify the approach and perform a mathematical analysis whereby we determine the sensitivity of the novel approach to mistakes in the estimation of the fraction of positive labels. On top of that, we conduct a number of experiments, both using generated and real life data. We compare the estimates of both the LL score and our newly introduced approach and show that the estimates using our approach are: (1) significantly closer to the true F_1 score, and (2) better at selecting the "best" model out of a set of models.

The rest of this paper is organized as follows. The formal problem description is given in Sect. 2. Related work is presented in Sect. 3, while our proposed approach is introduced in Sect. 4 together with the mathematical analysis of the approach. The experimental setup and accompanying results are described in Sect. 5 and 6 respectively. Finally, Sect. 7 concludes the paper.

2   Problem Formulation

Let us begin with formally specifying PU learning. Assume instances $i \in \mathcal{M}$ which are specified by their feature vector $\mathbf{x}_i \in \mathbb{R}^d$, corresponding label $y_i \in \{-1, 1\}$ and by the availability of the label $s_i \in \{0, 1\}$. Here, $\mathcal{M} := \{1, \ldots, M\}$ is the set of observations and d is the number of features. If, for an instance i, the label is available ($s_i = 1$), then it is always positive ($y_i = 1$). If the label is not available ($s_i = 0$), it can be either positive or negative. More specifically, let $\mathcal{P} \subseteq \mathcal{M}$ be the set of observations with a positive label. Let $\mathcal{S} \subseteq \mathcal{P}$ be the subset of observations for which the positive label is provided. Consequently, the labels of the observations in $\mathcal{U} := \mathcal{M} \backslash \mathcal{S}$ are not known.

An important assumption in most PU learning algorithms is that positive labeled instances are *Selected Completely At Random* among positive examples (SCAR assumption). This assumption lies at the basis of most PU learning algorithms [1]. Hence, $\mathcal{S}$ is a random subset of $\mathcal{P}$ under SCAR.

Using $\mathcal{S}$ and $\mathcal{U}$ we want to build a classifier f which can predict the label of the cases in $\mathcal{U}$, i.e. ideally $f(\mathbf{x}_i) = y_i$ for $i \in \mathcal{U}$. It should be stressed that, during the training and validation process, the final target of instances outside of $\mathcal{S}$ is not available. Therefore, learning should be done based on $\mathcal{S}$ combined with properties from the unlabeled data $\mathcal{U}$.

3   Literature Review

In this section, to provide an intuition of PU learning algorithms, we briefly introduce the commonly used two-step strategy. This is followed by metrics which estimate the performance of the resulting models.

3.1   PU Learning Algorithms: Two-Step Strategy

A well-known class of PU learning algorithms is the two-step strategy (cf. [7]). In step 1, a set of reliable negative instances is chosen from the unlabeled instances $\mathcal{U}$. It divides $\mathcal{U}$ into two sets: $\mathcal{N}_R$ and $\mathcal{U}\backslash\mathcal{N}_R$. In step 2, the algorithm iteratively adds more instances to $\mathcal{N}_R$, which are used as negative examples in the next iterations. This procedure is repeated until a convergence criterion is met or when no more instances are added to $\mathcal{N}_R$. There are several techniques for each of these steps. For example, the spy technique [8] and the Ricchio technique [6] are used for the first step. The EM algorithm [8] can be a natural choice for the second step. A deeper review about two-step techniques can be found in [7].

3.2   Performance Estimation

To select the classifier with the best generalizable performance, some evaluation is needed. In normal supervised learning, the F_1 score is a common performance measurement for binary classifiers. It is expressed as follows:

$$F_1 = 2 \cdot \frac{\text{recall} \cdot \text{precision}}{\text{recall} + \text{precision}},$$

with

$$\text{recall} = \frac{\text{TP}}{\text{TP} + \text{FN}} = \frac{\sum_{i \in \mathcal{M}} \mathbf{1}_{\{f(\mathbf{x}_i)=1, y_i=1\}}(i)}{\sum_{i \in \mathcal{M}} \mathbf{1}_{\{y_i=1\}}(i)} = \frac{P_1}{P}$$

$$\text{precision} = \frac{\text{TP}}{\text{TP} + \text{FP}} = \frac{P_1}{\sum_{i \in \mathcal{M}} \mathbf{1}_{\{f(\mathbf{x}_i)=1\}}(i)} = \frac{P_1}{M_1}.$$

Here, $\mathbf{1}_{\{\cdot\}}$ represents the indicator function. Moreover, $P := |\mathcal{P}|$ is the number of positive instances and P_1 is the number of positive instances which are also predicted to be 1, i.e. the number of true positives (TP). M_1 is the total number of observations which are predicted as positive.

In PU learning, the target label y_i is not available for unlabeled instances (with $s_i = 0$). Therefore, calculating the precision and recall is not directly possible. However, under the SCAR assumption, we expect the fraction of predicted positives in $\mathcal{S}$ to be the same as the fraction of predicted positives in $\mathcal{P}$:

$$\mathbf{E}\left(\frac{S_1}{S}\right) \overset{SCAR}{=} \frac{P_1}{P},$$

with S_1 the number of predicted positives in $\mathcal{S}$ and $S := |\mathcal{S}|$. Because of SCAR, the behavior of the classifier on $\mathcal{S}$ represents its behavior on $\mathcal{P}$. Hence, the recall can be estimated by

$$\overline{\text{rec}} = \frac{\sum_{i \in \mathcal{S}} \mathbf{1}_{\{f(\mathbf{x}_i)=1\}}(i)}{S} = \frac{S_1}{S}.$$

However, it is difficult to approximate the value of the precision, because it is less straightforward to obtain an estimate of P_1/M_1. This also means it is hard to estimate the F_1 score in PU learning. To solve this, multiple approaches exist, of which the LL score is commonly used. This score is given by

$$\text{LL} = \frac{\overline{\text{rec}}^2}{M_1/M} = \frac{S_1^2 \cdot M}{S^2 \cdot M_1}.$$

It can be directly calculated from a validation set, which contains positive and unlabeled examples. Moreover, it is shown that

$$\frac{\text{recall}^2}{M_1/M} = \frac{P_1^2 \cdot M}{P^2 \cdot M_1} = \frac{(P_1/M_1) \cdot (P_1/P)}{P/M} = \frac{\text{precision} \cdot \text{recall}}{P/M}.$$

Therefore, the LL score also has an estimation of the precision in its definition. It is claimed that the LL score has roughly the same behavior as the F_1 score: a high value of the LL score means both precision and recall are high, while a low value means that either recall or precision is low [5].

4   Estimating the F_1 Score

In this section, we present our approach to estimate the F_1 score in a PU learning problem. It is based on the assumption that we have an approximation of the fraction of positive instances that are labeled. Moreover, we analyze our approach mathematically.

4.1   Approach to Estimate F_1-score

First, we show how the precision can be estimated with the fraction ρ, defined as $\rho := S/P$. Under SCAR, $\mathbf{E}(S_1/\rho) \stackrel{SCAR}{=} P_1$, which yields

$$\overline{\text{prec}} = \frac{(1/\rho) \sum_{i \in \mathcal{S}} \mathbf{1}_{\{f(\mathbf{x}_i)=1\}}(i)}{\sum_{i \in \mathcal{M}} \mathbf{1}_{\{f(\mathbf{x}_i)=1\}}(i)} = \frac{S_1}{\rho \cdot M_1}.$$

The F_1 score can now be estimated by

$$\overline{F_1} := 2 \cdot \frac{\overline{\text{rec}} \cdot \overline{\text{prec}}}{\overline{\text{rec}} + \overline{\text{prec}}} = 2 \cdot \frac{(S_1/S) \cdot (S_1/(\rho \cdot M_1))}{(S_1/S) + (S_1/(\rho \cdot M_1))} = 2 \cdot \frac{S_1}{\rho \cdot M_1 + S},$$

while the actual F_1 score is given by

$$F_1 = 2 \cdot \frac{\rho \cdot P_1}{\rho \cdot M_1 + S}.$$

We are interested in how the approximated $\overline{F_1}$ differs from the actual F_1 score given a dataset and trained classifier f. Hence, we define the variable ΔF_1 as:

$$\Delta F_1 := \overline{F_1} - F_1 = 2 \cdot \frac{S_1 - \rho \cdot P_1}{\rho \cdot M_1 + S}.$$

The actual F_1 score is fixed, but $\overline{F_1}$ depends on which subset of $\mathcal{P}$ is chosen to be labeled. The number of predicted positive observations S_1 has a hypergeometric distribution [9] with P the population size, P_1 the number of 'success states' in the population and S the number of draws. An observation is 'successful' in this context if it is a true positive. Thus, $S_1 \sim \text{Hypergeometric}(P, P_1, S)$. Then,

$$\mathbf{E}(S_1) = S \cdot \frac{P_1}{P} = \rho P_1, \mathbf{Var}(S_1) = S \cdot \frac{P_1(P - P_1)(P - S)}{P^2(P - 1)} = \frac{\rho(1 - \rho)P_1(S - \rho P_1)}{S - \rho}.$$

We have for the approximated recall, precision and F_1 score:

$$\mathbf{E}(\overline{\text{rec}}) = \frac{\rho \cdot P_1}{S} = \text{recall}, \quad \mathbf{Var}(\overline{\text{rec}}) = \frac{\mathbf{Var}(S_1)}{S^2},$$

$$\mathbf{E}(\overline{\text{prec}}) = \frac{\rho \cdot P_1}{\rho \cdot M_1} = \text{precision}, \quad \mathbf{Var}(\overline{\text{prec}}) = \frac{\mathbf{Var}(S_1)}{\rho^2 M_1^2}$$

$$\mathbf{E}(\overline{F_1}) = 2 \cdot \frac{\rho \cdot P_1}{\rho \cdot M_1 + S} = F_1, \quad \mathbf{Var}(\overline{F_1}) = \frac{4 \cdot \mathbf{Var}(S_1)}{(\rho M_1 + S)^2}.$$

Since the expected values of the estimators are equal to the actual performance metrics, the estimators are unbiased.

4.2   Estimating ρ

Since $\rho = S/P$, and the size S of $\mathcal{S}$ is given, estimating P means estimating ρ. There are different approaches to estimate this value. We will not elaborate on this for the sake of brevity, but known approaches exploit domain knowledge or prior experiences with similar datasets. Or they use a classifier to make this estimate. In the remainder of the paper we use a classifier-based approach following [4] and also evaluate how well it works for real life cases.

4.3   Behavior Under Noisy ρ

Now, we analyse what the theoretical implications of a noisy ρ are on our estimators of the recall, precision and F_1 score. We assume that we do not know the real value of $\rho \in (0, 1]$. Let the random variable $\overline{\rho}$ indicate an estimator of ρ. In this case, ρ represents the probability that a positive observation is labeled. Since our estimator of the recall does not involve ρ, the distribution of $\overline{\text{rec}}$ remains the same when ρ is replaced by $\overline{\rho}$. However, the estimator of the precision does change:

$$\overline{\text{prec}}_{\overline{\rho}} = \frac{S_1}{\overline{\rho} \cdot M_1}. \quad \text{Consequently,} \quad \mathbf{E}(\overline{\text{prec}}_{\overline{\rho}}) = \frac{\mathbf{E}(S_1)}{M_1} \cdot \mathbf{E}\left(\frac{1}{\overline{\rho}}\right) = \frac{\rho P_1}{M_1} \cdot \mathbf{E}\left(\frac{1}{\overline{\rho}}\right)$$

$$\mathbf{Var}(\overline{\text{prec}}_{\overline{\rho}}) = \frac{1}{M_1^2} \left[\mathbf{E}(S_1^2) \cdot \mathbf{E}\left(\frac{1}{\overline{\rho}^2}\right) - \mathbf{E}(S_1)^2 \cdot \mathbf{E}\left(\frac{1}{\overline{\rho}}\right)^2 \right]$$

$$= \frac{\mathbf{Var}(S_1) \cdot \mathbf{E}\left(\frac{1}{\overline{\rho}^2}\right) + \rho^2 P_1^2 \cdot \mathbf{Var}\left(\frac{1}{\overline{\rho}}\right)}{M_1^2}.$$

This means $\overline{\text{prec}}_{\overline{\rho}}$ is an unbiased estimator only when $\mathbf{E}(1/\overline{\rho}) = 1/\rho$, which in general is not true. More specifically, consider the convex function $\varphi : (0,1] \to [1,\infty)$ given by $\varphi(x) = 1/x$. Hence, by Jensen's inequality, $\varphi(\mathbf{E}(X)) \le \mathbf{E}(\varphi(X))$ for random variable X and convex function φ. Thus, $1/\rho \le \mathbf{E}(1/\overline{\rho})$, and so

$$\mathbf{E}(\overline{\text{prec}}_{\overline{\rho}}) = \frac{\rho P_1}{M_1} \cdot \mathbf{E}\left(\frac{1}{\overline{\rho}}\right) \ge \frac{P_1}{M_1} = \text{precision}.$$

The approximated F_1 score with noisy ρ is given by

$$\overline{F_1}_{\overline{\rho}} := 2 \cdot \frac{S_1}{\overline{\rho} \cdot M_1 + S}.$$

The expected value of this estimator is at least equal to the actual F_1 score, which means it is biased. We show this again using Jensen's inequality and the convex function $\varphi : (0,1] \to (\frac{1}{M_1+S}, \frac{1}{S}]$ given by $\varphi(x) = \frac{1}{M_1 \cdot x + S}$. Now,

$$\mathbf{E}(\overline{F_1}_{\overline{\rho}}) = 2\mathbf{E}(S_1) \cdot \mathbf{E}\left(\frac{1}{\overline{\rho} \cdot M_1 + S}\right)$$

$$\ge 2\mathbf{E}(S_1) \cdot \frac{1}{\mathbf{E}(\overline{\rho}) \cdot M_1 + S} = 2 \cdot \rho \cdot P_1 \frac{1}{\rho \cdot M_1 + S} = F_1.$$

Consequently, when the fraction of labeled observations among the positive instances is deemed stochastic with an arbitrary distribution, then both the estimators of the precision and F_1 score are expected to overestimate.

5   Experimental Setup

In order to evaluate our approach we empirically compare it to the real F_1 score and to the behavior of the LL score on four different datasets [1]. For these datasets the ground truth for all instances is available.

5.1   Datasets and Setup

Generated Dataset. The first dataset, or actually set of datasets, is generated randomly. These datasets contain two features $X_1, X_2 \in [0,1]$ and points are generated uniformly random. In order to assign the points to one of the two classes ($y = 0$ or $y = 1$), their position is compared to a randomly generated line

---

[1] All code is available on Github: https://github.com/SEYED7037/PU-Learning-Estimating-F1-LOD2020-.

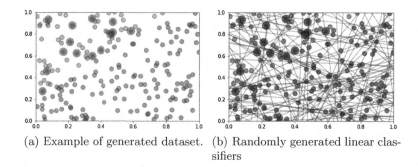

(a) Example of generated dataset. (b) Randomly generated linear classifiers

Fig. 1. Example of a generated dataset with accompanying classifiers. Positive labels are red and negatives are blue. Labeled samples are highlighted with a gray circle. (Color figure online)

$(X_1 - X_2 + 0.2 = 0)$. The classes are assigned to the points based on their position compared to this line. We then select a random sample (SCAR assumption) of size $\rho \cdot P$ of the positive examples $(y = 1)$ to act as $\mathcal{S}$ (with value $s = 1$) and take the rest as $\mathcal{U}$ (with $s = 0$). Figure 1a shows a randomly generated dataset, and Fig. 1b shows a set of linear classifiers which are generated randomly.

IRIS Dataset is a popular dataset in pattern recognition and machine learning literature [3]. It contains 3 flower classes (*setosa*, *versicolor* and *virginica*) of 50 instances each. There are 4 features available for each instance. By taking the last class (*virginica*) as positive and the two others as negative, it transfers to a binary classification problem. These two classes are not linearly separable. We again made a random fraction ρ of all positively labeled data points available as $\mathcal{S}$, the rest being $\mathcal{U}$. 4-D hyper planes were generated randomly to act as linear classifiers.

Heart Disease Dataset is well-known in pattern recognition literature [3]. The data contains both numerical and categorical features. The goal of models applied to this dataset is to predict the presence of heart disease in a patient. In our experiments, we trained random forest models with different numbers of estimators $(randint(1, 100))$ and maximum depth $(randint(1, 10))$.

Health Dataset was obtained from the VU University Medical Center and contains event logs of more than 300,000 patients. For more information about this dataset, please take a look at [10]. The goal is to identify certain types of patients based on their event log. Part of these patients are labeled as having kidney disease, others are labeled as diabetes, and the rest have another disease. For each disease, a fraction ρ of positive examples were randomly selected as labeled examples $\mathcal{S}$, while the rest were taken as unlabeled examples $\mathcal{U}$. Following [10] two features are present in the dataset to predict the label, namely $X_1, X_2 \in \mathbb{Z}$ which summarize the care paths of patients in a way that patients with that disease are optimally separable. A classifier is defined by a set of two thresholds, (θ_1, θ_2). An instance will be predicted as positive if $X_1 > \theta_1$ and $X_2 > \theta_2$.

5.2   Experimental Conditions and Performance Metrics

We compared our approach to the LL score based on two metrics: (1) distance to the real F_1 score; and (2) percentage of inversions. We compute the RMSE to measure the distance to the F_1 score. Computing the percentage of inversions, which is the key in showing that the right model was selected and thus our most important outcome, is a bit more difficult. The inversions were used to show how often the wrong model was selected based on either $\overline{F_1}$ or the LL score compared to the actual F_1 score. To this end, we took the different classifiers for each dataset and compared them pairwise. Each time a classifier that has a higher F_1 score compared to the other classifier is ranked lower we call this an *inversion*. Hence, we want to minimize the number of inversions. We compared the results using a Wilcoxon paired test to show possibly significant differences.

We conducted three types of experiments, namely: (1) empirically studying the assumptions and theoretical results of our approach; (2) evaluating the performance of the approach with the true value of ρ being available; and (3) evaluating the performance with noisy ρ. Each is explained in more detail below.

Empirical Evaluation of Assumptions. As has become clear, we make the assumption that ρ can be estimated. We have presented various approaches to estimate ρ, one of which involves a classifier g. This estimator is exactly correct if $g(x) = \Pr(s = 1|\mathbf{x})$ for all $\mathbf{x}$, but usually this condition does not hold in practice. To show the applicability of this technique (and how easily we can obtain the crucial ρ) we used the *IRIS data* and the *generated data* with different values of ρ and estimated the value of ρ using a trained classifier on the labeled data points. We conducted this experiment 100 times per value of ρ. To get more accurate results, we used the one-leave-out cross-validation technique.

Secondly, we evaluated another part of our approach, namely our result on the bounds. In this experiment, we again used the *generated* and *IRIS datasets* and took one randomly selected linear classifier with a real F_1 score of 0.44. We then took different values for ρ and for each value drew a random sample 100 times, thereby estimating the F_1 score using our approach. We used these results to compute the mean estimated value and the confidence bounds. We compared these to the bound following our mathematical result.

Performance Evaluation with Correct ρ. To evaluate the approach compared to the LL score, we first assume ρ to be known and correct. For these experiments, we selected the value of ρ ranging from 0 to 1 with increments of 0.01. For each setting of ρ for the *generated data* we generated 200 datasets and 100 random lines to act as classifiers. For the *IRIS* and the *health* dataset we generated 100 random classifiers. We measured the performance for both the deviation of the F_1 score and number of inversions.

Performance Evaluation with Noisy ρ. For the noisy ρ we varied the noise level and use the same experimental setup as presented under the *correct ρ* case. The noise level was varied from a 50% underestimation to a 100% overestimation. Due to the computational complexity, we only studied this part on the *generated data* and *IRIS* dataset and measure the percentage of inversions. Table 1 gives a brief overview of the datasets used for the various experiments.

Table 1. Datasets used for the various experiments.

Experiment	Generated data	Iris	Heart Disease	Health
Estimating ρ	X	X		
Evaluating bounds	X	X		
Perf. Eval. Correct ρ - F_1	X	X	X	X
Perf. Eval. Correct ρ - Inversions	X	X	X	X
Perf. Eval. Noisy ρ - Inversions	X	X		

6   Results

First, we report the results of the empirical evaluation of the assumptions followed by experiments in which the correct value of ρ was known. Then, we explore the cases where the value of ρ was noisy (either under- or overestimated).

6.1   Checking the Assumptions

Figure 2 shows the results on the estimation of ρ through our classifier including confidence bounds. We see that as ρ increases the variability of the estimation decreases, which makes sense as a small sample will make the estimation very sensitive to the sample drawn. However, it can be seen that estimations are very reasonable. We do not observe any obvious difference in the estimation behavior between the *generated dataset* (Fig. 2a) and the *IRIS dataset* (Fig. 2b).

Our second study about the underlying assumptions concerns our estimation of the bounds. Figure 3 shows the empirical results for various values of ρ, the empirical mean and bounds and the computed mean and bounds based on our mathematical results for both the *generated* and *IRIS datasets*. Results show that the two align very well for both datasets.

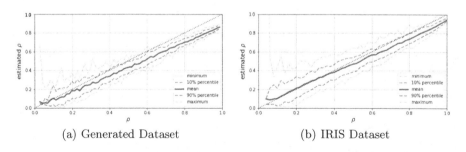

(a) Generated Dataset   (b) IRIS Dataset

Fig. 2. Estimating ρ using a classifier (see [4]). For each value of ρ, this experiment is conducted 50 times. Mean, minimum, maximum, 10 and 90 percentiles are reported.

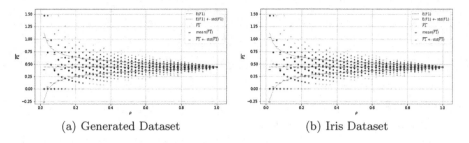

(a) Generated Dataset   (b) Iris Dataset

Fig. 3. Expected value and standard deviation of estimated $\overline{F_1}$ for different values of ρ. Each gray point shows $\overline{F_1}$ for one set of labeled data points. Points which overlap become darker. The empirical mean and bounds of $\overline{F_1}$ are shown by the blue and red dashed line respectively while the mean and bounds computed based on our mathematical result are shown by blue and red dashes respectively.

Table 2. RMSE of real F_1 vs. $\overline{F_1}$ and F_1 vs. LL score. For all cases, $\rho = 0.30$.

	Generated	IRIS	Heart disease	Health
F1	0.064	0.060	0.089	0.060
LL-score	0.772	0.420	0.344	0.623

6.2   Correct ρ

Let us move on to measuring the performance of our approach. We start by considering the case in which ρ was equal to the true value. Table 2 reports the RMSE for different datasets. The RMSE for our approach is much smaller compared to the LL score. This was also to be expected as the proposed score is an estimation of F1, while the LL score aims to approximate the behavior of the F_1 score and not necessarily its actual value. Most important to observe (as our aim is model selection and hyperparameter optimization) is that our estimated values are monotonically increasing with the true F_1 score. Therefore, our central metric is the number of inversions when performing model selection. Figure 4.a shows the results for the *generated dataset* for varying values of ρ. We see that as ρ increases the difference in performance between our approach and the LL score increases in favor of the approach we put forward. Also the confidence intervals become smaller as ρ increases. We also see that our approach never performs worse. Results of a paired Wilcoxon signed-rank test [11] show that for values of $\rho > 0.05$ the number of inversions caused by sorting classifiers based on the $\overline{F_1}$ score is significantly lower than those by the LL score. Moving on to the *IRIS dataset*, Fig. 4.b shows the average number of inversions for different values of ρ. Our approach is significantly better when $\rho > 0.02$. For the *heart disease dataset*, the Wilcoxon paired test shows a significant better performance for our approach for $\rho > 0.02$. Finally, for the real-life *health datasets* our approach is significantly better when $\rho > 0.08$ for kidney disorder and $\rho > 0.02$ for diabetes.

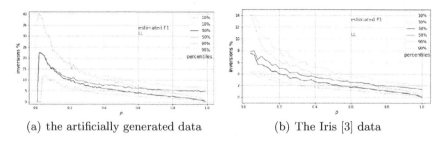

(a) the artificially generated data   (b) The Iris [3] data

Fig. 4. Number of inversions of both the LL score (red line) and the proposed $\overline{F_1}$ (blue line) including confidence bounds for different values of ρ.

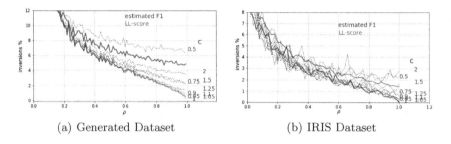

(a) Generated Dataset   (b) IRIS Dataset

Fig. 5. Effect of error in estimated ρ on the percentage of inversions.

6.3   Noisy ρ

In many cases, we do not know the exact value of ρ and might only be able to estimate it (see our first set of experiments). Figure 5 shows how under- and overestimations influence the number of inversions of our proposed approach for the *generated data* and the *IRIS dataset*. Here, the true value of ρ is multiplied with a value c. When considering the *generated dataset*, we see that only for a value of $c = 0.5$, i.e. an extreme underestimation of ρ, the proposed approach scores worse compared to the LL score. For the *IRIS dataset*, we see a similar pattern, except that also for a value of $c = 2$, i.e. a severe underestimation, our performance is worse. This shows that suffering from a bit of noise does not hamper our approach.

7   Conclusion

In this paper we have introduced a novel way of estimating the F_1 score to enable model selection and hyperparameter tuning in PU learning. This novel method is based on the assumption that an estimation can be made on the fraction of labeled positive cases. A mathematical analysis was performed to show the expected value of the estimation with respect to the real F_1 score. Also, we analyzed what the influence of stochasticity in ρ is on the estimations.

We showed that the estimators become biased when ρ is noisy, while they are unbiased when there is no noise.

Furthermore, we conducted experiments to evaluate our assumptions empirically, showing that the approach is practically applicable. On top, we have empirically compared our proposed approach to a well-known metric for model selection, namely the LL score. Results show that our approach (1) is closer to the true F_1 score, and (2) has fewer wrong selections of models (i.e. inversions) compared to the LL score for a variety of datasets. Both cases only hold for sufficiently large samples of training data, though the approach never performs worse. When considering wrongly estimating the fraction of positive labels we see that only severe underestimations hamper performance compared to the LL score. Our approach also brings advantages that the whole family of F scores can now be estimated.

References

1. Bekker, J., Davis, J.: Learning from positive and unlabeled data: A survey. arXiv preprint arXiv:1811.04820 (2018)
2. Denis, F., Gilleron, R., Tommasi, M.: Text classification from positive and unlabeled examples (2002)
3. Dua, D., Graff, C.: UCI machine learning repository (2017) http://archive.ics.uci.edu/ml
4. Elkan, C., Noto, K.: Learning classifiers from only positive and unlabeled data. In: Proceedings of the 14th ACM SIGKDD International Conference on Knowledge Discovery and Data Mining. pp. 213–220. ACM (2008)
5. Lee, W.S., Liu, B.: Learning with positive and unlabeled examples using weighted logistic regression. ICML. **3**, 448–455 (2003)
6. Li, X., Liu, B.: Learning to classify texts using positive and unlabeled data. IJCAI. **3**, 587–592 (2003)
7. Liu, B.: Web data mining: exploring hyperlinks, contents, and usage data. Springer Science & Business Media, Berlin (2007)
8. Liu, B., Lee, W.S., Yu, P.S., Li, X.: Partially supervised classification of text documents. In: ICML. vol. 2, pp. 387–394. Citeseer (2002)
9. Skala, M.: Hypergeometric tail inequalities: ending the insanity. arXiv preprint arXiv:1311.5939 (2013)
10. Tabatabaei, S.A., Lu, X., Hoogendoorn, M., Reijers, H.A.: Identifying patient groups based on frequent patterns of patient samples. arXiv preprint arXiv:1904.01863 (2019)
11. Wilcoxon, F.: Some rapid approximate statistical procedures. Ann. New York Acad. Sci. **52**(1), 808–814 (1950)
12. Zhao, Y., Kong, X., Philip, S.Y.: Positive and unlabeled learning for graph classification. In: 2011 IEEE 11th International Conference on Data Mining. pp. 962–971. IEEE (2011)

Dynamic Industry-Specific Lexicon Generation for Stock Market Forecast

Salvatore Carta[1], Sergio Consoli[2], Luca Piras[1],
Alessandro Sebastian Podda[1]([⊠]), and Diego Reforgiato Recupero[1]

[1] Department of Mathematics and Computer Science, University of Cagliari,
Cagliari, Italy
{salvatore,lucapiras,sebastianpodda,diego.reforgiato}@unica.it
[2] European Commission, Joint Research Center (DG-JRC),
Ispra, VA, Italy
sergio.consoli@ec.europa.eu

Abstract. Press releases represent a valuable resource for financial trading and have long been exploited by researchers for the development of automatic stock price predictors. We hereby propose an NLP-based approach to generate industry-specific lexicons from news documents, with the goal of dynamically capturing, on a daily basis, the correlation between words used in these documents and stock price fluctuations. Furthermore, we design a binary classification algorithm that leverages on our lexicons to predict the magnitude of future price changes, for individual companies. Then, we validate our approach through an experimental study conducted on three different industries of the Standard & Poor's 500 index, by processing press news published by globally renowned sources, and collected within the Dow Jones DNA dataset. Classification results let us quantify the mutual dependence between words and prices, and help us estimate the predictive power of our lexicons.

Keywords: Stock market forecast · Machine learning · Natural language processing · Financial technology.

1 Introduction

Seminal work in the financial literature [1] has shown that financial markets are "informationally efficient", meaning that stock prices reflect the information coming from exogenous factors, represented by events reported in the news. According to the *Adaptive Market Hypothesis* [2], which reconsiders the *Efficient Market Hypothesis* [1] in light of the behavioural economics, the excess return observed in stock market trading should be ascribed to information asymmetry: traders who are able to effectively access and mine information enjoy a competitive advantage. For these reasons, Text Mining, Natural Language Processing and Deep Learning techniques have been extensively explored by researchers facing the challenge of stock price forecasting [3,4].

© Springer Nature Switzerland AG 2020
G. Nicosia et al. (Eds.): LOD 2020, LNCS 12565, pp. 162–176, 2020.
https://doi.org/10.1007/978-3-030-64583-0_16

After a first phase, in which simple text-modeling methods such as bag-of-words and statistical measures were commonly employed, the beginning of the century was characterized by the development of more advanced methods, often based on Deep Learning [5,6]. This led to an exponential growth in the number of works related to financial technology and, more specifically, on NLP-based financial forecasting. Moreover, a further factor that contributed to this trend arises from the outbreak, from 2010 onwards, of social media platforms like Twitter[1] and StockTwits[2], with the generation of a vast amount of textual content that has made possible to improve existing classification techniques [7]. One of the areas that profited the most from this phenomenon is Sentiment Analysis [8], which exploits lexical resources to extract moods and opinions that have a relevant impact on the market [9,10].

Supported by the same motivation, we present a novel NLP-based approach whose goal is to identify potential correlations between words that appear in news stories and stock price movements. To do so, we define a process to automatically generate lexicons, by evaluating the impact of each term on the price fluctuations of a given time interval. Notably, the resulting lexicons are specialized on a specific *industry* or *business area* (e.g., Financial, Information Technology) and are dynamically updated to absorb the new information coming from the press (i.e., given an industry, we generate an updated version of its associated lexicon for each day on which the market is open).

Then, we design a predictive algorithm that matches such a lexicon with the most recent press articles dealing with a specific company, in order to forecast the *magnitude* of the stock price variation for that company in the following day, classified as *high* or *low*. From a general perspective, if the percentage of lexicon terms that appear in current news is similar to the percentage observed in previous news that were followed by significant price variations, then the algorithm predicts a significant price change for the following day.

It is noteworthy that having an estimate of the magnitude of the future variation, even without considering its direction, can still be a precious piece of information for the trader, as it represents an indicator associated to the volatility of the stock price, thus conveying a measure of risk. In this respect, it is comparable to the CBOE Volatility Index (VIX), a real-time market index that estimates the volatility expectation of a given stock in the following 30 days [11].

Finally, we experimentally validate our approach by considering a set of industries of the *Standard & Poor's 500* (S&P 500) index, relying on news documents delivered by globally renowned sources collected in the *Dow Jones DNA dataset*[3]. The results show that the selection process used to create the lexicons is effective at characterizing news that lead to significant stock price variations.

The original contributions of our paper can be summarized as follows:

– We propose a novel walk-forward approach to generate industry-specific lexicon;

---

[1] http://www.twitter.com.

[2] http://www.stocktwits.com.

[3] https://developer.dowjones.com/site/global/home/index.gsp.

- We design a classification algorithm that exploits the aforementioned lexicons to predict whether the market price variation for a given company will be high or low;
- We confirm the correlation between our lexicons and stock price variations of individual companies, through an experimental study on industries of the S&P 500 index.

The remainder of this paper is organized as follows. Section 2 describes the background work on financial forecasting based on Natural Language Processing techniques. Section 3 defines the forecasting problem we target in this paper. Then, our proposed approach is described within Sect. 4, where we show how we create the industry-specific lexicons and how we employ them to forecast the magnitude of stock price variations. We evaluate the performance of our method within Sect. 5, where we give details on the adopted methodology and datasets and then show the results, which indicate the effectiveness of the approach. Finally, in Sect. 6, we conclude the presentation and discuss possible future directions.

2   Related Work

Applications of NLP to financial forecasting have received increasing attention by researchers in the past two decades [12,13]. This phenomenon is due, on the one hand, to the spreading popularity of social media such as Twitter, StockTwits etc., that have led to a dramatic growth of the amount of user content and, on the other hand, to the development of techniques that are effective at automatically analyzing textual information, such as word-embeddings [14].

The work in the literature can be grouped according to the type of source. While a lot of investigation has been conducted on textual data coming from social media [15–17], we hereby focus our attention on studies that, similarly to ours, analyze data coming from press releases or company disclosures. These works give evidence that there exists correlation between the information contained in news stories and the stock prices, and it can be leveraged to make predictions. Hagenau et al. [18] exploit state-of-the-art text mining methods to detect expressive features to represent text, by including market information during the feature selection process; they show that such complex features significantly contribute to improve the accuracy of the classifier. Groth et al. [19] develop a risk management system based on a combination of intra-day stock prices and company disclosures, identifying those documents that have a significant impact on stock price volatility. Lavrenko et al. [20] define language models to represent patterns in news stories that are highly associated with price trends, in order to recommend articles that are likely to influence the behavior of the market.

Furthermore, several works aim at extracting events from unstructured data as a first step towards the prediction. For example in [21], authors train a neural tensor network to learn event embeddings, to be used as the input of a subsequent deep prediction stage. Conversely, Malik et al. [22] propose a combination of statistical classifiers and rules to extract events from expert-annotated press releases. Moreover, Carta et al. [23] retrieve data from Amazon products and their related prices over time; then, they augment these data with social media and Google Trends information, as exogenous features of an autoregressive and integrated moving average model; finally, they exploit the model to forecast products prices, showing the effectiveness of the approach.

To conclude our overview of the related work, we examine the integration of lexicons in the text-modeling phase and the forecasting algorithm. Indeed, one of the most common applications of lexicons is the extraction of the sentiment from documents [24–26]. Word connotations depend on the context in which they are used [27], hence computational intelligence systems require to adopt domain-specific lexicons. However, since the creation of tailor-made lexicons is still a challenging task, most works employ pre-existing public resources, such as LIWC[4], GI[5], Hu-Liu[6], which provide a binary classification for sentiment words (i.e., *positive* or *negative*) or ANEW[7] and SentiWordNet[8], which assign intensity scores to words [28,29]. Among the efforts done in the automatic generation of specialized resources, Oliveira et al. [30] propose a method to compute statistical metrics over StockTwits labeled messages, in order to create a financial lexicons, while, similarly to our approach, authors in [31] identify negative words in company-specific news articles with the goal of forecasting low firm earnings. To the best of our knowledge, no work in the literature has addressed the challenge of automatically creating a financial lexicon by assessing the value of terms based on the signal coming from stock price fluctuations (instead of common sentiment scores). Moreover, the novelty of our approach consists in the dynamics of the lexicon, which is industry-specific and is updated at a daily frequency to capture the information newly delivered on the press.

3   Problem Formulation

The problem we intend to tackle is to predict the magnitude of the variation, on a given day d, of the stock price of a company c; we denote this quantity as $\Delta_{d,c}$. It is defined as the relative increment – or decrement – of the *close* price (i.e., the price of the stock at the closing time of the market), registered on the day d with respect to the close price of the previous day $(d-1)$:

$$\Delta_{d,c} = \frac{\mid close_d - close_{(d-1)} \mid}{close_{(d-1)}} \tag{1}$$

Note that we consider the absolute value because we want to predict, as previously said, the magnitude of the variation, without taking into account its direction (positive or negative).

---

[4] http://liwc.wpengine.com/.

[5] http://www.wjh.harvard.edu/~inquirer/.

[6] https://www.cs.uic.edu/~liub/FBS/sentiment-analysis.html.

[7] https://csea.phhp.ufl.edu/media/anewmessage.html.

[8] http://sentiwordnet.isti.cnr.it/.

The input of our problem consists then of a set of companies, grouped into industries or business sectors. Besides, we can rely on a collection of news documents, which deal with one or more of such companies, and on the stock price time series of each of the companies belonging to the considered industries. In other words, we aim at solving a binary classification task in which: (i) each sample corresponds to a day-company pair (d, c) for which we have a set of news articles $N_{d-1,c}$ published on the day $(d-1)$ and involving the company c; (ii) the two classes are `high` and `low`, defined as follows:

$$Class(d, c) = \begin{cases} \texttt{high} & \text{if } |\Delta_{d,c}| > class\_threshold \\ \texttt{low} & \text{if } |\Delta_{d,c}| \leq class\_threshold \end{cases}$$

where *class_threshold* is a fixed value above which we consider the market variation as significant. The reasons behind the choice of a consistent value for *class_threshold* are explained in Sect. 5.

4   The Proposed Approach

Starting from the formulation given in Sect. 3, we propose an NLP-based approach that, first, builds a lexicon by searching the correlations between the stock prices and the words that appear on news articles, and, second, exploits such lexicons in order to predict the class of the future stock price variations for specific companies. The approach consists of two fundamental stages: (i) the generation of industry-specific lexicons; and, (ii) the class prediction algorithm. We now describe them in more details in the next paragraphs.

4.1   Generation of Industry-Specific Lexicons

The goal of the lexicon generation phase is to select the set of words that, more than any other, allow us to discriminate between days of high and low stock price variations for a specific industry in a given time interval. Intuitively, this is done by detecting the terms that, in a chosen period, show a significant effect on the price variations of individual companies belonging to the same industry. For this reason, we define our lexicons as *time-aware* and *industry-specific*. The decision of grouping the companies by industry relies on the following assumptions: (i) same words that appear in articles related to same industries are more likely to be used with the same connotation (i.e., phenomena such as polysemy are limited); (ii) stock prices of companies belonging to the same industry are more likely to have a similar behavior. In order to be able to capture the impact of events that occur day by day (and thus the effect of "new" words that show up in news articles reporting such events), we perform the lexicon creation in a dynamic way: we repeat its generation for every day on which the stock market is open, for each distinct industry.

Let I be an industry: for each day, we select all the news that are relevant for at least one company $c \in I$, published during the time frame $[d - \ell; d - 1]$ (with $\ell \geq 1$). For each news article in this set, we extract the text, consisting of the title, the snippet and the full body of the article, and then we perform some standard pre-processing techniques on it, such as stop-words removal, stemming and tokenization. In addition, we remove from the corpus all the words that appear too frequently and too infrequently, according to given tolerance thresholds.

Subsequently, we construct a document-term matrix, in which each row corresponds to a pair *(news, company)* and each column corresponds to a term, as obtained after the pre-processing. It is worth noting that the same news article might be relevant for more than one company of the underlying industry (e.g. the news might include references to both of them); in this case, the matrix will contain, for that news document, as many rows as the companies for which the document is relevant.

In the next step, we iterate over the rows of the matrix and, for each of them, we assign to each of its terms a value equal to the magnitude of the stock price variation (as defined in Eq. 1) registered, for the considered company, on the day after the article was published. Finally, each column is averaged (counting only non-zero entries), thus obtaining a list of terms, each associated to a score given by the average of the values assigned to them. We sort the terms by decreasing scores and select the first n. These are the ones associated to higher price variations and represent the time-aware, industry-specific lexicon that will be exploited in the prediction phase.

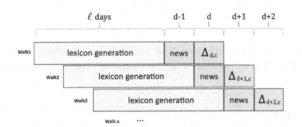

Fig. 1. Walk structure of the lexicon generation process.

4.2   Class Prediction Algorithm

As previously mentioned, the goal of our task is to predict, for a day d and a company c, if the stock price for c will have a large or a small variation $\Delta_{d,c}$. We only make predictions for companies for which there is at least one relevant news article published on the day $d - 1$. Intuitively, the algorithm is based on the assumption that the news followed by a high price variation will have a number of words in common with the lexicon which is higher compared to news followed by a low price variation. This is a direct consequence of the way in which the lexicon has been constructed. The prediction algorithm we have designed is composed of the following steps:

– for a given industry, we select all the articles associated to high and low variations, respectively, published during the time frame $[d - \ell; d - 1]$; for each of these two groups, after applying the same aforementioned pre-processing techniques, we calculate the percentage of words that the articles share with the lexicon created within the same time interval. Let $news\_match_{\mathtt{high}}$ and $news\_match_{\mathtt{low}}$ be these values.

– similarly, we select the news articles published on the day $d - 1$ relevant for a company c which belongs to the underlying industry and we calculate the percentage of words contained in the related lexicon. Let $current\_news\_match$ be this value.

– we choose a value included between $news\_match_{\mathtt{high}}$ and $news\_match_{\mathtt{low}}$ as $match\_threshold$.

– if $current\_news\_match > match\_threshold$, then we assign the class $\mathtt{high}$ to the sample; otherwise, we assign the class $\mathtt{low}$.

Figure 2 illustrates the flow chart of the stock variation prediction algorithm.

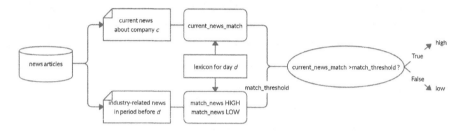

Fig. 2. Flow chart for the stock variation prediction algorithm. These steps refer to the prediction of a sample associated to the company c on a given day d.

5   Experimental Framework

5.1   Datasets

Dow Jones DNA. The Dow Jones "Data, News and Analytics" dataset[9] contains news articles from more than 33,000 authoritative global sources, including e.g. *The Wall Street Journal*, the *Dow Jones Newswires* and *The Washington Post*; it covers both print and online publications. The delivery frequency ranges from ultra-low latency newswires to daily, weekly, or monthly editions. The documents also cover a wide variety of topics, such as finance, business, current affairs and lifestyle. Every news item is enriched with a set of metadata providing information about the source, the time and place of the publication, the relevant companies and the topics involved.

---

[9] https://developer.dowjones.com/site/global/home/index.gsp.

S&P 500 Time Series. The second fundamental data exploited in our analysis consist of all the stock price *time series* of the companies included in the Standard & Poor 500 index (that measures the stock performance of 500 large companies listed on stock exchanges in the United States). Data are collected separately for each individual company at a daily frequency, and they include the *open*, *high*, *low* and *close* daily prices, along with the volume of operations, for each company. The dataset also provides information about the grouping of companies in sectors (e.g. Industrials, Healthcare, Information Technology, Communication Services, Financial, etc.).

5.2   Methodology

For our analysis, we collected from DNA all the news articles, in English language, published from 2015 to 2018 and relevant – according to the metadata field *company_codes_about* – for at least one company of one of these three industries: Information Technology, Finance and Industrials. We then grouped the news by industry, obtaining three groups, which respectively contain 3,623, 2,773 and 2,448 items. To align these documents with stock prices, we restricted the S&P 500 dataset to the same time span, for all the companies included in the aforementioned industries.

The overall objective of the experiments is to corroborate that there exists a correlation between the words that appear in news stories and stock price variations, but not necessarily to quantify such a correlation or to devise a predictor that is competitive with state-of-the-art methods. The experimental procedure that we implemented to validate our approach consists of two main sets of experiments:

- **Lexicon creation setup**: aimed at finding the best setup of parameters to select the terms that form the dynamic industry-based lexicons, evaluated on their ability to characterize news articles associated to high price variations;
- **Prediction algorithm evaluation**: performed to validate the effectiveness of our generated lexicons, by measuring the performance of the prediction algorithm with standard accuracy metrics.

5.3   Results

Lexicon Creation Setup. The first aspect we investigate is the choice of the method used to assign a score to each term contained in a news article, during the process of iterating through the document-term matrix as described in Sect. 4.1. We explored two alternatives:

- Δ: absolute value of the price variation on the day following the underlying news, as defined in Eq. 1;
- $\Delta \times w$: price variation weighted by the importance of the term for the underlying document as given by the *term frequency'inverse document frequency* (TFIDF) value, computed by considering all the articles employed to generate the lexicon.

For the evaluation of the best type of score, we need to introduce some notation. Let $news\_match_{high}$ be the percentage of words that news associated to high price variations share with the lexicon; analogously, $news\_match_{low}$ represents the same value for news followed by a low price variation. Then, let $match\_ratio$ be $news\_match_{high}$ / $news\_match_{low}$: the higher the $match\_ratio$, the more our lexicons are effective at characterizing news that have an impact on the stock prices. We measured $match\_ratio$ for every day of the interval chosen for our experimental framework and computed an average value.

Table 1. Comparison of the two different methods for assigning a score to the words contained in a news article. Different percentiles are considered, to select the words to include in the final lexicon; best value is denoted by *.

		Percentile				
		10	30	50	70	90
Δ	$news\_match_{high}$	73.5	61.2	40.2	17.3	3.5
	$news\_match_{low}$	70.4	52.3	29.2	9.9	1.2
	$match\_ratio$	1.05	1.18	1.43	2.19	6.42*
$\Delta \times w$	$news\_match_{high}$	73.3	60.8	45.2	29.6	12.8
	$news\_match_{low}$	71.1	53.6	37.3	23.1	9.4
	$match\_ratio$	1.03	1.14	1.22	1.32	1.53

The results for the two types of scores are showed in Table 1 (for space constraints, here we only show the results obtained on the Information Technology sector; however, the results for Finance and Industrials are consistent with the reported ones). The lexicon for day d is created using the news published within the 3 weeks before d, while the news on which $news\_match_{high}$ and $news\_match_{low}$ are computed are picked from within the 2 weeks before d; the $class\_threshold$ that defines the classes **high** and **low** is set to 0.005. We filtered out words that occur in less than 10 or in more than 80% of the documents.

Table 2. Comparison of different length choices of the time interval used for the lexicon creation, in weeks.

	Number of weeks						
	2	3	4	5	6	7	8
$news\_match_{high}$	3.4	3.5	3.6	3.4	3.3	2.8	2.5
$news\_match_{low}$	1.1	1.3	1.1	1.1	0.8	0.9	0.7
$match\_ratio$	6.29	7	9*	5.4	5.6	6.64	7.2

It is straightforward to notice that the method that scores the words only by the price variation leads to the creation of more effective lexicons, as the values of *match_ratio* are always much higher for Δ. The second parameter we want to observe here is the number of terms that are included in the final lexicon. We varied this value by selecting terms above the 10th, 30th, 50th, 70th and 90th percentile, respectively, from the ranking of words sorted in ascending order according to their scores. We can observe that *match_ratio* increases together with the percentile used to limit the number of words included in the lexicon; in other words, the smaller the lexicon, the better it discriminates between the two categories of news. Therefore, in the following experiments the score type is set to Δ and the percentile to 90.

The next aspect we want to explore is the length of the time interval within which the news used to generate the lexicon are collected (see Table 2). The ratios improve as we increase the number of weeks up to 4, but they degrade after this value. This behavior can be explained by considering the fact that the set of news articles that contribute to the lexicon should be sufficiently large for the lexicon to be representative; however, as we include news that are further away in time, the words have weaker correlations with the ones that appear within the 2 weeks on which *news_match*$_{high}$ and *news_match*$_{low}$ are computed. For this reason, we set the number of weeks to 4 for the prediction algorithm evaluation.

The resulting lexicons contain, on average, around 197 words for the Information Technology sector, 184 words for Financial and 157 words for Industrials. A typical lexicon presents, on the one hand, terms that are descriptive of the behaviour of the underlying market, such as *crisis*, *hedge*, *recovery* (here, terms are not stemmed for illustration purposes) and, on the other hand, words referring to specific events, for example *mexico*, *trump*.

Prediction Algorithm Evaluation. Once defined the best setup for the creation of the lexicons, we validate the effectiveness of the approach by testing the prediction algorithm. More precisely, we measure the performance on the three industries selected for our study separately, with a walk-forward strategy: for each industry, we make a prediction for every day between 2015 and 2018, for every company with at least one relevant news article published on the day before the prediction.

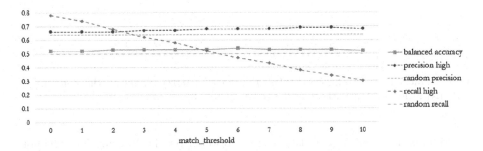

Fig. 3. Information Technology (*class_threshold* = 0.005). Comparison of our approach against a random classifier that predicts both classes with probability 0.5.

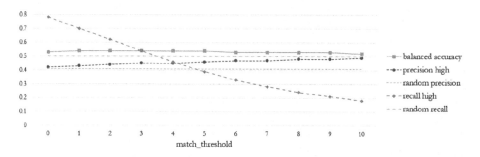

Fig. 4. Information Technology (*class_threshold* = 0.01). Same as the comparison reported in Fig. 3, but with *class_threshold* set to 0.01.

Through our experiments, we vary two parameters: the *class_threshold*, which separates the two classes, and the *match_threshold*, used by the algorithm to assign the predicted class to a sample. The different values for *match_threshold* are obtained as follows: we divide the interval between $news\_match_{high}$ and $news\_match_{low}$ in 10 steps, and we use each of these real numbers as a different value for *match_threshold*, so that class `high` is predicted with lower probability as the threshold increases (e.g., if $news\_match_{high} = 7.0$ and $news\_match_{low} = 2.0$, then *match_threshold* = 5 corresponds to the midpoint 4.5, *match_threshold* = 0 corresponds to the value 2.0 and *match_threshold* = 10 corresponds to 7.0). We measure the balanced accuracy on both classes and the precision and recall for the class `high`; these metrics are compared against a random classifier that predicts each class with a 50% probability, which constitutes our baseline. Please note that the precision of such a random classifier on a specific class is asymptotically equal to the coverage of that class over all the samples, while the recall tends to the value 0.5.

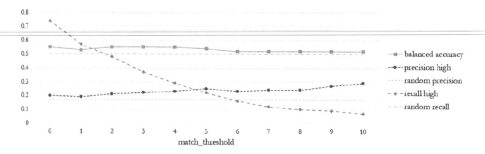

Fig. 5. Information Technology (*class_threshold* = 0.02). Same as the comparison reported in Fig. 3, but with *class_threshold* set to 0.02.

Figures 3, 4 and 5 show the results obtained on the Information Technology sector using as *class_threshold* the values 0.005, 0.01 and 0.02, respectively (please note that daily variations larger than 2% are quite rare in stock markets considered in our study, occurring in less than 17% of the cases). The balanced accuracy (which takes into account the unbalance between the two classes) is consistently above 0.5, indicating that the classifier performs globally better than random. Let us analyze the metrics on class **high** in detail. As expected, the class **high** becomes more rare as the threshold increases: its coverage (and, consequently, the precision of the random classifier) decreases from 0.64 to 0.41 to 0.17; interestingly though, our approach's improvements in precision against the baseline are more important in scenarios where the class **high** is more rare.

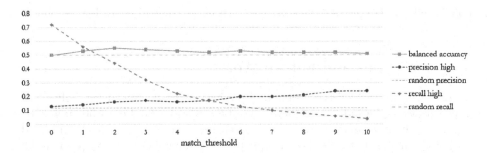

Fig. 6. Financial (*class_threshold* = 0.02). Same as comparison in Fig. 3, but on Financial business area, with *class_threshold* set to 0.02.

In other words, our approach is relatively more precise at detecting unusual, considerable variations in the stock prices. This suggests that the lexicons have a stronger correlation with the news published on the day $d - 1$ about company c in cases where the stock prices of c are subject to larger variations on day d. We also observe that precision values increase together with *match_threshold*. This comes, not surprisingly, at a cost of a significant degradation of recall; nevertheless, there are several combinations of *class_threshold* and *match_threshold* that lead to a significant improvement in both precision and recall (see, e.g., *class_threshold* = 0.02 with *match_threshold* = 1). The choice of the optimal setting, in this case, would heavily depend on how an hypothetical trading agent would integrate and exploit the classifier in order to make decisions. The experiments on Finance and Industrials are consistent with what we discussed so far and, in fact, the improvements of the classifier for these business areas are even more remarkable. For reasons of space constraint, we hereby show the details for both industries only for *class_threshold* = 0.02 (Fig. 6 and 7).

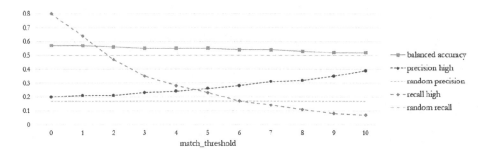

Fig. 7. Industrials (*class_threshold* = 0.02). Same as comparison in Fig. 3, but on Industrials business area, with *class_threshold* set to 0.02.

6   Conclusions and Future Work

In this work, we tackled the problem of predicting the magnitude of daily stock price variations, for individual companies of the Standard & Poor's 500 index, by exploiting the information given in press releases and news documents provided by global renowned sources. For this purpose, in previous Sect. 4 we presented a novel approach to generate industry-specific lexicons, by extracting terms from news documents that show a significant impact on the market. We also designed a predictive algorithm that, on a second stage, exploits our lexicons and incoming news to make forecasts about the future stock prices. To validate the approach, we carried out a set of experiments; the results have been reported and discussed in Sect. 5. They showed the capability of our lexicons of dynamically absorbing the information available on the press day by day and, as a consequence, their significant predictive power with respect to the price variations. Notably, although financial forecasting is still a challenging task, where accuracy results are typically much lower compared to other Machine Learning application fields (and usually very close to random accuracies), we noticed that our results showed a convincing improvement against the random predictor used as benchmark.

However, an actual limitation of our work is that, while the approach looks promising at extracting salient lexical properties from news sources, it does not take into account the semantic value of their content, which could potentially provide additional information to further improve the accuracy of the forecaster. Therefore, to integrate semantic features into our system, as a future enhancement, we intend to enrich the representation of terms and news of our method by exploiting state-of-the-art techniques such as *word embeddings*, and to leverage this enhanced model with advanced classifiers based on Deep Learning. Finally, we plan on enriching the data with more sources, starting from social media platforms.

References

1. Burton, G.M., Eugene, F.F.: Efficient capital markets: a review of theory and empirical work. J. Finance **25**(2), 383–417 (1970)

2. Andrew, W.L.: The adaptive markets hypothesis. J. Portfolio Manage. **30**(5), 15–29 (2004)
3. Barra, S., Carta, S.M., Corriga, A., Podda, A.S., Reforgiato Recupero, D.: Deep learning and time series-to-image encoding for financial forecasting. IEEE/CAA J. Automatica Sinica **7**(3), 683–692 (2020)
4. Carta, S., Corriga, A., Ferreira, A., Podda, A.S., Reforgiato Recupero, D.: A multi-layer and multi-ensemble stock trader using deep learning and deep reinforcement learning. In: Applied Intelligence, pp. 1–17 (2020)
5. Vargas, M.R., dos Anjos, C.E.M., Bichara, G.L.G., Evsukoff, A.G.: Deep learning for stock market prediction using technical indicators and financial news articles. In: 2018 International Joint Conference on Neural Networks (IJCNN), pp. 1–8 (2018)
6. Matsubara, T., Akita, R., Uehara, K.: Stock price prediction by deep neural generative model of news articles. IEICE Trans. Inf. Syst. **E101D**(4), 901–908 (2018)
7. Moro, G., Pasolini, R., Domeniconi, G., Pagliarani, A., Roli, A.: Prediction and trading of dow jones from twitter: a boosting text mining method with relevant tweets identification. Commun. Comput. Inf. Sci. **976**, 26–42 (2019)
8. Reforgiato Recupero, D., Consoli, S., Gangemi, A., Nuzzolese, A.G., Spampinato, D.: A semantic web based core engine to efficiently perform sentiment analysis. In: Presutti, V., Blomqvist, E., Troncy, R., Sack, H., Papadakis, I., Tordai, A. (eds.) ESWC 2014. LNCS, vol. 8798, pp. 245–248. Springer, Cham (2014). https://doi.org/10.1007/978-3-319-11955-7_28
9. Atzeni, M., Dridi, A., Reforgiato Recupero, D.: Using frame-based resources for sentiment analysis within the financial domain. Progress Artif. Intell. **7**(4), 273–294 (2018). https://doi.org/10.1007/s13748-018-0162-8
10. Dridi, A., Atzeni, M., Reforgiato Recupero, D.: Finenews: fine-grained semantic sentiment analysis on financial microblogs and news. Int. J. Mach. Learn. Cybernet. **10**(8), 2199–2207 (2019)
11. Corrado, C.J., Miller, T.W.: The forecast quality of cboe implied volatility indexes. J. Futures Markets: Futures, Options, and Other Derivative Products **25**(4), 339–373 (2005)
12. Xing, F.Z., Cambria, E., Welsch, R.E.: Natural language based financial forecasting: a survey. Artif. Intell. Rev. **50**(1), 49–73 (2017). https://doi.org/10.1007/s10462-017-9588-9
13. Fisher, I.E., Garnsey, M.R., Hughes, M.E.: Natural language processing in accounting, auditing and finance: a synthesis of the literature with a roadmap for future research. Intell. Syst. Account. Finance Manage. **23**(3), 157–214 (2016)
14. Mikolov, T., Sutskever, I., Chen, K., Corrado, G.S., Dean, J.: Distributed representations of words and phrases and their compositionality. In: Advances in Neural Information Processing Systems. pp. 3111–3119 (2013)
15. Bollen, J., Mao, H., Zeng, X.: Twitter mood predicts the stock market. J. Comput. Sci. **2**(1), 1–8 (2011)
16. Jianfeng, S., Arjun, M., Bing, L., Qing, L., Huayi, L., Xiaotie, D.: Exploiting topic based twitter sentiment for stock prediction. In: Proceedings of the 51st Annual Meeting of the Association for Computational Linguistics (Volume 2: Short Papers), pp. 24–29 (2013)
17. Nofer, M., Hinz, O.: Using twitter to predict the stock market. Bus. Inf. Syst. Eng. **57**(4), 229–242 (2015)
18. Hagenau, M., Liebmann, M., Neumann, D.: Automated news reading: stock price prediction based on financial news using context-capturing features. Decis. Support Syst. **55**(3), 685–697 (2013)

19. Groth, S.S., Muntermann, J.: An intraday market risk management approach based on textual analysis. Deci. Support Syst. **50**(4), 680–691 (2011)
20. Lavrenko, V., Schmill, M., Lawrie, D., Ogilvie, P., Jensen, D., Allan, J.: Language models for financial news recommendation. In: Proceedings of the ninth International Conference on Information and Knowledge Management. pp. 389–396 (2000)
21. Xiao, D., Yue, Z., Ting, L., Junwen, D.: Deep learning for event-driven stock prediction. In: Proceedings of the 24th International Conference on Artificial Intelligence, IJCAI 2015, pp. 2327–2333. AAAI Press (2015)
22. Hassan, H.M., Vikas, S.B., Huascar, F.: Accurate information extraction for quantitative financial events. In: Proceedings of the 20th ACM International Conference on Information and Knowledge Management. pp. 2497–2500 (2011)
23. Carta, S., et al.: Forecasting e-commerce products prices by combining an autoregressive integrated moving average (ARIMA) model and google trends data. Future Internet **11**(1), 5 (2019)
24. Taboada, M., Brooke, J., Tofiloski, M., Voll, K., Stede, M.: Lexicon-based methods for sentiment analysis. Comput. Ling. **37**(2), 267–307 (2011)
25. Sanjiv, R.D., Mike, Y.C.: Yahoo! for amazon: Sentiment extraction from small talk on the web. Manage. Sci. **53**(9), 1375–1388 (2007)
26. Moreno-Ortiz, A., Fernández-Cruz, J.: Identifying polarity in financial texts for sentiment analysis: a corpus-based approach. Procedia-Social Behav. Sci. **198**, 330–338 (2015)
27. Hamilton, W. L., Clark, K., Leskovec, J., Jurafsky, D.: Inducing domain-specific sentiment lexicons from unlabeled corpora. In: Proceedings of the Conference on Empirical Methods in Natural Language Processing. Conference on Empirical Methods in Natural Language Processing. vol. 2016, p. 595. NIH Public Access (2016)
28. Clayton, J.H., Eric, G.: Vader: A parsimonious rule-based model for sentiment analysis of social media text. In: Eighth International AAAI Conference on Weblogs and Social Media (2014)
29. Sahar, S., Nicholas, P., Dingding, W.: Financial sentiment lexicon analysis. In: 2018 IEEE 12th International Conference on Semantic Computing (ICSC). pp. 286–289. IEEE (2018)
30. Oliveira, N., Cortez, P., Areal, N.: Stock market sentiment lexicon acquisition using microblogging data and statistical measures. Decis. Support Syst. **85**, 62–73 (2016)
31. Tetlock, P.C., Saar-Tsechansky, M., Macskassy, S.: More than words: quantifying language to measure firms' fundamentals. J. Finance **63**(3), 1437–1467 (2008)

A Stochastic Optimization Model for Frequency Control and Energy Management in a Microgrid

Dhekra Bousnina[1(✉)], Welington de Oliveira[1], and Peter Pflaum[2]

[1] MINES ParisTech, PSL Research University, CMA - Centre de Mathématiques Appliquées, Sophia Antipolis, France
dhekra.bousnina@mines-paristech.fr
[2] Schneider-Electric Industries, Grenoble, France

Abstract. This paper presents an approach to develop an optimal control strategy for a Battery Energy Storage System (BESS) within a photovoltaic (PV)-battery microgrid in Finland. The BESS is used to assist the joint optimization of self-consumption and revenues provided by selling primary frequency reserves in the normal Frequency Containment Reserve (FCR) market. To participate in this frequency control market, a day-ahead planning of the reserve that will be held in the battery during the following day has to be submitted to the transmission system operator. To develop the optimal day-ahead reserve profile, the primary challenge we face is the random nature of the next-day frequency deviations, which we model as an AR (Autoregressive) process. We tackle these uncertainties by employing a two-stage stochastic programming formulation for the reserve planning and battery control optimization problem. Closed-loop simulations are then used to validate the energy management strategy, together with a Model-Predictive Controller (MPC) that is updated every 15 min to take into account an updated system state and new forecasts for load, PV production and electricity market prices. The results show that the considered stochastic programming approach significantly reduces energy costs and increases reliability when compared to a standard deterministic model.

Keywords: MPC · Frequency control · Stochastic optimization

Nomenclature

N	Number of scenarios for optimization
N'	Sample size for out-of-sample simulation
M	Sample size for closed-loop simulation
H	Number of time steps in prediction horizon
C_{TOU}	Time-of-Use (variable) energy price [\$]
P_{G}	Grid power consumption [kW]
C_{R}	Remuneration for provided reserve [\$/kW]

© Springer Nature Switzerland AG 2020
G. Nicosia et al. (Eds.): LOD 2020, LNCS 12565, pp. 177–189, 2020.
https://doi.org/10.1007/978-3-030-64583-0_17

R	Allocated frequency reserve
P_L	Load power
P_{pv}	PV power
P_B	Battery power
b	Stored energy in battery
τ	Time step
P_{Ch}	Battery charging power
P_{Disch}	Battery discharging power
η_{Ch}	Charging efficiency
η_{Disch}	Discharging efficiency
P_N	Nominal battery charging power
P_R	Frequency reserve power provided by the battery
P_{min}	Minimum battery power (i.e. max. discharging power)
P_{max}	Maximum battery charging power
b_{min}	Lower bound on energy stored in the battery
b_{max}	upper bound on energy stored in the battery
Δf	Normalized frequency deviation

1 Introduction

Historically in the electrical power systems, power production follows consumption. However, more recently, with the radical changes affecting the energy landscape, the production of fluctuating solar and wind energy started continuously increasing. In parallel to this increasing penetration of intermittent renewable energy sources, the European power systems are witnessing an increasing introduction of smart grids, along with a rise in the number of energy storage devices. On the other hand, intermittent renewable energies penetration also resulted in several challenges for the power grid, like the increasing need for frequency regulation to keep the frequency close to 50 Hz at all times: a short-term application to which some energy storage systems like Battery Energy Storage Systems (BESSs) are deemed to be very efficient [10]. Thanks to their fast ramp rate, BESS can play an important role in compensating deviations of the grid from the power balance. Frequency control is by no means the only application to which a BESS can be useful. Such systems can also be used, for instance, for arbitrage, i.e., for profiting from the variation of the power market prices to generate revenues by charging the battery during low price periods and discharging it during high price periods. Another interesting use of BESS is peak shaving, i.e., reducing peak power on the lines and therefore helping to avoid the need for grid reinforcements. BESSs can also be used for maintaining voltage on the distribution grid. In addition to all these applications, one of the major purposes behind installing BESSs remains the increase of self-consumption by storing excess of local power generation for later use.

It has been shown in many cases that investment returns from installing BESSs for only one of the purposes mentioned above remain too low due to the high costs [4]. However, using a battery to jointly deliver multiple services, like frequency regulation, together with self-consumption and arbitrage, can be much more profitable [12,14] and helps to increase return on investments. That is why the purpose of the present study is to jointly optimize self-consumption and revenues generated from selling Frequency Containment Reserves (FCRs) for a BESS within a microgrid. The 1.5 MW microgrid considered in this work, is composed of a PV generation system, a BESS and an electrical load, and is connected to the electricity grid. The objective of this work is to develop an optimal control strategy for the battery so that it can be used for multiple purposes: self-consumption within the microgrid and frequency regulation (FR) in the Finnish frequency reserve market. To participate in this market, a day-ahead bid of the reserve that will be held in the battery during the following day has to be submitted to the Transmission System Operator (TSO). In order to develop the optimal day-ahead reserve profile, the main challenge faced is the random nature of the frequency deviations. To tackle this problem of decision-making under uncertainty, we propose a two-stage stochastic programming approach for the day-ahead reserve planning problem and for the intraday battery control optimization problem. We compared the results of this approach with those of a classical deterministic formulation where uncertainties are ignored.

This work is organized as follows: Sect. 2.1 briefly describes the FCR market of the Nordic electrical system, and discusses some related works. The studied mathematical model for the day-ahead stochastic FCR problem is presented in Sect. 3, where we also describe the considered econometric model for generating future realizations of frequency deviation. The intraday MPC contemplated in this work is presented in Sect. 4, together with two essential simulation tools for assessing the benefits of the considered stochastic model. Section 5 presents a comparison between the stochastic and deterministic approaches in terms of energy costs and the quality of the obtained day-ahead reserve profiles.

2   Context of the Problem

2.1   Frequency Containment Reserve – FCR

The equilibrium between supply and demand is of paramount importance for a power grid. If consumption outweighs or falls behind production, the frequency of the grid diverges from its nominal value of 50 Hz. In this case, regulation is needed to bring the frequency back to its pre-defined value. In practice, this can be done by absorbing the surplus power from the grid, or by injecting the missing power when supply and demand are not balanced. To ensure this operation, referred to as frequency control or regulation, FCR are procured by the TSO.

In the Nordic system, TSOs agree with the obligations for maintaining these frequency control reserves. In Finland, where the TSO Fingrid is responsible for maintaining the balance between supply and demand, two types of FCRs are used for the constant control of frequency: normal (FCR-n) and disturbed (FCR-d) [11]. The BESS of the microgrid considered in this study will be used to provide frequency control in the FCR-n market. The remuneration for FCR-n capacity is of about 14.00 €/MW per hour that Fingrid shall pay to the reserve holder in compensation for the verified capacity. In this context, the main difficulty for the reserve holder is to determine the optimal day-ahead reserve plan to submit to the system operator such that the trade-off between using the battery for self-consumption and generating revenues from supplying frequency reserve is maximized. Besides this day-ahead planning problem, the reserve holder should solve intraday control problems to find the optimal control strategy for the battery in order to minimize the energy costs and to provide the committed FCR reserve. The main challenge comes from the uncertainty on the frequency deviation of the grid, which is a random vector. Therefore, our problem falls within the scope of stochastic programming [8].

2.2   Related Work

Many approaches have been proposed in the literature for both day-ahead planning and real-time control of PV-BESS [3]. One of the most efficient approaches is Model Predictive Control (MPC), which has been gaining popularity since the 1990s [5]. For instance, [13] proposes an MPC framework to develop a day-ahead planning and a real-time operation control for a PV-BESS. The study considered the use of a BESS for energy shifting, arbitrage and smoothing, but did not consider frequency regulation (FR). The work [3] revisits the same type of problem and proposes a chance-constrained optimization approach to account for the uncertainty in PV production. An MPC controller to achieve efficient operation of a BESS in an FR setting was investigated in [5]. To optimize the performance of the developed controller, the authors designed a frequency prediction method based on Grey theory, but they did not consider uncertainty on these predictions. Differently, [2] proposes a nested dynamic programming model to account for uncertainty on load demand, electricity price and frequency regulation signal while optimizing the use of a BESS for both FR and arbitrage. However, in contrast with the present work, [2] takes the FR capacity as a given parameter and not as a variable. More recently, [9] develops a stochastic model for jointly optimizing peak shaving and FR with a BESS. To deal with uncertainty on Frequency Deviation (FD) signal, the study considered a scenario-based model with only 10 FD scenarios selected from one-year historical data. In [6] a two-stage stochastic MPC framework is developed to determine real-time commitments in energy and FR markets for a BESS. Ledoit-Wolf covariance estimator is employed to generate load and price scenario profiles from historical data. The work does not consider FD scenarios. A multi-stage stochastic model is proposed in [4] for co-optimizing FR and self-consumption with a BESS. The authors take into account uncertainties on load and frequency deviation and split the problem into two distinct sub-problems: one based on Sample Average Approximation (SAA) for providing primary frequency control, and the other sub-problem handles self-consumption with a chance-constrained model. Our work builds on [4] by developing a stochastic MPC framework for the optimal planning and control of a PV-BESS microgrid. We formulate a two-stage stochastic model for the planning problem. FR scenarios are generated by an auto-regressive model, and out-of-sample simulations combined with MPC closed-loop simulations are performed to assess the quality of the computed solution. In the closed-loop simulations, the control problem takes into account, at every 15 min, the updated system state and forecasts together with the committed frequency regulation reserve plan (the first-stage decision) for optimizing the control for the microgrid.

3   A Stochastic Day-Ahead Optimization Model

The objective of the day-ahead optimization problem is to minimize the cost of electricity consumption and maximize the revenue provided by offering the frequency control service to Fingrid. In more specific terms, the frequency control capacity R_t, $t = 0, \ldots, H - 1$, is a profile with hourly time-steps: $H = 24$. Since the TSO requires a reserve plan with hourly sampling periods, we force the reserve R_t to be constant for one hour. This reserve plan is a *here-and-now* decision, meaning that it has to be made before the realization of the day-ahead uncertainties. In our problem, the uncertainties refer to the deviation of the normalized frequency (Δf_t, $t = 0, \ldots, H - 1$) in the next day, which can be positive or negative. If the delivered/received power differs from the required power $R_t \cdot \Delta f_t$ according to the previously committed day-ahead reserve R_t, then the microgrid operator has to pay penalties to the TSO. By considering a finite number N of scenarios $\Delta f^{(i)}$, $i = 1, \ldots, N$ for the frequency deviation, the proposed stochastic model consists of minimizing the expected costs of electricity consumption, and maximizing the revenue provided by offering the frequency control service:

$$\min \; - C_{\mathrm{R}} \sum_{t=0}^{H-1} R_t + \frac{1}{N} \sum_{i=1}^{N} \left[\sum_{t=0}^{H-1} C_{\mathrm{TOU},t} P_{\mathrm{G},t}^{(i)} \right] \tag{1a}$$

$$\text{s.t. } P_{\mathrm{G},t}^{(i)} = P_{\mathrm{L},t} + P_{\mathrm{B},t}^{(i)} + P_{\mathrm{pv},t} \qquad\qquad \forall t, i \tag{1b}$$

$$P_{\mathrm{B},t}^{(i)} = P_{\mathrm{N},t}^{(i)} + R_t \cdot \Delta f_t^{(i)} \qquad\qquad \forall t, i \tag{1c}$$

$$P_{\min} + R_t \le P_{\mathrm{N},t}^{(i)} \le P_{\max} - R_t \qquad\qquad \forall t, i \tag{1d}$$

$$P_{\mathrm{B},t}^{(i)} = P_{\mathrm{Ch},t}^{(i)} + P_{\mathrm{Disch},t}^{(i)} \qquad\qquad \forall t, i \tag{1e}$$

$$b_1^{(i)} = b_{\mathrm{init}} + \tau \left(P_{\mathrm{Ch},0}^{(i)} \eta_{\mathrm{Ch}} - P_{\mathrm{Disch},0}^{(i)} \frac{1}{\eta_{\mathrm{Disch}}} \right) \qquad\qquad \forall i \tag{1f}$$

$$b_{t+1}^{(i)} = b_t^{(i)} + \tau \left(P_{\mathrm{Ch},t}^{(i)} \eta_{\mathrm{Ch}} - P_{\mathrm{Disch},t}^{(i)} \frac{1}{\eta_{\mathrm{Disch}}} \right) \qquad\qquad \forall i, \forall t \ge 1 \tag{1g}$$

$$b_{\min} + 0.5 R_t \le b_t^{(i)} \le b_{\max} - 0.5 R_t \qquad\qquad \forall t, i \tag{1h}$$

$$R_t \ge 0 \qquad\qquad \forall t. \tag{1i}$$

The decision variables are $R, P_{\mathrm{B}}, P_{\mathrm{G}}, P_{\mathrm{N}}$ and b, and the time horizon is 24 h with time steps t of 15 min. In this formulation, constraint (1b) represents the power balance equation of the microgrid with P_L and P_{PV} being the forecasted load and PV production. In constraint (1c), we "virtually" divide the battery power into two parts: a quantity $R_t \cdot \Delta f_t^{(i)}$ that represents the (random) power provided by the battery for frequency control, and a nominal power P_N which must remain within the bounds in (1d). Constraints (1f) and (1g) correspond to the dynamic evolution of the battery state of charge (SoC), where η_{Ch} and η_{disch} denote respectively the battery charge and discharge efficiencies, and constraints (1d) and (1h) are used to force the battery to maintain energy and power reserves for frequency regulation. The multiplying factor 0.5 that appears in constraint (1h) illustrates a requirement set by the TSO for the operator to be able at each time step t to deliver the frequency reserve R_t for half an hour.

Note that (1) is a two-stage stochastic programming problem whose first-stage variable is the reserve profile R, and the remaining variables are the second-stage ones, which depend on the realization of the frequency deviation $\Delta f_t^{(i)}$. The given formulation, where the random vector Δf is represented by N scenarios $\Delta f^{(i)}$ with probability $1/N$, is known in the literature as *Sample Average Approximation* (SAA) [8]. For the Frequency Containment Reserve problem, the SAA model (1) is a Linear Programming (LP) problem whose dimension increases with the number N of scenarios. A deterministic version of (1) consists of taking $N = 1$ and $\Delta f^{(1)}$ a possible event of the random vector. A common approach is to ignore the uncertainties and replace the random vector Δf by its average, which is zero in our case: it amounts at making the unrealistic assumption that there will not be any frequency deviation Δf in the next day. In Sect. 5 such a deterministic approach is compared to the stochastic one in (1) by varying the number of scenarios N. The model employed for (frequency deviation Δf_t) scenario generation is now briefly discussed.

Scenarios for Frequency Deviation: The frequency deviation of the electricity grid can be modeled as a random process, which we approximate by finitely many scenarios. In this work, we employ an Autoregressive (AR) model to generate random scenarios of frequency deviation-profiles Δf, that we use in the optimization problem (1). To calibrate our AR model, we use historical frequency deviation data (from the TSO) that we re-sample at time steps of 15 min, the time step of the optimization model. Then the ARIMA(p, d, q) model in Python was used to fit a model with AR order $p = 1$, degree of differencing $d = 0$, and order of the Moving Average model $q = 0$. The resulting model can then be written in the following form:

$$X_t = c + \phi X_{t-1} + \epsilon_t, \qquad t = 0, 1, \ldots, H,$$

where c is a constant, ϕ is the AR coefficient and ϵ_t is a white noise.

4   MPC Controller and Simulation

The MPC controller's subproblem. The planning optimization problem (1) is solved once a day, in the evening (for example at 6 pm) to compute the day-ahead reserve profile R to be submitted to the TSO. Then, provided the TSO has accepted the offer, the reserve profile will be fixed and taken as a parameter by the intraday MPC controller. Thus, during the day itself, instead of solving the problem (1), a control optimization problem is solved at every time step to compute an optimal control strategy for the battery. Given a fixed reserve profile $R_t, t = 0, \ldots, H-1$, such a control problem is written at stage $\tau = 0, \ldots, H-1$ as follows, where $\Delta f_t^{(i)}, t = \tau, \ldots, \tau + H - 1$ is a fixed scenario (possibly updated with recent information):

$$Q_\tau^{(i)}(R) := \min \quad \sum_{t=\tau}^{H-1} \left[C_{\text{TOU},t} P_{\text{G},t}^{(i)} \right] \tag{2}$$

$$\text{s.t. } P_{\text{B},t}^{(i)} = P_{\text{N},t} + R_t \cdot \Delta f_t^{(i)} \qquad i \text{ fixed and } \forall t \geq \tau$$

$$(1\text{b}), (1\text{d}), (1\text{e}), (1\text{g}), (1\text{h}) \qquad i \text{ fixed and } \forall t \geq \tau.$$

The decision variables are $P_\text{B}, P_\text{G}, P_\text{N}$ and b, with the reserve profile R and scenario $\Delta f_t^{(i)}$ fixed. The so-called recourse function $Q_\tau^{(i)}(R)$ with $\tau = 0$ will be used to assess the quality of the reserve profile in an out-of-sample simulation.

Out-of-Sample Simulation. To assess the quality of the sample employed to define the model (1) we employ out-of-sample simulation [8]: we first generate a sample of N scenarios and compute a first-stage solution R^N of (1). Then, a second sample $\Delta \tilde{f}^{(j)}, j = 1, \ldots, N'$, with $N' > N$ scenarios is generated and the recourse values $Q_0^{(j)}(R^N)$ are computed for all $j = 1, \ldots, N'$. Note that this procedure amounts at solving N' individual LPs (2) and is, thus, easier than solving the larger LP (1) (where all the scenarios are jointly taken into consideration). Note that

$$\text{val}_{N'}^{\text{simul}}(R^N) := -C_\text{R} \sum_{t=0}^{H-1} R_t^N + \frac{1}{N'} \sum_{j=1}^{N} Q_0^{(j)}(R^N) \tag{3}$$

estimates the optimal value of the *true* stochastic problem, i.e., problem (1) with random vector Δf following its *true* continuous probability distribution. If the variations of $\text{val}_{N'}^{\text{simul}}(R^N)$ are robust w.r.t. the sample size N, meaning that if we change the sample of size N and recompute R^N by solving (1) the value $\text{val}_{N'}^{\text{simul}}(R^N)$ does not significantly change, we accept N as a reasonable size for the sample employed to define the two-stage stochastic problem (1). We care to mention that the higher is N, the better (by the Central Limit Theorem) is the decision R^N [8]. However, solving (1) for large N may be prohibitive in terms of CPU time, and decomposition techniques must come into play. We refer the interested reader to [7] and [1] for efficient decomposition techniques for two-stage stochastic programs. In this work, we solve (1) without decomposition by using the CPLEX LP solver.

5   Numerical Assessments

In this section, we consider two types of simulations: one to assess the quality of the considered sample of scenarios, and the other to simulate (via MPC) the microgrid operation along the day.

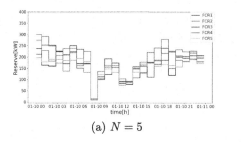

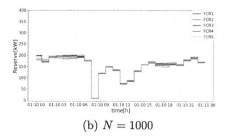

(a) $N = 5$ (b) $N = 1000$

Fig. 1. Five day-ahead reserve profiles obtained by solving (1) with different samples.

5.1 Sensitivity Analysis and Out-of-Sample Simulations

In order to evaluate whether the uncertainty has a considerable impact on our problem and to assess the impact of the number of scenarios on the computed frequency control capacity (i.e., R, the first-stage decision) we conducted the following sensitivity-analysis test. First, we considered five different samples with $N = 5$ scenarios (of frequency deviation) each. The scenarios were generated using the AR model described in Sect. 4. Then, for every sample, we solved the model (1) for obtaining five profiles of frequency reserve. These profiles are plotted in Fig. 1(a) (named $FCR1$ to $FCR5$), which shows that the scenario sample employed in (1) has a major impact on the solution of the problem. In other words, randomness plays an important role in (1), as expected. Having obtained very different profiles, which one shall be taken? The answer is none, because such profiles are too dependent of the considered samples, and therefore not reliable. In order to obtain a more reliable profile, we must increase the samples' size and verify if the associated reserve profiles are stable (in the sense that they do not differ much from each other). To this end, we perform the same analysis with five different samples, but now with $N = 1000$ scenarios each. As shown in Fig. 1(b), the computed reserve profiles are similar among them which means that $N = 1000$ allows to satisfactorily approximate the stochastic nature of the problem. Additional tests of all the scenario numbers ranging from 2 and 1000, in steps of 50, have shown that reserve profiles obtained with 400 to 600 scenarios are more "stable" than those obtained with 5 scenarios, and almost as stable as those yielded by the samples of $N = 1000$ scenarios. This kind of tests helps to determine the number N of scenario to be chosen so that a satisfactory trade-off between accuracy/stability of results and computational burden is achieved. For the case with $N = 5$ scenarios, the stochastic model was solved in less than 47 seconds. As for $N = 400$ and $N = 1000$, the CPU times were, respectively, 4712 seconds and 27367 seconds. The experiments were run on a computer with an Intel(R) Core(TM) i7-8650U CPU running at 1.90 GHz using 31.9 Go of RAM and running Windows 10.

To better assess the impact of the number of scenarios N on the solution and the optimal value of our stochastic problem we perform out-of-sample simulations, as described in Sect. 4. We proceeded as follows:

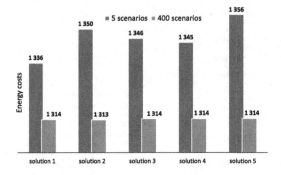

Fig. 2. Energy costs obtained from out-of sample simulations with 5 & 400 scenarios.

- problem (1) was solved with ten different samples; five samples of size $N = 5$ and five samples of size $N = 400$;
- another sample of $N' = 1000$ scenarios of frequency deviations was generated independently from the previous ten samples;
- for each of the ten obtained solutions: we fixed the first-stage solution (frequency reserve) R^N and solved N' LPs (2) by changing the scenarios;
- the simulated/averaged values $\mathtt{val}_{N'}^{\mathtt{simul}}(R^N)$ in (3) were computed to estimate the value of our *true* objective function (defined by the continuous stochastic process).

The ten simulated values are presented in Fig. 2. These values show that energy costs obtained with a number of scenarios $N = 400$ are lower and more stable (with variance equal to 0.04) than those obtained with $N = 5$, whose variance is 52.37.

5.2   Closed-Loop Simulations

In closed-loop simulations, we simulate one day of the microgrid operation using different layers of controllers as represented in Fig. 3. When the microgrid operator plans to participate in the FCR market for a given day, the planning optimization problem (1) is solved in the evening of the previous day. The obtained reserve plan R is then transmitted to the MPC control layer that will, at every 15 min of the following day, solve the control optimization problem (2) by taking into account *updated* system state and forecasts. Battery setpoints are then calculated and sent to the low-level control layer. The role of this controller is to measure real-time frequency deviations that take place at each time step and adjust the battery power in order to integrate the reserve power that should be delivered, according to the measured frequency deviations. Adjusted battery setpoints are then transmitted to the battery, and the new battery SoC is measured and given as input to the MPC control layer. This process is called closed-loop simulation. Some parameters and results obtained by performing closed-loop simulations are presented in Fig. 3.

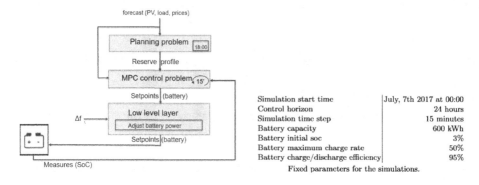

Simulation start time	July, 7th 2017 at 00:00
Control horizon	24 hours
Simulation time step	15 minutes
Battery capacity	600 kWh
Battery initial soc	3%
Battery maximum charge rate	50%
Battery charge/discharge efficiency	95%

Fixed parameters for the simulations.

Fig. 3. Different layers of the microgrid controller, parameters and results.

Deterministic Versus Stochastic Approach. To compare the results of the proposed stochastic approach with those of a classical deterministic method, we conducted the following test: (i) we generate $M = 40$ 24-h realizations of the frequency deviation; (ii) for each one of these scenarios, a closed-loop simulation is run for two different day-ahead reserve profiles R: one yielded by the stochastic model (1) with $N = 600$ scenarios, and the other given by the deterministic model, i.e., problem (1) with $N = 1$ and[1] $\Delta f = 0$; (iii) for each simulation, we compute the total energy TOU spendings of the day, taking into account the cost of energy consumption and the revenues gotten from the supplied FCR. The results in Fig. 4 show that not only the total costs obtained with the deterministic approach are in most of the cases higher than with the stochastic model, but they are also less stable: the "deterministic" costs have a more significant fluctuation.

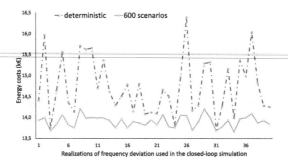

Fig. 4. Energy costs of stochastic approach with $N = 600$ × deterministic approach.

---

[1] Considering thus the average value of the frequency deviation time series and, therefore, assuming that there will be no frequency deviations.

Impact of the Sample Size on the Closed-Loop Simulation Results. The impact of the sample size on the out-of-sample simulation was investigated in Subsect. 5.1. In this subsection, the impact of the sample size (considered in (1)) on the closed-loop simulation is assessed. In the first test, closed-loop simulations are performed for five samples of $M = 20$ different realizations of frequency deviation each. For each realization, we perform closed-loop simulations with the day-ahead reserve profiles yielded by the stochastic model (1) defined with $N \in \{1, 5, 100, 600\}$ scenarios. Figure 5 presents, for each one of the five considered samples, the closed-loop simulation costs averaged over 20 different realizations. The results show that the higher the number of scenarios in (1), the lower the average costs of the closed-loop simulation. This is in accordance with the out-of-sample simulation of Subsect. 5.1, where the average planning cost was estimated. As we can see in Fig. 5a, the gain in terms of average costs is more pronounced when N increases from 1 to 5 and 100, while it is less remarkable from 100 to 600. This fact indicates that starting from a certain threshold, the gain in precision is not significant enough to justify the increase of the sample size in (1) and, thus, the increase of computational burden. In a second test, we take $M = 100$ different realizations of the frequency deviation resolving in the closed-loop simulation. The planning optimization problem (1) was solved with samples of size $N \in \{1, 5, 600\}$. Figure 5b shows the cumulative distribution functions of the M energy costs obtained in the closed-loop simulation. As the number of scenarios increases the cumulative distribution curve shifts to the left, and the spread between quantiles is reduced. This shows that not only the average cost diminishes as N increases (as already indicated in Fig. 5), but also the variance of the costs decreases.

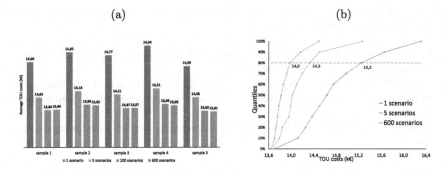

Fig. 5. (a): Impact of the number of scenarios on average energy costs; (b): Cumulative distribution of the energy costs of $M = 100$ closed-loop simulations

All in all, considering the uncertainty by using the stochastic model (1) allows reducing risks (cost fluctuations) and improving the predictability of the costs. Let us take, for instance, the quantile at 80%: we can see that if we use the deterministic model, the cost is lower than 15248 in 80% of the cases, whereas by using the stochastic model with $N = 600$ scenarios, this cost drops to 13974 in 80% of the cases. This measure is referred to as the Value at Risk at 80% (VaR): the lower it gets, the lower the exposure to risk.

6   Conclusion

This study investigated a microgrid energy management system and proposed a stochastic model to deal with the uncertain frequency deviations of the electricity grid while participating in the FCR market. We first described both deterministic and stochastic models for the reserve planning optimization problem and the battery control optimization problem. These models were numerically assessed by performing two types of tests: out-of-sample and closed-loop simulations. Comparison of the results of the classical deterministic model and the proposed stochastic approach showed that using stochastic programming to deal with the randomness of the frequency deviations results in reduced energy costs, reduced risks, and better predictability of the energy costs and revenues. For the problem of interest, the numerical results indicate that a sample with $N = 600$ scenarios generated by the AR model yields a satisfactory trade-off between precision and computational burden. As a byproduct of this study, a software that carries out scenario generation, two-stage stochastic optimization, out-of-sample, and closed-loop simulations, has been delivered for aiding decision making within a PV-BESS microgrid.

References

1. van Ackooij, W., de Oliveira, W., Song, Y.: Adaptive partition-based level decomposition methods for solving two-stage stochastic programs with fixed recourse. Inf. J. Comput. **30**(1), 57–70 (2017)
2. Cheng, B., Powell, W.B.: Co-optimizing battery storage for the frequency regulation and energy arbitrage using multi-scale dynamic programming. IEEE Trans. Smart Grid **9**(3), 1997–2005 (2016)
3. Conte, F., Massucco, S., Silvestro, F.: Day-ahead planning and real-time control of integrated PV-storage systems by stochastic optimization. IFAC-PapersOnLine **50**(1), 7717–7723 (2017)
4. Engels, J., Claessens, B., Deconinck, G.: Combined stochastic optimization of frequency control and self-consumption with a battery. IEEE Trans. Smart Grid **10**(2), 1971–1981 (2017)
5. Khalid, M., Savkin, A.V.: Model predictive control based efficient operation of battery energy storage system for primary frequency control. In: 2010 11th International Conference on Control Automation Robotics & Vision, pp. 2248–2252. IEEE (2010)
6. Kumar, R., Wenzel, M.J., Ellis, M.J., ElBsat, M.N., Drees, K.H., Zavala, V.M.: A stochastic MPC framework for stationary battery systems
7. de Oliveira, W., Sagastizábal, C., Scheimberg, S.: Inexact bundle methods for two-stage stochastic programming. SIAM J. Optim. **21**(2), 517–544 (2011)
8. Shapiro, A., Dentcheva, D., Ruszczyński, A.: Lectures on stochastic programming: modeling and theory. SIAM (2009)

9. Shi, Y., Xu, B., Wang, D., Zhang, B.: Using battery storage for peak shaving and frequency regulation: joint optimization for superlinear gains. IEEE Trans. Power Syst. **33**(3), 2882–2894 (2017)
10. Thorbergsson, E., Knap, V., Swierczynski, M., Stroe, D., Teodorescu, R.: Primary frequency regulation with li-ion battery based energy storage system-evaluation and comparison of different control strategies. In: Intelec 2013; 35th International Telecommunications Energy Conference, SMART POWER AND EFFICIENCY, pp. 1–6. VDE (2013)
11. Vilppo, O., Rautiainen, A., Rekola, J., Markkula, J., Vuorilehto, K., Järventausta, P.: Profitable multi-use of battery energy storage in outage mitigation and as frequency reserve. Int. Rev. Electr. Eng. **13**(3), 185–194 (2018)
12. Walawalkar, R., Apt, J., Mancini, R.: Economics of electric energy storage for energy arbitrage and regulation in New York. Energy Policy **35**(4), 2558–2568 (2007)
13. Wang, T., Kamath, H., Willard, S.: Control and optimization of grid-tied photovoltaic storage systems using model predictive control. IEEE Trans. Smart Grid **5**(2), 1010–1017 (2014)
14. White, C.D., Zhang, K.M.: Using vehicle-to-grid technology for frequency regulation and peak-load reduction. J. Power Sour. **196**(8), 3972–3980 (2011)

Using the GDELT Dataset to Analyse the Italian Sovereign Bond Market

Sergio Consoli[✉], Luca Tiozzo Pezzoli, and Elisa Tosetti

European Commission, Joint Research Centre, Directorate A-Strategy, Work
Programme and Resources, Scientific Development Unit, Via E. Fermi 2749,
21027 Ispra, VA, Italy
sergio.consoli@ec.europa.eu

Abstract. The Global Data on Events, Location, and Tone (GDELT) is
a real time large scale database of global human society for open research
which monitors worlds broadcast, print, and web news, creating a free
open platform for computing on the entire world's media. In this work,
we first describe a data crawler, which collects metadata of the GDELT
database in real-time and stores them in a big data management system based on Elasticsearch, a popular and efficient search engine relying
on the Lucene library. Then, by exploiting and engineering the detailed
information of each news encoded in GDELT, we build indicators capturing investor's emotions which are useful to analyse the sovereign bond
market in Italy. By using regression analysis and by exploiting the power
of Gradient Boosting models from machine learning, we find that the features extracted from GDELT improve the forecast of country government
yield spread, relative that of a baseline regression where only conventional regressors are included. The improvement in the fitting is particularly relevant during the period government crisis in May-December
2018.

Keywords: Government yield spread · Machine learning · Big data
management · Quantile regression · Feature Engineering · GDELT

1 Introduction

The explosion in computation and information technology experienced in the
past decade has made available vast amounts of data in various domains, that
has been referred to as *Big Data*. In Economics and Finance in particular, tap-
ping into these data brings research and business closer together, as data generated in ordinary economic activity can be used towards rapid-learning economic
systems, continuously improving and personalizing models. In this context, the
recent use of Data Science technologies for Economics and Finance is providing
mutual benefits to both scientists and professionals, improving forecasting and
nowcasting for several types of applications.

© The Author(s) 2020
G. Nicosia et al. (Eds.): LOD 2020, LNCS 12565, pp. 190–202, 2020.
https://doi.org/10.1007/978-3-030-64583-0_18

In particular, the recent surge in the government yield spreads in countries within the Euro area has originated an intense debate about the determinants and sources of risk of sovereign spreads. Traditionally, factors such as the credit-worthiness, the sovereign bond liquidity risk, and global risk aversion have been identified as the main factors having an impact on government yield spreads [2,20]. However, a recent literature has pointed at the important role of financial investor's sentiment in anticipating interest rates dynamics [17,25].

This paper exploits a novel, open source, news database known as *Global Database of Events, Language and Tone (GDELT)*[1] [15] to construct news-based financial indicators related to economic and political events for a set of Euro area countries. As described in Sect. 3.1, since the dimensions of the GDELT dataset make unfeasible the use of any relational database to perform an analysis in reasonable time, in Sect. 4.1 it is discussed the big data management infrastracture that we have used to host and interact with the data. Once GDELT data are crawled from the Web by the means of custom REST APIs[2], we efficiently transform and store them on our big data management system based on Elasticsearch, a popular and efficient NO-SQL search engine (Sect. 4.1).

Afterwards a feature engineering process is applied on the detailed information encoded in GDELT (Sect. 4.2) to select the most profitable variables which capture, among others, investor's emotions and popularity of news thematics, and that are useful to analyse the Italian sovereign bond market. In Sect. 4.3 we describe the Gradient Boosting machine we have used to analyse the Italian sovereign bond market. Our experimental analysis reported in Sect. 4.4 shows that the implemented machine learning model using the constructed GDELT indicators is useful to predict country government yield spread and financial instability, well aligned with previous studies in the literature.

2   Related Work

News articles represent a recent addition to the standard information used to model economic and financial variables. An early paper is [25] that uses sentiment from a column in the Wall Street Journal to show that high levels of pessimism are a relevant predictor of convergence of the stock prices towards their fundamental values. Following this early work, several other papers have tried to understand the role that news play in predicting, for instance, company news announcements, stock returns and volatility. For example recent works in finance exist on the application of semantic sentiment analysis from social media, financial microblogs, and news to improve predictions of the stock market (e.g. [1,7]). However these approaches generally suffer from a limited scope of the historical financial sources available. Recently, news have been also used in macroeconomics. For example, [13] looks at the informational content of the Federal Reserve statements and the guidance that these statements provide about the future evolution of monetary policy. Other papers ([26,27] and [23] among

---

[1] GDELT website: https://blog.gdeltproject.org/.
[2] See https://blog.gdeltproject.org/gdelt-2-0-our-global-world-in-realtime/.

others) use Latent Dirichlet allocation (LDA) to classify articles in topics and calculate simple measures of sentiment based on the topic classification. The goal of these papers is to extract a signal that could have some predictive content for measures of economic activity, such as GDP, unemployment and inflation [11]. Their results show that economic sentiment is a useful addition to the predictors that are commonly used to monitor and forecast the business cycle [7].

Machine learning approaches in the existing literature for controlling financial indexes measuring credit risk, liquidity risk and risk aversion include the works in [2,4,8,9,18], among others. Efforts to make machine learning models accepted within the economic modeling space have increased exponentially in recent years [19,24]. Among popular machine learning approaches, Gradient Boosting machines have been shown to be successful in various forecasting problems in Economics and Finance (see e.g. [5,6,16,28] among others).

3   Data

3.1   About GDELT

GDELT is the global database of events, location and tone that is maintained by Google [15]. It is an open Big Data platform on news collected at worldwide level, containing structured data mined from broadcast, print and web news sources in more than 65 languages. It connects people, organizations, quotes, locations, themes, and emotions associated with events happening across the world. It describes societal behavior through eye of the media, making it an ideal data source for measuring social factors and for testing our hypotheses. In terms of volume, GDELT analyses over 88 million articles a year and more than 150,000 news outlets. Its dimension is around 8 TB, growing 2TB each year. GDELT consists of two main datasets, the "Global Knowledge Graph (GKG)" and the "Events Table". For our study we have relied on the first dataset. GDELT's GKG captures what's happening around the world, what its context is, who's involved, and how the world is feeling about it; every single day. It provides English translation from the supported languages of the encoded information. In addition, the included themes are mapped into commonly used practitioners' topical taxonomies, such as the "World Bank (WB) Topical Ontology"[3], or into the GDELT built-in topical taxonomy. GDELT also measures thousands of emotional dimensions expressed by means of popular dictionaries in the literature, such as the "Harvard IV-4 Psychosocial Dictionary"[4], the "WordNet-Affect dictionary"[5], and the "Loughran and McDonald Sentiment Word Lists dictionary"[6], among others. For this application we use the GDELT GKG fields from

---

[3] https://vocabulary.worldbank.org/taxonomy.html.

[4] Harvard IV-4 Psychosocial Dictionary: http://www.wjh.harvard.edu/~inquirer/homecat.htm.

[5] WordNet-Affect dictionary: http://wndomains.fbk.eu/wnaffect.html.

[6] Loughran and McDonald Sentiment Word Lists: https://sraf.nd.edu/textual-analysis/resources/.

the World Bank Topical Ontology (i.e. WB themes), all emotional dimensions (GCAM), and the name of the journal outlet.

3.2 Yield Spread

We have extracted data from Bloomberg on the term-structure of government bond yields for Italy over the period 2 March 2015 to 31 August 2019. We calculate the sovereign spread for Italy against Germany as the difference between the Italian 10 year maturity bond yield minus the German counterpart. We also extract the standard level, slope and curvature factors of the term-structure using the Nelson and Siegel [21] procedure.

We estimate a model of credit spread forecast using convetional yield curve factors and the GDELT selected features, and compare with a classical model of credit spread forecast using only the level, slope and curvature as regressors.

4 Methods

4.1 Big Data Management

Massive unstructured datasets like GDELT require to be stored in specialized distributed file systems (DFS), joining together many computational nodes over a network, that are essential for building the data pipes that slice and aggregate this large amount of information. Among the most popular DFS platforms today, considering the huge number of unstructered documents coming from GDELT, we have used Elasticsearch [12,22] to store the data and interact with them. Elasticsearch is a popular and efficient efficient document-store which, instead of storing information as rows of columnar data like in classical relational databases, stores complex data structures that have been serialized as JSON documents. Being built on the Apache Lucene search library[7], it then provides real-time search and analytics for different types of structured or unstructured data.

Elasticsearch is built upon a distributed setting, allowing connecting multiple Elasticsearch nodes in a unique cluster. At the moment a document is stored, it is indexed and fully searchable in near real-time. An Elasticsearch index can be thought of as an optimized collection of documents and each document is a collection of fields, which are the key-value pairs that contain the stored data. An index is really just a logical grouping of one or more physical shards, where each shard is actually a self-contained index.

Elasticsearch is also schema-less, which means that documents can be indexed without explicitly specifying how to handle each of the different fields that might occur in a document. Elasticsearch provides a simple REST API for managing the created cluster and interacting with the stored documents. It is possible to submit API requests directly from the command line or through the Developer Console within the user web interface, referred to as *Kibana*[8]. The Elasticsearch

---

[7] https://lucene.apache.org/.

[8] Kibana, version 7.4: https://www.elastic.co/guide/en/kibana/7.4/.

REST APIs support structured queries, full text queries, and complex queries that combine the two, using Elasticsearch's JSON-style query language, referred to as *Query DSL*[9].

4.2 Feature Engineering

As GDELT adds news articles every fifteen minutes, each article is concisely represented as a row of a GDELT table in .csv format. The GKG table contains around 10 TB of data that need to be integrated and ingested as serialized JSON documents into our Elasticsearch framework. This involves applying the three usual steps of *Extract, Transform* and *Load* (ETL), that have to identify and overcome structural, syntactic, and semantic heterogeneity across the data. For this reason we have used the available World Bank Topical Ontology to understand the primary focus (theme) of each article and select the relevant news whose main themes are related to events concerning bond market investors. Such taxonomy is a classification schema for describing the World Bank's areas of expertise and knowledge domains representing the language used by domain experts. GDELT contains all themes discussed in an article as an entry in a single row. We separate these themes into separate entries.

We have extracted news information from GKG from a set of around 20 newspapers for Italy, published over the period March 2015 until end of August 2019. We rely on the Geographic Source Lookup file available on GDELT blog[10] in order to chose both generalist national newspapers with the widest circulation in that country, as well as specialized financial and economic outlets. Once collected the news data, we have mapped these to the relevant trading day. Specifically, we assign to a given trading day all the articles published during the opening hours of the bond market, namely between 9.00 am and 5.30 pm. Articles that have been published after the closure of the bond market or overnight are assigned to the following trading day.[11] Following [10], we assign the news published during weekends to Monday trading days, and omit articles published during holidays or in weekends preceding holidays.

Hence, we selected only articles such that the topics extracted by GDELT fall into one of the following WB themes of interest: *Macroeconomic Vulnerability and Debt*, and *Macroeconomic and Structural Policies*. We observe that articles can mention only briefly one of the selected topics and then focus on a totally different theme. To make sure that the main focus of the article is one of the selected WB topics, we have retained only news that contain in their text at least three keywords belonging to these themes. The aim is to select news that focus on topics relevant to the bond market, while excluding news that only briefly

---

[9] https://www.elastic.co/guide/en/elasticsearch/reference/current/query-dsl.html.

[10] See https://blog.gdeltproject.org/mapping-the-media-a-geographic-lookup-pf-gdelts-sources/.

[11] Since the GKG operates on the UTC time, https://blog.gdeltproject.org/new-gkg-2-0-article-metadata-fields/, we made a one-hour lag adjustment according to Italian time zone.

report macroeconomic, debt and structural policies issues. Finally, to obtain a pool of news that are not too heterogeneous in length, we have retained only articles that are at least 100 words long. After this selection procedure we obtain a total of 18,986 articles. From this large amount of information, we construct features counting the total number of words belonging to all WB themes and GCAMs detected each day. We also created the variables "Number of mentions" denoting the word count of each location mentioned in the selected news. By doing this we obtain a total of 2,978 GCAM, 1,996 Themes and 155 locations. Notice that, all of our features are expressed in terms of daily word count with the exception of the ANEW dictionary (v19) and the Hedenometer measure of happiness (v21) which are already provided as score values.

Once extracted the data from GKG, we have adopted a five step procedure to filter out features from news stories. In the first step we have applied a domain knowledge criteria and have retained a subset of 413 GCAM dictionaries that are potentially relevant for our analysis. Specifically we have extracted 31 dimensions of the General Inquirer Harvard IV psychosocial Dictionary, 61 dimensions of Roget's Thesaurus, 7 dimensions of the Martindale Regressive Imagery and 3 dimensions of the Affective Norms for English Words (ANEW) dictionary. The second step concerns the variability of the extracted features. In particular, we have retained variables with a standard deviation calculated over the full sample that is greater than 5 words and allowed a 10% of missing values on the total number of days. In addition, features that are missing at the beginning of the sample (more than 33% of the sample) have been excluded. In the final step we have performed a correlation analysis across the selected variables. We have normalized at first all features by number of daily articles. If the correlation between any two features is above 80% we give preference to the variable with less missing values, while if the number of missing values is identical and the two variables belong to the same category (i.e. both are themes or GCAMs), we randomly pick one of them. Finally, if the number of missing values is identical but the two variables belong to the same category, we consider the following order of priority: GCAM, WB themes, GDELT themes, locations. After this Feature Engineering procedure, we are left with a total of 45 variables, of which 9 are themes, 34 are GCAM, 2 are locations. A careful inspection among the selected topics reveals that WB themes such as Inflation, Government, Central Banks, Taxation and Policy have been selected by our procedure. These are important topics discussed in the news when considering interest rates issues. Moreover, features constructed and selected from GCAM dimensions such as optimism, pessimism or arousal are also inculed and allow us to explore the emotional state of the market.

4.3 Big Data Analytics

Classical economic models are not dynamically scalable to manage and maintain Big Data structures, like the GDELT one we are dealing with. A whole new set of big data analytics models and tools that are robust in high dimensions, like the ones from machine learning, are required [19,24]. In particular in our computational study we have chosen to rely on *Gradient Boosting* (GB) [14], a well-known machine-learning approach which has been shown to be successful in various modelling problems in Economics and Finance [5,6,16,28]. Gradient boosting is a machine learning technique for regression and classification problems, which produces a prediction model in the form of an ensemble of weak prediction models, typically decision trees. It builds the model in a stage-wise fashion like other boosting methods do, and it generalizes them by allowing optimization of an arbitrary differentiable loss function. That is, algorithms that optimize a cost function over function space by iteratively choosing a function (weak hypothesis) that points in the negative gradient direction.

In particular for our implementation we have used the $H2O^{12}$ library available for the R programming language. H2O is a scalable open-source machine learning platform that offers parallelized implementations of many supervised and unsupervised machine learning algorithms, including Gradient Boosting Machines.

In addition, to determine the optimal parameter values of our GB model, we have used a 10-fold cross-validation together with a grid search (or parameter sweep) procedure [3]. Grid search involves an exhaustive searching through a manually specified subset of the hyperparameter space of the learning algorithm, guided by some performance metric (like in our case minimizing the mean squared error mean). The main parameters to optimize in our GB model are the *maximum tree depth*, which indicates the the maximum possible depth of a tree in the model and is used to control over-fitting, as higher depth will allow model to learn relations very specific to a particular sample; and the *learning rate*, which determines the impact of each tree on the final outcome of the GB model. GB works by starting with an initial estimate which is updated using the output of each tree; the learning rate parameter controls then the magnitude of this change in the estimates. Therefore, lower values are generally preferred as they make the model robust to the specific characteristics of the tree and thus allowing it to generalize well. However lower values would require higher number of trees to model all the relations and will be computationally expensive.

To explore the hyperparameters space looking for optimal values of these parameters, the grid search procedure tests the GB model with values going from 1 to 10 (by steps of 1) for the maximum tree depth parameter and from 0.01 to 0.99 (by steps of 0.10) for the learning rate parameter, the best parameter values with respect to the mean squared error are produced as output. In the case that one of the produced parameter values reaches one of the related upper or lower bound, i.e. corner solutions, a greedy approach is iterated. The search boundaries of the specific parameter are perturbed, and the grid search procedure is restarted from the sub-optimal parameters coming from the previous estimation. The procedure halts when both produced parameter values fall inside the related search boundaries, giving these parameter values as output.

Although grid search does not provide an absolute guarantee that it will find the global optimal parameter values, in practice we have found it to work quite

---

[12] R Interface for the "H2O" Scalable Machine Learning Platform: https://cran.r-project.org/web/packages/h2o/index.html.

well, despite to be quite computationally expensive. In general grid searching is a widely adopted and accepted procedure for this kind of tuning tasks [3].

4.4   Experimental Analysis

The main objective of our empirical exercise is to assess the predictive power of GDELT selected features over and above the classical determinants for government credit spreads during stressed periods. We explore predictability for the 90th percentile of the credit spread distribution, since this is usually classified as a situation of financial distress (see [4] among others). Several studies in the literature have shown that during these periods, complex non-linear relationships among explanatory variables affect the behaviour of the output target which simple linear models are not able to capture. We account for that by using a GB model with a quantile distribution and we adopt a rolling window estimation technique where the first estimation sample starts at the beginning of March and ending in May 2017. We compare the forecasting perfomance of a GB model where classical determinants as well as selected GDELT features are included, with that of a GB model where only classical factors are considered. We measure the forecasting performance by calculating the absolute error for each forecast. We also assess the explanatory power of GDELT features over time by calculating the variable importance at each estimation, and explore the five most important variables at each rolling window estimation. This is in line with standard term structure literature, stating that from three to five factors are sufficient to explain the dynamics of yield spreads. We estimate the model on half of the sample and adopt a rolling window to generate one-step ahead forecasts. We use a 10-fold cross-validation for each estimation window, and apply the grid search procedure previously described in Sect. 4.3 to optimally find the hyperparameters of the GB model.

Figure 1 displays the time series of the Italian spread and the ratio between absolute forecast errors of the model augmented with GDELT features and that of the model with classical regressors only. Notice that, when the value of this ratio is below one, our augmented model performs better than the benchmark. It is interesting to observe that the performance of our augmented model improves considerably starting from the end of May 2018 when a period of political turmoil started in Italy. On the 29th of May, the Italian spread sharply rose reaching 250 basis point. Investors where particularly worried about the possibility of anti-euro government and not confident on the formation of a stable government. During these stressed events our GDELT features augmented model strongly outperforms the benchmark model with absolute ratios values well below one, i.e. 0.93 on May the 24th that is the minimum value across all the sample under analysis. This result emphasises the value added of news articles stories in forecasting the yield spreads dynamics during periods of financial distress. From June until November 2018, a series of discussions about deficit spending engagements and possible conflits with European fiscal rules continued to worry the markets. The spread strongly increased in October and November with values

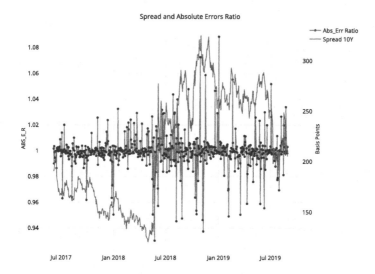

Fig. 1. Absolute errors ratios and spread

around 300 basis point. During this period our augmented model agains performs particularly well, with ratio around 0.95 with a minimum value of 0.94 on November the 9th. Since 2019, the italian political situation started to improve and the spread smothly decline, especially after the agreement with Brussels on buget deficit in December 2018. However, some events hit the Italian economy afterwards, such as the EU negative outlook and the European parliament elections which contributed to a temporary increase on interest rates.

Although in 2019 our aumented model did not perform as well as in 2018 in terms of absolute error ratios, it still consistently outperforms the benchmark model. From the analysis above we clearly observe three main sub-periods. The first pre-crisis period ranging form July 2017 till May 2018, the second one is the crisis period from June till December 2018, the third period from January 2019 till the end of the sample. We next analyse the contribution of each selected GDELT feature in the prediction of interest rates spread during periods of political stress, splitting the sample in the three identified subperiods.

Figure 2 shows the frequency of each variables appearing on the top five positions according to the Gradient Boosting Machine variable importance during (a) the pre-crisis period, (b) the crisis period, and (c) the post crisis period. It is interesting to observe that classical factors such as slope and curvature of spread yield curve are the most important variables during the pre-crisis period. However, we also observe that amongst the classical regressors, the level factor, which is pointed by the literature as the most important variable in explaining interest rates dynamics, is less important than two GCAM sentiment measures, namely the arousal form ANEW dictionary and Hate of Thesaurs. Figure 2(b) shows that, during crisis period, classical yield spread factors reduces considerably their predictive contribution. The most important variable is the Arousal

(a): Pre-Government crisis

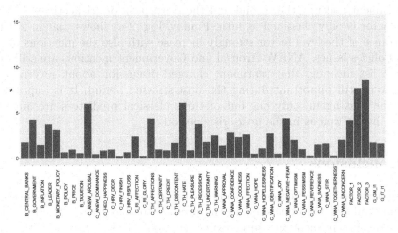

(b): Government crisis

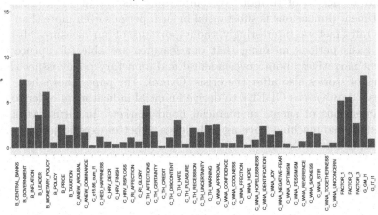

(c): Post-Government crisis

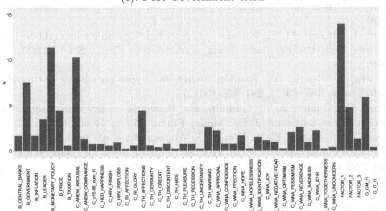

Fig. 2. Variable importance: percentage of times variables are top 5 classified

index of the ANEW dictionary. Interestingly, mentions of German locations and "Government" occupy the second and the third position, rispectively. Classical slope factor is only classified as fifth. Finally, Fig. 2(c) shows that in 2019, the importance of the level factor strongly increase with also the mentions of Monetary Policies issues. ANEW Arousal and Government mentions are still on the top five list meaning that sentiment charged discussion about governamental issues were still important during the post-stressed period. It is important to underline that again, only one out of three classical predictors are out of the top-five list as well as in the post-crisis period.

5   Conclusions

Our analysis is one of the first to study the behaviour of government yield spreads and financial portfolio decisions in the presence of classical yield curve factors and financial sentiment measures extracted from news. We believe that these new measures are able to capture and predict changes in interest rates dynamics especially in period of turmoil. We empirically show that the contribution of our sentiment dimensions is substantial in such periods even more than classical interest rates factors. Interestingly, the importance of our measures also extends in post-crisis periods meaning that our features are able to capture charged sentiment narratives about governmental and monetary policy issues that were the focus of stories also after the crisis. Overall, the paper shows how to use a large scale database as GDELT to derive financial indicators in order to capture future intentions of agents in financial bond markets. In future work we will adopt additional machine learning techniques to better exploit non-linear effects on the dependent variables.

References

1. Agrawal, S., Azar, P., Lo, A.W., Singh, T.: Momentum, mean-reversion and social media: evidence from StockTwits and Twitter. J. Portfolio Manag. **44**, 85–95 (2018)
2. Beber, A., Brandt, M.W., Kavajecz, K.A.: Flight-to-quality or flight-to-liquidity? Evidence from the Euro-area bond market. Rev. Finan. Stud. **22**(3), 925–957 (2009)
3. Bergstra, J., Bengio, Y.: Random search for hyper-parameter optimization. J. Mach. Learn. Res. **13**(1), 281–305 (2012)
4. Bernal, O., Gnabo, J.-Y., Guilmin, G.: Economic policy uncertainty and risk spillover in the Eurozone. J. Int. Money Finan. **65**(C), 24–451 (2016)
5. Chang, Y.-C., Chang, K.-H., Wu, G.-J.: Application of eXtreme gradient boosting trees in the construction of credit risk assessment models for financial institutions. Appl. Soft Comput. J. **73**, 914–920 (2018)
6. Deng, S., Wang, C., Wang, M., Sun, Z.: A gradient boosting decision tree approach for insider trading identification: an empirical model evaluation of china stock market. Appl. Soft Comput. J. **83**, 105652 (2019)

7. Dridi, A., Atzeni, M., Reforgiato Recupero, D.: FineNews: fine-grained semantic sentiment analysis on financial microblogs and news. Int. J. Mach. Learn. Cybernet. **10**(8), 2199–2207 (2018). https://doi.org/10.1007/s13042-018-0805-x

8. Favero, C., Pagano, M., von Thadden, E.-L.: How does liquidity affect government bond yields? J. Finan. Quant. Anal. **45**(1), 107–134 (2010)

9. Garcia, A.J., Gimeno, R.: Flight-to-liquidity flows in the Euro area sovereign debt crisis. Technical report, Banco de Espana Working Papers (2014)

10. Garcia, D.: Sentiment during recessions. J. Finan. **68**(3), 1267–1300 (2013)

11. Gentzkow, M., Kelly, B., Taddy, M.: Text as data. Journal of Economic Literature (2019, to appear)

12. Gormley, C., Tong, Z.: Elasticsearch: The definitive guide. O' Reilly Media, US (2015)

13. Hansen, S., McMahon, M.: Shocking language: understanding the macroeconomic effects of central bank communication. J. Int. Econ. **99**, S114–S133 (2016)

14. Hastie, T., Tibshirani, R., Friedman, J.: Additive models, trees, and related methods. The Elements of Statistical Learning. SSS, pp. 295–336. Springer, New York (2009). https://doi.org/10.1007/978-0-387-84858-7_9

15. Leetaru, K., Schrodt, P.A.: Gdelt: global data on events, location and tone, 1979–2012. Technical report, KOF Working Papers (2013)

16. Liu, J., Wu, C., Li, Y.: Improving financial distress prediction using financial network-based information and GA-based gradient boosting method. Comput. Econ. **53**(2), 851–872 (2019)

17. Loughran, T., McDonald, B.: When is a liability not a liability? Textual analysis, dictionaries and 10-ks. J. Finan. **66**(1), 35–65 (2011)

18. Manganelli, S., Wolswijk, G.: What drives spreads in the Euro area government bond markets? Econ. Policy **24**(58), 191–240 (2009)

19. Marwala, T.: Economic Modeling using Artificial Intelligence Methods. Springer-Verlag, London (2013). https://doi.org/10.1007/978-1-4471-5010-7

20. Monfort, A., Renne, J.-P.: Decomposing Euro-area sovereign spreads: credit and liquidity risks. Rev. Finan. **18**(6), 2103–2151 (2013)

21. Nelson, C., Siegel, A.F.: Parsimonious modeling of yield curves. J. Bus. **60**(4), 473–489 (1987)

22. Shah, N., Willick, D., Mago, V.: A framework for social media data analytics using elastic search and kibana. Wireless Networks (2018, in press)

23. Shapiro, A.H., Sudhof, M., Wilson, D.: Measuring news sentiment. Federal Reserve Bank of San Francisco Working Paper (2018)

24. Taddy, M.: Business Data Science: Combining Machine Learning and Economics to optimize, automate, and accelerate business decisions. McGraw-Hill, US (2019)

25. Tetlock, P.C.: Giving content to investor sentiment: the role of media in the stock market. J. Finan. **62**(3), 1139–1168 (2007)

26. Thorsrud, L.A.: Nowcasting using news topics. big data versus big bank. Norges Bank Working Paper (2016)

27. Thorsrud, L.A.: Words are the new numbers: a newsy coincident index of the business cycle. J. Bus. Econ. Stat. **38**(2), 1–17 (2018)

28. Yang, X., He, J., Lin, H., Zhang, Y.: Boosting exponential gradient strategy for online portfolio selection: an aggregating experts' advice method. Comput. Econ. **55**(1), 231–251 (2020)

Adjusted Measures for Feature Selection Stability for Data Sets with Similar Features

Andrea Bommert$^{(\boxtimes)}$ and Jörg Rahnenführer

TU Dortmund University, 44221 Dortmund, Germany
bommert@statistik.tu-dortmund.de

Abstract. For data sets with similar features, for example highly correlated features, most existing stability measures behave in an undesired way: They consider features that are almost identical but have different identifiers as different features. Existing adjusted stability measures, that is, stability measures that take into account the similarities between features, have major theoretical drawbacks. We introduce new adjusted stability measures that overcome these drawbacks. We compare them to each other and to existing stability measures based on both artificial and real sets of selected features. Based on the results, we suggest using one new stability measure that considers highly similar features as exchangeable.

Keywords: Feature selection stability · Stability measures · Similar features · Correlated features

1 Introduction

Feature selection is one of the most fundamental problems in data analysis, machine learning, and data mining. Recently, it has drawn increasing attention due to high-dimensional data sets emerging from many different fields. Especially in domains where the chosen features are subject to further experimental research, the stability of the feature selection is very important. Stable feature selection means that the set of selected features is robust with respect to different data sets from the same data generating distribution [7]. If for data sets from the same data generating process, very different sets of features are chosen, this questions not only the reliability of resulting models but could also lead to unnecessary expensive experimental research.

The evaluation of feature selection stability is an active area of research. Overviews of existing stability measures are given in [4] and [9]. The theoretical properties of different stability measures are studied in [10]. An extensive empirical comparison of stability measures is given in [3]. The research that has been done in various aspects related to stability assessment is reviewed in [1].

The source code for the experiments and analyses of this article is publicly available at https://github.com/bommert/adjusted-stability-measures.

© Springer Nature Switzerland AG 2020
G. Nicosia et al. (Eds.): LOD 2020, LNCS 12565, pp. 203–214, 2020.
https://doi.org/10.1007/978-3-030-64583-0_19

For data sets with similar features, the evaluation of feature selection stability is more difficult. An example for such data sets are gene expression data sets, where genes of the same biological processes are often highly positively correlated. The commonly used stability measures consider features that are almost identical but have different identifiers as different features. Only little research has been performed concerning the assessment of feature selection stability for data sets with similar features. Stability measures that take into account the similarities between features are defined in [15,16] and [17]. These measures, however, have major theoretical drawbacks. We call stability measures that consider the similarities between features "adjusted" stability measures.

In this paper, we introduce new adjusted stability measures. On both artificial and real sets of selected features, we compare them to each other and to existing stability measures and analyze their properties. The remainder of this paper is organized as follows: In Sect. 2, the concept of feature selection stability is explained in detail and adjusted stability measures are defined. The stability measures are compared in Sect. 3. Section 4 contains a summary of the findings and concluding remarks.

2   Concepts and Methods

In Subsect. 2.1, feature selection stability is explained and in Subsect. 2.2, measures for quantifying feature selection stability are introduced.

2.1   Feature Selection Stability

The stability of a feature selection algorithm is defined as the robustness of the set of selected features to different data sets from the same data generating distribution [7]. Stability quantifies how different training data sets affect the sets of chosen features. For similar data sets, a stable feature selection algorithm selects similar sets of features. An example for similar data sets could be data coming from different studies measuring the same features, possibly conducted at different places and times, as long as the assumption of the same underlying distribution is valid.

A lack of stability has three main reasons: too few observations, highly similar features and equivalent sets of features. Consider a group of data sets, for which the number of observations does not greatly exceed the number of features, from the same data generating process. The subsets of features with maximal predictive quality on the respective data sets often differ between these data sets. One reason is that there are features that seem beneficial for prediction, but that only help on the specific data set and not on new data from the same process. Selecting such features and including them in a predictive model typically causes over-fitting. Another reason is that there are features with similar predictive quality even though they are unrelated with respect to their content. Due to the small number of observations, chance has a large influence on which of these features has the highest predictive quality on each data set. The instability of feature selection resulting from both reasons is undesirable.

Regarding the case of highly similar and therefore almost identical features, it is likely that for some data sets, one feature is selected and for other data sets from the same process, another one of the similar features is chosen. As the features are almost identical, it makes sense to label this as stable because the feature selection algorithm always chooses a feature with the same information. Therefore, it is desirable to have a stability measure that takes into account the reason for the differences in the sets of chosen features. However, most existing stability measures treat both situations equally: if the identifiers of the chosen features are different, the feature selection is rated unstable.

Regarding the case of equivalent feature sets, for some data sets, there are different sets of features that contain exactly the same information. Finding all equivalent optimal subsets of features is an active field of research, see for example [13], and worst-case intractable. The selection of equivalent subsets of features is evaluated as unstable by all existing stability measures. Creating stability measures that can recognize equivalent sets of features is out of the scope of this paper.

2.2   Adjusted Stability Measures

For the definition of the stability measures, the following notation is used: Assume that there is a data generating process that generates observations of the p features $X_1, \ldots, X_p$. Further, assume that there are m data sets that are generated by this process. A feature selection method is applied to all data sets. Let $V_i \subseteq \{X_1, \ldots, X_p\}$, $i = 1, \ldots, m$, denote the set of chosen features for the i-th data set and $|V_i|$ the cardinality of this set. The feature selection stability is assessed based on the similarity of the sets $V_1, \ldots, V_m$. For all stability measures, large values correspond to high stability and small values to low stability.

Many existing stability measures that do not consider similarities between features assess the stability based on the pairwise scores $|V_i \cap V_j|$, see for example [3] and [10]. An example for an unadjusted stability measure is

$$\text{SMU} = \frac{2}{m(m-1)} \sum_{i=1}^{m-1} \sum_{j=i+1}^{m} \frac{|V_i \cap V_j| - \frac{|V_i| \cdot |V_j|}{p}}{\sqrt{|V_i| \cdot |V_j|} - \frac{|V_i| \cdot |V_j|}{p}}.$$

$\frac{|V_i| \cdot |V_j|}{p}$ is the expected value of $|V_i \cap V_j|$ if $|V_i|$ and $|V_j|$ features are chosen at random with equal selection probabilities. $\sqrt{|V_i| \cdot |V_j|}$ is an upper bound for $|V_i \cap V_j|$. Including it in the denominator makes 1 the maximum value of SMU. If many of the sets V_i and V_j have a large overlap, the feature selection is evaluated as rather stable. The basic idea of adjusted stability measures is to adjust the scores $|V_i \cap V_j|$ in a way that different but highly similar features count towards stability. Note that all of the following adjusted stability measures depend on a threshold θ. This threshold indicates how similar features have to be in order to be seen as exchangeable for stability assessment.

Zucknick et al. [17] extend the well known Jaccard index [6], considering the correlations between the features:

$$\text{SMZ} = \frac{2}{m(m-1)} \sum_{i=1}^{m-1} \sum_{j=i+1}^{m} \frac{|V_i \cap V_j| + C(V_i, V_j) + C(V_j, V_i)}{|V_i \cup V_j|} \quad \text{with}$$

$$C(V_i, V_j) = \sum_{x \in V_i} \frac{1}{|V_j|} \sum_{y \in V_j \setminus V_i} |\text{Cor}(x,y)| \, \mathbb{I}_{[\theta, \infty)} \left(|\text{Cor}(x,y)| \right).$$

$|\text{Cor}(x,y)|$ is the absolute Pearson correlation between x and y, $\theta \in [0,1]$ is a threshold, and $\mathbb{I}_S$ denotes the indicator function for a set S. One could generalize this stability measure by allowing arbitrary similarity values from the interval $[0,1]$ instead of the absolute correlations. A major drawback of this stability measure is that it is not corrected for chance. Correction for chance [10] means that the expected value of the stability measure for a random feature selection with equal selection probabilities for all features does not depend on the number of chosen features.

Zhang et al. [16] also present adjusted stability measures. Their scores are developed for the comparison of two gene lists. The scores they define are

$$\text{nPOGR}_{ij} = \frac{K + O_{ij} - E[K + O_{ij}]}{|V_i| - E[K + O_{ij}]}$$

with $ij \in \{12, 21\}$. K is defined as the number of genes that are included in both lists and regulated in the same direction. O_{ij} denotes the number of genes in list i that are not in list j but significantly positively correlated with at least one gene in list j. For each pair of gene lists, two stability scores are obtained.

Yu et al. [15] combine the two scores nPOGR_{ij} and nPOGR_{ji} into one score for the special case $|V_i| = |V_j|$:

$$\text{nPOGR} = \frac{K + \frac{O_{ij} + O_{ji}}{2} - E\left[K + \frac{O_{ij} + O_{ji}}{2}\right]}{|V_i| - E\left[K + \frac{O_{ij} + O_{ji}}{2}\right]}.$$

In this paper, we generalize this score to be applicable in the general context of feature selection with arbitrary feature sets $V_1, \ldots, V_m$ by

1. replacing the quantity K by $|V_i \cap V_j|$.
2. allowing the similarities between the features to be assessed by an arbitrary similarity measure instead of only considering significantly positive correlations, that is, replacing $\frac{O_{ij} + O_{ji}}{2}$ by $\frac{A(V_i, V_j) + A(V_j, V_i)}{2}$ with A defined below.
3. replacing $|V_i|$, which is the maximum value of $K + \frac{O_{ij} + O_{ji}}{2}$, by $\frac{|V_i| + |V_j|}{2}$, the maximum value of $|V_i \cap V_j| + \frac{A(V_i, V_j) + A(V_j, V_i)}{2}$.
4. calculating the average of the scores for all pairs $V_i, V_j, i < j$.

As a result, the stability measure

$$\text{SMY} = \frac{2}{m(m-1)} \sum_{i=1}^{m-1} \sum_{j=i+1}^{m} \frac{S_{\text{SMY}}(V_i, V_j) - E\left[S_{\text{SMY}}(V_i, V_j)\right]}{\frac{|V_i|+|V_j|}{2} - E\left[S_{\text{SMY}}(V_i, V_j)\right]}$$

with $\quad S_{\text{SMY}}(V_i, V_j) = |V_i \cap V_j| + \dfrac{A(V_i, V_j) + A(V_j, V_i)}{2}$

and $\quad A(V_i, V_j) = |\{x \in (V_i \setminus V_j) : \exists y \in (V_j \setminus V_i) \text{ with similarity}(x, y) \geq \theta\}|$

is obtained. E denotes the expected value for a random feature selection and can be assessed in the same way as described below for SMA. Similarity$(x, y) \in [0, 1]$ quantifies the similarity of the two features x and y and $\theta \in [0, 1]$ is a threshold.

In situations where V_i and V_j greatly differ in size and contain many similar features, the value of SMY may be misleading. Consider a scenario with $|V_i| \gg |V_j|$, $|V_i \cap V_j| = 0$, $A(V_i, V_j) = |V_i|$, and $A(V_j, V_i) = |V_j|$. In such situations, there are many features in the larger set that are similar to the same feature in the smaller set. Even though the sets V_i and V_j greatly differ with respect to feature redundancy and resulting effects for model building such as over-fitting, the stability score attains its maximum value.

To overcome this drawback, a new stability measure employing an adjustment $\text{Adj}(V_i, V_j)$ different from $\frac{A(V_i, V_j) + A(V_j, V_i)}{2}$ that fulfills

$$\max\left[|V_i \cap V_j| + \text{Adj}(V_i, V_j)\right] \leq \max\left[\left|\widetilde{V}_i \cap \widetilde{V}_j\right|\right] \text{ with } \left|\widetilde{V}_i\right| = |V_i| \text{ and } \left|\widetilde{V}_j\right| = |V_j|$$

is defined in this paper. This means that the adjusted score for V_i and V_j cannot exceed the value of $|\widetilde{V}_i \cap \widetilde{V}_j|$ that would be obtained if two sets $\widetilde{V}_i$ and $\widetilde{V}_j$ with $|\widetilde{V}_i| = |V_i|$ and $|\widetilde{V}_j| = |V_j|$ were chosen such that their overlap is maximal. This happens when $\widetilde{V}_i \subseteq \widetilde{V}_j$ or $\widetilde{V}_j \subseteq \widetilde{V}_i$. The resulting measure is

$$\text{SMA} = \frac{2}{m(m-1)} \sum_{i=1}^{m-1} \sum_{j=i+1}^{m} \frac{|V_i \cap V_j| + \text{Adj}(V_i, V_j) - E\left[|V_i \cap V_j| + \text{Adj}(V_i, V_j)\right]}{\text{UB}\left[|V_i \cap V_j|\right] - E\left[|V_i \cap V_j| + \text{Adj}(V_i, V_j)\right]}$$

with UB $\left[|V_i \cap V_j|\right]$ denoting an upper bound for $|V_i \cap V_j|$. The expected values $E\left[|V_i \cap V_j| + \text{Adj}(V_i, V_j)\right]$ cannot be calculated with a universal formula as they depend on the data specific similarity structure. However, they can be estimated by repeating the following Monte-Carlo-procedure N times: 1. Randomly draw sets $\widetilde{V}_i \subseteq \{X_1, \ldots, X_p\}$ and $\widetilde{V}_j \subseteq \{X_1, \ldots, X_p\}$, with $|\widetilde{V}_i| = |V_i|$, $|\widetilde{V}_j| = |V_j|$, and equal selection probabilities for all features. 2. Calculate the score $|\widetilde{V}_i \cap \widetilde{V}_j| + \text{Adj}(\widetilde{V}_i, \widetilde{V}_j)$. An estimate for the expected value $E\left[|V_i \cap V_j| + \text{Adj}(V_i, V_j)\right]$ is the average of the N scores.

Concerning the upper bounds UB $\left[|V_i \cap V_j|\right]$, $\min\{|V_i|, |V_j|\}$ is the tightest upper bound for $|V_i \cap V_j|$. However, this upper bound is not a good choice for UB $\left[|V_i \cap V_j|\right]$ because the stability measure could attain its maximum value for sets $V_i \subsetneq V_j$ or $V_j \subsetneq V_i$. To avoid it, UB $\left[|V_i \cap V_j|\right]$ must depend on both $|V_i|$ and $|V_j|$. Possible choices are for example $\frac{|V_i|+|V_j|}{2}$ or $\sqrt{|V_i| \cdot |V_j|}$. These choices are upper bounds for $|V_i \cap V_j|$ and they are met if and only if $V_i = V_j$. For $|V_i| \neq |V_j|$, the bounds differ and $\min\{|V_i|, |V_j|\} \leq \sqrt{|V_i| \cdot |V_j|} \leq \frac{|V_i| + |V_j|}{2}$ holds which makes $\sqrt{|V_i| \cdot |V_j|}$ more suitable. Therefore, $\mathrm{UB}\left[|V_i \cap V_j|\right] = \sqrt{|V_i| \cdot |V_j|}$ is used in this paper. If there are no similar features in the data set, SMA is identical to SMU, independent of the choice of adjustment. Four different adjustments are considered. We first define them and then give explanations for their construction.

$$\mathrm{Adj}_{\mathrm{MBM}}(V_i, V_j) = \text{size of maximum bipartite matching } (V_i \setminus V_j, V_j \setminus V_i)$$
$$\mathrm{Adj}_{\mathrm{Greedy}}(V_i, V_j) = \text{greedy choice of most similar pairs of features}$$
$$\text{determined by Algorithm 1 introduced on page 7}$$
$$\mathrm{Adj}_{\mathrm{Count}}(V_i, V_j) = \min\{A(V_i, V_j), A(V_j, V_i)\} \text{ with } A \text{ as defined for SMY}$$
$$\mathrm{Adj}_{\mathrm{Mean}}(V_i, V_j) = \min\{M(V_i, V_j), M(V_j, V_i)\} \text{ with}$$

$$M(V_i, V_j) = \sum_{x \in V_i \setminus V_j : |G_x^{ij}| > 0} \frac{1}{|G_x^{ij}|} \sum_{y \in G_x^{ij}} \text{similarity}(x, y) \text{ and}$$

$$G_x^{ij} = \{y \in V_j \setminus V_i : \text{similarity}(x, y) \geq \theta\}$$

The resulting four variants of SMA are named SMA-MBM, SMA-Greedy, SMA-Count and SMA-Mean. For the adjustment of SMA-MBM, a graph is constructed. In this graph, each feature of $(V_i \setminus V_j) \cup (V_j \setminus V_i)$ is represented by a vertex. Vertices $x \in V_i \setminus V_j$ and $y \in V_j \setminus V_i$ are connected by an edge, if and only if similarity$(x, y) \geq \theta$. An edge in the graph means that the corresponding features of the two connected vertices should be seen as exchangeable for stability assessment. A matching of a graph is defined as a subset of its edges such that none of the edges share a vertex [12, p. 63]. A maximum matching is a matching that contains as many edges as possible. The size of the maximum matching is the number of edges that are included in the maximum matching. The size of the maximum matching can be interpreted as the maximum number of features in $V_i \setminus V_j$ and $V_j \setminus V_i$ that should be seen as exchangeable for stability assessment with the restriction that each feature in $V_i \setminus V_j$ may only be seen as exchangeable with at most one feature in $V_j \setminus V_i$ and vice versa. There are no edges between vertices that both correspond to features of $V_i \setminus V_j$ or $V_j \setminus V_i$, so the graph is bipartite [12, p. 17]. For the calculation of a maximum matching for a bipartite graph, there exist specific algorithms [5].

The calculation of the maximum bipartite matching has the complexity $\mathcal{O}((\text{number of vertices} + \text{number of edges}) \cdot \sqrt{\text{number of vertices}})$ [5] and hence can be very time consuming. Therefore, a new greedy algorithm for choosing the most similar pairs of features is introduced in Algorithm 1. It is used to calculate the adjustment in SMA-Greedy. The return value of the algorithm is always smaller than or equal to the size of the maximum bipartite matching of the corresponding graph. The computational complexity of the algorithm is dominated by the sorting of the edges and hence is $\mathcal{O}(\text{number of edges} \cdot \log(\text{number of edges}))$.

```
1  size = 0
2  L_A = [X, Y, S] = list of tuples x ∈ V_i \ V_j, y ∈ V_j \ V_i, similarity(x, y) with
   similarity(x, y) ≥ θ, sorted decreasingly by similarity values
3  L_B = empty list
4  while length of L_A > 0 do
5    |  [x, y, s] = first tuple of L_A
6    |  add [x, y, s] to L_B
7    |  remove all tuples in L_A that contain x or y
8  end
9  return length of L_B
```

Algorithm 1: Greedy choice of the most similar pairs of features.

For SMA-Count, $A(V_i, V_j)$ is the number of features in V_i, that are not in V_j but that have a similar feature in $V_j \setminus V_i$. The minimum of $A(V_i, V_j)$ and $A(V_j, V_i)$ is used in order to guarantee that the adjusted score for V_i and V_j cannot exceed the value of $|\widetilde{V}_i \cap \widetilde{V}_j|$ that would be obtained if two sets $\widetilde{V}_i$ and $\widetilde{V}_j$ with $|\widetilde{V}_i| = |V_i|$ and $|\widetilde{V}_j| = |V_j|$ were chosen such that their overlap is maximal. $\min\{A(V_i, V_j), A(V_j, V_i)\}$ is always larger than or equal to the size of the maximum bipartite matching.

The adjustment of SMA-Mean is very similar to the one of SMA-Count. While $A(V_i, V_j)$ counts the number of features in $V_i \setminus V_j$, that have a similar feature in $V_j \setminus V_i$, $M(V_i, V_j)$ sums up the mean similarity values of the features in $V_i \setminus V_j$ to their similar features in $V_j \setminus V_i$. If there are no similarity values of features in $V_i \setminus V_j$ and $V_j \setminus V_i$ in the interval $[\theta, 1)$, the adjustments of SMA-Count and SMA-Mean are identical. Otherwise, the adjustment of SMA-Mean is smaller than the adjustment of SMA-Count.

3   Experiments and Results

The adjusted stability measures SMZ, SMY, SMA-Count, SMA-Mean, SMA-Greedy and SMA-MBM are compared to each other and to the unadjusted measure SMU. All calculations have been performed with the software R [11] using the package *stabm* [2] for calculating the stability measures and *batchtools* [8] for conducting the experiments on a high performance compute cluster.

3.1   Experimental Results on Artificial Feature Sets

First, a comparison in a situation with only 7 features is conducted. The advantage of this comparison is that all possible combinations of 2 subsets of features can be analyzed, as there are only $2^7 \cdot 2^7 = 16\,384$ possible combinations. For the adjusted and corrected measures SMY, SMA-Count, SMA-Mean, SMA-Greedy and SMA-MBM, the expected values of the pairwise scores are calculated exactly by considering all possible pairs of sets of the same cardinalities. The values of all stability measures presented in Subsect. 2.2 are calculated for all 16 384 possible combinations of 2 feature sets being selected from a total number of 7 features. Figure 1 displays the similarities between the 7 features used for this analysis. The threshold θ is set to $\theta = 0.9$, so there are 3 groups of similar features. Note that the similarity matrix is sufficient for calculating the stability measure for all pairs of possible combinations of 2 feature sets.

0.1	0.1	0.1	0.1	0.1	0.95	1
0.1	0.1	0.1	0.1	0.1	1	0.95
0.1	0.1	0.1	0.95	1	0.1	0.1
0.1	0.1	0.1	1	0.95	0.1	0.1
0.95	0.95	1	0.1	0.1	0.1	0.1
0.95	1	0.95	0.1	0.1	0.1	0.1
1	0.95	0.95	0.1	0.1	0.1	0.1

Similarity ≥ 0.9   No   Yes

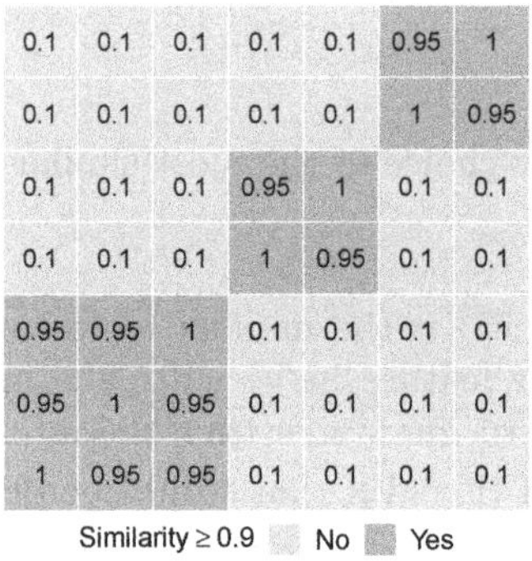

Fig. 1. Similarity matrix for the 7 features. Similarity values must be in $[0, 1]$.

To compare all stability measures, in Fig. 2, scatter plots of all pairs of stability measures are shown.   All adjusted measures differ strongly from the unadjusted stability measure SMU with respect to their stability assessment behavior. The adjusted stability measure SMZ, which is the only considered measure that is not corrected for chance, also differs strongly from all other stability measures. This demonstrates, that the missing correction has a large impact on the stability assessment behavior. SMA-Count, SMA-Mean, SMA-Greedy and SMA-MBM have almost identical values for all combinations. The values assigned by SMY and by the SMA variants are also quite similar. However, for combinations that obtain comparably large stability values by all of these measures, SMY often attains larger values than the SMA measures. These are combinations for which several features from the one set are mapped to the same feature of the other set, see the discussion in Subsect. 2.2. This undesired behavior of SMY occurs for large stability values. This is problematic because large stability values are what an optimizer is searching for when fitting models in a multi-criteria fashion taking into account the feature selection stability [3].

3.2   Experimental Results on Real Feature Sets

Now, the stability measures are compared based on feature sets that are selected for four real data sets with correlated features (OpenML [14] IDs 851, 41 163,

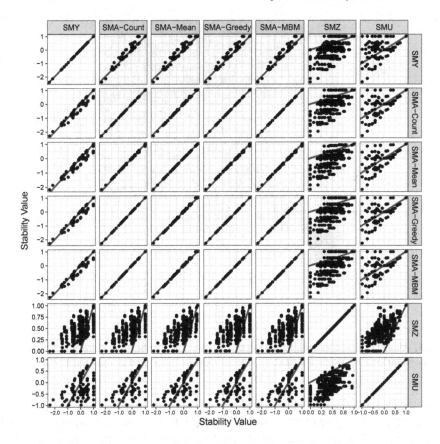

Fig. 2. Scatter plots of the stability values for the 16 384 combinations and all seven stability measures. The line in each plot indicates the identity.

1 484 and 1 458) with feature selection methods. The details of the feature selections are omitted here due to space constraints. Also, the focus is on the evaluation of the stability based on realistic feature sets resulting from real applications. To assess the similarity between features, the absolute Pearson correlation is employed for all adjusted stability measures. The threshold θ is set to $\theta = 0.9$ because in many fields, an absolute correlation of 0.9 or more is interpreted as a "strong" or even "very strong" association. For the adjusted and corrected measures SMY, SMA-Count, SMA-Mean, SMA-Greedy and SMA-MBM, the expected values of the pairwise scores are estimated based on $N = 10\,000$ replications. This value for N is suggested in [16] and has shown to provide a good compromise between convergence and run time in preliminary studies.

To analyze the similarities between the stability measures, Pearson correlations between all pairs of stability measures are calculated and averaged across data sets by calculating the arithmetic mean. Figure 3 displays the results. The adjusted and uncorrected stability measure SMZ differs most strongly from all

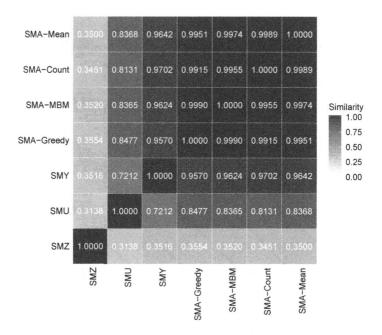

Fig. 3. Mean Pearson correlations between all pairs of the seven stability measures. The correlations between the stability measures are averaged across data sets.

other stability measures. The adjusted and corrected measures SMY, SMA-Count, SMA-Mean, SMA-Greedy and SMA-MBM assess the stability almost identically. The corrected and unadjusted measure SMU is more similar to this group than to SMZ. Here, SMU is much more similar to the corrected and adjusted stability measures SMY, SMA-Count, SMA-Mean, SMA-Greedy and SMA-MBM than in the previous subsection. The reason is that the real data sets considered here contain fewer similar features in comparison to the total number of features than in the artificial example in the previous subsection.

Now, the run times of the stability measures for realistic feature sets are compared. For SMU and SMZ, the run time is not an issue. For all of the considered data sets, they can be computed in less than one second. Figure 4 displays the run times for calculating the values of the adjusted and corrected stability measures. The run times of these measures are much longer than the run times of SMU and SMZ. The reason is that the expected values of the scores have to be estimated, which involves frequently repeated evaluation of the adjustments. For all data sets, SMY and SMA-Count require the least time for calculation among the adjusted and corrected measures. For most data sets, the calculation of SMA-Mean, SMA-Greedy and SMA-MBM takes much longer. For large data sets, the latter computation times are not acceptable.

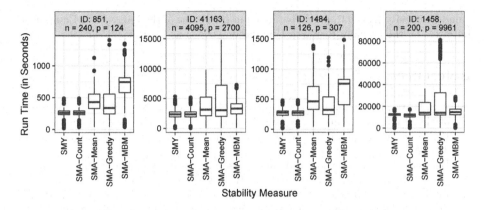

Fig. 4. Run times of the adjusted and corrected stability measures for four real data sets. n: number of observations, p: number of features.

4 Conclusions

For data sets with similar features, for example data sets with highly correlated features, the evaluation of feature selection stability is difficult. The commonly used stability measures consider features that are almost identical but have different identifiers as different features. This, however, is not desired because almost the same information is captured by the respective sets of features.

We have introduced and investigated new stability measures that take into account similarities between features ("adjusted" stability measures). We have compared them to existing stability measures based on both artificial and real sets of selected features. For the existing stability measures, drawbacks were explained and demonstrated.

For the newly proposed adjusted stability measure SMA, four variants were considered: SMA-Count, SMA-Mean, SMA-Greedy and SMA-MBM. They differ in the way they take into account similar features when evaluating the stability. Even though the adjustments for similar features are conceptually different for the four variants, the results are very similar both on artificial and on real sets of selected features. With respect to run time, the variant SMA-Count outperformed the others. Therefore, we conclude that SMA-Count should be used when evaluating the feature selection stability for data sets with similar features.

A promising future strategy is to employ SMA-Count when searching for models with high predictive accuracy, a small number of chosen features and a stable feature selection for data sets with similar features. To reach this goal, one can perform multi-criteria hyperparameter tuning with respect to the three criteria and assess the stability with SMA-Count.

Acknowledgements. This work was supported by German Research Foundation (DFG), Project RA 870/7-1 and Collaborative Research Center SFB 876, A3. We acknowledge the computing time provided on the Linux HPC cluster at TU Dortmund University (LiDO3), partially funded in the course of the Large-Scale Equipment Initiative by the German Research Foundation (DFG) as Project 271512359.

References

1. Awada, W., Khoshgoftaar, T.M., Dittman, D., Wald, R., Napolitano, A.: A review of the stability of feature selection techniques for bioinformatics data. In: 2012 IEEE International Conference on Information Reuse and Integration, pp. 356–363 (2012)
2. Bommert, A.: stabm: stability measures for feature selection (2019). https://CRAN.R-project.org/package=stabm, R package version 1.1.0
3. Bommert, A., Rahnenführer, J., Lang, M.: A multicriteria approach to find predictive and sparse models with stable feature selection for high-dimensional data. Comput. Math. Methods Med. **2017**, 7907163 (2017)
4. He, Z., Yu, W.: Stable feature selection for biomarker discovery. Comput. Biol. Chem. **34**(4), 215–225 (2010)
5. Hopcroft, J.E., Karp, R.M.: An $n^{5/2}$ algorithm for maximum matchings in bipartite graphs. SIAM J. Comput. **2**(4), 225–231 (1973)
6. Jaccard, P.: Étude comparative de la distribution florale dans une portion des Alpes et du Jura. Bulletin de la Société Vaudoise des Sciences Naturelles **37**, 547–579 (1901)
7. Kalousis, A., Prados, J., Hilario, M.: Stability of feature selection algorithms: a study on high-dimensional spaces. Knowl. Inf. Syst. **12**(1), 95–116 (2007)
8. Lang, M., Bischl, B., Surmann, D.: Batchtools: tools for R to work on batch systems. J. Open Source Softw. **2**(10), 135 (2017)
9. Lausser, L., Müssel, C., Maucher, M., Kestler, H.A.: Measuring and visualizing the stability of biomarker selection techniques. Comput. Statistics **28**(1), 51–65 (2013)
10. Nogueira, S.: Quantifying the Stability of Feature Selection. Ph.D. thesis, University of Manchester, United Kingdom (2018)
11. R Core Team: R: A Language and Environment for Statistical Computing. R Foundation for Statistical Computing, Vienna, Austria (2018). https://www.R-project.org/
12. Rahman, M.S.: Special classes of graphs. Basic Graph Theory. UTCS, pp. 111–133. Springer, Cham (2017). https://doi.org/10.1007/978-3-319-49475-3_9
13. Statnikov, A., Lytkin, N.I., Lemeire, J., Aliferis, C.F.: Algorithms for discovery of multiple markov boundaries. J. Mach. Learn. Res. **14**, 499–566 (2013)
14. Vanschoren, J., Van Rijn, J.N., Bischl, B., Torgo, L.: OpenML: Networked science in machine learning. ACM SIGKDD Explor. Newslett. **15**(2), 49–60 (2013)
15. Yu, L., Han, Y., Berens, M.E.: Stable gene selection from microarray data via sample weighting. IEEE/ACM Trans. Comput. Biol. Bioinf. **9**(1), 262–272 (2012)
16. Zhang, M., et al.: Evaluating reproducibility of differential expression discoveries in microarray studies by considering correlated molecular changes. Bioinformatics **25**(13), 1662–1668 (2009)
17. Zucknick, M., Richardson, S., Stronach, E.A.: Comparing the characteristics of gene expression profiles derived by univariate and multivariate classification methods. Stat. Appl. Genet. Molecular Biol. **7**(1), 7 (2008)

Reliable Solution of Multidimensional Stochastic Problems Using Metamodels

Marius Bommert[(✉)] and Günter Rudolph

TU Dortmund University, 44221 Dortmund, Germany
`marius.bommert@tu-dortmund.de`

Abstract. Multidimensional stochastic objective functions can be optimized by finding values in decision space for which the expected output is optimal and the uncertainty is minimal. We investigate the optimization of expensive stochastic black box functions $f : \mathbb{R}^a \times \mathbb{R}^b \to \mathbb{R}$ with controllable parameter $x \in \mathbb{R}^a$ and a b-dimensional random variable C. To estimate the expectation $\mathrm{E}(f(x,C))$ and the standard deviation $\mathrm{S}(f(x,C))$ as a measure of uncertainty with few evaluations of f, we use a metamodel of f. We compare an integration and a sampling approach for the estimation of $\mathrm{E}(f(x,C))$ and $\mathrm{S}(f(x,C))$. With the sampling approach, the runtime is much lower at almost no loss of quality.

Keywords: Optimization under uncertainty · Stochastic black box function · Metamodel

1 Introduction

Many optimization problems are influenced by random effects. The objective functions of such problems are stochastic. Stochastic functions can be denoted as $f(x,C)$ where x is a controllable parameter vector and C a random variable. In our previous work [3], we considered the case of expensive stochastic black box functions $f : \mathbb{R} \times \mathbb{R} \to \mathbb{R}$ with $x \in \mathbb{R}$ and C being one-dimensional. In this paper, we extend this work to higher dimensions of x and C. Independent of the dimensions of x and C, the goal is finding x-values for which f is optimal.

The evaluation of f in any point x does not yield a deterministic output. Instead, $f(x,C)$ is a random variable whose distribution depends on x. The expectation $\mathrm{E}(f(x,C))$ provides information about the central location of the distribution when f is evaluated in x. It describes the expected output of $f(x,C)$. If f is evaluated in x many times, the mean output value will be close to $\mathrm{E}(f(x,C))$ (law of large numbers). A single evaluation of f in x does not necessarily yield a value close to $\mathrm{E}(f(x,C))$. The variance $\mathrm{V}(f(x,C))$ gives the expected quadratic deviation of an evaluation of $f(x,C)$ from $\mathrm{E}(f(x,C))$. It

---

The source code for the experiments and analyses of this paper is publicly available at https://github.com/mariusbommert/LOD2020.

© Springer Nature Switzerland AG 2020
G. Nicosia et al. (Eds.): LOD 2020, LNCS 12565, pp. 215–226, 2020.
https://doi.org/10.1007/978-3-030-64583-0_20

is a measure of the spread of the distribution and can be used to assess the uncertainty about how far $f(x, C)$ will likely deviate from $E(f(x, C))$. As the variance uses a quadratic scale, the standard deviation $S(f(x, C)) = \sqrt{V(f(x, C))}$ is often used instead.

We optimize f by simultaneously optimizing the expectation $E(f(x, C))$ and the standard deviation $S(f(x, C))$. Depending on the context, $E(f(x, C))$ is minimized or maximized. $S(f(x, C))$ is always minimized to achieve small uncertainty. So, in order to optimize the single-objective stochastic function f, we optimize the biobjective deterministic function $(E(f(x, C)), S(f(x, C)))$:

$$E(f(x, C)) \to \min/\max! \quad \text{and} \quad S(f(x, C)) \to \min!.$$

Optimizing a stochastic black box function by maximizing the expectation and minimizing a risk measure is a common strategy in portfolio optimization. The risk is often assessed by the variance of f or other domain specific risk measures like the value at risk. The mean-variance model introduced by Markowitz [14] is a very popular approach in this field. For this model, expectation and variance are scalarized into a single objective function and then optimized. Portfolio optimization has also been analyzed as a multi-objective problem, see for example [15]. An overview of methods for optimizing stochastic functions is given in [9]. One approach for the optimization of stochastic functions presented in [9] also includes the simultaneous optimization of the expectation and a risk measure like the standard deviation. Methods for estimating expectation and standard deviation are not suggested by the authors.

Another field of research which deals with the optimization of stochastic functions is stochastic programming [11]. In this field, the expectation $E(f(x, C))$ is optimized for a stochastic function f. One approach for estimating $E(f(x, C))$ is sample average approximation [11]. For sample average approximation, samples $c_1, \ldots, c_n$ from C are drawn and the arithmetic mean of $f(x, c_i)$ is calculated. In contrast to our setting, f usually is known analytically and not an expensive function in this field.

In the field of optimizing stochastic black box functions f, many contributions exist. But, to the best of our knowledge, none of them optimize $E(f(x, C))$ and $S(f(x, C))$ simultaneously. Huang et al. [10] optimize the expectation $E(f(x, C))$ of a stochastic black box function f, using sequential Kriging optimization. Swamy [18] optimize the expectation $E(f(x, C))$ of a stochastic black box function f while controlling a risk measure using a probability constraint. In contrast to our work, both [10] and [18] do not optimize the standard deviation $S(f(x, C))$. In [10], the randomness in f is assumed to be additive noise, so $S(f(x, C))$ is assumed to be constant with respect to x. In [18], the risk measure is used for constraining the set of x-values for the optimization of $E(f(x, C))$.

In our previous work [3], we have shown that $E(f(x, C))$ and $S(f(x, C))$ can be estimated using a metamodel for f and numerical integration. This approach for estimating $E(f(x, C))$ and $S(f(x, C))$ is able to outperform the standard approach: moment estimation of $E(f(x, C))$ and $S(f(x, C))$. With our approach, more accurate estimates are obtained. Additionally, with our approach, it is possible to estimate $E(f(x, C))$ and $S(f(x, C))$ for any value $x \in \mathbb{R}$, while with the standard approach, estimates are only possible if f has been evaluated at least twice in the x-value of interest.

Analogously to our previous work [3], we assume that C is observable and that the class of its distribution is known. The assumption of C being observable holds in many scenarios. For example, in the newsvendor problem [1], which we consider in this paper, C describes the random demand for products. This demand can be measured even if it exceeds the supply by counting how many times the vendor is asked for the products after they have gone out of stock.

In this paper, we demonstrate that our approach [3] also works for the multi-dimensional setting. Also, we improve the runtime of our approach with almost no loss of quality. We achieve this by using sampling instead of numerical integration when estimating $E(f(x, C))$ and $S(f(x, C))$ based on a metamodel for f. This sampling approach is similar to the sample average approximation used in stochastic programming. In both sampling and sample average approximation, samples of C are drawn to estimate $E(f(x, C))$ or both $E(f(x, C))$ and $S(f(x, C))$, respectively. In contrast to sample average approximation, we use a metamodel of f to assess $f(x, c_i)$, because in our setting, f is a stochastic black box function.

The remainder of this article is organized as follows: In Sect. 2, basic concepts and methods are described. In Sect. 3, the estimation of $E(f(x, C))$ and $S(f(x, C))$ using a metamodel and numerical integration or sampling is explained. The design of our experiments for comparing the two approaches is presented in Sect. 4. In Sect. 5, the approaches are compared with respect to their quality and runtime. Concluding remarks and future work are presented in Sect. 6.

2   Concepts and Methods

In the following, Kriging, Pareto optimality and attainment functions are explained.

2.1   Kriging

Let $x_1, \ldots, x_n$ denote n points which are evaluated with a deterministic function f. Kriging is an interpolating method to build a metamodel $\hat{f}$ of f using the n given points. For that purpose, it is assumed that $f(x_1), \ldots, f(x_n)$ are realizations of a Gaussian random field. Numerical optimization is performed to fit the model to the data. For more information on Kriging see [17].

2.2   Pareto Set and Pareto Frontier

Let $f : X \to \mathbb{R}^m, f(x) = (f_1(x), \ldots, f_m(x))$ denote an objective function where all components should be minimized. $x \in X$ weakly dominates $y \in X$ (notation: $x \preceq y$ or $f(x) \preceq f(y)$) iff $\forall i \in \{1, \ldots, m\} : f_i(x) \leq f_i(y)$. The point x dominates the point y iff it weakly dominates it and $\exists i \in \{1, \ldots, m\} : f_i(x) < f_i(y)$. $x^\star \in X$ is Pareto optimal if there is no $x \in X$ which dominates it. The set of all Pareto optimal points is called Pareto set and the corresponding image is the Pareto frontier. For more information on Pareto frontiers and Pareto sets see [5].

2.3   Attainment Function

Let $\mathcal{X}_1, \ldots, \mathcal{X}_r$ denote r approximations of the same Pareto frontier resulting from r optimization runs. Each approximation can be seen as a realization of a random non dominated point set $\mathcal{X}^\star$. $\mathcal{X}^\star = \{X_1^\star, \ldots, X_N^\star\}$ is a random set of vectors in $\mathbb{R}^m$. The attainment function allows analyzing the distribution of this random set with respect to its location. A point $z \in \mathbb{R}^m$ is attained by the set $\mathcal{X}^\star$ iff

$$X_1^\star \preceq z \vee \ldots \vee X_N^\star \preceq z =: \mathcal{X}^\star \trianglelefteq z.$$

So, a point is attained by a set if at least one element of the set weakly dominates the point. For each $z \in \mathbb{R}^m$ the attainment function is defined as the probability that z is attained by $\mathcal{X}^\star$:

$$a(z) = P(\mathcal{X}^\star \trianglelefteq z).$$

The empirical attainment function estimates the attainment function. It is defined as

$$e(\mathcal{X}_1, \ldots, \mathcal{X}_r; z) = \frac{1}{r} \sum_{i=1}^{r} I(\mathcal{X}_i \trianglelefteq z)$$

with $I(\cdot) : \mathbb{R}^m \mapsto \{0, 1\}$ denoting the indicator function and $\mathcal{X}_1, \ldots, \mathcal{X}_r$ the r approximations of the true Pareto frontier.

For visualizing the empirical attainment function in the two dimensional case, contour lines are plotted. These lines display the tightest set of points which are attained for a given percentage of the r approximations of the true Pareto frontier. This and further information on the attainment function can be found in [7] and [8].

3   Estimating Expectation and Standard Deviation with Metamodels

In the following, methods for estimating $E(f(x, C))$ and $S(f(x, C))$ using a metamodel are explained.

3.1   General Idea

Let $f(x, C)$ be a stochastic function where $x = (x_1, \ldots, x_a) \in \mathbb{R}^a$ is a controllable parameter and C is a b-dimensional random variable. Our aim is optimizing $f(x, C)$ by simultaneous minimization of the expectation

$$E(f(x, C)) = \int_{-\infty}^{\infty} \cdots \int_{-\infty}^{\infty} f(x, c_1, \ldots, c_b) \cdot p_C(c_1, \ldots, c_b) dc_1 \ldots dc_b$$

and the standard deviation $S(f(x, C)) = \sqrt{V(f(x, C))}$ with

$$V(f(x, C)) = \int_{-\infty}^{\infty} \ldots \int_{-\infty}^{\infty} (f(x, c_1, \ldots, c_b) - E(f(x, C)))^2 \cdot p_C(c_1, \ldots c_b) dc_1 \ldots dc_b.$$

$p_C(c_1, \ldots, c_b)$ denotes the probability density function of C at the point $c = (c_1, \ldots, c_b)$. It is assumed that the distribution of C is continuous. For a discrete distribution, a summation over the support of C is needed instead of the integration. If the expectation should be maximized, this maximization problem can be transformed into a minimization problem by multiplying the expectation with -1.

3.2   Estimation of Expectation and Standard Deviation

For the estimation of $E(f(x, C))$ and $S(f(x, C))$, two different approaches are used. The first one is a generalization of the approach presented in [3]. We make this approach applicable in situations with a multidimensional parameter x and a multidimensional random variable C. The following integrals are calculated:

$$\hat{E}(f(x, C)) = \int_{l_b}^{u_b} \ldots \int_{l_1}^{u_1} \hat{f}(x, c_1, \ldots, c_b) \cdot \hat{p}_C(c_1, \ldots c_b) dc_1 \ldots dc_b$$

and $\hat{S}(f(x, C)) = \sqrt{\hat{V}(f(x, C))}$ with

$$\hat{V}(f(x, C)) = \int_{l_b}^{u_b} \ldots \int_{l_1}^{u_1} (\hat{f}(x, c_1, \ldots, c_b) - \hat{E}(f(x, C)))^2 \cdot \hat{p}_C(c_1, \ldots, c_b) dc_1 \ldots dc_b.$$

$l_1, \ldots, l_b$ and $u_1, \ldots, u_b$ denote the lower and upper integration limits and $\hat{p}_C$ the estimated probability density function of C. It is assumed that C is observable so that the probability density function p_C can be estimated. To build the metamodel $\hat{f}$, a data set with n values for x, c and $f(x, c)$ is needed. For the x-values, box-constraints $[x_{l_1}, x_{u_1}] \times \ldots \times [x_{l_a}, x_{u_a}]$ are used. These bounds are necessary to specify where to evaluate f and $\hat{f}$. The n values of $x \in [x_{l_1}, x_{u_1}] \times \ldots \times [x_{l_a}, x_{u_a}] \subset \mathbb{R}^a$ can be chosen in a controlled and smart way whereas the values $c_1, \ldots, c_n$ are given as realizations of the b-dimensional random variable C and hence cannot be chosen as desired.

The second method is estimating $E(f(x, C))$ and $S(f(x, C))$ by drawing samples $\tilde{c}_i = (\tilde{c}_{i1}, \ldots, \tilde{c}_{ib}), i = 1, \ldots, k$, of C and calculating the moment estimates

$$\hat{E}(f(x, C)) = \frac{1}{k} \sum_{i=1}^{k} \hat{f}(x, \tilde{c}_i)$$

and

$$\hat{S}(f(x,C)) = \sqrt{\frac{1}{k-1} \sum_{i=1}^{k} (\hat{f}(x,\tilde{c}_i) - \hat{E}(f(x,C)))^2}$$

with metamodel $\hat{f}$ as before. The parameters of the distribution are estimated using the given realizations $c_1, \ldots, c_n$. Then, the samples $\tilde{c}_1, \ldots, \tilde{c}_k$ are drawn from this estimated distribution.

4   Experiments

In this section, we describe the design of our experiments for analyzing and comparing the two approaches for estimating $E(f(x,C))$ and $S(f(x,C))$. Also, we explain some computational aspects.

4.1   Design of Experiments

The experiments are constructed as follows: We investigate an objective function $f : \mathbb{R}^2 \times \mathbb{R}^2 \to \mathbb{R}$. n x-values in the interval $[x_{l_1}, x_{u_1}] \times [x_{l_2}, x_{u_2}]$ are chosen using an improved latin hypercube sampling [2]. For n, the values $50, 100, 200, 500$ and $1\,000$ are used. For each x-value, a realization c of the two-dimensional random variable C is drawn. The objective function f is evaluated in each pair of $(x_i, c_i), i = 1, \ldots, n$. A Kriging model $\hat{f}$ is fitted to the generated data set. This model is used as metamodel for f. $E(f(x,C))$ and $S(f(x,C))$ are estimated with the methods *integration* and *sampling*. For *sampling*, $10\,000$ samples are drawn. The parameters of the distribution of C are estimated with moment estimates.

In this paper, we consider an inexpensive objective function in order to be able to perform a large number of experiments. However, we limit the budget of function evaluations like it is common for expensive black box optimization. The objective function that we consider is a two-dimensional generalization of the newsvendor problem [1]. The one-dimensional newsvendor problem is motivated by selling some product like newspapers which loses value very fast. It is not possible to sell the newspapers after a short period of time because they are outdated then. Given a stock of $x \in \mathbb{R}$ newspapers and a random demand C, the minimum of x and C is sold for a price p. The cost for purchase or production for x products is qx. So, the profit is given as

$$f(x,C) = p\min(x,C) - qx$$

with $p > q$. This version of the newsvendor model is generalized to the two-dimensional case in the following way:

$$f(x,C) = f(x_1, x_2, C_1, C_2) = p\min(x_1, C_1) - qx_1 + r\min(x_2, C_2) - sx_2.$$

This objective function describes the profit made based on two products with correlated demand. We use $p = 5, q = 3, r = 4, s = 2, x_1 \in [0,100]$ and

$x_2 \in [20, 120]$. For modeling the demand C, we use a two-dimensional normal distribution

$$C \sim \mathcal{N}\left(\begin{pmatrix} 50 \\ 70 \end{pmatrix}, \begin{pmatrix} 25 & -\frac{25}{3} \\ -\frac{25}{3} & 25 \end{pmatrix}\right).$$

The Pearson correlation in this scenario is $\mathrm{corr}(C_1, C_2) = -\frac{1}{3}$, so there is a weak negative association between C_1 and C_2. The profit is maximized and the uncertainty minimized. Therefore, $-\mathrm{E}(f(x, C))$ and $\mathrm{S}(f(x, C))$ are minimized. $\mathrm{E}(f(x, C))$ and $\mathrm{S}(f(x, C))$ are evaluated in a grid of the values $x_1 \in \{0, 1, \ldots, 100\}$ and $x_2 \in \{20, 21, \ldots, 120\}$ which results in 10 201 grid points. Figure 1 displays the true values $-\mathrm{E}(f(x, C))$ and $\mathrm{S}(f(x, C))$ for the considered objective function. The Pareto frontier is convex.

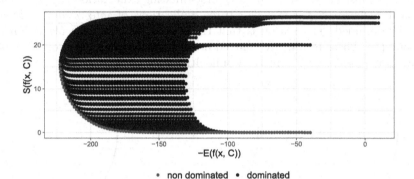

Fig. 1. Plot of $-\mathrm{E}(f(x, C))$ and $\mathrm{S}(f(x, C))$ for the considered objective function calculated in the 10 201 grid points. The Pareto frontier is displayed in red. (Color figure online)

To assess the quality of the approaches, $\hat{\mathrm{E}}(f(x, C))$ and $\hat{\mathrm{S}}(f(x, C))$ are evaluated in the grid points. To measure the overall quality of the estimation of $\mathrm{E}(f(x, C))$ and $\mathrm{S}(f(x, C))$, we calculate the mean Euclidean distance between the points $(\mathrm{E}(f(x, C)), \mathrm{S}(f(x, C)))$ and $(\hat{\mathrm{E}}(f(x, C)), \hat{\mathrm{S}}(f(x, C)))$ for all 10 201 grid points in which $\hat{\mathrm{E}}(f(x, C))$ and $\hat{\mathrm{S}}(f(x, C))$ have been evaluated:

$$\frac{1}{10201} \sum_{i=1}^{10201} \left\| (\mathrm{E}(f(x_i, C)), \mathrm{S}(f(x_i, C))) - (\hat{\mathrm{E}}(f(x_i, C)), \hat{\mathrm{S}}(f(x_i, C))) \right\|_2.$$

Small values are desirable. Because the results of the experiments are influenced by random effects, we repeat the experiments for each scenario 100 times.

Beyond the scope of this paper, we performed experiments with further covariance matrices. For these experiments, we obtained almost identical results.

4.2 Computational Aspects

For our experiments, we use the software R [16]. For generating a latin hypercube sampling the package lhs [6] is employed. As metamodel we apply Kriging which is implemented in the package `DiceKriging` [17]. For numerical integration the standard method `integrate` is used. `integrate` performs adaptive quadrature based on the QUADPACK routine dqags. To determine the Pareto frontiers, the package `ecr` [4] is employed. For generating the empirical attainment functions, the package `eaf` [13] is used. For performing the experiments on a high performance compute cluster, we rely on the package `batchtools` [12].

5   Evaluation of the Experiments

In this section, we analyze the quality of our approach for optimizing multidimensional stochastic black box functions based on metamodels. Also, we compare the methods *integration* and *sampling* for estimating expectation $E(f(x,C))$ and standard deviation $S(f(x,C))$. For three replications of *integration*, no results could be obtained. The reason is that the standard integration algorithm did not converge. These results are omitted in the following evaluation of the experiments. For *sampling*, no numerical difficulties were encountered.

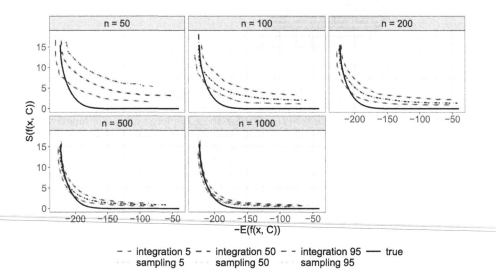

Fig. 2. Empirical attainment functions for the 5th, 50th and 95th percentile for *integration* and *sampling*. true: true Pareto frontier, *n*: number of observations for building the metamodel. (Color figure online)

First, we analyze how well our approach works in the multidimensional setting. Figure 2 displays empirical attainment functions for the optimization of $E(f(x,C))$ and $S(f(x,C))$. Separate attainment functions are shown for the methods *integration* and *sampling* and for the different numbers of observations for building the metamodels. The 5th, 50th and 95th percentile lines as well as the true Pareto frontier are displayed. The Pareto frontier estimates approach the

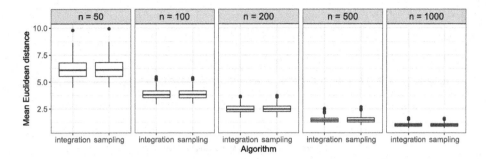

Fig. 3. Mean Euclidean distance between true and estimated expectation and standard deviation. n: number of observations for building the metamodel.

true Pareto frontier for increasing n. The more observations are used for creating the metamodel, the better estimates of the true Pareto frontier are obtained. Also, the 5th and 95th percentile lines, whose distance measures the variation of the Pareto frontier estimates within the 100 repetitions, are closer for increasing n. The more observations the metamodels are based on, the less variation there is in the estimated Pareto frontiers. For $n = 1\,000$, the estimated Pareto frontiers are quite close to the true Pareto frontier. In order to obtain Pareto frontier estimates with sufficient quality, metamodels based on at least $n = 500$ observations are required. The reason is that a Kriging model is only accurate if enough observations are used for building it. In our previous work [3], we found out that 100 observations are required for obtaining good results for one-dimensional x and C. Extrapolating from this, $n = 10\,000$ observations could be needed for a similar quality of results in the scenario of two-dimensional x and C. Because we consider expensive black-box functions, $10\,000$ function evaluations are far too many. Also, building Kriging models and predicting values of f is very time consuming for such a high number of points because the computational complexity of Kriging is $\mathcal{O}(n^3)$ [17]. Figure 3 shows the mean Euclidean distance between $(\mathrm{E}(f(x, C)), \mathrm{S}(f(x, C)))$ and $(\hat{\mathrm{E}}(f(x, C)), \hat{\mathrm{S}}(f(x, C)))$ for all $10\,201$ grid points x in which $\hat{\mathrm{E}}(f(x, C)$ and $\hat{\mathrm{S}}(f(x, C))$ have been evaluated. The mean Euclidean distance decreases when n increases. This means that the estimation of $\mathrm{E}(f(x, C))$ and $\mathrm{S}(f(x, C))$ improves not only for Pareto optimal x-values but for all x-values. In conclusion, a high quality of results can be achieved also if the function f has multidimensional input values x and C but more evaluations of f are needed than in the scenario of one-dimensional x and C.

Now, we compare the two methods *integration* and *sampling*. In Fig. 2, the empirical attainment functions for *integration* and *sampling* are almost identical. This means that both methods approximate the true Pareto frontier with nearly the same quality. Figure 4 displays the differences between the mean Euclidean distance for *integration* and *sampling*. More precisely, for each metamodel, the mean Euclidean distance obtained with *sampling* is substracted from the mean Euclidean distance obtained with *integration*. Because the same metamodels are used for both methods, the differences in mean Euclidean distance are only due

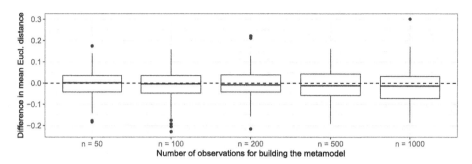

Fig. 4. Difference of mean Euclidean distance between true and estimated expectation and standard deviation between *integration* and *sampling*. n: number of observations for building the metamodel. Values greater than 0 mean that *sampling* is better and values smaller than 0 mean that *integration* is better.

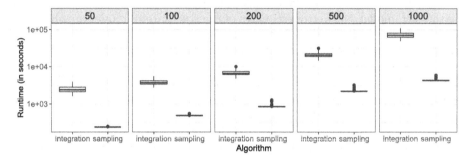

Fig. 5. Runtime for the estimation of $E(f(x, C))$ and $S(f(x, C))$ in seconds. n: number of observations for building the metamodel. The evaluation in the grid points is parallelized on 10 cores.

to the methods for estimating $E(f(x, C))$ and $S(f(x, C))$. The median values of the differences are close to zero and the boxes are small for all values of n. So, the overall quality of the estimates of $E(f(x, C))$ and $S(f(x, C))$ is almost identical for *integration* and *sampling* and never differs much. The quality of *integration* is a little bit higher, especially for $n = 1\,000$. But the deviations of the median values from zero are very small compared to the variation of differences in mean Euclidean distance, so they can be neglected. In conclusion, the mean Euclidean distance for *integration* and *sampling* is almost identical.

Figure 5 displays the runtimes of *integration* and *sampling*. The runtimes state the time needed for estimating $E(f(x, C))$ and $S(f(x, C))$ based on an existing metamodel $\hat{f}$. The runtimes are assessed on a high performance compute cluster with 10 cores in parallel. For both methods, the runtime increases for increasing n, because more complex Kriging models are used then. Estimating $E(f(x, C))$ and $S(f(x, C))$ with *integration* needs much more time than the estimation with *sampling*. In conclusion, the quality of both approaches is almost identical, but the runtime for *sampling* is much lower than for *integration*.

6   Conclusion

We analyzed the optimization of expensive stochastic black box functions $f(x, C)$ with $x \in \mathbb{R}^a$ and a b-dimensional random variable C. We transformed the problem of optimizing the stochastic function f into the biobjective optimization of the deterministic expectation $E(f(x, C))$ and standard deviation $S(f(x, C))$.

We estimate $E(f(x, C))$ and $S(f(x, C))$ by building a metamodel for f and using numerical integration or sampling. Compared to the standard approach of moment estimation of $E(f(x, C))$ and $S(f(x, C))$, our approach provides the advantage that $E(f(x, C))$ and $S(f(x, C))$ can be estimated in any x-value, regardless if f has been evaluated in x or not. Our approach enables us to estimate $E(f(x, C))$ and $S(f(x, C))$ with a high quality and yields a good approximation of the true Pareto frontier at the cost of requiring that C is observable.

In our previous work [3], we assumed that $a = b = 1$. In this paper, we generalized the approach to higher dimensions. We conducted a simulation study and found out that with the generalized approach, a good quality of results is obtained. Also, we presented another modification to the approach, using sampling instead of integration for estimating $E(f(x, C))$ and $S(f(x, C))$. With this modification, the runtime was heavily reduced while the quality of the estimates remained nearly the same. So, with the modifications presented in this paper, our approach of optimizing f is now applicable for higher dimensions of x and C, it is faster and it still provides a high quality of results.

Having heavily reduced the runtime of our approach in this paper enables us to consider even higher dimensions of x and C in future research. Another interesting aspect for future research is looking at the scenario that the distribution class of C is unknown.

Acknowledgments. The authors gratefully acknowledge the computing time provided on the Linux HPC cluster at TU Dortmund University (LiDO3), partially funded in the course of the Large-Scale Equipment Initiative by the German Research Foundation (DFG) as project 271512359.

References

1. Arrow, K.J., Harris, T., Marschak, J.: Optimal inventory policy. Econometrica **19**(3), 250–272 (1951). https://www.jstor.org/stable/1906813
2. Beachkofski, B.K., Grandhi, R.V.: Improved distributed hypercube sampling. In: 43rd AIAA/ASME/ASCE/AHS/ASC Structures, Structural Dynamics, and Materials Conference, p. 1274 (2002). https://doi.org/10.2514/6.2002-1274
3. Bommert, M., Rudolph, G.: Reliable biobjective solution of stochastic problems using metamodels. In: Deb, K., Goodman, E., Coello Coello, C.A., Klamroth, K., Miettinen, K., Mostaghim, S., Reed, P. (eds.) Evolutionary Multi-Criterion Optimization, pp. 581–592. Springer International Publishing, Cham (2019). https://doi.org/10.1007/978-3-030-12598-1_46
4. Bossek, J.: ecr 2.0: A modular framework for evolutionary computation in R. In: Proceedings of the Genetic and Evolutionary Computation Conference Companion, pp. 1187–1193. ACM (2017). https://doi.org/10.1145/3067695.3082470

5. Branke, J., Deb, K., Miettinen, K., Słowiński, R. (eds.): Multiobjective Optimization. LNCS, vol. 5252. Springer, Heidelberg (2008). https://doi.org/10.1007/978-3-540-88908-3

6. Carnell, R.: lhs: Latin Hypercube Samples (2019). https://CRAN.R-project.org/package=lhs, R package version 1.0.1

7. Fonseca, C.M., Guerreiro, A.P., López-Ibáñez, M., Paquete, L.: On the computation of the empirical attainment function. In: Takahashi, R.H.C., Deb, K., Wanner, E.F., Greco, S. (eds.) Evolutionary Multi-Criterion Optimization, pp. 106–120. Springer, Berlin, Heidelberg (2011). https://doi.org/10.1007/978-3-642-19893-9_8

8. Grunert da Fonseca, V., Fonseca, C.M., Hall, A.O.: Inferential performance assessment of stochastic optimisers and the attainment function. In: Zitzler, E., Deb, K., Thiele, L., Coello Coello, C.A., Corne, D. (eds.) Evolutionary Multi-Criterion Optimization, pp. 213–225. Springer, Berlin, Heidelberg (2001). https://doi.org/10.1007/3-540-44719-9_15

9. Gutjahr, W.J., Pichler, A.: Stochastic multi-objective optimization: a survey on non-scalarizing methods. Ann. Oper. Res. **236**(2), 475–499 (2013). https://doi.org/10.1007/s10479-013-1369-5

10. Huang, D., Allen, T.T., Notz, W.I., Zeng, N.: Global optimization of stochastic black-box systems via sequential kriging meta-models. J. Global Optim. **34**(3), 441–466 (2006). https://doi.org/10.1007/s10898-005-2454-3

11. Kleywegt, A.J., Shapiro, A., Homem-de-Mello, T.: The sample average approximation method for stochastic discrete optimization. SIAM J. Optim. **12**(2), 479–502 (2002). https://doi.org/10.1137/S1052623499363220

12. Lang, M., Bischl, B., Surmann, D.: batchtools: Tools for R to work on batch systems. J. Open Source Softw. **2**(10) (2017). https://doi.org/10.21105/joss.00135

13. López-Ibáñez, M., Paquete, L., Stützle, T.: Exploratory analysis of stochastic local search algorithms in biobjective optimization. In: Bartz-Beielstein, T., Chiarandini, M., Paquete, L., Preuss, M. (eds.) Experimental Methods for the Analysis of Optimization Algorithms, pp. 209–222. Springer, Heidelberg (2010). https://doi.org/10.1007/978-3-642-02538-9_9

14. Markowitz, H.: Portfolio selection. J. Finan. **7**(1), 77–91 (1952). https://doi.org/10.1111/j.1540-6261.1952.tb01525.x

15. Ponsich, A., Jaimes, A.L., Coello Coello, C.A.: A survey on multiobjective evolutionary algorithms for the solution of the portfolio optimization problem and other finance and economics applications. IEEE Trans. Evolution. Comput. **17**(3), 321–344 (2013). https://doi.org/10.1109/TEVC.2012.2196800

16. R Core Team: R: A Language and Environment for Statistical Computing. R Foundation for Statistical Computing, Vienna, Austria (2019). https://www.R-project.org/

17. Roustant, O., Ginsbourger, D., Deville, Y.: DiceKriging, DiceOptim: two R packages for the analysis of computer experiments by kriging-based metamodeling and optimization. J. Stat. Softw. **51**(1), 1–55 (2012). https://doi.org/10.18637/jss.v051.i01

18. Swamy, C.: Risk-averse stochastic optimization: probabilistically-constrained models and algorithms for black-box distributions. In: Proceedings of the twenty-second annual ACM-SIAM symposium on Discrete Algorithms, pp. 1627–1646. SIAM (2011). https://doi.org/10.1137/1.9781611973082.126

Understanding Production Process Productivity in the Glass Container Industry: A Big Data Approach

Maria Alexandra Oliveira$^{(\boxtimes)}$, Luís Guimarães$^{(\boxtimes)}$ ⓘ, and José Luís Borges$^{(\boxtimes)}$

Faculty of Engineering, University of Porto, 4200-465 Porto, Portugal
{alexandra.oliveira,lguimaraes,jlborges}@fe.up.pt

Abstract. It is becoming increasingly important to take advantage of Big Data in order to be able to understand industrial processes and improve their efficiency and effectiveness. This work presents an application on a glass container manufacturing plant, to detect and characterize patterns in the efficiency of the production process. Besides the inherent complexity of the pattern discovery task, the challenge is increased by the multivariate time series nature of the data. The main goal of this project, other than minimizing production losses, creating knowledge from data and therefore improving the company's overall efficiency, aims to contribute to literature in describing patterns on an univariate time series leveraging multivariate time series data, specially in manufacturing applications.

Keywords: Pattern discovery · Glass container production · Machine learning

1 Introduction

The current global market is marked by an extremely high competitiveness, caused mainly by the increase in supply and customer's expectations. Faced by this scenario, companies must strive to be more efficient and effective, in doing so quality and productivity play a major role. The capability to improve both quality and productivity of the delivered products or processes, will dictate the company's position in the market. Such effort can be measured by the enterprise's capacity to keep up with the evolution of the industry. With the rise of Industry 4.0, it became clear for organizations the importance of analysing the vast amount of data generated over the years by their core processes. The analysis of these large-size collections, known as Big Data, will enable enterprises to create knowledge from these collections and thus, gain insights to improve the quality and consequently the yield of their delivered products/services.

In fact, the availability of such amounts of data from various sources have been contributing to research in data mining (DM) applications. DM has been

---

Supported by FEUP-PRIME program and BA GLASS PORTUGAL.

G. Nicosia et al. (Eds.): LOD 2020, LNCS 12565, pp. 227–232, 2020.
https://doi.org/10.1007/978-3-030-64583-0_21

defined as a set of algorithms and techniques able to extract nontrivial information from large databases, with the goal of discovering trends, hidden patterns and relationships in the data. These techniques that DM resorts to, are a combination of machine learning, statistics and patterns recognition tools that allow to analyse large datasets [4]. In the particular context of manufacturing, the employment of DM tools has been gaining popularity for being able to discover interesting patterns and actionable knowledge, that can be subsequently exploited to improve a panoply of related areas such as, defect prevention and detection, quality control, production and decision support systems. [6]. The adoption of these techniques at the expense of more traditional multivariate statistical methods, is highly motivated by the inherent complexity and variability of production processes, the existence of non-linear relationships, the involvement of several factors and process stages with variable duration [8].

Manufacturing data is often available as time series, but the use of classical time series analysis methods, like autoregressive integrated moving average (ARIMA), becomes impractical due to not only the size and nature of the available datasets, but also due to the kind of knowledge that it is intended to extract or the information that is expected to estimate. For these reasons, efforts have been directed into the development of research in temporal data mining techniques. This variation of DM presents tools to deal with large sequential datasets and its the major advantage when compared with more conventional approaches for analysing time series data, is that its application is not limited to predictive and control tasks. Additionally, temporal data mining methods provide a more exploratory analysis since in many applications the correlations and causal relationships between variables are unknown. Besides, these techniques also allow to detect, very often unexpected, trends or patterns in the data that once interpreted, might be of great usefulness [4].

Pattern discovery is the most common task of temporal data mining, which consists of finding frequent and unknown patterns in a time-series without any previous information about their shapes and location [7]. This task is also known by motif discovery and anomaly detection. To address such task, distance-based clustering methods are one of the most preferred. However, the process of pattern discovery can be seriously compromised when it is needed to analyse large collections of historical data. Therefore, other techniques capable of dealing with this issue have been also adopted to find trends in a time series. [2] apply self-organizing Maps (SOM) to discover patterns from stock time-series and [5] presents an application in molecular biology using fuzzy c-means (FCM).

Motivated by a real-world case, the focus of this work is on detecting and explaining patterns in time series data from a glass container manufacturer. At this company, data has been collected and stored over the years, resulting in large amounts of datasets containing valuable information about the processes and products. Yet, these data are kept in silos corresponding to the different production stages. Up to now, few attempts have been made to either integrate all parts, or to correlate the information within an integrated approach. The dispersiveness of data increases the difficulty in replicating the best historical production cycles, and creates challenges in understanding quality defects causes. Bearing this in mind, we present a pattern discovery approach to identify trends in the process's production efficiency, based on the variables intervening in the glass container production. The algorithm is then tested on unseen data, in order to assess its performance and infer if it is suitable for monitoring and controlling tasks. Additionally, we also present a explainable model capable of identifying the root cause and explaining the patterns by assigning responsibilities to variables. Resulting from the application of the proposed work, we expect to decrease production losses, increase the availability of equipment, reduce the dependency on worker's experience, speed up the reaction to decreasing trends and provide an understanding of the causes of the identified patterns.

The work's contribution is twofold. First, we demonstrate a practical application on glass container industry, where little research can be found (see [1] and [3]), specially in the realm of pattern discovery. Secondly, it brings value to the company which acquires expertise on the factors responsible for the process's states and dexterity in the decision-making regarding where to intervene. As a result, the organization will be able to better dominate their processes and will be placed in a more favorable market positioning. This know-how can be further disseminated on the same industry and the methodology is extensible enough to receive other sources of data.

2   The Glass Container Production Process

The manufacturing process begins with the mixing of raw materials in the batch house. The mixture is then transported into the furnace where is melted up to 1500°C. After leaving the furnace, the liquefied glass goes through a conditioning process in order to assure the thermal homogeneity of the paste. Then, this glass paste ends its path in the feeders, where it is cut into gobs and distributed to a set of parallel independent section (IS) machines. Here, the container is formed by a molding process. After being given a shape, the container receives a heat treatment to reduce glass stress, followed by a coating treatment. These two steps are performed during the annealing process. The containers are then subjected to a strict quality control, performed by automated inspection machines capable of rejecting defective containers. Once they have been approved in the quality control process, the containers are packed on pallets at the end of the production lines. Figure 1 illustrates the manufacturing process stages.

Fig. 1. Glass manufacturing process.

3 Methodology

The main driver of the proposed methodology are the data we receive from the different production stages depicted in Fig. 1. For instance, between the batch house and feeders stages we have available information on the temperatures applied to homogenize the liquefied glass, as well as, the pressures of the valves releasing cold or hot air. Regarding IS Machines, the data concerns closing/opening timings of the grasp, cooling pressures and temperatures of the moulds, as well as maintenance interventions. In the inspection phase, we may find the total number of rejected and good containers at each timestamp, the type of defects detected and the corresponding section where the defective container came from.

Given the context in which this work is introduced, the complexity of the manufacturing process and the multivariate nature of the time series data, a great part of this work is dedicated to an exhaustive understanding of the process and variables involved, in order to conduct a proper integrated approach. Furthermore, as the dataset is directly built from raw data, many challenges are addressed before inputting the information into the pattern discovery algorithm:

- Based on the provided data, calculate the target variable to detect patterns on that better represents the process losses. In this procedure, attention must be paid to the temporal relationships.
- Handling missing data, since besides affecting the temporal dependencies, it is also difficult to extrapolate values for several variables.
- Selecting the variables relevant for the identification of the patterns of interest, considering that we intend to improve the algorithms' performance while not loosing information.

The processed data feeds the pattern discovery model, whose main function is to identify different trends during the production process. To perform this task, the series must be first divided resorting to a time-window. Subsequently, the resulting segments can be compared by applying a clustering method which allows to discriminate the different trends present in the time-series. As the segments are part of a time series from a process with a great variability and are of different sizes, the clustering algorithm is supported by a suitable similarity metric for this purpose. Once the clusters are well-defined, a classification model is trained to classify them based on the input variables (the variables intervening in the process) and tested on unseen data. The main purpose of the classification model is to allow explaining the clusters leveraging the input variables, so the knowledge extract from it can be generalized. The characterization of the different clusters, which represent the various patterns found in the time-series, explains mainly trends of different directions, either increasing or decreasing. Hence, the insights developed from this characterization will help to better directed the efforts, since the root cause of a particular pattern is identified. Figure 2 illustrates the proposed approach.

Fig. 2. Adopted methodology.

4   On Going Work

The experiments done so far, involve the production in one line and one product type during one month. Nevertheless, as we extend the period of the analysis, more than one product may be considered. In Fig. 3 are depicted two versions of the production efficiency evolution during one month. Figure 3a demonstrates the original time series. As we may observe from it, the process presents some fluctuations due to its inherent variability. For this reason, in order to better distinguish the trends and to facilitate the clustering task, the original time-series was smoothed using the Locally Weighted Scatterplot Smoothing (LOWESS) algorithm, applying a 3% of smoothing factor. Figure 3b illustrates this smoothed series where we can also distinguish the different patterns present in it, resulting

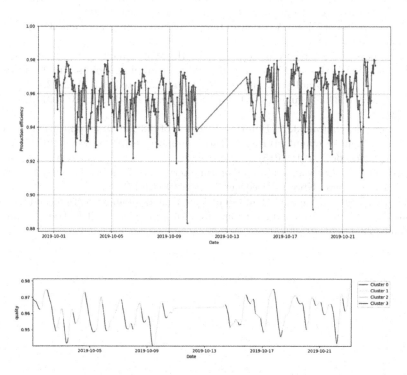

Fig. 3. (a) Original time series representing the production efficiency during one month. (b) Smoothed time-series highlighted by four different clusters.

from the application of the clustering algorithm. The straight line in both plots represents three days in a row without records.

The classification problem consists of classifying four classes based on the input variables. Besides, the multi-classification problem is very imbalanced due to the classes' distribution: 36.5% of class 0, 49% of class 1, 6.9% of class 2 and 7.6% of class 3. From the preliminary results, with 78% of accuracy, we could conclude that the variables impacting the most on the production efficiency are essentially the temperatures applied to the liquefied glass and the IS machines parameters.

References

1. Awaj, Y.M., Singh, A.P., Amedie, W.Y.: Quality improvement using statistical process control tools in glass bottles manufacturing company. Int. J. Quality Res. **7**(1), 107–126 (2013)
2. Fu, T.c., Chung, F., Ng, V., Luk, R.: Pattern discovery from stock time series using self-organizing maps. In: Workshop Notes of KDD2001 Workshop on Temporal Data Mining, pp. 26–29. Citeseer (2001)
3. Kuhnert, C., Bernard, T.: Extraction of optimal control patterns in industrial batch processes based on support vector machines. In: 2009 IEEE Control Applications, (CCA) & Intelligent Control, (ISIC), pp. 481–486. IEEE (2009)
4. Laxman, S., Sastry, P.S.: A survey of temporal data mining. Sadhana **31**(2), 173–198 (2006)
5. Möller-Levet, C.S., Klawonn, F., Cho, K.-H., Wolkenhauer, O.: Fuzzy clustering of short time-series and unevenly distributed sampling points. In: R. Berthold, M., Lenz, H.-J., Bradley, E., Kruse, R., Borgelt, C. (eds.) IDA 2003. LNCS, vol. 2810, pp. 330–340. Springer, Heidelberg (2003). https://doi.org/10.1007/978-3-540-45231-7_31
6. Rostami, H., Dantan, J.Y., Homri, L.: Review of data mining applications for quality assessment in manufacturing industry: support vector machines. Int. J. Metrol. Quality Eng. **6**(4), 401 (2015)
7. Torkamani, S., Lohweg, V.: Survey on time series motif discovery. Wiley Interdiscip. Rev. Data Mining Knowl. Discovery **7**(2), e1199 (2017)
8. Zhou, Z., Wen, C., Yang, C.: Fault detection using random projections and k-nearest neighbor rule for semiconductor manufacturing processes. IEEE Trans. Semicond. Manuf. **28**(1), 70–79 (2015)

Random Forest Parameterization for Earthquake Catalog Generation

David Llácer[1], Beatriz Otero[1(✉)], Rubén Tous[1], Marisol Monterrubio-Velasco[2], José Carlos Carrasco-Jiménez[2], and Otilio Rojas[2,3]

[1] Universitat Politécnica de Catalunya, Barcelona, Spain
botero@ac.upc.edu
[2] Barcelona Supercomputing Center, Barcelona, Spain
[3] Universidad Central de Venezuela, Facultad de Ciencias, Caracas, Venezuela

Abstract. An earthquake is the vibration pattern of the Earth's crust induced by the sliding of geological faults. They are usually recorded for later studies. However, strong earthquakes are rare, small-magnitude events may pass unnoticed and monitoring networks are limited in number and efficiency. Thus, earthquake catalog are incomplete and scarce, and researchers have developed simulators of such catalogs. In this work, we start from synthetic catalogs generated with the TREMOL-3D software. TREMOL-3D is a stochastic-based method to produce earthquake catalogs with different statistical patterns, depending on certain input parameters that mimics physical parameters. When an appropriate set of parameters are used, TREMOL-3D could generate synthetic catalogs with similar statistical properties observed in real catalogs. However, because of the size of the parameter space, a manual searching becomes unbearable. Therefore, aiming at increasing the efficiency of the parameter search, we here implement a Machine Learning approach based on Random Forest classification, for an automatic parameter screening. It has been implemented using the machine learning Python's library Sci-Kit Learn.

Keywords: Earthquakes · Synthetic catalogs · Machine learning · Random forest

1 Introduction

The study of earthquakes helps us to estimate their possible occurrence and observed magnitudes, and allow assessing potential actions to lessen their impact and terrible effects [3,10,12]. For instance, conventional seismic hazard analyses of a region start from a local earthquake catalog. Unfortunately, seismic catalogs cover a limited time span compared to some long recurrence earthquake intervals, and our monitoring and processing resources are still limited. Thus, these catalogs may present deficiencies. Earthquakes can be studied from either physical or statistical points of view [5]. The model proposed in [7], and named

© Springer Nature Switzerland AG 2020
G. Nicosia et al. (Eds.): LOD 2020, LNCS 12565, pp. 233–243, 2020.
https://doi.org/10.1007/978-3-030-64583-0_22

TREMOL, aims at generating synthetic earthquake catalogs by using the Fiber Bundle Model (FBM), which is a stochastic modeling technique that requires parameter tuning. Under appropiate parameterization, TREMOL reproduces an earthquake mainshock and the following aftershock sequence until the modeled seismicity ceases, as observed in nature. The collection of these events, with associated magnitudes, represents a TREMOL synthetic catalog. In [6], some machine learning (ML) techniques were developed to assist in the TREMOL screening parameter. These techniques use as a basis a real seismic catalog that record the full earthquake sequence and observed magnitudes, along with the epicentral and hipocentral coordinates, origin times, among other additional information. With this data, they fed a ML system implementing Support Vector Machine, Flexible Discriminant Analysis and Random Forest, in order to determine the best combination of TREMOL input parameters, and which statistical features of a catalog were more important to analyze synthetic and real sequences. In [6], the ML technique yielding the best results was Random Forest, that exhibits the lowest mean errors on the statistical features that allow comparing a synthetic to a real catalog. Thus, the work in [6] motivates our current implementation that extends the efforts for parameter screening with a more complex database. Following [6], we here use exclusively Random Forest to optimize TREMOL input parameters.

The main contribution of this work is the use of a more general dataset with respect to [6], and the development of a flexible and automatic tool based on the Random Forest classification technique for parameter searching. These topics are discussed in Sects. 2 and 3. Basically, the previously used data in [6] was generated with a TREMOL-2D version [7]. Simulations on this model take place on a 2D horizontal representation of the study region, where depth is omitted, and this fact may limit the modeling of some mainshock-aftershock sequences. A first 3D TREMOL prototype was developed and tested in [9], that shows potential improvements to solve complex mainshock-aftershock scenarios. Although, TREMOL-3D makes output seismicity more realistic, it also extends the number of parameters to be optimized, making more computationally demanding the parameter screening. Thus, in this work we start by developing a new Random Forest module for parameter screening, using as a basis the previous implementation in R language. We decide to switch to Python [11] for our new implementation, given that the new TREMOL-3D has also been programmed in Python and Julia [1]. Finally, we compare the TREMOL parameters found by an heuristic analysis in [9] with the automatically results found in this work. Details on the experimental framework and results are given in Sect. 4, and Sect. 5 states our final conclusions.

2   Data

Figure 1 shows some of the essential data recorded in a real seismic catalog. For each event, these data comprise occurrence time, hypocentral location, magnitude, among other seismological information. TREMOL models a mainshock-aftershock sequence and yields a similar catalog of synthetic earthquakes [9].

Table 1 lists some statistical parameters or features, associated to three fundamental empirical laws in Seismology, and commonly analyzed on earthquake catalogs [3,5]. The Gutenberg-Richter frequency-magnitude event distribution is given in terms of the linear regression parameters b_{value} and a_{value} values, and the largest and smallest magnitudes, M_{max} and M_{min}, respectively. The term M_{Bath} corresponds to the second largest event according to the Bath law, Tr_{max} estimates the rupture time of the mainshock, while the Omori's law predicts the aftershock decaying rate in terms of p_{Omori} and c_{Omori}. Among all these 8 features, c_{Omori}, given by the physical time span of the earthquake sequence, has been poorly predicted by TREMOL results in previous works [6,9]. The explanation may rely on the fact that TREMOL time formulation is dimensionless, and then c_{Omori} is low sensitive to input model parameters. The sensitivity and importance of remaining statistical features would highly depend on the modeled seismicity, and therefore, on the input TREMOL parameters.

Origin	Time	Lat. (N)	Lon. (E)	Depth (km)	Nsta	Gap	M_L	M_{L10}
01/10/95	07:51:16.12	23.66	121.41	15	12	170	5.23	4.77
02/23/95	05:19:12.15	24.22	121.66	18	25	139	6.33	5.60
03/24/95	04:13:52.14	24.62	121.86	79	24	180	5.81	5.16
04/03/95	11:54:47.98	23.95	122.31	16	25	230	5.92	5.30
04/24/95	10:04:01.52	24.65	121.65	63	28	60	5.44	5.05
06/25/95	06:59:08.74	24.58	121.69	43	36	84	6.50	5.89
07/07/95	03:04:48.29	23.88	121.10	15	28	49	5.61	5.08
07/14/95	16:52:47.89	24.36	121.76	6	23	178	5.70	4.98

Fig. 1. Example of a real earthquake catalog taken from [13].

Table 1. Relevant statistical features of an earthquake catalog and values associated to the 2010 El Mayor-Cucapah earthquake sequence.

Statistical features	Description	Value
b_{value}	Parameter of the Gutenberg-Richter law	0.96
M_{max}	The magnitude of the mainshock	7.2
M_{Bath}	Magnitude related to the Bath's law	5.7
Tr_{max}	Maximum rupture time duration	29
a_{value}	Parameter of the Gutenberg-Richter that is proportional to the seismicity rate	5.89
M_{min}	Minimum magnitude recorded in the synthetic catalogue	3
p_{Omori}	Exponent of the Omori-Utu law typically close to 10 It is a value that controls the decay rate of aftershock activity	1
c_{Omori}	Parameter of the Omori-Utsu law	0.5

In this work, we use as reference the real catalog corresponding to the Mw 7.2 *El Mayor-Cucapah* earthquake, a big event occurred in 2010 in Baja California,

Mexico, and whose descriptive statistical features are given on the last column of Table 1. The data is taken from [4], and stored in a *csv* file for our parameter screening implementation. On the other hand, Table 2 presents the input parameters of TREMOL, which are related to the initial conditions of the fault system and given in terms of load and roughness parameters. These parameters allow to configure the initial state and the TREMOL simulation environment. The statistical coherence of an output synthetic catalog is quantified in terms of the features in Table 1. Thus, a good input parameterization should allow TREMOL to produce a synthetic catalog with similar statistical features to the ones observed in the real catalog.

Table 2. Input paramaters of TREMOL.

Name	Description
S_a	Roughness size
$\vec{B}$	Vector that favors nucleation in Roughness
Θ	Load distribution percentage to diagonal cells
Π_{Asp}	Load distribution percentage of asperity cells
Π_{Faults}	Load distribution percentage of fault cells

3   Methodology

3.1   Data Processing

As real data hereafter we mean the 1D vector with the 8 statistical values associated to *El Mayor-Cucapah* earthquake in Table 1. A previous time consuming simulation stage based on TREMOL under a variety of input parameters with seismological significance, we obtain the training set for our parametric search engine. The resulting training set is given by 1680 shock simulations, each one with different combinations of input parameters. With this set, we train Random Forest and finally generate a prediction, that we later compare with the real data. As mentioned before, this comparison is performed in terms of the 8 statistical features that represent and aggregate all seismic events in a catalog. An example of this training data is given by the row shown in Table 3. As one can see, the training data is stored in a 1D vector with the same format as the real one, with an additional label. The first eight columns collect the statistical features for the candidate catalog, and the last column is a label representing the input parameter, that we have chosen to make the analysis for. For example, the label value given in Table 3 corresponds to the Roughness size (S_a) used in TREMOL simulations.

Table 3. Example of a training row data for Random Forest.

b_{value}	M_{Max}	M_{Bath}	Tr_{max}	a_{value}	M_{min}	p_{Omori}	c_{Omori}	$label_{input}$
0.94	7.0	5.1	32	5.46	3	1	0.5	5

3.2 Random Forest

Random Forest consists of a large number of individual decision trees that operate as an ensemble [2]. Decision trees are the building blocks of the Random Forest algorithm, which is a non-parametric supervised learning method used for classification or regression. Decision trees break down a dataset into smaller subsets, while an associated tree is incrementally developed in the process. The final result is a tree with decision nodes and leaf nodes. Each individual tree in the Random Forest splits a class prediction, and the class with the most votes becomes our model's prediction. The fundamental concept behind Random Forest is that a large number of trees, operating as a committee, will outperform any of the individual constituent models (see Fig. 2). The main reason for this attribute is that decision trees protect each other from their individual errors.

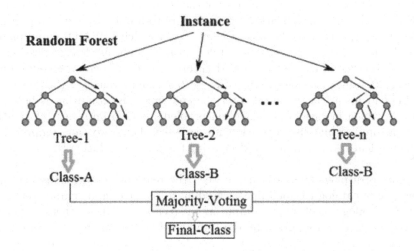

Fig. 2. Random Forest schema.

In this case we used a Random Forest classifier with 500 estimators. Random trees are also interesting because allows us to determine the importance of the features in order to see which parameters affects more when classifying. To compute the Random Forest, we used the Python library Sci-Kit Learn [8], an efficient group of tools for predictive data analysis. The input for the Random Forest algorithm is formed by two variables: a matrix (1680×8) with the eight statistical values of each row and the label (1680×1) we selected to make the analysis with one of the five labels. We fit the model with this data and generate the Random Forest. Then, with the real data, we obtain a prediction from the trained system, that is a label. So we take the rows of the synthetic data that have the same label as the prediction and calculate the mean, obtaining an approximation of the real data. We can see the different approximations depending on the input label we selected in Sect. 4.

4   Experiments

We perform several experiments according to the input parameter we want to analyze. Given that our methodology considers five different input parameters, we then use five labels in our dataset. We have trained the Random Forest algorithm with our entire dataset, and later test it using the real data. In this way, the algorithm outputted labels to establish the correspondence to real data. Finally, we take the mean column vector of the synthetic catalogs having this label as the definitive catalog. These outputs clearly depend on the considered input parameter. The accuracy of our simulation outputs is assessed through comparative plots between the synthetic and real catalogs, as shown in Fig. 3.

Furthermore, we can analyze the importance or sensitivity of any statistical feature with respect to each parameter of the Random Forest model, as given by Sci-Kit Learn. The quantification of this feature-to-input parameter importance is a key point of this study, because it allows to see which TREMOL input mainly affect the generation of a synthetic catalog with features, as close as possible, to the ones observed in the real case. In decision trees, every node is a condition of how to split values into a single feature, so that similar values of the dependent variable end up in the same set after the splitting. The condition is based on impurity, which in case of classification problems is Gini impurity/information gain (entropy), while the regression its variance. Thus, we can compute how much each feature contributes to decreasing the impurity, during the training of a tree.

Figure 3 displays comparisons between the real data (continuous line) and the synthetic catalog (dashed line) according to our five input parameters. The synthetic catalog was obtained computing the mean of all the synthetic data rows that contained the label given by the Random Forest algorithm when we fed it with the real data. In the X-axis we have the statistical features we mentioned in Sect. 2, and in the Y-axis we have their respective value. Furthermore, we can check the importance of each statistical feature with a parameter of the Random Forest model given by Sci-Kit Learn. Figure 4 plots this importance measure.

As we can observe, the most representative statistical feature is the b_{value}, followed by the M_{max} and the M_{Bath}. On the other hand, the feature with lower importance is the c_{Omori}, that actually has zero importance. We have to remark that the importance of each statistical feature was plotted for the input label that better approaches the real data, in this case the Π_{Asp} label.

The last step of our analysis, representing an important objective of this work, is knowing which are the input parameters of the earthquake simulator that

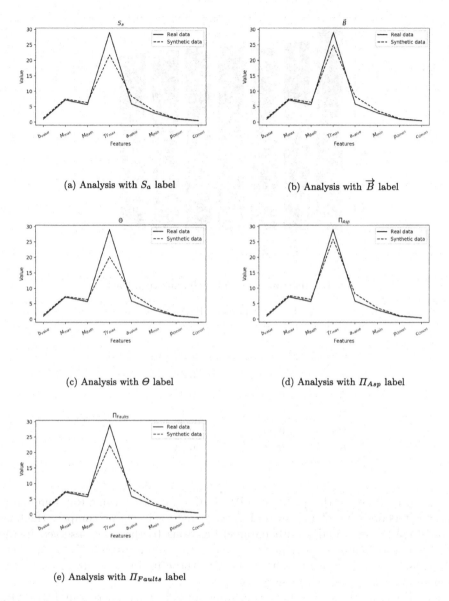

(a) Analysis with S_a label

(b) Analysis with $\vec{B}$ label

(c) Analysis with Θ label

(d) Analysis with Π_{Asp} label

(e) Analysis with Π_{Faults} label

Fig. 3. Comparison between real data and synthetic data based on each input parameter.

generates the best statistical catalog, with high similarity to the real data. To do so, we take the mean of the input parameters corresponding to the rows that contain the label predicted by the algorithm, and the resulting input parameters are shown in Table 4.

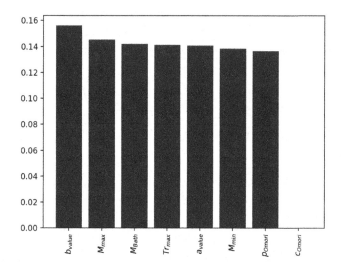

Fig. 4. Importance level of each statistical feature.

Table 4. Comparison of the Random Forest TREMOL parameters to the ones previously found by an heuristic analysis.

Parameter	Random Forest	Heuristic analysis
S_a	0.2	0.22
$\vec{B}$	0.7	0.9
Θ	0.1	0.1
Π_{Asp}	0.95	0.85
Π_{Faults}	0.84	0.75

In order to assess the quality of TREMOL results when using Random Forest parameters and those found by an heuristic analysis in [9], we compare the final statistical features under both parameterizations to the ones observed in the real mainshock-aftershock sequence. This comparison is given in Table 5, where differences on real features, with respect to those in Table 1, are given by the omission of the foreshock events.

In Table 5, we observe that using Random Forest parameters in TREMOL, the resulting statistical features improve with respect to the real ones. In particular, results are better for b_{value}, M_{Max}, Tr_{max}, and a_{value}. However, the remaining features stay close to those obtained by using as input parameters, the ones delivered by the heuristic analysis, with the exception of the last feature c_{Omori}.

It is worth noting that the heuristic analysis previously applied in [9] is not systematic and may take more time to arrive at similar parameter values than those provided by the Random Forest analysis. However, this heuristic analysis may also fail by delivering values far away from the optimal ones. The comparison

Table 5. The mean and standard deviation of the statistical features from 20 TREMOL executions considering both sets of input parameters given in Table 4. The real statistical features are also shown as reference.

Statistical feature	Real value	Heuristic mean	Heuristic deviation	Random Forest mean	Random Forest deviation
b_{value}	0.99	1.34	0.03	1.36	0.02
M_{max}	7.2	6.94	0.15	7.33	0.09
M_{Bath}	5.7	6.44	0.24	6.31	0.44
Tr_{max}	18.0	10.18	1.29	14.23	1.15
a_{value}	6.55	8.34	0.10	8.39	0.09
M_{min}	3.0	3.67	0.0	3.66	0.0
p_{Omori}	1.0	1.12	0.15	1.18	0.15
c_{Omori}	0.5	0.39	0.28	0.49	0.32

in Table 5 also allow the TREMOL sensitivity to the input parameters. Moreover, the differences between some real features and the corresponding synthetics help us to understand the possible improvements to our simulator. For example, in the case of M_{min} both sets of input parameters lead to larger values with respect to the real. This reflects the intrinsic TREMOL requirement of considering well refined meshes to be able to produce small magnitude earthquakes. Similarly, in the case of b_{value}, both TREMOL simulations shows higher values than the expected. These results could be also due to meshes too coarse, and then could be improved by higher mesh refinements. In summary, the collection results in Table 5 confirms that the Random Forest analysis is a proper tool to find TREMOL optimum parameters, and spend small amounts of computational time leading to higher accuracy than by heuristic analysis.

5   Conclusions

This work applies Random Forest classification for searching input parameters to the mainshock-aftershock simulator TREMOL. Under appropriate parameterization, TREMOL generates synthetic earthquake catalogs with a high statistical similarity to a real one. We also presents an importance measure of these input parameters on the statistical features that describe an earthquake catalog, and therefore serve as the similarity quantification. After training Random Forest classifier with different label values, where each label corresponding to an input TREMOL parameter, the label leading to the best approximation is Π_{Asp}. This label represents the percentage of the physical load distribution of asperity cells. Our results clearly support this important conclusion, in our study case of the Mw 7.2 *El Mayor-Cucapah* earthquake occurred in 2010.

We also would like to emphasize that during the Random Forest training, the most important feature was the b_{value}, followed by the M_{max}. Alternatively, we observed that the c_{Omori} feature has no a significant importance, compared to all other statistical features. Thus, this last feature might be omitted in future similar studies. Moreover, after comparing the heuristic values with those found in this work in our study case, we conclude that the Random Forest TREMOL parameters improved the similarity of the generated catalog with respect to the real one. These results also suggest that this ML parameterization, enable TREMOL for catalog generation with great similarities to other real seismicities.

As future works, we envision the development of a software package that embedded both the TREMOL simulator and the optimal parameter screening modules. In addition, we would like to explore the effectiveness of searching methods, such as Gradient Boosting or even Neural Networks, if the amount of simulation data is sufficient enough.

Acknowledgements. This work is partially supported by the Spanish Ministry of Economy and Competitivity under contract TIN2015-65316-P and by the Catalan Government through the programmes 2017-SGR-1414, 2017-SGR-962 and the RIS3CAT DRAC project 001-P-001723. Moreover, this project has received funding from the European Union's Horizon 2020 research and innovation programme under the Marie Sklodowska-Curie grant agreement No 777778 (MATHROCKS). The research leading to these results has received funding from the European Union's Horizon 2020 research and innovation programme under the ChEESE project, grant agreement No. 823844.

References

1. Bezanson, J., Edelman, A., Karpinski, S., Shah, V.B.: Julia: a fresh approach to numerical computing. SIAM Rev. **59**(1), 65–98 (2017). https://doi.org/10.1137/141000671
2. Breiman, L.: Random forests. Mach. Learn. **45**(1), 5–32 (2001). https://doi.org/10.1023/A:1010933404324
3. Burridge, R., Knopoff, L.: Model and theoretical seismicity. Bull. Seismol. Soc. Am. **57**(3), 341–371 (1967)
4. Hauksson, E., Stock, J., Hutton, K., Yang, W., Vidal-Villegas, J.A., Kanamor, H.: The 2010 mw 7.2 el Mayor-Cucapah earthquake sequence, Baja California, Mexico and Southernmost California, USA: Active seismotectonics along the Mexican Pacific margin. Pure Appl. Geophy. **168**, 1255–1277 (2011). https://doi.org/10.1007/s00024-010-0209-7
5. Kagan, Y.Y., Knopoff, L.: Stocastic synthesis of earthquake catalogs. J. Geophys. Res. Solid Earth **86**(B4), 2853–2862 (1981). https://doi.org/10.0129/JB086iB04p02853
6. Monterrubio-Velasco, M., Carrasco-Jiménez, J.C., Castillo-Reyes, O., Cucchietti, F., De la Puente, J.: A machine learning approach for parameter screening in earthquake simulation. In: 30th International Symposium on Computer Architecture and High Performance Computing, pp. 348–355 (2018). DOI: https://doi.org/10.1109/CAHPC.2018.8645865
7. Monterrubio-Velasco, M., Rodríguez-Pérez, Q., Zúñiga, R., Scholz, D., Aguilar-Meléndez, A., de la Puente, J.: A stochastic rupture earthquake code based on the fiber bundle model (TREMOL v0.1): application to mexican subduction earthquakes. Geosci. Model Dev. **12**(5), 1809–1831 (2019). https://doi.org/10.5194/gmd-12-1809-2019

8. Pedregosa, F., et al.: Scikit-learn: machine learning in Phyton. J. Mach. Learn. Res. **12**, 2825–2830 (2011)

9. Scholz, D.: Numerical simulations of stress transfer as a future alternative to classical Coulomb stress changes. Master's thesis, University College London, Department of Earth Sciences, London (2018)

10. Turcotte, D.L.: Seismicity and self-organized criticality. Phys. Earth Planet. Inter. **111**(3–4), 275–293 (1999). https://doi.org/10.1016/S0031-9201(98)00167-8

11. Van Rossum, G.: Python tutorial. Tech. Rep. CS-R9526, Centrum voor Wiskunde en Informatica (CWI), Amsterdam (1995)

12. Vázquez-Prada, M., González, A., Gómez, J.B., Pacheco, A.F.: A minimalist model of characteristic earthquakes. Nonlinear Process. Geophysi. **9**, 513–519 (2002). https://doi.org/10.5194/npg-9-513-2002

13. Wu, Y.M., Shin, T.C., Tsai, Y.B.: Quick and reliable determination of magnitude for seismic early warning. Bull. Seismol. Soc. Am. **88**(5), 1254–1259 (1998)

Convolutional Neural Network and Stochastic Variational Gaussian Process for Heating Load Forecasting

Federico Bianchi[(✉)], Pietro Tarocco, Alberto Castellini,
and Alessandro Farinelli

Department of Computer Science, Verona University,
Strada Le Grazie 15, 37134 Verona, Italy
{federico.bianchi,pietro.tarocco,alberto.castellini,
alessandro.farinelli}@univr.it

Abstract. Heating load forecasting is a key task for operational planning in district heating networks. In this work we present two advanced models for this purpose, namely a Convolutional Neural Network (CNN) and a Stochastic Variational Gaussian Process (SVGP). Both models are extensions of an autoregressive linear model available in the literature. The CNN outperforms the linear model in terms of 48-h prediction accuracy and its parameters are interpretable. The SVGP has performance comparable to the linear model but it intrinsically deals with prediction uncertainty, hence it provides both load estimations and confidence intervals. Models and performance are analyzed and compared on a real dataset of heating load collected in an Italian network.

Keywords: Heating load forecasting · Smart grids · Convolutional Neural Networks · Stochastic variational Gaussian processes · Model interpretability

1 Introduction

Energy management for smart grids gained strong interest from the artificial intelligence community [17]. A branch of smart grids concerns District Heating Networks (DHNs), centralized heating plants that provide heating to residential and commercial buildings through a network of pipes. In particular these measurements consider the temperature and the water flow rate. Accurate prediction of heating load plays a key role in energy production, supplying planning and energy saving, with economical and environmental benefits.

Data-driven forecasting [3–5] involves learning models of a variable of interest (i.e., the heating load in our case) from historical data of the same and other variables (e.g., meteorological or social factors) to predict future values of the variable of interest. Several methodologies are available in the literature for this

---

F. Bianchi and P. Tarocco—These authors contributed equally to this work.

© Springer Nature Switzerland AG 2020
G. Nicosia et al. (Eds.): LOD 2020, LNCS 12565, pp. 244–256, 2020.
https://doi.org/10.1007/978-3-030-64583-0_23

purpose. Autoregressive linear models [1,15] predict the target variable considering a linear combination of environmental and social factors (day of the week, calendar events). These models are usually simple to interpret but they have a quite rigid function form that can limit their performance in case of complex variable relationships. In [1] a multiple equation autoregressive linear approach is proposed, where the heating load of each pair (hour of the day, day of the week) is modeled independently, resulting in 168 equations.

Recurrent Neural Networks (RNN), in particular Long Short-Term Memory (LSTM) [10] and convolutional-LSTM [19], are among the most used methods for time series forecasting. The disadvantage of these models is that they require large training datasets to be learned and they are hardly interpretable. CNNs have been used for energy load forecasting and other problems related to time series forecasting [14]. What differentiates these works from ours is that we focus on the specific problem of heating load forecasting and provide a simple CNN model having good interpretability and better forecasting performance than available linear regression models.

Gaussian processes (GPs) [16] are other approaches used for time series forecasting. Their advantage is that they explicitly consider uncertainty in analyzed data, hence their predictions are equipped with both expected values and confidence intervals. On the other hand, exact learning of these models is very time consuming and unfeasible for large datasets, hence approximated training methods are used. Usage of GPs for time series analysis has recently gained interest [8,18], and applications to the energy forecasting domain are present in the literature [2,7,20]. The main differences between these approaches and our model is that we use stochastic variational Gaussian processes (SVGPs), which enable model training on a two-years dataset in few minutes using GPUs, and that our model uses specific variables for heating load forecasting.

In the following of this paper we propose two models for heating load forecasting, a CNN and a SVGP. Both models take inspiration from the autoregressive model proposed in [1] (see model called ARM_4) and extend it with specific features of CNN and SVGP, respectively. An analysis of the models is proposed with the aim to explain [6] how they extend the autoregressive model. Models are trained, tested and their performance are compared on a real dataset collected in a DHN located in Verona (Italy), containing hourly heating load produced by the plants in years 2016, 2017 and 2018. Novelties of this work are i) CNN outperforms a state-of-the-art autoregressive linear regression model, ii) SVGP has slightly lower performance but it provides useful confidence intervals on the prediction, iii) first step towards model interpretability. Comparison with other methods (e.g., LSTM and T-CNN) has been considered in some of our experiments but in this paper we only presented the two models with best results in terms of both performance and interpretability. The main contributions to the state-of-the-art are summarized in the following:

- a CNN and a SVGP model are proposed for heating load forecasting in DHNs;
- model parameters are analyzed and explained highlighting connections with parameters of the autoregressive linear model in the literature;
- model performance are tested and compared on a real-world dataset.

The rest of the manuscript is organized as follows. Section 2 presents the framework for model comparison, the dataset, the used methodologies and the performance measures. Result and performance are analyzed in Sect. 3. Conclusions and future directions are described in Sect. 4.

2   Materials and Methods

In this section we formalize the problem of heating load forecasting for DHNs, we describe the dataset composed by heating load and environmental variables. We finally introduce the modeling methodologies and performance measures we used to compare the models.

2.1   Problem Definition and System Overview

District heating networks are plants in which a power station, often through co-generation, produces heat and distributes it through a network of pipes connected to commercial and residential buildings. The heating load is collected by direct measurements performed in the plant. Forecasting methods are an important task for improving the process of planning, production and distribution of heating. In Fig. 1 an overview of data analysis framework is displayed. In the first phase, models are trained using real-world data. In the second phase models are tested on a different test set by performing all possible 48-h predictions of heating load. In the last phase model parameters and performance are analyzed and compared.

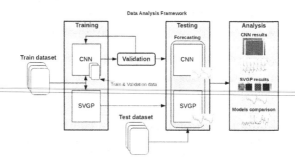

Fig. 1. Overview of the data analysis framework

2.2   Dataset

In the present work we use a real dataset provided by AGSM[1] an Italian utility company that manages a DHN in Verona. Data was collected from 01.01.2016

---

[1] https://www.agsm.it/.

to 21.04.2018 with hourly sampling interval including historical load l and fore-cast of weather variables like temperature T, relative humidity (R_H), rainfall (R), wind speed (W_S), wind direction (W_D). We first selected only observations belonging to time intervals in which the heating is on (this is regulated by law in intervals from 01.01.2016 to 11.05.2016, from 11.10.2016 to 14.05.2017, and from 16.10.2017 to 21.04.2018). We engineered new variables according to sim-ilar applications in the literature [15]. The complete list of variables is display in Table 1. Models are trained using data related to years 2016 and 2017 (10140 observations) and tested on 2018 data (2497 observations).

Table 1. List of variables used in the models.

Variable	Description	Variable	Description
l	Heating load [MW] (target)	W_S	Wind speed [m/s]
l_i	$i \in [1,7]$ Heating load i days ago	W_D	Wind direction [0,9], 9 = no wind
T	Temperature [°C]	R	Rainfall (1 = rain, 0 = no rain)
T^2	Square of T	H	Holiday (0 = false, 1 = true)
$T_{ma(7)}$	Moving avg of T last 7 days	h	Hour of the day [0, 23]
T_M	Maximum T of the day	d	Day of the week [1, 7]
T_M^2	Square of T_M of the day	w	Weekend (0 = false, 1 = true)
T_{Mp}	T_M of the previous day	Sa	Saturday (0 = false, 1 = true)
T_{Mp}^2	T_M^2 of previous day	Su	Sunday (0 = false, 1 = true)
R_H	Relative humidity [%]		

2.3 Convolutional Neural Network Model

CNNs are neural networks that use a linear mathematical operator called *con-volution* in at least one of their layers [10]. Each convolutional neuron takes two functions $x(t)$ and $k(t)$ as inputs and generates a new function $f(t)$ which is defined, when t is discrete, as $f(t) = (x * k)(t) = \sum_{i=-\infty}^{+\infty} x(i)k(t-i)$ where func-tion x is often referred to as *input*, function k as *kernel* (or *filter*) and function f as *feature map*. The kernel is learned by suitable algorithms to allow the net-work to approximate a function of interest, in our case study, the future values of heating load from past and present values of environmental and social factors. Working with time series data, we apply convolution over a single dimension, i.e., time t, hence our kernels are bidimensional matrices of parameters having one column for each input variable and one row for each time instant considered in the convolution.

The CNN presented in this work has a simple architecture which however outperforms the autoregressive linear model presented in [1], showing the strong capabilities of CNNs to forecast future values of heating load. The network archi-tecture is displayed in Fig. 2a. The input is a matrix having one column for each variable (22 columns in total) and one row for each time instant considered for predicting the heating load of the next hour (168 rows in total). The first layer is a CNN layer with five neurons. Each neuron performs a convolution of the input using a kernel of the same dimension of the input itself (i.e., 3697 weights are used in each kernel, bias included) and then applies a ReLU activation function (i.e., $ReLU(v) = max(v, 0)$) generating a single real value for each neuron. The five feature maps $f_0, \ldots, f_4$ thus obtained are then passed to a *dense* neuron, which computes their linear combination $y = w_0 f_0 + w_1 f_1 + w_2 f_2 + w_3 f_3 + w_4 f_4 + w_5$ where $w_i \in \mathbb{R}$ are the weights and the bias of the dense neuron. This operation can be seen as an extension of the linear model presented in [1] since that linear model has one autoregressive equation for each pair (weekday, hour), i.e., 168 equations with 20 variables for a total of 3360 parameters, while this CNN model is the composition of 5 convolutions made by kernels having a parameter for each pair *(variable, time instant)*, namely, 22 variables and 168 time instants for each of the 5 convolutions plus 6 parameters used to compose the convolutions, plus 6 biases, for a total of 18491 parameters. The CNN model is trained using a dataset containing weather variables T, R_H, W_S, W_D, R, T^2, $T_{ma(7)}$, T_M, T_M^2, T_{Mp}, T_{Mp}^2, historical heating load variables $l_i, 1 \le i \le 7$ and social factors H, h, d, w (see Table 1 for variable definitions).

Weight initialization for each layer is performed by *Xavier* normal initializer [9]. We trained the model using Keras[2], splitting further training dataset into train (6500 observations) and validation (3460 observations) to improve the model selection and avoid overfitting problem, typical of neural networks. The weights are learned by optimized gradient descent Adam [13], for 20 epochs using batch sizes of 32. Early stopping procedure monitors the training process, saving the best set of weights that minimize a loss function computed at the end of each epoch as mean squared error (MSE) over the validation dataset.

2.4   Stochastic Variational Gaussian Process Model

Let X be a finite set of input points $x_1, \ldots, x_n$ (they can be scalars or vectors), Gaussian processes assume the probability distribution of function values $p(f(x_1), \ldots, f(x_n))$ at those points to be jointly Gaussian, namely $\mathbf{x} \sim \mathcal{N}(\mathbf{m}, K)$ where matrix K is called the *covariance matrix* or *kernel* [16,18]. It has dimension $n \times n$, where n is the number of inputs of the training set. During model training the kernel matrix is filled in with covariance values between all possible pairs of inputs in the training sets. A key point of GP model design is the choice of kernels. We use *periodic kernels* $k_{\text{Periodic}}(x, x') = \sigma^2 \exp\left(-\frac{2\sin^2(\pi|x-x'|/p)}{\ell^2}\right)$ for modeling cyclical behaviors due to social factors and Radial Basis Function (RBF) kernels $k_{\text{RBF}}(x, x') = \sigma^2 \exp\left(-\frac{(x-x')^2}{2\ell^2}\right)$ for environmental factors, such as temperature. The parameters of periodic kernel are the output variance σ, the period length p and the length scale l. Those of RBF kernels are the output variance σ and the length scale l.

The posterior distribution of the function values on the testing locations f_* (i.e. load predictions) is jointly Gaussian distributed with the function values

---

[2] https://keras.io/.

f on the training locations. GP predictions provide both expected values and *confidence intervals*, which are extremely important in forecasting applications. The downside of this approach is its computational cost, the time and space complexity for training.

A solution to the complexity problem is approximate training methods. We propose, in particular, SVGPs that use stochastic optimization to scale GP training to large datasets. The main idea of this model is to select a set of datapoints called *inducing inputs* or pseudo inputs on which the covariance matrix is generated. The position of these points in the dataset is optimized together with the model parameters through gradient-based optimization with the aim to maximize the evidence lower bound (ELBO) [11], a lower bound of the log-marginal likelihood. Improving the ELBO improves the variational posterior approximation by minimizing the Kullback-Leibler divergence between the true posterior and the variational approximation. Since inducing inputs variational parameters and not model parameters, they can be optimized without risk of overfitting. We perform SVGP training by *Adamax* [13] with *predictive log likelihood* loss function [12]. *Batches* of 256 points are used and 500 *inducing inputs* are chosen from all input dimensions. Parameter optimization was iterated for 100 *epochs*. Finally, the *Cyclical Learning Rate* (CLR) [21] method is used to optimize the learning rate of the model during the training phase. Models are trained using *GPyTorch*[3] on GPUs provided by *Google Colab*[4].

The model proposed in this work uses 13 variables, namely, T, R_H, W_S, W_D, R, T^2, $T_{ma(7)}$, T_M, T_M^2, T_{Mp}, T_{Mp}^2, Sa and Su (see Table 1). For each of these variables we introduce an RBF kernel because future heating load values should be inferred from past load values having similar values for these variables. Then we introduce two periodic kernels for considering the daily and the weekly cycle of the heating load due to social factors. We finally compose the two periodic kernels by summing them (to consider both periodicities) and we multiply the result by the product of all the RBF kernels of environmental factors. We multiply kernels because the effect of multiplication is similar to the intersection (logical *and*) of data filters, hence we predict future heating loads considering more important past loads having similar values of all social and environmental factors. The final kernel is $k_F(\mathbf{x_1}, \mathbf{x_2}) = (k_{P\_24h}(\mathbf{x_1}, \mathbf{x_2}) + k_{P\_168h}(\mathbf{x_1}, \mathbf{x_2})) * \prod_{v \in V} k_{RBF\_v}(\mathbf{x_1}, \mathbf{x_2})$ where $k_{P\_24h}(\mathbf{x_1}, \mathbf{x_2})$ is the daily periodic kernel, $k_{P\_168h}(\mathbf{x_1}, \mathbf{x_2})$ is the weekly periodic kernel and $k_{RBF\_v}(\mathbf{x_1}, \mathbf{x_2})$ is the ard-version of the RBF kernel. We notice that in the proposed SVGP model, variables related to previous loads $l_i, 1 \leq i \leq 7$ are not used because the model intrinsically computes loads as a weighted sum of past loads corresponding to similar environmental and social conditions.

2.5   Performance Measure

Performance is evaluated by Root Mean Square Error (RMSE) on 48-h forecasting horizon. Given an observed time-series with n observations $y_1, \ldots, y_n$ and

---

[3] https://gpytorch.ai/.
[4] https://colab.research.google.com.

predictions $\hat{y}_1, \ldots, \hat{y}_n$, the formula is $RMSE = \sqrt{\frac{1}{N} \sum_{t=1}^{N} (\hat{y}_t - y_t)^2}$. Performance is evaluated on the overall test set, therefore we iterate the computation of the RMSE on a sliding window of 48-h, moving from the beginning to the end of the test set. For example, starting from the first point p_1 we forecast the next 48-h and compute the RMSE on the interval $[p_1, p_{48}]$, then we move to the next point p_2 repeating the previous step on the interval $[p_2, p_{49}]$, and so on. The measure thus obtained is called $(\overline{RMSE})$ in the following and it is the average RMSE over all 48-h predictions in the test set. The RMSE was computed on a 48-h basis for because of a specific application requirement.

3   Results

In this section we evaluate the proposed models, first analyzing the CNN and the SVGP independently, then for CNN we show some kernel parameters and how single CNN neuron signals are composed to generate the heating load prediction. For SVGP we display the kernels of a few single variables and their composition. Some details are provided to investigate the interpretability and the evaluation of models performance on test set.

3.1   CNN Model

The CNN model described in the previous section (Fig. 2a) computes the heating load as a weighted sum of five signals (i.e., $f_0, \ldots, f_4$) generated by convolution of the multivariate input signal. One of the five kernels used to perform the convolution, namely kernel 0, is displayed as a heatmap in Fig. 2c, where rows are time instants (i.e., index 168 corresponds to one hour before and 0 corresponds to 168 h before the current time), columns are variables and colors values of parameters.

Interestingly enough, temperature T and previews day load l_1 have value of parameters with a quite direct interpretation, although CNN are known to be hardly interpretable. In Fig. 2d we show the values of kernel for only these two variables in a line chart having time in the x-axis and value of parameter in the y-axis. Temperature parameters (blue line) are negative in 168 (i.e., one hour before the prediction instant) and they increase (from right to left) to about 0.05 moving towards 0 (i.e., one week before the prediction). These values show that the temperature in the previous two days have a negative impact on the heating load, and its informativeness becomes almost null after two days. This behaviour makes sense since we know that low temperatures imply high heating load to warm up buildings. Also the decrease of the absolute value of the parameter when moving back in time seems to make sense, since it means that recent temperatures are more informative than old one for the prediction. The parameters related to the load of the previous day (red line) have even more interesting behavior, with daily peaks that decrease from 168 to about 75 (from right to left) and then increase from 75 to 0. All peaks are positive because

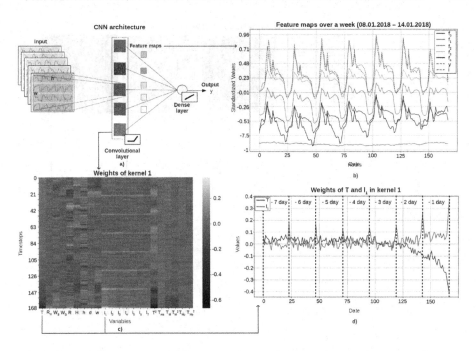

Fig. 2. a) CNN's architecture; b) Neuron outputs for one week of prediction; c) Kernel of neuron 1; d) Weights of T and l_1 variables in kernel 1. (Color figure online)

past load have a positive influence on future loads. The daily peaks show the social component of load, namely, to predict today's load at time $\bar{t}$ it is more informative the yesterday's load at $\bar{t}$ than loads at different hours of the day. This is because people usually warm up buildings differently in hours of the day. Moreover, the increase of the peak corresponding to indices 48 and 24 highlights the weekly pattern of the load, due to the fact that to predict heating load on day $\bar{d}$ (e.g., Sunday) it is more informative to use past loads observed in day $\bar{d}$ than in other days, because people use heating differently in days of the week.

Finally, the charts of Fig. 2b show the output of each of the 5 neurons of the convolutional layer (blue, green, gray, orange and black lines), the output of the network (red dashed line), and the true load (blue dashed line) for the week from 08.01.2018 to 14.01.2018. We first observe that feature map f_1 has an almost constant negative value, while feature map f_4 shows the typical load peaks more than others. Considering the weights of the dense layer, i.e., $w_0 = 0.900, w_1 = -0.561, w_2 = 0.221, w_3 = 0.928, w_4 = 0.298, w_5 = 0.030$, we notice that all of them but w_1 are positive, and w_0 and w_3 have higher absolute values, hence feature maps f_0 and f_3 have stronger influence to the final prediction than others. Finally, the load prediction y which is the weighted sum of convolution signals is very similar to the true load signal l.

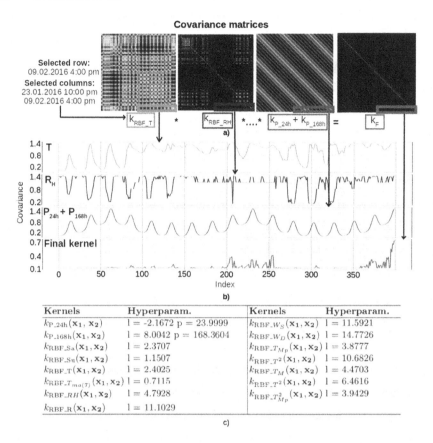

Fig. 3. a) SVGP kernels; b) Highlight on kernel parameters (similarity) of observations taken on 09/02/2016 at 4:00pm against observations taken from 23/01/2016 at 10:00pm to 09/02/2016 at 4:00pm; c) Kernel hyperparameters after training. (Color figure online)

3.2 SVGP Model

The SVGP model is based on a product of $n \times n$ kernels. Figure 3.a shows the final kernel on the right and some of the factor kernels in the center. They are all depicted as heatmaps in which yellow (and bright) to high parameter values and blue (and dark) corresponds to parameters close to zero. Notice that pictures show only a submatrix of each kernel and that each cell (i, j) of the final kernel contains the product of the corresponding (i, j) cells of all factor kernels. Patterns in the colors correspond to correlation patterns since each cell contains a similarity measure between two values. In Fig. 3b we explain them in the particular case of the row corresponding to 09.02.2016 at 4:00 pm. The x-axis value 400 corresponds to the same date and moving from 400 back to 0 the time decreases of one hour at each step. The day cycle is quite visible, since night temperatures are lower than the temperature of 09.02.2016 at 4:00 pm, hence the RBF kernel produces a smaller covariance value (see the periodical lower values in the chart). Similar pattern, although with differences, can be seen in the relative humidity kernel which is also RBF. Kernel $(k_{P\_24h} + k_{P\_168h})$ has a more stable trend with a clear daily period (with peaks at 4:00 pm) summed to a weekly period (with peaks on Tuesdays 09.02.2016). The corresponding values of the final kernel (red line) tend to be high only when all factor values are high. As the figure shows, recent values (close to 400) have high values since they are very important to predict the heating load of the next hour. The parameters (i.e., length scale and periodicity) of the various kernels are listed in Table 3c.

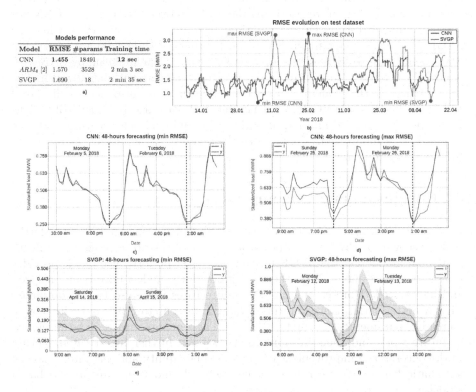

Fig. 4. a) Models performance; b) RMSE evolution of CNN and SVGP for each 48 h forecast performed in 2018. Each point (x, y) represents the RMSE (y) of a 48 h forecast starting at instant x; c-d-e-f) 48-h forecasting with minimum and maximum RMSE error for CNN and SVGP.

3.3 Model Comparison

The main properties of the two proposed models are listed in Fig. 4a, where also the autoregressive linear model presented in [1] is reported for comparison. The best performance is achieved by the CNN which however has a very high number of parameters. The SVGP model has a slightly worse performance than ARM but it has a small number of parameters and it provides confidence intervals on prediction, which is a key feature in some applications. The training time of the CNN is also low, because the network is very simple and the optimizer reaches good performance with a small number of training epochs. However, the training times of ARM and SVGP are also low for our application, since the model is required to be updated only every 24 h.

In Fig. 4b we show the trend of RMSE for the CNN (blue line) and the SVGP (red line) on the test set (year 2018). Each point (x, y) represents a RMSE (y) of a 48-h forecast starting at instant x. The blue and red points show the maximum and minimum RMSE of, respectively, the CNN and the SVGP model. CNN has the maximum RMSE on February 25th at 9:00 am with 3.295 MWh, while the SVGP model has it maximum RMSE on February 12th at 6:00 am with 3.184 MWh. Minimum RMSEs are achieved on February 5th at 10:00 am by CNN with 0.652 MWh, whereas for SVGP on April 14th at 9:00 am with 0.662 MWh. The 48-h predictions that generate these (best and worst) performance are displayed in Fig. 4c-d-e-f). For SVGP confidence intervals are also provided. Blue lines represent the true load and red lines the predicted load. The heating load here displayed is standardized, to guarantee the privacy of the dataset, as requested by the utility company.

4   Conclusion and Ongoing Work

CNN and SVGP models have been used to predict heating load in a real DHN. Results show that the CNN outperforms a state-of-the-art autoregressive linear regression model and the SVGP has slightly lower performance but it provides useful confidence intervals on the prediction. Both models have been analyzed and interpreted. More complex CNN architectures were also tested, obtaining slightly better performance in terms of $\overline{RMSE}$ (i.e., up to 1.397) but with a large increase of parameters (i.e., up to 121, 255). These architectures will be developed in future work and their interpretability will be further investigated. Future work concerns the improvement of model explainability and the integration of automatic feature engineering in neural network and Gaussian process based models.

Acknowledgments. The research has been partially supported by the projects "Dipartimenti di Eccellenza 2018–2022, funded by the Italian Ministry of Education, Universities and Research (MIUR), and "GHOTEM/CORE-WOOD, POR-FESR 2014–2020", funded by Regione del Veneto.

References

1. Bianchi, F., Castellini, A., Tarocco, P., Farinelli, A.: Load forecasting in district heating networks: model comparison on a real-world case study. In: Nicosia, G., Pardalos, P., Umeton, R., Giuffrida, G., Sciacca, V. (eds.) LOD 2019. LNCS, vol. 11943, pp. 553–565. Springer, Cham (2019). https://doi.org/10.1007/978-3-030-37599-7_46
2. Blum, M., Riedmiller, M.: Electricity demand forecasting using gaussian processes. In: Proceedings of 15th AAAIWS, pp. 10–13. AAAI Press (2013)
3. Castellini, A., et al.: Activity recognition for autonomous water drones based on unsupervised learning methods. In: Proceedings of 4th Italian Workshop on Artificial Intelligence and Robotics (AI*IA 2017), vol. 2054, pp. 16–21 (2018)
4. Castellini, A., Bicego, M., Masillo, F., Zuccotto, M., Farinelli, A.: Time series segmentation for state-model generation of autonomous aquatic drones: a systematic framework. Eng. Appl. Artif. Intell. **90**, 103499 (2020)
5. Castellini, A., Franco, G.: Bayesian clustering of multivariate immunological data. In: Nicosia, G., Pardalos, P., Giuffrida, G., Umeton, R., Sciacca, V. (eds.) LOD 2018. LNCS, vol. 11331, pp. 506–519. Springer, Heidelberg (2019). https://doi.org/10.1007/978-3-030-13709-0_43
6. Castellini, A. Masillo, F., Sartea, R., Farinelli, A.: eXplainable modeling (XM): data analysis for intelligent agents. In: Proceedings of 18th International Conference on Autonomous Agents and Multiagent Systems (AAMAS), pp. 2342–2344. IFAAMAS (2019)
7. Dahl, A., Bonilla, E.: Scalable Gaussian process models for solar power forecasting. In: Woon, W.L., Aung, Z., Kramer, O., Madnick, S. (eds.) DARE 2017. LNCS (LNAI), vol. 10691, pp. 94–106. Springer, Cham (2017). https://doi.org/10.1007/978-3-319-71643-5_9
8. Frigola-Alcalde, R.: Bayesian time series learning with Gaussian processes. Ph.D. thesis, University of Cambridge (2015)
9. Glorot, X., Bengio, Y.: Understanding the difficulty of training deep feedforward neural networks. In: AISTATS, pp. 249–256 (2010)
10. Goodfellow, I., Bengio, Y., Courville, A.: Deep Learning. MIT Press, Cambridge (2016)
11. Hensman, J., Matthews, A., Ghahramani, Z.: Scalable variational Gaussian process classification. In: 18th International Conference on Artificial Intelligence and Statistics (2015)
12. Jankowiak, M., Pleiss, G., Gardner, J.R.: Sparse Gaussian process regression beyond variational inference. CoRR (2019)
13. Kingma, D., Ba, J.: Adam: a method for stochastic optimization. CoRR (2014)
14. Koprinska, I., Wu, D., Wang, Z.: Convolutional neural networks for energy time series forecasting. In: 2018 International Joint Conference on Neural Nets (IJCNN), pp. 1–8 (2018)
15. Ramanathan, R., Engle, R., Granger, C.W.J., Vahid-Araghi, F., Brace, C.: Short-run forecast of electricity loads and peaks. Int. J. Forecast. **13**, 161–174 (1997)
16. Rasmussen, C.E., Williams, C.K.I.: Gaussian Processes for Machine Learning (Adaptive Computation and Machine Learning). The MIT Press, Cambridge (2005)
17. Raza, M., Khosravi, A.: A review on artificial intelligence based load demand forecasting techniques for smart grid and buildings. Ren. Sust. En. Rev. **50**, 1352–72 (2015)

18. Roberts, S., Osborne, M., Ebden, M., Reece, S., Gibson, N., Aigrain, S.: Gaussian processes for time-series modelling. Phil. Trans. Royal Soc. (Part A) **371**, 20110550 (2013)
19. Sainath, T., Vinyals, O., Senior, A., Sak, H.: Convolutional, long short-term memory, fully connected deep neural networks. In: ICASSP, p. 4580–4 (2015)
20. Shepero, M., Meer, D.V.D., Munkhammar, J., Widén, J.: Residential probabilistic load forecasting: a method using gaussian process designed for electric load data. Appl. Energy **218**, 159–172 (2018)
21. Smith, L.N.: No more pesky learning rate guessing games. CoRR (2015)

Explainable AI as a Social Microscope: A Case Study on Academic Performance

Anahit Sargsyan[1,2(✉)], Areg Karapetyan[1,3(✉)], Wei Lee Woon[1], and Aamena Alshamsi[1,4]

[1] Department of Computer Science, Masdar Institute, Khalifa University, Abu Dhabi, UAE
akarapetyan@masdar.ac.ae
[2] Division of Social Science, New York University Abu Dhabi, Abu Dhabi, UAE
as12831@nyu.edu
[3] Research Institute for Mathematical Sciences (RIMS), Kyoto University, Kyoto, Japan
[4] The MIT Media Lab, Massachusetts Institute of Technology, Cambridge, MA, USA

Abstract. Academic performance is perceived as a product of complex interactions between students' overall experience, personal characteristics and upbringing. Data science techniques, most commonly involving regression analysis and related approaches, serve as a viable means to explore this interplay. However, these tend to extract factors with wide-ranging impact, while *overlooking variations specific to individual students*. Focusing on each student's peculiarities is generally impossible with thousands or even hundreds of subjects, yet data mining methods might prove effective in devising more targeted approaches. For instance, subjects with shared characteristics can be assigned to clusters, which can then be examined separately with machine learning algorithms, thereby providing a more nuanced view of the factors affecting individuals in a particular group. In this context, we introduce a data science workflow allowing for fine-grained analysis of academic performance correlates that captures the *subtle differences in students' sensitivities to these factors*. Leveraging the Local Interpretable Model-Agnostic Explanations (LIME) algorithm from the toolbox of Explainable Artificial Intelligence (XAI) techniques, the proposed pipeline yields groups of students *having similar academic attainment indicators*, rather than similar features (e.g. familial background) as typically practiced in prior studies. As a proof-of-concept case study, a rich longitudinal dataset is selected to evaluate the effectiveness of the proposed approach versus a standard regression model.

Keywords: Explainable AI · LIME · Data science · Machine learning · Computational social science · Academic performance · GPA prediction

Electronic supplementary material The online version of this chapter (https://doi.org/10.1007/978-3-030-64583-0_24) contains supplementary material, which is available to authorized users.

© Springer Nature Switzerland AG 2020
G. Nicosia et al. (Eds.): LOD 2020, LNCS 12565, pp. 257–268, 2020.
https://doi.org/10.1007/978-3-030-64583-0_24

1   Introduction

With far-ranging consequences on young people's lives and careers, academic performance is susceptible to various types and forms of influence. It is often path-dependent, correlating with an individual's past performance [3,18]. Furthermore, factors with significant impact on academic performance can be specific to the subject in question, such as intelligence and determination [2,11,22], or exogenous, resulting from the social, emotional and socioeconomic environment in which the individual was raised [5,10,14]. Therefore, in general, investigation of academic performance predictors is attained through longitudinal studies [5,8].

Mainly, two directions are evidenced in this line of research. The first, followed in [6,21], relies on statistical models to measure the correlation of a few variables that were premised on prior results in the literature. The second, attended in [1, 12,13,17], resorts to data science and machine learning techniques for extracting informative predictors from large datasets with thousands of candidate features.

While both approaches recognize the uniqueness of the students' backgrounds, the effect of the correlates is still determined as an aggregate over the entire study population/group, leaving the *subtle variations between individuals* largely overlooked. In particular, the former assumes that all the subjects are impacted by the same set of selected correlates in the same manner, whereas the latter seeks to derive a predictive model with high accuracy that generalizes to the overall population. However, no two *subjects are identical*, and factors profoundly affecting one individual might have a merely negligible impact on another person, even under comparable circumstances.

Against this backdrop, we introduce a novel data science approach in which (i) the predictors of academic performance for each student *are identified and quantified* (ii) the study population is segmented into clusters based on the obtained values. The proposed pipeline allows the groups with *similar success indicators* to be analyzed collectively, which should enable their effects to reinforce each other and be more readily discoverable.

To quantify academic performance predictors specific to individual students, we avail of a recently developed XAI algorithm, known as LIME [16]. The outputs from LIME serve as "explanations", which are *localized* in that a unique explanation is generated for each subject. Though descriptive on an individual case basis, these explanations are intrinsically disassociated, and thus their direct interpretation (one by one) becomes intractable with a growing number of subjects. This paper presents an efficient solution by grouping the students according to LIME coefficients, as detailed in Sect. 3. Distinctively, under such clustering criterion, the subsequent analysis is explicitly centered at groups of students who *share similar academic performance correlates*, as opposed to the classical approach of clustering in the feature space (i.e., based on observable characteristics such as gender, familial background or financial class). In a sense, with this scheme in place, it proves possible for subjects with fairly diverse backgrounds and needs to be grouped together, provided they share common markers of academic attainment.

As one demonstration, the proposed approach is benchmarked on a longitudinal dataset, released for the Fragile Families Challenge (FFC) competition, against a standard regression model. The results reveal a striking difference in the depth of insights gained, with the devised pipeline featuring prominently. While intended as a proof-of-concept, this preliminary study unveils findings on academic performance indicators that could serve social and data science communities. Furthermore, the workflow proposed herein can potentially pave the way towards more efficient and targeted intervention strategies by providing insights that would be inconceivable to achieve with traditional methods.

2   Data and Pre-processing

2.1   FFC Dataset

The examined dataset stems from the Fragile Families and Child Wellbeing study that documented the lives of over 4000 births occurring between 1998 and 2000 in U.S. cities with at least 200,000 population. As such, the study was carried out in the form of questionnaire surveys and interviews with parents shortly after the children's birth, and when the infants were 1, 3, 5 and 9 years old (*overall five waves*). The elicited data covered essential temporal information on the children's attitudes, parenting behavior, demographic characteristics, to name a few (further details of the dataset can be consulted in [15]). The data was released within the scope of FFC competition which sought to predict 6 life outcomes, including Grade Point Average (GPA), based on these data records. Analysis of the overall results and ensuing findings of the competition are summarized in [19].

In total, the dataset comprises 4,242 rows (one per child) and 12,943 columns, including the unique numeric identifier. During FFC, however, only half of the data rows were released as a training set. Of these, 956 entries had GPA values missing, and therefore the final dataset analyzed in this study totalled 1,165 subjects, as appears in Fig. 6 in the Appendix.

2.2   Pre-processing and Feature Selection

Before proceeding with this step, we remark that it is stipulated exclusively by nature (e.g., missing values) and properties (e.g., dimensionality) of the dataset under study and per se is not a principal constituent of the developed pipeline. Indeed, it is tailored specifically for the FFC dataset and might very well be substituted by any other appropriate routine yielding a sufficiently informative feature subset (i.e., *with a decent predictive accuracy*) of reasonable cardinality. Thus, for clarity of exposition, the respective particulars are deferred to the Appendix.

The target subset of optimally descriptive features, as revealed through extensive experimentation and *validated by its predictive accuracy*[1], contained 65 fea-

---

[1] A mean squared error (MSE) of approximately 0.359 was achieved under 3-fold cross-validation (a result of comparable fidelity, submitted during the FFC, *secured a place in the top quartile* of the scoreboard).

tures, tabulated in Table 1 in [20]. This pool of features forms the input for the proceeding analysis laid out in Sect. 3. Figure 1 depicts the spread of these 65 features across the 5 waves (i.e., over the trajectory of children's lives). For each wave, the features are arranged into the following six categories: {*familial, financial, academic, social, personality, other*} and their respective counts are illustrated in Fig. 1 as a stacked bar chart.

As deduced from Fig. 1, the distribution of features in familial and financial categories is skewed towards the early span of children's lives. For the correlates falling in academic, social and personality categories, the opposite trend is evidenced.

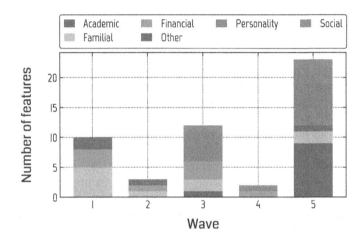

Fig. 1. The distribution of selected 65 features, categorized into 6 major factor types, among the 5 waves (i.e., over the course of the children's lives).

3   Comparative Analysis

This section contrasts the proposed approach against a conventional regression model, as illustrated in Fig. 2, and discusses the results.

We cast the problem as a classification task by discretizing the GPA scores into three classes, Low, Middle and Top, defined respectively by the following ranges: $[1, 2.5], (2.5, 3.25), [3.25, 4]$. Consequently, only the subjects falling into the Top and Low categories were retained (861 in total). The motivation is to steer the focus of classification algorithms towards the aspects discerning high and low performers. Indeed, the factors responsible for "borderline" performances are likely the ones with negligible impact, hence inferring them might obscure the results. On the other hand, omitting a large group of subjects could lead to the loss of pertinent data. Thus, the above thresholds were set according to the top and bottom 30% percentiles of GPA score records. This ensured a solid number of participant students while retaining a sizable gap between the two classes.

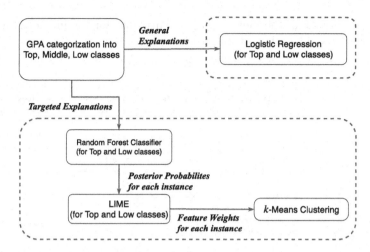

Fig. 2. Flowchart of the conducted comparative analysis including the featured methodology for obtaining targeted explanations.

3.1   General Indicators

Following the common practice of previous works, in this initial phase, we employed the logistic regression algorithm to screen the selected features that broadly correlate with academic performance. The subjects from the Top and Low categories were fit to the model and the resulting coefficients, under L1 regularization, are presented in Fig. 3.

As observed from Fig. 3, test grades, along with other early metrics of academic performance, are imperative, and so are the factors associated with the child's social background. In particular, the indicators in familial and financial categories appear to influence children's academic performance predominantly at an early age. Whereas the correlates associated with scholastic aptitude manifest their effect mostly at later stages of subjects' lives. These observations are consistent with prior findings in the literature, providing some measure of validation. Overall, the most influential predictors are listed below.

1. The two most important factors relate to the parents' education [4].
2. The *Peabody Picture Vocabulary Test* (PPVT) percentile rank correlated with academic performance. PPVT is a standardized test designed to measure an individual's vocabulary and comprehension and provide a quick estimate of verbal ability or scholastic aptitude. Another standard test's (known as Woodcock Johnson Test) percentile rank was identified as a significant indicator as well.
3. Interestingly, the fact that the father has been incarcerated[2] - a proxy for family support - affects children's performance negatively. Contrariwise, the

---

[2] Note that in Fig. 3 the positive correlation of this feature is due to the reversed order of values (i.e., the highest value indicates the father has not spent time in jail).

complexity/rank of the mother's job, a surrogate for the financial situation, conduced to enhanced academic performance.

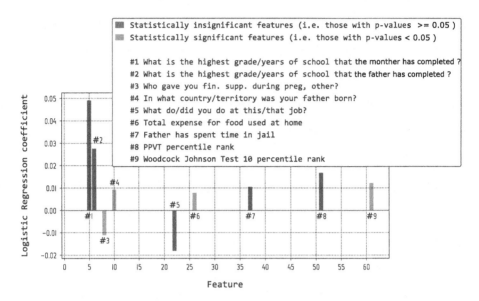

Fig. 3. Logistic Regression coefficients of the selected 65 features (ordered in a non-decreasing value of their waves) for **Top** and **Low** classes.

The emerging picture is compelling and multifaceted. On the one hand, test scores and academic aptitude occupy a central role, which is to be expected. Yet, there are indications that, beyond this, other features reflective of social and financial stability could also play a part, which strongly motivates the second, targeted part of this study.

3.2   Proposed Methodology: Targeted Indicators

While the insights highlighted in Sect. 3.1 were illuminating, they were extracted from the entire dataset, and the perspectives obtained were thus quite broad. To further extract targeted or localized indicators of academic success, we resorted to LIME [16]. For each instance, LIME produces a localized explanation of the classifier output by perturbing the feature values to generate a set of synthetic data points in the vicinity of the true instance. The posterior probability for each data point is estimated using the trained classifier, and a linear regression model is trained using the synthetic points as the inputs, and the posterior probabilities as the targets. The localized regression coefficients obtained in this way can then be interpreted as the *importance of a feature*, and are estimated separately for each subject.

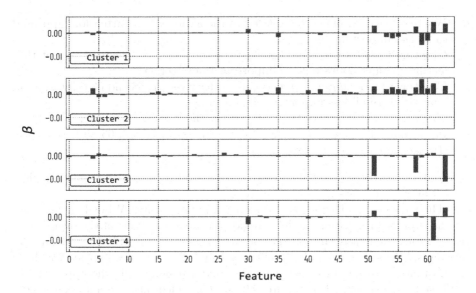

Fig. 4. Clustering of students in Top and Low classes based on LIME coefficients. The horizontal axis represents the selected 65 features sorted by their wave values in a nondecreasing order (i.e., over the course of children's lives). The vertical axis (β) depicts the deviations of the mean LIME coefficients for each cluster from the overall mean values of each LIME coefficient across the population.

This technique was adapted for the present context as follows. First, Random Forest classifier is trained on the data and is invoked to estimate the posterior probability for each instance. Then, LIME is applied to subjects falling into the Top and Low groups to produce feature weights specific to each subject, which are then clustered with the k-means algorithm. Each cluster is then characterized by the centroid of the LIME coefficients for the instances therein.

3.3   Results and Discussion

The k-means clustering results, with parameter $k = 4$ (i.e., four clusters), appear in Fig. 4. This choice of k enables a compact yet expressive representation of the underlying insights. In general, the higher the k, the more likely the groups are to resemble each other closely. On the other hand, when k is small, one might possibly overlook factors critical to some subset of subjects.

In Fig. 4, to emphasize the differences between clusters, we focus on each feature's relative weights. That is, a cluster is represented in terms of the difference between its centroid and population means. As the figure suggests, the characteristics of the subjects vary significantly between the clusters. In particular, the salient patterns observed were as follows:

- **Cluster 1** (*189 subjects*): The subjects in this cluster appear to be strongly influenced by features 51, 58, 61 and 63, which are all linked to test scores.

Whereas features 59 and 60, related to the ability to pay attention, appeared to be less important.

- **Cluster 2** (*141 subjects*): This cluster was similar to Cluster 1 except that features 59 and 60 were more important than average.
- **Cluster 3** (*297 subjects*): In this cluster, features 51, 58 and 63 (all test score related), were all less important than average.
- **Cluster 4** (*234 subjects*): Here, feature 61 (the Woodcock Johnson Test score) was significantly less important, while features 51, 58 and 63 all had slightly stronger impacts on performance.

Overall, the features with values close to zero are the ones with a uniform effect on all individuals regardless of a cluster.

These results are deeply compelling in a number of ways. While Fig. 3 (results of the standard regression model) provided a broad overview of the overall success factors, the relative importance of each of these factors differed substantially among subjects. For example, in clusters 1 and 2, test scores appeared to have a substantial impact on future academic performance, while the opposite was observed in cluster 3. Also, features 59 and 60, which measure a student's general attentiveness, seemed relatively peripheral in cluster 1 and crucial in cluster 2 (and close to the average in clusters 3 and 4).

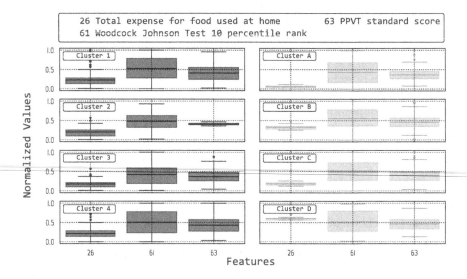

Fig. 5. Distributions of values of the selected numerical features within the clusters of students in Top and Low classes. Clusters 1 to 4 were produced with the proposed scheme (i.e., based on LIME coefficients), while clusters A to D with the conventional method (i.e., based on feature values).

Another interesting observation was the spread of feature values *within* clusters, which is depicted in Fig. 5 (representing three selected numerical features).

In particular, note the spread of values for feature 26, which is *the total expense for food used at home*. Apparently, the feature values in the LIME clusters exhibit a far greater range compared to the conventional clusters, which tend to group people with similar financial situations. For the other features, the results are slightly less straightforward, but LIME clusters 3 and 4, in particular, also exhibit a much broader spread of values. This underscores that clustering with the LIME coefficients does not merely group students based on their personal or social circumstances, but rather in terms of the factors which affect their future academic performance.

4   Future Work

The present findings indicate the potential effectiveness of the proposed methodology in analyzing causal relationships in datasets alike FFC. However, this work was intended as a proof-of-concept case study, leaving open several avenues for future investigations. In particular, below listed are several promising directions.

1. Analyze comparable data sets (e.g., The Millennium Cohort Study [7]), to test whether certain aspects of the observations are data-specific, or reflect true underlying patterns.
2. Perform a thorough sensitivity analysis on the chosen subset of features as well as incorporate XAI techniques alternative to LIME (e.g., SHapley Additive exPlanations [9]).
3. Apply the proposed approach to more diverse and larger datasets to demonstrate its generalizability to other use cases/study domains.

5   Concluding Remarks

In this study, a novel data science pipeline is proposed, which conduces the identification of the specific features associated with academic performance in different groups of students. A clustering algorithm was employed to group these subjects, then targeted success indicators were extracted from each of these groups and scrutinized. We note that the present findings rely on a technique (LIME), which was developed relatively recently and should be treated as preliminary. However, if and when superior methods are proposed, they can similarly be incorporated into the devised workflow.

The key point is that such localized models are vital if we are to obtain a more nuanced view of the actual success indicators for specific children and families. The findings suggest that the children of fragile families can be given the best chance of success through interventions that are tailored to their individual needs, e.g. in some families, a small home loan could be the difference between a star student and a dropout, whereas in others a free mentoring scheme might turn more valuable.

Appendix

This section details the employed pre-processing and feature selection steps, portrayed as a flowchart in Fig. 6.

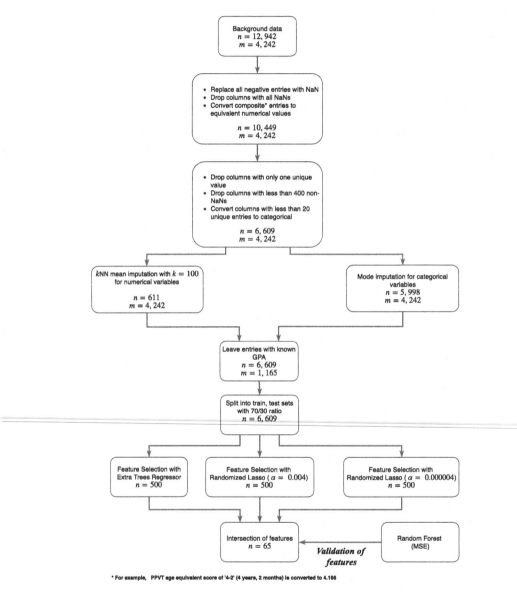

* For example, PPVT age equivalent score of '4-2' (4 years, 2 months) is converted to 4.166

Fig. 6. A step-by-step illustration of the performed pre-processing and feature selection routines, depicted as a flowchart, with n and m standing for the number of features and entries in the given step, respectively.

Due to the nature of the dataset under study, there were numerous instances of missing values where respondents either refused or were unavailable to answer. This was resolved through the judicious combination of data transformation, reduction, and imputation techniques, as listed below.

- All missing and negative values were replaced by NaN and the columns with 0 variance were removed. Then, only the columns having at least 400 non-NaN values were retained.
- Features with less than 20 unique values were treated as categorical and a simple median imputation was applied to replace the missing values.
- The variant of kNN (k-Nearest Neighbors) imputation algorithm, implemented in the Python package Fancyimpute, was leveraged, with value of 100 for the parameter k, to estimate the remaining NaN values.

The number of features in the resulting dataset was reduced from over $12,900$ to $6,609$. However, this number of covariates was still exceedingly high, necessitating a subsequent feature selection phase. An array of filter- and wrapper-based methods, including Principal Component Analysis, Lasso, and Gradient Boosting Regression, were attempted in search of the most informative feature subset of reasonable cardinality. These methods were applied to the extracted pool of $6,609$ features, both recursively and explicitly, and probed under diverse parameter settings. The acquired subsets were then evaluated for their predictive accuracy across various models trained, effectively providing a means of validation. In essence, the latter step intends to establish the *overall validity of the model* underlying the proceeding analysis, thereby solidifying credibility of the explanations derived therein.

The target subset of optimally descriptive features was obtained by the following means. Feature importances were estimated by the Extra Trees Regressor algorithm (with 500 estimators) and Randomized Lasso, and the top 500 features were retained from each. For the latter, two different values were considered for the regularization parameter α, namely 0.004 and 0.000004, thus resulting in two separate feature subsets. The intersection of these three subsets, containing 65 features, led to maximized GPA prediction accuracy. In particular, with the Random Forest algorithm, an MSE of approximately 0.359 was achieved under 3-fold cross-validation.

References

1. Asif, R., Merceron, A., Pathan, M.K.: Predicting student academic performance at degree level: a case study. Int. J. Intell. Syst. Appl. **7**(1), 49 (2014)
2. Colom, R., Escorial, S., Shih, P.C., Privado, J.: Fluid intelligence, memory span, and temperament difficulties predict academic performance of young adolescents. Pers. Individ. Differ. **42**(8), 1503–1514 (2007)
3. Coyle, T.R., Pillow, D.R.: Sat and ACT predict college GPA after removing G. Intelligence **36**(6), 719–729 (2008)
4. Ermisch, J., Francesconi, M.: Family matters: impacts of family background on educational attainments. Economica **68**(270), 137–156 (2001)

5. Graziano, P.A., Reavis, R.D., Keane, S.P., Calkins, S.D.: The role of emotion regulation in children's early academic success. J. Sch. Psychol. **45**(1), 3–19 (2007)
6. Jackson, L.A., Von Eye, A., Biocca, F.A., Barbatsis, G., Zhao, Y., Fitzgerald, H.E.: Does home internet use influence the academic performance of low-income children? Dev. Psychol. **42**(3), 429 (2006)
7. Joshi, H., Fitzsimons, E.: The millennium cohort study: the making of a multipurpose resource for social science and policy. Longit. Life Course Stud. **7**(4), 409–430 (2016)
8. Laidra, K., Pullmann, H., Allik, J.: Personality and intelligence as predictors of academic achievement: a cross-sectional study from elementary to secondary school. Pers. Individ. Differ. **42**(3), 441–451 (2007)
9. Lundberg, S.M., Lee, S.I.: A unified approach to interpreting model predictions. In: Advances in Neural Information Processing Systems, vol. 30, pp. 4765–4774. Curran Associates, Inc. (2017)
10. McLoyd, V.C.: Socioeconomic disadvantage and child development. Am. Psychol. **53**, 185 (1998)
11. Pajares, F., Hartley, J., Valiante, G.: Response format in writing self-efficacy assessment: greater discrimination increases prediction. Meas. Eval. Counsel. Dev. **33**(4), 214 (2001)
12. Pal, A.K., Pal, S.: Analysis and mining of educational data for predicting the performance of students. Int. J. Electron. Commun. Comput. Eng. **4**(5), 1560–1565 (2013)
13. Pandey, M., Sharma, V.K.: A decision tree algorithm pertaining to the student performance analysis and prediction. Int. J. Comput. Appl. **61**(13) (2013)
14. Pritchard, M.E., Wilson, G.S.: Using emotional and social factors to predict student success. J. Coll. Stud. Dev. **44**(1), 18–28 (2003)
15. Reichman, N.E., Teitler, J.O., Garfinkel, I., McLanahan, S.S.: Fragile families: sample and design. Child Youth Serv. Rev. **23**(4), 303–326 (2001)
16. Ribeiro, M.T., Singh, S., Guestrin, C.: Why should i trust you?: explaining the predictions of any classifier. In: Proceedings of the 22nd ACM SIGKDD International Conference on Knowledge Discovery and Data Mining, pp. 1135–1144. ACM (2016)
17. Romero, C., Ventura, S.: Educational data mining: a review of the state of the art. IEEE Trans. Syst. Man Cybern. Part C (Appl. Rev.) **40**(6), 601–618 (2010)
18. Salanova, M., Schaufeli, W., Martínez, I., Bresó, E.: How obstacles and facilitators predict academic performance: the mediating role of study burnout and engagement. Anxiety Stress Coping **23**(1), 53–70 (2010)
19. Salganik, M.J., Lundberg, I., Kindel, A.T., Ahearn, C.E., Al-Ghoneim, K., et al.: Measuring the predictability of life outcomes with a scientific mass collaboration. Proc. Natl. Acad. Sci. **117**(15), 8398–8403 (2020)
20. Sargsyan, A., Karapetyan, A., Woon, W.L., Alshamsi, A.: Explainable AI as a social microscope: a case study on academic performance. CoRR abs/1806.02615 (2020). http://arxiv.org/abs/1806.02615
21. Tillman, K.H.: Family structure pathways and academic disadvantage among adolescents in stepfamilies. Sociol. Inq. **77**(3), 383–424 (2007)
22. Tross, S.A., Harper, J.P., Osher, L.W., Kneidinger, L.M.: Not just the usual cast of characteristics: using personality to predict college performance and retention. J. Coll. Stud. Dev. **41**(3), 323 (2000)

Policy Feedback in Deep Reinforcement Learning to Exploit Expert Knowledge

Federico Espositi[(✉)] and Andrea Bonarini

AI and Robotics Lab, Dipartimento di Elettronica, Informazione e Bioingegneria,
Politecnico di Milano, Via Ponzio 34\5, 20133 Milano, MI, Italy
{federico.espositi,andrea.bonarini}@mail.polimi.it
http://airlab.deib.polimi.it/

Abstract. In Deep Reinforcement Learning (DRL), agents learn by sampling transitions from a batch of stored data called Experience Replay. In most DRL algorithms, the Experience Replay is filled by experiences gathered by the learning agent itself. However, agents that are trained completely Off-Policy, based on experiences gathered by behaviors that are completely decoupled from their own, cannot learn to improve their own policies. In general, the more algorithms train agents Off-Policy, the more they become prone to divergence. The main contribution of this research is the proposal of a novel learning framework called *Policy Feedback*, used both as a tool to leverage offline-collected expert experiences, and also as a general framework to improve the understanding of the issues behind *Off-Policy Learning*.

Keywords: Machine learning · Deep Reinforcement Learning · DDPG · Exploitation · Policy feedback

1 Introduction

It is a wide known fact that RL algorithms that combine Function Approximation, Bootstrapping and Off-Policy Learning, may not train an agent successfully. These three elements have been referred to as the *Deadly Triad* [4]. However, although many state-of-the art techniques such as DQN [7] and DDPG [6] are considered *Off-Policy* techniques, they can successfully train agents. Instead, *Batch Learning* techniques are unsuccessful unless the learning policy is constrained to be related to the exploratory one [3,5].

It was firstly suggested in [4] that the problem is that the concepts of *On* and *Off* Policy learning are not crisp, and thus, instability is obtained in an increasing way as the training procedure leans towards *complete off-policiness*. Though it seems that the On-Policy experiences are crucial for a successful training, their relative importance in the training process has not yet been characterized.

With this work we intend to propose the novel architecture of *Policy Feedback*, that allows to leverage both *on* and *off* policy learning with flexibility, providing a deeper understanding of the issues behind pure Off-Policy learning.

© Springer Nature Switzerland AG 2020
G. Nicosia et al. (Eds.): LOD 2020, LNCS 12565, pp. 269–272, 2020.
https://doi.org/10.1007/978-3-030-64583-0_25

The concept of *"feedback"* refers to the injection of transitions gathered by the learner itself alongside the expert's. The aim of this paper is to introduce and support the following *Feedback Hypothesis*:

"The most crucial problems in Off-Policy learning can be solved with the injection of an On-Policy feedback signal of any magnitude".

2   Design

We propose the *Policy Feedback* method, which is based on the DDPG algorithm [6], but the Experience Replay is composed of two separate buffers:

- The *Expert Memory* $\mathcal{M}_{\mathcal{E}}$, the fixed batch containing the expert transitions collected offline.
- The *Feedback Memory* $\mathcal{M}_{\mathcal{F}}$, a classical FIFO Experience Replay which is progressively filled with the most recent experiences gathered by the learner.

As the expert already provides highly fruitful experiences, there is no need for the agent's actions to be superimposed with random noise for exploration.

When building mini-batches for training, a single parameter, $P_{on\_p} \in [0,1]$ -probability of On-Policy-, controls the proportion of transitions belonging to the *Feedback Memory* Replay with respect to the *Expert*'s, thus controlling the magnitude of the equivalent feedback signal.

3   Experiments and Results

All the agent hyper-parameters were set as in the DDPG original paper [6].

Experiments were repeated for different values of P_{on_p} using the same Expert Replays $\mathcal{M}_{\mathcal{E}}$, for two environments of the gym library: *"LunarLanderContinuous-v2"* and *"Swimmer-v2"* [1]. Results are shown in Fig. 1.

In the full-Off-Policy case, with $P_{on\_p} = 0$, the agent is completely unable to improve its policy. Instead, with any tested value of $P_{on\_p} > 0$, learning was successful, and actually led to a considerable increase of the learning speed with respect to the DDPG baseline.

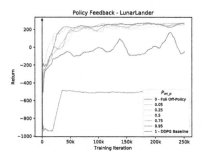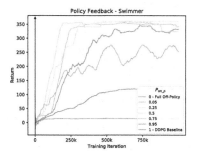

Fig. 1. Policy feedback for different values of feedback intensity P_{on_p}

4   Discussion

In Fig. 2, training losses of the neural Q functions are shown for the corresponding experimental results reported in Fig. 1.

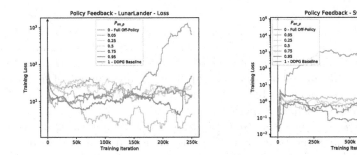

Fig. 2. Q-function training loss in policy feedback

It is clear from results reported in Fig. 1 that the On-Policy feedback signal is sufficient for stability.

The Generalization Over Preference Hypothesis

For the *LunarLander* case, results show that the low performance obtained when learning completely off-Policy seems to be decorrelated from Q-instability. We therefore assume that Q-divergence is not the only factor at play during Off-Policy learning, and propose to look for another one. In this study, the *Generalization Over Preference Hypothesis* is presented :

> "An issue with complete Off-Policy Learning with Function Approximation and Bootstrapping is the generalization of the estimated Q-function over regions of the state-action space that are not explored by the *Behavioral Policy*, but to which the training agent would be led to as a result of training".

Indeed, the Q function is learned from the transitions in the Experience Replay, and the policy is updated to choose actions which maximize the learned Q. We call these actions the *preference*. Due to function approximation, Q generalizes and may be artificially high in some regions, leading to *preferences* that are actually low-performing.

We argument that *Generalization Over Preference* is a crucial cause of low-performance of off-policy learning, and that it is only a subset of the *generalization* problem. Indeed, to solve the latter, the Q function would need to be learned accurately across the entire state-action space [2], while the *Generalization Over Preference* can be addressed by only adjusting the Q values over the current *preference*. This is exactly what is done in *Policy Feedback*, which always feeds training with experiences collected by the learner, and which are thus related precisely to the *preference*, that are therefore adjusted. As such, *Policy Feedback* provides an empirical confirmation of the *Generalization Over Preference hypothesis*

5   Conclusions

In this paper, the *Policy Feedback* framework has been introduced, with the intent to understand more deeply the issues related to *Off-Policy* learning and to characterize the relative importance of *On-Policy* samples coming from the agent's own policy. Results show that the injection of any amount of feedback from the agent's own policy is able to stabilize learning, allowing the agent to successfully leverage the expert experiences.

Policy Feedback leverages the *Generalization Over Preference* issue to its advantage, since it uses the transitions that the agent is directly led to by the training process (the *preference*) and immediately corrects for the inconsistencies.

As a result, *Policy Feedback* is able to solve both the problems of Off-Policy learning, and to exploit the available expert transitions.

These results support the *Feedback Hypothesis* introduced in Sect. 1.

References

1. Brockman, G., et al.: OpenAI gym. CoRR abs/1606.01540 (2016). arXiv: 1606.01540. http://arxiv.org/abs/1606.01540
2. de Bruin, T., et al.: Improved deep reinforcement learning for robotics through distribution-based experience retention. In: 2016 IEEE/RSJ International Conference on Intelligent Robots and Systems (IROS), pp. 3947–3952 (2016)
3. Fujimoto, S., Meger, D., Precup, D.: O-policy deep reinforcement learning without exploration. CoRR abs/1812.02900 (2018). arXiv: 1812.02900. http://arxiv.org/abs/1812.02900
4. van Hasselt, H., et al.: Deep reinforcement learning and the deadly triad. CoRR abs/1812.02648 (2018). arXiv: 1812.02648. http://arxiv.org/abs/1812.02648
5. Kumar, A., et al. Stabilizing O-policy Q-learning via bootstrapping error reduction (2019). arXiv: 1906.00949 [cs.LG]
6. Lillicrap, T.P., et al.: Continuous control with deep reinforcement learning (2015). arXiv: 1509.02971 [cs.LG]
7. Mnih, V., et al.: Human-level control through deep reinforcement learning. Nature **518**, 529–33 (2015). https://doi.org/10.1038/nature14236

Gradient Bias to Solve the Generalization Limit of Genetic Algorithms Through Hybridization with Reinforcement Learning

Federico Espositi[(✉)] and Andrea Bonarini

AI and Robotics Lab, Dipartimento di Elettronica, Informazione e Bioingegneria,
Politecnico di Milano, Via Ponzio 34\5, 20133 Milano, MI, Italy
{federico.espositi,andrea.bonarini}@mail.polimi.it
http://airlab.deib.polimi.it/

Abstract. Genetic Algorithms have recently been successfully applied to the Machine Learning framework, being able to train autonomous agents and proving to be valid alternatives to state-of-the-art Reinforcement Learning techniques. Their attractiveness relies on the simplicity of their formulation and stability of their procedure, making them an appealing choice for Machine Learning applications where the complexity and instability of Deep Reinforcement Learning techniques is still an issue. However, despite their apparent potential, the classic formulation of Genetic Algorithms is unable to solve Machine Learning problems in the presence of high variance of the fitness function, which is common in realistic applications.

To the best of our knowledge, the presented research is the first study about this limit, introduced as *the Generalization Limit of Genetic Algorithms*, which causes the solutions that Genetic Algorithms return to be not robust and in general low-performing. A solution is proposed based on the *Gradient Bias* effect, which is obtained by artificially injecting more robust individuals into the genetic population, therefore biasing the evolutionary process towards this type of solutions. This Gradient Bias effect is obtained by hybridising the Generic Algorithm with the Deep Reinforcement Learning technique DDPG, resulting in the Explorative-DDPG algorithm (*X*-DDPG).

X-DDPG will be shown to solve the Generalization Limit of its genetic component via Gradient Bias, while outperforming its DDPG baseline in terms of agent return and speed of learning.

Keywords: Machine Learning · Genetic Algorithms · Deep reinforcement learning · DDPG · Gradient bias · Distributed exploration

1 Introduction

In recent years, researchers have shown the ability of Genetic Algorithms (GA) to extend beyond rudimentary Machine Learning applications to the training of

G. Nicosia et al. (Eds.): LOD 2020, LNCS 12565, pp. 273–284, 2020.
https://doi.org/10.1007/978-3-030-64583-0_26

autonomous agents, for a number of different tasks. Their general formulation makes it possible to address very complex problems [4,9], even relating to the building blocks of intelligence [2,19], with minimal intervention of the designers, leaving the process free to extract from raw data as much information as required. In the last years, GAs have been used to evolve agents that obtained performance comparable or even superior to those trained with Deep Reinforcement Learning algorithms in a number of benchmark environments [17,20]. Indeed, the modern trend seems to be leaning to GAs to solve more and more complex tasks.

However, there is a structural limitation. We show that, if for fixed policies the tasks present stochasticity (in the policy and/or in the environment) which leads to trajectories that are subject to high variance, GAs are unable to evolve robust and highly-performing controllers. We call this problem the *Generalization Limit of Genetic Algorithms*. To the best of our knowledge, no previous study highlights these issues as a structural limitation of GAs that makes them incompatible with most realistic problems concerning autonomous agents.

The main issue resides in the fact that initialized controllers are strongly biased towards poorly performing and high variance individuals, and that such solutions are a basin of attraction for GAs, which prefer such type of controllers and evolve solutions with these undesired characteristics.

To solve this problem, we propose to hybridize the GA with a popular Reinforcement Learning algorithm, Deep Deterministic Policy Gradient (DDPG) [13]. Controllers obtained with Reinforcement Learning techniques are intrinsically more robust, since the learning objective itself is to increase the theoretically expected score over the environment stochasticity. By periodically injecting DDPG-trained agents into the genetic population, evolution is biased towards more robust solutions. We name this phenomenon the *Gradient Bias* effect. The resulting hybrid algorithm is called *Explorative DDPG* (*X*-DDPG), combining a GA with a distributed variant of DDPG called AE-DDPG [23]. It will be shown that X-DDPG is able to successfully face the *Generalization Limit* of its genetic component, while also outperforming its RL baseline AE-DDPG.

Experiments are performed on RL benchmarks offered by the OpenAI Gym platform [5].

This paper is organized as follows. In *Sect.* 2, the work related to this research is illustrated. *Section* 3 is dedicated to the *Generalization Limit of Genetic Algorithms*, introduced with experimental results and then discussed to provide a better understanding of the underlying issue. In *Sect.* 4, GAs are hybridized with AE-DDPG to create the $X-$DDPG algorithm, and the mutual benefits that each component receives from the other are discussed, focusing on the *Gradient Bias*.

2   Related Work

The central topic of this research are Genetic Algorithms, applied to the end-to-end training of autonomous agents, a field where they have gained increasing success in the last decades [1,4]. In the field of Evolutionary Robotics [9],

Artificial Neural Networks are used as robot controllers and trained by Evolutionary Strategies. Notable are the works on *Minimal Cognition* [19], robot locomotion [3], and symbol grounding [2]. They have also proved to be valid alternatives to Reinforcement Learning techniques for some tasks [17,20]. However, no previous study addresses the core topic presented in this paper, the *Generalization Limit of Genetic Algorithms*, the low performance of GAs in the presence of high variance of the fitness functions. Stochasticity in GAs has been widely addressed in literature [11,15,22], either by stating the general need for a reliable fitness estimation or the desire for a robust solution. Our contribution moves forward by stating that the selection mechanism of GAs is incompatible with high-variance fitness distributions, and that this limit is thus a structural problem of the algorithm.

To address the *Generalization Limit*, we propose to support the evolutionary process by *hybridization* with a RL algorithm: DDPG. The idea of hybridizing RL algorithms with GAs is not new, though in most cases these algorithms are built to have GAs support the RL process, either by influencing the agent network weights [6,18] or the training data [16]. In [7,12], the genetic component is used as a *distributed exploration mechanism*, where the transitions gathered when evaluating the population's new individuals are used to train the Reinforcement Learning agent using DDPG. The present research builds on similar ideas. In particular, [12] also includes the technique of injecting the DDPG-trained agent into the genetic population. This is the core idea behind the *Gradient Bias* effect introduced in this paper.

We present a hybrid algorithm, $X-$DDPG, which uses as RL component a *distributed exploration* algorithm, AE-DDPG [23]. While in regular DDPG training experiences are obtained only from the learning agent, in *distributed exploration* algorithms many copies of the centralized agent contribute to these experiences. This idea gave birth to techniques such as A3C [14], Apex [10], Impala [8] and AE-DDPG [23] itself. Each method differs from the others in the way the information gathered by the distributed explorers is used in the training process.

3   The Generalization Limit of Genetic Algorithms

The perspective of being able to train neural policies with GAs, without the need of computationally expensive and unstable backpropagation, seems ideal. However, no previous study explicitly addresses the structural issue which we introduce. The aim of this Section is to support the validity of the following statement, which we propose as the core of our contribution:

> *"Classical Genetic Algorithms cannot be employed effectively for agent training in the presence of high variance of the fitness function"*.

3.1   Experiments

Individuals of the population are agents controlled by neural policies, whose architectures follow that of the original DDPG algorithm [13]. Experiments are performed in two benchmark environments from the OpenAI-Gym library, *LunarLanderContinuous-v2* and *Swimmer-v2*[1]. The *scalar fitness* of each individual is the average score along 5 different episodes. Figure 1 shows the mean and best *scalar fitness* of the population during the evolutionary process.

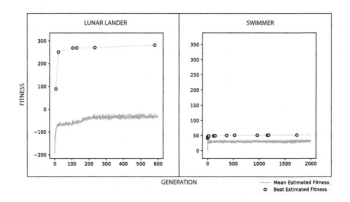

Fig. 1. *Mean and best assigned* scalar fitness *during evolution*

A complete evolutionary process was repeated in each environment 5 times, leading to similar early saturation of the fitness. The reason for this behavior may be caused by the very issue this research addresses.

Fitness Distribution. To monitor the expected value and variance of the true *fitness distributions* of individuals during evolution, they are estimated from data coming from 50 episode scores per-individual (Fig. 2). Note that the *scalar fitness* used in the evolutionary process is the average of 5 episode-wise scores instead.

We estimate the fitness distribution mean and spread of two groups of individuals, respectively:

- *Last generation population*: each individual of the last generation population. Figure 2a.
- *Best individuals in evolution history*: whenever an individual emerges with *scalar fitness* above the current best population fitness, we compute its *true fitness distribution*. Figure 2b.

It is clear that the *scalar fitness* is a high overestimation of the true expected value of the fitness distribution. Moreover, the variance is actually never reduced as learning proceeds. These results suggest that the algorithm is not actually finding solutions with highest performance.

---

[1] The structure and parameters of the algorithm can be found at https://github.com/AIRLab-POLIMI/GA-DRL, in the GA section.

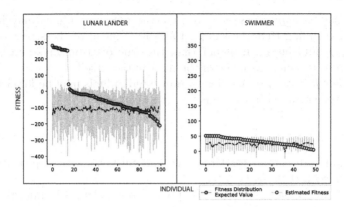

(a) *Fitness distribution and scalar fitness of each individual of the last generation population*

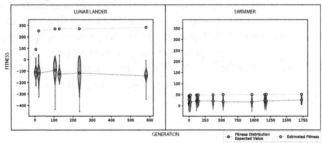

(b) *Fitness distribution and scalar fitness of individuals with highest scalar fitness in history*

Fig. 2. *Estimated* fitness distribution *expected value and spread compared to assigned* scalar fitness

3.2 Discussion

Multiple features of the classical Genetic Algorithm concur to generate the *Generalization Limit* in stochastic environments, and are illustrated in the following.

Evaluation

- Due to stochasticity, individual per-episode outcomes are not deterministic. Instead, each individual i has an associated *fitness distribution* f_i.
- The *scalar fitness* $\hat{f}_i$ of each individual, required for the evolutionary process, is only computed once as the algebraic mean over a limited number n of episodes.
- Computing an individual's *scalar fitness* from the outcome of n episodes is equivalent to sampling from the distribution of the algebraic mean, which is related to individual's own *fitness distribution* f_i (same expected value, variance rescaled by $\frac{1}{\sqrt{n}}$).

- As a result, *scalar fitnesses* $\hat{f}_i$ of high-variance individuals may be *overestimated*, meaning that the sampled algebraic mean of the scores over the n episodes would be much higher than the true distribution's expected value $\mathbb{E}_{f_i}$, due to the high variance of the distribution itself.
- The *scalar fitness* of robust/low-variance individuals may be lower than that of the overestimated individuals with high-variance.

Selection

- Individuals are chosen for Crossover and Survival with a probability that increases with their *scalar fitness* $\hat{f}_i$.
- Selected individuals will bias the propagation of genetic information, and therefore solution type, throughout evolution.
- The classical Genetic Algorithm uses only the individuals' *scalar fitnesses* for selection. There is no way to account for the true *fitness distribution*.
- Overestimated individuals present high *scalar fitness* but also high-variance, low-robustness *fitness distributions*. These overestimated individuals are therefore selected with high probability, but will transmit to future generations the actual low-performance genetic information of their distributions.
- *overestimated* individuals with high-variance distributions may be selected more frequently than *robust* individuals.

Initialized Population

- Only a portion of high-variance individuals will be *overestimated*.
- Initialized populations are usually mostly composed of high-variance individuals, providing the necessary initial bias for *overestimated* individuals to overrun the robust ones.

The problem is that Genetic Algorithms try to find solutions with the highest possible *scalar fitness*. When individuals are instead characterized by *fitness distributions*, individuals with high-variance are more likely to get higher *scalar fitness* than robust individuals do, because the latter, characterized by low variance, would have to also present high expected values to sample high values. Moreover, the initial population is mostly composed of individuals prone to *overestimation*, since random solutions are more likely to have high variance than high performance. As a result, high-variance solutions are a basin of attraction for classical GAs: the genetic search is biased towards these higher variance solutions, where the sampled fitness highly overestimates the true expected value.

4   *X*−DDPG

To confront the *Generalization Limit*, we propose a hybrid solution that combines the GA with an RL-based training process, with the aim of biasing the genetic search toward more robust solutions. We call this algorithm *Explorative-DDPG*.

4.1   Motivations

Addressing the Limits of Genetic Algorithms. When the amount of robust individuals is very low, the proportion of high-variance solutions will increase and eventually overrun the entire population. The problem could be solved by artificially injecting robust individuals in it. The problem is how to find a *source of robust individuals* without having to estimate their fitness distribution, which may in general be a very computationally expensive procedure.

Reinforcement Learning techniques find policies with the highest possible Q-value, which is the theoretical expected return, and, as a consequence, RL solutions are intrinsically robust.

We propose to support the genetic evolution with a parallel Reinforcement Learning process, used as a continuous source of more robust policies. Since RL agents are trained with backpropagation, we called this the *Gradient Bias* effect.

Addressing the Limits of Reinforcement Learning. GAs could also in turn support the RL process.

In most DRL algorithms, the Experience Replay is filled by experiences gathered only by the learning agent itself. This poor sample diversity may lead to early convergence, a problem addressed in literature with *distributed exploration* strategies. GAs could contribute to the distributed exploration, by inserting in the *Experience Replay* trajectories sampled by the genetic individuals when evaluated. This would provide a more diversified set of experiences available to the RL training.

These are the motivations that lead to the proposal of the novel algorithm $X-$DDPG.

4.2   The Algorithm

In $X-$DDPG, a distributed DDPG training and a Genetic evolution are executed in parallel. The two processes are mutually interacting.

- **Improve DDPG – Maximal Exploration:** The exploratory processes are now two. The DDPG agent learns from transitions stored in its Experience Replay, which is composed of two separate memories: one for the distributed AE-DDPG exploration, and the other for the Genetic Experiences.
- **Improve the GA – Gradient Bias:** At the beginning of each generation, a copy of the most recent DDPG agent is inserted in the population, and is subject to Evaluation and Selection. This mechanism *biases* the genetic search towards regions of more robust solutions, which the GA can traverse more efficiently than DDPG.

4.3   Experiments and Results

Experiments are performed on the two benchmark environments mentioned in Sect. 3. The structure of the neural policies and the DDPG training are the ones prescribed by the original DDPG paper [13][2].

X–DDPG and the Generalization Limit of Genetic Algorithms. We compare the *fitness distributions* of the best individuals in the evolution history (Fig. 3b) with those of the individuals from the last generation (Fig. 3a), for the two cases of regular GA and X−DDPG. Fitness distributions are estimated from 50 episode scores per-individual, with respect to the *scalar fitnesses*, which are only composed of 5.

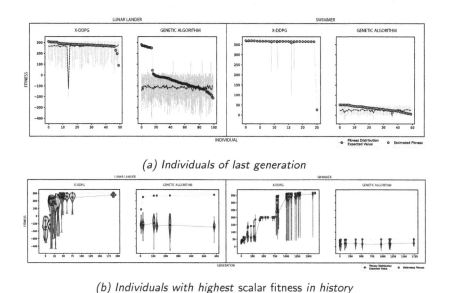

(a) Individuals of last generation

(b) Individuals with highest scalar fitness in history

Fig. 3. Estimated fitness distributions of individuals of last generation (3a) and individuals with highest scalar fitness (3b), computed over 50 episodes per-individual, with and without Gradient Bias

The best individuals throughout the evolution still show variability in their fitness distributions. However, their expected values are close to the maximum, around which most of the spread is concentrated, and thus, solutions are quite robust. Moreover, the *scalar fitness* is in general *not an overestimation* of the true expected value of the distribution. This is a very important property for a GA since, in general, fitness distributions are not computed during training, and designers must rely solely on the *scalar fitness* to monitor the evolutionary process and quality of the solutions.

---

[2] Further details on the algorithm configuration and parameters can be found at https://github.com/AIRLab-POLIMI/GA-DRL, in the X-DDPG section.

The sole *Gradient Bias* is able to make the evolutionary process not only feasible and reliable, but also capable of returning solutions with a quality comparable with the state of the art.

X–DDPG Outperforms RL Baselines. The algorithm uses as RL component the distributed AE-DDPG algorithm [23].

The performance of the X–DDPG gradient training is now compared to that of the pure AE-DDPG and regular DDPG. Results are reported in Fig. 4a, showing that the RL agent in X–DDPG outperforms agents trained by simple DDPG and AE-DDPG.

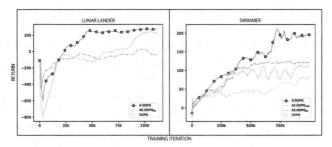

(a) *Comparing performance of agents trained by X − DDPG and its baselines AE-DDPG and DDPG*

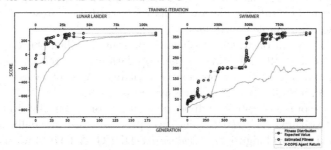

(b) *Comparison of the fitness of the best individuals in the history of the genetic search, showing both the* scalar fitness *and the* estimated distribution, *with the performance of the DDPG agent trained with X − DDPG*

Fig. 4. *Monitoring training agent performance*

The final results are a comparison of the performance of the solutions generated by the two components of X–DDPG, independently from each other. Figure 4b shows the performance of both its RL agent and the best individuals generated by its genetic evolution. The best individuals of the genetic search actually outperform the RL agents from the very first generations, reaching optimal regions. As a consequence, the overall algorithm obtains very high-performance solutions in a time even shorter than what required by the RL process, thanks to the GA, which is supported by the *Gradient Bias* provided by DDPG.

This is a very interesting result, as it shows that indeed GAs are very powerful tools to train autonomous agents, and that the *Generalization Limit* can be overcome to unleash their full potential. It is clear that genetic individuals do not only exploit the *Gradient Bias* in narrow regions surrounding the RL agent. Instead, this bias is just an initial tool, used to overcome the first barriers of high-variance solutions.

5   Conclusions

In this paper, a structural issue was presented, which makes Genetic Algorithms ineffective when the problem induces high variance of the fitness distributions. We called this the *Generalization Limit of Genetic Algorithms*, as the obtained controllers are unable to generalize their working behavior, resulting, in general, in this situation, in poor performance and unreliable learning.

In was shown that the problem arises from the interaction of GAs with the stochasticity of the trajectories. The latter leads to high variance *Fitness Distributions* of the controllers, while the first is prone to overestimation of the expected performance. When the variance is high, the overestimated fitness of higher variance individuals surpass that of the more robust, lower variance, individuals, and the overall population is biased towards the preservation of these high-variance, unreliable controllers.

In Sect. 4, a solution to this problem is presented, called the *Gradient Bias*. Indeed, by periodically injecting in the population more robust individuals, a threshold is reached where overestimated controllers cannot overrun the robust ones anymore, and are instead progressively discarded.

A reliable source of robust individuals comes from a parallel Reinforcement Learning algorithm, *AE-DDPG*. The two processes are mutually interacting, resulting in the hybrid algorithm *Explorative*-DDPG. *X*-DDPG was shown to be able to solve the *Generalization Limit* of the Genetic search, while also outperforming agents trained with pure AE-DDPG.

Genetic Algorithms are actually able to exploit regions with higher fitness more efficiently than DDPG, but only once such regions are approached by the population. This need of the genetic search to be biased outside lower performance regions may be the reason that led in the past to underestimate the potential of Genetic Algorithms.

5.1   Limits and Future Directions

Generality of the Statements. The presented results have been shown for a limited number of tasks. Our current research aim is to widen the range of environments to show the generality of both the issue and the solution that we presented. This will also include different incarnations of the Genetic Algorithm formulation.

A Pure Genetic Solution. Though promising, the *Gradient Bias* effect can only support the Genetic Algorithm when the GA is used to solve a problem formulated in an RL framework. Though it is argued by Sutton and Barto [21] that any learning problem can be formulated as such, it may still be an issue to always model a problem in this fashion. The ideal condition would be to solve the *Generalization Limit* with a variation of the pure GA. It has been shown in this paper that, once appropriately supported, GAs are more than capable to solve even complex problems, outperforming the more commonly used RL algorithms.

The Power of Hybridization. $X-$DDPG has been introduced in this paper mainly as a solution to the *Generalization Limit* of its genetic component, with the introduction of the *Gradient Bias* effect. However, in the environments under study, $X-$DDPG obtained state of the art results in terms of training iterations and final performance, by only affecting the training experiences. Future studies will focus on different types of DRL algorithms, applied to a larger and more diversified set of environments.

References

1. Bongard, J.: Morphological change in machines accelerates the evolution of robust behavior. Proc. Natl. Acad. Sci. **108**(4), 1234–1239 (2011). https://doi.org/10.1073/pnas.1015390108. ISSN 0027–8424. https://www.pnas.Org/content/108/4/1234
2. Bongard, J., Anetsberger, J.: Robots can groimd crowd-proposed symbols by forming theories of group mind. In: The 2019 Conference on Artificial Life, vol. 28, pp. 684–691 (2016). https://doi.org/10.1162/978-0-262-33936-0-chl09
3. Bongard, J., Pfeifer, R.: A method for isolating morphological effects on evolved behaviour, July 2003
4. Bongard, J., Zykov, V., Lipson, H.: Resilient machines through continuous self-modeling. Science **314**(5802), 1118–1121 (2006). https://doi.org/10.1126/science.1133687. ISSN 0036–8075. https://science.sciencemag.org/content/314/5802/1118
5. Brockman, G., et al.: OpenAI gym. CoRR abs/1606.01540 (2016). arXiv: 1606.01540. http://arxiv.org/abs/1606.01540
6. Chang, S., et al.: Genetic-gated networks for deep reinforcement learning, December 2018
7. Colas, C., Sigaud, O., Oudeyer, P.-Y.: GEP-PG: decoupling exploration and exploitation in deep reinforcement learning algorithms. CoRR abs/1802.05054 (2018). arXiv: 1802.05054. http://arxiv.org/abs/1802.05054
8. Espeholt, L., et al.: IMPALA: scalable distributed deep-RL with importance weighted actor-learner architectures. CoRR abs/1802.01561 (2018). arXiv: 1802.01561. http://arxiv.org/abs/1802.01561
9. Harvey, I., Husbands, P., Cliff, D.: Issues in evolutionary robotics (1992)
10. Horgan, D., et al.: Distributed prioritized experience replay. CoRR abs/1803.00933 (2018). arXiv: 1803.00933. http://arxiv.org/abs/1803.00933
11. Jin, Y., Branke, J.: Evolutionary optimization in uncertain environments- a survey. IEEE Trans. Evol. Comput. **9**(3), 303–317 (2005)

12. Khadka, S., Turner, K.: Evolution-guided policy gradient in reinforcement learning, November 2019
13. Lillicrap, T.P., et al.: Continuous control with deep reinforcement learning (2015). arXiv: 1509.02971 [cs.LG]
14. Mnih, V., et al.: Asjmchronous methods for deep reinforcement learning. CoRR abs/1602.01783 (2016). arXiv: 1602.01783. http://arxiv.org/ahs/1602.01783
15. Rakshit, P., Konar, A., Das, S.: Noisy evolutionary optimization algorithms-a comprehensive survey. Swarm Evol. Comput. **33**, 18–45 (2016). https://doi.org/10.1016/j.swevo.2016.09.002
16. Ramicic, M., Bonarini, A.: Selective perception as a mechanism to adapt agents to the environment: an evolutionary approach. IEEE Trans. Cogn. Dev. Syst. 1 (2019). https://doi.org/10.1109/TCDS.2019.2896306. ISSN 2379–8939
17. Salimans, T., et al.: Evolution strategies as a scalable alternative to reinforcement learning (2017). arXiv: 1703.03864 [stat.ML]
18. Sehgal, A., et al.: Deep reinforcement learning using genetic algorithm for parameter optimization. In: 2019 Third IEEE International Conference on Robotic Computing (IRC), pp. 596–601, February 2019
19. Slocum, A.C., et al.: Further experiments in the evolution of minimally cognitive behavior: from perceiving affordances to selective attention, pp. 430–439. MIT Press (2000)
20. Such, F.P., et al.: Deep neuroevolution: genetic algorithms are a competitive alternative for training deep neural networks for reinforcement learning. CoRR abs/1712.06567 (2017). arXiv: 1712.06567. http://cirxiv.org/abs/1712.06567
21. Sutton, R.S., Barto, A.G.: Reinforcement Learning: An Introduction. Second. The MIT Press, Cambridge (2018). http://incompleteideas.net/book/the-book-2nd.html
22. Yang, S., Ong, Y., Jin, Y.: Evolutionary Computation in Dynamic and Uncertain Environments, vol. 51 (2007). https://doi.org/10.1007/978-3-540-49774-5. ISBN 978-3-540-49772-1
23. Zhang, Z., et al.: Asynchronous episodic deep deterministic policy gradient: towards continuous control in computationally complex environments. CoRR abs/1903.00827 (2019). arXiv: 1903.00827. http://arxiv.org/abs/1903.00827

Relational Bayesian Model Averaging for Sentiment Analysis in Social Networks

Mauro Maria Baldi$^{(\boxtimes)}$ (ID), Elisabetta Fersini (ID), and Enza Messina (ID)

Department of Computer Science, University of Milano-Bicocca, Milan, Italy
{mauro.baldi,elisabetta.fersini,enza.messina}@unimib.it

Abstract. Nowadays, the exponential diffusion of information forces Machine Learning algorithms to take relations into account in addition to data, which are no longer independent. We propose a Bayesian ensemble learning methodology named Relational Bayesian Model Averaging (RBMA) which, in addition to a probabilistic ensemble voting, takes relations into account. We tested the RBMA on a benchmark dataset for Sentiment Analysis in social networks and we compared it with its previous non-relational variant and we show that the introduction of relations significantly improves the performance of classification. Moreover, we propose a model for making predictions when new data becomes available modifying and increasing the underneath graph of relations on which the RBMA was trained.

Keywords: Ensemble learning · Relational classifiers · Sentiment analysis

1 Introduction

Recent years have been characterized by an exponential diffusion of information and this trend is not going to cease. Nowadays, everything tends to be social and online [1]. This unavoidably leads to deal with relational data, where overall information is not given by just the values of data, but also by how data are linked. As an example, people in social networks publish messages and videos (data), but they also interact with other users (relations). Or, online corpora are huge collections of documents (data), but these documents are full of citations to other documents (relations). The importance of dealing with data and relations pushed researchers to propose a novel paradigm called approval network [2] which aims to model both constructualism and homophily. The ubiquitous presence of such networked data can reduce the effectiveness of classical Machine Learning algorithms. Classical Machine Learning algorithms might not be successful if exploited in relational datasets. In fact, these algorithms predict the category of new data assuming that all the instances in the dataset are independent. This assumption is clearly not true for relational datasets, in which the instances are linked through relations [3].

© Springer Nature Switzerland AG 2020
G. Nicosia et al. (Eds.): LOD 2020, LNCS 12565, pp. 285–296, 2020.
https://doi.org/10.1007/978-3-030-64583-0_27

Moreover, realistic classifiers are not error free. For the purpose of reducing the error of a classifier, a number of techniques known as Ensemble Learning are available in the literature [4]. These techniques consider the prediction of multiple classifiers on the same dataset. Although this approach improves the overall performance compared to those of a single classifier, it also has a drawback. In fact, the decision of each classifier making up the ensemble has the same weight compared to the remaining ones. In order to overcome this issue, a Bayesian Model Averaging (BMA) approach was proposed by Fersini et al. [5], which methodology aims to filter both data and model uncertainty.

The aim of this paper is threefold. First, we extend the theoretical aspects of the (non-relational) BMA to its relational version, namely the Relational Bayesian Model Averaging (RBMA). Second, we propose an initial comparison between non-relational and relational classifiers, the BMA and the RBMA on a benchmark dataset.

Finally, we propose a sentiment-analysis application of the RBMA where the sentiment of new people writing comments in social networks has to be inferred based on past reviews used to train the RBMA. This application is particularly complex for at least the following reasons: 1) data are not independent but related; 2) the underlying graph representing relations is not static but evolves over time.

This paper is organized as follows: in Sect. 2, we propose a brief state of the art of both relational classification and ensemble learning techniques. In Sect. 3, we introduce the background concept and we recall the basic model for training the BMA classifier. The same model is extended to the RBMA, which takes relations into account. In Sect. 4, we show how the RBMA can be used for making predictions. In particular, we present the sentiment-analysis application. Section 5 is devoted to the computational results. Finally, we conclude in Sect. 6.

2   Literature Review

The aim of this section is to provide a brief literature review on relational classifiers, ensemble learning techniques, together with their integrations. An early work about relational classification was proposed by Lu and Getoor [3], while more recent works can be found in [6–8]. Surveys about ensemble learning techniques can be found in [4,9,10]. A work integrating relational classification and ensemble learning algorithms can be found in [11], where the authors presented a methodology for dealing with heterogeneous, sparse and relational data. Recent works can be found in [12–14]. A tutorial and a review about a Bayesian approach for multiple classifiers can be found in [15,16], while an application for weather forecasts can be found in [17]. Finally, applications to Sentiment Analysis can be found in [5,18,19].

3   Background Concept and Model Training

Given a set of labels (or classes, or categories) $\mathcal{L}$, the aim of a classifier is to predict the class $l \in \mathcal{L}$ of a generic instance $\mathbf{x}$, having a dataset $\mathcal{D} = (\mathbf{X}, \mathbf{l})$ available, where $\mathbf{X}$ is a set of instances with their labels $\mathbf{l}$. Note that in a Natural Language Processing setting, matrix $\mathbf{X}$ can be represented by the tf-idf matrix [20]. An index $s \in S$ is associated to each instance in the dataset, where S denotes the set of instances. Thus, the s-th instance in the dataset has features $\mathbf{x}(s)$ and label $l(s)$, where $\mathbf{x}(s)$ is the s-th row of matrix $\mathbf{X}$ and $l(s)$ is the s-th row of column vector $\mathbf{l}$. For a non-relational classifier, the instances in the dataset are assumed to be independent. Vice versa, a relational classifier assumes the instances can be dependent and modeled through links or relations. Thus, the dataset $\mathcal{D}$ consists of the union of the features and labels plus a graph modeling the relations:

$$\mathcal{D} = (\mathbf{X}, \mathbf{l}, \mathcal{G}\{S, E\}), \tag{1}$$

where each node $s \in S$ corresponds to an instance in the dataset and each edge $e \in E$ represents a relation between two instances. As for a non-relational classifier, the aim of a relational classifier is to predict the label of a generic instance $\mathbf{x}$, but keeping relations into account in addition to the instances and the labels in the dataset.

The aim of the Bayesian Model Averaging is to provide predictions based on the results of independent classifiers. We name C the set of these single classifiers and $\mathcal{D}$ the dataset with features and labels. A single classifier $i \in C$ provides the probability $P(l(s) \mid i, \mathcal{D})$ for tuple s in dataset $\mathcal{D}$ to be classified with label $l(s)$. The Bayesian Model Averaging provides a weighted estimate based on the total probability theorem. In particular, the probability for tuple s to be labeled as $l(s)$ from the Bayesian Model Averaging is:

$$P(l(s) \mid C, \mathcal{D}) = \sum_{i \in C} P(l(s) \mid i, \mathcal{D})P(i \mid \mathcal{D}) \tag{2}$$

The posterior probability $P(i \mid \mathcal{D})$ can be computed through Bayes' theorem:

$$P(i \mid \mathcal{D}) = \frac{P(\mathcal{D} \mid i)P(i)}{P(\mathcal{D})} = \frac{P(\mathcal{D} \mid i)P(i)}{\sum\limits_{j \in C} P(\mathcal{D} \mid j)P(j)}, \tag{3}$$

where $P(i)$ is the prior probability for classifier $i \in C$ to vote and $P(\mathcal{D} \mid \cdot)$ is the model likelihood. The denominator in (3) is equal for each classifier $i \in C$ and so can be omitted. Moreover, each classifier taking part to the Bayesian Model Averaging has the same probability to vote, thus $P(i) = \frac{1}{|C|}$ and this constant term can also be omitted. Therefore, the label $l(s)$ predicted by the Bayesian Model Averaging for tuple s in dataset $\mathcal{D}$ is:

$$l^*(s) = \arg\max_{l(s)} P(l(s) \mid C, \mathcal{D}) = \arg\max_{l(s)} \sum_{i \in C} P(l(s) \mid i, \mathcal{D})P(\mathcal{D} \mid i). \tag{4}$$

The probability $P(\mathcal{D} \mid i)$ can be computed in an approximate manner as follows: first, a k-fold cross validation is performed over the dataset $\mathcal{D}$. After that, precision, recall and F1 measures are computed for each fold $l \in \{1, \ldots, k\}$ and

$P(\mathcal{D} \mid i)$ can be approximated as the average of the F1 measures in the various folds:

$$P(\mathcal{D} \mid i) \approx \frac{1}{k} \sum_{l=1}^{k} F1_{i,l} = \frac{1}{k} \sum_{l=1}^{k} \frac{2 \times P_{i,l} \times R_{i,l}}{P_{i,l} + R_{i,l}}, \tag{5}$$

where $P_{i,l}$, $R_{i,l}$ and $F1_{i,l}$ respectively denote the precision, the recall and the F1 measure of classifier $i \in C$ in fold $l \in \{1, \ldots, k\}$. Please, refer to [21] for a formal definition of precision, recall, F1 measure, and cross validation.

A relational classifier consists of a local (non-relational) classifier and another classifier which takes relations into account [22]. Relational classifiers are provided by toolkits such as NetKit [22]. If all the classifiers in set C are relational, we get the RBMA, still using formulae (2)–(5). These formulae also allow us to state the computational complexity of the RBMA. As seen in (3) and (5), it is strongly affected by the computational complexity of each classifier making up the RBMA because of probability $P(l(s) \mid i, \mathcal{D})$. Let $O(\mathcal{C})$ be the computational complexity of the hardest classifier to train in the RBMA, then for each fold we have $|C|$ classifier to train and $|C|$ F1-measures to compute according to (5). The latter operation requires a loop over the labels in the incumbent fold, which is $O(|\mathbf{l}|)$. Thus, the overall computational effort is:

$$O(k|C|(O(\mathcal{C}) + |\mathbf{l}|)). \tag{6}$$

4   RBMA: The Predictive Model

In Sect. 3, we presented a generic model with a detailed procedure for training the RBMA based on two fundamental results of probability theory: Bayes' theorem and the total probability theorem. But how can the RBMA be used for making predictions? In contrast to classical machine learning algorithms where the instances of both the training and the test set are independent, things become more complex when relations arise among data. Furthermore, in this section, we propose a testing procedure for making predictions when new unlabeled data becomes available with time, requiring to constantly update the underlying graph $\mathcal{G}\{S, E\}$ introduced in (1). In particular, a person leaving a comment becomes a node in dataset $\mathcal{D}$. However, comments and reviews can be directed to other people who previously wrote other comments. A classical example is when somebody comments a post on Facebook or Instagram including the tag of the person which is directed, or retweets a tweet on Twitter. According to the notation presented in Sect. 3, if a person $s_1 \in S$ writes a comment to another person $s_2 \in S$, we have the relation $(s_1, s_2) \in E$, where $E \in \mathcal{G}$. Information of person $s_1 \in S$ is gathered in vector $\mathbf{x}_{s_1}$, which is the s_1-th row of matrix $\mathbf{X} \in \mathcal{D}$.

A first group of people is initially selected with their comments and relations. This group forms the training set on which the RBMA is trained. This situation is symbolically depicted in Fig. 1, where, for the sake of simplicity, only three invented names and relations are given. Note that the weight w_{ij} on the arc

$(i, j) \in E$ represents the number of comments that person $s_i \in S$ wrote to person $s_j \in S$. Comments are aggregated and preprocessed according to Natural Language Techniques [20,21,23] and people in the training set are manually labeled as positive or negative. In particular, matrix $\mathbf{X}$ in (1) can represent the tf-idf matrix of the preprocessed and aggregated reviews of users. People labels, also known as polarities, represent the so-called sentiment of their reviews. In particular, people writing sentences of satisfaction and appreciation are labeled as positive, whilst people writing unhappy comments or complaints are labeled as negative.

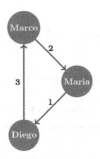

Fig. 1. The training set.

Figure 2 symbolically represents the moment when new data become available. For instance, new data can be available after a query to a web service. In particular, the graph on the left is the training set on which the RBMA was trained, while the graph on the right represent those people who wrote new comments and to whom they are directed, but which sentiment is unknown. According to the notation in Sect. 3, the dataset expression provided in (1) becomes:

$$\mathcal{D} = (\mathbf{X}_{TR}, \mathbf{l}_{TR}, \mathbf{X}_{TE}, \mathcal{G}\{S_{TR} \cup S_{TE}, E\}), \tag{7}$$

where $\mathbf{X}_{TR}$ is the tf-idf matrix of those people in the training set with their respective labels $\mathbf{l}_{TR}$, $\mathbf{X}_{TE}$ is the tf-idf matrix of those people in the test set, and the overall set of users S is given by the union of sets S_{TR} and S_{TE}. S_{TR} is the set of users in the training set, whilst S_{TE} is the set of users in the test set. Clearly, other matrix representations of information can be used in place of the tf-idf matrix. We do not write the set of the edges E as the union of the relations in the training set with those in the test set for a reason which we are going to explain with the help of Fig. 3. Figure 2 is also useful to introduce a concrete problem which has to be addressed when dealing with relational online data. In fact, in the example in Fig. 2, Elisa is a new user who writes three comments to Maria. But Maria is also in the training set because she has already written comments before the training of the RBMA. Hence, the situation depicted in Fig. 2 with the test set independent of the training set is not realistic. In fact,

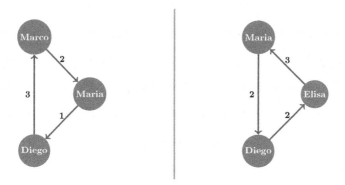

Fig. 2. The training and the test set separated.

not only data are relational, but the associated graph is dynamic and increases each time a query is performed.

We model this situation as suggested in Fig. 3, where the training and the test set are not separated but integrated through dummy edges. First, note that in Fig. 3 we added appropriate superscripts $_{TR}$ and $_{TE}$ according to (7) in order to emphasize the users respectively in the training and test set. Thus, in addition to the already existing arc linking Elisa with Maria in the test set, we add a dummy arc with the same weight from Elisa in the test set to Maria in the training set.

Moreover, we wish to point out a particular consideration one should keep into account. Again, we show the concept with an example. Maria writes three comments in total: one to Diego before the training procedure (i.e., the arc (Maria, Diego) in Fig. 1) and two to Diego after the training procedure and the query (i.e., the arc (Maria, Diego) in Fig. 2). We know Maria's sentiment because she belong to the group of people in the training set, but she is also present in the test set. For the latter group, the sentiment is unknown and will be predicted by the RBMA, Maria included. We assume that it is unlikely that someone opinion suddenly changes over time. Thus, in order to model this sentiment conservation, we add a dummy link from Maria in the test set to herself in the training set having the overall number of comments as weight. Links like this help the classifier to take previous sentiment into account. Similar dummy arcs can be traced as shown in Fig. 3.

We also wish to point out that in this model there are no links from users in the training set to users in the test set. The reason for this choice is for an easier management of the whole tf-idf matrix associated to the overall graph. Again, for the sake of simplicity, we show this updating procedure with the help of a toy example with the same users from Figs. 1 and 2. Still for simplicity, we are going to report tf matrices [21] rather than tf-idf ones. Let Table 1 represent the tf matrix $\mathbf{X}_{TR}$ of the users in the training set like in Fig. 1 and let Table 2 represent the tf matrix $\mathbf{X}_{TE}$ of the users in the training set like in Fig. 2. Unfortunately in principle the vocabularies in the two sets are not equal and thence it is not

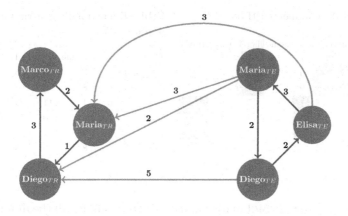

Fig. 3. The training and the test set integrated.

Table 1. The tf matrix for the training set.

Users	Car	Home	Train	Travel
Diego	7	5	0	0
Marco	0	8	4	0
Maria	0	1	11	5

Table 2. The tf matrix for the test set.

Users	Airplane	Home	Hotel	Train
Diego	2	6	5	0
Elisa	7	2	4	1
Maria	2	8	5	6

possible to concatenate the two tf matrices. However, we show this is possible after a simple manipulation.

In this example, the vocabulary of the training set is $V_{TR} = \{$car, home, train, travel$\}$ and the vocabulary of the test set is $V_{TE} = \{$airplane, home, hotel, train$\}$. First, it is necessary to join the two vocabularies in order to get $V = V_{TR} \cup V_{TE}$ $= \{$airplane, car, home, hotel, train and travel$\}$. At this point, we can pad with appropriate zero-column vectors the two matrices when necessary and get the merged matrix like in Table 3 where the additional zeros are highlighted in red and blue. This is the case for those words appearing in just one vocabulary. Note that words appearing in the test set but not in the training test do not have an impact on the output of any classifier because they correspond to a column of zeros in the training phase. Therefore, there is no need to include them in the merged matrix. Thus, we can eliminate these words from the overall vocabulary and from the merged matrix and get the final matrix in Table 4.

The labels for the users in the test set are still computed using (4), where $P(\mathcal{D}, |, i)$ is computed during the training procedure using (5) and $P(l(s) \mid i, \mathcal{D})$ is computed using the relational classifier $i \in C$ for all users $s \in S_{TE}$ on the overall dataset $\mathcal{D}$, this time given by:

$$\mathcal{D} = (\mathbf{X}, \mathbf{l}_{TR}, \mathcal{G}\{S_{TR} \cup S_{TE}, E\}), \tag{8}$$

Table 3. The merged tf matrix.

Users	airplane	car	home	hotel	train	travel
Diego$_{TR}$	0	7	5	0	0	0
Marco$_{TR}$	0	0	8	0	4	0
Maria$_{TR}$	0	0	1	0	11	5
Diego$_{TE}$	2	0	6	5	0	0
Elisa$_{TE}$	7	0	2	4	1	0
Maria$_{TE}$	2	0	8	5	6	0

Table 4. The final tf matrix for the overall dataset $\mathcal{D}$.

Users	Car	Home	Train	Travel
Diego$_{TR}$	7	5	0	0
Marco$_{TR}$	0	8	4	0
Maria$_{TR}$	0	1	11	5
Diego$_{TE}$	0	6	0	0
Elisa$_{TE}$	0	2	1	0
Maria$_{TE}$	0	8	6	0

where $\mathbf{X}$ is the overall tf-idf matrix and set E also includes all the dummy edges. The generalized testing procedure is presented in Algorithm 1.

To summarize, this model is able to deal with a dynamic graph avoiding to train the RBMA again if new data becomes available. This procedure allows to train the RBMA as shown in Sect. 3 only with the initial group of users. It is also able to model a change of opinion for an initial user. However, the addition of dummy arcs prevents to model unrealistic and abrupt changes of opinion.

5   Computational Results

Computational results were conducted on a benchmark dataset regarding Barack Obama [24]. This dataset consists of 159 messages (with positive and negative sentiments) written by 62 users on the Twitter social network. Moreover, there are also 267 relations among these users. Each relation is a tuple (x, y, w), where $w \in \mathbb{N}$ is the number of retweets of user $x \in S$ to user $y \in S$. Each user sentiment is given by the algebraic sum of the sentiment of her/his tweets. If this sum is zero, a positive sentiment is assigned to the user.

Computational tests were conducted in Python 3 on a workstation Intel Core TM i7 with 8 GB of RAM and Windows 10 as operationg system. The code uses NetKit [22], a powerful toolkit providing a set of relational classifiers working in three steps: 1) a non-relational (local) model, 2) a relational model and 3) a collective inferencing. NetKit relies on the Weka [20] machine learning toolkit. We used the following classifiers: naive Bayes, logistic and j48. We trained the BMA and the RBMA based on these classifiers. Table 5 shows a comparison among all these classifiers both in the non-relational and relational variant. The comparisons are reported in terms of the precisions (P), the recall (R) and the F1 measure (F1). The last three rows show the percentage of improvement between one relational classifier and its non-relational variant. In particular, given the figure of merit $m \in \{P, R, F1\}$, let $m_{i, R}$ be the result of the relational classifier $i \in C$ and let m_i be the result of the corresponding non-relational classifier for the same figure of merit, then the percentage improvement is given by:

$$\% \text{ improvement of classifier } i = 100 \times \frac{m_{i, R} - m_i}{m_i}. \tag{9}$$

Algorithm 1: The testing procedure

Data: the partial dataset $\mathcal{D} = (\mathbf{X}_{TR}, l_{TR}, \mathbf{X}_{TE}, \mathcal{G}\{S_{TR} \cup S_{TE}, E\})$
Data: the weight w_{ij} of edge $(i, j) \in E$
Data: the set of classifiers C
Data: the probabilities $P(\mathcal{D}, |, i)$, $\forall i \in C$ computed when training the RBMA
Data: the vocabulary of the training set V_{TR} sorted in alphabetical order
Data: the vocabulary of the test set V_{TE} sorted in alphabetical order
Result: the predicted label $l^*(s)$ $\forall s \in S_{TE}$
for i : $i_{TR} \in S_{TR} \wedge i_{TE} \in S_{TE}$ **do**
> **for** j : $j_{TE} \in S_{TE} \wedge (j_{TE}, i_{TE}) \in E$ **do**
>> $w_{j_{TE} i_{TR}} := w_{j_{TE} i_{TE}}$;
>> $E := E \cup \{(j_{TE}, i_{TR})\}$;
>
> **end**
> $w_{i_{TE} i_{TR}} = \sum_{j\,:\,(i_{TR}, j_{TR})\in E} w_{ij} + \sum_{j\,:\,(i_{TE}, j_{TR})\in E} w_{ij}$;
> $E := E \cup \{(i_{TE}, i_{TR})\}$;

end
$V := V_{TR} \cup V_{TE}$;
for w : $w \in V_{TE} \wedge w \notin V_{TR}$ **do**
> add a zero vector corresponding to the new word w in the tf-idf matrix $\mathbf{X}_{TR}$ respecting the alphabetical order;

end
for w : $w \in V_{TR} \wedge w \notin V_{TE}$ **do**
> add a zero vector corresponding to the new word w in the tf-idf matrix $\mathbf{X}_{TE}$ respecting the alphabetical order;

end
$\mathbf{X} := \left[\mathbf{X}_{TR}^T \, \mathbf{X}_{TE}^T\right]^T$;
$\mathcal{D} := (\mathbf{X}, l_{TR}, \mathcal{G}\{S_{TR} \cup S_{TE}, E\})$;
for $s \in S_{TE}$ **do**
> **for** $i \in C$ **do**
>> compute $P(l(s) \mid i, \mathcal{D})$;
>
> **end**
> $l^*(s) = \arg\max \sum_{i \in C} P(l(s) \mid i, \mathcal{D}) P(\mathcal{D} \mid i)$;

end

Table 1 reveals a number of properties. We can see that the BMA does not dominate the three classifiers in which it is rooted. This is obvious because, as its name says, the BMA performs a weighted mean based on each single classifier predictions according to (4). However, it is very helpful to increase the overall performance of the classifiers with respect to all the labels. For instance, the BMA provides an F1 measure for the positive labels of 0.625 which outperforms the F1 measure of the naive Bayes and the logistic classifier. However, the j48 classifier provides a greater F1 measure of 0.648. Nevertheless, for the negative labels, we find that the BMA outperforms the j48 classifier. Above all, we see that the RBMA outperforms all the classifiers with a F1 measure of 0.903 for both the labels and with a percentage improvement between 45% and 50%.

Table 5. Comparisons between relational and non-relational classifiers. For each line, we show in bold red and bold blue the best performance respectively for the negative and positive labels.

		Naive Bayes		Logistic		J48		BMA	
		neg	pos	neg	pos	neg	pos	neg	pos
Non-relational	P	0.593	0.571	0.571	0.593	0.636	0.575	0.621	**0.606**
	R	0.516	0.645	0.645	0.516	0.452	**0.742**	0.581	0.645
	F1	0.552	0.606	0.606	0.552	0.528	**0.648**	0.600	0.625
Relational	P	0.897	0.848	0.862	0.818	0.853	**0.929**	0.903	0.903
	R	0.839	**0.903**	0.806	0.871	0.935	0.839	0.903	**0.903**
	F1	0.867	0.875	0.833	0.843	0.892	0.881	0.903	**0.903**
% Improvement	P	51.265	48.511	50.963	37.943	34.119	**61.565**	45.411	49.010
	R	62.597	40.000	24.961	**68.798**	106.858	13.073	55.422	40.000
	F1	57.065	44.389	37.459	**52.717**	68.939	35.957	50.500	44.480

This shows the combined importance of relations and probabilistic ensemble voting in making predictions. We also computed the significance paired t-test (95% of confidence) on the results obtained by the RBMA against all the other approaches. The null hypothesis was rejected, denoting a statistical significance difference between the proposed approach and the state-of-the-art models.

6   Conclusions

Relations among data characterize real worlds applications. Thus, classical classifiers which assume instances of the dataset to be independent may lead to poorer performances if trained on a networked dataset because information contained in relations is lost. In this paper, we extended the Bayesian Model Averaging ensemble classifier to a more general setting which takes relations into account based on three classifiers: naive Bayes, logistic and j48. We found that the RBMA increases the average performance of classification basing its decisions on the predictions of each classifier. Then, we trained the RBMA considering the relational version of the aforementioned classifiers. We found a huge increase in performance for all the figures of merit, and this suggests that most relations bear a significant amount of information in a networked dataset. Finally, we presented a testing procedure for an updating graph able to model those settings where the sentiment of users writing new comments, posts, tweets or reviews has to be predicted based on past knowledge on which the RBMA was trained. This application is able to keep into account that users on which the RBMA was trained can leave new comments at a later moment, i.e., after the training procedure.

Our future perspectives consist in testing the Relational Bayesian Model Averaging with other datasets and in implementing an optimization algorithm which, starting with a set of classifiers making up the RBMA classifier, discards the least effective ones over the provided dataset.

Acknowledgements. This work has been partially funded by MISE (Ministero Italiano dello Sviluppo Economico) under the project "SMARTCAL – Smart Tourism in Calabria" (F/050142/01-03/x32).

Moreover, the authors are very grateful to Sofus A. Macskassy for his generosity in helping with the NetKit toolkit.

References

1. Weitz, S.: Search: How the Data Explosion Makes Us Smarter. GreenHouse Collection. Routledge, London (2014)
2. Fersini, E., Pozzi, F.A., Messina, E.: Approval network: a novel approach for sentiment analysis in social networks. World Wide Web **20**(4), 831–854 (2016). https://doi.org/10.1007/s11280-016-0419-8
3. Lu, Q., Getoor, L.: Link-based classification. In: ICML 2003: Proceedings of the Twentieth International Conference on International Conference on Machine Learning, pp. 496–503 (2003)
4. Rokach, L.: Ensemble-based classifiers. Artif. Intell. Rev. **33**(1), 1–39 (2010)
5. Fersini, E., Messina, E., Pozzi, F.A.: Sentiment analysis: Bayesian ensemble learning. Decis. Support Syst. **68**, 26–38 (2014)
6. Nickel, M., Murphy, K., Tresp, V., Gabrilovich, E.: A review of relational machine learning for knowledge graphs. Proc. IEEE **104**(1), 11–33 (2016)
7. Mooney, R.: Statistical relational learning and script induction for textual inference. Technical report AFRL-RI-RS-TR-2017-243, Air force research laboratory information dictorate (2017)
8. Ramanan, N., et al.: Structure learning for relational logistic regression: an ensemble approach. In: Proceedings of the Sixteenth International Conference on Principles of Knowledge Representation and Reasoning (KR 2018), pp. 661–662 (2018)
9. Gomes, H.M., Barddal, J.P., Enembreck, F., Bifet, A.: A survey on ensemble learning for data stream classification. ACM Computing Surveys **50**(2), 1–36 (2017). Article 23
10. Sagi, O.: L-Rokach: ensemble learning: a survey. WIREs Data Min. Knowl. Discov. (2018). https://doi.org/10.1002/widm.1249
11. Preisach, C., Schmidt-Thieme, L.: Ensembles of relational classifiers. Knowl. Inf. Syst. **14**(3), 249–272 (2008)
12. Li, Y., Zhong, S., Zhong, Q., Shi, K.: Lithium-ion battery state of health monitoring based on ensemble learning. IEEE Access **7**, 8754–8762 (2019)
13. Bablani, A., Edla, D.R., Tripathi, D., Kuppili, V.: An efficient concealed information test: EEG feature extraction and ensemble classification for lie identification. Mach. Vis. Appl. **30**(5), 813–832 (2019)
14. Alfred, R., Shin, K.K., Chin, K.O., Lau, H.K., Hijazi, M.H.A.: k-NN ensemble DARA approach to learning relational. In: Abawajy, J.H., Othman, M., Ghazali, R., Deris, M.M., Mahdin, H., Herawan, T. (eds.) Proceedings of the International Conference on Data Engineering 2015 (DaEng-2015). LNEE, vol. 520, pp. 203–212. Springer, Singapore (2019). https://doi.org/10.1007/978-981-13-1799-6_22
15. Hoeting, J.A., Madigan, D., Raftery, A.E., Volinsky, C.T.: Bayesian model averaging: a tutorial. Stat. Sci. **14**(4), 382–401 (1999)
16. Wasserman, L.: Bayesian model selection and model averaging. J. Math. Psychol. **44**, 92–107 (2000)

17. Raftery, A.E., Gneiting, T., Balabdaoui, F., Polakowski, M.: Using Bayesian model averaging to calibrate forecast ensembles. Mon. Weather Rev. **133**(5), 1155–1174 (2005)

18. Pozzi, F.A., Fersini, E., Messina, E.: Bayesian model averaging and model selection for polarity classification. In: Métais, E., Meziane, F., Saraee, M., Sugumaran, V., Vadera, S. (eds.) NLDB 2013. LNCS, vol. 7934, pp. 189–200. Springer, Heidelberg (2013). https://doi.org/10.1007/978-3-642-38824-8_16

19. Pozzi, F.A., Maccagnola, D., Fersini, E., Messina, E.: Enhance user-level sentiment analysis on microblogs with approval relations. In: Baldoni, M., Baroglio, C., Boella, G., Micalizio, R. (eds.) AI*IA 2013. LNCS (LNAI), vol. 8249, pp. 133–144. Springer, Cham (2013). https://doi.org/10.1007/978-3-319-03524-6_12

20. Witten, I.H., Frank, E., Hall, M.A., Pal, C.J.: Data Mining: Practical Machine Learning Tools and Techniques. Morgan Kaufmann, Burlington (2016)

21. Jurafsky, D., Martin, J.H.: Speech and Language Processing. Series in Artificial Intelligence. Prentice Hall, Upper Saddle River (2008)

22. Macskassy, S.A., Provost, F.: Classification in networked data: a toolkit and a univariate case study. J. Mach. Learn. Res. **8**, 935–983 (2007)

23. Swamynathan, M.: Mastering Machine Learning with Python in Six Steps: A Practical Implementation Guide to Predictive Data Analytics Using Python (2017)

24. Università degli Studî di Milano-Bicocca: The MIND laboratory. http://www.mind.disco.unimib.it/gallery/index.asp?cat=92&level=1&lang=en

Variance Loss in Variational Autoencoders

Andrea Asperti[✉]

Department of Informatics: Science and Engineering (DISI), University of Bologna,
Bologna, Italy
andrea.asperti@unibo.it

Abstract. In this article, we highlight what appears to be major issue of Variational Autoencoders (VAEs), evinced from an extensive experimentation with different networks architectures and datasets: the variance of generated data is significantly lower than that of training data. Since generative models are usually evaluated with metrics such as the Fréchet Inception Distance (FID) that compare the distributions of (features of) real versus generated images, the variance loss typically results in degraded scores. This problem is particularly relevant in a two stage setting [8], where a second VAE is used to sample in the latent space of the first VAE. The minor variance creates a mismatch between the actual distribution of latent variables and those generated by the second VAE, that hinders the beneficial effects of the second stage. Renormalizing the output of the second VAE towards the expected normal spherical distribution, we obtain a sudden burst in the quality of generated samples, as also testified in terms of FID.

1 Introduction

Since their introduction [19,21], Variational Autoencoders (VAEs) have rapidly become one of the most popular frameworks for generative modeling. Their appeal mostly derives from the strong probabilistic foundation; moreover, they are traditionally reputed for granting more stable training than Generative Adversarial Networks (GANs) [12].

However, the behaviour of Variational Autoencoders is still far from satisfactory, and there are a lot of well known theoretical and practical challenges that still hinder this generative paradigm. We may roughly identify four main (interrelated) topics that have been addressed so far:

balancing issue [3,5,7,8,15,18] a major problem of VAE is the difficulty to find a good compromise between sampling quality and reconstruction quality. The VAE loss function is a combination of two terms with somehow contrasting effects: the log-likelihood, aimed to reduce the reconstruction error, and the Kullback-Leibler divergence, acting as a regularizer of the latent space with the final purpose to improve generative sampling (see Sect. 2 for details). Finding a good balance between these components during training is a complex and delicate issue;

© Springer Nature Switzerland AG 2020
G. Nicosia et al. (Eds.): LOD 2020, LNCS 12565, pp. 297–308, 2020.
https://doi.org/10.1007/978-3-030-64583-0_28

variable collapse phenomenon [2,6,8,25,27]. The KL-divergence component of the VAE loss function typically induces a parsimonious use of latent variables, some of which may be altogether neglected by the decoder, possibly resulting in a under-exploitation of the network capacity; if this is a beneficial side effect or regularization (sparsity), or an issue to be solved (overpruning), it is still debated;

training issues VAE approximate expectations through sampling during training that could cause an increased variance in gradients [6,26]; this and other issues require some attention in the initialization, validation, and annealing of hyperparameters [4,5,15].

aggregate posterior vs. expected prior mismatch [1,8,11,18] even after a satisfactory convergence of training, there is no guarantee that the learned aggregated posterior distribution will match the latent prior. This may be due to the choice of an overly simplistic prior distribution; alternatively, the issue can e.g. be addressed by learning the actual distribution, either via a second VAE or by ex-post estimation by means of different techniques.

The main contribution of this article is to highlight an additional issue that, at the best of our knowledge, has never been pointed out so far: the variance of generated data is significantly lower than that of training data.

This resulted from a long series of experiments we did with a large variety of different architectures and datasets. The variance loss is systematic, although its extent may vary, and looks roughly proportional to the reconstruction loss.

The problem is relevant because generative models are traditionally evaluated with metrics such as the popular Fréchet Inception Distance (FID) that compare the distributions of (features of) real versus generated images: any bias in generated data usually results in a severe penalty in terms of FID score.

The variance loss is particularly serious in a two stage setting [8], where we use a second VAE to sample in the latent space of the first VAE. The reduced variance induces a mismatch between the actual distribution of latent variables and those generated by the second VAE, substantially hindering the beneficial effects of the second stage.

We address the issue by a simple renormalization of the generated data to match the expected variance (that should be 1, in case of a two stage VAE). This simple expedient, in combination with a new balancing technique for the VAE loss function discussed in a different article [3], are the basic ingredients that permitted us to get the *best FID scores* ever achieved with variational techniques over traditional datasets such as CIFAR-10 and CelebA.

The cause of the reduced variance is not easy to identify. A plausible explanation is the following. It is well known that, in presence of multimodal output, the mean square error objective typically results in blurriness, due to averaging (see [14]).

Variational Autoencoders are intrinsically multimodal, due to the sampling process during training, comporting averaging around the input data X in the data manifold, and finally resulting in the blurriness so typical of Variational Autoencoders [10]. The reduced variance is just a different facet of the same phenomenon: averaging on the data manifold eventually reduces the variance of data, due to Jensen's inequality.

The structure of the article is the following. Section 2 contains a short introduction to Variational Autoencoders from an operational perspective, focusing on the regularization effect of the Kullback-Leibler component of the loss function. In Sect. 3, we discuss the variance loss issue, relating it to a similar problem of Principal Component Analysis, and providing experimental evidence of the phenomenon. Section 4 is devoted to our approach to the variance loss, with experimental results on CIFAR-10 and CelebA, two of the most common datasets in the field of generative modeling. A summary of the content of the article and concluding remarks are given in Sect. 5.

2   Variational Autoencoders

A Variational Autoencoder is composed by an encoder computing an *inference* distribution $Q(z|X)$, and a decoder, computing the posterior probability $P(X|z)$. Supposing that $Q(z|X)$ has a Gaussian distribution $N(\mu_z(X), \sigma_z(X))$ (different for each data X), computing it amounts to compute its two first moments: so we expect the encoder to return the standard deviation $\sigma_z(X)$ in addition to the mean value $\mu_z(X)$.

During decoding, instead of starting the reconstruction from $\mu_z(X)$, we sample around this point with the computed standard deviation:

$$\hat{z} = \mu_z(X) + \sigma_z(X) * \delta$$

where δ is a random normal noise (see Fig. 1). This may be naively understood as a way to inject noise in the latent representation, with the aim to improve the robustness of the autoencoder; in fact, it has a much stronger theoretical foundation, well addressed in the literature (see e.g. [9]). Observe that sampling is outside the backpropagation flow; backpropagating the reconstruction error (typically, mean squared error), we correct the current estimation of $\sigma_z(X)$, along with the estimation of $\mu(X)$.

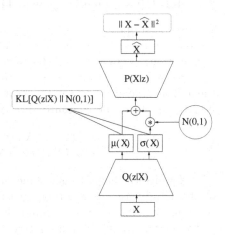

Fig. 1. VAE architecture

Without further constraints, $\sigma_z(X)$ would naturally collapse to 0: as a matter of fact, $\mu_z(X)$ is the expected encoding, and the autoencoder would have no reason to sample away from this value. The variational autoencoder adds an additional component to the loss function, preventing $Q(z|X)$ from collapsing to a dirac distribution: specifically, we try to bring each $Q(z|X)$ close to the prior $P(z)$ distribution by minimizing their Kullback-Leibler divergence $KL(Q(z|X)||P(z))$.

If we average this quantity on all input data, and expand KL-divergence in terms of entropy, we get:

$$
\begin{aligned}
&\mathbb{E}_X KL(Q(z|X)||P(z)) \\
&= -\mathbb{E}_X \mathcal{H}(Q(z|X)) + \mathbb{E}_X \mathcal{H}(Q(z|X), P(z)) \\
&= -\mathbb{E}_X \mathcal{H}(Q(z|X)) + \mathbb{E}_X \mathbb{E}_{z \sim Q(z|X)} log P(z) \\
&= -\mathbb{E}_X \mathcal{H}(Q(z|X)) + \mathbb{E}_{z \sim Q(z)} log P(z) \\
&= -\underbrace{\mathbb{E}_X \mathcal{H}(Q(z|X))}_{\substack{\text{Avg. Entropy} \\ \text{of } Q(z|X)}} + \underbrace{\mathcal{H}(Q(z), P(z))}_{\substack{\text{Cross-entropy} \\ \text{of } Q(X) \text{ vs } P(z)}}
\end{aligned}
\tag{1}
$$

By minimizing the cross-entropy between the distributions we are pushing $Q(z)$ towards $P(z)$. Simultaneously, we aim to augment the entropy of each $Q(z|X)$; assuming $Q(z|X)$ is Gaussian, this amounts to enlarge the variance, with the effect of improving the coverage of the latent space, essential for a good generative sampling. The price we have to pay is more overlapping, and hence more confusion, between the encoding of different datapoints, likely resulting in a worse reconstruction quality.

2.1 KL Divergence in Closed Form

We already supposed that $Q(X|z)$ has a Gaussian distribution $N(\mu_z(X), \sigma_z(X))$. Moreover, provided the decoder is sufficiently expressive, the shape of the prior distribution $P(z)$ can be arbitrary, and for simplicity it is usually assumed to be a normal distribution $P(z) = N(0, 1)$. The term $KL(Q(z|X)||P(z)$ is hence the KL-divergence between two Gaussian distributions $N(\mu_z(X), \sigma_z(X))$ and $N(1, 0)$ which can be computed in closed form:

$$
\begin{aligned}
&KL(N(\mu_z(X), \sigma_z(X)), N(0, 1)) = \\
&\tfrac{1}{2}(\mu_z(X)^2 + \sigma_z^2(X) - log(\sigma_z^2(X)) - 1)
\end{aligned}
\tag{2}
$$

The closed form helps to get some intuition on the way the regularizing effect of the KL-divergence is supposed to work. The quadratic penalty $\mu_z(X)^2$ is centering the latent space around the origin; moreover, under the assumption to fix the ratio between $\mu_z(X)$ and $\sigma_z(X)$ (rescaling is an easy operation for a neural network) it is easy to prove [1] that expression 2 has a minimum when $\mu_z(X)^2 + \sigma_z(X)^2 = 1$. So, we expect

$$
\mathbb{E}_X \mu(X) = 0
\tag{3}
$$

and also, assuming 3, and some further approximation (see [1] for details),

$$
\mathbb{E}_X \mu_z(X)^2 + \mathbb{E}_X \sigma_z^2(X) = 1
\tag{4}
$$

If we look at $Q(z) = \mathbb{E}_X Q(z|X)$ as a Gaussian Mixture Model (GMM) composed by a different Gaussian $Q(z|X)$ for each X, the two previous equations express the two moments of the GMM, confirming that they coincide with those of a normal prior. Equation 4, that we call *variance law*, provides a simple sanity check to ensure that the regularization effect of the KL-divergence is working as expected.

Of course, even if two first moments of the aggregated inference distribution $Q(z)$ are 0 and 1, it could still be very far from a Normal distribution. The possible mismatching between $Q(z)$ and the expected prior $P(z)$ is likely the most problematic aspect of VAEs since, as observed by several authors [1,16, 22], it could compromise the whole generative framework. Possible approaches consist in revising the VAE objective by encouraging the aggregated inference distribution to match $P(z)$ [23] or by exploiting more complex priors [4,17,24].

An interesting alternative addressed in [8] is that of training a second VAE to learn an accurate approximation of $Q(z)$; samples from a Normal distribution are first used to generate samples of $Q(z)$, that are then fed to the actual generator of data points. Similarly, in [11], the authors try to give an ex-post estimation of $Q(z)$, e.g. imposing a distribution with a sufficient complexity (they consider a combination of 10 Gaussians, reflecting the ten categories of MNIST and Cifar10).

These two works provide the current state of the art in generative frameworks based on variational techniques (hence, excluding models based on adversarial training), so we shall mostly compare with them.

3   The Variance Loss Issue

Autoencoders, and especially variational ones, seems to suffer from a systematic loss of variance of reconstructed/generated data with respect to source data. Suppose to have a training set X of n data, each one with m features, and let $\hat{X}$ be the corresponding set of reconstructed data. We measure the (mean) variance loss as the mean over data (that is over the first axis) of the differences of the variances of the features (i.e. over the second, default, axis):

$$\text{mean}(\text{var}(X) - \text{var}(\hat{X}))$$

Not only this quantity is always positive, but it is also approximately equal to the mean squared error (mse) between X and $\hat{X}$:

$$\text{mse}(X, \hat{X}) = \text{mean}((X - \hat{X})^2)$$

where the mean is here computed over all axes.

We observed the variance loss issue over a large variety of neural architectures and datasets. In particular cases, we can also give a theoretical explanation of the phenomenon, that looks strictly related to averaging. This is for instance the case of Principal Component Analysis (PCA), where the variance loss is precisely equal to the reconstruction error (it is well known that a *shallow* Autoencoder implements PCA, see e.g [13]).

Let us discuss this simple case first, since it helps to clarify the issue.

Principal component analysis (PCA) is a well know statistical procedure for dimensionality reduction. The idea is to project data in a lower dimensional space via an orthogonal linear transformation, choosing the system of coordinates that maximize the variance of data (principal components). These are easily computed as the vectors with the largest eigenvalues relative to the covariance matrix of the given dataset (centered around its mean points) (Fig. 2). Since

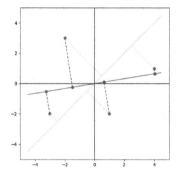

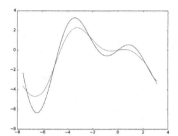

Fig. 2. The principal component is the green line. Projecting the red points on it, we maximize their variance or equivalently we minimize their quadratic distance. (Color figure online)

Fig. 3. The green line is a smoother version of the blue line, obtained by averaging values in a suitable neighborhood of each point. The two lines have the same mean; the mean squared error between them is 0.546, the variance loss is 2.648. (Color figure online)

the distance of each point from the origin is fixed, by the Pythagorean theorem, maximizing its variance is equivalent to minimize its quadratic error from the hyper-plane defined by the principal components. For the same reason, *the quadratic error of the reconstruction is equal to the sum of the variance errors of the components which have been neglected.*

This is a typical example of variance loss due to averaging. Since we want to renounce some components, the best we can do along them is to take the mean value. We entirely lose the variance along these directions, that is going to be paid in terms of reconstruction error.

3.1   General Case

We expect to have a similar phenomenon even with more expressive networks. The idea is expressed in Fig. 3. Think of the blue line as the real data manifold; due to averaging, the network reconstructs a smoother version of the input data, resulting in a significant loss in terms of variance.

The need for averaging may have several motivations: it could be caused by a dimensionality reduction, as in the case of PCA, but also, in the case of variational autoencoders, it could derive from the Gaussian sampling performed before reconstruction. Since the noise injected during sampling is completely unpredictable, the best the network can due is to reconstruct an "average image" corresponding to a portion of the latent space around the mean value $\mu_z(X)$, spanning an area proportional to the variance $\sigma_z(X)^2$.

In Fig. 4, we plot the relation between mean squared error (mse) and variance loss for *reconstructed images*, computed over a large variety of different neural architectures and datasets: the distribution is close to the diagonal. Typically, the variance loss for *generated images* is even greater. We must also account for a few pathological cases *not reported in the figure*, occurring with dense networks with very high capacity, and easily prone to overfitting. In this cases, mse is usually relatively high, while variance loss may drop to 0.

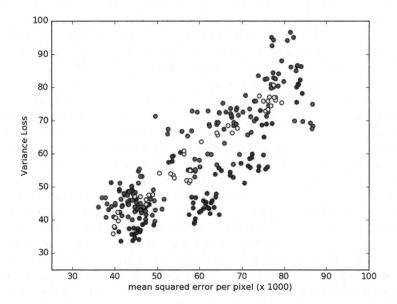

Fig. 4. Relation between mean squared error and variance loss. The different colors refer to different neural architectures: • (blue) Dense Networks; • (red) ResNet-like; • (green) Convolutional Networks; • Iterative Networks (DRAW-GQN-like) (Color figure online)

In the general, deep case, however, it is not easy to relate the variance loss to the mean squared error. We just discuss a few cases.

If for each data X, the reconstructed value $\hat{X}$ is comprised between X and its mean value μ, it is easy to prove that the mean squared error is a *lower bound* to the variance loss (the worse case is when $\hat{X} = \mu$, where the variance loss is just equal to the mean squared error, as in the PCA case).

Similarly, let X_p be an arbitrary permutation of elements of X and let $\hat{X} = (X + X_p)/2$. Then, the mean square distance between X and $\hat{X}$ is equal to the variance loss. However, the previous property does not generalize when we average over an arbitrary number of permutations; usually the mean squared error is lower than the quadratic distance between X and $\hat{X}$, but we can also get examples of the contrary.

We are still looking for a comfortable theoretical formulation of the property we are interested in.

4   Addressing the Variance Loss

As we explained in the introduction, the variance loss issue has a great practical relevance. Generative models are traditionally evaluated with metrics such as the popular Fréchet Inception Distance (FID) aimed to compare the distributions of real versus generated images trough a comparison of extracted features. In the case of FID, the considered features are inception features; inception is usually preferred over other models due to the limited amount of preprocessing performed on input images. As a consequence, a bias in generated data may easily result in a severe penalty in terms of FID score (see [20] for an extensive analysis of FID in relation to the training set).

The variance loss is particularly dangerous in a two stage setting [8], where a second VAE is used to sample in the latent space of the first VAE, in order to fix the possible mismatch between the aggregate inference distribution $Q(z)$ and the expected prior $P(z)$. The reduced variance induces a mismatch between the actual distribution of latent variables and those generated by the second VAE, hindering the beneficial effects of the second stage.

A simple way to address the variance loss issue consists in renormalizing generated data to match the actual variance of real data by applying a multiplicative scaling factor. We implemented this simple approach in a variant of ours of the two stage model of Dai and Wipf, based on a new balancing strategy between reconstruction loss and Kullback-Leibler described in [3]. We refer to this latter work for details about the structure of the network, hyperparameter configuration, and training settings, clearly outside the scope of this article. The code is available at https://github.com/asperti/BalancingVAE. In Fig. 5 we provide examples of randomly generated faces. Note the particularly sharp quality of the images, so unusual for variational approaches.

Both for CIFAR-10 and CelebA, the renormalization operation results in an improvement in terms of FID scores, particularly significant in the case of CelebA, as reported in Tables 1 and 2. At the best of our knowledge, these are the best generative results ever obtained for these datasets without relying on adversarial training. In the Tables, we compare our generative model with the original two-stage model in [8] and with the recent deterministic model in [11]; as we mentioned above, these approaches represent the state of the art for generative models not based on adversarial training. For our model, we provide scores with and without normalization. For each model, we give FID scores for reconstructed images (REC), images generated after the first stage (GEN-1), and images generated after the second stage (GEN-2). In the case of the deterministic

Fig. 5. Examples of **generated** faces. The resulting images do not show the blurred appearance so typical of variational approaches, significantly improving their perceptive quality.

Table 1. CIFAR-10: summary of results

Model	REC	GEN-1	GEN-2
RAE-l2 [11] (128 vars)	32.24±?	80.8±?	74.2±?
2S-VAE [8]		76.7 ± 0.8	72.9 ± 0.9
2S-VAE (ours)	53.8 ± 0.9	80.2 ± 1.3	69.8 ± 1.1
With normalization	53.5 ± 0.9	78.6 ± 1.2	**69.4 ± 1.0**

model [11], the "first stage" refers to sampling after fitting a Gaussian on the latent space, where the second stage refers to a more complex ex-post estimation of the latent space distribution via a GMM of ten Gaussians. The variance was computed over ten different trainings.

Table 2. CelebA: summary of results

Model	REC	GEN-1	GEN-2
RAE-SN [11]	36.0±?	44.7±?	40.9±?
2S-VAE [8]		60.5 ± 0.6	44.4 ± 0.7
2S-VAE (ours)	33.9 ± 0.8	43.6 ± 1.3	42.7 ± 1.0
With normalization	33.7 ± 0.8	42.7 ± 1.2	**38.6 ± 1.0**

Fig. 6. Faces with and without *latent space* re-normalization (right and left respectively). Images on the right have better contrasts and more definite contours.

In Fig. 6 we show the difference between faces generated from a same random seed with and without latent space re-normalization. We hope that the quality of images allows the reader to appreciate the improvement: renormalized images (on the right) have more precise contours, sharper contrasts and more definite details.

5   Conclusions

In this article, we stressed an interesting and important problem typical of autoencoders and especially of variational ones: the variance of generated data can be significantly lower than that of training data. We addressed the issue with a simple renormalization of generated data towards the expected moments of the data distribution, permitting us to obtain significant improvements in the quality of generated data, both in terms of perceptual assessment and FID score. On typical datasets such as CIFAR-10 and CelebA, this technique - in conjunction with a new balancing strategy between reconstruction error and Kullback-Leibler divergence - allowed us to get what seems to be the best generative results ever obtained without the use of adversarial training.

References

1. Asperti, A.: About generative aspects of variational autoencoders. In: Nicosia, G., Pardalos, P., Umeton, R., Giuffrida, G., Sciacca, V. (eds.) LOD 2019. LNCS, vol. 11943, pp. 71–82. Springer, Cham (2019). https://doi.org/10.1007/978-3-030-37599-7_7
2. Asperti, A.: Sparsity in variational autoencoders. In: Proceedings of the First International Conference on Advances in Signal Processing and Artificial Intelligence, ASPAI, Barcelona, Spain, 20–22 March 2019 (2019)
3. Asperti, A., Trentin, M.: Balancing reconstruction error and Kullback-Leibler divergence in Variational Autoencoders. CoRR, abs/2002.07514, February 2020
4. Bauer, M., Mnih, A.: Resampled priors for variational autoencoders. CoRR, abs/1810.11428 (2018)
5. Bowman, S.R., Vilnis, L., Vinyals, O., Dai, A.M., Józefowicz, R., Bengio, S.: Generating sentences from a continuous space. CoRR, abs/1511.06349 (2015)
6. Burda, Y., Grosse, R.B., Salakhutdinov, R.: Importance weighted autoencoders. CoRR, abs/1509.00519 (2015)
7. Burgess, C.P., et al.: Understanding disentangling in beta-vae, Nick Watters (2018)
8. Dai, B., Wipf, D.P.: Diagnosing and enhancing VAE models. In: Seventh International Conference on Learning Representations (ICLR 2019), 6–9 May, New Orleans (2019)
9. Doersch, C.: Tutorial on variational autoencoders. CoRR, abs/1606.05908 (2016)
10. Dosovitskiy, A., Brox, T.: Generating images with perceptual similarity metrics based on deep networks. In: Lee, D.D., Sugiyama, M., Luxburg, U., Guyon, I., Garnett, R. (eds.) Advances in Neural Information Processing Systems 29: Annual Conference on Neural Information Processing Systems 2016, 5–10 December 2016, Barcelona, Spain, pp. 658–666 (2016)
11. Ghosh, P., Sajjadi, M.S.M., Vergari, A., Black, M.J., Schölkopf, B.: From variational to deterministic autoencoders. CoRR, abs/1903.12436 (2019)
12. Goodfellow, I.J., et al.: Generative Adversarial Networks. ArXiv e-prints, June 2014
13. Goodfellow, I., Bengio, Y., Courville, A.: Deep Learning. MIT Press, Cambridge (2016). http://www.deeplearningbook.org
14. Goodfellow, I.J.: NIPS 2016 tutorial: Generative adversarial networks. CoRR, abs/1701.00160 (2017)
15. Higgins, I., et al.: beta-vae: Learning basic visual concepts with a constrained variational framework (2017)
16. Hoffman, M.D., Johnson, M.J.: Elbo surgery: yet another way to carve up the variational evidence lower bound. In: Workshop in Advances in Approximate Bayesian Inference, NIPS, vol. 1 (2016)
17. Kingma, D.P., et al.: Improving variational autoencoders with inverse autoregressive flow. In: Advances in Neural Information Processing Systems 29: Annual Conference on Neural Information Processing Systems 2016, 5–10 December 2016, Barcelona, Spain, pp. 4736–4744 (2016)

18. Kingma, D.P., Salimans, T., Welling, M.: Improving variational inference with inverse autoregressive flow. CoRR, abs/1606.04934 (2016)
19. Kingma, D.P., Max, W.: Auto-encoding variational bayes. In: 2nd International Conference on Learning Representations, ICLR 2014, Banff, AB, Canada, 14–16 April 2014, Conference Track Proceedings (2014)
20. Ravaglia, D.: Performance dei variational autoencoders in relazione al training set. Master's thesis, University of Bologna, school of Science, Session II (2020)
21. Rezende, D.J., Mohamed, S., Wierstra, D.: Stochastic backpropagation and approximate inference in deep generative models. In: Proceedings of the 31th International Conference on Machine Learning, ICML 2014, Beijing, China, 21–26 June 2014, volume 32 of JMLR Workshop and Conference Proceedings, pp. 1278–1286. JMLR.org (2014)
22. Rosca, M., Lakshminarayanan, B., Mohamed, S.: Distribution matching in variational inference (2018)
23. Tolstikhin, I.O., Bousquet, O., Gelly, S., Schölkopf, B.: Wasserstein auto-encoders. CoRR, abs/1711.01558 (2017)
24. Tomczak, J.M., Welling, M.: VAE with a vampprior. In: International Conference on Artificial Intelligence and Statistics, AISTATS 2018, 9–11 April 2018, Playa Blanca, Lanzarote, Canary Islands, Spain, pp. 1214–1223 (2018)
25. Trippe, B., Turner, R.: Overpruning in variational Bayesian neural networks. In: Advances in Approximate Bayesian Inference Workshop at NIPS 2017 (2018)
26. Tucker, G., Mnih, A., Maddison, C.J., Lawson, J., Sohl-Dickstein, J.: REBAR: low-variance, unbiased gradient estimates for discrete latent variable models. In: Guyon, I. (eds.) Advances in Neural Information Processing Systems 30: Annual Conference on Neural Information Processing Systems 2017, 4–9 December 2017, Long Beach, CA, USA, pp. 2627–2636 (2017)
27. Yeung, S., Kannan, A., Dauphin, Y., Li, F.-F.: Tackling over-pruning in variational autoencoders. CoRR, abs/1706.03643 (2017)

Wasserstein Embeddings for Nonnegative Matrix Factorization

Mickael Febrissy$^{(\boxtimes)}$ and Mohamed Nadif

Université de Paris, LIPADE, 75006 Paris, France
{mickael.febrissy,mohamed.nadif}@u-paris.fr

Abstract. In the field of document clustering (or dictionary learning), the fitting error called the Wasserstein (In this paper, we use "Wasserstein", "Earth Mover's", "Kantorovich–Rubinstein" interchangeably) distance showed some advantages for measuring the approximation of the original data. Further, It is able to capture redundant information, for instance synonyms in bag-of-words, which in practice cannot be retrieved using classical metrics. However, despite the use of smoothed approximation allowing faster computations, this distance suffers from its high computational cost and remains uneasy to handle with a substantial amount of data. To circumvent this issue, we propose a different scheme of NMF relying on the Kullback-Leibler divergence for the term approximating the original data and a regularization term consisting in the approximation of the Wasserstein embeddings in order to leverage more semantic relations. With experiments on benchmark datasets, the results show that our proposal achieves good clustering and support for visualizing the clusters.

1 Introduction

Nonnegative Matrix Factorization (NMF) provides a useful framework for learning and reducing in a smaller latent space of rank g, samples as well as features of a data matrix $X \in \mathbb{R}_+^{n \times d}$. The method consists in approximating X by the product of two low-dimensional matrices $Z \in \mathbb{R}_+^{n \times g}$ and $W \in \mathbb{R}_+^{d \times g}$, s.t. $X \approx ZW^\top$. Z will be seen as a coefficient matrix while W as the basis matrix. Text analysis or Dictionary Learning (where constraints are generally added to obtain sparsier coefficient vectors) are relevant applications of NMF. The Kullback-Leibler NMF proposed by Lee and Seung (1999), for instance, remains to this day quiet relevant for document clustering. Several other extensions of NMF aim to improve the quality of document clustering. We denote the orthogonal NMF (ONMF) (Ding et al. 2006) which consists in applying an orthogonal constraint on the coefficient matrix in order to deduce a clustering. Later, Yoo and Choi (2008) introduced another variant of ONNF using the Stiefield manifolds and Cai et al. (2010) proposed the Graph Regularized NMF (GNMF) where an affinity graph is constructed to encode the geometrical information. Further, 3 factors NMF/Nonnegative matrix Tri-factorization (NMTF) methods which aim to

© Springer Nature Switzerland AG 2020
G. Nicosia et al. (Eds.): LOD 2020, LNCS 12565, pp. 309–321, 2020.
https://doi.org/10.1007/978-3-030-64583-0_29

seek a decomposition of a data matrix X into 3 matrices have also been focusing on improving co-clustering performances (Labiod and Nadif 2011). Other approaches have been used combining data embedding and clustering (Allab et al. 2016). Nevertheless, despite all the notable efforts highlighting the potential of NMF for document clustering (Li and Ding 2018), these approaches still exhibit some limitations, namely they do not explicitly account for the semantic relationships between words as taken into account, for instance, by integrating a word embedding model into NMF (Ailem et al. 2017; Salah et al. 2017; 2018). Therefore, words having a common *meaning, synonyms* or more generally words that are about the same topic are not guaranteed to be mapped in the same direction within the latent space. This is simply due to the fact that words with similar meanings are not exactly used in similar documents. Consequently, the document embeddings resulting from the approximation are also not guaranteed to share all potential similarities when the documents are actually from the same topic. We illustrate our idea in the following example: Taking two groups of documents $group1 = \{$ "The professor is doing a lecture" (doc1), "The professor is giving a lesson" (doc2)$\}$ and $group2 = \{$ "The professor is on vacation in England" (doc3), "The students are on vacation in England" (doc4), "The students and their professor are on vacation in England" (doc5)$\}$ in terms of meaning but different regarding the words shared between each other. Considering a bag-of-word representation of these sentences (Table 1). In this example, we recognize that *lecture* and *lesson* are synonym. Nonetheless, if we compute the cosine similarity between those terms from X as shown in Fig. 1, they would not be related. In order to leverage this relation, we draw inspiration from several NMF algorithms (Sandler and Lindenbaum 2009, Rolet et al. 2016, Schmitz et al. 2018) aiming to overcome this issue (also regularly encountered in image processing) by using the Wasserstein distance (Villani 2008). This distance which aims to measure the gap between probability distributions/histograms is arguably less sensitive to these types of redundant representations. However, computing the distance between two histograms of dimension n is expensive and requires to solve a linear program in $\mathcal{O}(n^3 \log(n))$. Multiple works have shown that depending on the ground metric, it could be computed in $\mathcal{O}(n^2)$ time for instance using the L_1 ground distance (Ling and Okada 2007), or several orders of magnitude faster using threshold ground distances (Pele and Werman 2009). But in a NMF learning process using large-scale histograms for which the distance is computed between more than one pair, the process remains very expensive and time consuming. Therefore, to go further and take advantages of the Optimal Transport at a lower computational cost with NMF, we use the Wasserstein embeddings obtained from the Wasserstein distance computed between the two probability marginal of X. The model consists in transporting the weights of each marginals living in their respective simplex Δ_n and Δ_d into a respective lower dimensional simplex Δ_g using the data X and the lower dimensional factors data Z and W. Subsequently, a regularization of Z and W according to those embeddings is achieved. WE-NMF implies the computation of $g(n + d)$ Wasserstein parameters that can be stored inside two matrices $G \in \mathbb{R}_+^{g \times d}$ and $H \in \mathbb{R}_+^{g \times n}$. These parame-

Table 1. Document×term matrix.

	Professor	Lecture	Lesson	Vacation	Students	England
doc1	1	1	0	0	0	0
doc2	1	0	1	0	0	0
doc3	1	0	0	1	0	1
doc4	0	0	0	1	1	1
doc5	1	0	0	1	1	1

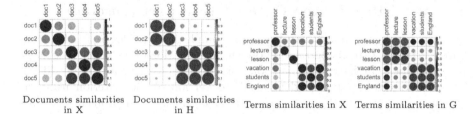

Documents similarities   Documents similarities   Terms similarities in X   Terms similarities in G
in X   in H

Fig. 1. Cosine similarity between documents or terms. The color and size indicate the binding force between the documents and the words in $X \in \mathbb{R}^{n \times d}$, $H \in \mathbb{R}^{g \times n}$ and $G \in \mathbb{R}^{g \times d}$.

ters deliver the optimal transportation for shifting the mass of documents (resp. terms) into the mass of the latent factors w_k with $k \in \{1, \ldots, g\}$ (resp. z_k). As shown in Fig. 1, we can see that this distance allows to highlight the relation between the two synonyms leading to a better understanding of relations between the documents citing those terms. Overall, we believe that this distance will be also relevant to leverage relations such as hyponyms (for instance: bus and car are hyponym of vehicle) which might be subject to reveal more proximities between documents.

The paper is organized as follow. In Sect. 2, we introduce the Optimal Transportation (OT) problem. In Sect. 3, we review its smooth approximation with the entropic regularization and the Cuturi sinkhorn-Knopp matrix scaling algorithm. In Sect. 4, we describe our model and algorithm. Section 5 presents our results obtained from several experiences on real world datasets.

2   Optimal Transport and Wasserstein Distance

Let $\mu = \sum_{i=1}^{n} a_i \delta_{x_i}$ be an empirical measure with a family of points $\mathcal{X} = (x_1, \ldots, x_n) \in \Omega^n$ and weights $a = (a_1, \ldots, a_n)$ living in the simplex Δ_n (where Ω is an arbitrary space and δ_{x_i} the Dirac unit mass on x_i). Let ν be another empirical measure with family $\mathcal{Y} = (y_1, \ldots, y_m) \in \Omega^m$ and weights $b = (b_1 \ldots, b_m)$ living in the simplex Δ_m. The Wasserstein distance between μ and ν, also known as the transportation problem is defined as the optimization of the following problem:

$$W_p(\mu, \nu) = \mathbf{p}(\boldsymbol{a}, \boldsymbol{b}, \boldsymbol{M}_{\mathcal{X}\mathcal{Y}}) = \min_{\boldsymbol{T} \in U(\boldsymbol{a}, \boldsymbol{b})} \langle \boldsymbol{T}, \boldsymbol{M}_{\mathcal{X}\mathcal{Y}} \rangle_F \tag{1}$$

where $U(\boldsymbol{a}, \boldsymbol{b})$ is the transportation polytope acting as the feasible set of all matrices $\boldsymbol{T} = (t_{ij}) \in \mathbb{R}_+^{n \times m}$ with the row and column marginals respectively equal to $\boldsymbol{a}$ and $\boldsymbol{b}$ s.t. $U(\boldsymbol{a}, \boldsymbol{b}) = \{\boldsymbol{T} \in \mathbb{R}_+^{n \times m} | \sum_i t_{ij} = a_i, \sum_j t_{ij} = b_j\}$, $\boldsymbol{M}_{\mathcal{X}\mathcal{Y}} = (m_{ij})$ is the matrix of pairwise distances (also called the cost parameter) between elements of $\mathcal{X}$ and $\mathcal{Y}$, $\mathbf{p}(\boldsymbol{a}, \boldsymbol{b}, \boldsymbol{M}_{\mathcal{X}\mathcal{Y}})$ is the Wasserstein distance in a form of the optimum of a linear program on $n \times m$ variables and parameter $\boldsymbol{a}, \boldsymbol{b}$ and $\boldsymbol{M}_{\mathcal{X}\mathcal{Y}}$; $\langle \boldsymbol{T}, \boldsymbol{M}_{\mathcal{X}\mathcal{Y}} \rangle_F = tr(\boldsymbol{T}^\top \boldsymbol{M}_{\mathcal{X}\mathcal{Y}}) = \sum_{i,j}^{n,m} t_{ij} m_{ij}$ is the Frobenius dot-product.

3   Cuturi Regularized Optimal Transport (Discrete)

$W_p(\mu, \nu)$ is a linear function with a cubic complexity $\mathcal{O}(n^3 \log(n))$ (when computed between two histograms of dimension n). Moreover, when n is large, $W_p(\mu, \nu)$ does not have a unique solution. In order to leverage these difficulties, Cuturi (2013) introduced a penalized version of the criterion using the Shannon's entropy which has for effects to smooth the linear problem and turns it into a strictly convex problem which can be solved faster. The regularized criterion $W_p^\lambda(\mu, \nu)$ takes the following form:

$$\mathbf{p}_\lambda(\boldsymbol{a}, \boldsymbol{b}, \boldsymbol{M}_{\mathcal{X}\mathcal{Y}}) = \min_{\boldsymbol{T} \in U(\boldsymbol{a}, \boldsymbol{b})} \langle \boldsymbol{T}, \boldsymbol{M}_{\mathcal{X}\mathcal{Y}} \rangle_F - \lambda^{-1} H(\boldsymbol{T}) \tag{2}$$

where $H(\boldsymbol{T}) = -\sum_{i,j}^{n,m} t_{ij} \log(t_{ij})$ is the Shannon's entropy and $\lambda \in [0, \infty]$ the regularization parameter. Depending on the value of λ, the smooth criterion converges toward the classical Wasserstein distance. If $\lambda \longrightarrow \infty$, $H(\boldsymbol{T})$ decreases and leans toward $W_p(\mu, \nu)$ (Deterministic coupling). In this case, $W_p^\lambda(\mu, \nu)$ becomes as or even more difficult to solve than the classical problem using a efficient linear solver. If $\lambda \longrightarrow 0$, $H(\boldsymbol{T})$ increases and pulls away $W_p^\lambda(\mu, \nu)$ from $W_p(\mu, \nu)$ (Independent coupling where μ and ν are assumed to be more independent). To solve $W_p^\lambda(\mu, \nu)$, $\boldsymbol{T}$ can be formulated as the solution of a scaling problem such as:

$$\boldsymbol{T} = diag(\boldsymbol{a}) \, \boldsymbol{K} \, diag(\boldsymbol{b}) \tag{3}$$

where $\boldsymbol{K}$ is the Gibbs kernel s.t $\boldsymbol{K} = e^{-\lambda \boldsymbol{M}_{\mathcal{X}\mathcal{Y}}}$. To obtain $\boldsymbol{T}$, solution of $W_p^\lambda(\mu, \nu)$, the Sinkhorn-Knopp's algorithm which has a complexity $\mathcal{O}(nm)$ is the most commonly used. It involves matrix/vector multiplications and converges with a speed of several orders of magnitude faster than the regular EMD (Earth Mover's Distance) solvers. A version of the algorithm adapted for the Wasserstein distance can be found in Cuturi (2013) as well as an updated version in Cuturi and Doucet (2014) which also solves the dual problem of Eq. (2). In the following, we will refer to this algorithm as SD for Sinkhorn Distance and its optimal solution for $\boldsymbol{T}$ as $\boldsymbol{T}^*$. It is also notable to note that Eq. (2) can be seen as a relative entropy and becomes a projection problem similar to the one encountered in NMF-KL; Eq. (2) is equivalent to $\min_{\boldsymbol{T} \in U(\boldsymbol{a}, \boldsymbol{b})} \mathcal{D}_{KL}(\boldsymbol{T}, \boldsymbol{K})$ where $\mathcal{D}_{KL}$ is the Kullback-Leibler divergence. The Sinkhorn-Knopp's algorithm can easily be adapted to matrix/matrix multiplications to allow the computation of the Wasserstein distance between one histogram and a set of histograms.

4   Wasserstein Embeddings NMF (WE-NMF)

Let $\boldsymbol{X} \in \mathbb{R}_+^{n \times d}$ be a document-term matrix. NMF using the Wasserstein distance as an error for approximating several histograms $\boldsymbol{x}_j \in \mathbb{R}_+^n$ in $\boldsymbol{X}$ can be stated as $\|\boldsymbol{X} - \boldsymbol{Z}\boldsymbol{W}^\top\| = \sum_j^d W_p(\boldsymbol{x}_j, [\boldsymbol{Z}\boldsymbol{W}^\top]_j)$ where $[\boldsymbol{Z}\boldsymbol{W}]_j \in \Delta_n$. This implies a number of d intermediate linear calculus of complexity $\mathcal{O}(n^3 \log(n))$ using W_p, or d matrix scaling problems of complexity $\mathcal{O}(n^2)$ using W_p^λ. Both methods, EMD-NMF Sandler and Lindenbaum (2009) and W-NMF Rolet et al. (2016) propose solutions to speed up the computational time. EMD-NMF uses the wavelet EMD approximation Shirdhonkar and Jacobs (2008) while W-NMF uses the Legendre-Fenchel conjugate of W_p^λ which has a closed form gradient and benefits from GPU parallelization. However with both methods, the overall computational time remains substantial for high dimensional data. With WE-NMF that we propose, we initiate a different approach aiming at reducing the amount of intermediate matrix scaling problems by considering only 4 histograms: the respective marginals of $\boldsymbol{X}$ of sizes d and n and their respective representations for the latent factors $\boldsymbol{z}_k \in \mathbb{R}_+^d$ and $\boldsymbol{w}_k \in \mathbb{R}_+^d$ of size g. Therefore, in WE-NMF we have the computation of two Wasserstein distances: $W_p^\lambda(\mu, \mu_{bis})$ (with the cost computed between the column vectors $\boldsymbol{x}_j$'s and the factors $\boldsymbol{z}_k$'s) and $W_p^\lambda(\nu, \nu_{bis})$ (with the cost computed between the row vectors $\boldsymbol{x}_i \in \mathbb{R}_+^d$ and the factors $\boldsymbol{w}_k$'s). Let $x_{i\cdot} = \sum_j x_{ij}$, $x_{\cdot j} = \sum_i x_{ij}$ and $\overline{Z} \in \{0,1\}^{n \times g}$ (resp. $\overline{W} \in \{0,1\}^{d \times g}$) be the classification matrix deduced from $\boldsymbol{Z}$ (resp. $\boldsymbol{W}$). We denote the respective weights for μ and μ_{bis} with $\boldsymbol{a} = (\frac{x_{1\cdot}}{n}, \dots, \frac{x_{n\cdot}}{n}) \in \Delta_n$ and $\boldsymbol{a}_{bis} = (\sum_i a_i \overline{z}_{i1}, \dots, \sum_i a_i \overline{z}_{ig}) \in \Delta_g$; the respective weights for ν and ν_{bis} with $\boldsymbol{b} = (\frac{x_{\cdot 1}}{d}, \dots, \frac{x_{\cdot d}}{d}) \in \Delta_d$ and $\boldsymbol{b}_{bis} = (\sum_j b_j \overline{w}_{j1}, \dots, \sum_j b_j \overline{w}_{jg}) \in \Delta_g$. Let $\boldsymbol{T}$ be the transportation matrix in the polytope $U(\boldsymbol{b}, \boldsymbol{b}_{bis})$ associated with the cost matrix $\boldsymbol{M}_{ZX} = [D(\boldsymbol{z}_k, \boldsymbol{x}_j)^p]_{kj} \in \mathbb{R}_+^{g \times d}$, $\boldsymbol{S}$ the transportation matrix in $U(\boldsymbol{a}, \boldsymbol{a}_{bis})$ associated with the cost matrix $\boldsymbol{M}_{WX} = [D(\boldsymbol{w}_k, \boldsymbol{x}_i)^p]_{ki} \in \mathbb{R}_+^{g \times n}$ and D is the ground metric. Thereby, we define the Wasserstein embedding matrices as:

$$\boldsymbol{G} \overset{def}{=} \boldsymbol{T} \odot \boldsymbol{M}_{ZX} \quad \text{and} \quad \boldsymbol{H} \overset{def}{=} \boldsymbol{S} \odot \boldsymbol{M}_{WX} \tag{4}$$

where $\odot$ refers to the Hadamard product. The parameter b_{bis} denotes the samples weights detained per each cluster of samples while a_{bis} denotes the features weights per cluster of features. Both are respectively updated at each iteration of the algorithm. In the sequel, we aim to solve the following problem which consists in minimizing the objective function $\mathcal{F}(\boldsymbol{Z}, \boldsymbol{W}, \boldsymbol{G}, \boldsymbol{H})$ taking the following form:

$$\min_{\boldsymbol{Z}, \boldsymbol{W}, \geq 0} \mathcal{D}_{KL}(\boldsymbol{X}, \boldsymbol{Z}\boldsymbol{W}^\top) + \gamma(\mathcal{D}_{KL}(\boldsymbol{G}, \boldsymbol{Z}^\top) + \mathcal{D}_{KL}(\boldsymbol{H}, \boldsymbol{W}^\top)) \tag{5}$$

where γ a regularization parameter $\in \mathbb{R}_+$. The minimization of the first two terms of (5) can be achieve through a set of multiplicative update rules. Let $\boldsymbol{\alpha} \in \mathbb{R}_+^{n \times g}$, $\boldsymbol{\beta} \in \mathbb{R}_+^{d \times g}$ and $\boldsymbol{\alpha} \in \mathbb{R}_+^{d \times g}$ be the Lagrange multipliers, the Lagrangian function $\mathcal{L}(\boldsymbol{Z}, \boldsymbol{W}, \boldsymbol{\alpha}, \boldsymbol{\beta})$ is equal to $\mathcal{F}(\boldsymbol{Z}, \boldsymbol{W}) + \boldsymbol{\alpha}\boldsymbol{Z} + \boldsymbol{\beta}\boldsymbol{W}$. The resulting gradients are $\nabla\mathcal{L}(Z_{ik}) = -\frac{\boldsymbol{X}}{\boldsymbol{Z}\boldsymbol{W}^\top}\boldsymbol{W} + \sum_j w_{jk} - \gamma\frac{\boldsymbol{H}^\top}{\boldsymbol{Z}} + \gamma + \boldsymbol{\alpha}$ and $\nabla\mathcal{L}(W_{jk}) = -\frac{\boldsymbol{X}^\top}{\boldsymbol{W}\boldsymbol{Z}^\top}\boldsymbol{Z} +$

$\sum_i z_{ik} + -\gamma \frac{G^\top}{W} + \gamma + \boldsymbol{\beta}$. Making use of the Kuhn Tucker conditions, we obtain: $\boldsymbol{Z} \odot \left(\frac{X}{ZW^\top} \boldsymbol{W} + \gamma \frac{H^\top}{Z}\right) - \boldsymbol{Z} \odot \left(\sum_j w_{jk} + \gamma\right) = 0$ and $\boldsymbol{W} \odot \left(\frac{X^\top}{WZ^\top} \boldsymbol{Z} + \gamma \frac{G^\top}{W}\right) - \boldsymbol{W} \odot \left(\sum_i z_{ik} + \gamma\right) = 0$. This leads to the following update rules:

$$Z = Z \odot \frac{\frac{X}{ZW^\top} W + \gamma \frac{H^\top}{Z}}{\sum_j w_{jk} + \gamma} \tag{6}$$

$$W = W \odot \frac{\frac{X^\top}{WZ^\top} Z + \gamma \frac{G^\top}{W}}{\sum_i z_{ik} + \gamma}. \tag{7}$$

Algorithm 1. Wasserstein Embeddings NMF (WE-NMF), $\mathcal{O}(ngd + gN)$.

Input: $\boldsymbol{X}$, γ, λ, p, $\boldsymbol{a} \in \Delta_n$, $\boldsymbol{b} \in \Delta_d$ and g.
Output: $\boldsymbol{Z}$, $\boldsymbol{W}$, $\boldsymbol{G}$, and $\boldsymbol{H}$.
Initialization: $\boldsymbol{Z} \leftarrow \boldsymbol{Z}^{(0)}$; $\boldsymbol{W} \leftarrow \boldsymbol{W}^{(0)}$
repeat
  1. $\boldsymbol{M}_{ZX} = \left[D(\boldsymbol{z}_k, \boldsymbol{x}_j)^p\right]_{kj}$, update $\boldsymbol{b}_{bis}$
  1′. $\boldsymbol{T} \leftarrow \boldsymbol{T}^*$ using $SD(\boldsymbol{M}_{ZX}, \lambda, \boldsymbol{b}, \boldsymbol{b}_{bis})$, $\boldsymbol{G} = \boldsymbol{T} \odot \boldsymbol{M}_{ZX}$
  2. $\boldsymbol{M}_{WX} = \left[D(\boldsymbol{w}_k, \boldsymbol{x}_i)^p\right]_{ki}$, update $\boldsymbol{a}_{bis}$
  2′. $\boldsymbol{S} \leftarrow \boldsymbol{S}^*$ using $SD(\boldsymbol{M}_{WX}, \lambda, \boldsymbol{a}, \boldsymbol{a}_{bis})$, $\boldsymbol{H} = \boldsymbol{S} \odot \boldsymbol{M}_{WX}$
  3. $\boldsymbol{Z} \leftarrow \boldsymbol{Z} \odot \frac{\frac{X}{ZW^\top} W + \gamma \frac{H^\top}{Z}}{\sum_j w_{jk} + \gamma}$
  4. $\boldsymbol{W} \leftarrow \boldsymbol{W} \odot \frac{\frac{X^\top}{WZ^\top} Z + \gamma \frac{G^\top}{W}}{\sum_i z_{ik} + \gamma}$
until convergence
  5. Normalize each $\boldsymbol{z}_k$ to unit-norm.
In this case, D is the cosine dissimilarity and $p = 2$. SD stands for Sinkhorn Distance. Note that step (1,1′) and step (2,2′) are independent and can be parallelized.

The convergence of WE-NMF is demonstrated. However, it was not included due to page limit.

In the worst case scenario the complexities of the multiplicative updates (6, 7) remain the same as for NMF which is $\mathcal{O}(ngd)$. However the computational cost of updates (4) becomes the main bottleneck as the complexity of WE-NMF depends directly on the complexity of the chosen algorithm used to compute $\boldsymbol{T}$ and $\boldsymbol{S}$. Using W_p^λ of complexity $\mathcal{O}(gd)$ for $\boldsymbol{G}$ and $\mathcal{O}(gn)$ for $\boldsymbol{H}$, the complexity for Eq. (4) is then $\mathcal{O}(gN)$ where $N = \max(d, n)$, leading to the overall complexity for one iteration at $\mathcal{O}(ngd + gN)$.

5   Experiments

To assess the performance of our model, we compare it with several NMF models commonly used for document clustering as well as Sinkhorn Distance/Earth Mover's Distance clustering methods. The list includes: orthogonal

NMF (ONMF) (Yoo and Choi 2008), Projective NMF (PNMF) (Yuan and Oja 2005), Graph Regularized NMF (GNMF) (Cai et al. 2010), Spherical K-means (Buchta et al. 2012), and Variational Wasserstein Clustering (VWC) (Mi et al. 2018), which is equivalent to a Wasserstein Spherical-Kmeans with the cosine dissimilarity as the ground metric. Moreover, WE-NMF collapses to the original NMF when $\gamma = 0$ which will be our baseline for comparing the direct gain of our model.

Five benchmarking document-term datasets highlighting several varieties of challenging situations were selected for these experiments. Their characteristics are displayed in Table 2. They differ in terms of amount of clusters, dimension, clusters balance (coefficient defined as the ratio of the number of documents in the smallest class to the number of documents in the largest class), degree of mixture of the different groups and sparsity (where nz indicates the percentage of non-zero values). We normalized each data matrix with TF-IDF and their respective documents to unit L_2-norm to remove the bias introduced by their length. Two measures widely used to quantify the clustering performance of an

Table 2. Datasets description (# denotes the cardinality).

Datasets	Characteristics				
	#Documents	#Words	#Clusters	nz (%)	Balance
CSTR	475	1000	4	3.40	0.399
CLASSIC4	7095	5896	4	0.59	0.323
NG5	4905	10167	5	0.92	0.943
NG20	18846	14390	20	0.59	0.628
SPORTS	8580	14870	7	0.86	0.0358

algorithm were employed, namely the Normalized Mutual Information (NMI) (Strehl and Ghosh 2002) and the adjusted Rand Index (ARI) (Hubert and Arabie 1985). Both criteria reach a value of 1 when the clustering is identical to the ground truth.

5.1   Settings

As defined earlier, matrices G and H can be recovered after minimizing problem 1 by using a Sinkhorn-Distance algorithm in order to obtain the optimal transportation matrices T and S. The results showcased here were obtained using algorithm 3 of Cuturi and Doucet (2014) (Smooth Primal and Dual Optima) in our algorithm. The number of clusters was set to the original number of classes for each dataset. The results of each respective algorithm were obtained over an average of 30 random runs. Their respective parameters (if required) were set according to the recommended settings; for instance $\gamma = 100$ for GNMF.

5.2   Other Optimal Transport Algorithms

Algorithms such as SO-TROT (Second Order Row-Tsallis regularized Optimal Transport), KL-TROT (Muzellec et al. 2017) and SAG (Stochastic Average Gradient) for discrete OT (Genevay et al. 2016) were also tested. In our model, S and T are quite small as the rank defined by the user remains low for clustering application. Thereby most of the time, the use of stochastic methods become unnecessary as conventional algorithms converge in a decent time. Nevertheless, they can become handy whether the user specifies a large amount of clusters. After testing, we denote very similar results with SAG for discrete OT compared to the ones obtained with SD. The results with SO-TROT and KL-TROT were similar on CSTR. The TROT distance is appealing since it generalizes Optimal transport and Cuturi (2013) approach as well as involving the escort distribution. Unfortunately, its very high computational cost makes unsuitable on larger dataset.

Ground Metric. We chose to retain the cosine dissimilarity as the ground metric function to map elements $(\boldsymbol{x}_j, \boldsymbol{z}_k)$ and $(\boldsymbol{x}_i, \boldsymbol{w}_k)$. Despite its limitations for advanced semantic relations, this measure is widely acknowledged as a referenced for text mining and remains relevant in most situations. Indeed, It is particularly appealing when we are dealing with large amount of directional sparse data; it does not affect therefore the computational cost of this dissimilarity.

λ **Setting.** As λ increases, the optimization problem (2) is expected to converge toward the classical Optimal Transportation distance. However, Cuturi (2013) has reported in his experiments that (2) tends *to hover* above the classical OT distance by about 10% and that practical value of λ were not necessarily the highest. Moreover, fixing λ has to be in regard to the order of magnitude of the cost matrix. While in our case, $m_{ij} \in [0,1]$, other continuous function may attribute larger values of m_{ij} which may results in constraining λ in a reduced interval to avoid numerical overflow. The regularization parameter γ has been studied across each data matrix along a range going from 10^{-5} to 10^3. Figure 2 showcases variations of NMI and ARI with WE-NMF according to the evolution of γ; taking $\gamma = 10$ seems to be good trade-off. NG5 and NG20 are the only datasets where $\gamma = 1$ will be better, however the performance at $\gamma = 10$ remains equal or superior to the one of NMF. For $\gamma > 10$, the algorithm fails due to numerical overflow.

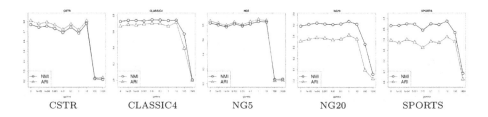

| CSTR | CLASSIC4 | NG5 | NG20 | SPORTS |

Fig. 2. Impact of the regularization parameter γ on WE-NMF (SD).

5.3   Empirical Results

Table 3 summarizes the different results. As we can see, WE-NMF provides better performances overall. Notice also that NMF-KL gives similar achievements in terms of ARI on CSTR and NG5. Acknowledging the abilities of the Wasserstein embeddings to improve the clustering performances, we decided to use them as supports for visualizing the data (samples and features) with respect to the clustering for the set of terms and the available original partition for the documents. The different groups are depicted in color. We observed that the embeddings matrices H and G respectively provide better representations for the actual document clusters and even more significant ones for the term clusters with a soaring separability. Figure 3 highlights different visualizations of the documents of each dataset using the UMAP (Uniform Manifold Approximation and Projection for Dimension Reduction) components (McInnes et al.2018) obtained on X, Z and $H^\top$, where the true classes are projected. Figure 4 shows similar projections for the set of terms where the depicted groups are the term clusters obtained from the solution with the best criterion according to NMF-KL (for $X^\top$ and W) and WE-NMF (for $G^\top$). In Fig. 3, UMAP does not always provide a meaningful visualization of the data samples x_i (neither the features x_j, see line 1 in Fig. 4). Several setups made according to the crucial parameters (min_dist and $n\_neighbors$) emphasized by the authors were tested with $n\_neighbors \in \{15, 80, 320\}$; neither of them was successful to circumvent these issues. Also, we observed that the use of different metrics such as cosine similarity instead of the euclidean distance did surprisingly not improved the visualization. Therefore, we conducted the rest of our experiments with the defaults parameters. CLASSIC4 and NG20 are the datasets with the highest sparsity rates which might be the reason leading UMAP to fail.

Table 3. Average NMI and ARI over different datasets.

Datasets	Metrics	NMF-KL	ONMF	PNMF	GNMF	S-Kmeans	VWC	WE-NMF(SD)
CSTR	NMI	0.77 ± 0.02	0.65 ± 0.05	0.66 ± 0.01	0.57 ± 0.08	0.76 ± 0.01	0.55 ± 0.03	$\mathbf{0.78} \pm 0.03$
	ARI	$\mathbf{0.81} \pm 0.02$	0.56 ± 0.04	0.56 ± 0.01	0.53 ± 0.11	0.79 ± 0.01	0.50 ± 0.03	$\mathbf{0.81} \pm 0.03$
CLASSIC4	NMI	0.72 ± 0.09	0.55 ± 0.09	0.59 ± 0.05	0.65 ± 0.04	0.60 ± 0.001	0.54 ± 0.02	$\mathbf{0.74} \pm 0.01$
	ARI	0.65 ± 0.10	0.39 ± 0.09	0.44 ± 0.01	0.49 ± 0.05	0.47 ± 0.001	0.45 ± 0.01	$\mathbf{0.72} \pm 0.01$
NG5	NMI	0.83 ± 0.01	0.65 ± 0.04	0.65 ± 0.05	0.63 ± 0.07	0.74 ± 0.03	0.68 ± 0.04	$\mathbf{0.84} \pm 0.01$
	ARI	$\mathbf{0.86} \pm 0.01$	0.48 ± 0.08	0.47 ± 0.09	0.62 ± 0.09	0.64 ± 0.07	0.68 ± 0.06	$\mathbf{0.86} \pm 0.01$
NG20	NMI	0.49 ± 0.01	0.44 ± 0.02	0.45 ± 0.02	0.50 ± 0.01	0.50 ± 0.01	0.41 ± 0.01	$\mathbf{0.51} \pm 0.01$
	ARI	0.35 ± 0.01	0.22 ± 0.02	0.24 ± 0.02	0.35 ± 0.05	0.31 ± 0.02	0.27 ± 0.01	$\mathbf{0.38} \pm 0.02$
SPORTS	NMI	0.53 ± 0.01	0.55 ± 0.02	0.56 ± 0.001	0.55 ± 0.001	$\mathbf{0.62} \pm 0.02$	0.55 ± 0.02	0.58 ± 0.01
	ARI	0.40 ± 0.01	0.28 ± 0.01	0.28 ± 0.001	0.28 ± 0.001	0.40 ± 0.04	0.39 ± 0.01	$\mathbf{0.43} \pm 0.01$

Furthermore, Fig. 3 shows in some cases that a better separability between document clusters can be observed from the Wasserstein Embeddings $H^\top$ of WE-NMF as opposed to NMF-KL factor Z. It is highlighted with CSTR and CLASSIC4 which in addition is the dataset where WE-NMF improves the most in terms of NMI and ARI. With NG20 and SPORTS were we also see major improvements, the document visualizations from $H^\top$ are substantially different from those of Z. In Fig. 4, although the true groups of terms are not available, our method seems to be more suitable by allowing representations with substantial clusters separability compared to what is given with NMF-KL factor W and $X^\top$. UMAP which builds its components according to the samples instead of the features does not provide any meaningful visualizations for the terms on the transposed data matrix $X^\top$ although it could detect groups of documents on X (Fig. 3). In practice, terms are more difficult to classify. They can appear in several contexts, be used in different topics or even be considered as noisy depending on the pre-processing applied to build the document-term matrix. From these observations, dimensionality reduction seems to be beneficial and WE-NMF which allows the decomposition matrices to be approximated by not only X but also by H and W might already gain an advantage for this type of data.

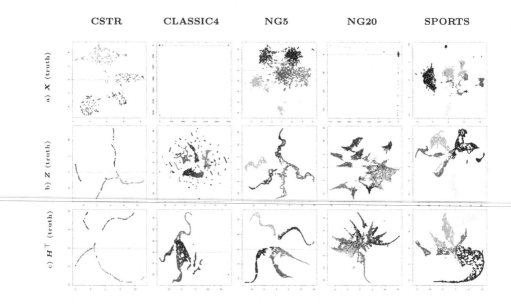

Fig. 3. Visualizations of true document classes by UMAP applied on a) X, b) Z obtained by NMF-KL, and c) $H^\top$ by WE-NMF.

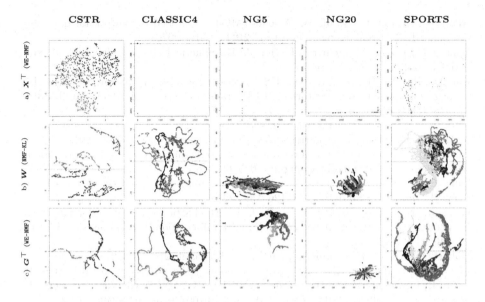

Fig. 4. Visualizations of term clusters by UMAP applied on a) $\boldsymbol{X}^\top$ using `WE-NMF` clusters, b) $\boldsymbol{W}$ using `NMF-KL` clusters, and c) $\boldsymbol{G}^\top$ using `WE-NMF` clusters.

6   Conclusion

We described a novel NMF model allowing to take into consideration semantic relationships between words such as synonyms and hyponyms (non linear relations) in order to boost the clustering performances of the Kullback-Leibler NMF. Moreover, we highlighted the visualization properties of this method which provides a good support for observing the data clustering which remaining fairly realistic to the true classification. On another hand, this method also provides a relevant representation of the terms which so far could not be well interpreted throughout NMF or conventional visualization methods.

Our approach has been successfully evaluated by comparing it with mainly NMF-type methods. It would be useful to also evaluate it with other methods tailored for sparse data and arising from other approaches based on the modularity criterion and von-Mises Fisher mixtures (Ailem et al. 2015, Salah and Nadif 2018, Affeldt et al. 2020).

References

Affeldt, S., Labiod, L., Nadif, M.: Ensemble block co-clustering: a unified framework for text data. In: CIKM (2020)

Ailem, M., Role, F., Nadif, M.: Co-clustering document-term matrices by direct maximization of graph modularity. In: CIKM, pp. 1807–1810 (2015)

Ailem, M., Salah, A., Nadif, M.: Non-negative matrix factorization meets word embedding. In: SIGIR, pp. 1081–1084 (2017)

Allab, K., Labiod, L., Nadif, M.: A semi-NMF-PCA unified framework for data clustering. IEEE Trans. Knowl. Data Eng. **29**(1), 2–16 (2016)

Buchta, C., Kober, M., Feinerer, I., Hornik, K.: Spherical k-means clustering. J. Stat. Softw. **50**(10), 1–22 (2012)

Cai, D., He, X., Han, J., Huang, T.S.: Graph regularized nonnegative matrix factorization for data representation. IEEE Trans. Pattern Anal. Mach. Intell. **33**(8), 1548–1560 (2010)

Cuturi, M.: Sinkhorn distances: lightspeed computation of optimal transport. In: Advances in Neural Information Processing Systems, pp. 2292–2300 (2013)

Cuturi, M., Doucet, A.: Fast computation of Wasserstein barycenters. In: International Conference on Machine Learning, pp. 685–693 (2014)

Ding, C., Li, T., Peng, W., Park, H.: Orthogonal nonnegative matrix t-factorizations for clustering. In: SIGKDD, pp. 126–135 (2006)

Genevay, A., Cuturi, M., Peyré, G., Bach, F.: Stochastic optimization for large-scale optimal transport. In: Advances in Neural Information Processing Systems, pp. 3440–3448 (2016)

Hubert, L., Arabie, P.: Comparing partitions. J. Classif. **2**(1), 193–218 (1985)

Labiod, L., Nadif, M.: Co-clustering under nonnegative matrix tri-factorization. In: Lu, B.-L., Zhang, L., Kwok, J. (eds.) ICONIP 2011. LNCS, vol. 7063, pp. 709–717. Springer, Heidelberg (2011). https://doi.org/10.1007/978-3-642-24958-7_82

Lee, D.D., Seung, H.S.: Learning the parts of objects by non-negative matrix factorization. Nature **401**(6755), 788 (1999)

Li, T., Ding, C.: Nonnegative matrix factorizations for clustering: a survey. In: Data Clustering, pP. 149–176. Chapman and Hall/CRC (2018)

Ling, H., Okada, K.: An efficient earth mover's distance algorithm for robust histogram comparison. IEEE Trans. Pattern Anal. Mach. Intell. **29**(5), 840–853 (2007)

McInnes, L., Healy, J., Melville, J.: Umap: Uniform manifold approximation and projection for dimension reduction. arXiv preprint arXiv:1802.03426 (2018)

Mi, L., Zhang, W., Gu, X., Wang, Y.: Variational Wasserstein clustering. In: Ferrari, V., Hebert, M., Sminchisescu, C., Weiss, Y. (eds.) ECCV 2018. LNCS, vol. 11219, pp. 336–352. Springer, Cham (2018). https://doi.org/10.1007/978-3-030-01267-0_20

Muzellec, B., Nock, R., Patrini, G., Nielsen, F.: Tsallis regularized optimal transport and ecological inference. In: AAAI (2017)

Pele, O., Werman, M.: Fast and robust earth mover's distances. In: CCV, pp. 460–467 (2009)

Rolet, A., Cuturi, M., Peyré, G.: Fast dictionary learning with a smoothed Wasserstein loss. In: Artificial Intelligence and Statistics, pp. 630–638 (2016)

Salah, A., Nadif, M.: Directional co-clustering. Adv. Data Anal. Classif. **13**(3), 591–620 (2018). https://doi.org/10.1007/s11634-018-0323-4

Salah, A., Ailem, M., Nadif, M.: A way to boost semi-NMF for document clustering. In: CIKM, pp. 2275–2278 (2017)

Salah, A., Ailem, M., Nadif, M.: Word co-occurrence regularized non-negative matrix tri-factorization for text data co-clustering. In: AAAI, pp. 3992–3999 (2018)

Sandler, R., Lindenbaum, M.: Nonnegative matrix factorization with earth mover's distance metric. In: CVPR, pp. 1873–1880 (2009)

Schmitz, M.A., et al.: Wasserstein dictionary learning: optimal transport-based unsupervised nonlinear dictionary learning. SIAM J. Imaging Sci. **11**(1), 643–678 (2018)

Shirdhonkar, S., Jacobs, D.W.: Approximate earth mover's distance in linear time. In: 2008 IEEE Conference on Computer Vision and Pattern Recognition, pp. 1–8. IEEE (2008)

Strehl, A., Ghosh, J.: Cluster ensembles–a knowledge reuse framework for combining multiple partitions. J. Mach. Learn. Res. **3**(Dec), 583–617 (2002)

Villani, C.: Optimal Transport: Old and New, vol. 338. Springer, Heidelberg (2008). https://doi.org/10.1007/978-3-540-71050-9

Yoo, J., Choi, S.: Orthogonal nonnegative matrix factorization: multiplicative updates on Stiefel manifolds. In: Fyfe, C., Kim, D., Lee, S.-Y., Yin, H. (eds.) IDEAL 2008. LNCS, vol. 5326, pp. 140–147. Springer, Heidelberg (2008). https://doi.org/10.1007/978-3-540-88906-9_18

Yuan, Z., Oja, E.: Projective nonnegative matrix factorization for image compression and feature extraction. In: Kalviainen, H., Parkkinen, J., Kaarna, A. (eds.) SCIA 2005. LNCS, vol. 3540, pp. 333–342. Springer, Heidelberg (2005). https://doi.org/10.1007/11499145_35

Investigating the Compositional Structure of Deep Neural Networks

Francesco Craighero[1], Fabrizio Angaroni[1], Alex Graudenzi[2(✉)], Fabio Stella[1], and Marco Antoniotti[1]

[1] Department of Informatics, Systems and Communication,
University of Milan-Bicocca, Milan, Italy
[2] Institute of Molecular Bioimaging and Physiology,
Consiglio Nazionale delle Ricerche (IBFM-CNR), Segrate, Italy
alex.graudenzi@unimib.it

Abstract. The current understanding of deep neural networks can only partially explain how input structure, network parameters and optimization algorithms jointly contribute to achieve the strong generalization power that is typically observed in many real-world applications. In order to improve the comprehension and interpretability of deep neural networks, we here introduce a novel theoretical framework based on the compositional structure of piecewise linear activation functions. By defining a direct acyclic graph representing the composition of activation patterns through the network layers, it is possible to characterize the instances of the input data with respect to both the predicted label and the specific linear transformation used to perform predictions. Preliminary tests on the MNIST dataset show that our method can group input instances with regard to their similarity in the internal representation of the neural network, providing an intuitive measure of input complexity.

Keywords: Deep learning · Interpretability · Piecewise-linear functions · Activation patterns

1 Introduction

Despite the extremely successful application of Deep Neural Networks (DNNs) to a broad range of distinct domains, many efforts are ongoing both to deepen their understanding and improve their interpretability [5,20]. This is particularly relevant when attempting to explain their generalization performances, which are typically achieved due to over-parameterized models [1,25].

To this end, many publications focus on the study of the *expressivity* of DNNs, i.e., how their architectural properties such as, e.g., depth or width, affect their performances [1,9,10,16,17,19,21]. These studies usually analyze DNNs

---

A. Graudenzi, F. Stella and M. Antoniotti—Co-senior authors.

G. Nicosia et al. (Eds.): LOD 2020, LNCS 12565, pp. 322–334, 2020.
https://doi.org/10.1007/978-3-030-64583-0_30

with piecewise-linear (PWL) activation functions, such as Rectified Linear Units (ReLUs [6]), which allow a simplified mathematical analysis of the feature space.

In particular, given a standard multinomial classification problem, it is possible to study how a given input dataset is processed in the internal representation of a ReLU DNN by analyzing the *activation patterns*, i.e., the sets of neurons that are null or positive for each instance of the dataset, in each layer of the network.

Each activation pattern uniquely defines a layer-specific activation (linear) region, i.e., the region of the input space which leads to the activation of the same pattern [9]; clearly, one or more instances can be mapped on the same activation region. Each instance will be then characterized by a specific trajectory through activation patterns in successive layers, as a result of the composition of multiple ReLUs. Accordingly, each instance will be mapped onto distinct activation regions in each layer. By analysing how different instances are characterized by common activation patterns and regions, it is possible to investigate how the input space is *folded* for any given dataset.

In particular, the so-called compositional structure [17] of the activation patterns can then be exploited to *interpret the representation* of the input data by a DNN, i.e., "to understand how data are represented and transformed throughout the network" [5]. This structure can be translated into an *Activation Pattern Direct acyclic graph* (APD), which we formally define in the following sections and that may represent a powerful tool to evaluate the expressivity of a DNN with respect to a specific dataset.

Accordingly, by analyzing how many distinct instances are mapped on shared sub-portions of the APD, i.e., belong to the same activation regions, it is possible to provide an intuitive measure of the input complexity, which can be then related to classification accuracy. We remark that the analysis of the relation between input data and the representation of DNNs is an active area of research in the sphere of explainable AI and covers topics such as, e.g., sample analysis [3,12,13].

In this work, we propose a new framework to quantitatively analyze the compositional structure of DNNs and, in particular:

1. we introduce and define the concept of Activation Pattern Direct acyclic graph (APD);
2. we describe a lightweight algorithm to cluster the instances of a dataset on the basis of their mapping on the APD;
3. we present an empirical analysis of the MNIST dataset [14], in which we show that the proposed clustering method on the APD could be employed to distinguish instances on which the network is more confident.

2   Related Work

The literature devoted to the study of *network expressivity* of ReLU DNNs is vast. Three topics are particularly relevant for the current work, namely: (*i*) the estimation of the upper-bound of the number of linear regions [16,17,21]; (*ii*) the analysis of the linear regions through input trajectories [19]; (*iii*) the analysis of other linear regions properties, such as their size or their average number [9,10].

Sample analysis is a timely research topic; in particular a variety of studies demonstrates how sampling instances by importance during training can improve learning. Again, to limit the scope of our investigation, we can distinguish four different sampling strategies: (*i*) curriculum learning [2], according to which it is preferable to start learning from easier to harder instances, also implemented in self-paced learning [13]; (*ii*) selecting only the hardest instances, e.g., the ones that induce the greater change in the parameters [12]; (*iii*) meta-learning, i.e., "learning to learn" [4]; (*iv*) favoring uncertain instances [3,24].

Given these premises, a first major challenge is the estimation of instance hardness/complexity. Accordingly, the choice of the right sampling strategy is essential and depends both on the task and on data type, e.g., (simple, noisy, ...).

In [24], the authors analyze the learning process by measuring the so-called *forgetting events* (defined formally in Definition 4). An instance is called *unforgettable* when no forgetting event occurs during training, otherwise it is called *forgettable*. The authors show that training a new model without unforgettable samples does not affect the accuracy. Similarly, two further studies show how to build an ensemble of DNNs by iteratively training a new network on a reduced version of the dataset. In [20] the authors iteratively mask the features that display the greatest input gradient. As a result, they define multiple models that make predictions based on "qualitatively different reasons", mainly to achieve greater explainability. In [23], the authors train each new network on a reduced version of the dataset, where "good" instances of the previous network are removed ("good" inputs are the ones with hidden features belonging to mostly correctly classified instances). To define the hidden features, the authors first cluster each hidden layer with k-means, and then characterize each instance with respect to the clusters of each layer to which it belongs. We here propose a similar approach, in which each instance is characterized by the path of linear regions (activation patterns) in each layer to which it belongs.

3   Methods

In this section we will formally define the Activation Pattern DAG (APD) and present a novel algorithm to cluster input instances on the basis of the respective path in the APD. In the following definitions, we will employ the notation used in [17], while we refer to [9] for an extensive formal description of activation patterns and activation regions.

3.1   Basic Definitions

Let $\mathcal{N}_\theta(x_0)$ be a *Feedforward Neural Network* (FNN) with input $x_0 \in \mathbb{R}^{n_0}$ and trainable parameters θ. Each layer h_l, for $l \in 1, \ldots, L$, is represented as a vector of dimension n_l, i.e., $h_l = [h_{l,1}, \ldots, h_{l,n_l}]^T$, where each component $h_{l,i}$ (i.e., a neuron or unit) is the composition of a linear preactivation function $f_{l,i}$ and a nonlinear activation function $g_{l,i}$, i.e. $h_{l,i} = g_{l,i} \circ f_{l,i}$.

Let x_l be the output of the l-th layer for $l = 1, \ldots, L$ and the input of the network for $l = 0$, then, we define $f_{l,i}(x_{l-1}) = W_l x_{l-1} + b_{l,i}$, where both $W_l \in \mathbb{R}^{n_{l-1}}$ and $b_{l,i} \in \mathbb{R}$ belong to the trainable parameters θ. Regarding activation functions, in this paper we will focus on piecewise linear activation functions. Thus, for the sake of simplicity, we define $g_{l,i}$ as a ReLU activation function, i.e., $g_{l,i}(x) = \max\{0, x\}$. When clear from the context, we will omit the second index of $f_{l,i}$ and $g_{l,i}$ to refer to the vector composed by all of them.

Finally, we can represent the FNN $\mathcal{N}_\theta$ as a function $\mathcal{N}_\theta : \mathbb{R}^{n_0} \to \mathbb{R}^{out}$ that can be decomposed as

$$\mathcal{N}_\theta(x) = f_{out} \circ h_L \circ \cdots \circ h_1(x), \tag{1}$$

where f_{out} is the output layer (e.g., softmax, sigmoid, ...).

3.2 From activation patterns to the APD

Given a FNN $\mathcal{N}_\theta$ and a dataset $\mathcal{D}$, we define the *activation pattern* of layer l given input $x \in \mathcal{D}$ as follows:

Definition 1 (Activation Pattern). *Let $\mathcal{N}_\theta(x_0)$ be the application of a FNN $\mathcal{N}$ with parameters θ on an input $x_0 \in \mathcal{D}$, with $\mathcal{D} \subseteq \mathbb{R}^{n_0}$. Then, by referring to x_{l-1} as the input to layer $l \in \{1, \ldots, L\}$, we can compute the activation pattern $A_l(x_0)$ of layer l on input x_0 as follows:*

$$A_l(x_0) = \{a_i \mid a_i = 1 \text{ if } h_{l,i}(x_{l-1}) > 0 \text{ else } a_i = 0, \; \forall i = 1, \ldots, n_l\}. \tag{2}$$

Thus, we can represent $A_l(x_0)$ as a vector in $\{0, 1\}^{n_l}$, i.e.:

$$A_l(x_0) = [a_1, a_2, \ldots, a_{n_l}]. \tag{3}$$

The above definition can be easily extended to other binary piecewise activations, e.g. Leaky-ReLUs [15], or to maxout activations [8], by using k-ary activation patterns due to the k thresholds. In Fig. 1 we show a simple example of a FNN $\mathcal{N}(x_0)$ and its activation patterns. In the following, we will represent generic activation patterns as a or a_i, and with $\text{layer}(a)$ we will refer to the layer corresponding to that pattern. In addition, we allow us to simplify the notation of A_l and refer to $A_l(\mathcal{X}_0)$ on $\mathcal{X}_0 \subseteq \mathbb{R}^{n_0}$ as $A_l(\mathcal{X}_0) = \bigcup_{x_0 \in \mathcal{X}_0} A_l(x_0)$.

Given an activation pattern $\hat{a}$, or a set of patterns $\mathcal{A}$ belonging to different layers, and a set of instances $\mathcal{X} \subseteq \mathcal{D}$, we call *activation region* the set composed by the instances in $\mathcal{X}$ that generate that activation pattern, or patterns, in their respective layers.

Definition 2 (Activation Region). *The activation region identified by an activation pattern $\hat{a}$ on an input subset $\mathcal{X} \subseteq \mathcal{D}$ is given by:*

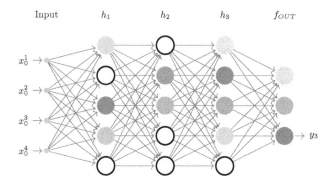

Fig. 1. The evaluation of the example neural FNN $\mathcal{N}$ on an input instance $x_0 = [x_0^1, x_0^2, x_0^3, x_0^4]$. Hidden units have different opacity depending on the module of the positive output, while the black border indicates output 0. In the last layer, the output label y_3 indicates the output unit with the largest value. In this example, we have: $A_1 = [1, 0, 1, 1, 0]$, $A_2 = [0, 1, 1, 0, 0]$, $A_3 = [1, 1, 1, 1, 0]$.

$$\mathcal{AR}(\hat{a}, \mathcal{X}) = \{x \in \mathcal{X} \mid A_l(x) = \hat{a}, \ l = \texttt{layer}(\hat{a})\}. \tag{4}$$

Given a set of activation patterns $\mathcal{A}$ belonging to different layers, i.e., $\forall a_i, a_j \in \mathcal{A} \ \texttt{layer}(a_i) \neq \texttt{layer}(a_j)$, we define their activation region as:

$$\mathcal{AR}(\mathcal{A}, \mathcal{X}) = \bigcap_{\hat{a} \in \mathcal{A}} \mathcal{AR}(\hat{a}, \mathcal{X}). \tag{5}$$

Given a dataset $\mathcal{D}$ and a FNN $\mathcal{N}_\theta$, we introduce the APD as the directed acyclic graph defined by all the activation patterns generated by instances in $\mathcal{D}$ and the way in which they are composed.

Definition 3 (Activation Patterns DAG). *Given a FNN $\mathcal{N}_\theta$ and a dataset $\mathcal{D} \subseteq \mathbb{R}^{n_0}$, the **A**ctivation **P**atterns **DAG** (**APD**) is a directed acyclic graph $APD_{\mathcal{N}_\theta}(\mathcal{D}) = (V, E)$, where:*

– *V is the set of vertices defined by*

$$V = \{1, ..., |\mathcal{A}|\},$$

where $\mathcal{A} = \bigcup_{l=1}^{L} A_l(\mathcal{D})$ is the set of all possible activation patterns and $|\mathcal{A}|$ is its cardinality. In addition, let $\texttt{patt} : V \to \mathcal{A}$ be a labelling function that associates each vertex to the corresponding activation pattern.

– *E is the set of edges defined by:*

$$E = \{(v_1, v_2) \in V \times V \mid \texttt{patt}(v_1) \text{ and } \texttt{patt}(v_2) \text{ are consecutive}\}, \tag{6}$$

where two patterns a_i, a_j are called consecutive if

$$\texttt{layer}(a_i) = l = \texttt{layer}(a_j) - 1$$

and there exists $x \in \mathcal{D}$ such that $A_l(x) = a_i$ and $A_{l+1}(x) = a_j$.

From Definition 3, it follows that the APD has the same depth of the corresponding FNN, and that a node at depth d in the APD corresponds to a pattern of the d-th layer in the FNN. In Fig. 2 we show a toy-example for the $APD_\mathcal{N}$ defined by the FNN $\mathcal{N}$ of Fig. 1, as generated on five example samples.

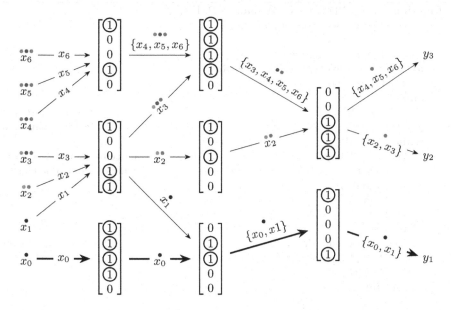

Fig. 2. The representation of $APD_\mathcal{N}(\{x_0, \ldots, x_6\})$, where x_0 is the input instance evaluated in Fig. 1. The thicker lines correspond to the edges generated by example x_0. In addition, on the right we represented the label predicted by the network. Finally, colored bullets over instances mark the splitting history (see Algorithm 1), from right to left, i.e., each new bullet identifies a new partition of the previous cluster.

3.3 Clustering the input dataset using the APD

As discussed in the previous sections, a given activation pattern defines an activation region in the input space, to which one or more instances are associated. Additionally, each activation pattern identifies the linear trasformation applied by the respective layer to all the instances in the corresponding activation region. Thus, activation patterns represent an abstraction on how a DNN elaborates and represents the input data. As a consequence, studying how patterns are shared between instances may provide useful information on how the learning process exploits symmetries in the feature space [17]. For example, in our toy-example of Fig. 2, the linear transformation defined by pattern $a = [1, 0, 0, 0, 1]$ of the third layer is common to both x_0 and x_1.

Similarly to [23], we here exploit the compositional structure of the APD to characterize each input instance on the basis of the trajectory through activation patterns in the distinct layers. The intuition is that the overlap among the trajectories of two instances on the APD is effective in assessing how the two instances are similarly transformed throughout the network.

Algorithm 1. Splitting algorithm.

---

function SPLIT(APD $G = (V, E)$, dataset $\mathcal{D}$, FNN $\mathcal{N}$)
    n.**pred**() ← predecessors of node $n \in V$
    L ← # layers of $\mathcal{N}$
    out ← dummy ending node
    for $v \in V$ s.t. `layer(patt(`v`))` $== L$ **do**
        E.**add**((v, out))
    end for
    $\mathcal{P}$ ← $\{(out, \mathcal{D})\})$ ▷ Current partition
    $\mathcal{F}$ ← $\emptyset$ ▷ Final partition
    while $\mathcal{P} \neq \emptyset$ **do**
        (n, $\mathcal{C}$) ← $\mathcal{P}$.**pop**()   ▷ Extract (current node, cluster)
        if n.**pred**() $== \emptyset \vee |\mathcal{C}| == 1$ **then**   ▷ Check if splittable cluster
            $\mathcal{F}$.**add**($\mathcal{C}$)
            break
        end if
        $\mathcal{S}$ ← $\emptyset$
        for $v \in n$.**pred**() **do**   ▷ Split current cluster
            $\mathcal{V}$ ← $\mathcal{AR}($`patt(`v`)`$, \mathcal{C})$
            $\mathcal{S}$.**add**($(v, \mathcal{V})$)
        end for
        $\mathcal{S}'$ ← $\{\mathcal{V} \mid (v, \mathcal{V}) \in \mathcal{S}\}$
        ig ← `InformationGain`$(\mathcal{C}, \mathcal{S}')$
        if $ig > 0$ **then**   ▷ Check splitting gain
            $\mathcal{P}$ ← $\mathcal{P} \cup \mathcal{S}$
        else
            $\mathcal{F}$.**add**($\mathcal{C}$)
        end if
    end while
    return $\mathcal{F}$
end function

Additionally, from Fig. 2 one can notice that some activation patterns are characterized by a decision boundary, such as activation pattern $[0, 0, 1, 1, 1]$ of the third layer, whereas some are not, such as activation pattern $[1, 0, 0, 1, 0]$ of the first layer. We will refer to the former as *unstable activation patterns* and to the latter as *stable*. In this respect, it would be interesting to test whether instances belonging to stable activation patterns are the ones on which the network is more confident.

Furthermore, in order to asses the confidence of the network on predictions, it might be effective to look for stable activation patterns belonging to many instances. The motivation is that we expect "learned" instances of the same class to be brought closer in the feature space by the first layers, resulting in stable patterns in layers closer to the output.

For example, in Fig. 1 instances x_4, x_5, x_6 activate the same activation patterns from the beginning, while instances x_1 and x_0 are folded in the same activation pattern after the first layer transformation.

To automatically identify similar instances, we defined a splitting algorithm, depicted in Algorithm 1. The goal is to cluster instances that share the same activation patterns and are classified with the same label, proceeding backwards from the output layer. The first partition of input data is performed by considering only the activation patterns of the last layer; if one of the identified clusters contains instances with distinct labels, it is split by considering which activation pattern they activate in the previous layer. Splitting is determined via information gain measure [18], since a decrease of entropy implies more homogeneous partitions.

In Fig. 2 colored bullets mark the splitting history of the 6 instances, following the clustering order from the output layer to the input layer. For example, the first partition is identified by cyan and blue color, i.e. $\{\{x_2, x_3, x_4, x_5, x_6\}, \{x_0, x_1\}\}$. Cluster $\{x_0, x_1\}$ is not split, because both instances are classified with y_1 label. Conversely, the other cluster is partitioned twice: the first splitting occurs when considering the second layer, as x_2 has a different activation pattern than the others and is classified with a different class; the same occurs at the first layer, this time between x_3 and the other instances. The final partition is the following $\{\{x_0, x_1\}, \{x_4, x_5, x_6\}, \{x_3\}, \{x_2\}\}$.

In the next section, we will present some preliminary results on how cluster size of the instances partition can be used to evaluate input similarity and hardness.

4   Results

We applied[1] the clustering algorithm discussed in the previous section on the MNIST dataset and tested it on ReLU networks with different architectures. In particular, we will show that instances included in larger clusters may be "easier" for the network, while errors and "hard" instances are usually included in small clusters. More in detail, we are looking for similar instances in the feature space, i.e. with the same activation pattern, that are classified with the same label, as this may be interpreted as a measure of "confidence" of the network in that specific composition of transformations. Finally, we will show a further experiment in which the information on clustering is used to reduce the number of instances in the training set by more than the 80%, showing an increase of the test error by less than 2%.

The results of Figs. 3, 4 and 5 were performed on all the 60 000 instances of the MNIST training set with a fixed learning rate of 0.001, 500 epochs, SGD as optimization algorithm and the following different architectures: (i) 32full: 5 layers with 32 neurons each; (ii) 16full: 5 layers with 16 neurons each; (iii) 32bottl: with 5 layers with 32, 16, 12, 10, 8 neurons each. The accuracy obtained on the MNIST dataset were, respectively: 98.3% for 32full, 97.2% for 32bottl and 95.8% for 16full.

---

[1] Code available at https://github.com/BIMIB-DISCo/ANNAPD.

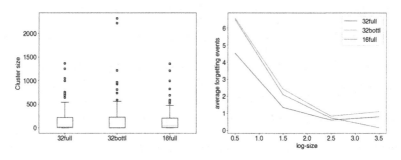

Fig. 3. (Left) Boxplots of cluster size distributions with different architectures. (Right) Average number of forgetting events against log-binned cluster size.

In Fig. 3 (left) the distribution of the sizes of the input partition obtained for different architectures is reported. The majority of the clusters are small (average size $\approx 4, 3, 5$ for `32full`, `16full` and `32bottl`, respectively), while even very large clusters (containing up to 2000 instances) are observed for all architectures. Bigger clusters are expected to contain a larger number of correctly classified instances, i.e., the instances on which the network is more "confident". To test our hypothesis, we analyzed the distribution of forgetting events by cluster size, where forgetting events are defined as follows:

Definition 4 (Forgetting event [24]**).** *Let x be an instance with label k and $pred_e(x)$ the predicted label of x at epoch e. A learning event at epoch e occurs when $pred_{e-1} \neq k$ and $pred_e = k$. A forgetting event at epoch e occurs when $pred_{e-1} = k$ and $pred_e \neq k$. If an instance has no forgetting event during the learning process, is called unforgettable, otherwise is a forgettable instance.*

In Fig. 3 (right) we display the average number of forgetting events with respect to (log-binned) cluster size. From the picture it seems to emerge that, for all architectures, the forgettable instances are grouped in the small clusters. This trend is confirmed by looking at the cumulative distributions of errors and forgetting events in Fig. 4.

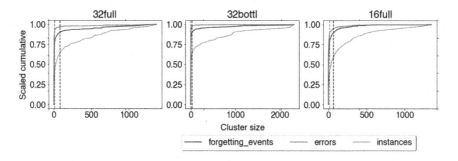

Fig. 4. Cumulative distribution of forgetting events and errors by (sorted) cluster size on three different architectures. In green the cumulative number of considered instances. All the three lines are normalized between 0 and 1. The vertical dotted lines represent where 90% of the respective cumulative is reached. (Color figure online)

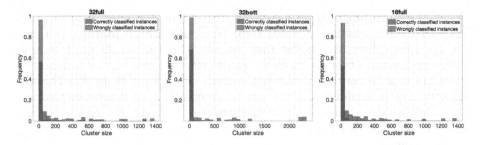

Fig. 5. Cluster size distribution for correctly and wrongly classified instances.

In Fig. 5 one can see the distribution of the cluster size with respect to either correctly and wrongly classified instances. Consistently with the other findings, wrongly classified instances are characterized by very small clusters (mostly singletons) for all architectures, whereas correctly classified instances are typically included in clusters with significantly larger size and a much higher variance. Again, this result would suggests the presence of a significant correlation between cluster size and the input hardness.

Lastly, in Table 1 are reported the results of the dataset reduction experiment, performed as follows: (i) the MNIST training set was split in 50 000 and 10 000 instances, for training and validation, respectively; (ii) the 32full architecture was trained with SGD, learning rate of 0.01 and early stopping [7]; (iii) the clustering algorithm was performed on the APD with the learned parameters; (iv) two more training sets were defined: one with only one representative for each cluster, and one with the same number of instances, but randomly selected; (v) training was performed again, with the same parameters as (ii).

The results confirmed the information extracted by the clustering procedure: with randomly selected instances, we decreased the generalization performances by more than 4%, while with selected instances the loss was only of 2%.

Table 1. Results for dataset reduction on MNIST. The table reports the mean test error and standard deviation of 5 seeds for three possible training sets: (i) all instances, (ii) one representative for each cluster and (iii) a random selection of the same reduced number of instances of (ii).

Trainset selection	Trainset size	Test error
All	50 000 inst.	$96, 37 \pm 0, 2$
Cluster representatives	8 871 inst.	$94, 64 \pm 0, 3$
Random	8 871 inst.	$92, 26 \pm 0, 4$

5 Conclusions and Future Developments

In this work we introduced the APD, a structure that represents the composition of piecewise linear functions defined by the layers of a ReLU network.

Additionally, we proposed an algorithm to partition the input dataset based both on the composition of linear transformations defined by the layers and the predicted labels. We showed that this partition can be efficiently used to group the instances that are similarly transformed by network. Furthermore, we speculate that the instances included in large clusters are those on which the network is more confident and that are better classified, an hypothesis that was confirmed by the preliminary tests performed on the MNIST dataset. Lastly, we reported preliminary results on how our clustering can be used to perform dataset reduction.

Clearly, these results were obtained on a small selection of the possible contributing factors. Further experiments will investigate the role of each variable, including depth and width of the network, learning rate and optimization algorithm. Moreover, the role of standard techniques such as Batch Normalization [11] and Dropout [22] on piecewise linear regions was recently investigated in [26]; therefore, we expect both techniques to have an influence on the topological properties of the APD. It would also be worth investigating the extension of the APD on Convolutional Neural Networks, by either studying the APD only on the fully connected part or by trying to extend the proposed structure also on filters with piecewise linear activations.

As a further future development, it will be interesting to analyze whether the APD can be effective also when approximating sigmoid or tanh activations with piecewise functions.

In conclusion, the APD represents a simple, but expressive tool, to study how DNNs learn data, motivated by geometrical studies on the properties of piecewise linear activation functions [16, 17].

References

1. Arpit, D., et al.: A closer look at memorization in deep networks". In: Proceedings of the 34th International Conference on Machine Learning, ICML 2017, pp. 233–242. JMLR.org, Sydney 6 Aug 2017
2. Bengio, Y., et al.: Curriculum learning. In: Proceedings of the 26th Annual International Conference on Machine Learning - ICML 2009. the 26th Annual International Conference, pp. 1–8. ACM Press, Montreal (2009)
3. Chang, H.-S., Learned-Miller, E., McCallum, A.: Active bias: training more accurate neural networks by emphasizing high variance samples. In: Guyon, I., et al. (ed.) Advances in Neural Information Processing Systems 30, pp. 1002–1012. Curran Associates Inc. (2017)
4. Fan, Y., et al.: Learning What Data to Learn, 27 February 2017. arXiv:1702.08635 [cs, stat]
5. Gilpin, L.H., et al.: Explaining explanations: an overview of interpretability of machine learning. In: 2018 IEEE 5th International Conference on Data Science and Advanced Analytics (DSAA), pp. 80–89, October 2018
6. Glorot, X., Bordes, A., Bengio, Y.: Deep sparse rectifier neural networks. In: Proceedings of the Fourteenth International Conference on Artificial Intelligence and Statistics, pp. 315–323 (2011)

7. Goodfellow, I., Bengio, Y., Courville, A.: Deep Learning, Chap. 7, pp. 246–252. MIT Press (2016)
8. Goodfellow, I.J., et al.: Maxout networks. In: Proceedings of the 30th International Conference on Machine Learning, ICML 2013, Atlanta, GA, USA, 16–21 June 2013, vol. 28, pp. 1319–1327. JMLRWorkshop and Conference Proceedings, JMLR.org (2013)
9. Hanin, B., Rolnick, D.: Deep ReLU networks have surprisingly few activation patterns. In: Wallach, H., et al. (ed.) Advances in Neural Information Processing Systems 32, pp. 359–368. Curran Associates Inc. (2019)
10. Hanin, B., Rolnick, D.: Complexity of linear regions in deep networks. In: Chaudhuri, K., Salakhutdinov, R. (eds.) Proceedings of the 36th International Conference on Machine Learning, ICML 2019, 9–15 June 2019, Long Beach, California, USA, vol. 97. Proceedings of Machine Learning Research. PMLR, pp. 2596–2604 (2019)
11. Ioffe, S., Szegedy, C.: Batch normalization: accelerating deep network training by reducing internal covariate shift. In: Bach, F.R., Blei, D.M. (eds.) Proceedings of the 32nd International Conference on Machine Learning, ICML 2015, Lille, France, 6–11 July 2015, vol. 37. JMLR Workshop and Conference Proceedings, pp. 448–456. JMLR.org (2015)
12. Katharopoulos, A., Fleuret, F.: Not all samples are created equal: arning with importance sampling. In: Dy, J.G., Krause, A. (eds.) Proceedings of the 35th International Conference on Machine Learning, ICML 2018, Stock-holmsm ässan, Stockholm, Sweden, 10–15 July 2018, vol. 80. Proceedings of Machine Learning Research. PMLR, 2018, pp. 2530–2539 (2018)
13. Kumar, M.P., Packer, B., Koller, D.: Self-paced learning for latent variable models. In: Lafferty, J.D., et al. (eds.) Advances in Neural Information Processing Systems 23, pp. 118–1197. Curran Associates Inc. (2010)
14. LeCun, Y., Cortes, C., Burges, C.J.: MNIST handwritten digit database. In: ATT Labs [Online], vol. 2 (2010). http://yann.lecun.com/exdb/mnist
15. Maas, A.L., Hannun, A.Y., Ng, A.Y.: Rectifier nonlinearities improve neural network acoustic models. In: Proceedings of ICML, vol. 30, p. 3 (2013)
16. Montúfar, G., et al.: On the number of linear regions of deep neural networks. In: Proceedings of the 27th International Conference on Neural Information Processing Systems - Volume 2. NIPS 2014, pp. 2924–2932. MIT Press, Montreal (2014)
17. Montufar, G.F., et al.: On the number of linear regions of deep neural networks. In: Ghahramani, Z., et al. (eds.) Advances in Neural Information Processing Systems 27, pp. 2924–2932. Curran Associates Inc. (2014)
18. Quinlan, J.R.: Induction of decision trees. Mach. Learn. 1(1), 81–106 (1986)
19. Raghu, M., et al.: On the expressive power of deep neural networks. In: Precup, D., Teh, Y.W. (eds.) Proceedings of the 34th International Conference on Machine Learning, ICML 2017, Sydney, NSW, Australia, 6–11 August 2017, vol. 70, pp. 2847–2854. Proceedings of Machine Learning Research. PMLR (2017)
20. Ross, A.S., Hughes, M.C., Doshi-Velez, F.: Right for the right reasons: training differentiable models by constraining their explanations. In: Proceedings of the Twenty-Sixth International Joint Conference on Artificial Intelligence. Twenty-Sixth International Joint Conference on Artificial Intelligence. Melbourne, Australia: International Joint Conferences on Artificial Intelligence Organization, pp. 2662–2670, August 2017
21. Serra, T., Ramalingam, S.: Empirical Bounds on Linear Regions of Deep Rectifier Networks. arXiv:1810.03370 [cs, math, stat], 24 January 2019
22. Srivastava, N., et al.: Dropout: a simple way to prevent neural networks from overfitting. J. Mach. Learn. Res. 15(56), 1929–1958 (2014)

23. Tao, S.: Deep neural network ensembles. In: Nicosia, G., Pardalos, P., Umeton, R., Giuffrida, G., Sciacca, V. (eds.) LOD 2019. LNCS, vol. 11943, pp. 1–12. Springer, Cham (2019). https://doi.org/10.1007/978-3-030-37599-7_1

24. Toneva, M., et al.: An empirical study of example forgetting during deep neural network learning. In: 7th International Conference on Learning Representations, ICLR 2019, New Orleans, LA, USA, 6–9 May 2019. OpenReview.net (2019)

25. Zhang, C., et al.: Understanding deep learning requires rethinking generalization. In: 5th International Conference on Learning Representations, ICLR 2017, Toulon, France, 24–26 April 2017, Conference Track Proceedings. OpenReview.net (2017)

26. Zhang, X., Wu, D.: Empirical studies on the properties of linear regions in deep neural networks. In: 8th International Conference on Learning Representations, ICLR 2020, Addis Ababa, Ethiopia, 26–30 April 2020. OpenReview.net (2020)

Optimal Scenario-Tree Selection for Multistage Stochastic Programming

Bruno G. Galuzzi[1]([⊠]) [ID], Enza Messina[1] [ID], Antonio Candelieri[1] [ID],
and Francesco Archetti[1,2] [ID]

[1] University of Milano-Bicocca, 20125 Milan, Italy
bruno.galuzzi@unimib.it
[2] Consorzio Milano-Ricerche, via Roberto Cozzi, 53, 20126 Milan, Italy

Abstract. We propose an algorithmic strategy for Multistage Stochastic Optimization, to learn a decision policy able to provide feasible and optimal decisions for every possible value of the random variables of the problem. The decision policy is built using a scenario-tree based solution combined with a regression model able to provide a decision also for those scenarios not included in the tree. For building an optimal policy, an iterative scenario generation procedure is used which selects through a Bayesian Optimization process the more informative scenario-tree. Some preliminary numerical tests show the validity of such an approach.

Keywords: Multistage stochastic programming · Scenario-tree generation · Supervised learning

1 Introduction

Multistage stochastic programming (MSSP) is an important framework for (linear) optimization problems characterized by making different decisions for the future under uncertainty of the problem data. The basic idea is to make some decision now and to take some corrective decision (recourse) in the future, after the revelation of the uncertainty. The uncertainty for a stochastic problem is revealed sequentially, in multiple stages $t = 1, \ldots, T$, as the realization of a stochastic process defined on some probability space. Therefore, at each stage t, the decision-maker takes the decision x_t using only the partial knowledge of stochastic process until stage t.

If the stochastic process has a small number of realizations (scenarios), then all the scenarios can be represented in a so-called scenario-tree. In such a case, the solution of the MSSP problem is obtained through the solution of a linear programming model, called deterministic equivalent, whose computational cost depends on the dimension of the scenario-tree. The situation is more complicated if the stochastic process takes a large (possibly infinite) number of realizations since the associated deterministic equivalent is typically hard to solve computationally in a reasonable time. In this case, we can approximate the stochastic process with a scenario-tree formed by a finite subset of scenarios.

© Springer Nature Switzerland AG 2020
G. Nicosia et al. (Eds.): LOD 2020, LNCS 12565, pp. 335–346, 2020.
https://doi.org/10.1007/978-3-030-64583-0_31

Methods for scenario-tree generation have been proposed over the past years in the literature [6, 8, 10–12, 16]. All these methods focus on the approximation of the stochastic process before solving the optimization problem and usually, for computational tractability, the number of scenarios in the tree is kept rather small. Then, apart from the first-stage decision, all the subsequent decisions are provided only for those values of the random parameters included in the tree. Nevertheless, the real-world realization of the stochastic process could not coincide with any of the scenarios represented by the tree. In this case, some of the decisions provided may not be optimal or even feasible. Therefore, a decision policy should be derived, which generalizes for any possible scenario. Such a policy can be built using interpolation [9] or regression techniques [3] and the derived decision policy will depend on the underlined scenario-tree.

In this work, the emphasis is on how to obtain a decision policy which is robust with respect to different scenario realizations. To this aim, we propose a sequential learning algorithm to iteratively generate the most informative scenarios to be included in the tree to approximate an optimal decision policy. At each iteration, a global optimization problem is solved where the search space consists of all possible scenarios and the objective function is a performance function able to measure both the feasibility and the optimality of the decisions associated to the estimated policy with respect to the underlying stochastic process. To keep the number of function evaluations small, we solve this problem using a Bayesian Optimization approach [1, 15].

The rest of the paper is organized as follows. In Sect. 2 we present the related literature. In Sect. 3 we illustrate how to derive a decision policy from a given scenario-tree. In Sect. 4 we describe the iterative scenario-generation procedure to obtain the optimal policy estimation. Section 5 shows some preliminary numerical experiments to validate our approach. Finally, in the last section, we derive some conclusions.

2 Related Works

Over the past years, a considerable number of books and articles have been written on stochastic optimization. However, stochastic optimization is a general term that includes over a dozen fragmented communities. Related to this, in [13], the author claimed that "*while deterministic optimization enjoys an almost universally accepted canonical form, stochastic optimization is a jungle of competing for notational systems and algorithmic strategies. This is especially problematic in the context of sequential stochastic optimization problems*".

One possible formulation of multistage stochastic optimization problem is given by stochastic programming, which is a framework for sequential decision making under uncertainty [2, 4, 17]. Multistage stochastic programming is generally solved building a scenario tree of possible realizations of the underlying stochastic process. How to generate scenarios is, therefore, a critical issue. Several scenario tree generation procedures have been proposed in the literature, among which: Monte Carlo methods [16], moment matching [8], integration quadrature [10], Hidden Markov models [11], optimal quantization [12], and scenario reduction methods [6].

Nevertheless, the solution obtained through this approximation does not consist in an optimal policy, i.e. they do not provide optimal decisions for all values of random parameters but only for those realization includes in the scenario tree. Few studies consider such a problem. In [7], the authors solved the scenario-tree deterministic program on a shrinking horizon dynamically in order to implement the state-1 decision recursively at every stage. However, a drawback of this approach is its computational cost.

In [3], the authors compute a decision policy for any possible scenario-tree through a supervised learning strategy. Moreover, since the learned policy depends on a given scenario-tree, the authors use a large number of randomly generated scenario-trees (with also a random branching structure) to obtain the associated policies and evaluate their quality through an out-of-sample procedure using Monte-Carlo simulation. Nevertheless, a random strategy for the generation of such policies is computationally expensive. Indeed, the generation of each policy requires the scenario-tree solution of the MSSP problem, and it could require the generation of a considerable number of trees for ensuring, with a certain probability, the quality of the selected decision policy.

Finally, in [9], the authors proposed a procedure to interpolate and extrapolate the scenario-tree decision to obtain a decision policy that can be implemented for any value of the random parameters at a little computational cost. They also provide some quality parameters to measure the quality of a decision, both in terms of feasibility and optimality. However, also in this case, the decision policy depends on the particular realization of the underlying decision tree.

We overcome this problem by implementing a sequential procedure that, at each step, generate the scenario to be included in the tree by maximizing the optimality and feasibility of the associated decision policy. The resulting policy is then more robust with respect to the possible realization of the underlying stochastic process.

In this respect, the proposed approach shares some connection with Reinforcement Learning (RL). Indeed, MSSP and RL are both sequential decision-making approaches. Nevertheless, they are very different in the way they handle learning and knowledge acquisition. Moreover, RL, as well as dynamic programming, stem from Markov Decision Processes and differently from MSSP they are based on the state model. The proposed method for iteratively adding scenario aimed at estimating a sequence approximated policies could be regarded as a sequential learning approach where knowledge from the underlying stochastic process is acquired iteratively by considering its value with respect to policy performance optimization computed on the true underlying process.

3 Methodology

3.1 Multistage Stochastic Programming

Let us consider a set of time stages $t = 1, \ldots, T$, and a stochastic process $\boldsymbol{\xi} = \{\xi_t\}_{t=1}^{T}$, where ξ_1 is supposed to be known and ξ_t, for $t > 1$, is a random variable with a specified probability distribution and whose value is revealed gradually over time. At stage t, a decision $x_t \in \mathbb{R}^{n_t}$ must be taken, which depends only on the information available up to time t, that is indicated by $\xi_{[t]} = (\xi_1, \ldots, \xi_t)$. Therefore, a generic T-stage stochastic programming problem can be expressed as follows:

$$v^* = \min_{x_1, \ldots, x_T} \left\{ c_1 x_1 + \mathbb{E}_\xi \left[\sum_{t=2}^{T} c_t(\xi_t) x_t(\xi_{[t]}) \right] \right\}, \tag{1}$$

subject to:

$$A_1 x_1 = b_1, \tag{2}$$

$$B_t(\xi_t)x_{t-1} + A_t(\xi_t)x_t = b_t(\xi_t), \forall t \in \{2, \dots, T\}, \tag{3}$$

$$x_t \geq 0, \forall t \in \{1, \dots, T\}, \tag{4}$$

where (A_t, B_t, b_t, c_t) is the set of coefficients at stage t, whose values depends on the value of ξ_t.

To solve the optimization problem (Eq. 1–4), the stochastic process can be discretized into a finite set of scenarios $\{\xi^s\}_{s=1}^S \subseteq \Omega$, each having a specific probability p^s, and organized in a convenient tree structure $\Im$, called scenario-tree. The path from the root node ξ_1 (the same for all the scenarios) to a leaf node corresponds to a single scenario ξ^s, and the event outcomes for scenarios that pass through the same intermediate nodes are identical for all stages up to that node (see Fig. 1).

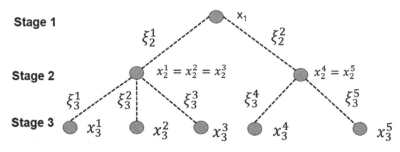

Fig. 1. Example of scenario-tree.

We can obtain an approximation of the solution of the MSSP problem by solving the so-called deterministic equivalent problem on a scenario-tree $\Im$, defined as follows:

$$\hat{v}^* = \min_{x_1, x_2^s, \dots, x_T^s} \left\{ c_1 x_1 + \sum_{s=1}^S p^s \left[c_{s,2} x_2^s + \dots + c_{s,T} x_T^s \right] \right\}, \tag{5}$$

subject to:

$$A_1 x_1 = b_1, \tag{6}$$

$$B_{s,t} x_{t-1}^s + A_{s,t} x_t^s = b_{s,t}, \quad \forall t \in \{2, \dots, T\} \text{ and } \forall s \in \{1, \dots S\}, \tag{7}$$

$$x_t^j \left(\xi_{[t]}^j \right) = x_t^k \left(\xi_{[t]}^k \right), \text{ if } \xi_{[t]}^j = \xi_{[t]}^k, \tag{8}$$

where $\hat{v}^*$ represents an approximation of v^* (Eq. 1), and $\xi_t^s = \{B_{s,t}, A_{s,t}, b_{s,t}, c_{s,t}\}$ is a specific scenario of the tree $\Im$ at stage t. The last set of constraints (Eq. 8) represent the non-anticipativity conditions, which means that every pair of scenarios have the same decision at stage t if they are indistinguishable up to stage t.

3.2 Inference of a Decision Policy from a Scenario-Tree Solution

Solving the MSSP problem through a scenario-tree approximation leads to a set of decisions associated with the observations represented by nodes of the scenario-tree. However, what happens in the real world is not necessarily coincident with any of the scenarios represented in the scenario-tree, and the decisions obtained by solving the MSSP could be either not optimal or not directly implementable because infeasible. For this reason, we need a strategy to obtain a decision (possibly optimal) for any possible scenario that could occur. It is here that machine learning regression techniques come into play.

Now suppose that we have a scenario-tree $\Im$ formed by S scenarios for the problem (Eq. 5–8). Let us consider, for each stage $t > 1$, the set of observation-decision pairs corresponding to the optimal solutions associated with each node at level t of a decision tree $\Im$:

$$S_{t,\Im} = \left\{ \xi^s_{[t],\Im}, x^{*s}_{t,\Im} \right\}^S_{s=1}, \tag{9}$$

where S is the number of scenarios and $x^{*s}_{2,\Im}, \ldots, x^{*s}_{T,\Im}$ are the optimal decisions associated with the scenario ξ^s of the tree $\Im$. We use each dataset $S_{t,\Im}$, to learn a regression model $\hat{x}_{t,\Im}(\cdot)$ able to infer a decision $x_{t,\Im}(\cdot)$ for every possible scenario. Therefore, the input space Ω_t corresponds to all the possible realizations of the stochastic process until stage t, that is $\xi_{[t]}$, whereas the output space is the decision variable x_t. The regression model $\hat{x}_{t,\Im}$ can be built using any possible regression technique [5]. For example, in [3], the authors use the Gaussian Process regressor, whereas in [9], the authors use the Nearest Neighbourhood.

In this work, we use of the Gaussian Process Regression [14]. In a nutshell, the function $x_t(\xi_{[t]})$ is described as a Gaussian Process $x_t(\xi_{[t]}) \sim \mathcal{GP}\left(m(\xi_{[t]}), k\left(\xi_{[t]}, \xi'_{[t]} \right) \right)$, where m and k are the mean and the kernel function, respectively. The mean function is usually assumed to be zero or constant, whereas the kernel function describes how much two points in the input space are closely related. The knowledge about the function is incorporated conditioning the joint Gaussian prior distribution on the observations provided by the dataset $S_{t,\Im}$.

Subsequently, we define a Machine-Learning (ML)-policy $\pi_\Im = \left(\pi_{1,\Im}, \ldots, \pi_{T,\Im} \right)$, as follows:

$$\pi_{1,\Im} = x^*_{1,\Im}, \tag{10}$$

$$\pi_{t,\Im}(\xi_{[t]}) = \hat{x}_{t,\Im}(\xi_{[t]}), \quad \forall t \in \{2, \ldots T\}, \tag{11}$$

where $\pi_{1,\Im}$ represents the first-stage decision (the same for all the scenarios).

3.3 Quality Evaluation of a Decision Policy

To evaluate a policy $\pi_\Im$ (Eq. 10–11) associated with a given scenario-tree $\Im$, we can assess its quality both in terms of objective function value and in terms of feasibility also, with respect to possible scenarios that have not been used for building the policy.

Let us assume to be able to generate a (possibly large) set of scenarios $V_\xi = \left\{\xi^s\right\}_{s=1}^{S} \subseteq \Omega$ from the underlying stochastic process. We can always assume that, for a sufficiently high value of S, such set represents a good approximation of the underlying stochastic process. Such a set V_ξ can be generated using Monte-Carlo Sampling Methods according to the stochastic process distribution.

In order to evaluate the quality (optimality and feasibility) of $\pi_\Im$, we consider the following quality indicators [9]:

$$F(\pi_\Im) = \frac{\left|V_\xi(\pi_\Im)\right|}{\left|V_\xi\right|}, \tag{12}$$

$$Q(\pi_\Im) = \left|V_\xi(\pi_\Im)\right|^{-1} \sum_{\xi^s \in V_\xi} C(\pi_\Im, \xi^s) \cdot 1_{V_\xi(\pi_\Im)}(\xi^s), \tag{13}$$

where $V_\xi(\pi_\Im) \subseteq V_\xi$ indicates the subset on which the policy $\pi_\Im$ provides a feasible decision with respect to V_ξ, and $C(\pi_\Im, \xi^s) = c_1\pi_{1,\Im} + \sum_{t=1}^{T} c_t \cdot \pi_{t,\Im}\left(\xi_{[t]}^s\right)$ is the specific cost of the policy $\pi_\Im$ for the scenario ξ^s, to be minimized. Note that Q is computed only on the subset $V_\xi(\pi_\Im)$.

4 Proposed Approach

At this point, we are able to generate a decision policy associated with a scenario-tree $\Im$ and to measure its quality through the indicators F and Q. We can aggregate these two indicators in a performance function as follows:

$$\alpha(\pi_\Im) = Q(\pi_\Im) + \beta(1 - F(\pi_\Im)), \tag{14}$$

where $\beta > 0$ is a parameter to be tuned. Obviously, if $F(\pi_\Im) \ll 1$, then $Q(\pi_\Im)$ could not represent any information about the optimal value v^* (Eq. 1). Indeed, it is important to have a correct balancing between these two terms using the parameter β. Now, if we have a set of different decision policies, we can rank them based on (Eq. 14). However, generating a large set of decision policies is computationally expensive. Indeed, each evaluation of the decision policy $\pi_\Im$ requires the computation of:

- the solution of the deterministic equivalent on the scenario-tree $\Im$ (Eq. 5–8),
- the associated policy $\pi_\Im$ through the set of regression models (Eq. 10–11),
- the value of the quality indicators Q and F (Eq. 12–13) over the set V_ξ.

Rather than generating a set of decision policies starting from different randomly generated scenario-trees as in [3], we propose to guide the generation of a sequence of "optimal" decision policies obtained iteratively by solving the following optimization problem:

$$\Im^* = \arg\min_{\Im \in \Sigma} \alpha(\pi_\Im), \tag{15}$$

where Σ represents the set of all possible scenario-trees that can be obtained by randomly selecting a set of scenarios from the stochastic process ξ.

The search space Σ is generally complex to formulate because it represents a high-dimensional space of branching structures. This is because we propose to use an iterative procedure to obtain an estimation of (Eq. 15). We build a scenario-tree, and its associated policy, by selecting, at each iteration, the scenario obtained by solving the following optimization problem:

$$\Im_n^* = \arg\min_{\Im \in \Sigma^n} \alpha(\pi_\Im), \tag{16}$$

where Σ^n represents the search space of all the scenario-tree formed by adding a scenario to $\Im_{n-1}^*$. Note that, in (Eq. 15) a new evaluation of the objective function α corresponds to test a decision policy based on new scenario-tree. In contrast, in (Eq. 16), the scenario-tree remains the same one except for only one scenario. Such a scenario, as shown in Fig. 2, can either be completely disjointed from the other scenarios of the tree or have a common partial path with other scenarios.

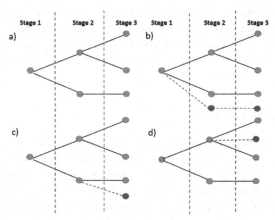

Fig. 2. An initial scenario-tree $\Im_3$, and three possible new scenario-trees $\Im_4$, formed by four scenarios: three scenarios from $\Im_3$ and a new scenario $\xi^4 = \left(\xi_1^4, \xi_2^4, \xi_3^4\right)$, which can be completely independent of the other scenarios (b), or with a partial common history with other scenarios, as in (c) and (d).

Applying this procedure for $n = 1..N$, we can obtain a scenario-tree $\Im_N$, whose N scenarios are not generated randomly, but are the results of N optimization procedures able to outperform a simple random generation of the scenario-tree. Moreover, adding only one scenario per iteration allows us to reduce the complexity of the optimization problem (Eq. 15). The decision-maker can stop the procedure after getting a decision policy characterized both by a high level of feasibility (e.g., $F \geq 0.95$) and a specific threshold value for Q.

5   Numerical Tests

In this section, we investigate the proposed methodology on a benchmark problem, that is a classical supply-chain optimization problem, called the Multi-Period Production and Inventory Planning Model[1].

We use the solver GUROBI[2] to solve the deterministic equivalent (Eq. 5–8) for a specific scenario-tree, and Bayesian Optimization (BO), an efficient global optimization strategy widely used for complex optimization problems, to solve (Eq. 16). For BO, we use the Gaussian Process as a surrogate model (Matern52 as kernel function), and the Expected Improvement as an acquisition function. The entire framework to pass from a specific scenario-tree $\Im$ to its corresponding decision policy has been implemented in MATLAB, using the Gaussian Process Regression to build the regression models (Eq. 10–11). More in detail, we use the *fitrgp* function from Statistics and Machine Learning Toolbox (Matern52 as kernel function). The set V_ξ, used to evaluate F and Q, is formed by 1000 scenarios. However, in the spirit of Machine Learning, we used only 800 of these scenarios (randomly selected) to evaluate the performance function during the iteration process. The other scenarios are used during the iteration procedure, to predict the unbiased value of F and Q on a new test set. We indicate these two subsets with V_ξ^{tr} (training set) and V_ξ^{val} (validation set). All the computations were performed on an Intel i7 CPU 16G RAM.

5.1   The Benchmark Problem

Let us consider a manufacturer that must produce a certain number of objects p_t (level of production) at a certain cost c_p, to satisfy a stochastic demand ξ_t, over a planning horizon $t = 1, \ldots, T$. If $p_t > \xi_t$, the extra-production can be used for the next stages. In this case, we assume that the manufacturer also has an additional cost c_s for stocking the rest of the products. We can express the minimization of the production costs as follows:

$$v^* = \min\left\{ c_p p_1 + c_s s_1 + \mathbb{E}_\xi\left[\sum_{t=2}^{T} \left(c_p p_t(\xi_{[t]}) + c_s s_t(\xi_{[t]}) \right) \right] \right\}, \tag{17}$$

subject to:

$$s_t = s_{t-1} + p_t - \xi_t, \ \forall\, t \in \{2, \ldots, T\}, \tag{18}$$

$$s_1 = s_0 + p_1 - \xi_1, \tag{19}$$

$$0 \leq p_t \leq p_{max}, \forall t \in \{2, \ldots, T\}, \tag{20}$$

$$0 \leq s_t \leq s_{max}, \ \forall\, t \in \{2, \ldots, T\}, \tag{21}$$

---

[1] https://www.aimms.com/english/developers/resources/examples/functional-examples/stocha stic-programming/.

[2] http://www.gurobi.com.

where s_t represents the amount of commodity stored from stage $t - 1$ for the next stage t. The data are set as follows: $s_0 = 1, p_{max} = 11, s_{max} = 15, c_p = 2.5$ and $c_s = 1$. Finally, we assume a time horizon of 4 stages, with $\xi_1 = 3$, and ξ_2, ξ_3, ξ_4 random variable following a uniform distribution in the interval $[1.5t, 4.5\ t]$. The values of the stochastic demands greatly affect both the total cost and the production decisions of the optimization problem. For example, if we solve the problem replacing the random variables with their expected values (i.e., $\boldsymbol{\xi} = (3, 6, 9, 12)$), we obtain a total cost of 73.5 and $p_1 = 2$, while if we replace the random variables with $\boldsymbol{\xi} = (3, 9, 13.5, 18)$, we obtain a cost of 130.25 and $p_1 = 9.5$. The true solution for the problem (Eq. 17–21) is about 95.5 and was estimated through scenario-tree characterized by a uniform branching factor $b = 11$ (corresponding to a scenario-tree formed by 1331 scenarios). Note that providing feasible decisions is an important issue of such a problem because, for $t \geq 2$, the maximum production p_t could be smaller than the actual demand ξ_t for a specific scenario.

5.2 Computational Results

We can build a decision policy $\pi = (\pi_1, \pi_2, \pi_3, \pi_4)$ from any scenario-tree solution related to (Eq. 17–21), as follows. A GP-regression model is built for the production policies $\pi_t(\xi_{[t]}) = \hat{p}_t(\xi_{[t]})$ with $t > 1$, whereas the value of the stocking quantities $\hat{s}_t(\xi_{[t]})$ are obtained by the constraints (Eq. 18).

To obtain the optimal decision policy for this problem, we use the iteration process explained in Sect. 4, starting from a tree formed by a single scenario $\boldsymbol{\xi}^1$, and increasing its size one scenario at a time, solving the optimization problem (Eq. 16) through BO, using 30 evaluations of the performance function (Eq. 14), at each iteration n. We set $\boldsymbol{\xi}^1 = (3, 6, 9, 12)$ (corresponding to the expected values for the stochastic demands). Then, at the end of each iteration, we add the scenario obtained to the current scenario-tree. We assume that the search space for the optimization problem is formed by the 1331 scenarios used to solve the problem in the previous subsection. In the performance function, we set $\beta = 1000$ to give more weight to F, whose value is between 0 and 1, than Q, whose values have an order of magnitude of 10^2.

Due to the randomness of the optimization process related to BO, we report the results on ten different runs. We summarize the computational results in the following boxplots. In Fig. 3, we report the boxplots comparing the values of the performance indicators, Q and F, and the value of the performance function, α, for the policies associated to the scenario-trees obtained through our procedure, for $n = 2, \ldots, 30$. The results are computed using the training test V_ξ^{tr}.

At the first iteration $n = 2$, we always obtain the same value of $Q = 119.4$ and $F = 0.928$. Indeed, the BO adds always the same scenario corresponding to $\boldsymbol{\xi} = (3, 9, 13.5, 18)$. The value of F increases from a mean value of 0.93 until a value of 0.99. This fact means that the associated decision policies, with the increase of n, can provide feasible decisions for almost all the scenarios of V_ξ^{tr}. We can note that the values of the standard deviations are smaller in the first iterations ($n \leq 10$) than in the next ones. This fact is because specific scenarios (most of the time situated at the limits of ranges) are always added in the first iterations to improve the value of the feasibility indicator F. Regarding the value of Q, we can note that there is no significant improvement at the first

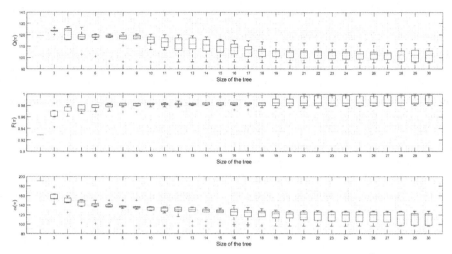

Fig. 3. Boxplots comparing the values of Q and F, and α on the training set V_ξ^{tr}, for the policies associated with the scenario-trees obtained through the iterative procedure, until $N = 30$.

iterations ($n \leq 10$) because the learning procedure focuses on the improvement of the feasibility indicator F, which has more influence than Q on the value of the performance function α. Subsequently, the value of Q starts to decrease, passing from a mean value of 119 to a value of 101 closer to the real value of the true solution. This fact means that the associated decision policies provide, with the increase of n, also optimal decisions for most of the scenarios of V_ξ^{tr}, only after the feasibility for most of the scenarios is assured.

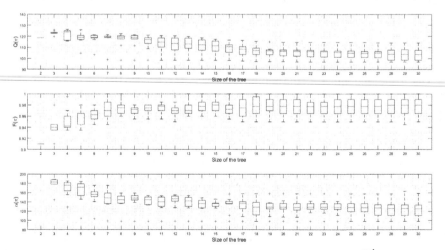

Fig. 4. Boxplots comparing the values of Q and F, and α on the validation set V_ξ^{val}, for the policies associated obtained through the iterative procedure, until $N = 30$.

Finally, to assure the robustness of the method, in Fig. 4, we report the value of F, Q and α computed on the validation set V_ξ^{val}. The results appear quite similar to the ones of Fig. 3, so the policies computed through our method seems to be able to provide feasible and optimal decisions also on new scenarios, not used to estimate the decision policy during the training process.

6 Conclusions

This work has presented a strategy to obtain an optimal decision policy for multistage stochastic programming, integrating elements from Machine Learning (sequential learning, Bayesian Optimization) and scenario-tree approximation. The search for an optimal policy was set as a global optimization problem, where the objective function is a performance function that measures both the feasibility and the optimality of the decision policy on a large set of new independent scenarios. To overcome the complexity of the search space formed by all the possible scenario-trees, we build an optimal scenario-tree using an iterative procedure, which selects, at each iteration, a scenario obtained by solving an optimization problem through Bayesian Optimization.

Preliminary computational results have been provided on a toy example showing the effectiveness of the procedure since we are able to reach good performance considering only a small number of scenarios. Future applications of such procedure will be made on different real-world application, involving multistage stochastic programming problems on supply-chain and financial domains. Nevertheless, further work is needed to validate the proposed procedure. Indeed, the goodness of the results depends on several settings. For example, the value of β (Eq. 14) depends on the considered problem and can influence the performance of the algorithm. Also, the type of regression model used to build decision policies is important and must be subject to more investigation.

References

1. Archetti, F., Candelieri, A.: Bayesian Optimization and Data Science. Springer, Cham (2019). https://doi.org/10.1007/978-3-030-24494-1
2. Birge, J.R., Louveaux, F.: Introduction to Stochastic Programming, 2nd edn. Springer, New York (2011). https://doi.org/10.1007/978-1-4614-0237-4
3. Defourny, B., Ernst, D., Wehenkel, L.: Scenario trees and policy selection for multistage stochastic programming using machine learning. INFORMS J. Comput. **25**(3), 488–501 (2013)
4. Dupacova, J.: Multistage stochastic programs: the state-of-the-art and selected bibliography. Kybernetika **31**(2), 151–174 (1995)
5. Hastie, T., Tibshirani, R., Friedman, J.: The Elements of Statistical Learning: Data Mining, Inference, and Prediction. Springer, New York (2009). https://doi.org/10.1007/978-0-387-84858-7
6. Heitsch, H., Römisch, W.: Scenario tree generation for multi-stage stochastic programs. In: Bertocchi, M., Consigli, G., Dempster, M. (eds.) Stochastic Optimization Methods in FINANCE and energy, vol. 163, pp. 313–341. Springer, New York (2011). https://doi.org/10.1007/978-1-4419-9586-5_14

7. Hilli, P., Pennanen, T.: Numerical study of discretizations of multistage stochastic programs. Kybernetika **44**(2), 185–204 (2008)
8. Høyland, K., Wallace, S.W.: Generating scenario trees for multistage decision problems. Manage. Sci. **47**(2), 295–307 (2001)
9. Keutchayan, J., Gendreau, M., Saucier, A.: Quality evaluation of scenario-tree generation methods for solving stochastic programming problems. CMS **14**(3), 333–365 (2017)
10. Leövey, H., Römisch, W.: Quasi-Monte Carlo methods for linear two-stage stochastic programming problems. Math. Program. **151**(1), 315–345 (2015). https://doi.org/10.1007/s10107-015-0898-x
11. Messina, E., Toscani, D.: Hidden Markov models for scenario generation. IMA J. Manage. Math. **19**(4), 379–401 (2008)
12. Pflug, G.C., Pichler, A.: Dynamic generation of scenario trees. Comput. Optim. Appl. **62**(3), 641–668 (2015). https://doi.org/10.1007/s10589-015-9758-0
13. Powell, W.B: Clearing the jungle of stochastic optimization. In: Bridging Data and Decisions, TutORials in Operations Research, pp. 109–137. INFORMS (2014)
14. Rasmussen, C.E., Williams, C.K.I.: Gaussian Processes for Machine Learning. MIT Press, Cambridge (2006)
15. Shahriari, B., Swersky, K., Wang, Z., Adams, R.P., De Freitas, N.: Taking the human out of the loop: a review of Bayesian optimization. Proc. IEEE **104**(1), 148–175 (2016)
16. Shapiro, A.: Monte Carlo sampling methods. Handb. Oper. Res. Manage. Sci. **10**, 353–425 (2003)
17. Shapiro, A., Dentcheva, D., Ruszczyński, A.: Lectures on Stochastic Programming: Modeling and Theory. Society for Industrial and Applied Mathematics (2014)

Deep 3D Convolution Neural Network
for Alzheimer's Detection

Charul Giri[1]([✉]), Morten Goodwin[1], and Ketil Oppedal[2]

[1] Centre for Artificial Intelligence Research (CAIR), University of Agder,
Kristiansand, Norway
{charug18,morten.goodwin}@uia.no
[2] Stavanger University Hospital, Stavanger, Norway
ketil.oppedal@sus.no

Abstract. One of the most well-known and complex applications of artificial intelligence (AI) is Alzheimer's detection, which lies in the field of medical imaging. The complexity in this task lies in the three-dimensional structure of the MRI scan images. In this paper, we propose to use 3D Convolutional Neural Networks (3D-CNN) for Alzheimer's detection. 3D-CNNs have been a popular choice for this task. The novelty in our paper lies in the fact that we use a deeper 3D-CNN consisting of 10 layers. Also, with effectively training our model consisting of Batch Normalization layers that provide a regularizing effect, we don't have to use any transfer learning. We also use the simple data augmentation technique of flipping. Our model is trained for binary classification that distinguishes between Alzheimer's and normal, as well as multiclass classification consisting of Alzheimer's, Mild Cognitive Impairment, and normal classes. We tested our model on the ADNI dataset and achieved 94.17% and 89.14% accuracy for binary classification and multiclass classification, respectively.

Keywords: Convolutional Neural Networks · Alzheimer's Detection · Medical Imaging · Deep Learning · MRI

1 Introduction

With all the advancement in medical technology, medical error is still a very common factor that contributes to 180,000 deaths every year as of 2008, reported by the US Department of Health and Human Services Office of the Inspector General. This fact makes medical error the third leading causes of deaths in the US according to [21]. Today, more than 50 million people suffer from dementia worldwide and with increasing life expectancy the number will increase. Alzheimer's disease (AD) is one of the most common forms contributing to 60%–70% of the cases (ref: https://www.who.int/news-room/fact-sheets/detail/dementia). AD is an irreversible neurodegenerative disease, in which the human brain cells involved in cognitive functioning are damaged and eventually die. Important symptoms are memory loss, reduced ability to learn new things and

© Springer Nature Switzerland AG 2020
G. Nicosia et al. (Eds.): LOD 2020, LNCS 12565, pp. 347–358, 2020.
https://doi.org/10.1007/978-3-030-64583-0_32

think. Additionally, orientation, comprehension, calculation, language and judgement are affected leading to the loss of the ability to perform everyday activities.

Diagnosing Alzheimer's disease in early stage is really difficult because it is thought to begin 20 years before the symptoms arrive [29]. The changes occurring in the brain during this stage are small and may be untraceable. According to [1], around 5.8 million people in US have Alzheimer's disease and the prediction goes to 13.8 million by 2050. Diagnosing AD is challenging and requires a thorough clinical assessment based on the patient's medical and family history, neuro-psychologic and - psychiatric testing, as well as blood and brain imaging exams such as CT-, PET- and MRI. Even though analysis of brain scans can point the doctors and researchers in the right direction, AD detection is challenging, since similar changes are common during aging as well. Computer-aided diagnostic technology has the potential to improve the challenging task of early and accurate detection of AD diagnosis. The latest research in deep learning has shown promising results in solving a large number of problems from various fields with a very high accuracy, including medical image analysis. Computer vision techniques and machine learning algorithms certainly have the potential to process medical images to efficiently diagnose Alzheimer's disease. The most successful type of model for medical image processing are Convolution Neural Networks (CNN).

In this study we have developed a CNN based pipeline to detect AD and classify it's different stages against the normal cognitive stage (NC), i.e, Mild cognitive impairment (MCI) and AD using MRI as input. MCI is a stage between normal age related forgetfulness due and the development of AD. Not all people with MCI necessarily develop AD, but many of them have a higher risk of developing AD. MRI of the brain can capture structural changes such as the decrease in size of the temporal and parietal lobes which are typically reduced in patients with AD.

CNN is a subclass of Artificial Neural Networks where features are extracted from data using feature maps or kernels that spatially share weights. These feature maps work towards finding distinct features in the images in order to distinguish them into different categories. This gives CNN the ability to extract features from MR images and to classify AD. The multiple building blocks of CNN such as convolution layer, pooling layer efficiently learn feature maps to gain spatial knowledge from image data. That's why CNN have an advantage over other deep learning algorithms in medical image analysis. The novelty of this research inheres within the use of 3D-CNNs to process whole MRI images instead of using MRI slices with 2D CNNS, giving us better performance.

2   Related Work

Several studies have been conducted in the recent years to develop a computer-aided diagnosis system for Alzheimer's detection. Traditional methods included researcher trying to handcraft features through voxel-based methods, ROI based methods, hippocampal shape and volume or patch-based methods.

[20] have attempted to compute region of interest (ROI) to detect AD. ROI is a section of image in which a binary mask is used to carry out various operations like filtering. [5] used voxel based morphometry (VBM) and MRI to investigate gray matter change in medial temporal structures and volume changes in several other brain regions.

[25] have analysed regional brain atrophy for example in the hippocampus to detect patterns of neuron death by segmenting different types of brain tissues such as grey-matter (GM), white-matter (WM) against cerebrospinal fluid (CSF) in the MRI. They segmented the images using watershed transformation algorithm [23] with marker image, and then calculating the shrinkage happened in the whole brain through Tissue Atrophy Ratio (AT) for early detection of AD. [17] has used an inherent structure-based multi-view learning (ISML) method in which they have extracted multi-view features based on multiple selected templates. They then employed a subclass clustering algorithm for feature selection in order to eliminate the redundant features. A SVM-based ensemble classifier is used to classify subjects into AD, MCI and NC.

[22] have used Hu moments invariants [13], calculating a set of seven invariant moments to extract features in the brain images (MRI) of all subjects. They also showed that normalizing these moments results in better feature extraction which makes it easier for the classifier to distinguish. The extracted features are then used as inputs to SVM and KNN classifiers to classify the subjects. They compared the classifiers, showing that SVM performed far better than KNN.

In [8] feature extraction is done by using ROI on three sMRI biomarkers, named as Voxel-based morphometry, Cortical and sub-cortical volume and Hippocampus volume. They used Principal component analysis (PCA) [4] for feature selection. PCA is a dimensionality reduction method simplifying a high dimension data into smaller dimension without losing the important patterns or trend in the data. Using PCA, they selected 61 features for the classification of AD. They studied three different classifiers: SVM, Random Forest and KNN, and evaluated their performance in two stages. First stage included individual features from s VBM-extracted ROI volumes, CSC-extracted feature volumes, and HV extracted features and second staged is evaluating classifiers using the combination of all 61 features. They concluded that SVM outperformed KNN and Random Forest in all cases.

Multi-modal data fusion using MRI and PET scans was proposed in [19]. They used stacked auto encoders and and a sigmoidal decoder to discover the synergy between MRI and PET scans for high level feature extraction with a softmax classifier. A zero-masking technique (SAE-ZEROMASK) is used in contrast to simple feature concatenation (SAE-CONCAT) technique. They randomly hide one modality and trained the hidden layers to reconstruct the multi-modal using inputs mixed with hidden modality.

However, SAE-CONCAT usually fails to captures the non-linear co-relation between two different modalities [26]. That's why authors in [26] proposed Multi-modal Stacked Deep Polynomial Networks algorithm (MM-SDPN) which uses multi-modalities like [19] but they have used two stages of SDPN to learn high-level features.

Various other machine learning algorithms have proven to be efficient when it comes to extracting high level features. Artificial neural networks were used by [9] for Nephropathy Detection and Classification. The drawback of using Feed Forward Neural Network (as usually called ANN) for computer vision is that they are computationally expensive. The number of learning parameters in ANN exponentially increase with respect to the size of the image. Thus to counter this problem, the use of convolution neural network (CNN) to automate feature learning in images has become popular because of their ability to generalize well to high dimensional data, without losing important patterns.

A 2-D CNN is presented by [2] where they used VGG16 [27] as a base model, and treated a 3-D MRI image as a stack of 2-D MRI slices. Other variants of CNNs have been used in the researches such as in [15] authors used a ROI focused 3-D CNN with multi-modality. Each modality and ROI region was assigned a dedicated pipeline of a CNN block, whose output was flattened. The flattened outputs are the extracted features from each modality and region of interests(ROI). These feature outputs were then concatenated, resulting into late data fusion and were passed to a softmax classifier. Problem with this approach is similar to what was described earlier, that while performing late data fusion using simple concatenation, it ignores the variance in the nature of multi-modalities and fails to learn the non-linear co-relation between modalities [26].

A few other research works employ pre-training 3-D CNN with auto encoders such as [24].They used sparse auto encoders for feature extraction and also compared the performance of 2-D CNN against 3-D CNN. Authors in [12] took a two stage approach where they first used a convolutional auto encoder in place of conventional unsupervised auto encoder to extract local features with possibly long voxel-wise signal vectors. These features are used to perform task-specific classification with a target-domain-adaptable 3D-CNN using transfer learning with Net2Net weight initialization. They later proposed [11] in which they trained the same model with deep supervision, which resultant in an improvement.

[6] also build a 3-D convolutional neural network for an end-to-end classification of subjects with AD. They added metadata (sex and age of subjects) to the first fully connected layer in their model. The downside of using metadata in the neural network is that the network will try to find the correlation based on the metadata that might be biased towards the predilection of meta-data, for e.g., older patient are more likely to be affected by Alzheimer's Disease, so the network might bias towards assigning older people to the Alzheimer's Class.

[30] studied various paradigms of 3-D CNNs like patch-level 3D CNN, ROI based 3D CNN, subject-level 3D CNN, along with exploring transfer learning using auto encoder pre-training and ImageNet pre-training. They also reviewed studies done on AD classification using Deep Learning from January 1990 to the 15th of January 2019, which proved very helpful in the proposed research.

3   Proposed Work

In this work we have addressed the problem of Alzheimer's Disease Detection, and proposed a novel architecture based on 3D convolution Networks. The

Table 1. Demographic data for 817 subjects from the ADNI database (STD - standard deviation))

Diagnosis	Subjects	Age($mean_{\pm std}$)	Gender (F/M)
AD	188	$75.36_{\pm 7.5}$	89/99
MCI	401	$74.84_{\pm 7.3}$	143/258
NC	228	$75.96_{\pm 5.0}$	110/118

advantage of using a 3D convolution Neural Network over 2D convolution Neural Network is that 3D CNN are able to extract features in 3D space. For example, in a video it can derive spatial information from 2 dimensions and as well as temporal information. In our case the MRI scans don't have the temporal dimension, but are 3D images. So by using a 3D CNN we can extract the spatial information from the three dimensional space.

We have performed binary classification between AD and NC (Normal Cognitive) and multi-class classification between AD, MCI and NC. We have used whole brain MRI scans on subject level for the network to focus on and eliminated the need of selecting ROI.

3.1 Dataset and Pre-processing

We collected the data from the Alzheimer's Disease Neuroimaging Initiative (ADNI). ADNI [14] is a collaboration for research in the progression of Alzheimer's Disease, started in 2004 as a private-public partnership among 20 companies. The data consists of 1075 1.5T Screening MRI records of 817 subjects from ADNI1 project. The statistics of subjects is shown in Table 1.

The MR images in the dataset don't have a standard size, so we downscaled the images to a resolution of 120 x 90 x 130 maintaining the axial view. We follow the common practice of downscaling for the deep learning model to process the large images.

Since Alzheimer's can start in any hemisphere of the brain, so it makes sense to augment the data by flipping horizontally. Having larger dataset is crucial for better performance in deep learning models. We only used left and right flip augmentation of the data. We have chosen axial view as it helps in avoiding the motion artifacts from eyeball which can appear in other views [7]. Table 1 gives a summary of the demographic information of the subjects studied in this paper.

3.2 Model

In this paper, we have used a 3-Dimensional Convolution Neural Network (3D CNN). As the name suggests 3D CNN performs convolution operation in 3 dimensions to extract features as opposed to a traditional 2D Convolution Neural Network, which works in only 2D space. Mathematically a 3D convolution operation in neural network is defined as follows:

Table 2. Model Architecture

Layer Name	Kernel size	No of kernels/neurons
Conv3D	5 × 5 × 5	32
BatchNorm	–	–
Conv3D	3 × 3 × 3	32
Conv3D	3 × 3 × 3	32
BatchNorm	–	–
MaxPool3D	2 × 2 × 2	–
Conv3D	3 × 3 × 3	64
Conv3D	3 × 3 × 3	64
BatchNorm	–	–
MaxPool3D	2 × 2 × 2	–
Conv3D	3 × 3 × 3	64
Conv3D	3 × 3 × 3	128
BatchNorm	–	–
MaxPool3D	2 × 2 × 2	–
Conv3D	3 × 3 × 3	64
Conv3D	3 × 3 × 3	128
BatchNorm	–	–
Conv3D	3 × 3 × 3	64
BatchNorm	–	–
MaxPool3D	2 × 2 × 2	–
Flatten	–	–
Dense	–	512
Dropout(0.1)	–	–
Dense	–	2

Given an input of size N,H,W,D,C_in where H = height, W = width, D = depth, C = no.of channels, N = batch size, the output of the convolution layer is produced as:

$$Out(N_i, C_{out_j}) = b(C_{out_j}) + \sum_{k=0}^{C_{in}-1} weight(C_{out_j}, k) * input(N_i, k) \qquad (1)$$

where * is the 3D cross correlation operation between two signals. The learnable kernels are $l \times l \times l$ matrices, which slide over the large input to detect relevant patterns creating new feature maps and convolving feature maps from previous layer. The architecture of the model is given in Table 2. Our model consists of 4 identical convolution blocks with 2 convolution layer stacked up on a batch normalization layer in each block. Every block goes through 3D maxpooling with a pool size of (2, 2, 2). Beside this, the model has an input block, last convolution block and a classifier block. The input block consists of one convolution layer and batch normalization layer. The last convolution block consists of 1 convolution layer followed by a batch normalization and maxpool layer respectively. The classifier block has one fully connected layer followed by a softmax classifier. All convolution blocks contain filters of size $3 \times 3 \times 3$, to learn the small details and patterns of Alzheimer's affected parts in the brain.

We have used relu activation function on all layers and categorical crossentropy cost function to optimize the loss, using Adam optimizer. ReLu doesn't activate the neurons with negative input values, which makes it is computationally very efficient over sigmoid and tanh function as it introduces sparsity.

3.3   Training

To train the model we have first initialized all the kernels with He normal initialization to achieve faster convergence. The model takes a batch size of 2 samples at a time. We have trained the model using 10-fold cross-validation to ensure that the model will perform well on all points of data, and not on just some random sets. To prevent the model from overfitting we have used L2 regularization on the last fully connected layer, which is followed by a dropout layer. To stop convolutional layer learning irrelevant features or over-fit to the features, we have used batch-normalization. The batch-normalization layer helps in reducing the co-variance shift of the hidden units in the neural network. It normalizes the output of the previous layer using the current batch statistics. In this way the distribution of the weights of the hidden units or kernels in our case can adapt to any change in the distribution of the data. Batch normalization layer also gave us some regularization effect as it scales the weights based on the batch mean and variance, which makes the weights stable. [16] showed that L2 regularizer has no regularizing effect when it is used along with Normalization, however it can influence the effective learning rate.

4   Experiments

We trained our model using k-fold cross validation with 10 folds. The data distribution of training, validation and testing is 80%, 10% and 10% respectively. The model performs two classification tasks: binary classification between AD and NC, and multiclass classification between AD, NC and MNC. We have seen in other studies that transfer learning using Net2Net and ImageNet has been used extensively before. But our model is able to learn the features for this classification task and achieve superior performance compared to the other models without transfer learning. The results of our approach are presented in Table 3 where we have compared it with other models.

We didn't use any kind of transfer learning because pre-trained models like ImageNet models don't contain relevant information for transfer learning of medical data classification. Other research works have used the weights of their pre-trained model from binary classification (e.g., AD and NC) to initialize the model for multi-class classification (e.g, AD, MCI and NC) or vice versa. The drawback of this approach is that if the same data is present for some classes in the next classification, then it will lead to data leakage in the model, causing biased transfer learning. To tackle this issue we initialized our model using He initialization during both classification tasks.

Table 3. Comparison with previous studies of Alzheimer's Detection using Deep Learning on ADNI Data. Accuracy metrics is used for comparison and represents binary classification between Alzheimer's Disease (AD) vs Normal Cognition (NC) and multi-class classification between Alzheimer's Disease (AD) vs Normal Cognitive (NC) Mild cognitive impairment (MCI)

Method	Modality	Dataset size	Accuracy	
			AD vs NC	AD vs MCI vs NC
[18]	FDG-PET	ADNI (339)	91.20	-
[28]	PET + MRI	ADNI (317)	91.10	-
[19]	MRI+PET	ADNI (331)	91.40	53.84
[3]	MRI	ADNI (815)	91.41	-
[6]	MRI	ADNI (841)	94.10	61.10
Our approach	MRI	ADNI (817)	94.17	89.14

We ran the model for 100 epochs for each fold. We experimented with different hyper-parameters like learning rate, drop out rate and regularization coefficient. We found the best set of parameters for the optimal performance of the model is an initial learning rate of 0.0001, drop out factor of 0.1 and regularization parameter as 0.001. We also used scheduled learning rate based on step decay to optimize the loss curve and avoid divergence. The learning rate is scheduled to drop by a factor of 0.1 every 40th epoch. As mentioned earlier we have used Batch normalization layer with Convolution Layers. We initially tried L2 regularizer. L2 regularizer when used along with Batch Normalization has no regularizing effect [16]. So we tried L2 regularizer and Batch Normalization separately, and found that the model performed better with Batch Normalization than with L2 regularizer, increasing the accuracy by 3% in binary classification and 2.34% in multiclass classification.

We thoroughly tested our model and compared it with previous Deep Learning based Alzheimer Detection model. [6] have used demographic information for the classification task. They merged the age and gender information as additional features in their network. Providing such information to the network can make it bias towards the correlation and pattern present in the information. And if the correlations are strong then the network will rely on these demographic features more than the features learned from the MRI images. For example it's a well know fact that eighty-one percent of people who have Alzheimer's disease are age 75 or older, and almost two-thirds of Americans with Alzheimer's are women [10]. If this statistics is provided then it can make the model biased towards a particular age or sex group.

To evaluate the performance of our model we have used accuracy (Acc), precision and recall, and F_2 score. Given True Positive, True Negative, False Positive and False Negative as TP, TN, FP, FN recall and precision is calculated as $Precision = TP/(TP + FP)$, $Recall = TP/(TP + FN)$. The F_2 score is calculated as $(5 \times Precision \times Recall)/(4Precision + Recall)$. We have used F_2 score because it gives more emphasis on false negatives making it a suitable option for medical experiments evaluation.

Evaluation Matrix. Table 4 presents the evaluation matrix averaged over 10 folds, of our model for the two classification tasks, and comparing the effect of l2 and batch normalization. The training vs validation loss and training vs validation accuracy of the average of the 10 folds is plotted in Fig. 1 for binary classification. It can be seen that after every 40 epochs the graph becomes smoother. It's due to the learning rate decay occurring every 40th epoch. The learning rate decreases by a factor of 0.1 here, helping the model to converge the optimization and making the learning more stable.

Table 4. Evaluation Matrix for the Deep 3D CNN Model (L2: L2 regularization, BN: Batch Normalization. Last row represents the configuration of our model)

Model	L2	BN	Binary Class				Multi Class			
			Acc(%)	F_2	Pre	Recall	Acc(%)	F_2	Pre	Recall
Deep 3D CNN	✓	✓	91.11	0.90	0.89	0.94	86.22	0.85	0.88	0.87
Deep 3D CNN	✓	✗	91.20	0.91	0.91	0.92	86.70	0.86	0.84	0.85
Deep 3D CNN	✗	✓	94.17	0.94	0.94	0.94	89.14	0.88	0.89	0.88

As can be seen from Fig. 2, the confusion matrix details the class-wise performance of our 3D CNN model on binary classification. During the evaluation of all the folds, we noted that the model has a very low false negative rate on its best fold, of 0.016 for AD class in binary classification, and 0.06 in multiclass classification, where it confuses a bit with MCI class. The average false negative rate is still very low at 0.04 for AD class in binary classification. These experiments show that our model has outperformed the other models with only single modality and training from scratch. The results also demonstrate the robustness and confidence of the Alzheimer's prediction by the proposed model.

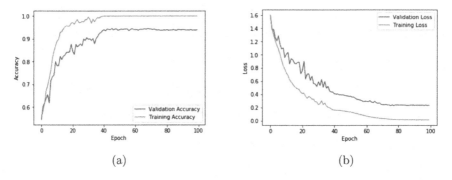

Fig. 1. (a) training vs validation accuracy (b) training vs validation loss

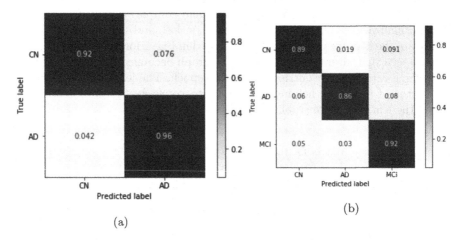

Fig. 2. 10 folds average of Confusion Matrix for (a) Binary Classification and (b) Multi-class Classification for model with best accuracy listed in Table 4

5   Conclusion and Future Work

In this paper we proposed a novel 3D CNN based classifier for Alzheimer's Detection. The model showed the ability to learn the relevant features on its own without the use of any transfer learning. From the results of multi-class classification it can be observed that the model has the potential of distinguishing between two very similar looking classes, in our case AD and MCI, and MCI and NC, with a very high confidence. We showed that proper initialization of model and fine parameter tuning leads to superior results. The future work involves experimenting with multiple modalities and exploring other feature extraction and feature selection algorithms so that the model will be able to learn more with small amount of data. The model can be improved in further studies by carrying out more sophisticated procedure for feature selection after feature extraction as just mentioned, and using deeply supervised learning. The next step could be to segment the biomarkers of Alzheimer's in the brain, and predicting the pattern of growth of the disease in a very early stage. We would also like to explore the machine learning models on MCI subjects to predict who might develop AD at a later stage.

References

1. 2019 alzheimer's disease facts and figures: Alzheimer's Dementia **15**(3), 321–387 (2019)
2. Convolutional neural network based Alzheimer's disease classification from magnetic resonance brain images. Cogn. Syst. Res. **57**, 147–159 (2019)
3. Aderghal, K., Benois-Pineau, J., Afdel, K., Catheline, G.: Fuseme: Classification of sMRI images by fusion of deep CNNs in 2D+E projections. In: CBMI 2017 (2017)
4. Andersen, A.H., Gash, D.M., Avison, M.J.: Principal component analysis of the dynamic response measured by fMRI: a generalized linear systems framework. Magnetic Resonance Imaging **17**(6), 795–815 (1999)
5. Bozzali, M., et al.: The contribution of voxel-based morphometry in staging patients with mild cognitive impairment. Neurology **67**(3), 453–460 (2006)
6. Esmaeilzadeh, S., Belivanis, D.I., Pohl, K.M., Adeli, E.: End-To-End Alzheimer's disease diagnosis and biomarker identification. In: Shi, Y., Suk, H.-I., Liu, M. (eds.) MLMI 2018. LNCS, vol. 11046, pp. 337–345. Springer, Cham (2018). https://doi.org/10.1007/978-3-030-00919-9_39
7. Goto, M., et al.: Utility of axial images in an early Alzheimer disease diagnosis support system (VSRAD). Nihon Hoshasen Gijutsu Gakkai zasshi **62**, 1339–1344 (2006)
8. Yubraj Gupta, et al.: Early diagnosis of Alzheimer's disease using combined features from voxel-based morphometry and cortical, subcortical, and hippocampus regions of MRI T1 brain images. PLOS ONE **14**(10), 1–30 (2019)
9. Hazarika, A.: A novel technique of neuropathy detection and classification by using artificial neural network (ANN) (2013)
10. Hebert, L.E., Weuve, J., Scherr, P.A., Evans, D.A.: Alzheimer disease in the United States (2010–2050) estimated using the 2010 census. Neurology **80**(19), 1778–1783 (2013)
11. Hosseini Asl, E.: Alzheimer's disease diagnostics by a 3D deeply supervised adaptable convolutional network. Front. Biosci. (Landmark edn.) **23**, 584–596 (2018)
12. Hosseini Asl, E., Keynton, R., El-Baz, A.: Alzheimer's disease diagnostics by adaptation of 3D convolutional network (2016)
13. Ming-Kuei, H.: Visual pattern recognition by moment invariants. IRE Trans. Inf. Theory **8**, 179–187 (1962)
14. Jack Jr., C.R., et al.: The Alzheimer's disease neuroimaging initiative (ADNI): MRI methods. J. Magnetic Resonance Imaging **27**(4), 685–691 (2008)
15. Khvostikov, A., Aderghal, K., Benois-Pineau, J., Krylov, A., Catheline, G.: 3D CNN-based classification using sMRI and MD-DTI images for Alzheimer disease studies (2018)
16. Laarhoven, T.: L2 regularization versus batch and weight normalization (2017)
17. Liu, M., Zhang, D., Adeli, E., Shen, D.: Inherent structure-based multiview learning with multitemplate feature representation for Alzheimer's disease diagnosis. IEEE Trans. Biomed. Eng. **63**(7), 1473–1482 (2016)

18. Liu, M., Cheng, D., Yan, W., Alzheimer's Disease Neuroimaging Initiative: Classification of Alzheimer's disease by combination of convolutional and recurrent neural networks using FDG-PET images. Front. Neuroinform. (2018)
19. Liu, S., et al.: Multimodal neuroimaging feature learning for multiclass diagnosis of Alzheimer's disease. IEEE Trans. Biomed. Eng. **62**(4), 1132–1140 (2015)
20. Arun, S., Latha, M.: Detection of ROI for classifying Alzheimer's disease using MR. Image of brain. Int. J. Innov. Technol. Exploring Eng. (IJITEE) 8 (2019)
21. Makary, M.A., Daniel, M.: Medical error—the third leading cause of death in the US. BMJ **353** (2016)
22. Mohammed, A., al azzo, F., Milanova, M.: Classification of Alzheimer disease based on normalized hu moment invariants and multiclassifier. Int. J. Adv. Comput. Sci. Appl. **8** (2017)
23. Najman, L., Schmitt, M.: Geodesic saliency of watershed contours and hierarchical segmentation. IEEE Trans. Pattern Anal. Mach. Intell. **18**(12), 1163–1173 (1996)
24. Payan, A., Montana, G.: Predicting Alzheimer's disease: a neuroimaging study with 3D convolutional neural networks. In: ICPRAM 2015–4th International Conference on Pattern Recognition Applications and Methods, Proceedings, 2 (2015)
25. Sadek, R.: Regional atrophy analysis of MRI for early detection of Alzheimer's disease. Int. J. Sig. Process. Image Process. Pattern Recogn. **6**, 50–58 (2013)
26. Shi, J., Zheng, X., Li, Y., Zhang, Q., Ying, S.: Multimodal neuroimaging feature learning with multimodal stacked deep polynomial networks for diagnosis of Alzheimer's disease. IEEE J. Biomed. Health Inform. **PP**, 1 (2017)
27. Simonyan, K., Zisserman, A.: Very deep convolutional networks for large-scale image recognition. arXiv:1409.1556 (2014)
28. Vu, T.D., Yang, H.-J., Nguyen, V.Q., Oh, A.-R., Kim, M.-S.: Multimodal learning using convolution neural network and sparse autoencoder. In: 2017 IEEE International Conference on Big Data and Smart Computing (BigComp), pp. 309–312, February 2017
29. Villemagne, V.L., et al.: Amyloid beta deposition, neurodegeneration, and cognitive decline in sporadic Alzheimer's disease: a prospective cohort study. Lancet Neurol. **12**(4), 357–367 (2013)
30. Wen, J., et al.: Convolutional neural networks for classification of Alzheimer's disease: overview and reproducible evaluation, April 2019

Combinatorial Reliability-Based Optimization of Nonlinear Finite Element Model Using an Artificial Neural Network-Based Approximation

Ondřej Slowik[(✉)] [iD], David Lehký[iD], and Drahomír Novák[iD]

Institute of Structural Mechanics, Faculty of Civil Engineering, Brno University of Technology, Brno, Czech Republic
slowik.o@fce.vutbr.cz

Abstract. The paper describes the reliability-based optimization of TT shaped precast roof girder produced in Austria. Extensive experimental studies on small specimens and small and full-scale beams have been performed to gain information on fracture mechanical behaviour of utilized concrete. Subsequently, the destructive shear tests under laboratory conditions were performed. Experiments helped to develop an accurate numerical model of the girder. The developed model was consequently used for advanced stochastic analysis of structural response followed by reliability-based optimization to maximize shear and bending capacity of the beam and minimize production cost under defined reliability constraints. The enormous computational requirements were significantly reduced by the utilization of artificial neural network-based approximations of the original nonlinear finite element model of optimized structure.

Keywords: Reliability-based optimization · Combinatorial optimization · Heuristic optimization · Artificial neural network · Double-loop reliability-based optimization · Prestressed concrete girder optimization · Stochastic analysis

1 Introduction

Reliability-based optimization (hereinafter RBO) represents a robust multidisciplinary approach for effective engineering design, ensuring both optimal and reliable structure at the output of the applied process. The concept itself appeared quite early in reliability engineering (see, e.g. [1, 2]). From those first pioneering works, the concept progressed from reliability-based to risk-based optimization approaches, e.g. [3], emphasizing robustness in structural optimization, e.g. [4]. The variety of RBO approaches have been evolved during the last decade. Most available procedures can be classified into three categories, namely, the double-loop (also known as two-level) approach, the single-loop (also known as mono-level) approach and the decoupled approach [5]. The double-loop algorithm structure is the most straightforward approach and consists of nesting the computation of the failure probability concerning the current design within

© Springer Nature Switzerland AG 2020
G. Nicosia et al. (Eds.): LOD 2020, LNCS 12565, pp. 359–371, 2020.
https://doi.org/10.1007/978-3-030-64583-0_33

the optimization loop [6]. It is also the most general approach applicable to all possible types of RBO tasks. This approach was utilized within the presented work.

Despite its great potential application of RBO within engineering practice, it is still sporadic and not very common. Most of the available applications come from fields where mass production of the final product is being planned (e.g. automotive industry [7]). Within civil engineering field, the RBO is often applied for optimization of precast concrete structures. For most practical civil engineering problems, deterministic optimization connected with partial safety factor-based reliability consideration [8] remains the usual choice due to its simplicity and low computational demands.

Optimization task defined within the following sections represents one of the outcomes from long term research cooperation between the team from Faculty of Civil Engineering at the Brno University of Technology (BUT) and the team from Institute for Structural Engineering at the University of Natural Resources and Life Sciences (BOKU) and industrial partner – Franz Oberndorfer GmbH & Co KG which is focusing on precast structural elements production in Austria.

Extensive experimental studies on small specimens and small and full-scale beams have been performed to comply with the required information for complex material laws as they are implemented in advanced probabilistic nonlinear numerical analyses. The experimental study of material parameters was conducted first [9]. The approach based on artificial neural network (hereinafter ANN) modelling [10] for advanced parameter identification was utilized. The following three fundamental parameters of concrete were identified: modulus of elasticity E_c, tensile strength f_t, and specific fracture energy G_f. The compressive strength of concrete f_c was measured by means of standard cubic compression tests.

Subsequently, the destructive shear tests under laboratory conditions were performed on ten scaled girders and proof loading test with full-size LDE7 roof girders [11]. Finally, the results of this research support in establishing robust deterministic and stochastic modelling techniques summarized in [12], which were also the basis for advanced statistical analysis of the shear capacity of LDE7 girders [13] and reliability-based optimization presented within this paper. Due to significant computational burden connected with evaluation of nonlinear finite element numerical models, the ANN-based surrogate models of LDE7 girder's performance were prepared and utilized in order to perform below-described optimization.

2   Optimized Structural Member

The LDE7 girder has a TT-shaped cross-section, a total length of 30.00 m and a height of 0.50 m at ends and 0.90 m in the middle of the beam. The width webs are 0.14 m. The slab has a dimension b/h of 3.00 m/0.07 m. The reinforcement and geometry of the beam are symmetrical according to middle cross-sectional and longitudinal plane. The girder is continuously prestressed to 1107,53 MPa via 32 (16 in each web) × 7-wire 1/2-inch strands with a wire quality of ST 1570/1770. The prestressing reinforcement is divided into four layers depending on the length of the isolated part of cables (the part where cables have no bond to the concrete). The lowest six cables (group S1) in each web are connected to the concrete at the whole length of the beam. The four cables in the second layer (group S2) of each web are isolated 2.00 m from both sides of the beam. The four cables in the third layer (group S3) of each web are isolated 4.00 m from both sides of the beam. The two cables in the fourth layer (group S4) of each web are isolated 5.60 m from both sides of the beam.

Thirteen stirrups with a diameter of 0.006 m (group R1) were mounted at 0.50 m to each other at the ends of the beam. Another 16 rebars with a diameter of 0.006 m (group R1) were installed in the middle of the girder. The concrete is strengthened in areas where tendons introduce prestressing force by 6 U shaped stirrups per each pair of tendons (group R2). The plate of the beam was equipped with orthogonal reinforcement of 0.008 m in diameter at a distance to each other of 0.20 m in the longitudinal and transverse directions (group R3). The lower reinforcement layer was anchored using four horizontal rebars in U-bolt shape with a diameter of 0.012 m (group R4) per side. Six reinforcement bars with a diameter of 0.014 m (group R5) are in the upper reinforcement layer of each web, and two reinforcement bars with a diameter of 0.02 m (group R6) are in the lower reinforcement layer and. Figure 1 shows detail of reinforcement groups and positions within the cross-section.

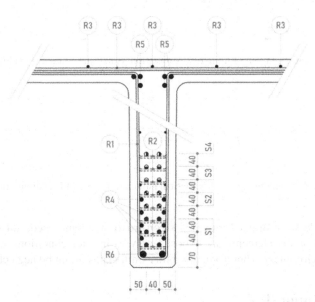

Fig. 1. Reinforcement layout within the cross-section of LDE7's web

The ATENA Science software package [14] was utilized for nonlinear finite element modelling of LDE7 roof girder. A comprehensive description of all parameters and settings utilized within the developed nonlinear numerical model is beyond the scope of this paper. Detailed information on performed numerical modelling is given in [12] and [13].

Two experimental setups have been modelled with the aim to capture the behaviour of the beam under two different failure conditions. Developed numerical models [13] have shown that girders fail in shear for loading situations with nonsymmetric force loading as well as for three-point bending (hereinafter 3PB) loading layout. At first, the modelled girder was loaded by displacement applied 4.125 m from support. Limit loading force at the peak of load versus deflection (hereinafter LD) diagram F_{max} was monitored outcome for above-described loading situation. Figure 2 displays the loading layout of numerical simulation using a nonsymmetric loading force.

Despite the shear failure of girder under 3PB loading layout the limit deflection defined for serviceability limit state as $l/200$ where l is the length of girder's span is reached before the shear failure of beam. Therefore, the monitored output for 3PB loading layout was loading force at the time when limit deflection occurs F_{wlim}. Figure 3 displays the loading layout of numerical simulation using 3PB loading scheme.

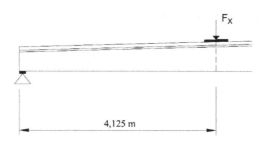

Fig. 2. Loading layout of numerical simulation using nonsymmetric loading force

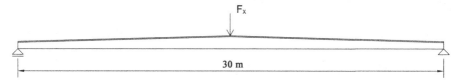

Fig. 3. Loading layout of numerical simulation using 3PB configuration

LDE7 girders are utilized mostly for constructions of mounted hall object roofs. In most applications elements are exposed only to interior conditions with the non-aggressive environment. Therefore, durability limit states might be neglected.

3 Optimization Task

Optimization procedure performed as final step of whole complex research reflects the producer's possibilities and production process from a practical and economical perspective. Considering the current product's established production procedures, related support utilities and materials ranging from design to marketing information, it would be too complicated to adjust the shape of the product, or it's reinforcement layout. Feasible optimization of LDE7 girder design should, therefore, focus on reinforcement diameters and utilized grade of concrete. Experimentally tested girders were cast only with the minimal amount of shear reinforcement following the idea of a manufacturer research department that prestressing of girder should ensure enough shear capacity. In contrast to that below described reliability-based optimization shown that increased amount of shear reinforcement might help to reduce diameters of reinforcement at other positions, increase shear capacity and utilize the full bending capacity of girder in combination with lower production cost.

The objectives for performed optimization were to maximize shear capacity F_{max} at ultimate limit state represented by nonlinear numerical model loaded according to the scheme in Fig. 2, to maximize bending capacity F_{wlim} represented by serviceability limit state corresponding to limit deflection 0.15 m during 3PB loading layout (see Fig. 3) and to minimize the overall price for utilized material.

3.1 Combinatorial Optimization Problem – Design Space Definition

Since construction materials and elements are defined by specified grades or dimensions, and it is not possible to purchase items out of available specifications, the optimization of real structures is often limited to a combinatorial problem. One of the above-declared optimization goals is to adjust the diameters of reinforcement groups and select a proper concrete grade to minimize price. Table 1 summarizes discrete options of reinforcement bars defined by diameters and corresponding unit prices while Table 2 summarizes available concrete grades and related prices. These data were provided by the manufacturer.

Table 1. Prices related to available diameters of conventional reinforcement

Diameter [mm]	6	8	10	12	14	16	20	26	30
Price [EUR/m]	0.19	0.21	0.52	0.65	0.83	1.13	1.68	2.72	3.63

Table 2 summarizes available grades of concrete which might be utilized for precast element production.

Table 2. Prices related to available concrete grades

Concrete grade	C 30/37	C 35/45	C 40/50	C 45/55	C 50/60
Price [EUR/m3]	64.46	68.06	74.98	79	83.93

Type of tendons might also be considered for optimization. Feasible cables, however, correspond to one of three types for the same price. Thus, no additional benefit might be expected by also varying type of tendons.

Concrete mixtures of grades C 50/60 and C 40/50 were subjected to fracture experiments in order to gain information on material parameters such as Young's modulus E, tensile strength f_t, fracture energy G_f, and compressive strength f_c (as summarized in [9]). Facing the fact that detailed statistical information on real values of mentioned parameters was known just for two tested mixtures, it was decided to utilize Model Code

2010 [15] recommendation for estimation of above-mentioned parameters mean values based on perceived characteristic compressive strength obtained during the standard cubic compression test.

For the stochastic simulation of material parameters within the inner loop of the RBO algorithm it was necessary to define the stochastic model of utilized concrete grades. Since no information on variability and probability density functions (hereinafter PDF) was available for 3 of 5 available concrete mixtures, these assumptions were made:

All mixtures do have the same values of coefficients of variation (hereinafter CoV) and same PDF. Utilized values of CoV corresponds to average CoV of parameters of mixtures C50/60 and C40/50 (obtained from experiments). Utilized PDFs corresponds to those of two known mixtures. Stochastic description of mixtures thus differs only by mean values of material parameters. Utilized CoVs and PDFs for concrete parameters are summarized in Table 3.

Table 3. Statistical distribution features of material concrete parameters

Parameter	CoV [%]	PDF
f_c	6	Lognormal (2par)
E_c	12	Lognormal (2par)
f_t	14	Lognormal (2par)
G_f	15	Lognormal (2par)

The value of prestressing force is determined based on selected grade according to the requirement that stress state within the concrete should not exceed the value of $0.45 f_{ck}$ in order to avoid nonlinear concrete creep behaviour evaluation (according to [8]). Stress state was evaluated for uniform loading situation (live load, weight of roof construction-0.3 kN/m^2 and snow load 1.14 kN/m^2) prescribed by producer. Prestressing force is thus entirely dependent on the value of compressive strength of concrete. Above described assumptions clearly render discrete optimization design space. The total sum of the available design options is listed in Table 4.

R1–R6 represent groups of reinforcement specified according to their purpose (see Fig. 1). Note that the number of options for a given reinforcement group reflects geometrical limitations of the construction. The total number of combinations might not appear significant compared to optimization problems within an infinite continuum. However, the computational time required to calculate one simulation represented by numerical model described in Sect. 2 is approximately 10 h, and two models for two separate loading conditions needs to be evaluated. The direct solution is not feasible since many simulations are needed for optimization using some of the standard optimization methods (such as Evolutionary algorithms). Note that the complexity is further increased by the need to perform a probabilistic assessment within the inner reliability loop. The final solution algorithm depicted in Fig. 5 utilizes 10 000 simulations to evaluate stochastic parameters (see Sect. 3.2) of both limit state configurations (see Fig. 2 and 3). That

Table 4. Total amount of possible combinations within optimization design space

Parameter	Min. value	Max. value	Options
f_c [MPa]	45	68	5
R1 (Orig. 6 mm)	6	10	3
R2 (Orig. 6 mm)	6	10	3
R3 (Orig. 8 mm)	6	10	3
R4 (Orig. 12 mm)	6	16	6
R5 (Orig. 14 mm)	6	26	8
R6 (Orig. 20 mm)	6	30	9
Combinations: 58320			

amount of simulations ensures stable point estimates of mean values and standard deviations of observed quantities (see Eq. 1 and 2). This statistical evaluation is performed for each simulation within outer optimization loop of double-loop RBO procedure (see Fig. 5). The total number of evaluations of defined nonlinear numerical model would be (in case of brute force solution) $58\ 320 \times 2 \times 10\ 000 = 1\ 166\ 400\ 000$. That would approximately correspond to $1\ 166\ 400\ 000 \times 10/12 = 972\ 000\ 000$ h of computation since 12 simulations might run in parallel on utilized hardware.

3.2 Stochastic Model

While above-described optimization design space is discrete, design space for statistical evaluation of observed quantities is naturally continuous and infinite. Since LDE7 roof girder is prefabricated, it can be utilized under various load conditions, and exact loading conditions are not defined at the time of production. Standard reliability evaluation requires knowledge of the statistical description of load actions. Thus, only semi probabilistic evaluation of design based on the separation of load actions and structural resistance, according to EC2 [8], might be performed for precast products. The natural uncertainties might be reflected by point estimates of mean value and standard deviation of structural response. For both above-mentioned loading setups (see Fig. 2 and Fig. 3) can be estimated mean values and standard deviations of observed quantities (F_{max} and F_{wlim}) based on small sample simulation using LHS mean method [16] and subsequent statistical evaluation of mean value:

$$\overline{\mu_g} \approx \frac{1}{N_{sim}} \sum_{N_{sim}}^{i=1} g(X_i) \tag{1}$$

and standard deviation:

$$\sigma \approx \sqrt{\frac{1}{N_{sim}-1} \sum_{i=1}^{N_{sim}} \left(g(X_i) - \overline{\mu_g}\right)^2} \tag{2}$$

where X_i is the vector of random variables and N_{sim} is the number of preformed simulations. For each simulation within the optimization-loop, it was necessary to evaluate statistics according to Eqs. (1) and (2) defined for separate loading conditions. For the simulation within the inner loop of optimization algorithm stochastic model, described in [13], was utilized. For stochastic simulation also a statistical correlation in the form of Spearman's correlation index was considered according to [17].

Computational requirements of direct RBO using original numerical models are enormous and direct approach is not feasible in this case. ANN surrogate models for both loading layouts might be created in order to perform RBO within a reasonable time. ANN structures require to be trained using a set of benchmark solutions. It is, therefore, necessary to simulate and evaluate the training set of realizations. The design vectors for the training set were generated within the discretized design space in order to focus simulation into points for optimization. Realizations were generated using a rectangular distribution with ranges defined in Table 4. Generated values were subsequently discretized using etalons in the form of Table 1 and Table 2. In order to avoid discrimination of limit categories (lower number of samples for a maximal and minimal category) the simulation boundaries depicted in Table 4 were extended by 10% of the arithmetical average of available discrete options. Performed simulations helped to perform sensitivity analysis and train two ANN approximations returning values of F_{max} and F_{wlim} based on input in the form of reduced design vector $\mathbf{X}_{ri.}$

3.3   Sensitivity Analysis

Created numerical models contain eighteen input parameters (twelve parameters of the stochastic model and six groups of optimized reinforcement diameters). It is beneficial to perform sensitivity analysis in order to reduce design space and focus ANN training to most important inputs. Sensitivity analysis in the form of Spearman's correlation index between input parameters and output quantity is the natural by-product of LHS simulation and does not require additional computational effort [18].

Sensitivity analysis showed that groups of reinforcement R2 – R5 do have an only negligible impact on girders performance. Thus, for those groups might be utilized reinforcement with a diameter of 6 mm. The ANN for F_{max} will use concrete material parameters and R1 group diameter as the input and second ANN for F_{wlim} will use losses of prestressing uncertainties, concrete material parameters and R6 group diameter as the input. Other parameters might be neglected within created approximations due to their small influence to observed outputs. ANN-based approximations and reduction of design space allowed to perform a simple heuristic evaluation of optimization loop. The utilization of ANN surrogate model allows to perform brute-force evaluation within reduced optimization design space corresponding to 135 discrete options and 10000 evaluations of surrogate model per each of those options.

3.4   The ANN Surrogate Models

An ANN is a parallel signal processing system composed of simple processing elements, called artificial neurons, which are interconnected by direct links (weighted connections). Such a system aims to perform parallel distributed processing to solve the desired computational task. One advantage of an ANN is its ability to adapt itself by changing its connection strengths or structure. For more details on ANN-based surrogate modelling, see [19]. Following the results of sensitivity analysis, ANN surrogate models for both considered limit states were created. Both ANNs are of a feed-forward multi-layered type. They consisted of one hidden layer with five nonlinear neurons (hyperbolic tangent transfer function) and an output layer with one output neuron (linear transfer function) which corresponds to one response value which is F_{max} for ULS and F_{wlim} in case of SLS, respectively. Both networks differ in the number of network inputs – there are five input parameters for ULS and seven for SLS, respectively. The ANN approximations were created using the training set of one hundred simulations of design vector $\mathbf{X}_i$ and corresponding responses evaluated for both limit states (for two loading layouts). Ten additional samples were utilized as a benchmark set in order to assess the performance of both ANN surrogates. Two hundred twenty evaluations of numerical models were performed in total. The total computational time necessary to evaluate the responses of training set was $220 \times 10/12 = 183,3$ h (approximately 8 days) since 12 simulations might run in parallel on utilized hardware. Figure 4 Shows agreement of simulated data with results obtained by ANN for F_{max}.

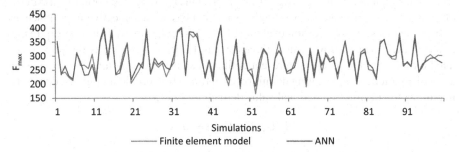

Fig. 4. Performance comparison ANN vs original finite element model

Surrogate model for F_{wlim} shows similar performance both on training as well as test data sets. A detailed description of utilized ANN structures is beyond the scope of this paper.

3.5 Optimization Problem, Decision-Making Criteria and Results

Performed optimization was constrained by the requirement for minimal performance for evaluated loading situations at the level of the current design. Since F_{max} and F_{wlim} are random variables, it is not possible to quantify performance only by mean value, but also variability in the form of standard deviation should be considered. For performance evaluation, the ratio between the mean value and standard deviation of observed quantities have been introduced in a similar way as Cornell's reliability index for reliability evaluation of problems with Gaussian distributions of both structural response and loads [20]. The values of such ratio have been calculated for current design and set as threshold values for constrained optimization. Constraint condition for shear capacity was prescribed as:

$$c_1 = \frac{\mu_{F_{max}}}{\sigma_{F_{max}}} \geq 5.661 \tag{3}$$

while constraint for bending capacity of LDE7 beam was defined as:

$$c_2 = \frac{\mu_{F_{wlim}}}{\sigma_{F_{wlim}}} \geq 5.969 \tag{4}$$

where utilized limit values come from statistical evaluation of the current performance of LDE7 girder. The values for both defined limit states were obtained using following statistics: $\mu_{F_{max}} = 237{,}79$, $\sigma_{F_{max}} = 42{,}005$ and $\mu_{F_{wlim}} = 202{,}05$, $\sigma_{F_{wlim}} = 33{,}849$, respectively [13]. Mean values and standard deviations are evaluated as point estimates using Eqs. (1) and (2). The described optimization problem is multicriterial one with multiple non-dominant solutions. In order to evaluate the performance of the given solution, it was necessary to define fitness function dependent at multiple monitored outputs:

$$F_i = w_1 \frac{c_{1i}}{\bar{c}_1} + w_2 \frac{c_{2i}}{\bar{c}_2} - w_3 \frac{c_{3i}}{\bar{c}_3} \tag{5}$$

where each member within the overall sum represents the weighted normalized criterion of optimality. w_1, w_2 and w_3 are weights specified based on subjective importance and user preferences. c_{1i} is criterion evaluated for i-th simulation according to Eq. (3), c_{2i} criterion evaluated for i-th simulation according to Eq. (4) and c_{3i} is the price calculated for i-th simulation, $\bar{c}_{1-3}$ represent arithmetical averages among the current generation of solutions. The resulting definition of optimization problem might be denoted as maximization of fitness functional value defined by Eq. (5):

$$F(C_1, C_2, C_3) \rightarrow max \tag{6}$$

where C_{1-3} are vectors of criteria evaluated in the current generation.

Above-defined combinatorial optimization task can be solved heuristically using algorithm depicted in Fig. 5. Simulation is performed within a design space in order to get the candidate solution. Simulated design vectors are discretized in the next step. Subsequently, dependent material properties are evaluated as a necessary input for the stochastic model within the inner loop along with the simultaneous evaluation of price. ANN surrogate models are utilized to process data simulated within the inner loop and to return approximate values of F_{max} and F_{wlim}. Those outputs are statistically processed, and the values of criterions c_1 and c_2 are calculated. If constraint conditions are not violated the solution is stored within the population of candidate solutions. The whole process repeats for every discrete solution. Fitness functional values are evaluated at the end of process and design vector with the highest value of fitness corresponds to Pareto-optimal solution dependent on values of weights w_1, w_2 and w_3. Since no dominant solution might be found for this type of problem, the different combinations of preferences (in the form of weights combinations) will generally lead to the different optimum. The heuristic solution allows to evaluate fitness function separately and to study multiple optimal designs without repeated execution of the whole task.

Multiple combinations of weights were tested in order to determine a set of Pareto-optimal solutions. It might be stated that for various combinations of weights, the determined optimum will most likely oscillate between three solutions summarized within

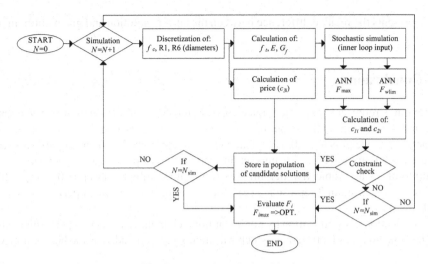

Fig. 5. Utilized RBO algorithm

Table 5. Optimal solutions for various combinations of weights

Variant	Concrete	R1 [mm]	R6 [mm]	w_1 (F_{max})	w_2 (F_{wlim})	w_3 (Price)
Optimum 1	C50/60	8	8	0.1	0.1	0.8
Optimum 2	C50/60	10	30	0.45	0.45	0.1
Optimum 3	C50/60	10	16	0.35	0.35	0.3

Table 5. Note that some combinations of weights might lead to different solutions, but they are limited to short and improbable intervals of weights selection.

Table 6 provides a comparison of optimal solutions performances. Improved performance is indicated by black color while the loss of performance is marked by red. Value

Table 6. Performance of identified optima

Quantity	Current	Optimum 1		Optimum 2		Optimum 3	
		Output	Δr	Output	Δr	Output	Δr
μ_{Fwlim} [kN]	202.05	**190.886**	−5.53%	215.983	6.90%	**198.524**	−1.75%
σ_{Fwlim} [kN]	33.849	31.899	5.76%	30.085	11.12%	31.558	6.77%
c_2	5.969	5.9841	0.25%	7.179	20.27%	6.291	5.39%
μ_{Fmax} [kN]	237.79	294.321	23.77%	357.87	50.50%	357.727	50.44%
σ_{Fmax} [kN]	42.005	24.818	40,92%	19.525	53.52%	19.248	54.18%
c_1	5.661	11.859	109.49%	18.329	223.77%	18.586	228.30%
c_3[EUR]	2126.70	1685.76	20.73%	**2230.43**	−4.88%	1910.02	10.19%

Δr represents the relative difference between the current solution and given optimum in % .

4 Conclusions

Optimum 1 represents extreme choice focused mostly on the reduction of price, and selected weights lead to the same solution as a single-objective definition of the task corresponding to weights $w_1 = 0$, $w_2 = 0$ and $w_3 = 1$. Optimum 2 is the most performance-focused solution; however, it brings an only negligible of price and corresponds to the solution of a single-objective definition with weight combination $w_1 = 0.5$, $w_2 = 0.5$ and $w_3 = 0$. Optimum 3 represent a reasonable compromise between performance and cost.

Verification of optimization by nonlinear finite element analysis was performed only on the benchmark of Optimum 3, which was finally recommended as the best optimized design.

The applicability of RBO for complex nonlinear finite element model was demonstrated within this paper. The modelled structure was optimized under multiple criteria of optimality, resulting in a set of non-dominant solutions. The results served as the base material for adjustment of LDE7 roof girder construction with the aim to reduce product price under the condition to improve or at least maintain current bearing capacity for two defined benchmark loading situations.

Acknowledgment. The authors would like to express their thanks for the support provided by the Czech Science Foundation (GAČR) Project RESUS No. 18-13212S and the project TAČR DELTA No. TF06000016.

References

1. Frangopol, D.M.: Interactive reliability based structural optimization. Comput. Struct. **19**(4), 559–563 (1984)
2. Li, W., Yang, L.: An effective optimization procedure based on structural reliability. Comput. Struct. **52**(5), 1061–1071 (1994)
3. Beck, A.T., Gomes, W.J.S.: A comparison of deterministic, reliability-based and risk-based structural optimization under uncertainty. Probab. Eng. Mech. **28**, 18–29 (2012)
4. Beyer, H.G., Sendhoff, B.: Robust optimization – a comprehensive survey. Comput. Methods Appl. Mech. Eng. **196**, 3190–3218 (2007)
5. Aoues, Y., Chateauneuf, A.: Benchmark study of numerical methods for reliability-based design optimization. Struct. Multidisc. Optim. **41**(2), 277–294 (2010)
6. Tu, J., Choi, K.K.: A new study on reliability-based design optimization. J. Mech. Des. (ASME) **121**(4), 557–564 (1999)
7. da Silva, G.A., Beck, A.T., Sigmund, O.: Stress-constrained topology optimization considering uniform manufacturing uncertainties. Comput. Methods Appl. Mech. Eng. **344**, 512–537 (2018). https://doi.org/10.1016/j.cma.2018.10.020
8. CEN: Eurocode 2: Design of Concrete Structures—Part 1-1: General Rules and Rules for Buildings. European Committee for Standardization – ECS (2004)

9. Strauss, A., Zimmermann, T., Lehký, D., Novák, D., Keršner, Z.: Stochastic fracture-mechanical parameters for the performance-based design of concrete structures. Struct. Concr. **15**(3), 380–394 (2014)
10. Lehký, D., Keršner, Z., Novák, D.: FraMePID-3PB software for material parameters identification using fracture test and inverse analysis. Adv. Eng. Softw. **72**, 147–154 (2014)
11. Stoerzel, J., Randl, N., Strauss, A.: Monitoring shear degradation of reinforced and pre-tensioned concrete members. In: IABSE Conference, Geneva (2015)
12. Strauss, A., Krug, B., Slowik, O., Novak, D.: Combined shear and flexure performance of prestressing concrete T-shaped beams: experiment and deterministic modelling. Struct. Concr. 1–20 (2017). https://doi.org/10.1002/suco.201700079
13. Slowik, O., Novák, D., Strauss, A., Krug, B.: Stochastic analysis of precast structural members failing in shear. In: Proceedings of 12th fib International Ph.D. Symposium in Civil Engineering, Prague, pp. 617–624 (2018). ISBN 978-80-01-06401-6
14. Červenka, V., Jendele, L., Červenka, J.: ATENA program documentation – part 1: theory, Červenka Consulting, Prague, Czech Republic (2019)
15. Taerwe, L., Matthys, S.: Fib Model Code for Concrete Structures 2010. Ernst & Sohn, Wiley, Berlin (2013)
16. McKay, M.D., Conover, W.J., Beckman, R.J.: A comparison of three methods for selecting values of input variables in the analysis of output from a computer code". Technometrics **21**, 239–245 (1979)
17. Strauss, A., Krug, B., Slowik, O., Novak, D.: Prestressed concrete roof girders: part I – deterministic and stochastic model. In: Proceedings of the Sixth International Symposium on Life-Cycle Civil Engineering (IALCCE 2018), vol. 1, pp. 510–517. CRC press, Taylor and Francis Group, London (2018). ISBN 9781138626331
18. Cheng, L., Zhenzhou, L., Leigang, Z.: New spearman correlation based sensitivity index and its unscented transformation solutions. J. Eng. Mech. **142**, 04015076 (2015). https://doi.org/10.1061/(ASCE)EM.1943-7889.0000988
19. Lehký, D., Šomodíková, M.: Reliability calculation of time-consuming problems using a small-sample artificial neural network-based response surface method. Neural Comput. Appl. **28**(6), 1249–1263 (2016). https://doi.org/10.1007/s00521-016-2485-3
20. Cornell, C.A.: A Probability-based structural code. ACI J. **66**, 974–985 (1969)

CMAC: Clustering Class Association Rules to Form a Compact and Meaningful Associative Classifier

Jamolbek Mattiev[1,3] and Branko Kavšek[1,2(✉)]

[1] University of Primorska, Koper, Slovenia
jamolbek.mattiev@famnit.upr.si, mattiev.jamolbek@urdu.uz,
branko.kavsek@upr.si
[2] Jožef Stefan Institute, Jamova cesta 39, 1000 Ljubljana, Slovenia
[3] Urgench State University, Urgench, Uzbekistan

Abstract. Huge amounts of data are being collected and analyzed nowadays. By using the popular rule-learning algorithms, the number of rules discovered on those big datasets can easily exceed thousands of rules. To produce compact and accurate classifiers, such rules have to be grouped and pruned, so that only a reasonable number of them are presented to the end user for inspection and further analysis. To solve this problem researchers have proposed several associative classification approaches that combine two important data mining techniques, namely, classification and association rule mining.

In this paper, we propose a new method that is able to reduce the number of class association rules produced by classical class association rule classifiers, while maintaining an accurate classification model that is comparable to the ones generated by state-of-the-art classification algorithms. More precisely, we propose a new associative classifier – CMAC, that uses agglomerative hierarchical clustering as a post-processing step to reduce the number of its rules.

Experimental results performed on selected datasets from the UCI ML repository show that CMAC is able to learn classifiers containing significantly less rules than state-of-the-art rule learning algorithms on datasets with larger number of examples. On the other hand, classification accuracy of the CMAC classifier is not significantly different from state-of-the-art rule-learners on most of the datasets. We can thus conclude that CMAC is able to learn compact (and meaningful) classifiers from "bigger" datasets, retaining an accuracy comparable to state-of-the-art rule learning algorithms.

Keywords: Frequent itemsets · Class association rules · Associative classification · Agglomerative hierarchical clustering

1 Introduction and Scientific Background

Huge amounts of data are being collected, stored and analyzed nowadays in many aspects of our everyday lives. Mining association rules from these data and reducing their num-

The original version of this chapter was revised: the allocations between authors and affiliations have been corrected. The correction to this chapter is available at
https://doi.org/10.1007/978-3-030-64583-0_64

bers is becoming a popular and important knowledge discovery technique [15]. Huge number of rules are being discovered in "real-life" datasets, which leads to combinatorial complexity and difficulty of interpretation. To overcome this problem, rules have to be pruned and/or clustered to reduce their number, thus leading to more compact and understandable models.

Association rule mining [1] aims to generate all existing rules in the database that satisfy some user-defined thresholds (e.g. minimum support and minimum confidence). Classification rule mining [4, 8, 23], on the other hand, tries to extract small sets of rules to form accurate and efficient models that are able to predict the class label of unknown objects. Associative Classification (AC) [6, 12, 17, 18] is a combination of these two important data mining techniques (classification and association rule mining) that aims to build accurate and efficient classifiers based on association rules. Research studies prove that AC methods can achieve higher accuracy than some of the traditional classification methods, although the efficiency of AC methods depends on the user-defined parameters such as minimum support and minimum confidence. Another group of approaches are clustering methods (unsupervised learning) studied in [16, 21, 22]. These clustering methods can be divided into two main parts: partitional and hierarchical clustering. In partitional clustering [24, 26], objects are grouped into disjoint clusters such that objects within the same cluster are more similar to each other than objects from another cluster. Hierarchical clustering [25], on the other hand, is a nested sequence of partitions. In the bottom-up approach, larger clusters are built by merging smaller clusters together, while the top-down approach starts with one big cluster containing all the objects and recursively divides it into smaller clusters.

In our research work, we propose a new associative classification method based on hierarchical agglomerative clustering. We define a new normalized direct distance to measures the similarities between class association rules (CARs), which we later use to cluster CARs in a bottom up hierarchical agglomerative fashion. Once CARs are clustered, we define a "representative" CAR within each cluster. Finally, just the representative CARs are chosen from each of the clusters to form a compact and understandable classification model.

We have performed experiments on 12 selected datasets from the UCI Machine Learning Database Repository [7] and compared the performance of our proposed method (CMAC) with 9 most popular (associative) classification algorithms (Naïve Bayes (NB) [2], Decision Table (DT) [13], OneR (1R) [11], PART (PT) [8], C4.5 [23], CBA [18], Random Forest (RF) [3], Simple Associative Classifier (SA) [20] and J&B [19]).

The rest of the paper is organized as follows. Section 2 highlights the related work. Problem statement and our goals are described in Sect. 3. Our proposed method is presented in Sect. 4. Section 5 describes the experimental settings, presents and discusses the experimental results. Conclusions and plans for future research are given in Sect. 6.

2   Related Work

The novelty in our proposed approach is in the way we select the strong class association rules, how we cluster then and how we choose the representative class association rule for each cluster. Other related research also deals with the notion of clustering class association rules, but all of them use different approaches. This section presents these related approaches to clustering class association rules and emphasizes the similarities and differences related to our proposed approach.

Clustering class association rules using the k-means clustering algorithm [5] first uses the "APRIORI" algorithm [1] to find class association rules. Then, it calculates a kind of interestingness measure for all rules and partitions the rules into disjoint clusters by using the k-means clustering method. Our proposed method uses the hierarchical agglomerative approach for clustering CARs instead of k-means and employs a different way of selecting "strong" CARs in the first place. The distance metric used by our approach during clustering is also different.

Another approach is distance-based clustering of association rules proposed in [9]. This approach uses a new normalized distance metric to group association rules. Based on the distances, agglomerative clustering is applied to cluster the rules. The rules are further embedded in a vector space with the use of multi-dimensional scaling and clustered using self-organizing maps. This method is very similar to ours, but we propose a new normalized distance metric based on direct distances between class association rules, whereas in [9] indirect measures are used based on CARs support and coverage.

Mining clusters with association rules [14] is another related approach. Here the rules are first generated using the APRIORI algorithm (like in [5]), but an "indirect" distance metric (based on coverage probabilities) is later used to find the similarities between rules. Rules are then clustered using a top-down hierarchical clustering method for finding clusters in a population of customers, where the list of products bought by the individual clients are given.

3   Problem Definition and Goals

We assume that a normal relational table is given with N examples (transactions). Each example is described by A distinct attributes and is classified into one of the M known classes. Since our algorithm supports just attributes of a nominal type (like the vast majority of association rule miners), we had to perform discretization on numeric attributes in some cases (the details about discretization are provided in Sect. 5). Goals and contributions of our proposed approach are the following:

1. Generate the strong class association rules that satisfiy some user-defined minimum support and minimum confidence constraints;
2. Propose a new normalized distance function based on the direct measures to find the similarities between two class association rules;
3. Cluster class association rules by using this normalized similarity measure and automatically determine the optimal number of clusters for each class value;
4. Define a way of extracting a representative CAR for each cluster to produce the final, compact and meaningful classifier;

4   Our Proposed Method – CMAC

Our approach – CMAC (Compact and Meaningful Associative Classifier) can be divided into 4 steps mentioned in previous section. Each of these 4 steps is presented in detail in the following 4 subsections.

4.1 Finding "Strong" Class Association Rules

We discuss how to discover the strong class association rules from frequent itemsets in this subsection. Generation of association rules (ARs) is usually split up into two main steps. First, minimum support is applied to find all frequent itemsets from the training dataset. Second, these frequent itemsets and the minimum confidence constraint are used to generate the strong association rules. Discovering of CARs also follows the same procedure. The only difference is that in the rule-generation part, the consequent (right-hand side) of the rule now contains only the class label (for CARs), while the consequent of a rule in AR-generation can include any frequent itemset.

In the first step, the APRIORI algorithm is used to find the frequent itemsets. The 'downward-closure' technique is used to accelerate the searching procedure by reducing the number of candidate itemsets at any level. APRIORI finds the 1-frequent itemset, then, 1-frequent itemsets are used to generate the 2-frequent itemsets and so on. If it finds any infrequent itemsets at any level, it stops, because, any infrequent itemsets cannot generate further frequent itemsets. APRIORI uses the 'downward-closure' property of frequent itemsets to reduce the time complexity of the algorithm.

After all frequent itemsets are generated from the training dataset, it is straightforward to generate the strong class association rules that satisfy both minimum support and minimum confidence constraints. Confidence of the rule can be computed by using Eq. (1).

$$confidence(A \rightarrow B) = \frac{support\_count(A \cup B)}{support\_count(A)} \tag{1}$$

Equation (1) is expressed in terms of support count of a frequent itemset, where A represents the antecedent (left-hand side of the rule) and B the consequent (right-hand side of the rule – class value in case of CARs) of a rule. Furthermore, $support\_count$ $(A \cup B)$ is the number of transactions that match the itemset $A \cup B$, and $support\_count$ (A) is the number of transactions that match the itemset A. Strong class association rules that satisfy the minimum confidence threshold can be generated based on Eq. (1), as follows:

- All nonempty subsets S are generated for each frequent itemsets L and a class label C;
- For every nonempty subset S of L, output the strong rule R in the form of "$S \rightarrow C$" if, $\frac{support\_count(R)}{support\_count(S)} \geq min\_conf$, where min_conf is the minimum confidence threshold;

4.2 The New "Direct" Distance Metric

Once we generate the class association rules (as described in Subsect. 4.1), our next goal is to cluster them. Since we intend to apply hierarchical agglomerative clustering, we must define a way of measuring the similarity between CARs, that is how far two rules are apart. Some candidate similarity measures have already been proposed in related research, but they all use indirect distance metrics for association rules (e.g. *absolute market basket difference, conditional market basket difference* and/or *conditional market basket log-likelihood* measures [9]). We compute the distance between two rules by ignoring the class label because we are clustering the rules belonging to the same class label. When we tried to apply these indirect distance measures to our proposed method, we got larger number of clusters that in turn resulted in bigger number or class association rules in the final classifiers. Therefore, we propose a new normalized direct distance measure that considers only the itemsets contained in the antecedent (left-hand side) of the rules.

Lets $R = \{r_1, r_2, \ldots, r_n\}$ be a set of class association rules found from a relational dataset D by the APRIORI algorithm that are defined by $A = \{a_1, a_2, \ldots, a_m\}$ distinct attributes classified into $C = \{c_1, c_2, \ldots, c_l\}$. Each rule is denoted as follows:

$$r_i = \{x_1, x_2, \ldots, x_k\} \rightarrow \{c_i\} \text{ where, } k \leq m, i = 1 \ldots l.$$

Given two rules $r, r' \in R$:

$$r = \{y_1, y_2, \ldots, y_s\} \rightarrow \{c_i\}$$
$$r' = \{z_1, z_2, \ldots, z_t\} \rightarrow \{c_i\}, \text{ where } s \leq m, t \leq m \text{ and } i = 1 \ldots l.$$

we compute the distance between rules r and r' as follows:

$$distance_q(r, r') = \begin{cases} 0, & \text{if } y_q = z_q \mid y_q = \emptyset \ \& \ z_q = \emptyset; \\ 1, & \text{if } y_q = \emptyset \ \& \ z_q \neq \emptyset \mid y_q \neq \emptyset \ \& \ z_q = \emptyset; \\ 2, & \text{if } y_q \neq z_q; \end{cases} \tag{2}$$

Equation (2) expresses how close two rules are one from another. If rules have similar items, then the distance function has a low value – a missing rule item is considered closer than a different rule item.

$$border = Max(s, t); \tag{3}$$

The *border* – Eq. (3) – is the length of the longest rule and it is used to normalize the distance metric. The actual distance between two rules is then calculated using Eq. (4), as follows:

$$d(r, r') = \sum_{i=1}^{border} distance_i / 2 \times border \tag{4}$$

4.3 Clustering

Clustering algorithms aim to group similar examples together, the examples in the same cluster should be similar and dissimilar from the examples in other clusters. Hierarchical clustering algorithms can be divided into 2 groups: top-down (*divisive hierarchical clustering*) and bottom-up (*agglomerative hierarchical clustering*).

Bottom-up algorithms initially, assume each example as a single cluster and then merge two closet clusters in every iteration until all clusters have been merged into a unique cluster that contains all examples. The resulting hierarchy of clusters is represented as a tree (or dendrogram). The root of the tree is the unique cluster that contains all the examples, the leaves are clusters with only one example.

Top-down approaches are the opposite of *agglomerative hierarchical clustering* methods. They consider all examples in single cluster and then they split the clusters into smaller parts until each example forms a cluster or until it satisfies some stopping condition.

We apply complete linkage method of agglomerative hierarchical clustering. In the complete linkage (farthest neighbor) method, the similarity of two clusters is the similarity of their most dissimilar examples, therefore, the distance between the farthest groups are taken as an intra cluster distance. We assume that we have given an $N \times N$ distance matrix D, containing all distances $d(i, j)$. The clusters are numbered 0, 1, ..., $(n - 1)$ and m represents the sequence number of a cluster. $L(k)$ is the level of the k-th clustering and the distance between two clusters $cl1$ and $cl2$ is defined as $d[cl1, cl2]$. Our agglomerative hierarchical clustering algorithm that uses complete linkage is outlined in Algorithm 1.

Algorithm 1: Agglomerative hierarchical clustering with complete linkage

---

1. **Initialization:** Each example is a unique cluster at level $L(0)=0$, sequence number $m=0$ and the optimal number of clusters (N) is identified. To get the intended number of clusters (N), the algorithm runs K times, where K=total number of clusters $- N$;
2. **Computation:** Find the most similar pair of clusters, $cl1$ and $cl2$ and merge them into a single cluster to form the next clustering sequence m. Increase the sequence number by one: $m=m+1$ and set the new level $L(m)=d(cl1, cl2)$;
3. **Update:** Update the distance matrix D, by removing the rows and columns corresponding to $cl1$ and $cl2$ and adding a new row and column corresponding to the new cluster. The distance between the new cluster ($cl1, cl2$) and some "old" cluster k is calculated as $d[cl1, cl2]=max\{d[k, cl1], d[k, cl2]\}$.
4. **Stopping condition:** if $m=K$ then stop, otherwise go to step 2.

When we cluster the rules, we need to find the number of clusters. We get the natural number of clusters by cutting the dendrogram at the point that represents the maximum distance between two consecutive cluster merges. The algorithm that identifies the natural number of clusters is presented in Algorithm 2.

The input to Algorithm 2 is a set of cluster distances that are calculated during the building of the dendrogram (the so-called cluster "heights"). The output is the natural number of clusters. In lines 1–3 the total number of clusters generated by hierarchical clustering is stored. Lines 4–7 outline the main part of the algorithm, *Opt_number_of_clusters* gets to the point where the difference between two consecutive cluster heights will be maximum. Since we start from *0, Opt_number_of_clusters* is equal to *(N-k)*. The last line returns the obtained result.

Algorithm 2: Computing the natural number of clusters

Input: a set of cluster heights.
Output: Optimal number of clusters.
1:   *Max_height_difference=cluster_height*[1]-*cluster_height*[0];
2:   *Opt_number_of_clusters*= 1;
3:   *N= cluster_height.length*;
4:   **for** (*k*=2; *k≤N*; *k*++) **do**
5:       **if** (*cluster_height*[*k*]-*cluster_height*[*k*-1]) > *Max_height_difference* **then**
6:           *Max_height_difference;*
7:           *Opt_number_of_cluster=N-k*;
8:   **return** *Opt_number_of_clusters*

4.4   Extracting the "Representative" CARs

Once we found all clusters, our final goal is to extract the "representative" CARs from each cluster to form our compact and meaningful associative classifier. Within each cluster we choose the CAR which is closer to the center of the cluster as a cluster representative CAR. In other words, the representative CAR must have the minimum average distance to all other CARs within a cluster. Algorithm 3 describes this procedure of selecting the representative CAR.

Algorithm 3: Extracting the representative class association rule

---

Input: a set of class association rules in the cluster.
Output: the "representative" class association rule in the cluster.
1:   *N=CARs.length*;
2:   *Fill*(distance, 0);
3:   *min_avg_distance=Integer.Max.value;*
4:   **for** (*i*=0; *i≤N*; *i*++) **do**
5:       **for** (*j*=0; *j≤N*; *j*++) **do**
6:           distance[*i*]=distance[*i*]+*distance*(CARs[*i*], CARs[*j*]);
7:       *avg_distance*=distance[*i*]/*N*;
8:       **if** (*avg_distance< min_avg_distance* **then**
9:           *min_avg_distance=avg_distance;*
10:          *representative_CAR_index=i;*
11: **return** CARs[*representative_CAR_index*];

In line 1 of Algorithm 3 the number of CARs within the cluster is stored. We use a distance array (line 2) to compute the distances from a selected CAR to all other CARs. The initial value of *min_avg_distance* (line 3) is initially set to the biggest integer. This variable is used to store the minimum average distance calculated in the double loop of the algorithm (lines 4–9) and represents the representative CAR that we seek.

Algorithm 4: Learning the CMAC associative classifier

Input: A training dataset D, *minimum support, minimum confidence.*
Output: CMAC associative classifier.
1: F=*frequent_itemsets*(D, *minsup*);
2: R=*genCARs*(F, *minconf*);
3: R=*sort*(R, *minsup*, *minconf*);
4: G=*Group*(R);
5: **for** (i=0; $i \leq$*number_of_class*; i++) **do**
6:      A=*distance*($G[i]$);
7:      *Cluster_heights*=*Hierarchical_clustering*(A, 1);
8:      N=*optimal_number_of_clusters*(*Cluster_heights*);
9:      *Cluster*=*Hierarchical_clustering*(A, N);
10:     **for** (j=0; $j \leq N$; j++) **do**
11:         Y=*representative_CAR*(*Cluster*[i]);
12:         *Associative_Clssifier*.add(Y);
13: **return** *Associative_Classifier*;

Finally, our proposed approach is represented in Algorithm 4. In lines 1–2 the strong class association rules that satisfy the user-specified minimum support (*minsup*) and minimum confidence (*minconf*) constraints are generated from the training dataset D by using the APRIORI algorithm. In line 3 these CARs are sorted by support and confidence in descending order according to the following criteria.

If R_1 and R_2 represent two CARs, R_1 is said to have higher rank than R_2 ($R_1 > R_2$):

- if and only if, $conf(R_1) > conf(R_2)$ or
- if $conf(R_1) = conf(R_2)$, but $supp(R_1) > supp(R_2)$ or
- if $conf(R_1) = conf(R_2)$ and $supp(R_1) = supp(R_2)$, but R_1 has fewer items in its left-hand side than R_2 or
- if all the parameters of the rules are equal, we choose any of them randomly.

CARs are grouped according to their class label in line 4. For each group of CARs (lines 5–12), the distance matrix (described in Subsect. 4.2) is constructed in line 6, hierarchical clustering algorithm (complete linkage method) computes the cluster heights (distances between clusters) by using the distance matrix (line 7) and these heights (distances) are used to find the optimal number of clusters (line 8). Then, we apply the hierarchical clustering algorithm again to identify the cluster of CARs (*Cluster* array stores the list of clustered CARs). In lines 10–12, the representative CAR is extracted (as described in Subsect. 4.4) for each cluster and added to our final classifier. The last line returns the final classifier.

Algorithm 5 depicts the procedure of using the CMAC classifier to find the class label of a (new) test example. For each rule in the CMAC classifier (line 1): if the rule can classify the example correctly, then, we increase the corresponding class count by one and store it (lines 2–3); if none of the rules can classify the new example correctly, then the algorithm returns the majority class value (lines 4–5); otherwise, it returns the majority class value of correctly classified rules (lines 6–7).

Algorithm 5: CMAC – classifying a (new) test example

Input: The CMAC classifier, the (new) test example
Output: Predicted class for the (new) test example
1: **for each** rule $y \in Classifier$ **do**
2:      **if** y classifies *test_example* **then**
3:          *class_count*[y.class]++;
4: **if** max(*class_count*)==0 **then**
5:      *predicted_class=majority_classifier*;
6: **else**
7:      *predicted_class*=max_index(*class_count*);
8: **return** *predicted_class*

---

5 Experimental Setting and Results

We tested our model on 12 real-life datasets taken from the UCI ML Repository. We evaluated our classifier (CMAC) by comparing it with 9 well-known classification algorithms on classification accuracy and the number of rules. All differences were tested for statistical significance by performing a paired t-test (with 95% significance threshold).

CMAC was ran with default parameters *minimum support = 1%* and *minimum confidence = 60%* (on some datasets, however, *minimum support* was lowered to *0.5%* or even *0.1%* to ensure enough CARs to be generated for each class value). For all other 8 rule learners we used their WEKA workbench [10] implementation with default parameters. Since association rule learning does not support numeric attributes, all numeric attributes (in all datasets) were pre-discretized with WEKA's "class-dependant" discretization method (that is able to automatically determine the number of bins for each numeric attribute).

Furthermore, all experimental results were produced by using a 10-times random split evaluation protocol – 2/3 of examples in each random split was used for learning the models, 1/3 was used to test them.

Experimental results are shown in Table 1 and Table 2. Table 1 illustrates classification accuracies (average values over the 10 random splits with standard deviations), while Table 2 contains the sizes of the classifiers (average number of rules over the 10 random splits, no standard deviations are shown here). **Bolded** results mean that the selected rule-learning algorithm significantly outperformed our CMAC algorithm, while underlined results mean that CMAC was significantly better. A "non-emphasized" text mean no significant difference has been detected in the comparison.

We can observe from Table 1 that CMAC achieves significantly higher accuracies than some rule-learners on some datasets, while on other datasets it achieves significantly lower accuracies or there is no significant difference. However, on average, the classification accuracies of CMAC are not much different than those of the other 9 rule-learners. CMAC's average accuracy of 80.3% is higher than that of algorithms NB, DT, 1R, C4.5 and lower than PT, RF, CBA, SA and J&B.

Interestingly, CMAC significantly outperforms all rule-learners on the "Breast Cancer" and "Hayes-root" datasets, while on the "Nursery" dataset it is outperformed by all

Table 1. Experimental comparison – classification accuracy

Dataset	#att	#cls	#recs	min supp	NB	DT	C4.5	PT	1R	RF	CBA	SA	J&B	CMAC
					Accuracy (standard deviation) (%)									
Breast Cancer	10	2	286	1%	73.0 (4.1)	71.1 (4.8)	70.9 (4.9)	69.2 (4.2)	67.5 (4.1)	70.0 (5.2)	74.6 (6.6)	80.1 (4.4)	78.8 (4.2)	81.5 (4.1)
Balance	7	2	432	1%	89.3 (1.2)	66.0 (2.3)	66.5 (2.9)	74.5 (2.9)	60.6 (1.5)	76.3 (3.1)	76.3 (1.9)	74.0 (1.7)	72.1 (1.5)	72.8 (3.2)
Car.Evn	7	4	1728	1%	84.8 (0.9)	86.1 (1.1)	89.5 (1.5)	95.0 (1.5)	70.0 (0.1)	93.4 (1.1)	91.1 (2.4)	86.2 (2.8)	88.2 (2.1)	84.8 (1.7)
Vote	17	2	435	1%	88.9 (2.1)	94.4 (0.8)	94.1 (1.4)	94.4 (2.6)	95.4 (1.1)	94.9 (1.6)	93.1 (2.0)	94.1 (2.3)	92.4 (1.6)	92.4 (1.8)
Tic-Tac-Toe	10	2	958	1%	69.9 (1.9)	74.1 (2.4)	84.7 (3.2)	89.3 (2.8)	69.7 (2.3)	94.1 (2.0)	100.0 (0.0)	93.6 (2.4)	98.4 (1.1)	87.3 (0.7)
Nursery	9	5	12960	1%	90.4 (0.4)	94.0 (0.2)	96.2 (0.4)	98.7 (0.4)	71.0 (0.5)	98.2 (0.2)	92.1 (0.4)	91.0 (0.9)	89.6 (1.1)	88.2 (1.0)
Hayes-root	6	3	160	0.1%	79.7 (7.9)	53.4 (4.8)	76.0 (4.2)	73.3 (7.7)	54.3 (4.5)	78.8 (5.6)	77.0 (9.8)	73.2 (4.9)	80.2 (4.7)	80.2 (4.3)
Lymp	19	4	148	1%	85.0 (4.9)	76.7 (4.3)	80.1 (5.2)	81.4 (5.5)	74.7 (5.2)	83.5 (4.9)	81.3 (4.5)	74.3 (4.9)	80.6 (4.1)	78.5 (4.5)
Spect.H	23	2	267	0.1%	77.8 (3.4)	78.8 (1.9)	79.3 (3.3)	80.2 (2.6)	79.4 (0.4)	82.7 (2.7)	83.4 (1.0)	83.5 (1.0)	80.1 (0.9)	81.4 (1.2)
Abalone	9	3	4177	1%	58.1 (0.7)	61.8 (1.5)	62.3 (1.2)	62.3 (1.1)	60.2 (1.0)	61.3 (0.6)	61.1 (1.0)	61.0 (0.9)	60.8 (0.9)	60.7 (1.1)
Adult	15	2	45221	0.5%	81.3 (0.3)	82.2 (0.3)	82.9 (0.2)	82.6 (0.2)	77.3 (0.3)	82.2 (0.3)	81.8 (0.4)	81.1 (0.9)	80.9 (1.4)	81.7 (2.1)
Insuranc	7	3	1338	1%	76.1 (1.4)	75.7 (1.6)	75.8 (1.4)	75.0 (1.8)	54.1 (0.6)	73.4 (1.6)	75.4 (2.0)	74.5 (1.6)	76.3 (0.7)	73.8 (0.8)
Average accuracy (standard deviation) (%)					79.5 (2.1)	76.2 (1.9)	79.9 (2.6)	81.3 (2.4)	70.0 (1.6)	82.4 (2.1)	82.3 (2.3)	80.5 (2.1)	81.5 (2.0)	80.3 (2.1)

other algorithms (except for 1R). Standard deviations of accuracy results decrease with increasing number of examples in a dataset, which is an expected behavior.

Since Naïve Bayes (NB) and Random forest (RF) do not explicitly produce rules in their classification models and OneR (1R) always contains one rule, we omitted them from the results in Table 2.

We can observe from Table 2, that CMAC significantly outperforms all other rule-learners on almost all datasets (except DT and PT on "Balance", "Hayes-root" and "Spect.H"; C4.5, CBA and J&B on "Spect.H"; C4.5 on "Balance" and J&B on "Tic-Tac-Toe") and it produces classifiers that have on average far less rules than those produced by the other 6 rule-learning methods included in the comparison.

Table 2. Experimental comparison – size of the classifier (number of rules)

Dataset	#attr	#cls	#recs	DT	C4.5	PT	CBA	SA	J&B	CMAC
Breast Cancer	10	2	286	14	13	12	41	20	51	8
Balance	5	3	625	25	31	21	83	45	84	39
Car.Evn	7	4	1728	342	100	66	35	160	37	31
Vote	17	2	435	19	13	6	18	30	13	6
Tic-Tac-Toe	10	2	958	137	73	37	19	60	14	18
Nursery	9	5	12960	810	288	161	122	175	113	74
Hayes-root	6	3	160	12	20	11	34	50	37	20
Lymp	19	4	148	15	13	10	16	42	37	5
Spect.H	23	2	267	2	9	12	2	50	12	15
Abalone	9	3	4177	60	49	71	131	155	67	14
Adult	15	2	45221	1656	302	545	71	110	86	12
Insurance	7	3	1338	48	21	49	84	62	52	18
Average no. of rules				263	76	82	51	80	51	21

6   Conclusion and Future Work

Experimental evaluations show that we could achieve our intended goal in this research to produce a compact and meaningful classifier by exhaustively searching the entire example space using constraints and clustering. Our CMAC classifier was able to reduce the number of classification rules while maintaining a classification accuracy that was comparable to state-of-the art rule-learning classification algorithms. Moreover, we showed in the experiments that CMAC was able to reduce the number of rules in the classifier by 2–4 times on average compared to the other rule-learners, while this ratio is even bigger on datasets with higher number of examples.

The main drawback of our proposed method (CMAC) is its time efficiency. In future work we plan to parallelize CMAC to bring its time complexity at least a bit closer to state-of-the-art "divide-and-conquer" rule-learning algorithms.

We also plan to combine direct and indirect measures to define a new distance metric for clustering and also plan to investigate a new method of extracting the representative CAR based on example coverage to improve the overall coverage and accuracy while still maintaining less number of classification rules.

Acknowledgement. The authors gratefully acknowledge the European Commission for funding the InnoRenew CoE project (Grant Agreement #739574) under the Horizon2020 Widespread-Teaming program and the Republic of Slovenia (Investment funding of the Republic of Slovenia and the European Union of the European Regional Development Fund). Jamolbek Mattiev is also funded for his Ph.D. by the "El-Yurt-Umidi" foundation under the Cabinet of Ministers of the Republic of Uzbekistan.

References

1. Agrawal, R., Srikant, R.: Fast algorithms for mining association rules. In: Bocca, J.B., Jarke, M., Zaniolo, C. (eds.) Proceedings of the 20th International Conference on Very Large Data Bases, VLDB 1994, Chile, pp. 487–499 (1994)
2. Baralis, E., Cagliero, L., Garza, P.: A novel pattern-based Bayesian classifier. IEEE Trans. Knowl. Data Eng. **25**(12), 2780–2795 (2013)
3. Breiman, L.: Random forests. Mach. Learn. **45**(1), 5–32 (2001)
4. Cohen, W.W.: Fast effective rule induction. In: Prieditis, A., Russel, S.J. (eds.) Proceedings of the Twelfth International Conference on Machine Learning, ICML 1995, California, pp. 115–123 (1995)
5. Dahbi, A., Mouhir, M., Balouki, Y., Gadi, T.: Classification of association rules based on K-means algorithm. In: Mohajir, M.E., Chahhou, M., Achhab, M.A., Mohajir, B.E. (eds.) 4th IEEE International Colloquium on Information Science and Technology, Tangier, Morocco, pp. 300–305 (2016)
6. Deng, H., Runger, G., Tuv, E., Bannister, W.: CBC: an associative classifier with a small number of rules. Decis. Support Syst. **50**(1), 163–170 (2014)
7. Dua, D., Graff, C.: UCI Machine Learning Repository. University of California, Irvine (2019)
8. Frank, E., Witten, I.: Generating accurate rule sets without global optimization. In: Shavlik, J.W. (eds.) Fifteenth International Conference on Machine Learning, USA, pp. 144–151 (1998)
9. Gupta, K.G., Strehl, A., Ghosh, J.: Distance based clustering of association rules. In: Proceedings of Artificial Neural Networks in Engineering Conference, USA, pp. 759–764 (1999)
10. Hall, M., Frank, E., Holmes, G., Pfahringer, B., Reutemann, P., Witten, I.H.: The WEKA data mining software: an update. SIGKDD Explor. **11**(1) (2009)
11. Holte, R.: Very simple classification rules perform well on most commonly used datasets. Mach. Learn. **11**(1), 63–91 (1993)
12. Hu, L.-Y., Hu, Y.-H., Tsai, C.-F., Wang, J.-S., Huang, M.-W.: Building an associative classifier with multiple minimum supports. SpringerPlus **5**(1), 1–19 (2016). https://doi.org/10.1186/s40 064-016-2153-1
13. Kohavi, R.: The power of decision tables. In: Lavrač, N., Wrobel, S. (eds.) 8th European Conference on Machine Learning, Crete, Greece, pp. 174–189 (1995)
14. Kosters, W.A., Marchiori, E., Oerlemans, A.A.J.: Mining clusters with association rules. In: Hand, D.J., Kok, J.N., Berthold, M.R. (eds.) IDA 1999. LNCS, vol. 1642, pp. 39–50. Springer, Heidelberg (1999). https://doi.org/10.1007/3-540-48412-4_4
15. Lent, B., Swami, A., Widom, J.: Clustering association rules. In: Gray, A., Larson, P. (eds.) Proceedings of the Thirteenth International Conference on Data Engineering, England, pp. 220–231 (1997)
16. Kaufman, L., Rousseeuw, P.J.: Finding Groups in Data: An Introduction to Cluster Analysis. Wiley, Hoboken (1990)
17. Khairan, D.R.: New associative classification method based on rule pruning for classification of datasets. IEEE Access **7**, 157783–157795 (2019)
18. Liu, B., Hsu, W., Ma, Y.: Integrating classification and association rule mining. In: Agrawal, R., Stolorz, P. (eds.) Proceedings of the 4th International Conference on Knowledge Discovery and Data Mining, New York, USA, pp. 80–86 (1998)
19. Mattiev, J., Kavšek, B.: A compact and understandable associative classifier based on overall coverage. Procedia Comput. Sci. **170**, 1161–1167 (2020)

20. Mattiev, J., Kavšek, B.: Simple and accurate classification method based on class association rules performs well on well-known datasets. In: Nicosia, G., Pardalos, P., Umeton, R., Giuffrida, G., Sciacca, V. (eds.) LOD 2019. LNCS, vol. 11943, pp. 192–204. Springer, Cham (2019). https://doi.org/10.1007/978-3-030-37599-7_17
21. Zait, M., Messatfa, H.: A comparative study of clustering methods. Future Gener. Comput. Syst. **13**(2–3), 149–159 (1997)
22. Phipps, A., Lawrence, J.H.: An Overview of Combinatorial Data Analysis. Clustering and Classification, pp. 5–63. World Scientific, New Jersey (1996)
23. Quinlan, J.: C4.5: programs for machine learning. Mach. Learn. **16**(3), pp. 235–240 (1993)
24. Ng, T.R., Han, J.: Efficient and effective clustering methods for spatial data mining. In: Bocca, J.B., Jarke, M., Zaniolo, C. (eds.) Proceedings of the 20th Conference on Very Large Data Bases (VLDB), Santiago, Chile, pp. 144–155 (1994)
25. Theodoridis, S., Koutroumbas, K.: Hierarchical algorithms. Pattern Recogn. **4**(13), 653–700 (2009)
26. Zhang, T., Ramakrishnan, R., Livny, M.: BIRCH: an efficient data clustering method for very large databases. In: Widom, J. (eds.) Proceedings of the 1996 ACM-SIGMOD International Conference on Management of Data, Montreal, Canada, pp. 103–114 (1996)

GPU Accelerated Data Preparation
for Limit Order Book Modeling

Viktor Burján[(✉)] and Bálint Gyires-Tóth[(✉)]

Department of Telecommunications and Media Informatics,
Budapest University of Technology and Economics, Budapest, Hungary
burjan.viktor@gmail.com, toth.b@tmit.bme.hu

Abstract. Financial processes are frequently explained by econometric models, however, data-driven approaches may outperform the analytical models with adequate amount and quality data and algorithms. In the case of today's state-of-the-art deep learning methods the more data leads to better models. However, even if the model is trained on massively parallel hardware, the preprocessing of a large amount of data is usually still done in a traditional way (e.g. few hundreds of threads on Central Processing Unit, CPU).

In this paper, we propose a GPU accelerated pipeline, which assesses the burden of time taken with data preparation for machine learning in financial applications. With the reduced time, it enables its user to experiment with multiple parameter setups in much less time. The pipeline processes and models a specific type of financial data – limit order books – on massively parallel hardware. The pipeline handles data collection, order book preprocessing, data normalisation, and batching into training samples, which can be used for training deep neural networks and inference. Time comparisons of baseline and optimized approaches are part of this paper.

Keywords: Financial modeling · Limit order book · Deep learning · Massively parallel

1 Introduction

A limit order book (LOB) contains the actual buy and sell limit orders that exists on a financial exchange at a given time.[1] Every day on major exchanges, like New York Stock Exchange, the volume of clients' orders can be tremendous, and each of them has to be processed and recorded in the order book. Deep learning models are capable of representation learning and modeling jointly [2],thus, other

---

[1] Such an order book does not exist for the so called dark pool, because the orders are typically not published in dark pools [1].

© Springer Nature Switzerland AG 2020
G. Nicosia et al. (Eds.): LOD 2020, LNCS 12565, pp. 385–397, 2020.
https://doi.org/10.1007/978-3-030-64583-0_35

machine learning methods (which requires roboust feature engineering) can be outperformed in case a large amount of data is available. This makes the application of deep learning-based models is particularly reasonable for LOB data. The parameter and hyperparameter optimization of deep learning models are typically performed on Graphic Processing Units (GPUs). Still, nowadays deep learning solutions preprocess the data mostly on the CPU (Central Processing Unit) and transfer it to GPUs afterwards. Finding the optimal deep learning model involves not only massive hyperparameter tuning but a large number of trials with different data representations as well. Therefore, CPU based preprocessing introduces a critical computational overhead, which can severely slow down the complete modeling process. In this work, a GPU accelerated pipeline is proposed for LOB data. The pipeline preprocesses the orders and converts the data into a format that can be used in the training and inference phases of deep neural networks. The pipeline can process the individual orders that are executed on the exchange, extract the state of the LOB with a constant time difference in-between, then batch these data samples together, normalise them, feed them into a deep neural network and initiate the training process. When training is finished, the resulted model can be used for inference and evaluation. In order to define output values for supervised training, further steps are included in the pipeline. A typical step is calculating features of the limit order book (mid-price or volume weighted average price, for example) for future timesteps, and labelling the input data based on these data. The main goal of the proposed method is to reduce the computational overhead of hyperparameter-tuning for finding the optimal model and make inference faster by moving data operations to GPUs. As data preprocessing is among the first steps in the modeling pipeline, if the representation of the order book for modeling is changed, the whole process must be rerun. Thus, it is critical, that not only the model training and evaluation, but the data loading and preprocessing are as fast as possible. The remaining part of this paper is organized as follows. The current section provides a brief overview of the concepts used in this work. Section 2 explores previous works done in this domain. The proposed method is presented in Sect. 3 in details, while Sect. 4 evaluates the performance. Finally, the possible applications are discussed and conclusions are drawn.

1.1   The Limit Order Book

The LOB consists of limit orders. Limit orders are defined by clients who want to open a long or short position for a given amount of a selected asset (e.g. currency pair), for a minimum or maximum price, respectively. The book of limit orders has two parts: bids (buy orders) and asks (sell orders). The bid and ask orders contain the price levels and volumes for which people want to buy and sell an asset. The two sides of the LOB are stored sorted by price - the lowest price of asks and the highest price of bids are stored at the first index. This makes accessing the first few elements quicker, as these orders are the most relevant ones. If the highest bid price equals or is more than the lowest ask price (so, the two parts of the order book overlap) the exchange immediately matches the bid and ask orders, and after execution, the bid-ask spread is formed.

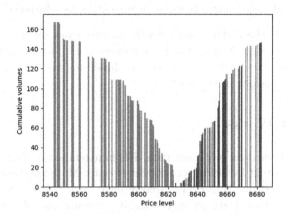

Fig. 1. Example of a limit order book snapshot with depth limited to 100. The asymmetry is due to the higher number of ask orders on the same price levels. The data is taken from the Coinbase Pro cryptocurrency exchange.

The bid with the largest price at a given t time: $b(t)$, while the ask with the smallest price at a given t time: $a(t)$.

Using these notations, the bid-ask spread is defined as

$$spread = a(t) - b(t) \tag{1}$$

The middle-price (also referenced as mid-price) is:

$$\frac{a(t) - b(t)}{2} \tag{2}$$

which value is within the spread.

Feature engineering can produce further order book representations that are used as inputs or outputs of machine learning methods. A commonly used representation is the Volume Weighted Average Price (VWAP). This feature considers not just the price of the orders, but also their side:

$$\frac{\sum price * volume}{\sum volume} \tag{3}$$

It is sometimes considered as a better approximation of the real value of an asset - orders with bigger volume have bigger impact on the calculated price.

A 'snapshot' of the order book contains the exchange's orders, ordered by price at a given t time, represented as

$$a_{depth}, a_{depth-1}, ..., a_1, b_1, b_2, ..., b_{depth-1}, b_{depth}, \qquad (4)$$

where *depth* is the order book's depth. In this work we use same depth on both bid and ask sides. A snapshot is visualised in Fig. 1. The orders which are closer to the mid-price are more relevant, as these may get executed before other orders which are deeper in the book.

2   Previous Works

As the volume of the data increases, data driven machine learning algorithms can exploit deeper context in the larger amount of data. Modeling financial processes has quite a long history, and there already have been many attempts to model markets with machine learning methods. [3] presents one of the most famous models for option pricing, written in 1973, still used today by financial institutions. Also, there have been multiple works on limit order books specifically. [4] examines traders behaviour with respect to the order book size and shape, while e.g. [5] is about modelling order books using statistical approaches.

Multiple data sources can be considered for price prediction of financial assets. These include, for example LOB data and sentiment of different social media sites, such as Twitter. [6] observes Twitter media to track cryptocurrency news, and also explores machine learning methods to predict the number of future mentions. [7] targets to use Twitter and Telegram data to identify 'pump and dump' schemes used by scammers to gain profits.

As limit order books are used in exchanges, many research aim to outperform the market by utilizing LOB data. The most common deep learning models for predicting the price movement are the convolutional neural networks (CNNs) and Long Short-Term Memory (LSTM) networks. [8] emphasizes the time series-like properties of the order book states, and uses specific data preprocessing steps, accordingly.

A benchmark LOB dataset is published [9], however, it is not widely spread yet. It provides normalised time-series data of 10 days from the NASDAQ exchange, which researchers can use for predicting the future mid-price of the limit order book. The LOBSTER dataset provides data from Nasdaq [10] and is frequently used to train data driven models. [11] shows a reinforcement learning based algorithm, which is used for cryptocurrency market making. The authors define 2 types of orders: market and limit, and approach the problem in a market maker's point of view. They experiment with multiple reward functions, such as Trade completion or P&L (Profit and Loss). They train deep learning agents using data of the Coinbase cryptocurrency exchange, which is collected through Websocket connection. [12] gives a comprehensive comparison between LSTM, CNN and fully-connected approaches. They have used a wider time-scale, the data are collected from the S&P index between 1950 and 2016.

3   Proposed Methods

In this section, we introduce the proposed GPU accelerated pipeline for LOB modeling. A high-level overview is shown on Fig. 2. The components of the pipeline are executed sequentially. When a computation is done, the result is saved to a centralised database, which is accessable by every component. The next step in the pipeline just grabs the needed data from this data store.

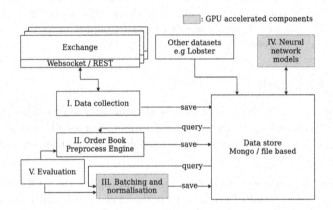

Fig. 2. Block diagram of the proposed GPU accelerated pipeline

The data collection component is used to store the data of the exchange(s) in a database. The Order Book Preprocess Engine is responsible for applying the update messages on a locally maintained LOB, and save snapshots with a fixed time delay between them. Batching and normalisation create normalised data, which is then mapped into discrete intervals and summed up, creating the output. The resulting data can be used as training data for deep neural networks. The evaluation is the last step of the pipeline, with which the performance of the preprocessing components can be measured. All of the components are detailed in the next sections.

3.1   Data Collection

The data format should be universal - no exchange or asset specific features should be included in the process. However, to collect a large enough dataset for training and performance evaluation, a data collection tool is a part of the pipeline, that can save every update that's happening on an exchange.

The data collection tool saves the following information from an exchange:

- Snapshot: The tool downloads the current state of the exchange periodically - the length of this period can be supplied when starting the process. The saved data is organised so the snapshots can be queried rapidly by dates.
- Updates: updates that are executed on the exchange and have effect on the order book.

The framework supports multiple assets when (down)loading data, due to the supposed universal features - training with multiple assets may increase the model accuracy [13,14]. As the number of LOB updates in an exchange can be extremely high, compression is used to store the data.

3.2   Data Preparation

The data preparation consists of two steps; the first one is the Order Book Preprocess Engine, which executes specific orders on a locally maintained order book, while the second component is responsible for normalization and batching.

Order Book Preprocess Engine. This step applies the LOB updates onto the snapshots, as it would be applied on the exchange itself, thus, creating a time series of LOB snapshots. As this part of the pre-processing purely relies on the previous state of the order book, and needs to be done sequentially, it cannot be efficiently parallelised and transferred to GPUs.

In this paper, only limit orders are considered – market orders are not the scope of the current work, as these are executed instantly. For the Order Book Preprocess Engine, the following order typesare defined that can change the state of the LOB:

- **Open:** A limit order has been opened by a client.
- **Done:** A limit order has either been fulfilled or cancelled by the exchange; it must be removed from the book.
- **Match:** Two limit orders have been matched, either completely or partially. If the match is partial, the volume of the corresponding orders will be changed. If the match is complete, both orders will be removed.
- **Change:** The client has chosen to change the properties of the order - the price and/or the volume.

There could be additional types of orders on different exchanges, indeed.

The goal of this first phase is to create snapshots of the exchange, by a fixed time gap between them. Each order in the stored book has the following properties:

- **Price:** The ask or bid price of the order.
- **Volume:** The volume of the asset which the client would like to sell/buy.
- **Id:** A unique identifier of the order.

There are only two parameters for this process - **time gap between generated snapshots** and **LOB depth**. We would like to emphasize that this component does not yet use GPU acceleration, however, it only needs to be run for the whole dataset once. By supplying adequate values for the input parameters - a small enough time frame, and a large enough LOB depth - the normalisation component afterwards can choose to ignore some of the snapshots, or the orders which are too deep in the book.

Batching and Normalisation. The input of this component is n pieces of snapshots provided by the previous steps. The component can be tore down to smaller pieces - it creates fixed-size or logarithmic intervals, scales the prices of the orders in a snapshot, maps these into the intervals and calculates cumulated sum of the volumes on each side of the LOB. The process is described in details below.

1. Splitting the input data into batches: n consecutive snapshots are concatenated (batched) with a rolling window. The number of snapshots in each batch, and the index of the considered snapshots (every 1st, 2nd, 3rd, etc.) are parameters. These batches can also be interpreted as matrices, which look like the following:

$$X_1 = \begin{bmatrix} a(t-k)_d & a(t-k)_{d-1} & \cdots & a(t-k)_1 & b(t-k)_1 & b(t-k)_2 & \cdots & b(t-k)_d \\ a(t-k+1)_d & a(t-k+1)_{d-1} & \cdots & a(t-k+1)_1 & b(t-k+1)_1 & b(t-k+1)_2 & \cdots & b(t-k+1)_d \\ & & & \cdots & & & \\ a(t)_d & a(t)_{d-1} & \cdots & a(t)_1 & b(t)_1 & b(t)_2 & \cdots & b(t)_d \end{bmatrix} \quad (5)$$

$$X_2 = \begin{bmatrix} a(t-k+m)_d & \cdots & a(t-k+m)_1 & b(t-k+m)_1 & \cdots & b(t-k+m)_d \\ a(t-k+m+1)_d & \cdots & a(t-k+m+1)_1 & b(t-k+m+1)_1 & \cdots & b(t-k+m+1)_d \\ & & \cdots & & & \\ a(t+m)_d & \cdots & a(t+m)_1 & b(t+m)_1 & b(t+m)_2 & \cdots & b(t+m)_d \end{bmatrix} \quad (6)$$

$$X_{l-1} = \begin{bmatrix} a(t-k+l*m)_d & \cdots & a(t-k+l*m)_1 & b(t-k+l*m)_1 & \cdots & b(t-k+l*m)_d \\ a(t-k+l*m+1)_d & \cdots & a(t-k+l*m+1)_1 & b(t-k+l*m+1)_1 & \cdots & b(t-k+l*m+1)_d \\ & & \cdots & & & \\ a(t+l*m)_d & \cdots & a(t+l*m)_1 & b(t+l*m)_1 & \cdots & b(t+l*m)_d \end{bmatrix} \quad (7)$$

In these matrices, t is the timestamp on the last snapshot of the first batch. d is the order book depth, m shows how many snapshots we skip when generating the next batch, and k is the length of a batch. X_i refers to the actual batch,

2. Normalising each batch independently: the middle price of the last snapshot in the batch is extracted for each, and all the prices in a batch are divided by this value. This also means, for the last snapshot of the batch, the bid prices will always have a value between 0 and 1.0, while the normalised asks will always be larger than 1. However, this limitation does not apply to the previous snapshots of the batch, as the price moves move up or down. This scaling is necessary, because we intend to use the pipeline for multiple assets. However, one asset could be traded in a price range with magnitudes of difference compared to other assets. Applying the proposed normalisation causes prices of any asset to be scaled close to 1. This will make the input for the neural network model quasi standardised.

3. Determining price levels for the 0-th order interpolation. In this step, the discrete values are calculated, which will be used for representing an interval of prices levels in the order book. These will contain the sum of volumes available in each price range at the end of processing.

The pipeline offers two methods to create these: linear (the intervals has the same size) and logarithmic (the intervals closer to the mid-price are smaller). The logarithmic intervals have a benefit over the linear ones: choosing linear intervals that are nor too big or too small needs optimization, as there could be larger price jumps - considering volatile assets and periods. These jumps could cause the values to be outside the intervals. With logarithmic intervals, this problem can be decreased, as the intervals on both sides can collect a much greater range of price values. Furthermore, prices around the spread are the most important ones, as discussed above – which are represented more detailed in the logarmithmic case.

One interval can be notated as:

$$[r_{lower}; r_{upper})$$ (8)

where *lower* and *upper* are the limits. In this step, multiple of these intervals are created, and stored in a list:

$$[r_{1;lower}; r_{1;upper}), [r_{2;lower}; r_{2;upper}), ... [r_{n;lower}, r_{n;upper})$$ (9)

The lower bound of the $k + 1$th interval always equals to the upper bound of the kth interval.

The selected intervals are re-used through the application, in the next steps.

4. 0-th order interpolation and summation. For each order in a snapshot the corresponding interval is defined (the price of the order is between the limits of the interval). This is done by iterating over the intervals, and verifying the price of the orders:

$$r_{i;lower} \leq price < r_{i,upper}$$ (10)

where i is the index of the actual interval. The volumes of orders which should be put into the same interval are summed. At the end of this step, the output contains the sum of volumes for each interval.

5. Iterating through the intervals which contain the volumes and calculating the accumulated volumes, separately for bid and ask side. At the beginning of this step, the summed volume for each interval is known. After the step is done, the value in each interval tells what is the volume that can be bought/sold below/above the lower/upper price level of the interval. When this part of the processing is done, the output for each snapshot could be plotted as in Fig. 1. This step can also be parallelized for each batch. An example to a completely processed batch can be seen on Fig. 3a and 3b.

6. Generating labels for the batches. This step is required for supervised learning algorithms to define a target variable for every batch. In the scope of this work, the mid-price and VWAP properties of the order book have been chosen to be

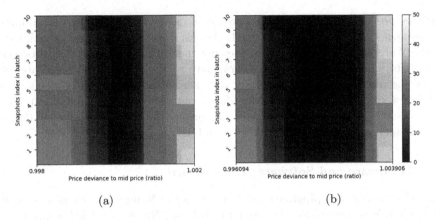

(a) (b)

Fig. 3. An example of the preprocessed LOB batch of 200 intervals with 30 depth. (a) was generated with linear scaling, while (b) uses logarithmic scale. Data taken from Coinbase Pro, 29. 10. 2019

used as labels, however, it can be arbitrary. Labels are generated according to the snapshots after the batch and scaled by the mid-price (or VWAP) of the last snapshot in the batch. If a regression model is used, then the value itself is the target variable. If the goal is to have a classification model, the target value can be clustered e.g., into 3 labels:

- Upward, downward and sliding movement. When calculating these, an α parameter is considered, which is a percentage value. If the average of the next snapshots is greater than $(1+\alpha) * midprice$ it is considered upward. If it is less than $(1+\alpha) * midprice$ it is downward, else the batch is labelled as a sideway movement.

The steps above result in a set of batches, where each batch contains n snapshots and corresponding output label(s). When displaying a batch, the relative price movement over time can be discovered.

4   Evaluation and Results

In order to measure the speed enhancements of the proposed methods, experiments with different data sizes and hardware accelerators have been carried out. The evaluation was done with data taken with the data collection tool. The data was collected with Coinbase Pro API. Coinbase exchange provides REST support for polling snapshots. For evaluation, a snapshot has been saved each 5 min, while the updates of the book are provided via Websocket streams.

4.1   Used Data

The data used for performance measurement was collected from the Coinbase exchange, using 4 currency pairs - BTC-USD, BTC-EUR, ETH-USD, XRP-USD, starting in September, 2019. For the test, we have limited the number of updates processed - the algorithm would stop after the supplied "n" number of updates - "n" is shown in Table 1. As the data contains discrepancies due to the Websocket connection, this does not mean that the number of processed snapshots and updates will always translate to the same number of batches - hence the number of batches slightly vary across the measurements - e.g. 2088 batches were generated for 100000 updates, but only 9941 for 1 million updates. The reason why we did not consider batches as the base limitation, is that in real-life scenarios these discrepancies would also happen.

4.2 Hardware and Software Architecture

The GPU accelerated pipeline extensively uses the Numpy and Numba packages - the CUDA accelerated code was written using Numba's CUDA support. To compare the performance the pipeline was executed on different GPUs (single GPU in each run) for the same amount of raw data. The deep learning models were created with Keras. The data collection, preparation and trainings were run in a containerized environment, using Docker.

The hardware units used in the tests:

- GPU 1: NVidia Titan V, 5120 CUDA cores, 12 GB memory.
- GPU 2: NVidia Titan Xp, 3840 CUDA cores, 12 GB memory.
- GPU 3: NVidia Titan X, 3072 CUDA cores, 12 GB DDR5 memory.
- CPU: Intel Core i7-6850K CPU, 3.60 GHz

4.3 Execution Time

In the first phase of the pipeline – the Order Book Preprocess Engine – the updates are sequential, which makes its execution in parallel a difficult problem. Because of that, the measurements have been broken into 2 parts: the snapshot generation and the parallellised normalisation with other steps.

Table 1 shows the measurements taken on the Preproces Engine component, while Table 2 measures the components that apply the normalisation on the output of the Preprocess Engine. For comparison, the CPU column runs the same algorithms, but using plain Python with libraries such as Numpy. In the 'CPU + LLVM' measurements, we utilise the optimization techniques of the Numba library. This project uses a just-in-time compiler to compile certain steps of the process - the same steps which are accelerated by CUDA when running on a GPU. We have chosen the CPU + LLVM method as a comparison base - the GPU performance is compared to this one. Some experiments were made on using parallel execution models of the Numba library (such as prange) - which did not bring significant performance improvements. Also, it's worth to note, that a server-grade CPU, which uses many more cores compared to the used one could speed up the results when executing parallelly.

Table 1. Performance of the Order Book Preprocess Engine component. Data taken from the Coinbase Exchange. The time delay between the snapshots is 1 s.

Pipeline performance		
Batches generated	Time taken [seconds]	Nr. of processed updates
2088	8.4 s	100000
9941	18.4 s	1000000
90086	169.58 s	10000000
504880	792.01 s	57.35 million

Table 2. Performance measurements of preparing the data after the Order Book Preprocess Engine. (LOB depth: 100, batch size: 10).

Data used	Pipeline performance [seconds]				
	CPU	CPU + LLVM	GPU 1	GPU 2	GPU 3
2070 batches	129.93	2.08	2	1.7	1.71
9941 batches	641.84	3.9	2.59	2.6	3.2
504880 batches	19404.38	137.237	58.76	61.77	101,86

4.4 Inference

To be able to integrate the solution into real-world trading environment, the pipeline should be able to process the stream of order book data and execute the steps even for a single batch as fast as possible. To simulate the inference phase time measurements were carried out with only a few snapshots. In these experiments 500 updates were used, from which 29 snapshots were created. The first phase - applying these updates and creating the next snapshots - took 7 seconds to generate 29 snapshots (0.15 s/snapshot). Normalising with CPU took 2.06 s (0.071 s/snapshot), while it took around 1.70, 1.95, 1.59 on the Titan V/Xp/X GPUs. The processing took 0.058/0.067/0.054 s/snapshot on each of them. This means there is 18% improvement when running on Titan V, and 22% on the Titan X).

5   Application and Conclusions

The most important feature of the GPU accelerated pipeline is that with the performance gain multiple parameter setups can be tested in less time, when compared with CPU preprocessing. The solution helps researchers to train deep learning models with several kinds of data representation and labelling methods and parameters - and select the method which yields the best accuracy. In absence of a GPU, the LLVM compiler can also improve on processing the data. The proposed GPU accelerated pipeline uses structured input data in a way that's not specific to any data source. As a result, it could be used for any exchange, which provides an interface to their order book data. Financial institutions or individual clients on the exchange could benefit from the proposed pipeline, as it helps them make data processing and normalisation much faster by running on the GPU instead of a CPU. An additional benefit of the pipeline is its modularity - the components can be used independently of each other. In future research, the pipeline can be performance-tested with multiple hyperparameter set-ups, and the GPU execution can be further investigated using profilers. The pipeline can also be the base of a real-time, GPU accelerated trading bot.

The solution is available at https://github.com/jazzisnice/LOBpipeline.

Acknowledgment. The research presented in this paper, carried out by BME was supported by the Ministry of Innovation and the National Research, Development and Innovation Office within the framework of the Artificial Intelligence National Laboratory Programme, by the NRDI Fund based on the charter of bolster issued by the NRDI Office under the auspices of the Ministry for Innovation and Technology, by the European Union, co-financed by the European Social Fund (EFOP-3.6.2-16-2017-00013, Thematic Fundamental Research Collaborations Grounding Innovation in Informatics and Infocommunications), by János Bolyai Research Scholarship of the Hungarian Academy of Sciences and by Doctoral Research Scholarship of Ministry of Human Resources (ÚNKP-20-5-BME-210) in the scope of New National Excellence Program. We gratefully acknowledge the support of NVIDIA Corporation with the donation of the Titan V GPU used for this research.

References

1. Ganchev, K., Kearns, M., Nevmyvaka, Y., Vaughan, J.W.: Censored exploration and the dark pool problem (2012)
2. LeCun, Y., Bengio, Y., Hinton, G.: Deep learning. Nature **521**(7553), 436–444 (2015)
3. Black, F., Scholes, M.: The pricing of options and corporate liabilities. J. Polit. Econ. **81**(3), 637–654 (1973)
4. Ranaldo, A.: Order aggressiveness in limit order book markets. J. Financ. Mark. **7**, 53–74 (2004)
5. Cont, R., Stoikov, S., Talreja, R.: A stochastic model for order book dynamics. Oper. Res. **58**, 549–563 (2010)
6. Beck, J., et al.: Sensing social media signals for cryptocurrency news. In: Companion Proceedings of the 2019 World Wide Web Conference on '- WWW 2019 (2019)
7. Mirtaheri, M., Abu-El-Haija, S., Morstatter, F., Steeg, GV., Galstyan, A.: Identifying and analyzing cryptocurrency manipulations in social media (2019)
8. Tsantekidis, A., Passalis, N., Tefas, A., Kanniainen, J., Gabbouj, M., Iosifidis, A.: Using deep learning for price prediction by exploiting stationary limit order book features (2018)
9. Ntakaris, A., Magris, M., Kanniainen, J., Gabbouj, M., Iosifidis, A.: Benchmark dataset for mid-price forecasting of limit order book data with machine learning methods. J. Forecast. **37**(8), 852–866 (2018)
10. Bibinger, M., Neely, C., Winkelmann, L.: Estimation of the discontinuous leverage effect: evidence from the NASDAQ order book. In: Federal Reserve Bank of St. Louis, Working Papers, April 2017

11. Wei, H., Wang, Y., Mangu, L., Decker, K.: Model-based reinforcement learning for predictions and control for limit order books (2019)
12. Di Persio, L., Honchar, O.: Artificial neural networks architectures for stock price prediction: comparisons and applications. Int. J. Circ. Syst. Signal Process. **10**, 403–413 (2016)
13. Sirignano, J., Cont, R.: Universal features of price formation in financial markets: perspectives from deep learning (2018)
14. Zhang, Z., Zohren, S., Roberts, S.: DeepLOB: deep convolutional neural networks for limit order books. IEEE Trans. Signal Process. **67**(11), 3001–3012 (2019)

Can Big Data Help to Predict Conditional Stock Market Volatility? An Application to Brexit

Vittorio Bellini[1], Massimo Guidolin[1,2], and Manuela Pedio[1,2,3(✉)]

[1] Bocconi Univesity, Via Roentgen 1, 20136 Milan, Italy
manuela.pedio@unibocconi.it
[2] Baffi Carefin Centre, Via Roentgen 1, 20136 Milan, Italy
[3] Bicocca University, Piazza dell'Ateneo Nuovo, 1, 20126 Milan, Italy

Abstract. Nowadays, investors making financial decisions have access to plentiful information. In this paper, we investigate whether it is possible to use the information retrievable from a large number of online press articles to construct variables that help to predict the (conditional) stock market volatility. In particular, we use the Brexit referendum as an experiment and compare the predictive performance of a traditional GARCH model for the conditional variance of the FTSE 100 Index with a number of alternative specifications where measures of the tone and the proliferation of news related Brexit is used to augment the standard GARCH model. We find that most of the proposed news-augmented models outperform a traditional GARCH model both in-sample and out of sample.

Keywords: Big data · Sentiment analysis · Conditional variance · GARCH models · Forecasting · Brexit

1 Introduction

According to the Efficient-Market Hypothesis (EMH) formulated by Fama (1970) [9], the price of financial assets should reflect all the available information. However, this hypothesis, which postulates that the investors act as rational, profit maximizing agents, has been recently put into discussion by the advocates of Behavioral Finance, who emphasize the role of human psychology in the process of price formation (see, e.g., Barberis and Thaler, 2003 [1]). As a consequence, in the last twenty years a new strand of literature has emerged, which studies the relationship between the sentiment of investors – expressed through news and posts in the media – and the returns on financial securities. Following this literature, in this paper we investigate whether variables that measure the extent and the tone of press coverage of certain events that are perceived as particularly relevant by the financial market operators may help to forecast (conditional) stock market volatility when used to augment a standard Generalized Autoregressive Conditional Heteroskedasticity (GARCH) model.

© Springer Nature Switzerland AG 2020
G. Nicosia et al. (Eds.): LOD 2020, LNCS 12565, pp. 398–409, 2020.
https://doi.org/10.1007/978-3-030-64583-0_36

In particular, the event on which we focus is the Brexit vote that took place in June 2016 along with the intense political debate that preceded it and the uncertainty that followed it. Therefore, we forecast the variance of open-to-close returns on the FTSE100 index during the period October 1, 2015–October 31, 2018. We evaluate the forecasting performance of news/sentiment-augmented models against a standard GARCH benchmark both in-and out-of-sample (OOS) using a variety of standard performance measures.

In the perspective of showing the importance of big data sentiment in forecasting the volatility of asset returns, the Brexit vote represents a highly relevant case for several reasons. Indeed, Brexit has undoubtfully raised concerns among investors: Dhingra, Ottaviano, Sampson, and Van Reenen (2016) [7], among the others, estimate a perspective, strong negative economic impact for the United Kingdom in terms of reduction of their gross domestic product and of foreign investment inflows. Therefore, in this paper we argue in favor of identifying the uncertainty regarding the outcome of the referendum first, and the doubts concerning the exit strategy later, as clear factors of long-term tension, both in the economic and social perspectives. The impact of Brexit on the volatility of the stock market is confirmed by Raddant and Karimi (2016) [14], who have examined the effects of the results of the Brexit vote on the correlation and volatility of the European financial markets, represented by five major equity indices. They show that the outcome of the Brexit vote led to a general increase in volatility, especially in the United Kingdom and in Italy, although the effect only lasted for three weeks. Belke, Dubova, and Osowski (2018) [2] have proposed a political uncertainty index directly linked to the political dispute process around Brexit in the year preceding the consultation and used a Granger-causality test to show that part of the volatility was transmitted from the political uncertainty regarding Brexit to the financial and currency markets.

The use of measures of news coverage (and their tone) to predict market returns and volatility has been largely debated in the literature although the results are yet inconclusive. In his seminal paper, Tetlock (2007) [16] first constructed an index of pessimism characterizing traders by analyzing the semantics used in the short comments contained in the popular column "Abreast of the Market" in the Wall Street Journal. He finds that high media pessimism predicts downward pressure on market prices followed by a reversion to fundamentals. Similarly, Chen, De, Hu, and Hwang (2014) [6] have investigated the extent to which opinions transmitted through social media platforms may predict future stock market returns and report that the views expressed in the articles are able to forecast stock returns and earnings surprises.

Guidolin, Orlov, and Pedio (2017) [11] use the volume of internet searches (or, alternatively, its growth rate) retrieved from Google Trends as a proxy for investors– interest in a specific asset to forecast the returns of the UK, French, German, and US stock and bond markets and find that such heuristic models perform better than the average of all the alternative predictive models. Blasco, Corredor and Santamaria (2005) [3] study what kind of information affects close-to-open and open-to-close returns, the volatilities, and the volume of actively traded securities in the Spanish market. Their study is close to ours as they also propose several specifications of the GARCH model that accounts for the asymmetric effects of the news on the stock market (conditional) volatility. However, they are not interested in the predictive accuracy of their models but only in the explanatory power of the news-related variables.[1]

Our study contributes to this literature in several ways. First, to the best of our knowledge, we are the first to employ the extensive information coming from the *Global Database of Events, Language, and Tone* (GDELT) to forecast stock return volatility. This is an enormous data repository collecting and analyzing the textual content of news coming from hundreds of thousands different sources in 100 different languages. The unique features of this database allow us to gain an unprecedented coverage of the topic and to quantify both the growth of the attention devoted to Brexit and the tone attributed to the event by the commentators. Second, we propose and evaluate several extensions of the traditional GARCH model that allows the news-related variable to enter in the prediction of future conditional volatility in asymmetric ways.

The empirical results support the usefulness of news-related variables to predict the conditional variance of UK stock returns during the Brexit period. In particular, the tone of media coverage of the Brexit is a significant additional variable when used to augment a standard GARCH model. Conversely, the level of media attention to the topic – measured by the growth rate of the articles where the word Brexit appears – is also significant, but the predictive performance is better than the one of the benchmark only when the variance of returns is high.

The rest of the paper is structured as follows. Section 2 describes the data used in the analysis and the alternative models that we propose to forecast conditional volatility. Section 3 discusses the empirical results. Section 4 concludes.

2   Data and Methodology

2.1   The News Related Variables

The main novelty of our paper is the use of GDELT as the main source of data. This dataset is based on a set of natural language processing algorithms capable of analyzing aggregated data in the form of texts from hundreds of thousands of news sources, mostly online but also printed, and disseminated by news agencies, broadcasters and publishers, in more than 100 different languages. In particular, for the purposes of our analysis, we have employed the *Global Knowledge*

---

[1] There are also a few studies that have analyzed the impact of news concerning specific events on stock market volatility. For instance, Lutz (2014) [13] study the interaction between the content, the tone, and the extent of the media coverage concerning the uncertainty of unconventional monetary policies and the stock volatility. He shows that shocks to monetary policy, combined with situations of high uncertainty and general pessimism among traders and in the media, lead to an increase in volatility higher than the one that is predicted by traditional models.

Graph tool (GKG), which is a large collection of online articles classified by the project algorithms based on the "keywords" contained in the text.[2] In addition, a sentiment analysis algorithm is used to process the whole text of the news and to attribute a "positive" or "negative" tone to it. The tone is summarized by a number which is computed as the difference between the "Positive Score" and the "Negative Score", i.e., the percentage of the words of the article with a positive emotional connotation vs. those with a negative connotation, respectively. In this sense, a positive tone indicates that the author attributes positive implications to the event being discussed while the opposite applies to a negative tone. As we are interested in the media coverage of Brexit, we extract from the dataset all the records that include the word "Brexit". Our sample period spans the period October 1, 2015–October 31, 2018.[3]

The data so extracted from the GDELT database consist of more than half a million articles written during our sample, with peaks of more than 10,000 articles in the days close to the referendum, and a stable and negative tone on average. Based on these data, we construct two different indicators that we use to augment the standard GARCH model: the first variable, *Art growth*, captures the rate of growth in the attention devoted to the topic, irrespective to its tone; conversely, the *tone* variable is a sentiment indicator. In particular, the tone variable is constructed as the weighted average of the tone scores of each *nameset i* published at time *t*, assuming as weights the total number of articles containing *nameset i*:

$$tone_t = \frac{\sum_{i=1}^{I} tone_{i,t} \times \omega_{i,t}}{\sum_{i=1}^{I} \omega_{i,t}}, \tag{1}$$

where $tone_{i,t}$ represents the tone score of the *nameset i* published on the date *t*, and ω_i is the number of articles containing the *nameset i* (corresponding to the GKG field indicated as "NUMARTS" in the GDELT database, see Appendix A). Thanks to this methodology, we attribute higher importance to the tone of the news that have raised more attention in terms of origination of articles and number of citations. Figure 1 shows the trend of the tone variable over the sample period.

---

[2] GKG identifies each article in a unique way with a set of keywords, referred to as the *nameset* of the article, i.e., a summary containing topic, actors, places, etc. More information can be found in Appendix A posted at http://dx.doi.org/10.17632/h4rtr9h5kd.1).

[3] Even though according to the Oxford English Dictionary (2017) the word "Brexit" can be traced back to May 2012, we start our analysis on October 1, 2015 because this date corresponds to the implementation of the GKG tool. This is reasonable because a closer media coverage of the topic has begun in proximity of the Referendum (June 2016).

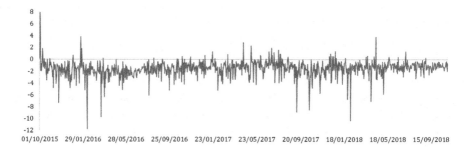

Fig. 1. The figure shows the evolution of the average tone of daily media coverage represented by articles that mention the word "Brexit". The global tone score is computed using a weighting scheme based on the number of sources that have mentioned or shared them.

The growth variable (henceforth, *Art Growth*), is the linear growth of the number of articles concerning Brexit with respect to the previous day.[4] Figure 2 shows the dynamics of this variable over the sample. To make the scale of all variables close and facilitate smooth running of the estimation algorithms, both Tone and Art Growth have been divided by 100. Additional details about the GDELT dataset together with a sample of data can be found in Appendix A (available at http://dx.doi.org/10.17632/h4rtr9h5kd.1).

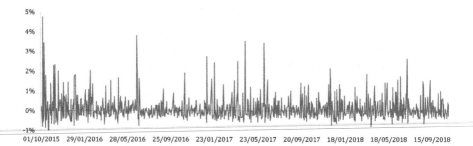

Fig. 2. The figure shows the evolution of the growth rate of the number of published articles containing the word "Brexit". The number of articles is the sum of the number of original contributions that have been written and published by a primary source, and the number of times that they have been mentioned/republished/shared from other news sources.

---

[4] To simplify the construction of the series, we have considered all articles published throughout the day, without limiting ourselves to the opening hours of the trading venues. However, this does not represent a serious limitation to our analysis as the majority of the articles considered are typically published during the opening hours of the relevant market. Additional details concerning the construction of the series are available upon request from the Authors.

2.2   The News Augmented GARCH Models

The variables growth and tone are used to augment a standard GARCH model (see Bollerslev, 1986 [4]) for the variance of the open-to-close return on the FTSE100 index.[5] The first specification that we propose – constructed by adding an exogenous variable (similar to Engle, 2002 [8]) to the standard GARCH – is:

$$R_{t+1} = \sigma_{t+1|t}z_{t+1} \qquad z_{t+1} \ i.i.d. \ N(0,1) \tag{2}$$

$$\sigma^2_{t+1|t} = \omega + \alpha R^2_t + \beta \sigma^2_{t|t-1} + \gamma x^2_{i,t+1} \qquad i = 1,2 \tag{3}$$

where R_t is the (daily) log return defined as $R_t \equiv \ln\left(\frac{P_t}{P_{t-1}}\right)$, with P_t being the value of the FTSE 100 index at time t, z_t is a normally distributed random shock and $\sigma^2_{t|t-1}$ is the conditional variance at time t given the information set at time $t-1$, $x_{i=1,t} = tone_t$, while $x_{i=2,t} = Artgrowth_t$; ω, α, β and γ are the estimated parameters.

The second extension that we propose concerns the sign of the news-related variables, as both tone and Art growth can display negative values. Considering this, it is appropriate to ask whether this information deserves to be studied (indeed, in Eq. 3 we lose the information regarding the sign by squaring the variables). As far as the series of the tone of media coverage is concerned, this means assuming that, given two news with the same impact in terms of absolute value, the most pessimistic news has a greater impact on the return variance than the most optimistic news. This idea is in line with the findings of Campbell and Hentschel (1992) [5], who argue that the most relevant news (both positive and negative) have a larger impact on the volatility than less relevant news and that important negative news show an even greater multiplicative effect. As far as the Art growth series is concerned, justifying the use of the squared variable in Eq. 3 is even more problematic: indeed, this means assuming that a reduction in the number of articles covering the topic under examination contributes to an increase in volatility of the same intensity as a similar increase in the number of articles. To overcome these limitations, we also estimate the following GARCH *Asymmetric News Dummy* (that we shall call GARCHAND):

$$\sigma^2_{t+1|t} = \omega + \alpha R^2_t + \beta \sigma^2_{t|t-1} + \gamma(1 + d_1\theta)\,tone^2_{t+1} \tag{4}$$

$$\sigma^2_{t+1|t} = \omega + \alpha R^2_t + \beta \sigma^2_{t|t-1} + \gamma d_2 Artgrowth^2_{t+1}, \tag{5}$$

where the conditional mean of returns is modeled as in Eq. 2 and d_1 (d_2) is a dummy variable that is equal to one if *tone* (*Art growth*) is negative (positive); in Eq. 4, we impose $\theta > 0$ so that the model forecasts a greater conditional variance when the average tone score of the day assumes a negative value, while in Eq. 5

---

[5] In our analysis, we restrict ourselves to GARCH $(1,1)$ and its extensions; as often discussed in the literature (see, e.g., Hansen and Lunde, 2005 [12]) adding lags of past squared errors and/or the predicted variance tends not improve much the forecasting performance of the model.

we impose $\gamma > 0$ so that only a positive growth in the number of articles may lead to an increase in predicted variance.[6]

Finally, it is also legitimate to check whether the information deriving from the exogenous variables has always the same impact, irrespective of the current level of market volatility. In fact, we may hypothesize that, in an environment of high volatility, some news reported with a very positive or very negative tone may trigger an even more marked increase in volatility vs. lukewarm reporting of news; a similar reasoning applies to an increase in media attention and/or diffusion of the topic. Following this idea, we propose a GARCH model augmented by a dummy variable that is linked not to intrinsic characteristics of the exogenous variable (sign), but to a "threshold" parameter observed with reference to the predicted conditional variance in the previous period. This model that we shall call GARCH *News Dummy* (GARCHND) has the following specification:

$$\sigma_{t+1|t}^2 = \omega + \alpha R_t^2 + \beta \sigma_{t|t-1}^2 + \gamma d_3 x_{i,t+1}^2 \, i = 1, 2, \tag{6}$$

where $x_{i=1} = tone_{t+1}$ and $x_{i=2} = Art\ growth_{t+1}$, d_3 is a dummy variable which is positive if $\sigma_{t|t-1}^2 \geq \kappa$, and $\sigma_{t|t-1}^2$ represents the daily conditional variance forecast for the previous period; the conditional mean of returns is modeled as in (2). The threshold κ is set alternatively to 10% and 20%. An example of the code performing the estimation of the models is available in Appendix B. Additional details are available upon request from the Authors. Notably, the use of exogenous variables based on news referring to the same day of prediction of the conditional variance is consistent with Blasco et al. (2005) [3] and Shi, Ho, and Liu (2011) [15]. This particular choice may raise some skepticism, given that the objective of our application is to forecast the conditional variance (volatility) in real time. However, most of the press articles tend to be released in the morning. As we observe the news variable at the end of the day, this has two consequences. On the one hand, the variable at time $t - 1$ will contain information that has already been largely processed by the market participants and incorporated into the market prices. On the other hand, it would be completely plausible for a trader or a risk manager to implement the model that we propose at some point during the day (for instance, after mid-day) to forecast the conditional volatility at the end of the day. We deem our exercise to represent a good proxy for this situation.

3   Empirical Results

The main goal of our analysis is to assess whether the news-augmented models described in Sect. 2 outperform a simple GARCH model in terms of forecasting accuracy both in-sample and (especially) out-of-sample (OOS). As far as the OOS performance is concerned, our experiment consists of setting up a genuine OOS exercise: model parameters are estimated over a October 1, 2015–June 30,

---

[6] The specification in (4) is similar to the GJR-GARCH proposed by Glosten, Jagannathan, and Runkle (1993) [10] with the difference that the asymmetric effect concerns the exogenous variable and the not past errors.

2018 estimation period, while the OOS evaluation period goes from July 1, 2018 to October 30, 2018. In order to evaluate the predictive performance, we use a set of standard performance measurement statistics and perform systematic comparisons among alternative models. More specifically, we compute the Root Mean Squared Error (RMSE), the Mean Absolute Percentage Error (MAPE) and the Relative MAPE (RMAPE), besides applying Diebold-Mariano (DM) tests. Notably, the variance at time t is a latent variable that is not observed even after time t; therefore, we use the realized variance (computed at a 10-min frequency, without sub-sampling) published by the Oxford-Man Institute as a proxy for the unobserved variance. The realized volatility over the sample period is plotted in Fig. 3.

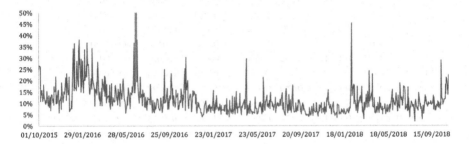

Fig. 3. The figure shows the dynamics of daily realized volatility of the returns on the London FTSE100 index, annualized considering 252 trading days and calculated from the realized variance estimated by the Oxford-Man Institute at a 10-min frequency, without sub-sampling. To enhance readability, the series has been cut at 50%.

3.1 In-Sample Forecasting Results

Table 1 presents the estimation results within the estimation sample for the models presented in Sect. 2. We note that all the models, with the exception of the GARCHND based on the "tone" series and with a threshold level $\kappa = 20\%$, are characterized by slightly lower mean squared residuals than a standard GARCH $(1,1)$ benchmark; in particular the model with the lowest in-sample forecast errors is the GARCHND based on the *Art Growth* series and with a threshold level $\kappa = 10\%$. It is important to emphasize the results of the estimation of γ (and θ in the case of Eq. 4), a coefficient that determines the weight of the information coming from the exogenous variables. Interestingly, γ is positive and significant in GARCHX models, in GARCHND models when $\kappa = 10\%$ and in the GARCHAND model with the growth rate of the number of articles used as an exogenous variable. As far as the hypotheses of a stronger effect on the variance coming from the negative tone of the news, the GARCHAND model including Tone fails to confirm the significance of the coefficient γ when inserted in an asymmetric structure defined by the coefficient θ, as in Eq. 4. Therefore, there is evidence that the information coming from media coverage tone is significant in predicting the variance of FTSE100 returns only if its square is considered symmetrically, either always (GARCHX) or when the predicted variance from

Table 1. The table shows the estimation results. The standard errors are in parentheses. The individual coefficient significance is emphasized according to the p-values of a standard (Wald) t-test for size levels equal to 1%, 5%, and 10% (***, ** and * respectively). In-sample fitting results which are superior to a standard GARCH model are boldfaced.

	RMSE	MAPE	ω	α	β	γ	θ
GARCH	0.0343	1.0471	0.0003	0.1072***	0.8485***		
			(0.0555)	(0.0186)	(0.0140)		
GARCHX Tone	**0.0343**	**1.0389**	0.0002	0.1117***	0.8132***	0.8239**	
			(0.0540)	(0.0250)	(0.0333)	(0.3770)	
GARCHX Growth	**0.0342**	**1.0399**	0.0003	0.1152***	0.8226***	2.1632*	
			(0.0559)	(0.0214)	(0.0187)	(1.2681)	
GARCHND Tone ($\kappa = 10$)	**0.0343**	**0.9979**	0.0003	0.0909***	0.8257***	1.0157***	
			(0.0645)	(0.0185)	(0.0151)	(0.2840)	
GARCHND Tone ($\kappa = 20$)	0.0343	1.0557	0.0003	0.1144***	0.8313***	0.5384	
			(0.0544)	(0.0211)	(0.0155)	(1.1148)	
GARCHND Growth ($\kappa = 10$)	**0.0342**	**0.9805**	0.0003	0.0947***	0.8443***	3.2991	
			(0.0573)	(0.0129)	(0.0084)	(1.8006)	
GARCHND Growth ($\kappa = 20$)	**0.0341**	**1.0472**	0.0003	0.1076***	0.8399***	4.3001	
			(0.0509)	(0.0193)	(0.0148)	(9.3667)	
GARCHAND Tone	**0.0343**	**1.0373**	0.0002	0.1110***	0.8153***	0.6885	0.1522
			(0.0537)	(0.0248)	(0.0336)	(4.3212)	(7.1158)
GARCHAND Growth	**0.0342**	**1.0375**	0.0003	0.1131***	0.8258***	2.2847**	
			(0.0558)	(0.0210)	(0.0179)	(1.3314)	

the previous period had exceeded 10% (GARCHND). The information resulting from the growth rate of the number of articles is significant for the same models mentioned above, including the case where an asymmetric impact on the variance as specified in Eq. 5 is considered. In conclusion, it seems that both the tone and the growth rate of the articles affect the prediction of stock market variance, but the attention received by news seems to be more significant when the volatility in the equity market is already high since the previous period.

3.2   Out-of-Sample Forecasting Results

Table 2 presents the OOS forecast performance results from the various models. As far as the tone, sentiment variable is concerned, the GARCHX and GARCHND models with $\kappa = 10\%$ continue to outperform the standard GARCH benchmark. Indeed, both these models outperform the GARCH benchmark, in terms of all the predictive accuracy statistics; this empirical evidence is confirmed by the results from the Diebold-Mariano test: the value of the statistics are significant at a 1% confidence level for the absolute loss function and at 5% for the quadratic loss function at least in the case of the GARCHND model.

On the contrary, most of the models augmented by the "attention" variable (namely, the rate of growth in the number of the articles concerning Brexit) seems not to outperform the GARCH $(1, 1)$ benchmark OOS. Indeed, while the

Table 2. The RMAPE and Diebold-Mariano (DM) test assume the GARCH $(1,1)$ model as a benchmark. The Diebold-Mariano statistics are reported for two alternative loss functions, i.e., the quadratic loss (DM SQ) and absolute value (DM AV). Significance at the size levels of 1%, 5%, and 10% is denoted by ***, ** and *, respectively. A negative value indicates a predictive performance superior to that of the benchmark model. Forecasting performances superior to the GARCH benchmark are boldfaced.

	RMSE	MAPE	RMAPE	DM SQ	DM AV
GARCH	0.0048	1.5137			
GARCHX Tone	**0.0047**	**1.2179**	**80.38%**	−0.7894	**−3.8885*****
GARCHX Art Growth	0.0048	**1.4792**	**97.64%**	0.8625	−0.4340
GARCHND Tone ($\kappa = 10$)	**0.0047**	**1.3253**	**87.47%**	**−1.5442****	**−4.3912*****
GARCHND Tone ($\kappa = 20$)	0.0048	**1.4992**	**98.95%**	0.1482	−0.6973
GARCHND Art Growth ($\kappa = 10$)	0.0048	**1.5150**	**99.99%**	0.9436	0.0980
GARCHND Art Growth ($\kappa = 20$)	**0.0048**	**1.4916**	**98.45%**	−1.2285	**−3.5342*****
GARCHAND Tone	**0.0047**	**1.2240**	**80.79%**	−0.8101	**−3.9346*****
GARCHAND Art Growth	0.0048	**1.4797**	**97.66%**	0.8486	−0.4675

GARCHX, the GARCHND with $\kappa = 10\%$, and GARCHAND show a strong in-sample predictive performance and they fail to significantly improve over the OOS performance of the GARCH model, as shown by the Diebold-Mariano tests. The only exception is represented by GARCHND with $\kappa = 20\%$, which significantly outperforms the GARCH benchmark in terms of OOS predictive accuracy. Therefore, our OOS analysis shows that, in contrast with the conclusions drawn from the in-sample estimation, there is empirical evidence that the growth rate of the articles concerning Brexit adds significant information in contexts of higher volatility. Moreover, despite the estimated γ turns out to be not significantly different from zero, the model is the one with the lowest RMSE values among all the alternatives assessed in the estimation stage. It should be considered that the lack of significance of γ may depend on the fact that there are very few observations for volatility higher than 20%, which leads to a high standard error for γ. However, the choice of the best performing model depends on the loss function that we decide to adopt. As far as the RMSE is considered, the GARCHND model based on the tone variable and with $\kappa = 10\%$ is the one with the lowest value of the statistic (and therefore the best one in terms of OOS predictive accuracy). However, as far as relative performance measures (namely, the RMAPE) are considered, a simple GARCHX that includes the tone as an exogenous variable without accounting for any asymmetric effects outperforms all the other models (including the standard GARCH benchmark).

Overall, it seems that the addition of a variable reflecting the sentiment surrounding Brexit, as captured by the tone of the media coverage, improves the prediction of the variance of FTSE 100 returns both in-sample and OOS. However, adding a variable that captures the growth of the media coverage concerning Brexit, despite showing some predictive power in sample, does not improve the forecasting ability of a simple GARCH $(1,1)$ model when considered OOS.

4   Conclusion

In this paper, we have proposed a set of alternative augmented GARCH models that include variables measuring the extent and the tone of the press coverage of a specific event, the so-called Brexit, which is deemed to have had a great impact on stock market operators, thus impacting equity volatility. We have benchmarked the predictive performance of these models vs. a standard GARCH. With respect to the existing literature, we have studied the increases in predictive accuracy due to the inclusion of sentiment/attention indices specifically related to a single socio-political issue, such as the exit process of the UK from the European Union. Differently from us, previous research has alternatively focused on: 1) the general tone and level of media coverage of equity markets as a whole, unrelated to any specific political or macroeconomic event; 2) the tone and level of firm-specific media coverage, referring to individual listed companies, or news concerning deviations of firm performance results from consensus. As far as the construction of the exogenous variables is concerned, to the best of our knowledge we are the first to tap into the extensive information collected by GDELT to forecast stock return volatility. We find that a few of the innovative models that include the media coverage and tone series outperform GARCH in three different specifications: GARCHX, GARCHND with $\kappa = 10\%$ and GARCHAND. The only model based on the series of the growth rates of the number of articles that persistently leads to a stronger performance than GARCH is GARCHND with $\kappa = 20\%$. From these results, we learn that knowing the tone with which an author reports the news significantly improves the forecast of the variance when such information is entirely considered (GARCHX), when it is only considered above a certain level of previously predicted variance (GARCHND), and when asymmetric effect driven by the sign (negative) of the tone (GARCHAND) are accounted for. Moreover, the growth rate in the number of articles is only significant in high variance contexts (GARCHND). A few extensions of our paper appear to be possible. As far as the GARCHND model is concerned, the dummy variable is only activated for conditions that refer to earlier predictions from the model. However, it would be interesting to activate the dummies based on the levels of effective realized variance on the previous day, which appears to be immediately measurable after the markets close. This would mean linking the activation of the exogenous variable to the same measure chosen as a proxy for the variable to be predicted.

References

1. Barberis, N., Thaler, R.: A survey of behavioral finance. In: Handbook of the Economics of Finance, vol. 1, pp. 1053–1128 (2003). https://doi.org/10.1515/9781400829125-004
2. Belke, A., Dubova, I., Osowski, T.: Policy uncertainty and international financial markets: the case of Brexit. Appl. Econ. **50**, 1–19 (2018). https://doi.org/10.1080/00036846.2018.14361520

3. Blasco, N., Corredor, P., Del Rio, C., Santamaria, R.: Bad news and Dow Jones make the Spanish stocks go round. Eur. J. Oper. Res. **163**, 253–275 (2005). https://doi.org/10.1016/j.ejor.2004.01.001
4. Bollerslev, T.: Generalized autoregressive conditional heteroskedasticity. J. Econometrics **31**, 307–327 (1986)
5. Campbell, J.Y., Hentschel, L.: No news is good news: an asymmetric model of changing volatility in stock returns. J. Finan. Econ. **31**, 281–318 (1992). https://doi.org/10.1016/0304-405X(92)90037-X
6. Chen, H., De, P., Hu, Y.J., Hwang, B.H.: Wisdom of crowds: the value of stock opinions transmitted through social media. Rev. Finan. Stud. **27**, 1367–1403 (2014). https://doi.org/10.2139/ssrn.1807265
7. Dhingra, S., Ottaviano, G., Sampson, T., Van Reenen, J.: The impact of Brexit on foreign investment in the UK. Working Paper, London School of Economics (2016)
8. Engle, R.: New frontiers for ARCH models. J. Appl. Econometrics **17**, 425–446 (2002). https://doi.org/10.1002/jae.683
9. Fama, E.: Efficient capital markets: a review of theory and empirical work. J. Finan. **25**, 383–417 (1970). https://doi.org/10.1111/j.1540-6261.1970.tb00518.x
10. Glosten, L.R., Jagannathan, R., Runkle, D.E.: On the relation between the expected value and the volatility of the nominal excess return on stocks. J. Finan. **48**, 1779–1801 (1993)
11. Guidolin, M., Orlov, A.G., Pedio, M.: How good can heuristic-based forecasts be? A comparative performance of econometric and heuristic models for UK and US asset returns. Quant. Finan. **18**, 139–169 (2017). https://doi.org/10.1080/14697688.2017.1351619
12. Hansen, P.R., Lunde, A.: A forecast comparison of volatility models: does anything beat a GARCH (1, 1)? J. Appl. Econometrics **20**, 873–889 (2005). https://doi.org/10.2139/ssrn.264571
13. Lutz, C.: Unconventional monetary policy, media uncertainty, and expected stock market volatility. Working Paper, Copenhagen Business School (2014)
14. Raddant, M., Karimi, F.: Cascades in real interbank markets. Comput. Econ. **47**, 49–66 (2016)
15. Shi, Y., Ho, K. Y., Liu, W. M. R.: Foreign exchange volatility, media coverage, and the mixture of distributions hypothesis: evidence from the Chinese renminbi currency. In: 19th International Congress on Modelling and Simulation, Perth, Australia (2011)
16. Tetlock, P.C.: Giving content to investor sentiment: the role of media in the stock market. J. Finan. **62**, 1139–1168 (2007). https://doi.org/10.2139/ssrn.685145

Importance Weighting of Diagnostic Trouble Codes for Anomaly Detection

Diogo R. Ferreira[1(✉)], Teresa Scholz[2], and Rune Prytz[2]

[1] IST, University of Lisbon, Lisbon, Portugal
`diogo.ferreira@tecnico.ulisboa.pt`
[2] Stratio Automotive, Lisbon, Portugal
`rune@stratioautomotive.com`

Abstract. Diagnostic Trouble Codes (DTCs) allow monitoring a wide range of fault conditions in heavy trucks. Ideally, a perfectly healthy vehicle should run without any active DTCs; in practice, vehicles often run with some active DTCs even though this does not pose a threat to their normal operation. When a DTC becomes active, it is therefore unclear whether it should be ignored or considered as a serious issue. Recent approaches in machine learning, such as training Variational Autoencoders (VAEs) for anomaly detection, do not help in this respect, for a number of reasons that we discuss based on actual experiments. In particular, a VAE tends to learn that a frequently active DTC is of no importance, when in fact it should not be dismissed completely; instead, such DTC should be assigned a relative weight that is smaller but still non-negligible when compared to other, more serious DTCs. In this work, we present an approach to measure the relative weights of active DTCs in a way that allows the analyst to prioritize them and focus on the most important ones. The approach is based on the concept of binary cross-entropy, and finds application not only in the analysis of DTCs from a single vehicle, but also in monitoring active DTCs across an entire fleet.

Keywords: Automotive industry · Anomaly detection · Variational Autoencoders · Binary cross-entropy

1 Introduction

In the automotive industry, Diagnostic Trouble Codes (DTCs) [1] are a standard way to monitor the health of a vehicle. This applies not only to automobiles that people use routinely for their transportation needs, but also to other classes of road vehicles, namely heavy-duty trucks.

Basically, a DTC is an alarm that is activated under certain operating conditions. For example, in a diesel engine it is important to maintain a certain fuel pressure. Since the fuel goes through several filters, pumps and injectors, a low fuel pressure may cause damage to these components and ultimately to the engine itself. Therefore, if the fuel pressure drops below a certain threshold, a special DTC for that purpose will become active.

© Springer Nature Switzerland AG 2020
G. Nicosia et al. (Eds.): LOD 2020, LNCS 12565, pp. 410–421, 2020.
https://doi.org/10.1007/978-3-030-64583-0_37

However, the fact that a DTC is active does not indicate, by itself, whether the issue is serious or not. For example, the fuel pressure may be momentarily low due to the ambient temperature, such as a cold start in an early morning with freezing temperatures. On the other hand, it could be caused by a fuel leak, and this would be much more serious, requiring repair.

An assumption that is often made in practice is that the importance of an issue correlates inversely with its frequency. For a vehicle that starts every morning in cold temperatures, the low fuel pressure DTC may be frequently active without representing a serious issue, whereas for another vehicle a single, unusual activation of the same DTC may require immediate attention.

This makes it possible to train machine learning models – such as Variational Autoencoders (VAEs) [2] – for anomaly detection based on data from a single vehicle, but it is difficult to generalize the results to other vehicles. In particular, training a model on a vehicle and then applying it to detect anomalies on another vehicle will not yield good results.

This problem is further exacerbated by the fact that companies have fleets comprising heterogeneous vehicles (e.g. trucks from different brands, different models, different years, etc.). When monitoring such a fleet, it becomes very difficult to determine which DTCs should be the focus of attention.

In this work, we apply anomaly detection to prioritize DTCs according to their importance, where this importance must be learned from data. The method should be able to determine the importance of each DTC not only in the context of a single vehicle, but also across multiple vehicles in order to assist the analyst in identifying which vehicles and issues should be at the top of the list when it comes to potential anomalies. An initial attempt at achieving this with a VAE eventually leads us to define a proper metric for importance weighting of DTCs.

2   Analysis of a Sample Vehicle

Figure 1 shows the DTCs for one particular vehicle in a fleet comprising about 40 trucks. The DTCs are being plotted over the period of one year, with a 60-s resolution. This particular vehicle had a relatively large number of DTCs that were active at least once during that period.

In Fig. 1 there are several different kinds of behaviors, from DTCs that appear to be randomly triggered on and off in quick succession, to DTCs that are active only once for a very brief period of time (short spikes). However, the most interesting DTCs are those that, being usually inactive, become active and stay that way for an extended period of time, such as several weeks or even a few months. This is what happens around May 2019, when the vehicle starts accumulating DTCs that were otherwise inactive and which, when considered altogether, point to one or more serious issues.

The repair history for this particular vehicle shows that it had a breakdown due to "overheating" (no further info provided) in mid-May. In Fig. 1, the fact that several active DTCs are being turned off (or reset) simultaneously in mid-May is consistent with a repair at that point in time.

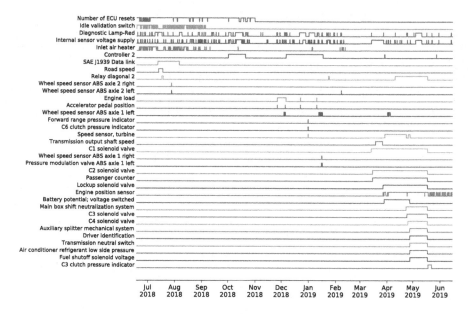

Fig. 1. DTCs for a sample vehicle over a period of one year.

An anomaly detection model trained on the first half of this one-year period should be able to pick up this anomaly in the second half. However, in the absence of labeled data, it might be that this anomalous behavior ends up in the training data for the model, which will then learn it as usual behavior and fail to recognize it as an anomaly in the future. The following section illustrates this point by means of an actual experiment.

3 Anomaly Detection with VAEs

Variational Autoencoders (VAEs) have been extensively used for anomaly detection, usually in combination with other methods, such as recurrent neural networks, to capture the evolution of time series data, e.g. [3–7]. In its basic form, anomaly detection relies on the ability of a VAE to reconstruct a given input signal [8]. When the VAE struggles to reproduce the input signal, this is an indication that it has not seen such behavior before, and therefore such behavior is classified as an anomaly.

While any neural network that produces an output of the same type and shape as its input can considered to be an autoencoder, the variational autoencoder has the distinctive feature of encoding the input as a probability distribution, and then generating the output by sampling from such distribution. As a result, the output is a stochastic approximation of the input, but such approximation should be sufficiently good for the training data that the model has already seen; it is only for data that the model has not seen before that the reconstruction error will be large.

Figure 2 shows the general structure of a VAE, illustrating the fact that the encoder learns the parameters of a latent distribution (μ and σ of a bivariate normal distribution, in this case), while the decoder produces the output based on samples drawn from that distribution.

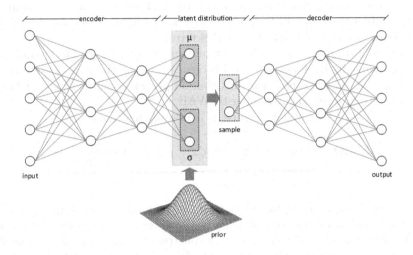

Fig. 2. General structure of a variational autoencoder.

Furthermore, the parameters of the latent distribution are not to be learned freely, as this could possibly lead to overfitting to the training data. Instead, a prior distribution is usually imposed (a standard bivariate normal, in this case), which provides some (tunable) degree of regularization.

The VAE is trained by minimizing both the reconstruction error and the Kullback-Leibler divergence of the latent distribution with respect to the prior, which is equivalent to maximizing the well-know evidence lower bound (ELBO) [9].

3.1 Formal Background

Let x be the input that the VAE is trying to reconstruct, and let z be the latent representation of that input, which the encoder is trying to learn by adjusting its network weights θ. In effect, what the encoder is learning is to approximate the posterior distribution $p(z|x)$ with a variational approximation $q_\theta(z|x)$, where θ are the variational parameters of such approximation.

On the other hand, the decoder is learning to generate x from the latent representation z by adjusting its network weights ϕ. In other words, the decoder is learning the likelihood distribution of data $p(x|z)$ with a parametrized approximation that is usually denoted by $p_\phi(x|z)$.

With some mathematical manipulations, it is possible to show that:

$$\log p(x) = \underbrace{\int q_\theta(z|x) \log \frac{p(x,z)}{q_\theta(z|x)} dz}_{\text{ELBO}} + \underbrace{\int q_\theta(z|x) \log \frac{q_\theta(z|x)}{p(z|x)} dz}_{D_{\mathrm{KL}}(q_\theta(z|x)\|p(z|x))} \qquad (1)$$

Since $\log p(x)$ is constant w.r.t. $q_\theta(z|x)$, and since the Kullback-Leibler divergence in the second term is non-negative, the first term on the right-hand side is effectively a lower bound for the log-evidence $\log p(x)$. Hence the designation of evidence lower bound (ELBO) for this term.

Furthermore, since $\log p(x)$ is constant, maximizing the ELBO is equivalent to minimizing the Kullback-Leibler divergence between the variational approximation $q_\theta(z|x)$ and the posterior distribution $p(z|x)$, which is exactly what we would like to achieve when training the VAE.

With some further manipulations, it can be shown that:

$$\text{ELBO} = \underbrace{\int q_\theta(z|x) \log p(x|z) dz}_{\mathbb{E}_{q_\theta(z|x)}[\log p(x|z)]} - \underbrace{\int q_\theta(z|x) \log \frac{q_\theta(z|x)}{p(z)} dz}_{D_{\mathrm{KL}}(q_\theta(z|x)\|p(z))} \qquad (2)$$

Therefore, maximizing the ELBO can be achieved by maximizing the first term and minimizing the second one. The first term is maximized by training the decoder to learn $p(x|z)$ in order to reconstruct x from z as good as possible, while the second term is minimized by training the encoder to keep the latent distribution as close as possible to a given prior $p(z)$.

In this context, the second term in Eq. (2) can be viewed as a regularization term that precludes the latent distribution from overfitting the training data. A simple choice for the prior distribution $p(z)$ is, for example, a multivariate standard normal with an identity covariance matrix.

3.2 Implementation

To detect anomalies in DTC data, we implemented a VAE to be applied to DTC signals such as those depicted in Fig. 1. For this sample vehicle, we found 62 DTCs that were active at least once over that one-year period. Therefore, we designed a VAE with layer dimensions that were commensurate with the dimensionality of the input data. A sequence of dense layers encodes the input data into a 4-dimensional latent distribution, and another sequence of dense layers decodes the output from a sample of that distribution.

An implementation model for the VAE is shown in Fig. 3, and it uses both deterministic and probabilistic layers. The deterministic layers (e.g. Dense) come from the base TensorFlow library,[1] while the probabilistic layers (such as MultivariateNormalTriL) come from TensorFlow Probability.[2] Both kinds of layers can be used seamlessly in the same model.

---

[1] https://github.com/tensorflow/tensorflow.
[2] https://github.com/tensorflow/probability.

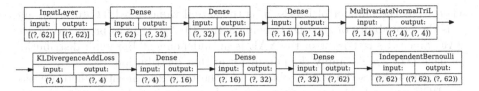

Fig. 3. Implementation of a VAE with deterministic and probabilistic layers.

In particular, `MultivariateNormalTriL` represents a multivariate normal distribution that is parametrized by a mean vector and the lower triangular part of the covariance matrix (since the matrix is symmetric, there is no need to specify the upper part). For an n-dimensional normal distribution, the total number of parameters is $n+n(n+1)/2 = 14$ for $n = 4$.

An interesting feature of `MultivariateNormalTriL` (and this applies to other layers representing probability distributions) is that it receives as input a tensor with the distribution parameters and produces as ouput a sample drawn from a distribution with those parameters.

The same principle applies to the very last layer in Fig. 3, where we use `IndependentBernoulli` to generate a binary random variable for each DTC. Rather than outputting the probability of a DTC being active, we sample from a Bernoulli distribution in order to produce an actual value (0 or 1). However, the main advantage of using a probabilistic layer here is that we can train the VAE by maximizing the log probability of its output (or by minimizing the negative of such log probability, which is the same).

A third probabilistic layer, `KLDivergenceAddLoss`, is a pass-through layer that adds a Kullback-Leibler divergence penalty to the loss function. This penalty represents the second term in Eq. (2) and is computed with respect to a prior distribution which, in our case, is a 4-dimensional independent normal.

The remaining layers in Fig. 3 are deterministic dense layers to implement the compression done by the encoder (top row, from 62 to 14 units) and the decompression by the decoder (bottom row, from 4 back to 62 units). The question mark (?) is a placeholder for the batch size to be used during training.

3.3   Anomaly Detection

To test the VAE in an anomaly detection setting, we split the data (Fig. 1) in two halves, where one half is used for training and the other is used for testing. In a first experiment, we train on Jul–Dec 2018 and test on Jan–Jun 2019. In a second experiment, we do the opposite by swapping the train and test sets.

To provide a single measure for anomaly detection, we use the binary cross-entropy (BCE) to assess the accuracy of the VAE in reconstructing the DTC signals at each point in time. Specifically, the BCE is given by:

$$\text{BCE} = -\frac{1}{N} \sum_{i=1}^{N} [y_i \log p_i + (1 - y_i) \log(1 - p_i)] \tag{3}$$

where N is the number of DTCs, y_i is the actual value of a DTC (0 or 1), and p_i is the probability of such DTC being active ($0 < p_i < 1$) as predicted by the VAE. Since the last layer of the VAE is an independent Bernoulli distribution for each DTC, p_i is simply the parameter of that distribution.

Figure 4 shows the results obtained in the two experiments, where we plot the BCE across both the training set and the test set. In the first plot, we see that the VAE has learned to reproduce the training data almost perfectly, regardless of what it contains. As for the test data, the VAE clearly recognizes the anomalous behavior around May 2019 (cf. Fig. 1).

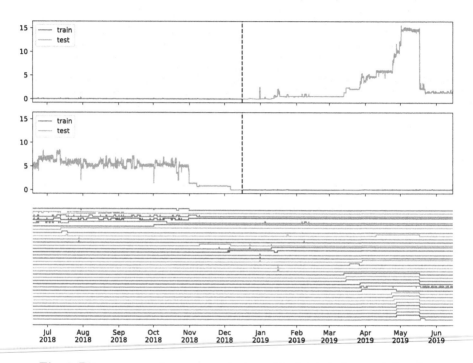

Fig. 4. Binary cross-entropy across train and test sets in two experiments.

In the second plot, where the training set is Jan–Jun 2019, again the VAE learns to reproduce the training data almost perfectly, regardless of the anomalous behavior that it contains. It is then in Jul–Dec 2018 that the VAE recognizes some unusual behavior, especially at the beginning of that period.

This points to a fundamental issue with this approach. Since the VAE learns to reproduce whatever training data it is given, any anomalous behavior must be labeled in order to split the training and test data in a way that the training set does not contain the kind of anomalies that need to be detected.

Furthermore, an analysis of multiple vehicles reveals that a model trained on one vehicle is not directly transferable to other vehicles, as they often display significantly different behavior in terms of active DTCs, even for vehicles with the same specifications (e.g. same brand, model and year). This makes it difficult to apply anomaly detection across a heterogeneous fleet of vehicles.

4   Importance Weighting of DTCs

In the previous section, we used binary cross-entropy (BCE) as an anomaly score, and with this scoring function we showed that a VAE can learn to reproduce whatever training data it is provided, even if the training data contains actual anomalies. In this section, we leave the VAE aside, but the concept of BCE is still useful to come up with an anomaly score that can be applied directly to DTC data, even without training a model.

According to Eq. (3), the BCE will be low either when $y_i = 1$ and $p_i \rightarrow 1$ (the second term under the summation will be zero) or when $y_i = 0$ and $p_i \rightarrow 0$ (the first term will be zero). Conversely, the BCE will be high either when $y_i = 1$ and $p_i \rightarrow 0$, or when $y_i = 0$ and $p_i \rightarrow 1$. In other words, the BCE will be low whenever y_i and p_i agree, and it will be high whenever y_i and p_i disagree.

However, when applied to DTC data, the two situations $y_i = 0$ and $y_i = 1$ should be treated differently. It is correct to say that when $y_i = 0$ (DTC is off) and $p_i \rightarrow 0$, the anomaly score should be low; as it is correct to say that when $y_i = 1$ (DTC is on) and $p_i \rightarrow 0$, the score should be high, because the DTC is active and this could point to an anomaly, regardless of the value of p_i.

Now, the BCE specifies that when $y_i = 1$ and $p_i \rightarrow 1$ the anomaly score is low, which should not happen, because an active DTC always points to a potential issue. Similarly, when $y_i = 0$ and $p_i \rightarrow 1$ the anomaly score is high, when it should not be, because the DTC is inactive, regardless of the value of p_i.

If a DTC is active ($y_i = 1$) this should always point to an anomaly, albeit a small anomaly if $p_i \rightarrow 1$, and a large anomaly if $p_i \rightarrow 0$. On the other hand, if a DTC is inactive ($y_i = 0$) this should indicate no anomaly, regardless of p_i.

Therefore, we introduce a different anomaly score, which is obtained by removing the second term in Eq. (3), and which we denote by loss L:

$$L = -\frac{1}{N} \sum_{i=1}^{N} y_i \log p_i \qquad (4)$$

According to this formula, if a DTC is inactive ($y_i = 0$) then the corresponding term is zero, and the DTC does not contribute to the anomaly score. On the other hand, if a DTC is active ($y_i = 1$) then the corresponding term is non-zero, and the DTC contributes with a weight of ($-\log p_i$) to the anomaly score.

If $p_i \rightarrow 0$ then the weight of the DTC can become very large, but in practice this does not occur, for two main reasons. First, in the extreme case when $p_i = 0$ the weight of a DTC would become infinite, but such DTC will never turn on, so $y_i = 0$. Second, a DTC that is active for only one minute in an entire year has a weight of $\left(-\log \frac{1}{365 \times 24 \times 60}\right) \approx 13.2$, which is of the same order of magnitude as the maximum values observed in the first plot of Fig. 4.

To avoid training a model, we take the probability p_i to be the fraction of time that a DTC was active during a certain period of time. For example, from the data in Fig. 1 it is possible to compute p_i for each DTC over the period of one year. We then plug these p_i values together with the actual data y_i into Eq. (4) to compute the anomaly score across the same time span.

Figure 5 shows the results, where the anomaly score is plotted at the top with the DTC signals right below, for comparison.

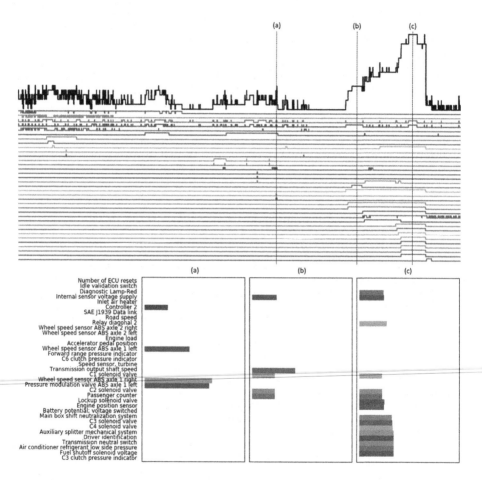

Fig. 5. Importance weights of DTCs at three points in time (a), (b) and (c).

In addition, by means of colored horizontal bars, Fig. 5 illustrates the contribution of each DTC to the anomaly score at three points in time: point (a) represents a brief spike in a few DTCs; point (b) represents the beginning of a long period of anomalous behavior; and point (c) represents the flat top where the anomaly score reaches its peak.

The weight $(-\log p_i)$ can therefore be regarded as a measure of the relative importance of each active DTC with respect to the remaining DTCs. In the midst of several active DTCs, the ones with a larger weight will point to more unusual events, as the examples in Fig. 5 illustrate.

This importance measure can be used not only across DTCs of a single vehicle, but also across DTCs from multiple vehicles simultaneously, provided that the p_i values for all DTCs are evaluated based the same period of time for all vehicles. This makes it possible to prioritize DTCs across an entire fleet and identify vehicles that possibly require immediate attention.

Being able to quickly identify vehicles with issues or issues within vehicles can significantly reduce the amount of data that an analyst must go through in order to find or confirm those issues. The approach can also be used as a first step towards building labeled datasets for training other models.

5   Related Work

The analysis of DTC data is central to many anomaly detection approaches in the automotive industry. For example, in [10] the authors propose a framework where DTC data plays two different roles. First, they focus on the detection of anomalous repairs by training a model to learn the appropriate repairs for given issues, where issues are characterized by DTC data. This requires the availability of both repair data and DTC data for training purposes. Second, they use decision trees to infer the operating conditions under which DTCs are activated; this can be used to check if a DTC is being activated according to the correct preconditions.

In more recent works, the focus has been shifting towards the use of machine learning for predictive maintenance. For example, in [11] the authors use random forests to predict which components are about to fail based on DTC data. Again, not only the DTC data but also the repair data are needed in order to build a labeled dataset for training. An interesting aspect of this work is that the authors extract features from DTC data instead of working with DTCs directly. Most of the extracted features are based on aggregated counts within a certain observation window. The authors then go into a detailed study of how the size of this window affects the prediction results.

The use of unsupervised learning techniques (with at least part of the data being unlabeled) has also been gaining ground. For example, in [12] the author focuses on DTC activations which occur in rare operating conditions (called operating modes). Using a clustering algorithm (DBSCAN) and also with the aid of domain experts, the author identifies clusters of operating modes where the DTC activations predominantly occur. The outliers to those clusters are labeled as anomalies. Then, using classifiers based on support vector machines (SVMs), it becomes possible to determine whether a new DTC activation takes place under a typical or atypical operating mode. The underlying assumption is that if the DTC was activated in a rare operating mode, this may point to an uncommon fault which could take longer or be more expensive to repair.

A more fully unsupervised approach is adopted in [13], where the authors use different kinds of models, including autoencoders, for fault detection across a fleet of city buses. A separate autoencoder is used to reconstruct the raw signals from each vehicle. Also, the autoencoder is trained on a chunk of data (e.g. on a daily basis) so there are several trained models for the same vehicle at different points in time. All of these models from each and every vehicle are then compared by assessing their performance on a reference dataset which, ideally, should contain only typical behavior. The models that deviate the most, not only from the reference dataset but also from a "consensus" that is obtained from all vehicles, point to potential faults. In their experiments, the authors have shown that these deviations correlate well with the repair history.

The relation of this brief survey of selected works to the present paper can be described as follows:

- When using supervised learning for anomaly detection, as in [10], this requires the availability of labeled data, for example repair data that can be used to label the anomalies in DTC data. In some cases, such as [13], the repair history is available but is not used for training; rather, it is used for validating the results. This is the same approach that we adopted here.
- While it is possible to apply feature extraction over the training data, as in [11], deep learning models such as convolutional networks and autoencoders should be able to perform feature extraction themselves, and this was one of the reasons for using autoencoders alongside with linear models in [13]. This was also our motivation for experimenting with VAEs.
- Anomaly detection can be performed at different levels of abstraction, either from raw signals [13], or from DTC data [10], or from feature engineering over DTC data [11]. Here we have found DTC data to be a good compromise between overly detailed raw data and overly simplistic features, with the additional benefit of being able to work with binary distributions.
- Eventually, we introduced a simple anomaly score loosely inspired by the concept of binary cross-entropy, which provided similar results to a VAE but without having to train such model. This could be the first stage in a multi-step approach towards obtaining labeled data for training classifiers in a subsequent stage. Such multi-step approach would be somewhat similar to [12], where our anomaly score could be used to label the outliers.

6   Conclusion

In the automotive industry, there are large amounts of data that can be used to train different kinds of models. One of the major challenges is making use of unlabeled data for anomaly detection. While a VAE is certainly capable of capturing normal behavior and signaling anomalies, some care must be taken to avoid including anomalies in the training data. In practice, this is harder than it seems because even healthy vehicles often run with some issues.

In addition, the problem of performing anomaly detection across a heterogeneous fleet does not have a one-fits-all solution, since each vehicle requires a separately trained model. In this work, we proposed an anomaly score that was based on our previous experiments with VAEs. As is often the case, a simpler solution turns out to be the more effective in practice. The anomaly score and importance weighting of DTCs are being implemented in a monitoring platform that allows an operator to prioritize issues across their fleet.

References

1. Walter, E., Walter, R.: Diagnostic trouble codes (DTCs). In: Data Acquisition from Light-Duty Vehicles Using OBD and CAN, pp. 97–108. SAE International (2018)
2. Kingma, D.P., Welling, M.: Auto-encoding variational Bayes. In: International Conference on Learning Representations (ICLR) (2014)
3. Suh, S., Chae, D.H., Kang, H., Choi, S.: Echo-state conditional variational autoencoder for anomaly detection. In: International Joint Conference on Neural Networks (IJCNN), pp. 1015–1022, July 2016
4. Guo, Y., Liao, W., Wang, Q., Yu, L., Ji, T., Li, P.: Multidimensional time series anomaly detection: a GRU-based gaussian mixture variational autoencoder approach. In: 10th Asian Conference on Machine Learning, PMLR, vol. 95, pp. 97–112 (2018)
5. Park, D., Hoshi, Y., Kemp, C.C.: A multimodal anomaly detector for robot-assisted feeding using an LSTM-based variational autoencoder. IEEE Robot. Autom. Lett. **3**(3), 1544–1551 (2018)
6. Kawachi, Y., Koizumi, Y., Harada, N.: Complementary set variational autoencoder for supervised anomaly detection. In: IEEE International Conference on Acoustics, Speech and Signal Processing (ICASSP), pp. 2366–2370 (2018)
7. Pereira, J., Silveira, M.: Unsupervised anomaly detection in energy time series data using variational recurrent autoencoders with attention. In: 17th IEEE International Conference on Machine Learning and Applications (ICMLA), pp. 1275–1282 (2018)
8. An, J., Cho, S.: Variational autoencoder based anomaly detection using reconstruction probability. Technical report, SNU Data Mining Center (2015)
9. Hoffman, M.D., Johnson, M.J.: ELBO surgery: yet another way to carve up the variational evidence lower bound. In: Advances in Approximate Bayesian Inference (NIPS Workshop) (2016)
10. Singh, S., Pinion, C., Subramania, H.S.: Data-driven framework for detecting anomalies in field failure data. In: IEEE Aerospace Conference, pp. 1–14 (2011)
11. Pirasteh, P., et al.: Interactive feature extraction for diagnostic trouble codes in predictive maintenance: a case study from automotive domain. In: Proceedings of the Workshop on Interactive Data Mining (WIDM). ACM (2019)
12. Theissler, A.: Multi-class novelty detection in diagnostic trouble codes from repair shops. In: 15th IEEE International Conference on Industrial Informatics (INDIN), pp. 1043–1049 (2017)
13. Rögnvaldsson, T., Nowaczyk, S., Byttner, S., Prytz, R., Svensson, M.: Self-monitoring for maintenance of vehicle fleets. Data Mining Knowl. Disc. **32**(2), 344–384 (2017)

Identifying Key miRNA–mRNA Regulatory Modules in Cancer Using Sparse Multivariate Factor Regression

Milad Mokhtaridoost[1] and Mehmet Gönen[2,3,4(✉)] [ID]

[1] Graduate School of Sciences and Engineering,
Koç University, 34450 İstanbul, Turkey
`mmokhtaridoost15@ku.edu.tr`
[2] Department of Industrial Engineering, Koç University, 34450 İstanbul, Turkey
`mehmetgonen@ku.edu.tr`
[3] School of Medicine, Koç University, 34450 İstanbul, Turkey
[4] Department of Biomedical Engineering, School of Medicine,
Oregon Health & Science University, Portland, OR 97239, USA

Abstract. The interactions between microRNAs (miRNAs) and messenger RNAs (mRNAs) are known to have a major effect on the formation and progression of cancer. In this study, we identified regulatory modules of 32 cancer types using a sparse multivariate factor regression model on matched miRNA and mRNA expression profiles of more than 9,000 primary tumors. We used an algorithm that decomposes the coefficient matrix into two low-rank matrices with separate sparsity-inducing penalty terms on each. The first matrix linearly transforms the predictors to a set of latent factors, and the second one regresses the responses using these factors. Our solution significantly outperformed another decomposition-based approach in terms of normalized root mean squared error in all 32 cohorts. We demonstrated the biological relevance of our results by performing survival and gene set enrichment analyses. The validation of overall results indicated that our solution is highly efficient for identifying key miRNA–mRNA regulatory modules.

Keywords: Cancer biology · Machine learning · MicroRNAs · Messenger RNAs · Optimization · Regulatory modules

1 Introduction

Cancer has become one of the most important causes of death worldwide. Despite the significant improvements in cancer therapies, the molecular mechanisms of cancer have not yet been fully understood [19]. Messenger RNAs (mRNAs) are a large family of RNA molecules that carry genetic information from DNA to the ribosome, where they specify the amino acid sequence of the protein products of gene expression. A microRNA (miRNA) is a small non-coding RNA molecule found in plants, animals, and some viruses that functions in mRNA silencing and

© Springer Nature Switzerland AG 2020
G. Nicosia et al. (Eds.): LOD 2020, LNCS 12565, pp. 422–433, 2020.
https://doi.org/10.1007/978-3-030-64583-0_38

post-transcriptional regulation of gene expression through translational repression or mRNA degradation. The detection of miRNAs and considerable advances in the characterization of this gene group have shown that these small non-coding RNAs are an important family of regulatory RNAs [4]. miRNAs in the genomic region are participating in the regulation of pathways that are involved in the formation and progression of many types of tumors [2]. As a result, understanding RNA regulation in cancer has attracted a lot of attention.

A considerable amount of literature in large-scale genomic studies has focused on multivariate regression algorithms to model complex relationships (e.g. miRNA–mRNA regulatory modules). In high-dimensional regression applications where the numbers of predictors and/or responses far exceed the sample size, it is standard to impose some low-dimensional structure on the coefficient matrix to decrease the effect of the curse of high dimensionality.

1.1   Literature Review

To impose low-dimensional structure, several researches have used group Lasso or block-sparse multiple regression by imposing mixed ℓ_1/ℓ_γ norms ($\gamma > 1$) on the coefficient matrix. There are some particular examples for ℓ_1/ℓ_γ norm family, namely, ℓ_1/ℓ_∞ [11] and ℓ_1/ℓ_2 [12]. Another approach is to impose a constraint on the rank of the coefficient matrix instead of its entries, which is known as linear factor regression. Linear factor regression methods usually define a penalty function on the rank of the coefficient matrix, leading to low-rank solutions [14]. Some popular methods including principal components regression can be cast into linear factor regression formulation [9]. Several studies attempted to write the coefficient matrix as the multiplication of two smaller matrices [1,6], which can be interpreted as projecting the input data matrix into a low-dimensional space using the first matrix and performing linear regression in this projected space using the second matrix. Thus, the ranks of these two matrices are smaller than or equal to the dimensionality of the projected space, and the rank of the coefficient matrix cannot be greater than this value.

miRNAs have critical roles in gene regulation. In identifying miRNA–mRNA regulatory modules, similar to many high dimensional biology applications such as studies involving gene expression data, the number of miRNAs (i.e. predictors) and mRNAs (i.e. responses) are considerably greater than the sample size. Moreover, an acceptable assumption has been established in this sense that relationship between miRNAs and mRNAs is sparse (i.e. a small subset of miRNAs affect a small subset of mRNAs). To fit models on such datasets, regularized or penalized methods are needed, and feature selection or feature extraction is often the most important aspect of the analysis. To find the unknown complex interactions between miRNAs and mRNAs using expression profiles, a large and growing body of literature was developed using data mining approaches.

Several studies tried to identify miRNA–mRNA regulatory modules using miRNA and mRNA expression profiles [8,10,15–18]. [8] formulated an iterative algorithm based on singular value decomposition (SVD) with a thresholding step named T-SVD regression. [10] presented a two-stage method to identify miRNA–mRNA regulatory modules. In the first step, they extended the single output sparse group Lasso regression model towards multi-output regression scenario to learn the regulatory network between miRNAs and mRNAs. In the second step, to identify regulatory modules from the learned network, they proposed an ℓ_0-penalized SVD model. They reported empirical results on a breast cancer cohort. [15] introduced a computational method to find the miRNAs associated with each mRNA using rule learning. [17] proposed a machine learning technique to integrate the three different dataset sources, namely, gene–gene and DNA–gene interaction networks, and miRNA and mRNA expression profiles in an integrated non-negative matrix factorization model. They applied the proposed method to a dataset of ovarian cancer. To integrate other data sources (i.e. methylation) into miRNA–mRNA regulatory module identification, [16] and [18] applied linear regression and joint matrix factorization models, respectively.

1.2   Our Contributions

Among the aforementioned computational studies about inferring miRNA–mRNA regulatory modules, there is not a single study with a comprehensive set of experiments on several diseases. In this study, we implemented a multivariate regression method, which is called the Sparse Multivariate Factor Regression (SMFR) [5], to identify miRNA–mRNA regulatory modules from matched expression profiles. We tested SMFR, to the best of our knowledge, on the largest collection of diseases by performing computational experiments on 32 cancer cohorts.

To show the biological relevance of the identified modules, we performed survival and gene set enrichment analyses. To validate the identified modules, we applied two strategies: (i) We checked the literature to find supporting evidence for the top five miRNAs in the corresponding cancer type. (ii) We checked all identified miRNA–mRNA interactions to find whether they were reported in the miRTArBase database [3]. The results showed that a large portion of the identified modules have supporting evidence in the existing literature. To see the predictive performance of SMFR, we compared the normalized root mean squared error (NRMSE) of SMFR with another decomposition-based algorithm (i.e. T-SVD regression). The average NRMSE values of both algorithms showed that SMFR achieved considerably lower error in comparison to T-SVD regression. Our results showed that our computational approach is highly efficient in identifying key miRNA–mRNA regulatory modules.

2   Problem Definition

We addressed the problem of finding cancer-specific miRNA–mRNA regulatory modules from matched miRNA and mRNA expression profiles of tumors. We formulated this problem as a linear regression model that finds mRNA expression levels using miRNA expression levels. For this problem, we are given a training set that contains matched miRNA and mRNA expression profiles of N tumors.

The training set can be represented as $\mathcal{D} = \{(\boldsymbol{x}_i, \boldsymbol{y}_i)\}_{i=1}^N$, where $\boldsymbol{x}_i \in \mathbb{R}^D$ and $\boldsymbol{y}_i \in \mathbb{R}^T$ denote the miRNA and mRNA expression profiles of tumor i, respectively. The miRNA and mRNA expression profiles of tumors can be shown as $\mathbf{X}$ and $\mathbf{Y}$, respectively, which are of size $N \times D$ and $N \times T$.

3 Method

To find miRNA–mRNA regulatory modules using $\mathcal{D}$, we formulated a linear regression problem under the assumption that the columns of $\mathbf{X}$ and $\mathbf{Y}$ are centered (i.e. zero mean columns) and normalized (i.e. unit standard deviation columns) as $\mathbf{Y} = \mathbf{XW} + \mathbf{E}$, where $\mathbf{W} \in \mathbb{R}^{D \times T}$ is the matrix of regression coefficients, and $\mathbf{E} \in \mathbb{R}^{N \times T}$ is the matrix of error terms. The error terms in $\mathbf{E}$ are assumed to be independent and identically distributed Gaussian random variables with zero mean and σ^2 variance.

Without any explicit constraint on $\mathbf{W}$, this problem is equivalent to solving T separate univariate regression problems. We assumed that $\mathbf{W}$ has a low-rank structure, so that the regression equation becomes $\mathbf{Y} = \mathbf{XW}_\mathcal{X}\mathbf{W}_\mathcal{Y} + \mathbf{E}$, where $\mathbf{W}_\mathcal{X} \in \mathbb{R}^{D \times R}$ projects D-dimensional miRNA profiles into an R-dimensional space, and $\mathbf{W}_\mathcal{Y} \in \mathbb{R}^{R \times T}$ performs linear regression in this projected space.

This low-rank assumption has two main motivations: (i) We can reduce the number of regression coefficients that we need to learn from $(D \times T)$ to $(D \times R + R \times T)$. Considering that the numbers of miRNAs and mRNAs in our experiments are in the orders of hundreds and thousands, respectively, this reduction is quite significant when R is in the order of tens. (ii) We can write this low-rank approximation as the summation of rank-one matrices, where each rank-one matrix can be used as a distinct miRNA–mRNA regulatory module. For example, the first column of $\mathbf{W}_\mathcal{X}$ and the first row of $\mathbf{W}_\mathcal{Y}$ correspond to the first miRNA–mRNA regulatory module. This low-rank decomposition approach is illustrated in Fig. 1.

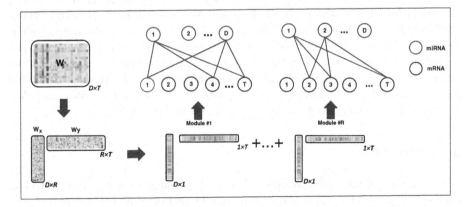

Fig. 1. Overview of our low-rank decomposition approach for identifying miRNA–mRNA regulatory modules.

However, this low-rank structure does not guarantee sparsity in the regression coefficients. We know that each miRNA–mRNA regulatory module contains a small list of miRNAs that regulate the expression levels of a limited number of mRNAs. To be able to extract knowledge from $\mathbf{W}_\mathcal{X}$ and $\mathbf{W}_\mathcal{Y}$ matrices, we need to enforce sparsity on their columns and rows, respectively. We used SMFR to be able to enforce these sparsity constraints and to learn the number of regulatory modules (i.e. R) during regression modeling [5].

3.1 Sparse Multivariate Factor Regression

[5] suggested to use an elastic net penalty on $\mathbf{W}_\mathcal{X}$ and an ℓ_1 penalty on $\mathbf{W}_\mathcal{Y}$, thus the general formulation for SMFR can be written as

$$\text{minimize} \quad \frac{1}{2}\|\mathbf{Y} - \mathbf{X}\mathbf{W}_\mathcal{X}\mathbf{W}_\mathcal{Y}\|_F^2 + \overbrace{\lambda_1\|\mathbf{W}_\mathcal{X}\|_{1,1} + \lambda_2\|\mathbf{W}_\mathcal{X}\|_F^2}^{\mathcal{L}_\mathcal{X}(\mathbf{W}_\mathcal{X})} + \overbrace{\lambda_3\|\mathbf{W}_\mathcal{Y}\|_{1,1}}^{\mathcal{L}_\mathcal{Y}(\mathbf{W}_\mathcal{Y})}$$

$$\text{with respect to } \mathbf{W}_\mathcal{X} \in \mathbb{R}^{D \times R}, \mathbf{W}_\mathcal{Y} \in \mathbb{R}^{R \times T},$$

(1)

where $\|\cdot\|_F$ and $\|\cdot\|_{1,1}$ indicate Frobenius and element-wise ℓ_1 norms, respectively. $\mathcal{L}_\mathcal{X}(\mathbf{W}_\mathcal{X})$ and $\mathcal{L}_\mathcal{Y}(\mathbf{W}_\mathcal{Y})$ are the regularization functions defined on $\mathbf{W}_\mathcal{X}$ and $\mathbf{W}_\mathcal{Y}$, respectively, to restrict the search space and to enforce some structure on the coefficients such as sparsity. $\lambda_1 \in \mathbb{R}_+$ and $\lambda_2 \in \mathbb{R}_+$ are the coefficients of the elastic net penalty on $\mathbf{W}_\mathcal{X}$, and $\lambda_3 \in \mathbb{R}_+$ is the coefficient of the ℓ_1 penalty on $\mathbf{W}_\mathcal{Y}$. Instead of determining the dimensionality of the projected space using cross-validation, they developed an iterative algorithm to obtain $\mathbf{W}_\mathcal{X}$ and $\mathbf{W}_\mathcal{Y}$ in full rank. By doing so, the columns of $\mathbf{W}_\mathcal{X}$ and the rows of $\mathbf{W}_\mathcal{Y}$ would be linearly independent, leading to better interpretation at the end.

The algorithm starts with an initial upper bound on the rank (i.e. $R = R_U$) and solves the optimization problem (1) using an alternative strategy since the objective function is not jointly convex with respect to $\mathbf{W}_\mathcal{X}$ and $\mathbf{W}_\mathcal{Y}$, but it is convex with respect to $\mathbf{W}_\mathcal{X}$ if we fix $\mathbf{W}_\mathcal{Y}$ or vice versa. After convergence, the algorithm checks whether both $\mathbf{W}_\mathcal{X}$ and $\mathbf{W}_\mathcal{Y}$ are full rank. If both of them are not full rank, they reduce R parameter by one and solve the optimization problem (1) again. To update $\mathbf{W}_\mathcal{X}$ and $\mathbf{W}_\mathcal{Y}$ in each iteration, gradient descent algorithm was used, and Prox-Linear update with extrapolation was applied to accelerate the convergence of gradient descent method [13]. At the end of this iterative procedure, the algorithm is guaranteed to have full rank $\mathbf{W}_\mathcal{X}$ and $\mathbf{W}_\mathcal{Y}$ matrices. This optimization strategy is detailed in Algorithm 1.

4   Materials

In this study, we used the data provided by the Cancer Genome Atlas (TCGA) consortium in the Genomic Data Commons Data Portal at https://portal.gdc. cancer.gov, which includes genomic characterizations and clinical information of more than 10,000 cancer patients for 33 different cancer types. All cohorts except glioblastoma multiforme (GBM) included both miRNA and mRNA expression profiles. That is why we performed computational experiments on 32 cancer types. We used the most recent data freeze (i.e. Data Release 24 – May 7, 2020) to obtain miRNA, mRNA, and clinical data files. We also did not include metastatic tumors in our computational analyses since their underlying biology would be very different than primary tumors.

Algorithm 1. SMFR via Alternating Minimization

Input: $\mathbf{X} \in \mathbb{R}^{N \times D}$, $\mathbf{Y} \in \mathbb{R}^{N \times T}$, $\lambda_1 \in \mathbb{R}_+$, $\lambda_2 \in \mathbb{R}_+$, $\lambda_3 \in \mathbb{R}_+$, $R_U \in \mathbb{Z}_{++}$
Output: $\mathbf{W}_{\mathcal{X}}^\star$, $\mathbf{W}_{\mathcal{Y}}^\star$, $R^\star$
$R \leftarrow R_U$
while true **do**
   $t \leftarrow 0$
   $\mathbf{W}_{\mathcal{X}}^{(t)} \leftarrow$ a random matrix from $\mathbb{R}^{D \times R}$
   while optimization problem (1) not converged **do**
      $\mathbf{W}_{\mathcal{Y}}^{(t+1)} \leftarrow$ update $\mathbf{W}_{\mathcal{Y}}$ with $\mathbf{W}_{\mathcal{X}}$ fixed at $\mathbf{W}_{\mathcal{X}}^{(t)}$
      $\mathbf{W}_{\mathcal{X}}^{(t+1)} \leftarrow$ update $\mathbf{W}_{\mathcal{X}}$ with $\mathbf{W}_{\mathcal{Y}}$ fixed at $\mathbf{W}_{\mathcal{Y}}^{(t+1)}$
      $t \leftarrow t + 1$
   end while
   if rank$(\mathbf{W}_{\mathcal{X}}^{(t+1)}) < R$ **or** rank$(\mathbf{W}_{\mathcal{Y}}^{(t+1)}) < R$ **then**
      $R \leftarrow R - 1$
   else
      break
   end if
end while
$\mathbf{W}_{\mathcal{X}}^\star \leftarrow \mathbf{W}_{\mathcal{X}}^{(t+1)}$
$\mathbf{W}_{\mathcal{Y}}^\star \leftarrow \mathbf{W}_{\mathcal{Y}}^{(t+1)}$
$R^\star \leftarrow R$

For miRNA and mRNA expression profiles, we downloaded BCGSC miRNA Profiling and HTSeq-FPKM files of all primary tumors preprocessed by the unified miRNAseq and RNA-Seq pipeline of TCGA consortium, respectively. At the end, we obtained 9,975 BCGSC miRNA Profiling files and 9,755 HTSeq-FPKM files in total.

For our computational experiments, we needed matched miRNA and mRNA expression profiles. Hence, we discarded primary tumors with just one expression profile from further analysis, which resulted in a total sample size of 9,541 for 32 cancer types. The details about the numbers of samples can be seen in Supplementary Table S1. Our Supplementary Material can be publicly accessed at https://drive.google.com/open?id=1uZi9f8rqv6F85sD8HvBQwIH7N-iFYeSl.

For each cancer type, we included miRNAs and mRNAs that were expressed in at least 50% of the tumors in the analysis. The original miRNA and mRNA profiles of tumors contain expression values for 1,881 miRNAs and 19,814 mRNAs. However, after filtering, these numbers reduced to around 600 miRNAs and 17,000 mRNAs on the average. The numbers of miRNAs and mRNAs (i.e. D and T) included in each cancer type can be seen in Supplementary Table S1.

After finding miRNA–mRNA regulatory modules, we performed survival analyses to evaluate the biological relevance of these modules. For this purpose, we also downloaded Clinical Supplement files of all patients. We then extracted their survival characteristics (i.e. days to last follow-up or days to last known alive for alive patients and days to death for dead patients).

To find the biological relevance of the results, we downloaded the Hallmark gene set collection from the Molecular Signatures Database (MSigDB), which is publicly available at https://www.gsea-msigdb.org/gsea/msigdb. This collection presents information about groups of genes which are similar or dependent in terms of their functions.

5   Results and Discussion

5.1   Experimental Settings

To determine reasonable values for the regularization parameters, i.e. λ_1, λ_2, and λ_3, we considered six possible values (i.e. 20, 40, 60, 80, 100, and 120) for each of them and trained the algorithm on the datasets that have less than 100 primary tumors with available miRNA and mRNA expression profiles (i.e. ACC, CHOL, DLBC, KICH, LAML, MESO, SKCM, UCS, and UVM). We first calculated NRMSE for all of them, then we selected the triplet of λ_1, λ_2, and λ_3 with the minimum average NRMSE. The selected values of λ_1, λ_2, and λ_3 were 40, 80, and 100, respectively. We considered $R_U = 10$ as the upper bound on the number of latent factors.

In our experiments, we used the ratio $|f^{(t+1)} - f^{(t)}|/f^{(t)}$ as the convergence criterion, where $f^{(t+1)}$ and $f^{(t)}$ are the values of objective function of optimization problem (1) in the current and previous iterations, respectively. We stopped the algorithm if $|f^{(t+1)} - f^{(t)}|/f^{(t)} < \epsilon$, where we set $\epsilon = 10^{-6}$. We also limited the maximum number of iterations to 10^5, so if the algorithm did not converge in 10^5 iterations, we stopped the algorithm.

After the algorithm stops, we need to assign weights to the identified regulatory modules. Our initial assumption was $\mathbf{W} = \mathbf{W}_{\mathcal{X}}\mathbf{W}_{\mathcal{Y}}$. To put all modules into the same scale, we normalized each row of $\mathbf{W}_{\mathcal{X}}$ and each column of $\mathbf{W}_{\mathcal{Y}}$ to unit norm. The low rank decomposition now becomes $\mathbf{W} = \widetilde{\mathbf{W}}_{\mathcal{X}}\mathbf{D}_{\mathcal{X}}\mathbf{D}_{\mathcal{Y}}\widetilde{\mathbf{W}}_{\mathcal{Y}}$, where $\mathbf{D}_{\mathcal{X}}$ and $\mathbf{D}_{\mathcal{Y}}$ are $R \times R$ diagonal matrices. The relative importance weights to the regulatory modules are assigned based on the diagonal entries of $\mathbf{D}_{\mathcal{X}}\mathbf{D}_{\mathcal{Y}}$. We then ranked the regulatory modules from the largest weight to the smallest weight by ordering the rows of $\mathbf{W}_{\mathcal{X}}$ and the columns of $\mathbf{W}_{\mathcal{Y}}$. The regulatory module with the largest diagonal entry has the largest contribution (i.e. the first regulatory module).

Table 1. Predictive performance values of SMFR and T-SVD regression algorithm on 32 datasets in terms of average NRMSE over mRNAs.

Cohort	T-SVD		SMFR		Cohort	T-SVD		SMFR	
	$R^\star$	NRMSE	$R^\star$	NRMSE		$R^\star$	NRMSE	$R^\star$	NRMSE
ACC	10	0.9829	8	0.7438	LUSC	10	0.9581	10	0.7994
BLCA	10	0.9807	10	0.7837	MESO	10	0.9689	8	0.7609
BRCA	10	0.9373	10	0.8093	OV	10	0.9385	10	0.8173
CESC	10	0.9549	10	0.7861	PAAD	10	0.9274	10	0.6710
CHOL	10	0.9460	6	0.9623	PCPG	10	0.9369	10	0.7132
COAD	10	0.9255	10	0.7438	PRAD	10	0.9417	10	0.7233
DLBC	10	0.9450	4	0.7675	READ	10	0.9854	10	0.7224
ESCA	10	0.9331	10	0.7628	SARC	10	0.9090	10	0.7665
HNSC	10	0.9487	10	0.7675	SKCM	10	0.9366	10	0.7281
KICH	10	0.9923	10	0.6607	STAD	10	0.9442	10	0.7664
KIRC	10	0.9393	10	0.7147	TGCT	10	0.8411	10	0.6107
KIRP	10	0.9305	10	0.7007	THCA	10	0.9438	10	0.6712
LAML	10	0.9803	10	0.7188	THYM	10	0.9578	10	0.5775
LGG	10	1.0918	10	0.6752	UCEC	10	0.9588	10	0.7679
LIHC	10	0.9369	10	0.7290	UCS	10	0.9612	9	0.9775
LUAD	10	0.9253	10	0.7850	UVM	10	0.8019	10	0.5988

5.2   Predictive Performance Comparison

To evaluate the predictive performance of SMFR algorithm, we calculated the NRMSE between observed and predicted expression values for each mRNA. We also applied T-SVD regression on the same data using tsvd R package [7]. We reported the average NRMSE value over mRNAs for each cancer type in Table 1 for both SMFR and T-SVD regression algorithms. In our implementation, we observed that some of the cancer types (mostly cohorts with very few samples) used less than 10 regulatory modules (i.e. $R^\star < 10$). However, most of the cohorts (i.e. 27 out of 32) used all 10 modules to capture the regulatory relationships between miRNAs and mRNAs. The predictive performance values showed that, in comparison to T-SVD regression, SMFR algorithm was able to explain a higher proportion of the variance on all cohorts, e.g. SMFR explains 22% and 20% of variance on BLCA and BRCA cohorts, respectively, whereas T-SVD regression explains 2% and 7% of variance cohort on the same cohorts.

5.3   Extracting Key miRNA–mRNA Regulatory Modules

To check whether SMFR algorithm finds meaningful miRNA–mRNA regulatory modules, we identified selected miRNAs and mRNAs for each module. We thresholded the rows of $\widetilde{\mathbf{W}}_{\mathcal{X}}$ and the columns of $\widetilde{\mathbf{W}}_{y}$ as follows: a miRNA is considered to be selected if the magnitude of its weight is larger than one over square root of the total number of miRNAs included, and an mRNA is considered to be selected if the magnitude of its weight is larger than one over square root of the total number of mRNAs included. We then clustered the patients into two groups using k-means clustering on the expression values of the selected mRNAs only. We finally presented the expression values of selected miRNAs and mRNAs together with the clustering assignment of patients in heat maps of selected miRNAs and mRNAs for all identified modules. Figure 2 shows an example heat map of selected miRNAs and mRNAs for KIRC cohort. In this heat map, red colors show over-expression (i.e. higher than the population mean) and blue colors show under-expression (i.e. lower than the population mean). The font sizes for the names of miRNAs and mRNAs are proportional to the magnitudes of their weights inferred by SMFR algorithm.

We then investigated the functional importance of identified miRNA–mRNA regulatory modules by checking whether there is a significant survival difference between the two groups of patients found by the clustering algorithm using the log-rank test. We checked 305 miRNA–mRNA regulatory modules in total (Table 1), and, in only 77 of them, we observed a significant survival difference between two groups (i.e. p-value < 0.05 in the log-rank test). Figure 2 shows an example Kaplan-Meier survival curve for KIRC cohort.

We calculated enrichment of 77 identified modules with significant survival differences in Hallmark gene sets collection using the hypergeometric test. We found at least one enriched gene set in Hallmark collection with an FDR-corrected q-value < 0.05 in 72 out of 77 (93.51%) regulatory modules. These results show the power of our method in grouping genes that participate in the same biological processes or gene set. For 72 key regulatory modules, we reported the miRNA and mRNA expression heat maps and Kaplan-Meier survival curves in Supplementary Figs. S1–S72.

5.4 Literature Validation of Identified miRNA–mRNA Regulatory Modules

We checked the literature to find supporting evidence for 72 key miRNA–mRNA regulatory modules identified by SMFR algorithm. We searched the PubMed repository at https://pubmed.ncbi.nlm.nih.gov using the top five miRNA names in terms of the magnitude of their weights and cancer name. We were able to find supporting evidence for 55 out of 72 (76.39%) regulatory modules. The publications that provide supporting evidence for identified modules by SMFR are listed in Supplementary Table S2.

We used miRNA–mRNA interactions in the miRTArBase database [3], which provides a comprehensive list for experimentally validated miRNA–target interactions, to validate the miRNA–mRNA interactions identified by our implementation. We found that 220 miRNA–mRNA interactions identified in our modules are also reported in the miRTarBase database, and these interactions are listed in Supplementary Table S3.

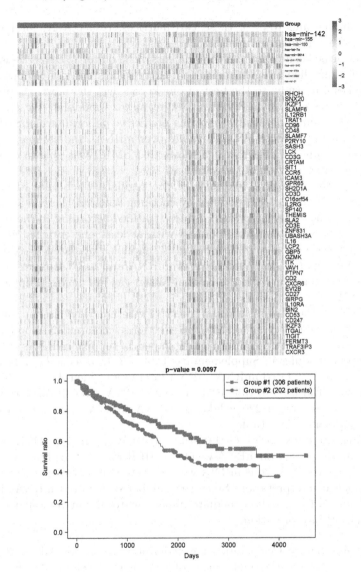

Fig. 2. Regulatory module #8 identified for KIRC cohort. (top) Heat map of expression profiles of top 10 miRNAs and top 50 mRNAs in this module. (bottom) Kaplan-Meier curves of two patient groups stratified according to this module.

6 Conclusions

MicroRNAs are important regulators of gene expression by directly or indirectly targeting mRNAs, and they are also shown to be involved in the formation and progression of cancer. In this study, we applied SMFR algorithm which decomposes the coefficient matrix using two low-rank matrices and enforces sparsity on the columns and rows of these two matrices, respectively, for identifying miRNA–mRNA regulatory modules.

We performed computational experiments on 32 out of 33 cancer cohorts from TCGA with matched miRNA and mRNA expression profiles of primary tumors, which makes our study, to the best of our knowledge, the most comprehensive computational study on miRNA–mRNA regulatory modules. We showed the effectiveness of our approach by reporting predictive performance of SMFR algorithm in terms of average NRMSE values over mRNAs and comparing against T-SVD regression in Table 1, which shows that SMFR was able to explain significantly a larger portion of the variance.

To show the biological relevance of the regulatory modules identified by SMFR, we first filtered the modules by checking whether there is a significant survival difference between patient groups stratified according to the regulatory modules. Next, we filtered the modules with at least one enriched gene set in hallmark collection. We were able to find 72 modules validated by two filtering analyses. We reported the heat maps of miRNA and mRNA expression values for 72 modules together with the Kaplan-Meier survival curves (Supplementary Figs. S1–S72).

To validate the results, we first checked whether there is any supporting evidence from the literature to show the effect of selected miRNAs in the underlying cancer. We reported the supporting papers of top five miRNAs of 55 out of 72 regulatory modules in Supplementary Table S2. In a second validation approach, we found validated miRNA–mRNA interactions identified by SMFR that matched with the largest validated miRNA–mRNA interaction database (i.e. miRTarBase). We found 220 validated miRNA–mRNA interactions that have listed in Supplementary Table S3.

We envision an extension of this study towards modifying SMFR algorithm to obtain more sparsity on mRNA weights. SMFR was able to obtain very sparse coefficients for miRNAs, but mRNA weights were not sparse enough to make a clear biological interpretation. Since the number of involved mRNAs is in the order of tens of thousands, it is quite important to obtain very sparse weights for the sake of interpretation.

Acknowledgements. Computational experiments were performed on the OHSU Exacloud high performance computing cluster. Mehmet Gönen was supported by the Turkish Academy of Sciences (TÜBA-GEBİP; The Young Scientist Award Program) and the Science Academy of Turkey (BAGEP; The Young Scientist Award Program).

References

1. Ajana, S., Acar, N., Bretillon, L., Hejblum, B.P., Jacqmin-Gadda, H., Delcourt, C.: BLISAR study group: benefits of dimension reduction in penalized regression methods for high-dimensional grouped data: a case study in low sample size. Bioinformatics **35**(19), 3628–3634 (2019)
2. Calin, G.A., Croce, C.M.: MicroRNA signatures in human cancers. Nat. Rev. Cancer **6**(11), 857–866 (2006)

3. Chou, C.H., et al.: miRTarBase update 2018: a resource for experimentally validated microRNA-target interactions. Nucleic Acids Res. **46**(D1), D296–D302 (2018)
4. He, L., Hannon, G.J.: MicroRNAs: small RNAs with a big role in gene regulation. Nat. Rev. Genet. **5**(7), 522–531 (2004)
5. Kharratzadeh, M., Coates, M.: Sparse multivariate factor regression. In: Proceedings of the IEEE Statistical Signal Processing Workshop, pp. 1–5 (2016)
6. Kumar, A., Daume III, H.: Learning task grouping and overlap in multi-task learning. In: Proceedings of the 29th International Conference on Machine Learning, pp. 1383–1390 (2012)
7. Ma, X., Xiao, L., Wong, W.H.: tsvd: Thresholding-based SVD for multivariate reduced rank regression, R package version 1.4 (2015)
8. Ma, X., Xiao, L., Wong, W.H.: Learning regulatory programs by threshold SVD regression. Proc. Natl. Acad. Sci. U.S.A. **111**(44), 15675–15680 (2014)
9. Massy, W.F.: Principal components regression in exploratory statistical research. J. Am. Stat. Assoc. **60**(309), 234–256 (1965)
10. Min, W., Liu, J., Luo, F., Zhang, S.: A novel two-stage method for identifying microRNA-gene regulatory modules in breast cancer. In: Proceedings of the IEEE International Conference on Bioinformatics and Biomedicine, pp. 151–156 (2015)
11. Negahban, S.N., Wainwright, M.J.: Simultaneous support recovery in high dimensions: Benefits and perils of block ℓ_1/ℓ_∞-regularization. IEEE Trans. Inf. Theory **57**(6), 1161–1168 (2011)
12. Obozinski, G., Wainwright, M.J., Jordan, M.I.: Support union recovery in high-dimensional multivariate regression. Ann. Stat. **39**(1), 1–47 (2011)
13. Parikh, N., Boyd, S.: Proximal algorithms. Found. Trends Opt. **1**(3), 127–239 (2014)
14. Pourahmadi, M.: High-Dimensional Covariance Estimation. Wiley, New York (2013)
15. Tran, D.H., Satou, K., Ho, T.B.: Finding microRNA regulatory modules in human genome using rule induction. BMC Bioinform. **9**(Suppl 12), S5 (2008)
16. Yang, D., et al.: Integrated analyses identify a master microRNA regulatory network for the mesenchymal subtype in serous ovarian cancer. Cancer Cell **23**(2), 186–199 (2013)
17. Zhang, S., Li, Q., Liu, J., Zhou, X.J.: A novel computational framework for simultaneous integration of multiple types of genomic data to identify microRNA-gene regulatory modules. Bioinformatics **27**(13), i401–i409 (2011)
18. Zhang, S., Liu, C.C., Li, W., Shen, H., Laird, P.W., Zhou, X.J.: Discovery of multidimensional modules by integrative analysis of cancer genomic data. Nucleic Acids Res. **40**(19), 9379–9391 (2012)
19. Zhang, Y., et al.: Genome-wide identification of the essential protein-coding genes and long non-coding RNAs for human pan-cancer. Bioinformatics **35**(21), 4344–4349 (2019)

A Krill Herd Algorithm for the Multiobjective Energy Reduction Multi-Depot Vehicle Routing Problem

Emmanouela Rapanaki[✉], Iraklis - Dimitrios Psychas, Magdalene Marinaki,
Nikolaos Matsatsinis, and Yannis Marinakis

School of Production Engineering and Management, Technical University of Crete,
Chania, Greece
emmarap@hotmail.com, ipsychas102@gmail.com, magda@dssl.tuc.gr,
{nikos,marinakis}@ergasya.tuc.gr

Abstract. Krill Herd algorithm is a powerful and relatively new Swarm
Intelligence Algorithm that has been applied in a number of different
kind of optimization problems since the time that it was published. In
recent years there is a growing number of optimization models that are
trying to reduce the energy consumption in routing problems. In this
paper, a new variant of Krill Herd algorithm, the Parallel Multi-Start
Non-dominated Sorting Krill Herd algorithm (PMS-KH), is proposed
for the solution of a Vehicle Routing Problem variant, the Multiobjective Energy Reduction Multi-Depot Vehicle Routing Problem (MERMD-
VRP). Four different models are proposed where the distances between
the customers and between the customers and the depots are either symmetric or asymmetric and the customers have either demand or pickup.
The algorithm is compared with four other multiobjective algorithms,
the Parallel Multi-Start Non-dominated Sorting Artificial Bee Colony
(PMS-ABC), the Parallel Multi-Start Non-dominated Sorting Differen-
tial Evolution (PMS-NSDE), the Parallel Multi-Start Non-dominated
Sorting Particle Swarm Optimization (PMS-NSPSO) and the Parallel
Multi-Start Non-dominated Sorting Genetic Algorithm II (PMS-NSGA
II) in a number of benchmark instances, giving very satisfactory results.

Keywords: Vehicle routing problem · Krill Herd · NSGA II · DE ·
PSO · ABC · VNS

1 Introduction

The **Vehicle Routing Problem** is one of the most famous optimization problems and its main goal is the design of the best routes in order the selected (or
calculated) vehicles to serve the set of customers with the best possible way based
on the selected criteria that each different variant of the problem includes. As the
interest of the researchers and of the decision makers in industries for the solution of different variants of the Vehicle Routing Problem continuously increases,

© Springer Nature Switzerland AG 2020
G. Nicosia et al. (Eds.): LOD 2020, LNCS 12565, pp. 434–447, 2020.
https://doi.org/10.1007/978-3-030-64583-0_39

a number of more realistic and, thus, more complicated variants/formulations of the Vehicle Routing Problem and, in many cases with more than one objective functions, have been proposed and a number of more sophisticated algorithms [5] are used for their solutions [13]. There is a growing number of papers for solving multi-depot vehicle routing problems [6] or energy vehicle routing problems [1,4,17]. However, there are few papers for the solution of Multi-Depot vehicle routing problem with fuel consumption [3].

In this research, the formulation of the problem combines more than one depot, the possibility of pickups and deliveries and the simultaneous reduction of the fuel consumption in symmetric and, in the more realistic, asymmetric cases. In the first objective function (OF1), the total travel time is minimized, while in the second objective function the Fuel Consumption (FC) taking into account the traveled distance and the load of the vehicle is, also, minimized. In the second objective function we study two different scenarios, in the first scenario, we have deliveries in the customers (OF2) and, in the other, we have pickups from the customers (OF3). Also, we study a symmetric case, where except of the load and the traveled distance, we consider that there are no other route parameters or that there are perfect route conditions, and an asymmetric case, where we take into account parameters of real life, such as weather conditions or uphill and downhill routes [11,12]. Thus, we produce four different problems, in the first one we have the minimization of the OF1 and OF2 objective functions in the symmetric case, in the second one we have the minimization of the OF1 and OF3 in the symmetric case and in the other two scenarios, we have the minimization of the corresponding combinations of objective functions in the asymmetric case. In recent years a number of evolutionary, swarm intelligence and, in general, nature inspired algorithms have been proposed for the solution of a Vehicle Routing Problem both when one or more objectives are considered.

In the following the three objective functions and the constraints of the problem are presented [8,9,12]:

$$\min OF1 = \sum_{i=I_1}^{n} \sum_{j=1}^{n} \sum_{\kappa=1}^{m} (t_{ij}^{\kappa} + s_j^{\kappa}) x_{ij}^{\kappa} \tag{1}$$

$$\min OF2 = \sum_{h=I_1}^{I_\pi} \sum_{j=2}^{n} \sum_{\kappa=1}^{m} c_{hj} x_{hj}^{\kappa} (1 + \frac{y_{hj}^{\kappa}}{Q}) r_{hj}$$

$$+ \sum_{i=2}^{n} \sum_{j=I_1}^{n} \sum_{\kappa=1}^{m} c_{ij} x_{ij}^{\kappa} (1 + \frac{y_{i-1,i}^{\kappa} - D_i}{Q}) r_{ij} \tag{2}$$

$$\min OF3 = \sum_{h=I_1}^{I_\pi} \sum_{j=2}^{n} \sum_{\kappa=1}^{m} c_{hj} x_{hj}^\kappa r_{hj}$$

$$+ \sum_{i=2}^{n} \sum_{j=I_1}^{n} \sum_{\kappa=1}^{m} c_{ij} x_{ij}^\kappa \left(1 + \frac{y_{i-1,i}^\kappa + D_i}{Q}\right) r_{ij} \tag{3}$$

$$\sum_{j=I_1}^{n} \sum_{\kappa=1}^{m} x_{ij}^\kappa = 1, i = I_1, \cdots, n \tag{4}$$

$$\sum_{i=I_1}^{n} \sum_{\kappa=1}^{m} x_{ij}^\kappa = 1, j = I_1, \cdots, n \tag{5}$$

$$\sum_{j=I_1}^{n} x_{ij}^\kappa - \sum_{j=I_1}^{n} x_{ji}^\kappa = 0, i = I_1, \cdots, n, \kappa = 1, \cdots, m \tag{6}$$

$$\sum_{j=I_1, j\neq i}^{n} y_{ji}^\kappa - \sum_{j=I_1, j\neq i}^{n} y_{ij}^\kappa = D_i, i = I_1, \cdots, n, \kappa = 1, \cdots, m, \; for \; deliveries \tag{7}$$

$$\sum_{j=I_1, j\neq i}^{n} y_{ij}^\kappa - \sum_{j=I_1, j\neq i}^{n} y_{ji}^\kappa = D_i, \; i = I_1, \cdots, n, \; \kappa = 1, \cdots, m, \; for \; pick-ups \tag{8}$$

$$Q x_{ij}^\kappa \geq y_{ij}^\kappa, i, j = I_1, \cdots, n, \kappa = 1, \cdots, m \tag{9}$$

$$\sum_{i,j \in S} x_{ij\kappa} \leq |S| - 1, \; for \; all \; S \subseteq \{2, \cdots, n\}, \kappa = 1, \cdots, m \tag{10}$$

$$x_{ij}^\kappa = \begin{cases} 1, & if (i,j) belongs \; to \; the \; route \\ 0, & otherwise \end{cases} \tag{11}$$

The objective function (1) is used for the minimization of the time needed to travel between two customers or a customer and the depot (where t_{ij}^κ is the time needed to visit customer j immediately after customer i using vehicle κ, s_j^κ is the service time of customer j using vehicle κ, n is the number of nodes ($\{I_1, I_2, ... I_\pi, 2, 3, ..., n\}$), m is the number of homogeneous vehicles and the depots are a subset $\Pi = \{I_1, I_2, ... I_\pi\}$ of the set of the n nodes denoted by $i = j = I_1, I_2, ... I_\pi$ (π is the number of homogeneous depots)). The second objective function (2) is used for the minimization of the **Route based Fuel Consumption** (RFC) in the case that the vehicle performs only deliveries taking into account real life route parameters (weather conditions or uphills and downhills or driver's behavior - The parameter r_{ij} corresponds to the route parameters from the node i to the node j and it is always a positive number). The third objective function (3) is used for the minimization of the **Route based Fuel Consumption** (RFC) in the case that the vehicle performs only pick-ups in its route. Constraints (4) and (5) represent that each customer must be visited only by one vehicle; constraints (6) ensure that each vehicle that arrives at a node must leave from that node also. Constraints (7) and (8) indicate that the reduced (if it concerns deliveries) or increased (if it concerns pick-ups) load (cargo) of the vehicle after it visits a node is equal to the demand of that node. Constraints (9) are used to limit the maximum load carried by the vehicle and to force y_{ij}^{κ} to be equal to zero when $x_{ij}^{\kappa} = 0$, constraints (10) are the tour elimination constraints while constraints (11) ensure that only one vehicle will visit each customer. It should be noted that the problems solved in this paper are symmetric (where $r_{ij} = 1$, $\forall(i, j)$) or asymmetric (where $r_{ij} \neq r_{ji}$, $\forall(i, j)$).

In this paper, a variant of the Krill Herd (KH) algorithm, suitable for the Multiobjective Vehicle Routing Problems, the Parallel Multi-Start Non-dominated Sorting Krill Herd Algorithm (PMS-KH), is proposed for the solution of the Multiobjective Energy Reduction Multi-Depot Vehicle Routing Problem (MEMDVRP). The KH algorithm is based on simulating the herding behavior of krill individuals, using a Langrangean model [2]. The original Krill Herd algorithm was proposed for the solution of continuous optimization problems and the proposed algorithm was modified properly in order to be used for the solution of the studied problem. The algorithm is compared with four other evolutionary algorithms, the Parallel Multi-Start Non-dominated Sorting Particle Swarm Optimization (PMS-NSPSO) [10], the Parallel Multi-Start Non-dominated Sorting Differential Evolution (PMS-NSDE) [8], the Parallel Multi-Start Non-dominated Sorting Artificial Bee Colony (PMS-ABC) [12] and the Parallel Multi-Start Non-dominated Sorting Genetic Algorithm II (PMS-NSGA II) [9]. This paper is organized into three sections. In the next section, the proposed algorithm is described, while, in Sect. 3 the computational results are presented. The last section provides concluding remarks and future research.

2   Parallel Multi-Start Non-dominated Sorting Krill Herd Algorithm (PMS-KH)

Krill Herd algorithm (KH) is a relatively new swarm intelligence algorithm [2,14–16] that simulates the herding behavior of krill individuals. The algorithm was proposed, initially, for the solution of global optimization problems but it can be used with the suitable modifications for the solution of any other problem [16]. The algorithm consists of three movements that simulate the hunting of the krills and the communication with each other, thus, an equation is produced that is consisted of three different movements, the foraging action, the movement influenced by other krill and the physical diffusion [14]. The equations of the Krill Herd algorithm are [2,14–16]:

$$\frac{dX_i}{dt} = F_i + N_i + D_i \tag{12}$$

$$F_i = V_f \beta_i + \omega_f F_i^{old} \tag{13}$$

$$\beta_i = \beta_i^{food} + \beta_i^{best} \tag{14}$$

$$N_i^{new} = N^{max}\alpha_i + \omega_n N_i^{old} \tag{15}$$

$$D_i = D^{max}\delta \tag{16}$$

$$X_i(t + \Delta t) = X_i(t) + \Delta t \frac{dX_i}{dt} \tag{17}$$

where F_i, N_i, and D_i denote the foraging motion, the motion influenced by other krill, and the physical diffusion of the krill i, respectively. The F_i simulates the current food location and the information of previous location (Eq. 13), V_f is the foraging speed, ω_f is the inertia weight of the foraging motion in $(0,1)$, F_i^{old} is the last foraging motion. The N_i movement (Eq. 15) is estimated by the target effect, the local effect and the repulsive effect. The N^{max} is the maximum induced speed, ω_n is the inertia weight of the second motion, N_i^{old} is the last motion influenced by other krill. The physical diffusion (Eq. 16) is a random process where D^{max} is the maximum diffusion speed and δ is a random number. Finally, the time-relied position from time t to $t + \Delta t$ is given by Eq. 17.

In the Parallel Multi-Start Non-dominated Sorting Krill Herd algorithm (PMS-KH), initially, a set of krills (possible solutions) (X) are randomly placed in the solution space. As the studied problem is a variant of the Vehicle Routing Problem, the solution should represent a number of routes that vehicles follow. However, as the equations of the Krill Herd algorithm are suitable for global optimization problems, we transformed suitably the initial solutions. Then, we used the equations of the KH algorithm, described previously, and we transformed back the solutions in order to calculate the objective functions. In every iteration, the non-dominated solutions are found.

As the use of the equations for the calculation of the KH algorithm, could produce some inefficient solutions (due to the transformation of the solutions from continuous values to discrete values and vice versa), it was decided to add another phase for the calculation of the new positions in the algorithm in order to take advantage of possible good new and old positions in the whole set of solutions. Thus, the solutions of the last two iterations (iteration it and $it + 1$) are combined in a new vector and, then, the members of the new vector are sorted using the *rank* and the *crowding distance* as in the NSGA II algorithm as it was modified in [7]. The first W solutions of the new vector are the produced solutions of the iteration $it+1$. The distribution of the new solutions is performed based on the values of the *rank* and the *crowding distance*. The new solutions are evaluated by each objective function separately. A Variable Neighborhood Search (VNS) algorithm is applied to the solutions with both the vns_{max} and the

$local_{max}$ equal to 10 [7]. The personal best solution of each krill is found by using the following observation. If a solution in iteration $it + 1$ dominates its previous best solution of the iteration it, then, the previous best solution is replaced by the current solution. On the other hand if the previous best solution dominates the current solution, then, the previous best solution remains the same. Finally, if these two solutions are not dominated between them, then, the previous best solution is not replaced. It should be noted that the non-dominated solutions are not deleted from the Pareto front and, thus, the good solutions will not be disappeared from the population. In the next iterations, in order to insert a new solution in the Pareto front archive, there are two possibilities. First, the new solution is non-dominated with respect to the contents of the archive and secondly it dominates any solution in the archive. In both cases, all the dominated solutions, that already belong to the archive, have to be deleted from the archive. At the end of each iteration, from the non-dominated solutions from all populations the Total Pareto Front is updated considering the non-dominated solutions of the last *population*. The algorithm terminates when a predefined maximum number of iterations reached.

3   Computational Results

The whole algorithmic approach was implemented in Visual C++ and was tested on a set of benchmark instances. The data were created according Rapanaki et al. [11,12] research and the size of each benchmark instance is equal to 100 nodes. Also, the algorithms were tested for ten instances for the two objective functions (OF1-OF2 or OF1-OF3) five times [11]. For the comparison of the four algorithms, we use four different evaluation measures: the range to which the front spreads out (M_k), the number of solutions (L), the combination of the spread and distribution of the solutions (Δ) and the coverage measure, where the fraction of solutions in one algorithm that are weakly dominated by one or more solutions of the other algorithm is calculated (C). It is preferred the L, M_k and C measures to be as larger as possible and the Δ value to be as smaller as possible (for analytical presentation of the measures please see [8]). The results of the proposed algorithm are compared with the results of the Parallel Multi-Start Non-dominated Sorting Particle Swarm Optimization (PMS-NSPSO) [10], the Parallel Multi-Start Non-dominated Sorting Differential Evolution (PMS-NSDE) algorithm [8], the Parallel Multi-Start Non-dominated Sorting Artificial Bee Colony (PMS-ABC) algorithm [12] and the Parallel Multi-Start Non-dominated Sorting Genetic Algorithm II (PMS-NSGA II) algorithm [9].

In Tables 1, 2, 3, 4, 5, 6, 7 and in Fig. 1, the computational results of the proposed algorithm are presented. Initially, in Tables 1 and 2 the computational results of the best execution of the proposed algorithm are compared with the computational results of the other four algorithms in the three evaluation measures in the asymmetric and symmetric cases, respectively. As in the proposed

Table 1. Average results and best runs of all algorithms used in the comparisons in the asymmetric problems.

	Algorithms	Asymmetric Delivery FCVRP			Asymmetric Pick-up FCVRP		
		L	M_k	Δ	L	M_k	Δ
A-B-CD	PMS-KH	46.00(57)	603.32(594.16)	0.63(0.60)	46.60(53)	591.77(593.94)	0.64(**0.59**)
	PMS-NSPSO	46.40(53)	598.41(592.84)	0.70(0.68)	47.00(56)	597.42(609.09)	0.66(0.62)
	PMS-NSGA II	56.40(**62**)	592.33(598.84)	0.61(0.54)	59.80(**63**)	598.92(608.39)	0.61(0.65)
	PMS-NSDE	50.00(59)	598.17(**604.03**)	0.61(**0.53**)	46.00(49)	598.81(605.87)	0.61(0.62)
	PMS-ABC	25.20(26)	519.39(537.22)	0.68(0.69)	30.20(31)	594.50(**615.83**)	0.72(0.70)
A-C-BD	PMS-KH	48.20(51)	595.72(**613.14**)	0.65(**0.62**)	44.60(39)	592.00(603.58)	0.61(0.62)
	PMS-NSPSO	47.60(51)	600.33(611.96)	0.68(0.68)	47.80(53)	601.70(595.13)	0.64(**0.59**)
	PMS-NSGA II	61.80(**72**)	594.92(602.67)	0.61(0.68)	58.80(**56**)	604.25(**613.08**)	0.59(0.64)
	PMS-NSDE	49.40(56)	594.19(603.86)	0.63(0.64)	44.40(50)	597.25(612.67)	0.63(0.62)
	PMS-ABC	26.40(30)	514.57(523.10)	0.80(0.94)	29.60(33)	599.64(609.91)	0.72(0.73)
A-D-BE	PMS-KH	47.40(47)	602.83(**612.39**)	0.65(**0.55**)	42.40(45)	600.29(615.68)	0.64(0.71)
	PMS-NSPSO	48.80(52)	592.57(606.60)	0.64(0.62)	46.00(48)	608.80(**620.73**)	0.66(0.65)
	PMS-NSGA II	54.20(**54**)	601.74(597.52)	0.61(**0.55**)	54.60(**63**)	602.82(610.88)	0.61(0.60)
	PMS-NSDE	46.80(51)	591.33(585.05)	0.68(0.58)	46.20(53)	594.36(608.66)	0.64(**0.58**)
	PMS-ABC	21.20(25)	513.52(533.70)	0.79(0.84)	30.40(33)	604.45(614.92)	0.80(0.85)
A-E-BD	PMS-KH	46.00(51)	597.29(**607.78**)	0.65(0.65)	43.80(50)	586.47(594.49)	0.61(0.54)
	PMS-NSPSO	48.00(**58**)	586.44(595.07)	0.65(0.63)	49.00(**60**)	597.10(610.34)	0.67(0.69)
	PMS-NSGA II	57.20(56)	595.30(604.81)	0.58(**0.56**)	56.40(**60**)	591.73(**615.11**)	0.58(0.53)
	PMS-NSDE	53.80(**58**)	589.49(595.89)	0.66(0.66)	45.40(49)	598.88(578.64)	0.64(**0.52**)
	PMS-ABC	24.40(30)	515.43(519.05)	0.71(0.64)	30.80(28)	597.77(603.00)	0.80(0.72)
B-C-AD	PMS-KH	42.40(49)	580.80(**604.86**)	0.69(0.70)	41.80(45)	596.24(605.06)	0.68(0.58)
	PMS-NSPSO	42.00(45)	587.45(587.39)	0.66(0.69)	43.20(38)	601.86(**616.66**)	0.67(**0.57**)
	PMS-NSGA II	51.20(**64**)	596.11(602.55)	0.61(0.60)	53.60(**60**)	602.63(609.71)	0.63(0.65)
	PMS-NSDE	42.20(47)	593.19(586.51)	0.63(**0.56**)	44.00(47)	596.09(594.45)	0.69(0.61)
	PMS-ABC	19.60(23)	479.34(507.12)	0.87(0.78)	25.60(31)	590.78(585.40)	0.79(0.76)
B-D-AC	PMS-KH	43.60(30)	595.59(601.52)	0.65(**0.53**)	43.40(46)	581.21(602.56)	0.65(0.74)
	PMS-NSPSO	42.00(53)	589.78(591.38)	0.62(0.66)	42.20(**50**)	581.03(589.53)	0.68(0.75)
	PMS-NSGA II	54.60(**63**)	593.43(**618.00**)	0.61(**0.53**)	51.40(46)	587.73(**606.69**)	0.60(**0.51**)
	PMS-NSDE	43.20(48)	591.59(611.63)	0.71(0.72)	44.80(**50**)	592.20(586.71)	0.66(0.58)
	PMS-ABC	21.00(26)	508.53(514.14)	0.72(0.69)	24.40(25)	590.85(602.09)	0.72(0.67)
B-E-AD	PMS-KH	47.60(**58**)	592.51(**614.57**)	0.66(0.73)	42.40(48)	588.98(575.90)	0.60(**0.55**)
	PMS-NSPSO	42.00(45)	584.42(543.81)	0.59(**0.54**)	48.00(47)	601.43(611.90)	0.65(0.62)
	PMS-NSGA II	50.80(57)	597.52(599.73)	0.56(0.56)	50.20(**55**)	594.89(600.52)	0.60(0.58)
	PMS-NSDE	41.80(47)	596.38(611.30)	0.66(0.61)	41.00(46)	598.64(**622.57**)	0.66(0.66)
	PMS-ABC	23.20(26)	489.52(510.33)	0.76(0.86)	29.40(34)	598.01(599.91)	0.77(0.73)
C-D-AE	PMS-KH	48.80(49)	587.41(593.82)	0.64(0.60)	44.40(42)	591.21(600.51)	0.63(**0.53**)
	PMS-NSPSO	45.00(**57**)	593.92(593.63)	0.67(0.66)	44.00(36)	592.59(602.11)	0.67(0.55)
	PMS-NSGA II	53.60(49)	597.24(589.81)	0.58(**0.53**)	51.80(**59**)	597.78(610.64)	0.63(0.59)
	PMS-NSDE	42.40(42)	594.55(**610.84**)	0.63(0.60)	44.40(51)	594.87(591.47)	0.70(0.71)
	PMS-ABC	25.40(30)	524.53(520.80)	0.82(0.76)	28.60(27)	598.07(**614.45**)	0.72(0.63)
C-E-AB	PMS-KH	46.40(**51**)	588.58(**608.94**)	0.63(0.62)	46.00(48)	591.57(583.06)	0.65(0.65)
	PMS-NSPSO	45.00(48)	586.58(577.86)	0.66(**0.54**)	46.80(41)	588.78(601.76)	0.63(**0.53**)
	PMS-NSGA II	55.00(47)	592.95(602.86)	0.61(0.57)	51.00(**65**)	595.58(608.99)	0.65(0.74)
	PMS-NSDE	39.40(47)	592.59(569.13)	0.67(0.63)	47.20(47)	593.76(**612.40**)	0.63(0.57)
	PMS-ABC	25.80(33)	520.43(520.96)	0.81(0.85)	26.00(27)	594.23(597.77)	0.84(0.76)
D-E-BC	PMS-KH	51.80(**56**)	580.24(**595.11**)	0.68(0.63)	43.00(44)	574.96(580.66)	0.68(0.71)
	PMS-NSPSO	42.60(50)	571.71(568.50)	0.67(0.69)	39.60(44)	578.42(573.72)	0.60(0.68)
	PMS-NSGA II	43.60(48)	526.74(503.05)	0.62(**0.53**)	50.80(**56**)	581.08(**602.31**)	0.66(0.69)
	PMS-NSDE	42.00(44)	582.50(589.22)	0.68(0.64)	43.60(52)	581.38(588.35)	0.60(0.64)
	PMS-ABC	23.20(23)	510.71(514.93)	0.70(0.60)	28.00(30)	581.88(577.52)	0.77(**0.63**)

Table 2. Average Results and Best Runs of all algorithms used in the comparisons in the symmetric problems.

	Algorithms	Symmetric Delivery FCVRP			Symmetric Pick-up FCVRP		
		L	M_k	Δ	L	M_k	Δ
A-B	PMS-KH	49.40(51)	606.19(**621.97**)	0.72(0.68)	49.80(59)	603.44(618.82)	0.65(0.73)
	PMS-NSPSO	47.80(52)	613.92(610.29)	0.68(0.62)	51.60(54)	607.75(602.86)	0.69(0.62)
	PMS-NSGA II	56.40(**61**)	602.89(603.41)	0.66(0.62)	58.60(**79**)	596.88(605.27)	0.66(**0.60**)
	PMS-NSDE	44.60(45)	605.73(596.04)	0.67(**0.55**)	44.60(46)	602.96(**622.10**)	0.68(0.62)
	PMS-ABC	29.00(31)	603.23(610.84)	0.76(0.69)	21.40(25)	600.76(611.89)	0.75(0.69)
A-C	PMS-KH	50.20(53)	608.63(**617.01**)	0.67(0.59)	49.80(58)	613.38(**617.90**)	0.67(**0.58**)
	PMS-NSPSO	50.20(53)	604.15(615.43)	0.66(0.67)	47.40(53)	604.96(608.06)	0.68(0.72)
	PMS-NSGA II	62.60(**66**)	609.36(611.80)	0.63(**0.54**)	56.60(**63**)	606.55(615.56)	0.64(0.61)
	PMS-NSDE	51.40(57)	596.73(602.61)	0.65(0.69)	49.20(44)	603.51(607.51)	0.65(0.61)
	PMS-ABC	32.20(36)	602.04(602.18)	0.83(0.70)	22.20(25)	599.57(601.56)	0.79(0.76)
A-D	PMS-KH	47.80(49)	585.82(598.02)	0.68(0.62)	44.40(46)	585.13(**596.17**)	0.69(0.68)
	PMS-NSPSO	48.40(54)	577.25(575.63)	0.62(**0.61**)	49.00(58)	586.10(594.99)	0.63(0.64)
	PMS-NSGA II	54.40(**57**)	592.85(587.50)	0.67(0.63)	58.60(**66**)	580.64(582.86)	0.66(0.60)
	PMS-NSDE	46.20(40)	581.70(591.46)	0.66(0.67)	47.00(47)	579.02(577.96)	0.66(**0.58**)
	PMS-ABC	29.20(35)	584.48(**610.10**)	0.84(0.88)	25.20(26)	583.36(582.98)	0.69(0.59)
A-E	PMS-KH	48.60(**53**)	604.35(**616.24**)	0.66(0.66)	43.80(52)	603.55(**619.68**)	0.71(0.64)
	PMS-NSPSO	48.00(47)	595.33(602.24)	0.68(0.69)	41.60(46)	592.88(612.23)	0.68(0.81)
	PMS-NSGA II	51.20(51)	598.85(608.34)	0.62(**0.60**)	61.00(**74**)	598.56(598.90)	0.65(0.59)
	PMS-NSDE	44.60(49)	604.13(608.11)	0.68(0.63)	43.80(43)	602.41(605.73)	0.62(**0.58**)
	PMS-ABC	35.00(36)	598.62(596.71)	0.83(0.74)	20.80(27)	596.34(603.89)	0.72(0.72)
B-C	PMS-KH	41.80(48)	580.63(591.97)	0.70(0.67)	43.60(49)	592.56(603.54)	0.69(0.73)
	PMS-NSPSO	41.00(51)	589.04(587.03)	0.68(0.62)	47.40(49)	595.11(605.91)	0.67(0.64)
	PMS-NSGA II	58.80(55)	591.49(602.55)	0.65(**0.55**)	56.20(**61**)	597.28(590.74)	0.65(**0.61**)
	PMS-NSDE	49.80(**60**)	589.78(**610.81**)	0.66(0.56)	42.00(49)	589.82(**613.07**)	0.72(0.64)
	PMS-ABC	24.00(28)	554.94(590.73)	0.83(0.95)	24.20(26)	595.01(604.70)	0.76(0.69)
B-D	PMS-KH	48.20(54)	592.58(608.00)	0.64(0.65)	44.40(52)	575.50(589.47)	0.65(0.69)
	PMS-NSPSO	43.40(55)	594.80(598.42)	0.65(**0.54**)	43.20(49)	591.07(600.78)	0.64(**0.54**)
	PMS-NSGA II	58.60(**56**)	595.91(**609.20**)	0.64(0.63)	55.80(**54**)	592.55(**608.67**)	0.65(0.66)
	PMS-NSDE	43.20(45)	595.65(595.43)	0.69(0.60)	41.20(**54**)	582.58(600.87)	0.70(0.68)
	PMS-ABC	28.40(30)	589.34(598.79)	0.81(0.93)	20.40(24)	583.20(600.10)	0.69(0.69)
B-E	PMS-KH	39.80(57)	592.71(598.83)	0.65(0.71)	44.60(56)	594.89(609.97)	0.69(0.62)
	PMS-NSPSO	49.40(52)	606.42(590.66)	0.60(**0.53**)	41.40(40)	604.84(**618.80**)	0.68(0.63)
	PMS-NSGA II	57.60(**59**)	603.48(607.87)	0.61(0.60)	58.60(**63**)	603.78(618.11)	0.63(**0.55**)
	PMS-NSDE	47.80(45)	603.11(609.40)	0.67(0.60)	44.40(46)	580.23(573.63)	0.66(0.58)
	PMS-ABC	28.40(31)	594.76(**614.93**)	0.85(0.78)	17.60(19)	602.44(606.64)	0.72(0.67)
C-D	PMS-KH	49.20(**62**)	592.77(603.59)	0.67(0.69)	43.40(45)	583.62(587.21)	0.69(**0.55**)
	PMS-NSPSO	48.80(46)	584.51(602.85)	0.64(**0.59**)	46.60(56)	586.85(594.19)	0.67(0.73)
	PMS-NSGA II	56.60(59)	587.97(**604.68**)	0.63(0.64)	51.20(**57**)	586.43(587.41)	0.59(0.60)
	PMS-NSDE	42.80(51)	577.40(594.58)	0.65(0.61)	46.80(51)	586.32(573.07)	0.66(0.57)
	PMS-ABC	29.00(32)	586.84(585.57)	0.76(0.62)	19.40(22)	583.38(**606.27**)	0.61(0.57)
C-E	PMS-KH	44.80(55)	602.00(596.86)	0.69(0.64)	53.00(54)	602.97(618.94)	0.69(0.67)
	PMS-NSPSO	48.00(51)	598.45(604.81)	0.64(0.65)	43.20(33)	599.71(609.57)	0.67(**0.59**)
	PMS-NSGA II	60.40(**63**)	599.00(592.77)	0.64(**0.58**)	60.40(**61**)	607.48(613.53)	0.65(0.60)
	PMS-NSDE	49.40(53)	594.95(611.88)	0.73(0.75)	52.20(60)	601.83(615.53)	0.67(0.72)
	PMS-ABC	33.20(29)	603.84(**616.56**)	0.80(0.68)	22.40(23)	600.35(**619.16**)	0.71(0.65)
D-E	PMS-KH	51.00(53)	604.44(612.04)	0.67(0.63)	52.00(**60**)	605.05(611.73)	0.69(0.70)
	PMS-NSPSO	49.20(**57**)	606.90(**620.98**)	0.71(0.80)	48.00(55)	609.46(617.27)	0.69(0.70)
	PMS-NSGA II	60.20(52)	601.63(610.06)	0.67(0.66)	59.00(53)	606.62(617.50)	0.63(**0.57**)
	PMS-NSDE	49.80(46)	604.82(619.96)	0.66(**0.55**)	49.00(54)	615.41(**622.74**)	0.70(0.64)
	PMS-ABC	31.80(33)	604.43(599.06)	0.77(0.68)	24.00(26)	608.29(620.04)	0.71(0.58)

Table 3. Results of the C measure for the four algorithms in ten instances when the Asymmetric Delivery problem using objective functions OF1-OF2 is solved

OF1-OF2	Multiobjective Asymmetric Delivery Route based Fuel Consumption VRP										
A-B-CD	NSPSO	NSDE	NSGA II	ABC	KH	B-D-AC	NSPSO	NSDE	NSGA II	ABC	KH
NSPSO	-	0.25	0.82	0	0.49	NSPSO	-	0.63	0.89	0.08	0.43
NSDE	0.55	-	0.90	0.12	0.71	NSDE	0.30	-	0.63	0	0.23
NSGA II	0.04	0.08	-	0	0.09	NSGA II	0.02	0.17	-	0	0.10
ABC	0.87	0.69	0.85	-	0.84	ABC	0.66	0.96	0.83	-	0.86
KH	0.25	0.15	0.82	0.04	-	KH	0.43	0.60	0.70	0	-
A-C-BD	NSPSO	NSDE	NSGA II	ABC	KH	B-E-AD	NSPSO	NSDE	NSGA II	ABC	KH
NSPSO	-	0.59	0.78	0.37	0.47	NSPSO	-	0.15	0.75	0.04	0.45
NSDE	0.39	-	0.83	0.17	0.31	NSDE	0.73	-	0.93	0.23	0.69
NSGA II	0.02	0.04	-	0	0.10	NSGA II	0.07	0	-	0	0.03
ABC	0.63	0.66	0.69	-	0.67	ABC	0.82	0.38	0.77	-	0.74
KH	0.45	0.55	0.78	0.1	-	KH	0.49	0.30	0.86	0.04	-
A-D-BE	NSPSO	NSDE	NSGA II	ABC	KH	C-D-AE	NSPSO	NSDE	NSGA II	ABC	KH
NSPSO	-	0.16	0.74	0	0.32	NSPSO	-	0.38	0.82	0	0.37
NSDE	0.79	-	0.93	0.16	0.66	NSDE	0.65	-	0.82	0.1	0.53
NSGA II	0.15	0.02	-	0	0	NSGA II	0.07	0.07	-	0	0.16
ABC	0.83	0.57	0.93	-	0.83	ABC	0.88	0.77	0.88	-	0.92
KH	0.65	0.29	0.91	0.08	-	KH	0.54	0.36	0.76	0	-
A-E-BD	NSPSO	NSDE	NSGA II	ABC	KH	C-E-AB	NSPSO	NSDE	NSGA II	ABC	KH
NSPSO	-	0.21	0.89	0.03	0.12	NSPSO	-	0.45	0.79	0.21	0.61
NSDE	0.55	-	0.91	0.1	0.29	NSDE	0.48	-	0.81	0.15	0.67
NSGA II	0.03	0	-	0	0	NSGA II	0.13	0.13	-	0.09	0.20
ABC	0.76	0.67	0.87	-	0.57	ABC	0.73	0.64	0.79	-	0.76
KH	0.71	0.69	0.84	0.2	-	KH	0.25	0.26	0.68	0.09	-
B-C-AD	NSPSO	NSDE	NSGA II	ABC	KH	D-E-BC	NSPSO	NSDE	NSGA II	ABC	KH
NSPSO	-	0.28	0.86	0.13	0.27	NSPSO	-	0.16	0.38	0.17	0.32
NSDE	0.38	-	0.94	0.04	0.43	NSDE	0.50	-	0.40	0.30	0.57
NSGA II	0.04	0	-	0	0.04	NSGA II	0.56	0.64	-	0.52	0.68
ABC	0.82	0.72	0.78	-	0.78	ABC	0.82	0.68	0.52	-	0.68
KH	0.64	0.53	0.94	0.04	-	KH	0.44	0.27	0.38	0.13	-

algorithm, the algorithms used in the comparisons, were executed five times and the computational results of the best execution were used for the comparisons. In Tables 3, 4, 5 and 6 the comparisons in the four different problems for the coverage measure are presented. Finally, in Table 7 a comparison of all algorithms in all measures is presented. In Fig. 1 four selected representative Pareto fronts of the four algorithms are given.

Generally, based on all Tables and Figures we can say that the proposed algorithm gave very competitive results compared to four well known algorithms from the literature. The proposed algorithm compared to the PMS-ABC algorithm gives superior results in three out four measures and inferior results in the Coverage measure. However, with all other algorithms used in the comparisons, the proposed algorithm gives superior results in Coverage measure and competitive

Table 4. Results of the C measure for the four algorithms in ten instances when the Symmetric Delivery problem using objective functions OF1-OF2 is solved

OF1-OF2	Multiobjective Symmetric Delivery Route based Fuel Consumption VRP										
A-B	NSPSO	NSDE	NSGA II	ABC	KH	B-D	NSPSO	NSDE	NSGA II	ABC	KH
NSPSO	-	0.33	0.97	0.06	0.57	NSPSO	-	0.67	0.96	0.27	0.30
NSDE	0.42	-	0.97	0.13	0.53	NSDE	0.22	-	0.84	0.06	0.19
NSGA II	0.02	0	-	0.03	0.12	NSGA II	0	0.07	-	0	0
ABC	0.81	0.69	0.95	-	0.78	ABC	0.65	0.87	0.93	-	0.63
KH	0.33	0.27	0.90	0.09	-	KH	0.55	0.76	0.98	0.37	-
A-C	NSPSO	NSDE	NSGA II	ABC	KH	B-E	NSPSO	NSDE	NSGA II	ABC	KH
NSPSO	-	0.54	0.97	0.11	0.43	NSPSO	-	0.38	0.78	0.03	0.35
NSDE	0.34	-	0.98	0.08	0.21	NSDE	0.54	-	0.90	0.19	0.39
NSGA II	0	0	-	0	0	NSGA II	0.10	0.07	-	0.03	0.02
ABC	0.83	0.86	1.00	-	0.85	ABC	0.90	0.69	0.85	-	0.75
KH	0.49	0.63	1.00	0.05	-	KH	0.50	0.64	0.92	0.16	-
A-D	NSPSO	NSDE	NSGA II	ABC	KH	C-D	NSPSO	NSDE	NSGA II	ABC	KH
NSPSO	-	0.35	0.96	0.2	0.43	NSPSO	-	0.35	0.81	0.03	0.40
NSDE	0.59	-	0.82	0.17	0.43	NSDE	0.35	-	0.90	0.03	0.34
NSGA II	0	0.10	-	0.14	0.06	NSGA II	0.07	0.06	-	0	0
ABC	0.70	0.60	0.84	-	0.67	ABC	0.87	0.90	1.00	-	0.79
KH	0.43	0.33	0.86	0.17	-	KH	0.43	0.51	0.98	0.16	-
A-E	NSPSO	NSDE	NSGA II	ABC	KH	C-E	NSPSO	NSDE	NSGA II	ABC	KH
NSPSO	-	0.39	0.92	0	0.15	NSPSO	-	0.42	0.81	0.17	0.47
NSDE	0.60	-	0.86	0	0.28	NSDE	0.53	-	0.86	0.28	0.47
NSGA II	0.06	0.06	-	0	0	NSGA II	0.20	0.15	-	0.14	0.07
ABC	0.96	0.88	0.98	-	0.72	ABC	0.82	0.62	0.87	-	0.76
KH	0.72	0.65	0.98	0.11	-	KH	0.49	0.26	0.86	0.14	-
B-C	NSPSO	NSDE	NSGA II	ABC	KH	D-E	NSPSO	NSDE	NSGA II	ABC	KH
NSPSO	-	0.38	0.78	0.11	0.25	NSPSO	-	0.35	0.92	0.15	0.42
NSDE	0.63	-	0.91	0.11	0.17	NSDE	0.51	-	0.98	0.33	0.53
NSGA II	0.12	0.10	-	0.14	0.02	NSGA II	0.02	0	-	0.03	0.02
ABC	0.82	0.82	0.82	-	0.82	ABC	0.70	0.50	0.96	-	0.66
KH	0.57	0.57	0.84	0.07	-	KH	0.58	0.26	0.94	0.15	-

results in all other measures. The PMS-ABC algorithm gives superior results in the Coverage measure compared to all other algorithms and inferior results compared to all other algorithms in the other three measures. The best results in all other measures are produced by PMS-NSGA II algorithm with slightly better results from the three other algorithms (PMS-PSO, PMS-DE and PMS-KH) and much better results than the PMS-ABC algorithm. The PMS-NSGA II could be the most efficient algorithm if its results in the Coverage measure were competitive with the results of any of the other four algorithms. Also, as it can be seen

Table 5. Results of the C measure for the four algorithms in ten instances, when an Asymmetric Pick-up problem using objective functions OF1-OF3 is solved

OF1-OF3	Multiobjective Asymmetric Pick-up Route based Fuel Consumption VRP										
A-B-CD	NSPSO	NSDE	NSGA II	ABC	KH	B-D-AC	NSPSO	NSDE	NSGA II	ABC	KH
NSPSO	-	0.41	0.67	0.13	0.47	NSPSO	-	0.48	0.96	0.12	0.52
NSDE	0.43	-	0.76	0.26	0.53	NSDE	0.38	-	0.98	0.08	0.57
NSGA II	0.09	0.14	-	0.16	0.13	NSGA II	0	0	-	0	0.02
ABC	0.70	0.63	0.75	-	0.71	ABC	0.92	0.94	1.00	-	0.89
KH	0.39	0.20	0.73	0.26	-	KH	0.44	0.42	0.93	0.08	-
A-C-BD	NSPSO	NSDE	NSGA II	ABC	KH	B-E-AD	NSPSO	NSDE	NSGA II	ABC	KH
NSPSO	-	0.40	0.91	0.09	0.38	NSPSO	-	0.50	0.75	0.12	0.54
NSDE	0.49	-	0.82	0.27	0.38	NSDE	0.49	-	0.84	0.18	0.48
NSGA II	0.02	0.06	-	0.03	0.08	NSGA II	0.11	0.11	-	0.08	0.10
ABC	0.70	0.50	0.88	-	0.59	ABC	0.70	0.78	0.84	-	0.81
KH	0.57	0.38	0.82	0.18	-	KH	0.45	0.37	0.87	0.05	-
A-D-BE	NSPSO	NSDE	NSGA II	ABC	KH	C-D-AE	NSPSO	NSDE	NSGA II	ABC	KH
NSPSO	-	0.25	0.87	0.09	0.33	NSPSO	-	0.55	0.90	0.07	0.40
NSDE	0.63	-	0.87	0.36	0.33	NSDE	0.19	-	0.83	0.03	0.40
NSGA II	0.08	0.02	-	0.15	0.04	NSGA II	0.03	0.02	-	0.07	0.05
ABC	0.73	0.64	0.89	-	0.62	ABC	0.83	0.92	0.88	-	0.83
KH	0.58	0.62	0.92	0.39	-	KH	0.42	0.61	0.81	0.18	-
A-E-BD	NSPSO	NSDE	NSGA II	ABC	KH	C-E-AB	NSPSO	NSDE	NSGA II	ABC	KH
NSPSO	-	0.22	0.80	0.21	0.30	NSPSO	-	0.40	0.85	0.26	0.40
NSDE	0.65	-	0.82	0.29	0.58	NSDE	0.32	-	0.86	0.19	0.27
NSGA II	0.08	0.10	-	0.11	0.08	NSGA II	0.10	0.04	-	0	0
ABC	0.70	0.47	0.77	-	0.60	ABC	0.61	0.74	0.98	-	0.66
KH	0.57	0.43	0.85	0.29	-	KH	0.63	0.57	0.92	0.15	-
B-C-AD	NSPSO	NSDE	NSGA II	ABC	KH	D-E-BC	NSPSO	NSDE	NSGA II	ABC	KH
NSPSO	-	0.74	0.90	0.16	0.51	NSPSO	-	0.50	0.77	0.30	0.52
NSDE	0.18	-	0.68	0.06	0.24	NSDE	0.36	-	0.73	0.17	0.36
NSGA II	0	0.17	-	0.03	0.09	NSGA II	0.09	0.17	-	0.13	0.11
ABC	0.58	0.87	0.93	-	0.69	ABC	0.64	0.69	0.82	-	0.64
KH	0.32	0.57	0.87	0.19	-	KH	0.20	0.50	0.73	0.20	-

from Figure (1), the Pareto fronts produced by PMS-NSGA II are inferior from the ones of all the other algorithms. In total, the proposed algorithm (PMS-KH) is the most stable algorithm taking into account all the measures used in the comparisons.

Table 6. Results of the C measure for the four algorithms in ten instances when a Symmetric Pick-up problem using objective functions OF1-OF3 is solved

OF1-OF3	Multiobjective Symmetric Pick-up Route based Fuel Consumption VRP										
A-B	NSPSO	NSDE	NSGA II	ABC	KH	**B-D**	NSPSO	NSDE	NSGA II	ABC	KH
NSPSO	-	0.48	1.00	0.32	0.61	NSPSO	-	0.35	0.94	0.13	0.56
NSDE	0.31	-	1.00	0.28	0.61	NSDE	0.51	-	0.89	0.21	0.48
NSGA II	0	0	-	0	0.02	NSGA II	0.02	0.02	-	0.08	0.04
ABC	0.54	0.43	0.95	-	0.61	ABC	0.71	0.59	0.87	-	0.65
KH	0.24	0.24	0.95	0.20	-	KH	0.37	0.37	0.89	0.12	-
A-C	NSPSO	NSDE	NSGA II	ABC	KH	**B-E**	NSPSO	NSDE	NSGA II	ABC	KH
NSPSO	-	0.30	0.84	0.12	0.21	NSPSO	-	0.50	0.68	0	0.20
NSDE	0.75	-	0.90	0.32	0.38	NSDE	0.43	-	0.71	0	0.29
NSGA II	0.08	0	-	0	0	NSGA II	0.15	0.11	-	0	0.02
ABC	0.89	0.59	0.95	-	0.53	ABC	0.95	0.89	0.92	-	0.82
KH	0.70	0.39	0.95	0.24	-	KH	0.48	0.61	0.95	0.12	-
A-D	NSPSO	NSDE	NSGA II	ABC	KH	**C-D**	NSPSO	NSDE	NSGA II	ABC	KH
NSPSO	-	0.23	0.86	0.19	0.24	NSPSO	-	0.65	0.88	0.23	0.29
NSDE	0.62	-	0.94	0.35	0.48	NSDE	0.13	-	0.81	0.04	0.02
NSGA II	0.07	0.04	-	0.15	0.09	NSGA II	0.07	0.14	-	0.14	0.04
ABC	0.78	0.49	0.94	-	0.65	ABC	0.52	0.72	0.84	-	0.58
KH	0.52	0.17	0.89	0.27	-	KH	0.46	0.78	0.88	0.27	-
A-E	NSPSO	NSDE	NSGA II	ABC	KH	**C-E**	NSPSO	NSDE	NSGA II	ABC	KH
NSPSO	-	0.49	0.85	0.18	0.44	NSPSO	-	0.58	0.85	0.26	0.50
NSDE	0.30	-	0.85	0.18	0.44	NSDE	0.36	-	0.92	0.43	0.28
NSGA II	0.09	0.12	-	0.18	0.06	NSGA II	0.03	0.03	-	0	0.02
ABC	0.54	0.60	0.86	-	0.52	ABC	0.58	0.58	0.93	-	0.37
KH	0.28	0.47	0.93	0.37	-	KH	0.42	0.50	0.95	0.43	-
B-C	NSPSO	NSDE	NSGA II	ABC	KH	**D-E**	NSPSO	NSDE	NSGA II	ABC	KH
NSPSO	-	0.31	0.95	0.27	0.49	NSPSO	-	0.33	0.89	0.19	0.40
NSDE	0.57	-	1.00	0.42	0.53	NSDE	0.49	-	0.94	0.53	0.42
NSGA II	0	0	-	0	0.06	NSGA II	0.02	0	-	0.03	0.02
ABC	0.67	0.40	0.98	-	0.63	ABC	0.67	0.43	0.92	-	0.52
KH	0.37	0.45	0.90	0.19	-	KH	0.44	0.43	0.92	0.50	-

Table 7. Comparisons of all algorithms in all measures

C measure							L	M_k	Δ
	PMS-ABC	PMS-NSPSO	PMS-NSDE	PMS-NSGA II	PMS-KH				
PMS-ABC		40	38	39	39	PMS-ABC	25.91	573.03	0.76
PMS-NSPSO	0		16	39	16	PMS-NSPSO	45.92	594.76	0.66
PMS-NSDE	2	24		39	19	PMS-NSDE	45.70	594.21	0.66
PMS-NSGA II	1	1	1		1	PMS-NSGA II	55.74	595.15	0.63
PMS-KH	1	24	21	39		PMS-KH	46.16	593.73	0.66

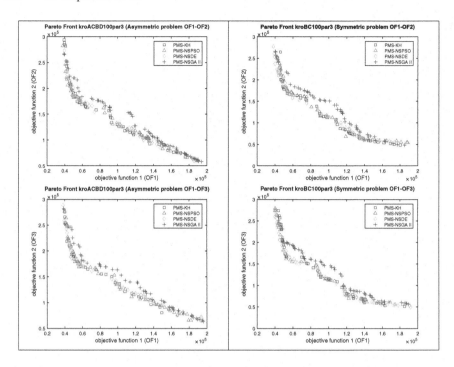

Fig. 1. Pareto fronts of the four algorithms for the instances "A-C-BD" and "B-C".

4 Conclusions and Future Research

In this paper, we proposed an algorithm (PMS-KH) for solving four newly formulated multiobjective fuel consumption multi-depot vehicle routing problems (symmetric and asymmetric pick-up and symmetric and asymmetric delivery cases). The proposed algorithm was compared with other four algorithms, the PMS-ABC, the PMS-NPSO, the PMS-NSDE and PMS-NSGA II. In general, in the four different problems, the PMS-KH algorithm performs slightly better than the other four algorithms in most measures, as we analyzed in the Computational Results section. As expected, the behavior of the algorithms was slightly different when a symmetric and an asymmetric problem was solved. Our future research will be, mainly, focused on PMS-KH algorithm in other multiobjective combinatorial optimization problems.

References

1. Demir, E., Bektaş, T., Laporte, G.: A review of recent research on green road freight transportation. Eur. J. Oper. Res. **237**(3), 775–793 (2014)
2. Gandomi, A.H., Alavi, A.H.: Krill Herd: a new bio-inspired optimization algorithm. Commun. Nonlinear Sci. Numer. Simul. **17**(12), 4831–4845 (2012)
3. Li, J., Wang, R., Li, T., Lu, Z., Pardalos, P.: Benefit analysis of shared depot resources for multi-depot vehicle routing problem with fuel consumption. Transp. Res. Part D: Transp. Environ. **59**, 417–432 (2018)

4. Lin, C., Choy, K.L., Ho, G.T.S., Chung, S.H., Lam, H.Y.: Survey of green vehicle routing problem: past and future trends. Expert Syst. Appl. **41**(4), 1118–1138 (2014)

5. Marti, R., Pardalos, P.M., Resende, M.G.: Handbook of Heuristics. Springer (2018). ISBN 978-3-319-07123-7

6. Montoya-Torres, J.R., Franco, J.L., Isaza, S.N., Jimenez, H.F., Herazo-Padilla, N.: A literature review on the vehicle routing problem with multiple depots. Comput. Ind. Eng. **79**, 115–129 (2015)

7. Psychas, I.-D., Marinaki, M., Marinakis, Y.: A parallel multi-start NSGA II algorithm for multiobjective energy reduction vehicle routing problem. In: Gaspar-Cunha, A., Henggeler Antunes, C., Coello, C.C. (eds.) EMO 2015. LNCS, vol. 9018, pp. 336–350. Springer, Cham (2015). https://doi.org/10.1007/978-3-319-15934-8_23

8. Psychas, I.D., Marinaki, M., Marinakis, Y., Migdalas, A.: Non-dominated sorting differential evolution algorithm for the minimization of route based fuel consumption multiobjective vehicle routing problems. Energy Syst. **8**, 785–814 (2016)

9. Psychas, I.-D., Marinaki, M., Marinakis, Y., Migdalas, A.: Minimizing the fuel consumption of a multiobjective vehicle routing problem using the parallel multi-start NSGA II algorithm. In: Kalyagin, V.A., Koldanov, P.A., Pardalos, P.M. (eds.) Models, Algorithms and Technologies for Network Analysis. SPMS, vol. 156, pp. 69–88. Springer, Cham (2016). https://doi.org/10.1007/978-3-319-29608-1_5

10. Psychas, I.-D., Marinaki, M., Marinakis, Y., Migdalas, A.: Parallel multi-start non-dominated sorting particle swarm optimization algorithms for the minimization of the route-based fuel consumption of multiobjective vehicle routing problems. In: Butenko, S., Pardalos, P.M., Shylo, V. (eds.) Optimization Methods and Applications. SOIA, vol. 130, pp. 425–456. Springer, Cham (2017). https://doi.org/10.1007/978-3-319-68640-0_20

11. Rapanaki, E., Psychas, I.D., Marinaki, M., Marinakis, Y., Migdalas, A.: A clonal selection algorithm for multiobjective energy reduction multi-depot vehicle routing problem. In: Nicosia, G., Pardalos, P., Giuffrida, G., Umeton, R., Sciacca, V. (eds.) Machine Learning, Optimization, and Data Science LOD 2018. LNCS, vol. 11331, pp. 381–393. Springer, Cham (2019). https://doi.org/10.1007/978-3-030-13709-0_32

12. Rapanaki, E., Psychas, I.-D., Marinaki, M., Marinakis, Y.: An artificial bee colony algorithm for the multiobjective energy reduction multi-depot vehicle routing problem. In: Matsatsinis, N.F., Marinakis, Y., Pardalos, P. (eds.) LION 2019. LNCS, vol. 11968, pp. 208–223. Springer, Cham (2020). https://doi.org/10.1007/978-3-030-38629-0_17

13. Toth, P., Vigo, D.: Vehicle Routing: Problems, Methods and Applications, 2nd edn. MOS-SIAM Series on Optimization, SIAM, Philadelphia (2014)

14. Wang, G.-G., Guo, L., Gandomi, A.H., Hao, G.-S., Wang, H.: Chaotic Krill Herd algorithm. Inf. Sci. **274**, 17–34 (2014)

15. Wang, G.-G., Gandomi, A.H., Alavi, A.H.: Stud krill herd algorithm. Neurocomputing **128**, 363–370 (2014)

16. Wang, G.-G., Gandomi, A.H., Alavi, A.H., Gong, D.: A comprehensive review of krill herd algorithm: variants, hybrids and applications. Artif. Intell. Rev. **51**(1), 119–148 (2017). https://doi.org/10.1007/s10462-017-9559-1

17. Xiao, Y., Zhao, Q., Kaku, I., Xu, Y.: Development of a fuel consumption optimization model for the capacitated vehicle routing problem. Comput. Oper. Res. **39**(7), 1419–1431 (2012)

Optimal Broadcast Strategy
in Homogeneous Point-to-Point Networks

Franco Robledo$^{(\boxtimes)}$, Pablo Rodríguez-Bocca, and Pablo Romero

Instituto de Computación, INCO Facultad de Ingeniería,
Universidad de la República, Montevideo, Uruguay
{frobledo,prbocca,promero}@fing.edu.uy

Abstract. In this paper we address a fundamental combinatorial optimization problem in communication systems. A fully-connected system is modeled by a complete graph, where all nodes have identical capacities. A message is owned by a singleton. If he/she decides to forward the message simultaneously to several nodes, he/she will take longer, with respect to a one-to-one forwarding scheme. The only rule in this communication system is that a message can be forwarded by a node that owns the message. The *makespan* is the time when the message is broadcasted to all the nodes. The problem under study is to select the communication strategy that minimizes both the makespan and the average waiting time among all the nodes. A previous study claims that a sequential or *one-to-one* forwarding scheme minimizes the average waiting time, but they do not offer a proof. Here, a formal proof is included. Furthermore, we show that the sequential strategy minimizes the makespan as well. A discussion of potential applications is also included.

Keywords: Communication system · Forwarding · Waiting time · Makespan

1 Motivation

The Internet is supported by the client-server architecture, where users connect with a specific server to download data. This architecture has some benefits. The service is both simple and highly predictable. However, the server infrastructure is not scalable when demand is increased. A natural idea to overcome this scalability issue is to consider content popularity, where the most popular contents can be shared by the users. The server invites users to communicate and offer those files which are normally replicated in the network. An abstraction of this concept is accomplished with peer-to-peer systems (P2P for short). They are self-organized virtual communities developed on the Internet Infrastructure, where users, called peers, share resources (content, bandwidth, CPU-time, memory) to others, basically because they have common interests. From a game-theoretic point of view, cooperation is better than competition. From an engineering point of view, we understand that the power of user-cooperation

© Springer Nature Switzerland AG 2020
G. Nicosia et al. (Eds.): LOD 2020, LNCS 12565, pp. 448–457, 2020.
https://doi.org/10.1007/978-3-030-64583-0_40

in P2P systems is maximized, but the real-life design is jeopardized by other factors. Indeed, broadband resources are better exploited with cooperation. The altruistic behaviour in P2P networks is achieved with incentives, using a *give-to-get* concept [2, 14]. Nevertheless, the design of a resilient P2P network has several challenges. Indeed, the Internet access infrastructure is usually asymmetric, hindering peer exchange; peers arrive and depart the system when they wish [15]; free-riders exploit network resources but do not contribute with the system; a failure in the underlying network usually damage the P2P service; there is an explicit trade-off between the full knowledge of the network (topology, peers resources) and payload, which directly impacts in the throughput and network performance.

The main purpose of this paper is to understand the best forwarding schemes in an ideal abstract setting, and how different forwarding schemes are used in real-life systems. Even though we motivate this paper by P2P systems, our main result apply to several communication systems, such as scheduling in parallel unrelated machines, social networks and content delivery networks. Our work is inspired by a fundamental problem posed for the first time by Qiu and Srikant, where they state that *it should be clear that a good strategy* is the one-to-one forwarding scheme [13]. Even though the authors study the service capacity of a file sharing peer-to-peer system, its formulation is general enough. For practical purposes they find a closed formula for the average waiting time following a one-to-one forwarding scheme when the population N is a power of two. In [9], a formal proof that the one-to-one forwarding scheme achieves the minimum waiting time is included, when the population is a power of two. Here, we formally prove that it is not only good, but also optimal, for both makespan and waiting time measures. The result holds for an arbitrary population size. We remark that the optimum forwarding scheme rarely appears in real-life systems. We discuss this phenomenon showing the gap between this theoretical result and real-life implementations of communication systems. The main contributions of this paper are two-fold:

1. The best forwarding scheme in complete homogeneous communication networks is found.
2. The gap between the best theoretical forwarding scheme and real-life implementations is discussed.

This paper is organized as follows. The problem under study is presented in Sect. 2. The mathematical analysis provides a full solution of the problem, which is derived in Sect. 3. The gap between theory and real-life applications is considered in Sect. 4. Section 5 contains the main conclusions and trends for future work.

2   Problem

We are given a full network composed by N peers with identical capacity b (in bits per second), and a message with size M (measured in bits). A single node that belongs to the network owns the message, and at time $t_1 = 0$ he/she forwards the message to one or several peers belonging to the network.

Let us denote $\tau = M/b$ the time-slot following one-to-one forwarding time. If some peer sends the message to c other peers, it will take $c\tau$ seconds to perform the forwarding task. Let us denote $0 = t_1 \leq t_2 \leq \ldots \leq t_N$ the corresponding completion times of the N peers in this cooperative system. The *makespan* is t_N, while the *average waiting time*, $\bar{t}$, is the average over the set $\{t_1, \ldots, t_N\}$. Clearly, $\bar{t} \leq t_N$.

In a *one-to-many* forwarding scheme, every peer selects a fixed number c of peers to forward the message. In general, in a simultaneous forwarding scheme there is some peer i that, at time t_i, simultaneously forwards the message to more than one peer. In contrast, the only remaining strategy is a sequential or *one-to-one* forwarding strategy.

The goal in the Minimum Point-to-Point Makespan (MPTPM) is to minimize t_N:

$$\min_{s \in \mathcal{S}} \max_{1 \leq i \leq N} t_i, \tag{1}$$

being s a member belonging to the family of forwarding strategies $\mathcal{S}$, where each node decides to send the message either to one or to multiple peers, and may decide a delay to start the transmission as well.

An analogous problem is the Minimum Point-to-Point Waiting Time (MPT-PWT), where the goal is to minimize $\bar{t}$ among all possible forwarding strategies:

$$\min_{s \in \mathcal{S}} \frac{1}{N} \sum_{i=1}^{N} t_i, \tag{2}$$

In a *one-to-many* forwarding scheme, every peer selects a fixed number c of peers to forward the message. In general, in a simultaneous forwarding scheme there is some peer i that, at time t_i, simultaneously forwards the message to more than one peer. In contrast, the only remaining strategy is a sequential or *one-to-one* forwarding strategy. A peer can decide to delay the transmission as well.

Here we formally prove that the one-to-one forwarding strategy is optimal for both the MPTPM and MPTPWT. For short, we will use $n = \lceil log_2(N) \rceil$ and $n_c = \lceil log_c(N) \rceil$.

3   Solution

A straight calculation provides the makespan and average waiting time in the one-to-one forwarding scheme:

Lemma 1. *The makespan in the one-to-one forwarding scheme is $n\tau$.*

Proof. The message is fully owned by 2^i peers at time $i\tau$, for $i = 1, \ldots, n - 1$. The remaining $N - 2^{n-1}$ peers receive the message at time $n\tau$.   ∎

Lemma 2. *The average waiting time in the one-to-one forwarding scheme is $\bar{t} = \frac{\tau}{N}(nN - 2^n + 1)$.*

Proof.

$$
\begin{aligned}
\bar{t} &= \frac{1}{N}[\sum_{i=1}^{n-1} 2^{i-1}i\tau + (N - 2^{n-1})\tau] \\
&= \frac{\tau}{N}[(n2^{n-1} - 2^n + 1) + (N - 2^{n-1})n] \\
&= \frac{\tau}{N}(nN - 2^n + 1).
\end{aligned}
$$

∎

Since $n = \lceil log_2(N) \rceil$, we conclude that both makespan and average waiting time grow logarithmically with the population of the system and linearly with respect to the time-slot τ when the one-to-one strategy is considered. Let us contrast the result with a one-to-many strategy in what follows, where each peer forwards the message to $c - 1$ different peers, for some $c > 2$.

Lemma 3. *The makespan in the one-to-many forwarding scheme of type $c - 1$ is $n_c(c - 1)\tau$.*

Proof. By the definition of one-to-many forwarding schemes of type $c - 1$, there are $N_i = c^{i-1}(c - 1)$ peers whose completion time is $T_i = i(c - 1)\tau$. Therefore, the message is fully owned by c^i peers at time T_i, for $i = 1, \ldots, n_c - 1$. The remaining $N - c^{n_c - 1}$ peers receive the message at time $T_{n_c} = n_c(c - 1)\tau$.   ∎

Lemma 4. *The average waiting time in a one-to-many forwarding scheme of type $c - 1$ is $\bar{t}_{c-1} = \frac{\tau}{N}[n_c(c - 1)N - c^{n_c} + 1]$.*

Proof.

$$
\begin{aligned}
\bar{t}_{c-1} &= \frac{1}{N}[\sum_{i=1}^{n_c - 1} N_i T_i + (N - c^{n_c - 1})T_{n_c} \\
&= \frac{1}{N}[\sum_{i=1}^{n_c - 1} c^{i-1}(c - 1)i(c - 1)\tau + (N - c^{n_c - 1})T_{n_c}] \\
&= \frac{\tau}{N}[n_c c^{n_c - 1}(c - 1) - (c^{n_c} - 1) + (N - c^{n_c - 1})T_{n_c}] \\
&= \frac{\tau}{N}[n_c(c - 1)N - c^{n_c} + 1].
\end{aligned}
$$

∎

On one hand, we can check that $\bar{t}_{c-1}$ equals $\bar{t}$ when $c = 2$, as expected. In fact, the one-to-one strategy is the one-to-many if $c - 1 = 1$. On the other, $\bar{t}_{c-1} > \bar{t}$ for every $c > 2$. The makespan is studied first:

Lemma 5. *The makespan in the one-to-one strategy is never greater than in the one-to-many strategy.*

Proof. If $c = 2$ we see that $n_2 = n$, so $(c - 1)n_c = n$. It suffices to prove that $(c - 1)n_c \geq n$ for any $c \geq 3$, being $n = \lceil log_2(N) \rceil$ and $n_c = \lceil log_c(N) \rceil$:

$$(c - 1)n_c \geq log_c(N^{c-1}) = (c - 1)log_c(2)log_2(N)$$
$$= log_c(2^{c-1})log_2(N) > log_2(N);$$

where the last inequality follows from the fact that $2^{c-1} > c$ whenever $c \geq 3$. Since $(c-1)n_c$ is an integer, we obtain that $(c-1)n_c \geq \lceil log_2(N) \rceil$, and the result follows. ∎

A technical lemma will be used in the main result:

Lemma 6. *Given two partitions of $N = \sum_{i=1}^{m} x_i = \sum_{i=1}^{m} y_i$ such that $x_i \geq y_i \geq 0, \forall i = 1,\ldots,m - 1$ and $0 \leq x_m < y_m$. Consider an arrange of times $0 \leq t_1 \leq t_2 \leq \ldots \leq t_m$, and a partition for each t_i, $t_i = \sum_{j=1}^{m_i} t_{ij}$, where $0 \leq t_{ij} \leq t_i$. Given any related partition of x_i, $x_i = \sum_{j=1}^{m_i} x_{ij}$, then $\overline{W}_x = \frac{1}{N}\sum_{i=1}^{m}\sum_{j=1}^{m_i} x_{ij}t_{ij}$ is strictly lower than $\overline{W}_y = \frac{1}{N}\sum_{i=1}^{m} y_i t_i$.*

Proof.

$$\overline{W}_x = \frac{1}{N}\sum_{i=1}^{m}\sum_{j=1}^{m_i} x_{ij}t_{ij} < \frac{1}{N}\sum_{i=1}^{m} x_i t_i$$

$$= \frac{1}{N}\left(\sum_{i=1}^{m-1} x_i t_i + x_m t_m\right)$$

$$= \frac{1}{N}\left(\sum_{i=1}^{m-1}(x_i - y_i)t_i + \sum_{i=1}^{m-1} y_i t_i - y_m t_m + x_m t_m\right)$$

$$= \overline{W}_y + \sum_{i=1}^{m-1}(x_i - y_i)t_i - (y_m - x_m)t_m$$

$$= \overline{W}_y + \sum_{i=1}^{m-1}(x_i - y_i)t_i - \left(\sum_{i=1}^{m}(x_i - y_i)\right)t_m < \overline{W}_y.$$

∎

In words, if more peers own the message at any time t_i using strategy x instead of y ($x_i \geq y_i, i = 1,\ldots,m - 1$) and the population is constant (N is constant, so $x_m < y_m$), then x outperforms y in terms of average waiting time.

Lemma 7. *The average waiting time in the one-to-one strategy is never greater than in the one-to-many strategy.*

Proof. When $t_i = i(c-1)\tau$ we know that c^i peers own the message following the one-to-many forwarding scheme of type $c-1$ versus $2^{i(c-1)}$ following the one-to-one forwarding scheme. By Lemma 6, it suffices to prove that $2^{i(c-1)} \geq c^i$ whenever $c \geq 3$. Taking logarithms on both sides yields $c - 1 \geq log_2(c)$, which holds for all $c \geq 2$. ∎

In order to illustrate the previous results, Figs. 1 and 2 present the makespan and average waiting times respectively for one-to-one and one-to-many strategies in representative cases.

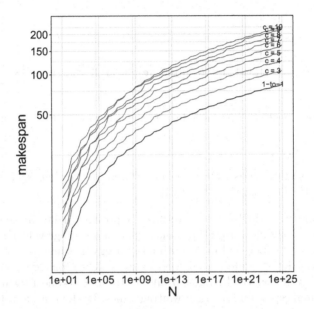

Fig. 1. Makespan as a function of the size of the system N, for $c \in \{1, \ldots, 10\}$).

Lemma 8 (Local Replacement). *If we are given a strategy where some peer x forwards the message to k new peers in a given time-slot $[t, t+T]$, and there exists an alternative strategy where x forwards the message to $k' > k$ peers in the same time-slot, then the local replacement for the alternative strategy in x reduces both the makespan and average waiting time if all the k' nodes behave as in the original strategy.*

Proof. During the specific time-slot $[t, t+T]$, the message is fully owned by more peers. By Lemma 6, the local replacement has lower average waiting time. Analogously, there are more successors of x, so they feed more peers and the makespan is lower as well. ∎

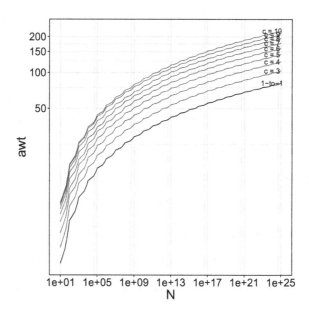

Fig. 2. Average waiting time as a function of the size of the system N, for $c \in \{1, \ldots, 10\}$).

Theorem 1 (Main Result). *The one-to-one forwarding scheme is optimal for both the MPTPM and the MPTPWT.*

Proof. If some peer deliberately produces a positive delay in the forwarding, there is a corresponding shift in both makespan and average waiting time. Therefore, delays are not included in an optimal strategy. If some peer x forwards the message to $c - 1 > 1$ nodes, we can consider a local replacement into the one-to-one strategy for x. By Lemmas 5 and 7, the one-to-one forwarding scheme offers lower makespan and average waiting times. By Lemma 8, a local replacement improves both measures. A local replacement is conducted in every node that forwards the message to many nodes. The result is a one-to-one forwarding scheme. ∎

It is worth to mention that a historical problem from telephonic services is the Minimum Broadcast Time or MBT for short [5]. In the MBT, we are given a simple graph, and a target node which owns the message. The goal is to select a forwarding one-to-one strategy, in order to minimize the broadcast time (i.e., the makespan). Observe that we studied complete networks. However, the makespan in an arbitrary simple (non-complete) graph is extremely challenging. In fact, the problem is formally known as the Minimum Broadcast Time (MBT), and it belongs to the class of $\mathcal{NP}$-Complete problems [6]. The MBT is equivalent to find a spanning tree rooted at the holder of the message with the minimum makespan. The hardness of the MBT promotes the development of metaheuristics. In particular, a Greedy randomized heuristic is already available in the literature, together with an efficient Integer Linear Programming (ILP) formulation for the MBT [4]. In the following result, we consider the MPTPM for general, for non-complete networks, here called MPTPMNC:

Theorem 2. *Finding the optimal strategy for the MPTPMNC belongs to the class of $\mathcal{NP}$-Hard problems.*

Proof. Consider the MPTPMNC in general, for an arbitrary graph G and target node v. By an iterative application of Lemma 8, a one-to-many strategy can by replaced by a one-to-one strategy, minimizing the makespan. Then, the globally optimum solution for the MPTPMNC is precisely the globally optimum solution for the MBT. Then, the MPTPMNC is at least as hard as the MBT. Since the latter belongs to the class of $\mathcal{NP}$-Hard problems [6], the result follows. ∎

4   Discussion

As far as we know, the MPTPWT was posed for the first time by Yang and de Veciana [13]. The authors study the service capacity of a file sharing peer-to-peer system, and the problem under study serves as a fluid model for replication. They literally state that *it should be clear that a good strategy* is the one-to-one forwarding scheme. For practical purposes they find a closed formula for the average waiting time following a one-to-one forwarding scheme when N is a power of two. In [9], a formal proof of Lemma 7 is provided when the population is a power of two. Here, we formally prove that it is not only good, but also optimal, for both makespan and waiting time measures (this is, for the MPTPM as well). The result holds for an arbitrary population size.

Theorem 1 is counterintuitive, and could be used in several fields of knowledge. For instance, the earliest-finish-time in the context of parallel computing systems is precisely our makespan, and forwarding strategies are identified with a formal scheduling on this machines [1, 11]. The main goal in a Content Delivery Network is to minimize the delivery time, which is strictly related with makespan and average waiting time [7, 8, 12]. The time needed to distribute information in a social network, or a virus by an epidemic, are one of the main factors studied in these disciplines [3, 10]. Several real networks use one-to-many forwarding schemes. This fact suggests that in practice at least one assumption does not hold. First, we remark that full connectivity holds in overlay networks, but does not hold in most real-life scenarios, such as social networks. Second, there is no matching between modelling and reality when identical capacity is assumed. Last but not least, in an information-centric network the behaviour of nodes could be affected with information.

5   Conclusions and Trends for Future Work

In this paper we show that a one-to-one forwarding scheme provides both the lowest makespan and average waiting time, under complete homogeneous networks. The merit of this strategy was suggested by previous authors in the context of peer-to-peer systems for average waiting times. Forwarding schemes have a direct implication in many different contexts, such as scheduling in parallel unrelated machines, social networks and content delivery networks. As a future work, we would like to extend our analysis to incomplete graphs with heterogeneous and dynamic nodes. Observe that the optimum forwarding scheme is completely deterministic. Furthermore, we would like to better understand the gap between the theoretical predictions from this paper and real-life applications such as social networks and cellular systems.

Acknowledgment. This work is partially supported by MATHAMSUD 19-MATH-03 Raredep, *Rare events analysis in multi-component systems with dependent components*, STICAMSUD 19-STIC-01 ACCON, *Algorithms for the capacity crunch problem in optical networks* and Fondo Clemente Estable *Teoría y Construcción de Redes de Máxima Confiabilidad*.

References

1. Casavant, T.L., Kuhl, J.G.: A taxonomy of scheduling in general-purpose distributed computing systems. IEEE Trans. Softw. Eng. **14**(2), 141–154 (1988)
2. Cohen, B.: Incentives Build Robustness in BitTorrent. vol. 1, 1–5 May 2003. https://www.bramcohen.com
3. Daley, D.J., Gani, J., Gani, J.M.: Epidemic Modelling: An Introduction, vol. 15. Cambridge University Press (2001)
4. de Sousa, A., Gallo, G., Gutierrez, S., Robledo, F., Rodríguez-Bocca, P., Romero, P.: Heuristics for the minimum broadcast time. Electron. Notes Discrete Math. **69**, 165–172 (2018). Joint EURO/ALIO International Conference 2018 on Applied Combinatorial Optimization (EURO/ALIO 2018)
5. Farley, A.M.: Broadcast time in communication networks. SIAM J. Appl. Math. **39**(2), 385–390 (1980)
6. Garey, M.R., Johnson, D.S.: Computers and Intractability: A Guide to the Theory of NP-Completeness. W. H. Freeman & Company, New York (1979)
7. Liu, Y., Guo, Y., Liang, C.: A survey on peer-to-peer video streaming systems. Peer-to-Peer Netw. Appl. **1**(1), 18–28 (2008)
8. Qiu, D., Srikant, R.: Modeling and performance analysis of bittorrent-like peer-to-peer networks. In: Proceedings of the 2004 Conference on Applications, Technologies, Architectures, and Protocols for Computer Communications, SIGCOMM 2004, pp. 367–378. ACM, New York (2004)
9. Romero, P.: Mathematical analysis of scheduling policies in peer-to-peer video streaming networks. Ph.D. thesis, Universidad de la República, Montevideo, Uruguay, November 2012
10. Saito, K., Nakano, R., Kimura, M.: Prediction of information diffusion probabilities for independent cascade model. In: Lovrek, I., Howlett, R.J., Jain, L.C. (eds.) KES 2008. LNCS (LNAI), vol. 5179, pp. 67–75. Springer, Heidelberg (2008). https://doi.org/10.1007/978-3-540-85567-5_9
11. Topcuoglu, H., Hariri, S., Wu, M.-Y.: Performance-effective and low-complexity task scheduling for heterogeneous computing. IEEE Trans. Parallel Distrib. Syst. **13**(3), 260–274 (2002)
12. Vakali, A., Pallis, G.: Content delivery networks: status and trends. IEEE Internet Comput. **7**(6), 68–74 (2003)

13. Yang, X., de Veciana, G.: Service capacity of peer to peer networks. In: Twenty-Third Annual Joint Conference of the IEEE Computer and Communications Societies (INFOCOM 2004), vol. 4, pp. 2242–2252 (2004)
14. Yu, B., Singh, M.P.: Incentive Mechanisms for Peer-to-Peer Systems, pp. 77–88. Springer, Heidelberg (2005)
15. Zhou, Y., Chen, L., Yang, C., Chiu, D.M.: Video popularity dynamics and its implication for replication. IEEE Trans. Multimedia **17**(8), 1273–1285 (2015)

Structural and Functional Representativity of GANs for Data Generation in Sequential Decision Making

Ali el Hassouni[1,2(✉)] [iD], Mark Hoogendoorn[1] [iD], A. E. Eiben[1] [iD],
and Vesa Muhonen[2] [iD]

[1] Department of CS, Vrije Universiteit Amsterdam, Amsterdam, The Netherlands
{a.el.hassouni,m.hoogendoorn,g.eiben}@vu.nl
[2] Data Science and Analytics, Mobiquity Inc., Amsterdam, The Netherlands
{aelhassouni,v.muhonen}@mobiquityinc.com

Abstract. In many sequential decision making problems progress is predominantly based on artificial data sets. This can be attributed to insufficient access to real data. Here we propose to mitigate this by using generative adversarial networks (GANs) to generate representative data sets from real data. Specifically, we investigate how GANs can generate training data for reinforcement learning (RL) problems. We distinguish structural properties (does the generated data follow the distribution of the original data), functional properties (is there a difference between the evaluation of policies for generated and real life data), and show that with a relatively small number of data points (a few thousand) we can train GANs that generate representative data for classical control RL environments.

1 Introduction

Machine learning has proven to be widely and efficaciously applicable in the real-world. Many of these applications rely on vast amounts of data. However, a great deal of real-world problems are sequential decision making problems in nature where long-term dynamics of an environment must be taken into account [19]. Typically, these problems are solved by RL using training data consisting of sequences of observations, actions and rewards. RL as a field has made groundbreaking advancements possible in many domains [4,5,15,23,24]. Generally, this progress is predominantly noticeable in artificial domains. State-of-the-art results in RL are obtained with agents that learn to play games in simulation environments such as OpenAI Gym [4] and Atari [15].

The application of RL to real-world scenarios has proven to be hard [6]. For many real-world applications, it is not possible to develop policies directly but instead these need to be learned from logs generated by a different behavioral policy [6]. In these cases, we are training the new policy to improve upon an existing policy. To learn and evaluate effective and optimal policies, large sets

G. Nicosia et al. (Eds.): LOD 2020, LNCS 12565, pp. 458–471, 2020.
https://doi.org/10.1007/978-3-030-64583-0_41

of logged experiences are required [6, 11]. Such data should be sufficiently representative for the problem at hand. Unfortunately, poor coverage of experiences from the real-world is often observed, leading to insufficient representative data. It can be an onerous task to accumulate a wide range of historical logs generated by a particular behaviour policy.

Generative models are not exempt from the limitations posed by the insufficiency of representative training data. However, recent developments in GANs [9] prove that these models are effective at synthesizing data at an unprecedented realism. We study the use of GANs to synthesize experiences based on a limited amount of data to help overcome the problems with RL training data sketched above. Consequently, we pose our main research question: **is generative modeling able to generate representative training data by synthesizing sequences of experiences for RL tasks?**

In order to answer this question, we need means to operationalize the term *representative*. We therefore introduce a framework to evaluate this property of the data. Within this framework, we introduce two properties of data representativity - the *structural* and *functional* representativity properties. To measure the structural representativity, we compare the synthesized data to the original sample and perform statistical evaluations to quantify the structural representativity of the generative model. As a measure for functional representativity we compare the difference in outcome of applying off-policy evaluation on real and generated data. We experiment with a number of RL problems taken from the OpenAI classic control suite [4]: CartPole and LunarLandar, and provide different amounts of original data to the GAN and study the impact on each of the criteria identified above.

This paper is organized as follows. First, we will discuss related work in Sect. 2, followed by the explanation of our method in Sect. 3. In Sect 4 we present our experimental setup. We present the results in Sect 5. Section 6 concludes the paper.

2   Related Work

Data representativity can be viewed from several perspectives - the ability to generate the right data for the tasks at hand (i.e. structural representativity), and the effectiveness of the synthesized data at improving this task (i.e. functional representativity). We address these aspects one by one and we refer to each part in the related work section.

Challenges in Real-World Reinforcement Learning. State-of-the-art developments in RL are frequently obtained with agents that learn to play games [5, 23, 24] in simulation environments such as OpenAI Gym and Atari. Although much work has been done in the area, limited work has been done in the area intersecting generative models and RL for domains with practical constraints. Furthermore, research has shown that applications of RL in the real-world still have challenges to overcome. [Dulac-Arnold, et al. 2019] mention that training off-line on data logged from a different behavior policy and learning on a set of limited samples as some of the challenges to be overcome [6].

GANs for Sequential Data. RL techniques [19,24] are ideally suited for sequential decision making problems [10]. In domains such as health care, work has just begun to explore computational approaches. Typically, there is lack of data in real-life applications which sometimes makes deep learning driven RL techniques unsuitable [6]. Recently, work started exploring the use of GANs [9] to generate data that can be utilized to resolve the issue with lack of data.

GANs in Reinforcement Learning. [Tseng et al. 2017] employed deep RL techniques to develop automated radiation protocols for patients with lung cancer using historical treatment plans [22]. They use a GAN to learn the characteristics of the patients from a relatively small dataset. Secondly, they use a deep neural network to reconstruct an artificial environment for radiotherapy. The obtained model estimates the underlying Markov Decision Process for adaptation of personalized radiotherapy patients' treatment courses [22]. Finally, they use a deep Q-network to choose the optimal dose during treatments. They showed that these techniques that partially rely on GANs are feasible and promising for achieving results that are similar to those achieved by clinicians. Contrary to our work, the GAN was not systematically tested for its ability to generate representative data which could pose safety concerns in the domain at hand.

[Liu et al. 2019] combined RL and GANs in such a manner that a self-improving process is built upon a policy improvement operator. The agent is developed iteratively to imitate behaviors that are generated by the operator (cf. [14]). [Antoniou et al. 2017] demonstrated that conditional GANs can be employed for data augmentation. This approach takes data from a source domain and learns to take any data item and generalise it to generate other within-class data items (cf. [1]). The main focus of this work is the computer vision domain. Worth mentioning is that the GANs were not systematically tested for their ability to generate representative data.

3 Method

In this section, we introduce a framework with properties to evaluate the suitability of generated data using GANs for RL problems. We distinguish between structural and functional properties. First, we introduce some formal terminology based on Markov Decision Processes (MDP).

3.1 Preliminaries

Let M be a task in the real-world that can be modeled as an MDP. Define M as $\langle S, A, T, R \rangle$. S is a finite state space. The set of actions that can be selected at each time step t is given by A. $T :: S \times A \times S \rightarrow [0,1]$ is a probabilistic transition function over the states in S where selecting an action $a \in A$ while in $s \in S$ at time t leads to the transition to the next state $s' \in S$ at $t+1$.

$R :: S \times A \to \mathbb{R}$ is the reward function that outputs a scalar $r = R(s, a)$ to each state $s \in S$ and action $a \in A$. States $s \in S$ consist of features denoted by the feature vector representation $\psi(s) = \langle \psi_1(s), \psi_2(s), \ldots, \psi_n(s) \rangle^\top$ of the state $s \in S$. Problems that are considered Markovian can be modeled using RL algorithms. The goal in RL is to learn a policy π that determines which action $a \in A$ to take in a state $s \in S$, $\pi :: S \to A$. The action $a = \pi(s)$ will lead to the transition to a new state s' with a scalar reward $r = R(s, a)$ being obtained. This transition denotes an experience $\langle s, a, r, s' \rangle$. We define a sequence of multiple experiences following each other in time as a trace. A trace is denoted by $\zeta : \langle s, a, r, s', a', r', s'', a'', r'', \ldots \rangle$. Multiple traces observed over time form a data set $Z \in \langle \zeta_1, \ldots \zeta_k \rangle$.

3.2   Reinforcement Learning

The goal in RL is to find a policy π^* out of all possible policies $\Pi :: S \times A \to$ [0,1] that select actions such that the sum of future rewards (at any time t) is maximized. The value of taking action $a \in A$ in state s of policy π, where $\pi(s) = a$, is:

$$Q^\pi(s, a) = E_\pi\{\sum_{k=0}^{K} \gamma^k r^{t+k+1} | s^t = s, a^t = a\} \tag{1}$$

where γ is defined as the discount factor that weights rewards occurring in the future, and s^t and a^t are states and actions occurring at time t. Define $Q(s, a)$ as the expected long-term value of being in state s and selecting action a. Taking the best action a in each possible state $s \in S$, a policy can be derived from the Q-function, i.e.

$$\pi'(s) = \arg\max_{a \in A} Q^\pi(s, a), \ \forall s \in S \tag{2}$$

Off-Policy Policy Evaluation (OPE). Similar to testing models in the traditional machine learning setting, it is important to perform counterfactual policy evaluation of RL policies before deploying them in the real-world. In this setting, there is a dataset of traces Z generated by a behaviour policy π_b, and a fixed evaluation policy π_e whose value-function Q^{π_e} we want to estimate using the data set Z. The goal is to learn an estimator $\hat{Q}^{\pi_e}(Z)$ of Q^{π_e} such that the mean squared error is minimized [21]:

$$MSE(\hat{Q}^{\pi_e}(Z), Q^{\pi_e}) := E\{(\hat{Q}^{\pi_e}(Z) - Q^{\pi_e})^2\} \tag{3}$$

3.3   Generating Experiences Using GANs

When employing RL techniques to learn a good estimate of the value-function $Q^\pi(s, a)$, a large amount of traces ζ are required [6]. The availability of a data set Z that is representative is crucial for both off-policy learning and evaluation. We employ generative models to demonstrate that experiences $\zeta : \langle s, a, r, s' \rangle$

can be synthesized using GANs. In this work, we focus on generating traces of experiences with a horizon length of one time step. In many decision making problems this is considered sufficient.

Traditional GAN. GANs are a class of algorithms with two neural networks, a generator, and a discriminator, that are competing in a zero-sum game [9]. The generator network, denoted by G, generates experiences $\zeta_G = G(z; \Theta^{(G)})$. Here z denotes random noise and $\Theta^{(G)}$ the weights of the network. The adversary of the generator, the discriminator network D, has to distinguish between experiences Z^{π_b} sampled from the training data and experiences Z^G generated by the generative network. The discriminator generates a probability, denoted by $D(\zeta; \Theta^{(D)})$, indicating whether the experience is a real example drawn from the training data or whether it is an experience generated by the generator network. Here $\Theta^{(D)}$ denotes the weights of the network.

The process of learning in GANs can be formulated as a zero-sum game. Each network aims at maximizing its payoff function. The discriminator payoff function is defined as follows:

$$v(\theta^{(G)}, \theta^{(D)}) = \underset{\zeta \sim p_{data}}{\mathbb{E}} \log D(\zeta) + \underset{\zeta \sim p_{model}}{\mathbb{E}} \log(1 - D(\zeta)) \tag{4}$$

where p_{data} denotes the distribution under which the training data was sampled and p_{model} the data sampled from the generator. The generator payoff function on the other hand is denoted by $-v(\theta^{(G)}, \theta^{(D)})$, so that at the moment of convergence of the GAN:

$$g^{\star} = \underset{G}{\operatorname{argmin}} \max_{D} v(G, D) \tag{5}$$

Improved Stability of Learning. As an alternative, the Wasserstein GAN was introduced by [2] to solve problems encountered during training of the traditional GAN. WGAN improves the stability of learning by solving problems related to mode collapse. Furthermore, the WGAN provides learning curves that are useful for performing hyperparameter optimziation. To be able to approximate the distribution of the experiences ζ from the data set Z, WGAN replaces the discriminator network with a critic that outputs a metric of the realness of the generated experience instead of predicting a probability of the generated experiences being real of fake. To further increase the chance of training generators that learn to generate experiences that are very realistic, architectural features and training procedures explained in [12,16,18,20] were used when possible.

3.4   Evaluation Framework

Below, we present the components of our framework to evaluate the representativity of the data generated by the GAN.

Structural Representativity Property. Using a generator $G(z; \Theta^{(G)})$ we can synthesize a set of experiences $Z \in \langle \zeta_1, \dots \zeta_k \rangle$ each of them representing a trace $\langle s, a, r, s' \rangle$. The states and next states are represented by the vector $\psi(s)$ consisting of the features $\langle \psi_1(s), \psi_2(s), \ldots, \psi_n(s) \rangle^\top$. To evaluate the structural representativity of the experiences Z^G generated by a generative model G we use the Kullback-Leibler (KL) [13] divergence metric. The KL-divergence is defined as follows:

$$V_{KL}(p_{\pi_b} || q_{G(z;\Theta^{(G)})}, c) := \sum_{i=1}^{N} p(c_{\pi_b}) \cdot \log \frac{p(c_{\pi_b})}{q(c_{G(z;\Theta^{(G)})}})} \tag{6}$$

where N denotes the number of experiences and $c \in \langle \psi_1(s), \psi_2(s), \ldots, \psi_n(s), a, r, \psi_1(s'), \psi_2(s'), \ldots, \psi_n(s') \rangle^\top$. This metric helps quantify how much information one may have lost when using the GAN to synthesize experiences similar to the experiences generated under π_b.

Functional Representativity Property. We employ the weighted doubly robust estimator [3] for off-policy evaluation to quantify the functional properties of our representativity framework. This estimator allows us to measure the difference between the evaluation of policies for generated and real life data. Ideally, these values would be identical, showing that there is no difference in usage of real or artificial data. The weighted doubly robust estimator is defined as follows.

$$V_{DR} := \hat{Q}(s) + \rho(r - \hat{R}(s, a)) \tag{7}$$

where:

$$\rho := \frac{\pi_e(a|s)}{\pi_b(a|s)} \tag{8}$$

and

$$\hat{V}(s) := \sum_a \pi_e(a|s) \hat{R}(s, a) \tag{9}$$

Here, $\hat{R}(s, a)$ represents an estimator for the reward function R, r the observed reward and $\hat{Q}(s)$ an estimator for expected future reward for being in state s. The doubly robust estimator is an unbiased estimator of $\hat{Q}_{\pi_e}(s)$ that is known to achieve accurate empirical and theoretical results. Contrary to the importance sampling estimator [17] that only relies on the unbiased estimates, the doubly robust estimator relies on an approximate model of the MDP M to decrease the variance.

4   Experimental Setup

Here we explain the experimental setup, including more detailed research questions, and the way in which we go about answering them.

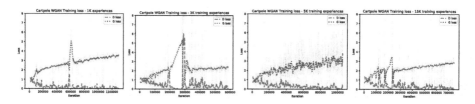

Fig. 1. WGAN training loss for CartPole with training data sizes 1K, 3K, 5K, and 15K.

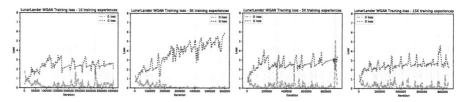

Fig. 2. WGAN training loss for LunarLander with training data sizes 1K, 3K, 5K, and 15K.

4.1 Specific Research Questions

Given our framework, we try to answer the main research question: **is generative modeling able to generate representative training data by synthesizing sequences of experiences for RL tasks?** and the following specific questions:

- RQ1: Can we use GANs to learn to generate one-step horizon sequential data suitable for RL tasks from relatively small samples?
- RQ2: Do generative models we developed for synthesizing sequential data suitable for RL tasks prove to be structurally representative?
- RQ3: Do generative models we developed for synthesizing sequential data suitable for RL tasks prove to be functionally representative?

4.2 Evaluation Environments

We evaluate our proposed method on two problems from the commonly used OpenAI classic control suite [4]. We selected CartPole and LunarLander because they resemble real-world control problems. In the CartPole environment, an observation represents the state of the pole consisting of 4 dimensions $\psi(s) = \langle \psi_1(s), \psi_2(s), \psi_4(s), \psi_4(s) \rangle^\top$. This environment has 2 actions, moving left or right. In the LunarLander environment, an observation consists of 8 features $\psi(s) = \langle \psi_1(s), \psi_2(s), \ldots, \psi_8(s) \rangle^\top$ where $s \in S$ represents the state of the ship. This environment has 4 actions, doing nothing, fire left orientation engine, fire main engine and fire right orientation engine.

4.3   Behaviour Policies and Datasets

The original data sets in all our experiments are sampled from the environments CartPole [4] and LunarLander [4]. The behaviour policies π_b used to generate experiences from these environments are trained using deep reinforcement learning whereby online policy training was performed using the RL framework Horizon [7]. For the CartPole environment we use the discrete DQN algorithm and for the LunarLander environment we use the continuous action DQN algorithm. For details about the implementation we refer to [7] as we reused their architecture and other hyperparameter settings. As a dataset, we generate around 20000 experiences under the behaviour policy π_b for each of the two environments Cartpole and Lunarlander. From each dataset we create subsets with 1000, 3000, 5000, and 15000 experiences each. These datasets will be used for training the GANs.

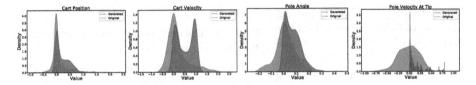

Fig. 3. Original (blue) and WGAN generated (green) state features for CartPole with training data size 1K. (Color figure online)

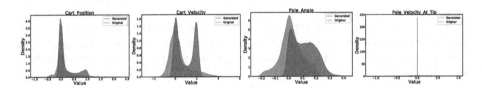

Fig. 4. Original (blue) and WGAN generated (green) state features for CartPole with training data size 3K. (Color figure online)

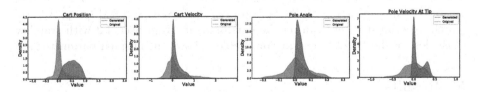

Fig. 5. Original (blue) and WGAN generated (green) state features for CartPole with training data size 5K. (Color figure online)

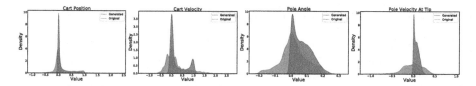

Fig. 6. Original (blue) and WGAN generated (green) state features for CartPole with training data size 15K. (Color figure online)

4.4 GAN Architecture and Hyperparameters

Considering the range of real-world applications our framework is applicable to, it is important that the GANs have a robust architecture and show stability of learning on the datasets from our case studies. We want to avoid to situation where every single GAN needs a specific architecture and hyperparameter tuning.

Wasserstein GAN. Regarding the type of GAN, we opted for the Wasserstein GAN as described in Sect. 3.3 for its stability during training and relative ease of performing hyperparameter tuning. Furthermore, given the fact that the data to be generated are experiences consisting of a state, action, reward and next state values, a generic feed forward neural network architecture is selected. Our architecture consists of input units, a hidden layer with hidden and bias units and output units for both the generator and the discriminator. We performed a grid search for the number of hidden units in the hidden layer for values between 50 and 1000 with step size 50 using the dataset Z_{π_b}. We chose the size of the input of the generator equal to the size of the experience to be generated.

Table 1. Optimal parameters generator and discriminator networks.

Network	Generator	Discriminator
Hidden layers	1	1
Hidden neurons	750	500
Optimizer	Adam	Adam
Learning rate	1e−5	1e−5
Initialization	Xavier	Xavier

Optimal Parameters. Based on literature we opted for Xavier initialization [8] of the weights for its benefits on the speed of learning and stability. We performed a grid search for the learning rate on the values 1e−1, 1e−2, 1e−3, 1e−4, and 1e−5. We experimented with the optimizers Adam, SGD and RMSprop. All grid search experiments were run on the environments from Sect. 4.2 with stability of the loss of the WGAN as criterion. Table 1 shows the optimal parameters.

5 Results

Below we present the results related to the four research questions we have identified.

5.1   Generating One-Step Horizon Sequential Data

Figures 1 and 2 show the training loss of the GANs for varying sizes of data for CartPole and LunarLander. We let the models train until the learning starts stabilizing. In a few occasions we can observe simultaneous spikes in the generator loss and drops in the discriminator loss and vice versa.

We generate experiences using the generator networks and compare the distribution of the synthesized experiences with the data that was used for training. Figs. 3, 4, 5, and 6 demonstrate such comparisons for the state features of the CartPole environment. As the size of data the GANs are trained on increases, the distribution of the generated data moves closer to the distribution of the original data. For the state feature Pole Velocity At Tip the GANs seem to have difficulties capturing the distribution completely while for the other features this seems less of a problem.

5.2   Structural Representativity

We use the KL-divergence metric to quantify the structural representativity of GANs when generating sequential data. Figures 7 and 8 provide graphs of the different features representing the state, action, next state and reward in the generated experiences. We compare these features to the same features from the original dataset. In the ideal situation the KL-divergence should approach 0 as quickly as possible. This would mean that the GAN needs a small amount of data to learn a good representation of the data.

CartPole. In Fig. 7 we see that for the CartPole environment it takes around 4K experiences for the KL-divergence metric to get closer to the original data. The steepest improvement takes place betwwen 1K and 3K training experiences. With 15K experiences the KL-divergence metric starts approaching 0. Similar to what we have seen in the previous section, the feature Pole Velociy At Tip seems to have the highest KL-divergence metric across all features after 15K training experiences. This seems to be a logical outcome because this feature seems to have a relatively high variance. In Fig. 8 we see a similar behaviour for most features from the next state. The KL-divergence for the action and reward approaches 0 very rapidly.

LunarLander. In Fig. 7 we see that for the LunarLander environment it takes around 15K experiences for the KL-divergence metric to get closer to 0. For some features such as the Velocity values and Angle values, we see from the KL-divergence that the distribution of the generated data shifts away from the distribution of the original dataset. These features are represented by distributions with means centered around 0 and low standard deviations. In Fig. 8 we see a similar behaviour for most features from the next state. The KL-divergence for a few features is constant however. The KL-divergence for the reward approaches 0 after 15K experiences while the action features seem to improve but are not approaching 0 at the same rate as the reward.

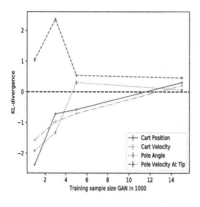
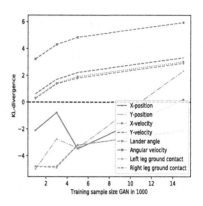

Fig. 7. KL-divergence values per state feature for CartPole (left) and LunarLandar (right).

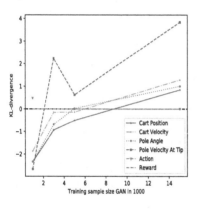
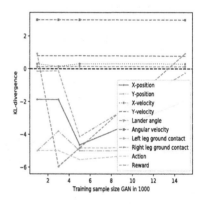

Fig. 8. KL-divergence values per action, next state feature and reward for Cartpole (left) and LunarLandar (right).

5.3   Functional Representativity

As a measure for functional representativity of the GANS, we compare the difference in outcome of applying off-policy evaluation on real and generated data. For each GAN trained on the varying data sizes, we obtain the generator network and generate experiences to be used for off-policy evaluation. For each GAN that we train on the different data sizes, we generate experiences and perform off-policy evaluation until we see that the off-policy evaluation metric has stabilized (1K generated experiences without change). The value we observe after the stabilization is used to calculate one MSE value. We repeat this for all 4 sample sizes and we obtain the graph seen in Fig. 9.

CartPole. Figure 9 shows the CartPole MSE of the evaluation reward compared to the perfect policy. With a perfect policy, a reward of 1 is obtained during each experience. We observe an MSE value of almost 0.25 when using $1K$ experiences to train the GAN (and the same amount of generated experiences to evaluate it). With $3K$ experiences we see a large drop in the MSE to a value of around 0.09. After this, the MSE gradually decreases to 0.075 for a data size of $5K$ and 0.033 for a data size of $15K$.

LunarLander. Figure 9 shows the LunarLander MSE of the evaluation reward compared to the policy from the logged data. With this behaviour policy, we could calculate that an average reward of 0.255 was obtained during each experience. We observe an MSE value of 0.065 when using $1K$ experiences to train the GAN (and the same amount of generated experiences to evaluate it). With $3K$ experiences we see a significant increase in the MSE to a value of around 0.14. After this, the MSE gradually decreases to 0.065 for a data size of $5K$ and 0.062 for a data size of $15K$. The MSE for the GAN trained on $1K$ experiences seems to be an out-

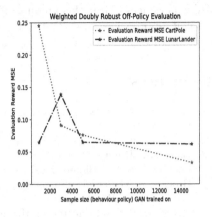

Fig. 9. Off-policy evaluation Mean Squared Error for CartPole and LunarLander using the weighted doubly robust evaluator.

lier as the off-policy evaluation value was very close to 0. As we could see from the KL-divergence metric in Fig. 7, the GAN trained on $1K$ experiences does not seem to capture the dataset well yet. Inspection of the generated data leads to the same conclusion where several generated values are still very close to zero. With this information and KL-divergence values we can safely treat the MSE for the GAN trained on $1K$ datapoints as an outlier.

6   Conclusion

In this paper, we have introduced a framework of properties to evaluate the suitability of synthesized sequences of experiences for RL tasks using GANs. We have introduced a framework of properties to evaluate the suitability of the generated data. Here we distinguished structural and functional properties. We tested our methods on standard RL benchmarks that have similarities to real-world problems. We obtained historical logs from a behaviour policy from literature and trained GANs on varying training data sizes. Using our framework we showed that GANs start generating representative training data with just a few thousand experiences.

To this end, we used the KL-divergence metric and showed that for the CartPole environment the KL-divergence starts approaching 0 after training on $3K$ to $5K$ experiences. For the LunarLander environment –a significantly more complex problem– we see similar behaviour, although it takes a slightly larger sample of training data for the KL-divergence to start approaching 0 for this environment.

We used the weighted doubly robust off-policy evaluation and relied on the MSE of the off-policy evaluation using the generated data. Ideally, the MSE value would approach zero. With the CartPole environment, we demonstrated that the MSE of the evaluations using generated data shows a steep drop in MSE with just $3K$ experiences. After this amount of data the MSE keeps decreasing gradually until it approaches zero. With the LunarLander environment we empirically demonstrate the scalability of our framework.

References

1. Antoniou, A., Storkey, A., Edwards, H.: Data augmentation generative adversarial networks (2017)
2. Arjovsky, M., Chintala, S., Bottou, L.: Wasserstein Gan (2017)
3. Bang, H., Robins, J.M.: Doubly robust estimation in missing data and causal inference models. Biometrics **61**(4), 962–973 (2005)
4. Brockman, G., et al.: Openai gym. CoRR abs/1606.01540 (2016). http://arxiv.org/abs/1606.01540
5. Silver, D., et al.: Mastering the game of go with deep neural networks and tree search. Nature **529**, 484–503 (2016). http://www.nature.com/nature/journal/v529/n7587/full/nature16961.html
6. Dulac-Arnold, G., Mankowitz, D., Hester, T.: Challenges of real-world reinforcement learning (2019)
7. Gauci, J., et al.: Horizon: Facebook's open source applied reinforcement learning platform (2018)
8. Glorot, X., Bengio, Y.: Understanding the difficulty of training deep feedforward neural networks. In: Proceedings of the Thirteenth International Conference on Artificial Intelligence and Statistics, pp. 249–256 (2010)
9. Goodfellow, I., et al.: Generative adversarial nets. In: Advances in Neural Information Processing Systems, pp. 2672–2680 (2014)
10. Hoogendoorn, M., Funk, B.: Machine Learning for the Quantified Self: On the Art of Learning from Sensory Data. Springer, Heidelberg (2017). https://doi.org/10.1007/978-3-319-66308-1
11. Jiang, N., Li, L.: Doubly robust off-policy value evaluation for reinforcement learning. arXiv preprint arXiv:1511.03722 (2015)
12. Kodali, N., Abernethy, J., Hays, J., Kira, Z.: On convergence and stability of GANs (2017)
13. Kullback, S., Leibler, R.A.: On information and sufficiency. Ann. Math. Stat. **22**(1), 79–86 (1951). https://doi.org/10.1214/aoms/1177729694
14. Liu, Y., Zeng, Y., Chen, Y., Tang, J., Pan, Y.: Self-improving generative adversarial reinforcement learning. In: Proceedings of the 18th International Conference on Autonomous Agents and MultiAgent AAMAS 2019, Systems, pp. 52–60. International Foundation for Autonomous Agents and Multiagent Systems, Richland (2019). http://dl.acm.org/citation.cfm?id=3306127.3331673
15. Mnih, V., et al.: Human-level control through deep reinforcement learning. Nature **518**(7540), 529–533 (2015). https://doi.org/10.1038/nature14236
16. Nagarajan, V., Kolter, J.Z.: Gradient descent GAN optimization is locally stable. In: Advances in Neural Information Processing Systems, vol. 30
17. Precup, D., Sutton, R.S., Dasgupta, S.: Off-policy temporal-difference learning with function approximation. In: ICML, pp. 417–424 (2001)

18. Salimans, T., Goodfellow, I., Zaremba, W., Cheung, V., Radford, A., Chen, X.: Improved techniques for training GANs (2016)
19. Sutton, R.S., Barto, A.G.: Reinforcement Learning: An Introduction, 2nd edn. MIT Press, Cambridge (2018)
20. Thanh-Tung, H., Tran, T., Venkatesh, S.: Improving generalization and stability of generative adversarial networks (2019)
21. Thomas, P., Brunskill, E.: Data-efficient off-policy policy evaluation for reinforcement learning. In: International Conference on Machine Learning, pp. 2139–2148 (2016)
22. Tseng, H.H., Luo, Y., Cui, S., Chien, J.T., Ten Haken, R.K., Naqa, I.E.: Deep reinforcement learning for automated radiation adaptation in lung cancer. Med. Phys. **44**(12), 6690–6705 (2017)
23. Vinyals, O., et al.: Grandmaster level in starcraft ii using multi-agent reinforcement learning. Nature **575**, 350–354 (2019)
24. Wiering, M., van Otterlo, M. (eds.): Reinforcement Learning: State of the Art. Springer, Heidelberg (2012). https://doi.org/10.1007/978-3-642-27645-3

Driving Subscriptions Through User Behavior Modeling and Prediction at Bloomberg Media

Rishab Gupta[✉], Rohit Parimi, Zachary Weed, Pranav Kundra, Pramod Koneru, and Pratap Koritala

Bloomberg LP, New York City, NY 10022, USA
{rgupta296,rparimi}@bloomberg.net

Abstract. In May 2018, Bloomberg Media launched a digital subscriptions business with a paywall. This meant that users would now be blocked after reading a fixed number of articles per month on Bloomberg.com and the Bloomberg mobile application. For the launch, all users were allowed to read 10 articles for free per month and were offered the same introductory price for subscription. Although we were able to get a large number of subscribers immediately after launch with this "one size fits-all" approach, the conversion rate steadily dropped in the next few months (the initial spike in subscriptions can be attributed to the purchase of subscription by our loyal users). In order to overcome this problem and to keep our subscription conversion numbers high, our team leveraged user engagement data and built a subscription propensity machine learning model to intelligently target users with the right introductory offers. Furthermore, our team used the engagement data trends across months along with the subscription propensity scores to optimize the paywall height for each user, thus maximizing revenue.

In this talk, we will present details about the challenges we faced and approaches we considered to implement the subscription propensity model. We will also talk about how we combined this model with user engagement trends to dynamically calculate the paywall height for each user for every month. Finally, we will share details on how we integrated and used the subscription propensity and dynamic paywall techniques in production, along with results that demonstrate the success of these approaches in increasing Bloomberg Media's subscription revenue.

Keywords: Machine learning · Data science · User behavior modeling and prediction · Subscription propensity · Elastic paywall · A/B testing

1 Introduction

At the time of the launch of Bloomberg Media's subscription business, all users of Bloomberg's web and mobile applications were given the same introductory price and paywall height. Paywall height is the number of articles a user is allowed to read for free per month. However, this is not an optimal approach as content consumption patterns and price sensitivities are different across users and geographical regions. For example, some users are avid news readers and can be considered our core audience whereas some others

© Springer Nature Switzerland AG 2020
G. Nicosia et al. (Eds.): LOD 2020, LNCS 12565, pp. 472–476, 2020.
https://doi.org/10.1007/978-3-030-64583-0_42

are casual readers. Therefore, it is important for us to identify and categorize our users so that we can market the right introductory price to them and maximize subscription revenue. Similarly, having a static paywall height for all users is also not optimal; if the paywall height is too large, we will only be able to block a small percentage of our audience, thereby not providing an incentive to subscribe leading to a loss in subscription revenue and if the paywall height is too small, most users may be prone to leave our site and not come back as it affects their news reading experience. This leads to a loss in Ad revenue as we lose a lot of page views by blocking content and from users who never return. Therefore, the right thing to do is nurture a user's engagement so that we can target them at an opportune moment while ensuring the Ad revenue is not negatively affected.

In order to target users with appropriate introductory offers, we decided to build a subscription propensity machine learning model which classifies our users into two buckets: users with high propensity to subscribe and users with low propensity to subscribe. The input to the model is user clickstream data which contains information about user's interactions with our content along with other metadata. The output is the propensity score (or the probability to subscribe) for that user. More details about this problem and our solution are explained in Sect. 2.

Once we had subscription propensities for users, our goal is to optimize the paywall height for each user such that we maximize subscription and Ad revenue combined. To achieve this, we make use of the subscription propensity scores along with user engagement trends in previous months to compute a paywall height for the current month for each user. More details about the problem and its solution are provided in Sect. 3.

These two optimizations led to a statistically significant increase in conversion rates and led to a total increase in revenue. The subscription propensity model is at the core of the solution. We will talk about some of the A/B test results and our insights in Sect. 4.

Some of the most important yet frequently ignored details when using machine learning models in the real-world are details about data pipelines (data gathering, cleaning and filtering, GDPR handling, and feature generation) and continuous model evaluation and validation in the production environment. Although these details are not part of this paper, we plan to present these details as part of our presentation.

2 Subscription Propensity Model

The input data for our subscription propensity model is user click stream data. This data can provide a lot of information about a user's behavior on our platforms. We can look at their activity with respect to article interactions, which can be broken down by page types. We also get information about the country of access along with device information such as browser and operating system. More details about this will be shared during our talk.

The class label for this problem was constructed by using active subscribers as positive examples and all other users as negative examples. This gave rise to the class imbalance problem, one of the most common challenges when working with real-world problems. For example, in our dataset, the subscribers constituted only about 0.025% of all users. An important consideration when working with imbalanced datasets is selecting the right evaluation metric. Model accuracy is not an insightful metric to use because the model will have 99.975% accuracy if it predicted every user as a non-subscriber. Therefore, to evaluate our model, we used area under precision-recall curve (AUPRC) along with precision and recall values for the positive and the negative classes.

We experimented with multiple models like logistic regression, neural nets, and decision trees and used k-fold cross-validation to tune hyper-parameters for each of these models. For our problem, Logistic Regression performed the best and is currently used in production. More details on feature engineering, experimentation, and periodic model evaluation in production will be shared during our presentation.

The goal of a classifier is to predict whether a data point belongs to a positive class or a negative class. However, instead of using the predictions from the classifier directly in production, we decided to use the probability scores given by the classifier as the subscription propensity score. These probability scores represent a user's likelihood to subscribe. This allows us to selectively target, for example, high propensity users who are not yet subscribers.

We used the results of the subscription propensity model, more precisely the propensity scores, to run A/B tests on Bloomberg's website. In these A/B tests, we experimented with different introductory offers on targets of users bucketed according to their propensity scores. Each A/B test was run for roughly a month and at the end of the month, the variation that resulted in statistically significant increase in conversions as compared to control was declared the winner and was made the new control for subsequent A/B test. Details about our A/B test set-up and results for the same will be shared in our presentation.

3   Elastic Paywall Optimization

The elastic paywall problem was aimed to determine the optimum paywall height for each user. This problem is central to business revenue since it controls the trade-off between subscriptions and ad revenue. As mentioned earlier, blocking users with a paywall can lead to higher subscription conversions with a loss in ad revenue and vice-versa.

Subscription propensity is a valuable signal since a high value indicates that a user is likely to subscribe and they should be targeted accordingly. This can be treated as a representative for subscription revenue. However, it cannot be used in isolation to decide the opportune moment to affect said targeting. As mentioned earlier, a naive approach of aggressively blocking high propensity users leads to higher bounce rates which has repercussions on both ads and subscription revenue. While we want to increase subscription conversion rates, we need to be mindful that we do not do so at the risk of impacting ad revenue This calls for the need of additional signals to be used in combination with subscription propensity. A key signal that we identified is user engagement trends, which can serve as a representative for ads revenue, since more on-site engagement leads to higher ad revenue.

The subscription propensity and engagement trends thus provide us with vital insight into both aspects of the ads versus subscription revenue trade-off. In order for a paywall to be truly elastic, it is necessary to consider the prior paywall height instead of starting fresh each time. The previous paywall height for users can be factored into the calculations to achieve this. The diagram below highlights the factors that go into the elastic paywall height calculation engine. This paywall height is then applied to users for the coming month. As a part of our talk, we will talk about how we solved the elastic paywall optimization problem using these key metrics, our method of testing and some of the observed results (Fig 1).

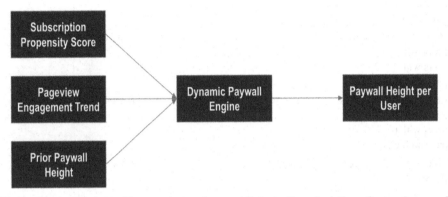

Fig. 1. Block diagram to illustrate inputs that contribute to the calculation of a user's paywall height for the month

4 Results

We use an in-house A/B testing solution to run experiments on our website. This allows us to create custom target audience segments, set up experiments for those targets and measure experiment performance. Some A/B test results for the subscription propensity model are as follows (Table 1):

Table 1. A/B test results for the subscription propensity model

User segment	Conversion lift	Revenue lift	Statistical significance
High propensity users with 2+ paywall hits	57%	27%	98.3%
Low propensity users - Segment 1	125%	186%	98.8%
Low propensity users - Segment 2	144%	150%	97.8%

Similarly, we ran A/B tests to experiment with the elastic paywall. We tested a few variants of the elastic paywall with respect to how the paywall adjusted itself depending

Table 2. A/B test results for Elastic Paywall

Segment	Conversion Lift (vs. height of 10)	Revenue Lift	Change in ad revenue	Statistical significance	Change in paywall stop rate
High Propensity users with 0 paywall hits	+108%	+20%	+4%	99.6%	+114%
High Propensity users with 1+ paywall hits	+114%	+25%	+2%	99.9%	+36%
Low Propensity users with 0 paywall hits	+33%	+12%	+0%	96.4%	+140%
Low Propensity users with 1+ paywall hits	+108%	+22%	-12%	95.2%	+21%

on subscription propensity and article engagement. Following are some key results we observed (Table 2):

The propensity experiment numbers clearly demonstrate the success of using subscription propensity to target users effectively with different introductory offers. The elastic paywall numbers show that some variants were more successful than others in increasing conversions and overall revenue. One key thing to note is that there wasn't a major impact on ad revenue by this effort while there was a lift in conversions and the paywall stop rate increased across the board. These efforts allowed Bloomberg Media to exceed its year-end subscriptions goal by about 40% and were a big success story.

Automatic Setting of DNN Hyper-Parameters by Mixing Bayesian Optimization and Tuning Rules

Michele Fraccaroli[1]([⊠]) [iD], Evelina Lamma[1] [iD], and Fabrizio Riguzzi[2] [iD]

[1] DE - Department of Engineering, University of Ferrara,
Via Saragat 1, 44122 Ferrara, Italy
{michele.fraccaroli,evelina.lamma}@unife.it
[2] DMI - Department of Mathematics and Computer Science,
University of Ferrara, Via Saragat 1, 44122 Ferrara, Italy
fabrizio.riguzzi@unife.it

Abstract. Deep learning techniques play an increasingly important role in industrial and research environments due to their outstanding results. However, the large number of hyper-parameters to be set may lead to errors if they are set manually. The state-of-the-art hyper-parameters tuning methods are grid search, random search, and Bayesian Optimization. The first two methods are expensive because they try, respectively, all possible combinations and random combinations of hyper-parameters. Bayesian Optimization, instead, builds a surrogate model of the objective function, quantifies the uncertainty in the surrogate using Gaussian Process Regression and uses an acquisition function to decide where to sample the new set of hyper-parameters. This work faces the field of Hyper-Parameters Optimization (HPO). The aim is to improve Bayesian Optimization applied to Deep Neural Networks. For this goal, we build a new algorithm for evaluating and analyzing the results of the network on the training and validation sets and use a set of tuning rules to add new hyper-parameters and/or to reduce the hyper-parameter search space to select a better combination.

1 Introduction

Deep Neural Networks (DNNs) provide outstanding results in many fields but, unfortunately, they are also very sensitive to the tuning of their hyper-parameters. Automatic hyper-parameters optimization algorithms have recently shown good performance and, in some cases, results comparable with human experts [3]. Tuning a big DNN is computationally expensive and some experts still perform manual tuning of the hyper-parameters. To do this, one can refer to some tricks [15] to determine the best set of hyper-parameter values to use for obtaining good performance from the network. This work aims to combine the automatic approach with these tricks used in the manual approach but in a fully-automated integration for drive the training of DNNs, automatizing the

© Springer Nature Switzerland AG 2020
G. Nicosia et al. (Eds.): LOD 2020, LNCS 12565, pp. 477–488, 2020.
https://doi.org/10.1007/978-3-030-64583-0_43

choice of HPs and analyzing the performance of each training experiment to obtain a network with better performance. Heuristics and Bayesian Optimization for parameter tuning have already been used in other application contexts other than Deep Learning, e.g., using heuristics to shrink the size of a parameter space for Self-Adapting Numerical Software (SANS) systems [6], using rollout heuristics that work well with Bayesian optimization when a finite budget of total evaluations is prescribed [14] or using domain-specific knowledge to reduce the dimensionality of Bayesian optimization to tune a robot's locomotion controllers [16]. For the automatic approach, we use a Bayesian Optimization [5] because it limits the evaluations of the objective function (training and validation of the neural network in our case) by spending more time in choosing the next set of hyper-parameter values to try. We aim at improving Bayesian Optimization [5] by integrating it with a set of tuning rules. These rules have the purpose of reducing the hyper-parameter space or adding new hyper-parameters to set. In this way, we constrain the Bayesian approach to select the hyper-parameters in a restricted space and add new parameters without any human intervention. In this way, we can avoid typical problems of the neural networks like overfitting and underfitting, and automatically drive the learning process to good results.

After discussing Bayesian Optimization (Sect. 2), we introduce the approach used to analyze the behaviour of the networks, the execution pipeline, the performed analysis and the tuning rules used to fix the detected problems (Sect. 3, Sect. 3.1, Sect. 3.2, and Sect. 3.3). Experimental results with different networks and different datasets are described in Sect. 4. All experiments are performed on benchmark datasets, MNIST and CIFAR-10 in particular.

2   Bayesian Optimization

In this section, we shortly review the state-of-the-art of hyper-parameter optimization for DNNs, i.e., grid search, random search and, with special attention, Bayesian Optimization. Grid Search applies exhaustive research (also called *brute-force search*) through a manually specified hyper-parameter space. This search is guided by some specified performance metrics. Grid Search tests the learning algorithm with every combination of the hyper-parameters and, thanks to its *brute-force search*, guarantees to find the optimal configuration. But, of course, this method suffers from the *curse of dimensionality*. The problem of Grid Search applied to DNNs is that we have a huge amount of hyper-parameters to set and the evaluating phase of the algorithm is very expensive. This definitely raises a time problem.

Random Search uses the same hyper-parameter space of Grid Search, but replaces the *brute-force search* with *random search*. It is, therefore, possible that this approach will not find results as accurate as Grid Search, but it requires less computational time while finding a reasonably good model in most cases [2].

Bayesian Optimization (BO) is a probabilistic model-based approach for optimizing objective functions (f) which are very expensive or slow to evaluate [5]. The key idea of this approach is to limit the evaluation phase of the objective function by spending more time to choosing the next set of hyper-parameters' values to try. In particular, this method is focused on solving the problem of finding the maximum of an expensive function $f : X \rightarrow \mathbb{R}$,

$$argmax_{x \in X} f(x) \tag{1}$$

where X is the hyper-parameter space while can be seen as a hyper-cube where each dimension is a hyper-parameter. BO builds a surrogate model of the objective function and quantifies the uncertainty in this surrogate using a regression model (e.g., *Gaussian Process Regression*, see next section). Then it uses an acquisition function to decide where to sample the next set of hyper-parameter values [7]. The two main components of Bayesian Optimization are: the *statistical model* (regression model) and the *acquisition function* for deciding the next sampling. The statistical model provides a posterior probability distribution that describes potential values of the objective function at candidate points. As the number of observations grows, the posterior distribution improves and the algorithm becomes more certain of which regions in hyper-parameter space are worth to be explored and which are not. The acquisition function is used to find the point where the objective function is supposed to be maximal [12]. This function defines a balance between *Exploration* (new areas in the parameter space) and *Exploitation* (picking values in areas that are already known to have favorable values); the aim of the *Exploration* phase is to select samples that shrink the search space as much as possible, and the aim of the *Exploitation* phase is to focus on the reduced search space and to select samples close to the optimum [10]. Figure 1 (a) shows the behaviour of this algorithm.

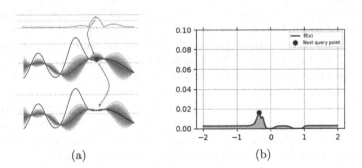

(a)   (b)

Fig. 1. (a) BO behaviour. The figure shows the objective function (black line), mean (light purple line), covariance (purple halo) and the samplings (black dots). The top of the figure shows the Expected Improvement function (blue line) and his maximum (yellow dot). The next sampled point will be exactly the maximum of the activation function (red dot). The bottom of the figure shows the new statistical model [4]. (b) Example of Expected Improvement with the sampling of the next point (blue dot) generated with Scikit-Optimize library. (Color figure online)

A Gaussian Process (GP) is a stochastic process and a powerful prior distribution on functions. Any finite set of N points $\{x_n \in X\}_{n=1}^{N}$ induces a multivariate Gaussian distribution on $\mathbb{R}^N$. GP is completely determined by the mean function $m : X \to \mathbb{R}$ and the covariance function $K : X \times X \to \mathbb{R}$ [17]. For a complete and more precise overview of the Gaussian Process and its application to Machine Learning, see [17].

The acquisition function is used to decide where to sample the new set of hyper-parameters' values, Fig. 1(b) shows. In the literature, there are many types of acquisition function. Usually, only three are used: *Probability of Improvement, Expected Improvement* and *GP Upper Confidence Bound* (GP-UCB) [12] [18]. The acquisition function used in this work is *Expected Improvement* [11]. The idea behind this acquisition function is to define a non-negative expected improvement over the best previously observed target value (previous best value) at a given point x. Expected Improvement is formalized as follows:

$$EI_{y^*}(x) = \int_{-\infty}^{y^*} (y^* - y)p(y|x)\,dy \qquad (2)$$

where y^* is the actual best result of the objective function, x is the new proposed set of hyper-parameters' values, y is the actual value of the objective function using x as hyper-parameters' values and $p(y|x)$ is the surrogate probability model expressing the probability of y given x.

3   Network Behaviour Analysis

By automatically analyzing the behaviour of the network, it is possible to identify problems that DNNs could have after the training phase (e.g., overfitting, underfitting, etc). Bayesian Optimization only works with a single metric (validation loss or accuracy, training loss or accuracy) and is not able to find problems like overfitting or fluctuating loss. When these problems are diagnosed, the proposed algorithm aims to shrink the hyper-parameters' value space or update the network architecture to drive the training to a better solution. Many types of analyses can be performed on the network results, e.g., analyzing the relation between training and validation loss or the accuracy, the trend of accuracy or loss during the training phase in terms of direction and form of the trend. For example: if a significant difference between the training loss and validation loss is found, and the validation loss is greater then the training loss, the diagnosis is that the network has a high variance and overfitting occurred.

3.1   Execution Pipeline

The software structure is composed by four main modules: *training* module (NN in Fig. 2), *Diagnosis* module, *Tuning Rules* module and the *Search Space* driven by the *controller* as can be seen in Fig. 2. Controller rules the whole process. The main loop of the Algorithm 1 uses the Controller for running the *Diagnosis* module, the *Tuning Rules* module and BO for performing the optimization. BO is used for choose the hyper-parameters' value and is forced to do so from a restricted search space.

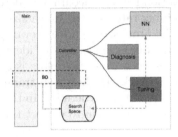

Fig. 2. The structure of the software with the main modules.

Algorithm 1 encapsulates the whole process (called *DNN-Tuner*). With S, N, *Params* and n we refer respectively to the hyper-parameters' value search space, initial neural network definition, a value picked from the search space and the number of cycles of the algorithm. R, H, and M are the results of the evaluation of the trained network, the history of the loss and accuracy in both training and validation phases and the trained model of the network, respectively. $Ckpt$ is a checkpoint of the Bayesian Algorithm that can be used to restore the state of the Bayesian Optimization. *Issues*, *NewM* and are a list of issues found by the *Diagnosis* module, the new model, respectively. All the modules are implemented in Python. The DNN and BO are implemented with Keras and Scikit-Optimize respectively.

Algorithm 1. DNN-Tuner

Require: S, N, n
  $Params \leftarrow Randomization(S)$
  $M \leftarrow BuildModel(P, N)$
  $R, H \leftarrow Training(Params, M)$
  $Ckpt \leftarrow None$
  while $n > 0$ **do**
    $Issues \leftarrow Diagnosis(H, R)$
    $NewM, S \leftarrow Tuning(Issues, M, S)$
    if $NewM \neq M$ **then**
      $Params \leftarrow Randomization(S)$
      $R, H \leftarrow Training(Params, NewM)$
    else
      $Ckpt \leftarrow BO(S, NewM, Ckpt)$
    end if
    $n \leftarrow (n - 1)$
  end while

3.2 Diagnosis Module

We have performed some proof-of-concept analyses. DNN-Tuner analyses the correlation between training and validation accuracy and loss to detect overfitting. It also analyses validation loss and accuracy to detect underfitting, and the trend of loss to fix the learning rate and the batch size.

For the first analysis, the proposed algorithm evaluates the difference of the accuracy in the training and validation phases to identify any gap as shown by Fig. 3. An important gap between these two plots is a clear sign of overfitting since this means that the network classifies well training data but not so well validation data, that is, it is unable to generalize. In this case, the algorithm can diagnose overfitting and then take action to correct this behaviour.

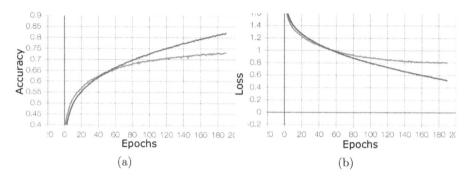

Fig. 3. An example of accuracy (a) and loss (b) in the training (red line) and validation phase (light blue line). (Color figure online)

For detecting underfitting, DNN-Tuner analyses the results of loss in both phases and estimates the amount of loss. The loss value that can be used as a threshold to diagnose underfitting is variable and depends on the problem. Then, a threshold is imposed and, if this threshold is exceeded, DNN-Tuner will diagnose the underfitting.

The analysis of the trend of the loss is useful for detecting the effects of the learning rate and batch size on the network. Given the trend, the proposed algorithm can check whether the loss is too noisy, and whether the direction of the loss is descending, ascending or constant. In response to this diagnosis, the algorithm can attempt to set the batch size and learning rate correctly. These adjustments are performed by applying rules called *Tuning Rules* (Sect. 3.3).

3.3 Tuning Rules

Tuning rules are the actions applied to the network or to the hyper-parameters' value search space in response to the issues identified with the *Diagnosis* module. Then, for each issue found by the diagnosis module, a tuning rule is applied and

Table 1. Rules applied on each issue.

Issue	Tuning rules
Overfitting	Regularization
	Batch normalization
Underfitting	More neurons
Fluctuating loss	Increase of batch size
Increasing loss trend	Decrease of learning rate

a new search space and a new model are derived from the previous ones. In the context of this work, the rules applied for each issue are shown Table 1.

In order to prevent overfitting, the tuning module applies Batch Normalization [9] and L2 regularization [13]. For underfitting, it tries to increase the learning capacity of the network by removing small values from the space of the number of neurons in the convolutional and the fully connected layers. In this way, it drives the optimization to choose higher values for the number of neurons. If DNN-Tuner finds the trend of the loss to be fluctuating, it forces the optimization to choose larger batch size values.

The amount of oscillation in the loss is related to the batch size. When the batch size is 1, the oscillation will be relatively high and the loss will be noisy. When the batch size is the complete dataset, the oscillation will be minimal because each update of the gradient should improve the loss function monotonically (unless the learning rate is set too high). The last rule of Table 1 tries to fix a growing loss. When the loss grows or remains constant at too high values, this means that the learning rate is probably too high and the learning algorithm is unable to find the minimum because it takes too large steps. When the algorithm detects this behaviour, it removes large values from the learning rate search space. Tuning is applied following the IF-THEN paradigm as shown in Algorithm 2, where $NewS$ is the altered search space.

Algorithm 2. Tuning

Require: $Issues, M, S$
  for I in $Issue$ **do**
    if $I =$ "$overfitting$" **then**
      $NewM, NewS \leftarrow Apply([Regularization, Batch Normalization], M, S)$
    else if $I =$ "$underfitting$" **then**
      $NewM, NewS \leftarrow Apply(More neurons, M, S)$
    ...
    end if
  end for
  return $NewM, NewS$

---

484 M. Fraccaroli et al.

4 Experiments

We compare *DNN-Tuner* with standard Bayesian Optimization implemented with the library Scikit-Optimize. Experiments were performed on CIFAR-10 and MNIST dataset. For each experiment, a cluster with NVIDIA GPU K80 and 8-cores Intel Xeon E5-2630 v3 was used. Due to the hardware settings, there is a time limit of 8 h for each the experiment.

The initial state of the DNN is composed of two blocks consisting of two Convolutional layers with ReLU as activation followed by a MaxPooling layer. At the end of the second block, there is a Dropout layer followed by a chain of two Fully Connected layers separated by a Dropout layer. The initial hyper-parameters to be set by the algorithm are the number of the neurons in the Convolutional layers, the values of the Dropout layers, the learning rate and the batch size. The size of the search space depends on hyper-parameter. The domains of hyperparameters are as follows. The size of the first and second convolutional layers are between 16 and 48 and between 64 and128 respectively. The domains of the rate of dropout for the first and second convolutional layers are $[0.002, 0.3]$ and $[0.03, 0.5]$ respectively. The size of the fully connected layes is between 256 and 512. The domain of the learning rate is $[10^{-5}, 10^{-1}]$.

Unlike Bayesian Optimization, our algorithm can update the network structure and the hyper-parameters to be set. For example, when our algorithm detects overfitting, it updates the network structure by adding L2 regularization at each Convolutional layers (kernel regularization), Batch Normalization after the activations and adds L2 regularization parameters to the hyper-parameters to be set. Both DNN-Tuner and BO start with the same neural network and the same hyper-parameters to tune and the same domains for hyperparameters.

Figures 4, 5, 6, 7 and 8 show accuracy and loss in training and validation phases of both algorithms (DNN-Tuner and BO). Both algorithms have completed seven optimization cycles. The neural network has been trained for 200 epochs and the dataset used are CIFAR-10 and MNIST.

Figures 4 show the difference in training and validation at the first cycle of optimization of DNN-Tuner and Bayesian Optimization respectively. You can observe that the trends are noisy because, in the first cycle, the hyper-parameters' values are chosen randomly.

Figures 5 show the difference in training and validation at the fourth cycle of optimization of DNN-Tuner and Bayesian Optimization respectively. You can observe that the trends of BO continue to be noisy, while the DNN-Tuner trends become less fluctuating with an accuracy of ∼0.75 and a loss of ∼0.8 on the validation set.

Figures 6 show the difference in training and validation at the seventh and last cycle of optimization of DNN-Tuner and Bayesian Optimization respectively. You can observe that the trends of BO still continues to be noisy, while the DNN-Tuner trends reach an accuracy of ∼0.83 and a loss of ∼2 on the validation set.

Figures 7 and 8 show the difference of training and validation at the seventh and last cycle of optimization of DNN-Tuner and Bayesian Optimization on the MNIST dataset, also with 200 training epochs. We can see that DNN-Tuner,

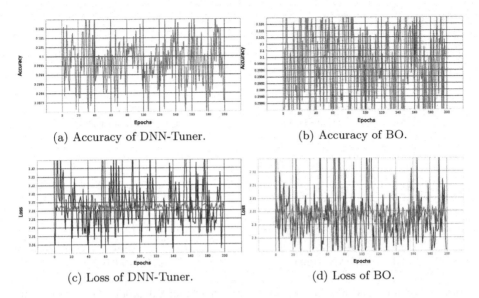

(a) Accuracy of DNN-Tuner.   (b) Accuracy of BO.

(c) Loss of DNN-Tuner.   (d) Loss of BO.

Fig. 4. Accuracy and loss in training (orange line) and validation (blue line) in the first step of both algorithms (DNN-Tuner and Bayesian Optimization). (Color figure online)

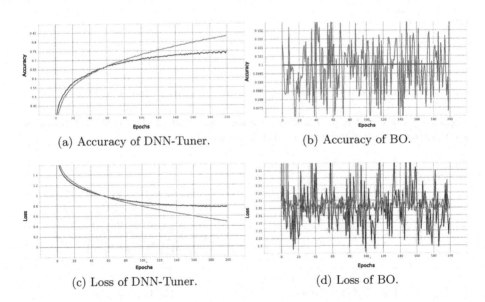

(a) Accuracy of DNN-Tuner.   (b) Accuracy of BO.

(c) Loss of DNN-Tuner.   (d) Loss of BO.

Fig. 5. Accuracy and loss in training (orange line) and validation (blue line) in the fourth step of both algorithms. (Color figure online)

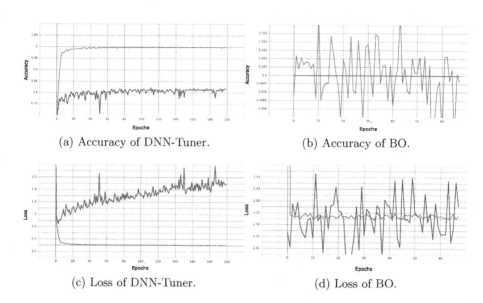

(a) Accuracy of DNN-Tuner.

(b) Accuracy of BO.

(c) Loss of DNN-Tuner.

(d) Loss of BO.

Fig. 6. Accuracy and loss in training (orange line) and validation (blue line) in the seventh step of both algorithms. (Color figure online)

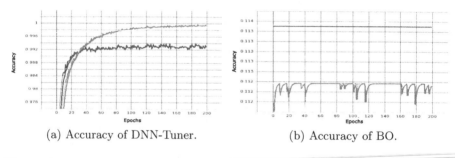

(a) Accuracy of DNN-Tuner.

(b) Accuracy of BO.

Fig. 7. Accuracy and loss in training (orange line) and validation (blue line) of the seventh step of both algorithms. (Color figure online)

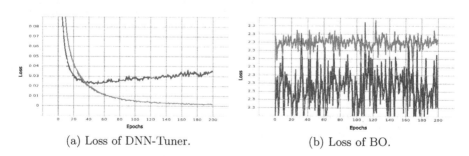

(a) Loss of DNN-Tuner.

(b) Loss of BO.

Fig. 8. Accuracy and loss in training (orange line) and validation (blue line) of the seventh step of both algorithms. (Color figure online)

reaches an accuracy ~ 0.993 and a loss of ~ 0.04 in the validation set unlike BO which reaches ~ 0.114 and ~ 2.3. In terms of time, for the seven cycles of optimization with 200 training epochs on MNIST dataset: DNN-Tuner employed 4.75 h while Bayesian Optimization 5.74 h.

5   Conclusion and Future Work

DNNs have achieved extraordinary results but the number of hyper-parameters to be set has led to new challenges in the field of Artificial Intelligence. Given this problem and the ever-larger size of the networks, the standard approaches of hyper-parameters tuning like Grid and Random Search are no longer able to provide satisfactory results in an acceptable time. Inspired by the Bayesian Optimization (BO) technique, we have inserted expert knowledge into the tuning process, comparing our algorithm with BO in terms of quality of the results and time. The experiments show that it is possible to combine these approaches to obtain a better optimization of DNNs. The new algorithm implements tuning rules that work on the network structure and on the hyper-parameter' search space to drive the training of the network towards better results. Some experiments were performed over standard datasets to demonstrate the improvement provided by DNN-Tuner. Future works will be focused on the implementation of the tuning rules with symbolic languages (either abductive as in [8], or probabilistic as in [1]) in order to achieve higher modularity, and interpretability (as well as explainability) for them.

Acknowledgements. This work was also supported by the National Group of Computing Science (GNCS-INDAM). The first author is supported by a scholarship funded by Emilia-Romagna region, under POR FSE 2014-2020 program.

References

1. Alberti, M., Cota, G., Riguzzi, F., Zese, R.: Probabilistic logical inference on the web. In: Adorni, G., Cagnoni, S., Gori, M., Maratea, M. (eds.) AI*IA 2016. LNCS (LNAI), vol. 10037, pp. 351–363. Springer, Cham (2016). https://doi.org/10.1007/978-3-319-49130-1_26
2. Bergstra, J., Bengio, Y.: Random search for hyper-parameter optimization. J. Mach. Learn. Res. **13**(Feb), 281–305 (2012)
3. Bergstra, J.S., Bardenet, R., Bengio, Y., Kégl, B.: Algorithms for hyper-parameter optimization. In: Advances in Neural Information Processing Systems, pp. 2546–2554 (2011)
4. Wikimedia Commons: File:gpparbayesanimationsmall.gif – wikimedia commons, the free media repository (2019). https://commons.wikimedia.org/w/index.php?title=File:GpParBayesAnimationSmall.gif&oldid=380025760. Accessed 16 Jan 2020
5. Dewancker, I., McCourt, M., Clark, S.: Bayesian optimization primer (2015)
6. Dongarra, J., Eijkhout, V.: Self-adapting numerical software and automatic tuning of heuristics. In: Sloot, P.M.A., Abramson, D., Bogdanov, A.V., Gorbachev, Y.E., Dongarra, J.J., Zomaya, A.Y. (eds.) ICCS 2003. LNCS, vol. 2660, pp. 759–767. Springer, Heidelberg (2003). https://doi.org/10.1007/3-540-44864-0_78

7. Frazier, P.I.: A tutorial on bayesian optimization. arXiv preprint arXiv:1807.02811 (2018)
8. Gavanelli, M., Lamma, E., Riguzzi, F., Bellodi, E., Zese, R., Cota, G.: An abductive framework for datalog ± ontologies. In: Proceedings of the Technical Communications of the 31st International Conference on Logic Programming (ICLP 2015), Cork, Ireland, 31 August–4 September 2015. CEUR Workshop Proceedings, vol. 1459, pp. 128–143. CEUR-WS.org (2015)
9. Ioffe, S., Szegedy, C.: Batch normalization: accelerating deep network training by reducing internal covariate shift. CoRR abs/1502.03167 (2015). http://arxiv.org/abs/1502.03167
10. Jalali, A., Azimi, J., Fern, X.Z.: Exploration vs exploitation in Bayesian optimization. CoRR abs/1204.0047 (2012). http://arxiv.org/abs/1204.0047
11. Jones, D.R., Schonlau, M., Welch, W.J.: Efficient global optimization of expensive black-box functions. J. Glob. Optim. **13**(4), 455–492 (1998)
12. Korichi, R., Guillemot, M., Heuséle, C.: Tuning neural network hyperparameters through Bayesian optimization and application to cosmetic formulation data. In: ORASIS 2019 (2019)
13. van Laarhoven, T.: L2 regularization versus batch and weight normalization. CoRR abs/1706.05350 (2017). http://arxiv.org/abs/1706.05350
14. Lam, R., Willcox, K., Wolpert, D.H.: Bayesian optimization with a finite budget: an approximate dynamic programming approach. In: Advances in Neural Information Processing Systems, pp. 883–891 (2016)
15. Montavon, G., Orr, G.B., Müller, K.-R. (eds.): Neural Networks: Tricks of the Trade. LNCS, vol. 7700. Springer, Heidelberg (2012). https://doi.org/10.1007/978-3-642-35289-8
16. Rai, A., Antonova, R., Song, S., Martin, W., Geyer, H., Atkeson, C.: Bayesian optimization using domain knowledge on the ATRIAS biped. In: 2018 IEEE International Conference on Robotics and Automation (ICRA), pp. 1771–1778. IEEE (2018)
17. Rasmussen, C.E.: Gaussian processes in machine learning. In: Bousquet, O., von Luxburg, U., Rätsch, G. (eds.) ML 2003. LNCS (LNAI), vol. 3176, pp. 63–71. Springer, Heidelberg (2004). https://doi.org/10.1007/978-3-540-28650-9_4
18. Snoek, J., Larochelle, H., Adams, R.P.: Practical Bayesian optimization of machine learning algorithms. In: Advances in Neural Information Processing Systems, pp. 2951–2959 (2012)

A General Approach for Risk Controlled Trading Based on Machine Learning and Statistical Arbitrage

Salvatore Carta[ID], Diego Reforgiato Recupero[(⊠)][ID], Roberto Saia[ID], and Maria Madalina Stanciu[ID]

Department of Mathematics and Computer Science, University of Cagliari, Cagliari, Italy
{salvatore,diego.reforgiato,roberto.saia,madalina.stanciu}@unica.it

Abstract. Nowadays, machine learning usage has gained significant interest in financial time series prediction, hence being a promise land for financial applications such as algorithmic trading. In this setting, this paper proposes a general approach based on an ensemble of regression algorithms and dynamic asset selection applied to the well-known statistical arbitrage trading strategy. Several extremely heterogeneous state-of-the-art machine learning algorithms, exploiting different feature selection processes in input, are used as base components of the ensemble, which is in charge to forecast the return of each of the considered stocks. Before being used as an input to the arbitrage mechanism, the final ranking of the assets takes also into account a quality assurance mechanism that prunes the stocks with poor forecasting accuracy in the previous periods. The approach has a general application for any risk balanced trading strategy aiming to exploit different financial assets. It was evaluated implementing an intra-day trading statistical arbitrage on the stocks of the S&P500 index. Our approach outperforms each single base regressor we adopted, which we considered as baselines. More important, it also outperforms Buy-and-hold of S&P500 Index, both during financial turmoil such as the global financial crisis, and also during the massive market growth in the recent years.

Keywords: Stock market forecast · Machine learning · Statistical arbitrage · Ensemble learning

1 Introduction

In financial investing, the general goal is to dynamically allocate a set of assets to maximize the returns over time and minimize risk simultaneously. A very well-known financial trading strategy is statistical arbitrage, or StatArb for short, which evolved out of pairs trading strategy [15], where stocks are paired based on fundamental or market similarities [20]. In pairs intra-day trading, when one stock of the pair over-performs the other, the stock is sold short with the

© Springer Nature Switzerland AG 2020
G. Nicosia et al. (Eds.): LOD 2020, LNCS 12565, pp. 489–503, 2020.
https://doi.org/10.1007/978-3-030-64583-0_44

expectation that its price will *drop* when the positions are closed. Similarly, the under-performer is bought with the expectation that its price will *climb* when positions are closed. The same concept applies to the StatArb strategy, except that it extends at portfolio level with more stocks [34]. Furthermore, the portfolio construction is automated and comprises two phases: (i) the *scoring* phase, where each stock is assigned to a relevance score, with high scores indicating stocks that should be held long and low scores indicating stocks that are candidates for short operations; and (ii) the *risk reduction* phase, where the stocks are combined to eliminate, or at least significantly reduce the risk factor [4, 28].

Financial investors that use StatArb strategy face the important challenge to correctly identify pairs of assets that exhibit similar behaviour, also determining the point in time when such assets' prices start moving away from each-other. As such, researchers have expended unremitting efforts on investigating novel approaches to tackle the asset choice problem and developed a wide range of *statistical tools* for the matter: distance based [20], co-integration approach [42], and models based on stochastic spread [26]. As previously noted in the literature [23], these tools exhibit a drawback as they rely solely on statistical relationship of a pair at the price level, and lack forecasting component. Moreover, if a divergence between stocks in a pair is observed, then it is *assumed* that the prices must converge in the future and positions are closed only when the equilibrium is reached, an event that is not accurately determined in time.

At the same time, the rapid growth of market integration yielded massive amounts of data in the finance industry, which promotes the study of advanced data analysis tools. By the same token, considering that StatArb is performed at portfolio level (hence a large number of assets is involved), the strategy needs to be implemented in an automated fashion. As such, cutting-edge analytical techniques and machine learning algorithms use has grown [22]. However, incorporating machine learning algorithms comes with its own set of drawbacks as the financial data contains a large amount of noise, jump and movement, leading to highly non-stationary time series that are thought to be highly unpredictable [35], thus deteriorating the forecasting performances. One successful alternative to mitigate the noise present in the data has already been proven to be ensemble methods. In literature, they demonstrated superior predictive performance compared to individual forecasting algorithms and hence their notorious success in different domains such as credit scoring [11] or sentiment analysis [3, 37, 38]. Furthermore, in literature, it has been proved that the employment of heterogeneous ensembles for forecasting outperforms homogeneous ones [9, 31]. When mentioning the forecasting, there are two different tasks that can be targeted: classification and regression. In literature, we can find several implementations of StatArb that use classification [30, 40] and this has always been proved easier to solve than the regression [39]. Although regression in the context of financial predictions poses more challenges [18, 33], it allows for a more *granular ranking*, without reference to any balance point. As such, in this paper we propose a general approach for risk-controlled trading based on machine learning and StatArb. The approach employs an ensemble of regressors and provides three levels of heterogeneous features:

1. Its components consist of any number of state-of-the-art machine learning and statistical models.
2. We train our models with information pertaining to constituents of financial time series with a diversified feature set, considering not only lagged daily prices return, but also a series of technical indicators.
3. We consider diversified models such as the ones that use as training either data from individual companies or companies in the same industry.

Finally, in our approach, after the assets have been ranked in descending order, we propose the use of a *dynamic asset selection*, which looks at the past and influences the ranking by removing stocks with bad past behavior. Then, the strategy buys (performing long operations) the flop k stocks and sells (performing short operations) the top k stocks.

In this paper, we also propose one possible instance of our approach that has been configured for intra-day operations and on the well-known S&P500 Index. The regressors we have employed for such an instance are the following state-of-the-art machine learning algorithms, Random Forests (RF), Light Gradient Boosted trees (LGB), Support Vector Regressors (SVR), and the widely known statistical model, ARIMA. ARIMA models are known to be robust and efficient for short-term prediction when employed to model economical and financial time series [1, 17] even more than the most popular ANNs techniques [32, 36].

To validate the configuration we have chosen for our instance, we evaluate its performance from both return and risk performance perspectives. The comparisons against Buy-and-Hold strategy of S&P500 Index and individual regressors that we adopted in our instance, lucidly illustrate its superiority in performing the forecast.

In summary, the contributions of this paper are the following:

1. We propose a general approach for risk-controlled trading based on machine learning and StatArb.
2. We defined the problem as a regression of price returns, instead of a classification one.
3. Our approach can be easily implemented using different types of assets.
4. We propose an ensemble methodology for StatArb, tackling the ensemble construction from three different perspectives:
   - *model diversity*, by using machine learning algorithms and even statistical algorithms;
   - *data diversity*, by considering lagged price returns and technical indicators so to enrich the data used by models;
   - *method diversity*, by simultaneously training single models across several assets (*i.e.*, models per industries) and, conversely, models for each stock.
5. We develop a dynamic asset selection based on models' most recent prediction performance that keeps the ranking of an asset if the past predictions of its return trend exceed a pre-determined behavior.

6. We provide a possible instance of our approach for intra-day trading with four kinds of regressors (machine learning algorithms and statistical models) for StatArb within the S&P500 Index.
7. We carried out a performance evaluation of our instance and its results outperform baseline methods on the S&P500 Index for intra-day trading.

The remaining of this paper is organized as it follows. Section 2 briefly describes relevant related work in the literature. Section 3 introduces the problem we are facing whereas Sect. 4 includes the architecture of the proposed general approach and the instance we have generated. All the features that we have used are described within Sect. 5. Section 6 details the regressors that we have been considered in the ensemble of our instance. Section 7 describes the proposed ensemble methodology and how we have aggregated the results of the single components. The dynamic asset selection approach is illustrated in Sect. 8. Section 9 discusses the experiments we have carried out and Sect. 10 ends the paper.

2   Related Work

The literature dealing with applications on machine learning and neural networks in finance is presented and analyzed in several works [2,10,12,22]. The work in [23] proposes a StatArb system that entails three phases: forecasting, ranking and trading. For the forecasting phase, the authors propose the use of an Elman recurrent neural network to perform weekly predictions and anticipate return spreads between any two securities in the portfolio. Next, a multi-criteria decision-making method is considered to outrank stocks based on their weekly predictions. Lastly, trading signals are generated for top k and bottom k stocks. This approach considers constituents of S&P100 Index on a period spanning from 1992 to 2006. Although this approach also considers regression, it lacks scalability as its application is limited to 100 stocks, and in case of broader indexes such as S&P500 or Russell 1000, would become computationally intractable. In [40], deep neural networks were used and standardized cumulative returns were considered as features. Following the approach proposed by [40], in [30] the authors construct a similar classification problem using cumulative returns as input features and employ models like deep neural networks, random forests, gradient boosted trees and three of their ensembles. The authors validate their study using $S\&P500$ Index constituents on a period ranging from 1992 to 2015, with trading frequency of one day. Later, the authors extend their work in [19] by using a Long Short-Term Memory network for the same prediction task. This enhanced approach outperforms memory-free classification methods. However, as the authors note, the out-performance is registered from 1992 to 2009, whereas from 2010 the excess return fluctuates around zero. The ensemble proposed in this work is used to tackle a classification problem whereas ours aims at solving a more difficult regression problem. In [29], the authors take a different approach for predicting returns of S&P500, where the used features are stock tweets information. The aim is to unveil how the textual data reflects in stocks' future returns. For this goal, they use factorization matrix and support vector machines. The proposed system performs prediction in a 20 min frequency over a two years period: from January 2014 to December 2015. The selection of flop and top stocks is made at the formation period based on the algorithms performance evaluation (*i.e.* lowest root relative squared error) and trading signals are generated based on Bollinger bands. The authors state that their factorization machines approach yields positive results even after transaction costs. In contrast to previously presented studies, in this work we consider the trading performance of an ensemble of diversified regression techniques that considers diverse models and data. Additionally, our approach includes in the pipeline a dynamic asset selection within the *risk reduction* phase, in order to avoid bad past stocks performances that jeopardize future trading. Such a heterogeneous setup is important to deal with the uncertain behavior of the market, as richer models and complementary information are used in the process. Moreover, the proposed approach can be regarded as generic as it can be instantiated with a huge number of configurations: number and types of regressors, market type (e.g. intra-day), selected features (e.g. lagged returns, technical indicators), number of assets to buy or sell (choice for k).

3   Problem Formulation

The problem tackled by our general approach consists of an algorithmic trading task in the context of StatArb that leverages machine learning to identify possible sources of profit and balance risk at the same time. The StatArb technique consists of three steps: forecasting, ranking, and trading.

- *Forecasting* - We tackle StatArb as a regression problem, investigating the potential of forecasting price returns for each of the assets in a pre-selected asset collection S, on a target trading day d.
- *Ranking* - Based on the anticipated price returns for the assets, we rank them in descending order. We *balance the risk* incurred by inaccurate predictions by pruning the "bad" assets based on their past behavior. This dynamical asset pruning yields a reorganized ranking of the assets.
- *Trading* - Having the trading desirability given by ranking in the previous stage, we issue trading signals for the top k and flop k stocks.

4   The Proposed Approach

Figure 1 depicts the architecture for the general approach for risk controlled trading we propose in this paper. Once the set of assets to work with has been selected, first we collect raw financial information for each asset s_i in the preselected asset collection S. We split our raw data in study periods, composed of training (in-sample data, used for training models) and trading (test) sets, which are non-overlapping. This procedure is a well-known validation procedure for time-series data-sets [16], known as walk-forward strategy. Figure 2 illustrates

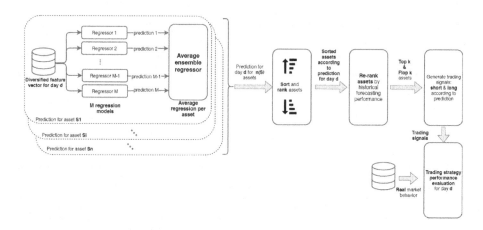

Fig. 1. Architecture of the proposed general approach for risk controlled trading

such a procedure. For each study period and each asset s_i, we generate the diversified feature set denoted by $\mathcal{F}_{d-1}^{s_i}$, using information available prior to the target date d. For in sample period we also generate the label $y_d^{s_i}$. The feature set it used as input to each regressor m in our regressors pool $\mathcal{M}$. The *forecast* is then performed using test data, where each trained model makes its prediction, $o_d^{s_i,m}$ for day d and stock s_i. Then, their results are averaged by a given ensemble method, to obtain a final output $o_d^{s_i,ENS} = \frac{\sum_{m \in \mathcal{M}} o_d^{s,m}}{n(\mathcal{M})}$. Next, we *sort assets* in descending order. That means that we will find at the top assets whose prices are expected to increase, and at the bottom assets whose prices will drop. Assets at the top and at the bottom of our sorting represent the most suitable candidates for trading. After the ranking is performed, we introduce the *dynamic asset selection* step: from this pool of assets, we discard those that do not satisfy a prediction accuracy higher than a given threshold ε in a past trading period, rearranging the ranking accordingly. The next step consists of selecting the top k (winners) and flop k (losers) assets and issue the corresponding trading signals: k long signals for the top k stocks and k short signals for the bottom k stocks. These selections are repeated for every day d

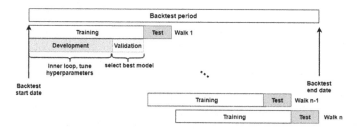

Fig. 2. Illustration of walk-forward procedure

in the trading period. Finally, we evaluate the performance of our architecture by means of back-testing strategy [4]. As mentioned in the introduction we have instantiated one example out of our general approach by using as pool of assets the stocks within the S&P500 Index [19, 30], the trading session to be intra-day. Also, we fixed the number of pairs to be traded to $k = 5$, based on the findings in similar works [19, 30] where higher k values leads to a decrease in portfolio performance both in terms of returns and risks. The set of features $\mathcal{F}$ and the regressors will be described, respectively, in the next two sections.

5   Feature Engineering

As already mentioned, our dataset of reference for the instance we propose is the S&P500 Index. Therefore we have collected the information for all the stocks that have been listed, at least once, as constituents of it in a period from January 2003 to January 2016.

For each stock, we have available daily raw financial information such as *Open Price*, *High Price* in the day, *Close Price*, *Low Price* in the day, and *Volume* of stocks traded during the day. Based on this information, we have created two different kinds of features:

i. **Lagged daily price returns (LR):** historical price returns are the set of features most used in financial studies. For a given trading day d, in the lag $[d - \Delta d, d - 1]$, we compute the $LR_{d,\Delta d}$ as follows:

$$LR_{d,\Delta d} = \frac{closePrice_{d-\Delta d} - openPrice_{d-\Delta d}}{openPrice_{d-\Delta d}}, \tag{1}$$

We have set $\Delta d \in \{1, \ldots, 10\}$, thus having for each day d 10 different lagged price returns shown as it follows:

$$[LR^{s_i}_{d-10}, LR^{s_i}_{d-9}, LR^{s_i}_{d-8}, LR^{s_i}_{d-7}, LR^{s_i}_{d-6}, LR^{s_i}_{d-5}, LR^{s_i}_{d-4}, LR^{s_i}_{d-3}, LR^{s_i}_{d-2}, LR^{s_i}_{d-1}]$$

The target value associated to this feature vector is the intra-day price return for d.

ii. **Technical Indicators (TI):** following [25], we use a set of technical indicators summarized in Table 1. We opted for this set of features as we are interested in predicting the price movement range and also its direction. Each of the technical indicators has different insights of the stock price movement.

For this second type of feature we built the following vector:

$$[EMA(10), \%K, ROC, RSI, AccDO, MACD, \%R, Disp(5), Disp(10)]$$

Similarly as for the **LR** feature vector, the associated target value (label) is the intra-day price return for the current day.

Table 1. Selected technical indicators and their acronyms throughout this paper.

Name of technical indicator	Formula
Exponential Moving Average ($EMA(10)$)	$(C_t \times a) + (EMA_{t-1} \times (1-a))$ where $a = 2/(n+1)$
Stochastic %K (%K)	$\%K = \frac{(C_t - LL_{t-n})}{(HH_{t-n} - LL_{t-n})} \times 100$
Price rate of change (ROC)	$\frac{C_t - C_{t-n}}{C_{t-n}} \times 100$
Relative Strength Index (RSI)	$100 - \frac{100}{1 + (U/T_n)}$
Accumulation Distribution Oscillator (AccDO)	$\frac{(C_t - LL_{t-n}) - (HH_{t-n} - C_t)}{HH_{t-n} - LL_{t-n}} \times V$
Moving Average Convergence - Divergence (MACD)	$EMA_{12}(t) - EMA_{26}(t)$
Williams %R	$\frac{HH_{t-n} - C_t}{HH_{t-n} - LL_{t-n}} \times 100$
Disparity 5 (Disp (5))	$\frac{C_t}{MA_5} \times 100$
Disparity 10 (Disp (10))	$\frac{C_t}{MA_{10}} \times 100$

C_t is the closing price at time t, L_t the low price at time t, H_t high price at time t, LL_{t-n} lowest low in the last $t - n$ days, HH_{t-n} highest high in the last $t - n$ days, MA_t the simple moving average of t days, U represents the total gain in the last n days and T_n represents the total loss in last n days

6   Baselines

In the proposed instance of our general approach we considered the following three different state-of-the-art machine learning models, and the widely known statistical model, ARIMA. We based our choice to employ such models on the following criteria: (i) robustness to noisy data and over-fitting. (ii) diversity amongst models in the final ensemble, and (iii) adoption of such models in the scientific community for similar tasks.

Light Gradient Boosting (LGB) is a relatively new Gradient Boosting Decision Tree algorithm, proposed in [27], which has been successfully employed in multiple tasks not only for classification and regression but also for ranking. LGB applies iteratively weak learners (decision trees) to re-weighted versions of the training data [21]. After each boosting iteration, the results of the prediction are evaluated according to a decision function and data samples are re-weighted in order to focus on examples with higher loss in previous steps. This method grows the trees by applying the leaf-wise (or breadth-first) strategy until the maximum depth is reached, thus making this algorithm more prone to over-fitting. To control this behavior we defined the maximum depth levels of the tree, *max_depth*, to 8. We chose to vary the *num_leaves* parameter in the set $[70, 80, 100]$, achieving a balance between a conservative model and a good generalization. The feature selection is restricted by a parameter *colsample_by_tree* set at 0.8 of the total number of features, which can be thought as a regularization parameter. The work in [21] suggests a learning rate lower than 0.1, so we set it to 0.01 to account for a better generalization over the data set.

Random Forests (RF) belong to a category of ensemble learning algorithms introduced in [8]. This learning method is the extension of traditional decision trees techniques where random forests are composed of many deep de-correlated decision trees. Such a de-correlation is achieved by bagging and by random feature selection. These two techniques make this algorithm robust to noise and outliers. In the case of RF, the larger the size of the forest (the number of trees), the better the convergence of the generalization error. But a higher number of trees or a higher depth of each tree induces computations costs, therefore a trade-off must be made between the number of trees in the forest and the improvement in learning after each tree is added to the forest. We opt to vary the number of trees by ranging *n_estimators* from 50 to 500 with a 25 increment. We based our choice on the work of [24]. Random feature selection operations substantially reduce trees bias, thus we set *min_samples_leaf* to 3 of the total number of features in a leaf. The learning rate is set to 0.01.

Support Vector Regressors (SVR) were proposed initially as supervised learning model in classification, and later revised for regression in [41]. Given the set of training data the goal is to find a function that deviates from actual data by a value no greater than ε for each training point, and at the same time is as flat as possible. It extends least-square regression by considering an ε-insensitive loss function. Further, to avoid over-fitting of the training data, the concept of regularization is usually applied. An SVR thus solves an optimization problem that involves two parameters: the regularization parameter (referred to as C) and the error sensitivity parameter (referred to as ε). C, the regularization cost, controls the trade off between model complexity and the number of non-separable samples. A lower C will encourage a larger margin, whereas higher C values lead to hard margin [41]. Thus, we set our search space in $\{8, 10, 12\}$. Parameter ε controls the width of the ε-insensitive zone, and is used to fit the training data. A too high value leads to flat estimates, whereas a too small value is not appropriate for large or noisy data-sets. Therefore, we set it to 0.1. In this study, we selected the radial basis function (RBF) as kernel. The work in [13] suggests that the γ value of the kernel function should vary together with C, and higher values of C require higher values for gamma too. Therefore, we set a smaller search space in $\{0.01, 0.5\}$.

ARIMA model was first introduced by [7], and has been ever-since one of the most popular statistical methods used for time-series forecasting. The algorithm captures a suite of different time-dependent structures in time series. As its acronym indicates $ARIMA(p, d, q)$ comprises three parts: *autoregression model* that uses the dependencies between an observation and a number of lagged observations (p); *integration differencing* of observations with different degree, to make the time series stationary; and *Moving Average model* that accounts the dependency between observations and the residual error terms when a moving average model is used to the lagged observations (q). We chose the lag order

$p \in \{1,5\}$, the degree of differencing $d \in \{1,5\}$, the size of the moving average window $q \in \{0,5\}$.

7   Ensemble

In the last section we have described the regressors that are included in the ensemble of the instance we proposed in this paper alongside with the parameters space used for each of them. Besides features mentioned in Sect. 5, and parameters intrinsic to each of forecasting models mentioned in Sect. 6, we also considered: – a model for each stock $s_i \in S$ in the training period, – a model for each industry by grouping stocks by their industry sector as given by the Global Industry Classification Standard (GICS). This was encouraged by previous work [20], where some portfolios were restricted to only include stocks from the same industry. Moreover, usually companies in the same industry tend to have similar behavior and exhibit some sort of correlation in their stock prices movement. As such, our training and model selection procedure is composed of three steps. As illustrated in Fig. 2, for each walk and each asset (stock):

- We split the training portion of the data-set into development and validation sets;
- Each type of model has been trained on the development subset. For the training of each regressor, we used an inner cross-validation with 10 folds to find the optimal hype-parameters. Consequently, to forecast the return of each asset, we created 4 models: 2 models (per industry) using TI or LR as features, that use data of all assets associated to that industry, and, in turn, forecast one asset at a time; 2 models (per asset) using TI and LR, that use data of a single asset. Then, using the validation set, we compute the MSE between the forecast and the ground truth, and choose the best model out of the four, per each asset for that walk;
- Finally the best model found at the previous step is trained on the full training set and tested on the test set.

During each walk and for each stock, LGB, RF, SVM, and ARIMA predictions are averaged to obtain the ensemble forecast.

8   Dynamic Asset Selection

We propose a stock pruning mechanism by performing a *dynamic asset selection* strategy. For a stock $s_i \in S$, given its past forecastings $o_t^{s_i,ENS}$, and also its past real values $y_d^{s_i}$ in a predefined look-back period T, we compute a modified version of the mean directional accuracy [5,6] as follows:

$$MDA_{s_i,T,d} = \frac{1}{T} \sum_{t=d-1}^{d-T-1} 1_{sgn(o_t^{s_i,ENS})==sgn(y_t^s)}, \qquad (2)$$

where d is the current trading day, T is the look-back length and $\mathbf{1}_P$ is the indicator function that converts any logical proposition P into a number that is 1 if the proposition is satisfied, and 0 otherwise, $sgn(\cdot)$ is the sign function. The $MDA_{s,T,d}$ metric compares the forecasted direction (upward or downward) with the realized direction, providing the probability that the forecasting model can detect the correct direction of returns for a stock s_i on a given timespan T prior to day d. Such a component introduces a new step in the StatArb pipeline: after the forecast is done, we rank the companies by their forecasted daily price returns. From this pool of companies, we discard those that do not satisfy a prediction accuracy higher than a given threshold ε in a past trading period, rearranging the ranking accordingly. The proposed dynamic asset selection strategy requires a series of parameters: the accuracy threshold ε, and rolling window length related to the past trading period, T. We made these choices based on findings in [14] where the authors noticed that MDA can efficiently capture the inter-dependence between asset returns and their volatility (hence forecastability) when using intermediate return horizons, e.g. two months. The threshold value has been set to $\varepsilon = 0.5$ as advised in [23] for a similar scenario.

9   Experimental Framework

We conducted the experiments on the S&P500 Index dataset focusing on data from January 2003 to January 2016. We considered four years for training (that is why our tests begin from March 2007)[1] and approximately one year for trading (or testing). We compared our approach (ensemble with the dynamic asset selection, ENS-DS), against the ensemble without the dynamic asset selection (ENS) and against each single regressor and the well known Buy&Hold passive investment strategy, known to be representative in finance communities [30].

The metrics we have used for comparison are: (i) return (cumulative, annual and mean daily); (ii) Sharpe ratios; and (iii) Maximum drawdown. Return defines the amount that the returns on assets have gained or lost over the indicated period of time. The Sharpe ratio (SR) measures the reward-to-risk ratio of a portfolio strategy, and is defined as excess return per unit of risk measured in standard deviations. The Maximum drawdown (MaxDD) is the maximum amount of wealth reduction that a cumulative return has produced from its maximum value over time. The results are summarized in Table 2. According to the cumulative return development over time in Table 2, the ENS strategy outperforms all the other non-ensemble models. Its daily returns is almost ten times the level of the Buy&Hold and up to three times the return of some individual regressors, (e.g., RF). Moreover, compared to the simple average ensemble, the ENS-DS approach (with $T = 40$) has a performance increase of 5% points.

Besides the return, in terms of risk exposure, the MaxDD offers an outlook on how sustainable an investment loss can be (lower is better). Also for this metric we notice the better performance of ENS-DS compared to the Buy&Hold strategy and each other baseline. The ENS-DS strategy produces a MaxDD of

---

[1] There are 21 trading days in one month.

Table 2. Results of the StatArb strategy over a period between March 2007 to January 2016

Method	Cumulative return (%)	Annual return (%)	Daily return (%)	MaxDD (%)	SR
LGB	157.52	20.13	0.071	38.52	1.08
RF	78.476	7.89	0.035	24.15	0.5
SVM	160.56	20.13	0.072	32.35	0.1
ARIMA	108.62	12.82	0.049	42.05	0.64
S& P500 Buy-and-Hold	37.36	3.02	0.013	45.42	0.15
ENS	250.95	31.30	0.113	14.42	1.76
ENS-DS, $T = 40$ days	**263.99**	**36.6**	**0.119**	**11.5**	**2.01**

11.5% that is less than one fourth of the Buy-and-Hold strategy(45%). Finally, it can be noticed that SR started from 1.76 for the simple ensemble and turned into 2.01 for the proposed ENS-DS, beating all the other baselines.

10 Conclusions and Future Work

In order to provide insights about efficient stock trading, in this paper we proposed a general approach for risk controlled trading based on machine learning and statistical arbitrage. The forecast is performed by an ensemble of regression algorithms and a dynamic asset selection strategy that prunes assets if they had a decreasing performance in the past period. As the proposed approach is general as all of its components, we created an instance out of it where we focused on the S&P500 Index, using the statistical arbitrage as a trading strategy. Moreover, we propose to forecast intra-day returns using an ensemble of Light Gradient Boosting, Random Forests, Support Vector Machines and ARIMA. We also proposed a set of heterogeneous features that can be used to train the models. By performing a walk-forward procedure, for each company and walk we tested all the combinations of features and internal parameters of each regressor to select the best model for each of them. The ensemble decision has been performed for each walk and company by averaging the forecast of each regressor. Our experiments showed that our ensemble strategy with the dynamic asset selection reaches significant returns of 0.119% per day, or 36.6% per year. As future work we are already working on the application of our approach in other markets and comparisons with different baselines. Further directions where we are headed include enriching the current approach with new types of assets, exogenous variables and the employment of deep neural networks.

Acknowledgments. The research performed in this paper has been supported by the "Bando "Aiuti per progetti di Ricerca e Sviluppo"—POR FESR 2014-2020—Asse 1, Azione 1.1.3, Strategy 2- Program 3, Project AlmostAnOracle - AI and Big Data Algorithms for Financial Time Series Forecasting".

References

1. Ariyo, A.A., Adewumi, A.O., Ayo, C.K.: Stock price prediction using the Arima model. In: 2014 UKSim-AMSS 16th International Conference on Computer Modelling and Simulation, pp. 106–112 (2014). https://doi.org/10.1109/UKSim.2014.67

2. Atsalakis, G.S., Valavanis, K.P.: Surveying stock market forecasting techniques - Part II: soft computing methods. ESWA **36**(3), 5932–5941 (2009). https://doi.org/10.1016/J.ESWA.2008.07.006

3. Atzeni, M., Recupero, D.R.: Multi-domain sentiment analysis with mimicked and polarized word embeddings for human-robot interaction. FGCS (2019). https://doi.org/10.1016/j.future.2019.10.012. http://www.sciencedirect.com/science/article/pii/S0167739X19309719

4. Avellaneda, M., Lee, J.H.: Statistical arbitrage in the us equities market. Quan. Finan. **10**(7), 761–782 (2010). https://doi.org/10.1080/14697680903124632

5. Bergmeir, C., Hyndman, R.J., Koo, B.: A note on the validity of cross-validation for evaluating autoregressive time series prediction. Comput. Stat. Data Anal. **120**, 70–83 (2018). https://doi.org/10.1016/j.csda.2017.11.003

6. Blaskowitz, O.J., Herwartz, H.: Adaptive forecasting of the EURIBOR swap term structure (2009)

7. Box, G.E.P., Jenkins, G.: Time Series Analysis, Forecasting and Control. Holden-Day Inc., San Francisco (1990)

8. Breiman, L.: Random forests. Mach. Learn. **45**(1), 5–32 (2001). https://doi.org/10.1023/A:1010933404324

9. Brown, G., Wyatt, J.L., Tiňo, P.: Managing diversity in regression ensembles. J. Mach. Learn. Res. **6**, 1621–1650 (2005)

10. Carta, S., Corriga, A., Ferreira, A., Recupero, D.R., Saia, R.: A holistic auto-configurable ensemble machine learning strategy for financial trading. Computation **7**(4), 67 (2019)

11. Carta, S., Ferreira, A., Recupero, D.R., Saia, M., Saia, R.: A combined entropy-based approach for a proactive credit scoring. Eng. Appl. Artif. Intell. **87**, 103292 (2020). https://doi.org/10.1016/j.engappai.2019.103292

12. Cavalcante, R.C., Brasileiro, R.C., Souza, V.L., Nobrega, J.P., Oliveira, A.L.: Computational intelligence and financial markets: a survey and future directions. Expert Syst. Appl. **55**, 194–211 (2016). https://doi.org/10.1016/J.ESWA.2016.02.006

13. Chalimourda, A., Schölkopf, B., Smola, A.J.: Experimentally optimal ν in support vector regression for different noise models and parameter settings. Neural Netw. **17**(1), 127–141 (2004). https://doi.org/10.1016/S0893-6080(03)00209-0

14. Christoffersen, P.F., Diebold, F.X.: How relevant is volatility forecasting for financial risk management? Rev. Econ. Stat. **82**(1), 12–22 (2000). https://doi.org/10.1162/003465300558597

15. Damghani, B.M.: The non-misleading value of inferred correlation: an introduction to the cointelation model. Wilmott **2013**(67), 50–61 (2013). https://doi.org/10.1002/wilm.10252

16. Dawid, A.P.: Present position and potential developments: some personal views statistical theory the prequential approach. J. R. Stat. Soc.: Ser. A (Gener.) **147**(2), 278–290 (1984)

17. Devezas, T.: Principles of Forecasting. A Handbook for Researchers and Practitioners: J. Scott Armstrong. Kluwer Academic Publishers, Norwell (2001). xii and 849 p. ISBN 0-7923-7930-6 (hardbound); us$190. Technol. Forecast. Soc. Change, **69**(3), 313–316 (2002). https://doi.org/10.1016/S0040-1625(02)00180-4

18. Enke, D., Thawornwong, S.: The use of data mining and neural networks for forecasting stock market returns. Expert Syst. Appl. **29**(4), 927–940 (2005). https://doi.org/10.1016/J.ESWA.2005.06.024. https://www.sciencedirect.com/science/article/pii/S0957417405001156?via%3Dihub

19. Fischer, T., Krauss, C.: Deep learning with long short-term memory networks for financial market predictions. Eur. J. Oper. Res. **270**(2), 654–669 (2018). https://doi.org/10.1016/J.EJOR.2017.11.054

20. Gatev, E., Goetzmann, W.N., Rouwenhorst, K.G.: Pairs trading: performance of a relative-value arbitrage rule. Rev. Finan. Stud. **19**(3), 797–827 (2006). https://doi.org/10.1093/rfs/hhj020

21. Hastie, T., Tibshirani, R., Friedman, J.: The Elements of Statistical Learning: Data Mining, Inference, and Prediction. Springer Series in Statistics, 2nd edn. Springer, Heidelberg (2009). https://doi.org/10.1007/978-0-387-84858-7

22. Henrique, B.M., Sobreiro, V.A., Kimura, H.: Literature review: machine learning techniques applied to financial market prediction. Expert Syst. Appl. **124**, 226–251 (2019). https://doi.org/10.1016/J.ESWA.2019.01.012

23. Huck, N.: Pairs selection and outranking: an application to the S&P 100 index. Eur. J. Oper. Res. **196**(2), 819–825 (2009). https://doi.org/10.1016/j.ejor.2008.03.025

24. Huck, N.: Large data sets and machine learning: applications to statistical arbitrage. Eur. J. Oper. Res. **278**(1), 330–342 (2019). https://doi.org/10.1016/J.EJOR.2019.04.013

25. Kara, Y., Acar Boyacioglu, M., Baykan, Ö.K.: Predicting direction of stock price index movement using artificial neural networks and support vector machines: the sample of the Istanbul Stock Exchange. Expert Syst. Appl. **38**(5), 5311–5319 (2011). https://doi.org/10.1016/J.ESWA.2010.10.027

26. Kaufman, C., Lang, D.T.: Pairs trading. In: Data Science in R: A Case Studies Approach to Computational Reasoning and Problem Solving, pp. 241–308 (2015). https://doi.org/10.1201/b18325

27. Ke, G., et al.: LightGBM: a highly efficient gradient boosting decision tree. In: Guyon, I., et al. (eds.) Advances in Neural Information Processing Systems, vol. 30, pp. 3146–3154. Curran Associates, Inc. (2017)

28. Khandani, A.E., Lo, A.W.: What happened to the quants in august 2007? Evidence from factors and transactions data. J. Finan. Mark. **14**(1), 1–46 (2011). https://doi.org/10.1016/j.finmar.2010.07.005

29. Knoll, J., Stübinger, J., Grottke, M.: Exploiting social media with higher-order factorization machines: statistical arbitrage on high-frequency data of the S&P 500. Quan. Finan. **19**(4), 571–585 (2019). http://www.scopus.com

30. Krauss, C., Do, X.A., Huck, N.: Deep neural networks, gradient-boosted trees, random forests: statistical arbitrage on the S&P 500. Eur. J. Oper. Res. **259**(2), 689–702 (2017). https://doi.org/10.1016/J.EJOR.2016.10.031

31. Large, J., Lines, J., Bagnall, A.: The heterogeneous ensembles of standard classification algorithms (HESCA): the whole is greater than the sum of its parts (2017)

32. Lee, K.J., Yoo, S., Jin, J.J.: Neural network model vs. Sarima model in forecasting Korean stock price index (KOSPI) (2007)

33. Leung, M.T., Daouk, H., Chen, A.S.: Forecasting stock indices: a comparison of classification and level estimation models. Int. J. Forecast. **16**(2), 173–190 (2000). https://doi.org/10.1016/S0169-2070(99)00048-5. http://www.sciencedirect.com/science/article/pii/S0169207099000485

34. Lo, A.W.: Hedge Funds: An Analytic Perspective (Revised and Expanded Edition), Student edn. Princeton University Press, Princeton (2010)

35. Lo, A., Hasanhodzic, J.: The Evolution of Technical Analysis: Financial Prediction from Babylonian Tablets to Bloomberg Terminals. Wiley, Bloomberg (2011)
36. Merh, N., Saxena, V.P., Pardasani, K.R.: A comparison between hybrid approaches of ANN and ARIMA for Indian stock trend forecasting (2010)
37. Recupero, D., Dragoni, M., Presutti, V.: ESWC 15 challenge on concept-level sentiment analysis. Commun. Comput. Inf. Sci. **548**, 211–222 (2015). https://doi.org/10.1007/978-3-319-25518-7_18. Cited By 17
38. Reforgiato Recupero, D., Cambria, E.: ESWC 14 challenge on concept-level sentiment analysis. Commun. Comput. Inf. Sci. **475**, 3–20 (2014). https://doi.org/10.1007/978-3-319-12024-9_1. Cited By 17
39. Sutherland, I., Jung, Y., Lee, G.: Statistical arbitrage on the KOSPI 200: an exploratory analysis of classification and prediction machine learning algorithms for day trading. J. Econ. Int. Bus. Manag. **6**(1), 10–19 (2018)
40. Takeuchi, L.: Applying deep learning to enhance momentum trading strategies in stocks (2013)
41. Vapnik, V.N.: An overview of statistical learning theory. IEEE Trans. Neural Netw. **10**(5), 988–999 (1999). https://doi.org/10.1109/72.788640
42. Vidyamurthy, G.: Pairs Trading : Quantitative Methods and Analysis. Wiley, Hoboken (2004)

Privacy Preserving Deep Learning Framework in Fog Computing

Norma Gutiérrez, Eva Rodríguez, Sergi Mus, Beatriz Otero[⊠], and Ramón Canal

Computer Architecture Department, Universitat Politècnica de Catalunya (UPC),
Barcelona, Spain
{norma,evar,smus,botero,rcanal}@ac.upc.edu

Abstract. Nowadays, the widespread use of mobile devices has raised serious cybersecurity challenges. Mobile services and applications use deep learning (DL) models for the modelling, classification and recognition of complex data, such as images, audio, video or text. Users benefit from the wide range of services and applications offered by these devices but pay an enormous price, the privacy of their personal data. Mobile services collect all different types of users' data, including sensitive personal data, photos, videos, clinical data, banking data, etc. All this data is pooled to the Cloud to train global DL models, and big companies benefit from all the collected users' data, posing obvious serious privacy issues.

This paper proposes a privacy preserving framework for Fog computing environments, which adopts a distributed deep learning approach. Internet of Things (IoT) end nodes never reveals their sensitivity to the Cloud server; instead, they share a fraction of the users' data, blurred with Gaussian noise, with a nearby Fog node. The DL methods considered in this work are Multilayer Perceptron (MLP) and Convolutional Neural Networks (CNN), and for both cases the accuracy is similar to the centralized and privacy violating approach, obtaining the best results for the CNN model.

Keywords: Cyber-attacks · Privacy · Deep learning · Machine learning · Security · Fog computing · IoT

1 Introduction

Nowadays, mobile devices have become essential in people's daily lives. Users benefit from a wide range of services and applications that makes use of DL methods with high accuracy on the modelling, classification and recognition of complex data, as images, audio, video or text. These DL-based services and applications include recommendation systems, speech and image recognition, target advertising, health monitoring, etc. Most of these services are for free, but they collect high sensitive users' data, such as personal users' data, photos, videos, and even users' banking data. The widespread usage of these services and applications has raised important cybersecurity challenges, especially privacy issues for users' personal data. Individuals are concerned about their collected data being used for unintended purposes, for example the data collected by e-healthcare

© Springer Nature Switzerland AG 2020
G. Nicosia et al. (Eds.): LOD 2020, LNCS 12565, pp. 504–515, 2020.
https://doi.org/10.1007/978-3-030-64583-0_45

systems for real-time diagnosis and medical consultancy, or even the data collected by face recognition applications that can be later used for targeted social advertising. The principal beneficiaries are companies that collect massive amounts of data from their users, since the success of existing DL techniques highly depends on the amount of available data for training the network. The centralized Cloud approach does not respect the privacy and confidentiality of users' data and also, it is staring to result inefficient due to limited computation resources and bandwidth. This has led to the adoption of distributed DL [1] to overcome response delays, computation bottlenecks, as well as privacy issues preventing users from pooling their data to a central server. On the one hand, the exchange of information with the server poses evident privacy issues. On the other hand, DL trained models incorporate essential information of its training data sets, and then it is not complicated to extract sensitive information from DL classifiers.

Fog computing brings Cloud Computing closer to the physical world of smart things. It was Cisco who initially conceived Fog computing as a Cloud computing extension to the edge of an enterprise network. Continuous innovations in hardware and software have made possible transferring computation to the edge of the networks. Currently, Cloud based architectures are used to process and store IoT devices' generated data, but fog computing paradigm is a promising solution to scale and optimize IoT infrastructures. Fog architectures consist of three layers, unlike cloud-based solutions that consider two layers, introducing a fog computing layer that can be divided into multiple abstraction sub-layers, between the Cloud and the IoT end devices. Edge nodes will provide computing, storage, communication, control and security services overcoming centralized Cloud based approaches issues. Fog computing considers distributed DL, where IoT devices, as well as Fog devices perform part of the learning process. IoT devices can conduct part of the learning process thanks to lightweight DL libraries developed by industrial companies, as TensorFlow Lite [2]. However, it is not realistic to suppose that all IoT devices are equipped with substantial computational resources, since most of them have low-spec chips. Moreover, end users would not wish to share their local models for privacy concerns. All this led us to devise a framework that considers DL models running on Fog devices, not in end devices. IoT end devices will send a fraction of their data, blurred with Gaussian noise, to nearby fog nodes, where DL algorithms will be executed.

Lot of work has been done in the privacy area, mainly based on differential privacy, introduced by Dwork [3] in 2006. Differential privacy guarantees that the models or algorithms learnt from participants released information, but without participants' data being included. Then, it will not be possible to determine the specific participant information used in computation. Differential privacy is used in DL frameworks to avoid disclosure of private information. The algorithms proposed in [4, 5] were based on a differentially private version of stochastic gradient descent (SGD). Hitaj et al. [6] continued this work, training DL structures locally and only sharing a subset of the parameters, obfuscated via differential privacy. But all these works do not consider a distributed DL architecture, they consider a central server to build a global model which combines all the users' data or provides all the parameters to end users.

In this paper, we focus on privacy and confidentiality concerns proposing a DL privacy preservation framework for IoT. This framework preserves users' privacy performing distributed DL. In this way, IoT nodes never reveal their sensitive data to the Cloud server, they only share a fraction their sensitive data, blurred with Gaussian noise, with a nearby Fog node, which subsequently computes the gradient of the users' data and updates a fraction of them to the Cloud. To evaluate the results obtained the framework is compared with centralized privacy-violating solutions, considering both MLP and CNN models, since they are the most widely used for DL privacy preservation.

2   Related Work

Centralized deep learning poses serious privacy threats, since users pool their data in a central server that train a global model, which combines data from all participants. First works that address privacy issues, proposed a privacy preserving model. Shokri and Shmatikov [1] propose a distributed training technique to preserve users' privacy based on selective stochastic gradient descent (SGD). The authors take advantage of the fact that optimization algorithms used in DL can be parallelized and executed asynchronously. Then, users can train their own datasets independently and only share a small subset of the key parameters of their models. This methodology works for all different types of DL models guaranteeing users privacy, while maintaining accuracy levels. The solution was evaluated using MLP and CNN models and two datasets MNIST [7] and SVHN [8], typically used for image classification. For the MNIST dataset, the system achieved an accuracy of 99.14% when participants share the 10% of their data, close to the accuracy (99.17%) obtained for a centralized privacy violating model approach. While for the SVHN dataset, the accuracy was of 93.12%. Pong et al. [9] continued this work improving the system in two ways: first, data is not leaked to the server, and second, homomorphic encryption is used providing more security to the system without compromising its accuracy.

A similar approach has been undertaken in Fog computing environments. Lyu et al. [10] defined a Fog embedded privacy preserving deep learning framework (FPPDL) to reduce computation and communication costs while preserving privacy. Privacy is preserved by a two-level protection mechanism, which first uses Random Projection (RP) to protect privacy, perturbing original data, but preserving certain statistical characteristics of the original data, and then Fog nodes train differentially fog-level models applying Differentially Private SGD. The framework was implemented using MLP with two hidden layers, using ReLU activations, and it was tested using three different datasets typically used in image classification: MNIST [7], SVHN [8] and multiview multicamera dataset [11]. The solution achieved an accuracy of 93.31% for the MNIST dataset and 84.27% for the SVHN dataset, almost 10% lower than for the centralized framework, but reducing communication and computation costs significantly.

The work presented in this paper proposes a framework that preserves users' privacy performing distributed DL in Fog computing, exploring two models: MLP [12] and CNN [13], since they are the models most commonly used for privacy preservation [1, 4, 9, 10, 14]. The framework is evaluated using a digits' dataset. Decentralized protection techniques are applied to this dataset on the Fog to classify the digit, preserving its original value correctly. The information used in this dataset can be related to the telephone number of an individual of the digits of his DNI.

The framework proposed in this paper differs from [1, 9] because we only share the gradients with the Fog nodes, and we inject Gaussian noise to improve users' privacy. And from [10] because we also consider the CNN model, not just MLP, obtaining better results for CNN. The framework proposed in this paper achieves better results for CNN. Another key aspect, that differentiates our framework from all previous works [1, 9, 10], is that we perform a validation process in the server when updating the mean gradients, which lead our framework to obtain better accuracy than in previous works.

3   Framework

This section describes the privacy preserving DL framework proposed in this paper. A three-level framework is proposed as depicted in Fig. 1, which consists of numerous IoT devices at the bottom level (V devices), or end nodes, Fog nodes in the middle level (N fog nodes) and the Cloud server at the top level.

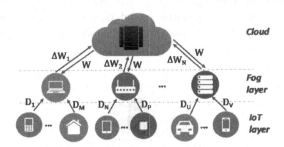

Fig. 1. Fog-embedded architecture.

The framework in this architecture has been designed to preserve users' privacy. Therefore the IoT end nodes never reveal their data to the Cloud server by only sharing part of their data with the nearby fog node. Moreover, Gaussian noise has been injected to provide higher privacy to users' sensitive data. The devices at the end nodes contain all the data to train the model. Each device selects a specified portion of the dataset and preprocesses the data. This preprocessing consists of loading the raw data from the dataset, separate it into labels and 3D images, which will later be flattened in the model. Then, a selected amount of data will be stored in a pickle which will be sent to the fog node. Once the data is selected and preprocessed, it is sent to its' associated fog node. Each fog node has a selected number of end nodes associated with it; in our experiments, a total of 6 end nodes were selected for each fog node. Once all the preprocessed data from the different IoT devices is in the fog node, it downloads a copy of the latest version of the model from the Cloud server. Initially, the model weights have a random initialization ($\bar{W}$, or random weight vector). When the fog node receives the data and the model it starts the training process with the specified epochs. In order to share as little as possible to the Cloud, once each epoch is completed only the gradient between the initial weights and the new weights, $\overrightarrow{\Delta W}$, is passed to the Cloud, thus preserving the client's privacy by only passing a little portion of the total results.

The Cloud server validates the received updates with a portion of the reserved dataset called validation data which is stored in the Cloud server, and stores them in $\overrightarrow{W}$ as $\overrightarrow{W} = \overrightarrow{W} + \overrightarrow{\Delta W}$. After the framework is collaboratively trained, each Fog node can test and evaluate its data independently.

The privacy preserving framework is compared with a centralized privacy violating Cloud architecture (see Fig. 2), where all the users' data is uploaded to the server, and the global model is trained using the same metrics as before. That is to say, devices, or end nodes, only upload raw data to the Cloud, who is responsible for the preprocessing, training, and validation of the model. In the fog-embedded architecture, not only the data is preprocessed in the end nodes, but the fog nodes train the model leaving the Cloud server with one task, to validate and update the model to be accessible for other fog nodes. We have chosen MLP and CNN for the fog-embedded architecture since these are the best architectures and most commonly used to analyze the selected problem. It is noteworthy that the accuracy results obtained for the privacy preserving framework proposed in this paper are similar to the achieved by the centralized framework, as detailed in the next section. That is, privacy is preserved achieving similar accuracy results. In Sect. 5, it will be shown that the models have less accuracy in the fog-embedded architecture. However, privacy is much higher than the centralized architecture, which violates end-users' privacy.

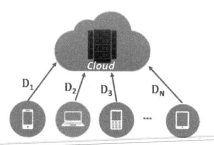

Fig. 2. Centralized architecture.

4   Performance Evaluation

4.1   Dataset

We evaluate our system on The Street View House Number (SVHN) dataset [8] that contains 73.257 digit images of 32×32 pixels for training, and 26.032 for testing. The objective is to classify the input as one of ten possible digits in the integers $\{0, 1, ..., 9\}$, so the size of the output layer is ten. From these samples, we used a total of 70.000 for training and 10.000 for validation. The validation samples reside in the Cloud server, whereas the training samples come from the different end nodes that each fog node has associated. In the fog-embedded architecture, we consider balanced data partitions for three, five, and seven fog nodes and its end nodes. Each fog node connects with six end nodes, and then each fog-level model is trained on 23.300, 14.000, and 10.000 samples depending if we are running three, five or seven fog nodes. This way, each end node owns 5%, 3.3%, and 2.4% data of the entire training. Decreasing the percentage of data that each end node provides to the fog node preserves the client's privacy since less data is demanded as the number of fog nodes is increased. Table 1 summarizes training and test examples used in this work for centralized and fog-embedded architectures.

Table 1. Size of training and test SVHN dataset.

Architecture	Training	Testing
Centralized	70.000	10.000
Fog-embedded (with 3 fog nodes)	23.300 per fog node Each fog node has 6 clients, and each end node has 3.900 training samples	10.000
Fog-embedded (with 5 fog nodes)	14.000 per fog node Each fog node has 6 clients, and each end node has 2330 training samples	10.000
Fog-embedded (with 7 fog nodes)	10.000 per fog node Each fog node has 6 clients, and each end node has 1.700 training samples	10.000

4.2 Description Models

This work proposes a framework that preserves users' privacy performing distributed DL in Fog computing using two models: Multilayer Perceptron (MLP) and Convolutional Neural Network (CNN).

The MLP architecture consists of an input layer, one or more hidden layers and an output layer. In our model, we have a simple multilayer perceptron with two hidden layers with 256 and 128 units using ReLU activations, and a final SoftMax layer with 10 units. The exact description of the model used is below.

```
model.Sequential([
  model.Flatten(),
  model.Dense(1024, ReLU),
  model.Dense(1024 -> 256, ReLU),
  model.Dense(256 -> 128, ReLU),
  model.Dense(128-> 10, SoftMax)
])
```

The CNN architecture consists of three different types of layers: convolutional pooling and classification. In the implementation, the CNN model consists of a simple Convolutional Neural Network with also, two hidden layers with 128 and 64 units respectively, since our main goal was to maintain the structure of the model in both cases as similar as possible. In our hidden layers, each layer consists of a convolutional layer and a pooling layer. The model also included a SoftMax layer with 10 units. The detailed model is in the diagram below.

```
model.Sequential([
  model. Conv2D(256, ReLU),
  model.MaxPooling2D(pool_size),
  model.Conv2D(256 -> 128, ReLU),
  model.MaxPooling2D(pool_size),
  model.Conv2D(128-> 64, ReLU),
  model.MaxPooling2D(pool_size),
  model. Flatten(),
  model.Dense(64- > 10, SoftMax)
])
```

Finally, for both implementations, other relevant parameters were considered as SGD (Stochastic gradient descent) with a learning rate of 0.1 and a categorical cross entropy loss.

4.3   Experiments

The network has been trained using a 1.6 GHz Dual-Core Intel Core i5 with 8 GB of memory and an Intel UHD Graphics 617 1536 MB, using Tensorflow version 2.0.0 with Python 3.7.4 version and Numpy library. Experiments using MLP and CNN were executed in a Cloud environment, specifically Microsoft Azure. It was used the virtual machine STANDARD_D14_V2. We used a callback to register all the statistics of each execution, enabling us to compare each model easily and storing all the data in a logs folder.

Moreover, we normalized the dataset by subtracting the average and dividing by the standard deviation of data training samples. After that, we changed the data format converting the images into Numpy arrays and stored them a file. Finally, in the fog-embedded architecture we applied 1% of Gaussian noise to the samples to preserve their privacy.

As stated before each data end-node sends a different portion of the dataset to the fog node and the fog node trains the model a stated number of epochs. In MLP a total of 100 epochs were chosen, and in CNN, a total of 20 epochs were chosen. These numbers are a good compromise between an optimal convergence and the improvement of the accuracy since if the number of epochs was to be increased; the accuracy progressed poorly and consumed many resources.

Once the model is trained, only the mean value of each weight of every epoch is passed to the Cloud server allowing thus, to preserve the user's privacy, since no raw data was sent, only a statistical value of the weight of each epoch. Then, when the Cloud server has received all the mean values from each fog-node, updates the model, as explained in Sect. 3. Finally, for each experiment in both architectures we have conducted 50 simulations and the results obtained were shown in the next section.

4.4 Results

During the training process, a Tensor board callback was used, which enabled to monitor the accuracy and loss over each epoch (see Fig. 3 and Fig. 4). In this way, the networks can be compared. After network training, we determine which architectures have major benefits, and those were kept while the underperforming ones were rejected. For both DL methods the mean-gradient was uploader for the server.

Experiments using the centralized architecture model and MLP reached an accuracy of 88.92% after 100 epochs (see Fig. 3), while the accuracy for CNN was 94.08% with only 20 epochs (see Fig. 4).

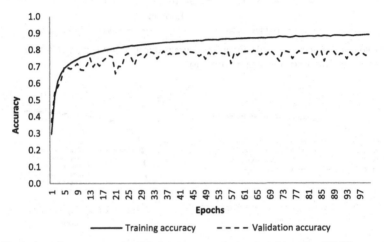

Fig. 3. Evolution of accuracy and validation accuracy during training using MLP in a centralized architecture (non-privacy data preserving).

Figures 5 and 6 show the results obtained for the experiments conducted on the fog-embedded architecture for both DL models using the SVHN dataset. As before, figures show the accuracy achieved in the training and validation processes. However, in this case the validation accuracy values are determined by getting the averaged value over all fog nodes. The graphs in both figures (MLP and CNN) show that accuracy for the fog-embedded architecture is slightly improved compared to the centralized model (see Figs. 3 and 4).

For all experiments in each scenario the percentage difference in loss between the training and validation dataset is smaller than 1%. This indicates that no overfitting takes place in any case.

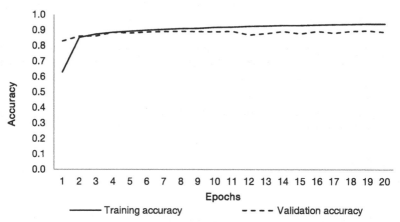

Fig. 4. Evolution of accuracy and validation accuracy during training using CNN in a centralized cloud architecture (non-privacy data preserving).

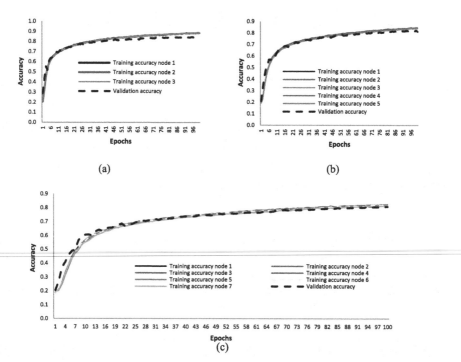

Fig. 5. Evolution of accuracy and validation accuracy during training using MLP in a fog- embedded architecture (privacy data preserving): (a) for 3 fog nodes; (b) for 5 fog nodes; (c) for 7 fog nodes.

Table 2 summarizes the obtained results in all experiments, where we can appreciate the difference between both architectures and models for validation and training. In this table, we observed the average values of the validation process to fog-embedded

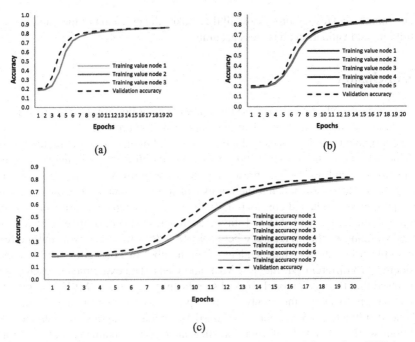

Fig. 6. Evolution of accuracy and validation accuracy during training using CNN in a fog-embedded architecture (privacy data preserving): (a) for 3 fog nodes; (b) for 5 fog nodes; (c) for 7 fog nodes.

Table 2. Summary of the training and validation accuracy average with 100 epochs (to MLP architecture) and 20 epochs (to CNN architecture)

Architecture	Framework			
	MLP		CNN	
	Training	Validation	Training	Validation
Centralized	0,8892	0,7841	0,9408	0,8878
Fog-embedded (with 3 fog nodes)	0,8837	0,8357	0,8606	0,8624
Fog-embedded (with 5 fog nodes)	0,8426	0,8120	0,8370	0,8500
Fog-embedded (with 7 fog nodes)	0,8205	0,8092	0,7923	0,8104

architecture is slightly close (or equal in the order of the tenths) to the training values to the same architecture in both models. This is due to in the fog-embedded architecture we have distributed 10.000 validation samples, and this amount of samples is very similar to the training samples. Usually, we have a more significant amount of training samples than validation samples, such as in the centralized model. Also, Figs. 5 and 6 show the effect to increase the fog nodes number influences the decrease in precision. This is mainly because increasing the number of fog nodes decreases the number of samples for the training and testing processes, all this makes the precision of the model found in the training and validation phase less accurate.

5   Conclusion

This paper proposed a privacy preserving framework for Fog computing environments, which adopted a distributed deep learning approach. The proposed framework consists of three levels, with IoT devices at the bottom level, Fog nodes in the middle level and the Cloud server at the top level. Fog nodes trained efficiently the network combining fractions of end nodes data, preserving in this way users' privacy. Each fog node was connected to six clients, and the number of fog nodes varied between 3, 5 and 7. The experiments conducted considered 70.000 training samples, distributed uniformly between all the clients, and obtained similar accuracies to the centralized Cloud training privacy violating method. Accuracy between 80%–86% was achieved, and the results were similar to each other independent of the number of Fog nodes. The framework considered two different DL models: MLP and CNN. The experiments results carried out showed that the CNN model worked significantly better than MLP, being the most remarkable improvement the number of the epochs needed in each architecture, since CNN needed five times fewer epochs than MLP to achieve the same convergence.

Future work will include a formal statistical analysis the obtained results, and further word will be conducted for improving the framework exploring new ways to update the model with a portion of the gradients, and users' privacy protection mechanism will be improved applying Random Projection to the original data.

Acknowledgements. This work is partially supported by Generalitat de Catalunya under the SGR program (2017-SGR-962) and the RIS3CAT DRAC project (001-P-001723).

References

1. Shokri, R., Shmatikov, V.: Privacy-preserving deep learning. In: Proceedings of the 22nd ACM SIGSAC Conference on Computer and Communications Security, pp. 1310–1321, October 2015
2. TensorFlow Lite. https://www.tensorflow.org/lite
3. Dwork, C.: Differential privacy. In: Proceedings of the International Colloquium on Automata, Languages and Programming (ICALP), pp. 1–12 (2006)
4. Abadi, M., et al.: Deep learning with differential privacy. In: Proceedings of the 2016 ACM SIGSAC Conference on Computer and Communications Security, pp. 308–318, October 2016
5. Dwork, C., Roth, A.: The algorithmic foundations of differential privacy. J. Found. Trends Theor. Comput. Sci. **9**(3–4), 211–407 (2014)
6. Hitaj, B., Ateniese, G., Perez-Cruz, F.: Deep models under the GAN: Information leakage from collaborative deep learning. In: Proceedings of the ACM SIGSAC Conference on Computer and Communications Security (CCS), pp. 603–618, November 2017
7. LeCun, Y., Bottou, L., Bengio, Y., Haffner, P.: Gradient-based learning applied to document recognition. Proc. IEEE **86**(11), 2278–2324 (1998)

8. Netzer, Y., Wang, T., Coates, A., Bissacco, A., Wu, B., Ng, A.: Reading digits in natural images with unsupervised feature learning. In: NIPS 2011 Workshop on Deep Learning and Unsupervised Feature Learning, vol. 2011, no. 2, p. 5 (2011)

9. Phong, L.T., Aono, Y., Hayashi, T., Wang, L., Moriai, S.: Privacy-preserving deep learning: revisited and enhanced. In: Batten, L., Kim, D.S., Zhang, X., Li, G. (eds.) ATIS 2017. CCIS, vol. 719, pp. 100–110. Springer, Singapore (2017). https://doi.org/10.1007/978-981-10-542 1-1_9

10. Lyu, L., Bezdek, J.C., He, X., Jin, J.: Fog-embedded deep learning for the internet of things. IEEE Trans. Industr. Inf. **15**(7), 4206–4215 (2019)

11. Roig, G., Boix, X., Shitrit, H.B., Fua, P.: Conditional random fields for multi-camera object detection. In: Proceedings of the IEEE International Conference Computer Vision (ICCV), pp. 563–570, November 2011

12. Roopak, M., Yun Tian, G., Chambers, J.: Deep learning models for cyber security in IoT networks. In: 2019 IEEE 9th Annual Computing and Communication Workshop and Conference (CCWC), pp. 452–457 (2019)

13. LeCun, Y., Bengio, Y.: Convolutional networks for images, speech, and time series. In: Arbib, M.A. (ed.) The handbook of Brain Theory and Neural Networks, pp. 255–258 (1998)

14. Osia, S.A., Shamsabadi, A.S., Taheri, A., Rabiee, H.R., Haddadi, H.: Private and scalable personal data analytics using hybrid Edge-to-Cloud deep learning. Computer **51**(5), 42–49 (2018)

On the Reliability of Dynamical Stochastic Binary Systems

Guido Lagos[1,2(✉)] and Pablo Romero[3(✉)]

[1] Departamento de Ingeniería Industrial, Universidad de Santiago de Chile,
Santiago, Chile
guido.lagos@usach.cl
[2] Facultad de Ingeniería y Ciencias, Universidad Adolfo Ibáñez, Santiago, Chile
[3] Facultad de Ingeniería, Universidad de la República, Montevideo, Uruguay
promero@fing.edu.uy

Abstract. In system reliability analysis, the goal is to understand the correct operation of a multi-component on-off system, i.e., each component can be either working or not, and each component fails randomly. The *reliability* of a system is the probability of correct operation. Since the reliability evaluation is a hard problem, the scientific literature offers both efficient reliability estimations and exact exponential-time evaluation methods.

In this work, the concept of Dynamical Stochastic Binary Systems (DSBS) is introduced. Samaniego signature provides a method to find the reliability and Mean-Time-to-Failure of a DSBS. However, we formally prove that the computation of Samaniego signature belongs to the hierarchy of $\#\mathcal{P}$-Complete problems.

The interplay between static and dynamic models is here studied. Two methodologies for the reliability evaluation are presented. A discussion of its applications to structural reliability and analysis of dependent failures in the novel setting of DSBS is also included.

Keywords: Network optimization · Computational complexity · Reliability · Samaniego signature · Crude Monte Carlo

1 Introduction

Reliability analysis is a growing research field, given its paramount importance in the understanding of communication networks, power grids, transportation systems, to name a few real-life complex systems. The goal is to find the probability of correct operation of a multi-component on-off system, subject to random failures. Practical applications include system optimization and diagnostics, in order to predict potential failures of critical systems.

In the network reliability setting, we are given a simple graph where the links fail independently, with identical probability. The connectedness probability of the random graph is known as the *all-terminal reliability*. Pioneer works

© Springer Nature Switzerland AG 2020
G. Nicosia et al. (Eds.): LOD 2020, LNCS 12565, pp. 516–527, 2020.
https://doi.org/10.1007/978-3-030-64583-0_46

from Ball and Provan formally prove that the all-terminal reliability evaluation belongs to the class of NP-Hard problems [3].

As a corollary, the scientific literature on network reliability offers exact (exponential-time) methods [19] and efficient (polynomial-time) statistical estimations [9], as well as exact efficient methods for special graphs. The reader is invited to consult the excellent monograph authored by Charles Colbourn [10], and its applications in network optimization [5].

More recent works try to extend the analysis to a more general class of stochastic binary systems (SBS). Naturally, the reliability evaluation is also hard, since it subsumes the all-terminal reliability. In [7], the concept of separable systems is introduced. Combining duality [15] and large deviation theory, the authors provide reliability bounds for general SBS. A first attempt of reliability maximization for general SBS is proposed in [8], where the goal is to select the uniformly most-reliable sub-system subject to budget constraints.

Independently, Samaniego proposed a signature-based method to compare the performance of coherent systems under specific assumptions on the failures, considering dynamic lifetimes on the models [18]. Here, we formally prove that the computation of Samaniego signature belongs to the hierarchy of $\#P$-Complete problems. Hence, Samaniego method has its limitations. Then, we promote Monte Carlo-based methods for the reliability evaluation of Dynamical SBS. The main contributions of this paper are the following:

1. The concept of a Dynamical Stochastic Binary System is introduced.
2. We formally prove that the computation of Samaniego signature belongs to the class of $\#P$-Complete problems.
3. The interplay between static and dynamical systems is explored.
4. A pointwise estimation of the Reliability of static models and Mean Time to Failure (MTTF) of dynamical systems is proposed, comparing Crude Monte Carlo versus Samaniego signature.
5. A proof of concept illustrates the benefits and shortcomings of both approaches.
6. A final discussion is oriented toward the applicability of system reliability to optimization.

The document is organized in the following manner. In Sect. 2, the concept of Dynamical Stochastic Binary Systems (DSBS) is introduced. Section 3 explores the interplay between the static model of SBS and the novel setting of DSBS, inspired by Samaniego signature. A pointwise methodology to find the MTTF is presented in Sect. 4. A proof-of-concept is presented in Sect. 5, where the performance of Crude Monte Carlo (CMC) and Samaniego methods is studied under systems that accept closed-form reliability evaluation. A discussion of extension of our proposals and potential applications to system optimization is included in Sect. 6. Section 7 presents concluding remarks and trends for future work.

2   Dynamical Stochastic Binary Systems

The following terminology is adopted from [2].

Definition 1 (Stochastic Binary System (SBS)).
An SBS is a triad (S, p, ϕ):

- *$S = \{1, \ldots, n\}$ is a ground set of components,*
- *$p = (p_1, \ldots, p_n) \in [0, 1]^n$ has their elementary reliabilities, i.e., component i is on with marginal probability p_i, and*
- *$\phi : \{0, 1\}^n \to \{0, 1\}$ is the structure function.*

Definition 2 (Reliability/Unreliability).
Let $\mathcal{S} = (S, p, \phi)$ be an SBS, and consider a random vector $X = (X_1, \ldots, X_n)$ with independent coordinates governed by Bernoulli random variables such that $P(X_i = 1) = p_i$. The reliability of $\mathcal{S}$ is the probability of correct operation:

$$r_{\mathcal{S}} = P(\phi(X) = 1) = E(\phi(X)). \tag{1}$$

The unreliability of $\mathcal{S}$ is $q_{\mathcal{S}} = 1 - r_{\mathcal{S}}$.

Definition 3 (Pathsets/Cutsets).
Let $\mathcal{S} = (S, p, \phi)$ be an SBS.
A state $x \in \{0, 1\}^n$ is a pathset (resp. cutset) if $\phi(x) = 1$ (resp., if $\phi(x) = 0$).

Most real-life systems present a *well-behavior* of its failures, in the sense that if a system is already non-operational, after new failures the system is necessarily non-operational again.

Definition 4 (Stochastic Monotone Binary System).
The triad $\mathcal{S} = (S, p, \phi)$ is an SMBS if the structure function $\phi : \{0, 1\}^n \to \{0, 1\}$ is monotone, $\phi(0_n) = 0$ and $\phi(1_n) = 1$, where 0_n and 1_n stand for the states where all the components are non-operational (resp. operational).

Given $x, y \in \{0, 1\}^n$, we consider the order of the cartesian product of the natural set. Then, $(x_1, \ldots, x_n) \le (y_1, \ldots, y_n)$ if and only if $x_i \le y_i$ for all $i = 1, \ldots, n$. Furthermore, $x < y$ if $x \le y$ and $x \ne y$.

Definition 5 (Minpaths/Mincuts).
Let $\mathcal{S} = (S, p, \phi)$ be an SMBS. A pathset x is a minpath if $\phi(y) = 0$ for all y such that $y < x$. A cutset y is a mincut if $\phi(x) = 1$ for all x such that $y < x$.

Let us identify a component with its corresponding index in the ordered set $\{0, 1\}^n$. We denote $x(u = 1)$ (or $x(u = 0)$) to the state x where the component in position u is set to 1 (resp. u is set to 0).

Definition 6 (Essential/Irrelevant Components).
Let $\mathcal{S} = (S, p, \phi)$ be a stochastic binary system. A component $u \in S$ is essential if the system is always down when u fails. The component u is irrelevant if $\phi(x(u = 0)) = \phi(x(u = 1))$ for all $x \in \{0, 1\}^n$.

An SBS is *homogeneous* if its elementary reliabilities are identical. The following examples provide a picture of the wide scope of applicability of SBS:

1. All-Terminal Reliability: the ground set is precisely the links of a simple graph. The system is up if the resulting random graph is connected.
2. K-Terminal Reliability: in the same random graph, the system is up if some distinguished node-set K, called terminals, belong to the same connected component.
3. Source-Terminal Reliability model: the previous model with $K = \{s, t\}$.
4. Diameter Constrained Reliability: a diameter constraint d is added to the K-Terminal Reliability. The system is up if every pair of terminals are connected by paths whose length is not greater than the diameter [6].
5. k-out-of-n: the system is up if and only if there are at least k components in operational state out of n equally reliable components.
6. Consecutive k-out-of-n: the k operational components must be contiguous in the ordered set $S = \{1, \ldots, n\}$.
7. Hamiltonian model: Given a simple graph G, the system is up if the resulting random graph has a Hamiltonian tour.

Samaniego in his monograph [18] deals with *coherent* systems:

Definition 7 (Coherent Systems).
An SMBS is coherent if it has no irrelevant components, and furthermore, $\phi(0_n) = 0$, $\phi(1_n) = 1$.

A first shortcoming is that the recognition of irrelevant components in an SMBS is not trivial [16]:

Theorem 1. *The recognition of an irrelevant component in an arbitrary SMBS belongs to the class of NP-Hard problems.*

Proof. Consider Hamiltonian model with $p = 1/2$ for all its links. A fixed link $e = (x, y) \in E$ is irrelevant iff there is no Hamiltonian path in $G - \{e\}$ whose terminals are precisely x and y. Since Hamiltonian path is $\mathcal{NP}$-Complete [12], the result follows. $\qquad\square$

Theorem 1 is pessimistic; however, the analysis of signatures can be employed using irrelevant components in general SMBS. In the following, we define separable systems and Dynamical SBS, as a particular stochastic process ruled by a given structure [7]. First, observe that an arbitrary structure ϕ can be understood as $0 - 1$ labels of the vertices of an n-cube Q_n in the Euclidean space, this is, $\phi : Q_n \to \{0, 1\}$. Let us denote $Q_n^+ = \phi^{-1}(1)$ and $Q_n^- = \phi^{-1}(0)$ to the pathsets and cutsets, respectively.

Definition 8 (Separable System).
The system S is separable if the pathsets Q_n^+ and cutsets Q_n^- can be separable by a hyperplane. Formally, $S = (S, p, \phi)$ is separable if there exists real numbers $\alpha_1, \ldots, \alpha_n$ and α such that $\sum_{i=1}^{n} \alpha_i x_i \geq \alpha$ for all pathset $x = (x_1, \ldots, x_n)$. We must replace $\geq$ by $<$ for all cutsets.

Here we introduce the following concept:

Definition 9 (Dynamical Stochastic Binary System).

A DSBS is a continuous-time stochastic process $\{(X_1(t), \ldots, X_n(t))\}_t$ equipped with a monotone structure $\phi : Q_n \rightarrow \{0, 1\}$, such that the domain of the random vector $(X_1 \ldots, X_n)$ is the state-space Q_n, the process starts at $t = 0$ in the unit-vector 1_n, and the process is finished in the absorbing state $Q_n^- = \phi^{-1}(0)$.

Observe that we considered the quotient of the set Q_n^- as a *single-state* of all the cutsets in the corresponding SBS. Furthermore, the model accepts single or multiple failures on a given instant, as well as repairs, since the transitions can occur between arbitrary pathsets of Q_n. We will measure the performance of a DSBS by means of the traditional MTTF:

Definition 10 (Mean-Time-To-Failure (MTTF)).

Consider a stochastic process $(Y_t)_{t \geq 0}$ and a hitting-set Q^-, that consists of the union of the failure states. The MTTF is the expected value of the random variable $T = \inf\{t \geq 0 : Y_t \in Q^-\}$.

A special family of DSBS will be considered:

Definition 11 (Associated DSBS).

Let $\mathcal{S} = (S, \phi, p)$ be an SMBS. Its associated DSBS is the stochastic process $\{(X_1(t), \ldots, X_n(t))\}$ with independent exponential lifetimes for each component $X_i \sim Exp(\lambda_i)$, being $\lambda_i = -Ln(p_i)$, that starts in the special state $X_i(0) = 1$ for all $i = 1, \ldots, n$.

Our goal is to study the interplay between SBS and its associated SBS. Furthermore, to discuss its impact in optimization and trends for future work in this promising field, which has a vast number of applications.

3   Interplay Between SBS and DSBS

Consider a homogeneous SMBS given by $\mathcal{S} = (S, p, \phi)$. Let us partition the pathsets using the following family: $\mathcal{A}_k = \{x \in \{0, 1\}^n : \phi(x) = 1, \sum_{i=1}^n x_i = k\}$. Clearly, $\cup_{k=0}^n \mathcal{A}_k = Q_n^+ = \phi^{-1}(1)$, is a partition of the pathsets. Further, let us denote $F_k = |\mathcal{A}_k|$, and $\boldsymbol{F} = (F_0, \ldots, F_n)$.

Proposition 1. *In a homogeneous SMBS, the reliability $R_{\mathcal{S}}(p)$ is a polynomial, and can be expressed as follows:*

$$R_{\mathcal{S}}(p) = \sum_{k=0}^n F_k p^k (1-p)^{n-k} \tag{2}$$

Proof. We will partition the pathsets using the sets $\{F_i\}_{i=0,...,n}$ and use the definition of system reliability:

$$R_{\mathcal{S}}(p) = P(\phi(X) = 1) = \sum_{x \in Q_n^+} P(X = x)$$

$$= \sum_{k=0}^{n} \sum_{x \in \mathcal{A}_k} P(X = x) = \sum_{k=0}^{n} F_k P(X = x)$$

$$= \sum_{k=0}^{n} F_k p^k (1 - p)^{n-k},$$

where the last step uses the fact that the failures on the components are independent, and all the pathsets belonging to $\mathcal{A}_k$ have probability $p^k(1 - p)^{n-k}$. $\square$

Consider the DSBS $\mathcal{D}$ associated to a homogeneous SMBS. Observe that the lifetimes of all the components are governed by an identical exponential distribution $F(t)$, with rate $\lambda = -Ln(p)$. Let us sort the component lifetimes in increasing order: $X_1^* < X_2^* \ldots < X_n^*$. Consider the hitting-set $Q_n^- = \phi^{-1}(0)$, and the random variable T that represents the *system's lifetime*, this is, the time until some cutset is reached. Naturally, $T = Y_j^*$ for some $j \in \{1, \ldots, n\}$.

Definition 12 (Signature). *The signature of an SMBS is the vector* $s = (s_1, \ldots, s_n)$ *such that* $s_i = P(T = X_i^*)$.

The following result can be obtained using the fact that the event $T = X_j^*$ holds almost surely for some $j \in \{1, \ldots, n\}$, and conditional probabilities [14]:

Proposition 2.

$$P(T > t) = \sum_{i=1}^{n} s_i \sum_{j=0}^{i-1} \binom{n}{j} (F(t))^j (1 - F(t))^{n-j} \tag{3}$$

There is a strict relation between $P(T > t)$ and the reliability of an SBS. Indeed, observe that the rate λ is fixed in such a way that $F(1) = 1 - e^{-\lambda} = p$. Then:

$$P(T > 1) = R_{\mathcal{S}} \tag{4}$$

Furthermore, the following relation between the coefficients hold:

Lemma 1.

$$f_k = \sum_{i > n-k} s_i, \tag{5}$$

being $f_k = \frac{F_k}{\binom{n}{k}}$ *the proportion of operational states with exactly* k *operational components.*

Proof. Equation (4) establishes that the polynomials from (2) and (3) must be identical. Comparing their coefficients in Bernstein basis $\{p^i(1-p)^{n-i}\}_{i=0,\ldots,n}$, $F_k = f_k\binom{n}{k}$ must be equal to $\binom{n}{k}\sum_{i>n-k} s_i$. $\square$

Consider the vector with the proportion of operational states, or f-vector: $f = (f_0, \ldots, f_n)$.

Theorem 2. *The computation of a signature of an arbitrary DSBS belongs to the class of #$\mathcal{P}$-Complete problems.*

Proof. Lemma 1 shows that the signature s of an associated DSBS is immediately recovered with the vector f; and vice-versa. Since Ball and Provan formally proved that counting the f-vector belongs to the class of #$\mathcal{P}$-Complete problems [3], the result follows. $\square$

Theorem 2 promotes heuristics or simulation techniques, either for the reliability estimation of an SBS or the MTTF of a DSBS.

4 Reliability Evaluation

Monte Carlo is a noteworthy computational tool for simulation. In a macroscopic point of view, the idea is to faithfully simulate a complex system (or a part of it), and consider N independent experiments of that simulation, in order to determine the performance of the system (or subsystem) and assist decisions. It has been widely applied to a great diversity of problems, including numerical integration, counting, discrete-event (and rare event) simulation. The reader can find a generous variety of applications in [11], and a thorough analysis of rare event simulation using Monte Carlo methods in [17]. In the classical or Crude Monte Carlo (CMC), N independent replicas $X_1, \ldots, X_N$ of a stochastic system with finite mean $E(X)$ are carried-out. By Kolmogorov's strong law, the average $\overline{X}_N$ is unbiased, and converges almost surely to $E(X)$. Its variance (and its mean square error) is $Var(\overline{X}_N) = \frac{Var(X)}{N}$. Figures 1 and 2 contain a pseudocode for CMC and Samaniego methods. Note that CMC is here applied to the static model, while *Samaniego* considers a dynamical system, and hence can produce a simultaneous estimation of both the reliability of an SBS (or DSBS), and MTTF parameter for the DSBS associated to a given SBS. Here, exponential lifetimes are selected for *Samaniego* method. In the following, both CMC and *Samaniego* methods are respectively detailed.

Crude Monte Carlo (CMC) receives the structure ϕ, a homogeneous probability of operation p for all the components, sampling size N, and returns the reliability estimation $\mathcal{R}_{CMC}$ for a static SBS. The rationale is just an averaging among observations of the system. A *Sum* is initialized as zero in Line 1. In a *double-for* loop of Lines 2–7, we will generate N independent replicas of the system, each with a vector of size n. In the block of Lines 3–5, the components are set to 1 with probability p. If the resulting observation is operational, the variable *Sum* is updated correspondingly in Line 6. The last *for*-loop is concluded with a global *Sum*, that counts all the operational observations (Line 7). The average is found as the global *Sum* divided by the sampling size N in Line 8. Finally, this average $\mathcal{R}_{CMC}$ is returned as the reliability estimation of the static SBS in Line 9.

Algorithm 1 $\mathcal{R}_{CMC} = CMC(\phi, p, N)$

1: $Sum \leftarrow 0$
2: **for** $i = 1$ **to** N **do**
3:      **for** $j = 1$ **to** n **do**
4:         $X_j^i \leftarrow (Random < p)$ $\{X_j^i = 1$ with probability $p.\}$
5:      **end for**
6:      $Sum \leftarrow Sum + \phi(X_1^i, \ldots, X_n^i)$
7: **end for**
8: $\mathcal{R}_{CMC} \leftarrow Sum/N$
9: **return** $\mathcal{R}_{CMC}$

Fig. 1. Pseudocode for CMC method.

Algorithm 2 $(\hat{MTTF}, \mathcal{R}_S) = Samaniego(\phi, p, N)$

1: $\mathit{s} \leftarrow 0$
2: $Sum_{MTTF} \leftarrow 0$
3: $\lambda \leftarrow -Ln(p)$
4: **for** $i = 1$ **to** N **do**
5:      **for** $j = 1$ **to** n **do**
6:         $L_j^i \leftarrow Exponential(\lambda)$
7:      **end for**
8:      $\mathit{s} \leftarrow \mathit{s} + CriticalIndex(\phi, (L_1^i, \ldots, L_n^i))$ $\{$Add 1 to critical index$\}$
9:      $Sum_{MTTF} \leftarrow Sum_{MTTF} + Lifetime(\phi, (L_1^i, \ldots, L_n^i))$ $\{$Add i-th lifetime$\}$
10: **end for**
11: $\mathit{s} \leftarrow \mathit{s}/N$
12: $\mathit{f} \leftarrow 0$
13: **for** $i = 1$ **to** $n - 1$ **do**
14:     **for** $j = i + 1$ **to** n **do**
15:        $f_i \leftarrow f_i + s_j$
16:     **end for**
17: **end for**
18: $Sum_R \leftarrow 0$
19: **for** $j = 0$ **to** $n - 1$ **do**
20:     $Sum_R \leftarrow Sum_R + f_j \binom{n}{j}(1-p)^j p^{n-j}$
21: **end for**
22: $\mathcal{R}_S \leftarrow Sum_R/N$
23: $\hat{MTTF} \leftarrow Sum_{MTTF}/N$
24: **return** $(\hat{MTTF}, \mathcal{R}_S)$

Fig. 2. Pseudocode for $Samaniego$ method.

Let us proceed with the details of $Samaniego$ (see Fig. 2). It can serve both for a static SBS and to estimate the reliability and MTTF of a Dynamical SBS with independent failures on its components (in particular, to the DSBS associated to a given SBS). The rationale is to pick independent lifetimes of the components, and determine the *critical component* from each observation, such that the system goes down exactly when it fails. The signature, MTTF and exponential rate are initialized in Lines 1–3. The rate λ is chosen in such a way that $P(T > 1)$ is precisely the reliability of the static SBS. In the *double-for* loop of Lines 4–10, the exponential lifetimes are generated, in a similar fashion as CMC. Function *CriticalIndex* finds the first component (index) such that the system is down, using the lifetimes for each component $(L_1^i, \ldots, L_n^i)$. In the block of Lines 11–17, the signature vector s is first estimated, and then the f-vector is estimated using Lemma 1. The last block of Lines 18–22 provides the reliability estimation, using Expression (3). In Line 23, the estimation for the MTTF takes place, and it is precisely the averaging of the system lifetime. Finally, both estimations $M\hat{T}TF$ and $\mathcal{R}_S$ are returned in Line 25.

5   Proof-of-Concept

In order to illustrate Samaniego and CMC approaches, we use the same sampling size of $N = 10^7$ independent vectors, picking samples in each component from an exponential distribution. A simple averaging among the operational systems is the estimator $\mathcal{R}_{CMC}$ for CMC, while the proportion of critical components s is first found, and then Expression (3) is considered as an alternative reliability estimation $\mathcal{R}_S$ (see Figs. 1 and 2 for details). Both techniques are developed in the following elementary systems, where a closed-form for the reliability is straight:

1. $\mathcal{S}_1$: Homogeneous 5-out-of-10 System, with $p = 1/2$, and
2. $\mathcal{S}_2$: Homogeneous Consecutive 3-out-of-10 System, with $p = 1/2$.

The exact reliability $R_{\mathcal{S}_1}$ is $P(Y \geq 5) = 1 - F_B(4)$, being $B \sim Bin(10, 1/2)$. Therefore, $R_{\mathcal{S}_1} = 0,623$. The recursive relation $a_{n+3} = a_{n+2} + a_{n+1} + a_n + 2^n$ subject to $a_1 = a_2 = 1, a_3 = 1$, counts the number of binary words which have at least three consecutive ones, and $a_{10} = 520$, so $R_{\mathcal{S}_\epsilon} = 520/2^{10} = 0.5078$. Table 1 shows the CPU-time and reliability estimations for both systems, using CMC and Samaniego methods.

Table 1. Estimations for $R_{\mathcal{S}_1}$ and $R_{\mathcal{S}_2}$ with $p = 1/2$.

Model	CPU Time $\mathcal{S}_1$ (s)	CPU Time $\mathcal{S}_2$ (s)	$R'_{\mathcal{S}_1}$	$R'_{\mathcal{S}_\epsilon}$	$M\hat{T}TF_1$	$M\hat{T}TF_2$
CMC	23.79	231.53	0.6229	0.5078	–	–
Samaniego	54.70	278.24	0.6230	0.5079	1.2202	1.4194

The reader can appreciate that both methods are adequate for small instances. In fact, these methods are statistically consistent, and unbiased for the reliability. The MTTF estimation exceeds the unit, in accordance with the fact that both probabilities exceed one-half (since more than half of the samples survive the unit-time).

By means of the central limit theorem, the main drawback can be appreciated: the convergence rate is slow for highly reliable systems, which makes both methods inaccurate under large values of p, and promotes the study of rare event simulation. Samaniego method is computationally more demanding than CMC; however, it also provides the MTTF estimation at the same time. The simplest CMC for static models works for both instances.

As an evidence of the difficulties with rare event scenarios, identical homogeneous systems S_1 and S_2 are considered in a second stage, but with failure probability $q = 1 - p = 10^{-6}$ in their components. The new results are summarized in Table 2.

Table 2. Estimations for R_{S_1} and R_{S_2} with $q = 1 - p = 10^{-6}$.

Model	CPU Time S_1 (s)	CPU Time S_2 (s)	R'_{S_1}	R'_{S_ϵ}	$M\hat{T}TF_1$	$M\hat{T}TF_2$
CMC	12.37	242.84	1	1	–	–
Samaniego	32.56	230.49	1	1	8.46E05	9.83E05

From Table 2, we can appreciate that both methods become useless under a rare event scenario. The reliability is close to the unit, but all the observations were operational under both methods. For a better understanding of the results shown in Table 2, consider an independent identically distributed vector $X_1, \ldots, X_N$ where its coordinates are Bernoulli variables with failure probability $q = 1 - p$. We want to estimate the unreliability $Q_S = 1 - R = E(1 - \phi(X))$, which is the same for practical purposes. The central limit theorem states that the radius of the confidence interval with level $1 - \alpha$ for $E(1 - \phi(X))$ using the averaging $\overline{\phi(X)}_N$ is approximately:

$$\epsilon = \frac{z_{\alpha/2}\sqrt{Q_S(1 - Q_S)}}{\sqrt{N}}. \tag{6}$$

The relative error $RE(Q_S)$ is the ratio between half the confidence interval and the correct target value Q_S.

$$RE(Q_S) = \frac{z_{\alpha/2}\sqrt{(1 - Q_S)}}{\sqrt{Q_S}\sqrt{N}}. \tag{7}$$

Observe that RE tends to infinity when the rarity of the event Q_S tends to zero. Then the simplest method, Crude Monte Carlo, does not have the Bounded Relative Error (BRE) property. This is the central problem of rare event simulation. The reader is invited to consult the excellent book on rare event simulation [17] for further details.

6   Discussion

From Expression (2), if a system maximizes simultaneously all the coefficients $\{F_k\}_{k=1,\ldots,n}$, then it is the optimal (most-reliable) system. In the special all-terminal reliability model, some works offer new graph with the maximum reliability, called *uniformly most-reliable graphs* [1]. Recent works study a topological network optimization under variable costs and dependent link failures [4]. Since the problem is $\mathcal{NP}$-Hard, the authors considered Sample Average Approximation (SAA) method [13], which consists in the application of CMC to the objective, and a smart link selection, that maximizes the number of operational configurations. This methodology can be clearly extended to the design of general Stochastic Binary Systems. The analysis of Dynamical SBS is more complex, but Markov Chain and Monte Carlo-based methods are useful. It is worth to remark that if the system is highly reliable and the failure is a rare event, CMC is not suitable, and there are tools already developed such as Importance Sampling [17].

7   Concluding Remarks

The concept of Dynamical Stochastic Binary System (DSBS) is introduced. In general, a DSBS is a stochastic process of multi-components subject to (possibly different and or dependent) failures and repairs. Here, we studied the interplay between the static model and dynamism, considering independent identically distributed random lifetimes for its components. Essentially, both problems can be understood as hard counting problems in the homogeneous system (with identical failures). Then, we illustrated the static concept of Crude Monte Carlo and the dynamic Samaniego signature (based on random lifetimes on the components). Both methods are suitable for small systems. Highly reliable systems should be analyzed in a different way.

As future work, we want to design Dynamical SBS trying to maximize the system reliability, meeting budget constraints. We would like to include realism to the model, by means of dependent failures on its components (using Marshal-Olkin copula or alternatives). There are already realistic models in the literature, and they could be generalized under the setting of DSBS.

Acknowledgment. This work is partially supported by Project ANII FCE_1_2019_1_156693 *Teoría y Construcción de Redes de* Máxima Confiabilidad, MATHAMSUD 19-MATH-03 *Rare events analysis in multi-component systems with dependent components* and STIC-AMSUD ACCON *Algorithms for the capacity crunch problem in optical networks*. Guido Lagos acknowledges the financial support of FONDECYT grant 3180767.

References

1. Archer, K., Graves, C., Milan, D.: Classes of uniformly most reliable graphs for all-terminal reliability. Discret. Appl. Math. **267**, 12–29 (2019)

2. Ball, M.O.: Computational complexity of network reliability analysis: an overview. IEEE Trans. Reliab. **35**(3), 230–239 (1986)
3. Ball, M.O., Scott Provan, J.: The complexity of counting cuts and of computing the probability that a graph is connected. SIAM J. Comput. **12**, 777–788 (1983)
4. Barrera, J., Cancela, H., Moreno, E.: Topological optimization of reliable networks under dependent failures. Oper. Res. Lett. **43**(2), 132–136 (2015)
5. Boesch, F.T., Satyanarayana, A., Suffel, C.L.: A survey of some network reliability analysis and synthesis results. Networks **54**(2), 99–107 (2009)
6. Canale, E., Cancela, H., Robledo, F., Romero, P., Sartor, P.: Diameter constrained reliability: complexity, distinguished topologies and asymptotic behavior. Networks **66**(4), 296–305 (2015)
7. Cancela, H., Ferreira, G., Guerberoff, G., Robledo, F., Romero, P.: Building reliability bounds in stochastic binary systems. In: 10th International Workshop on Resilient Networks Design and Modeling, pp. 1–7, August 2018
8. Cancela, H., Guerberoff, G., Robledo, F., Romero, P.: Reliability maximization in stochastic binary systems. In: 21st Conference on Innovation in Clouds, Internet and Networks and Workshops (ICIN), pp. 1–7, February 2018
9. Cancela, H., El Khadiri, M., Rubino, G.: A new simulation method based on the RVR principle for the rare event network reliability problem. Ann. OR **196**(1), 111–136 (2012)
10. Colbourn, C.J.: The Combinatorics of Network Reliability. Oxford University Press Inc., New York (1987)
11. Fishman, G.: Monte Carlo. Springer Series in Operations Research and Financial Engineering. Springer, Heidelberg (1996)
12. Garey, M.R., Johnson, D.S.: Computers and Intractability: A Guide to the Theory of NP-Completeness. W. H. Freeman and Company, New York (1979)
13. Kleywegt, A.J., Shapiro, A., Homem-de-Mello, T.: The sample average approximation method for stochastic discrete optimization. SIAM J. Optim. **12**(2), 479–502 (2002)
14. Lindqvist, B.H., Samaniego, F.J., Wang, N.: Preservation of the mean residual life order for coherent and mixed systems. J. Appl. Probab. **56**(1), 153–173 (2019)
15. Romero, P.: Duality in stochastic binary systems. In: 2016 8th International Workshop on Resilient Networks Design and Modeling (RNDM), pp. 85–91, September 2016
16. Romero, P.: Challenges in stochastic binary systems. In: 11th International Workshop on Resilient Networks Design and Modeling, pp. 1–8, October 2019
17. Rubino, G., Tuffin, B.: Rare Event Simulation Using Monte Carlo Methods. Wiley, Hoboken (2009)
18. Samaniego, F.J.: On closure of the IFR class under formation of coherent systems. IEEE Trans. Reliab. **R-34**(1), 69–72 (1985)
19. Satyanarayana, A., Chang, M.K.: Network reliability and the factoring theorem. Networks **13**(1), 107–120 (1983)

A Fast Genetic Algorithm for the Max Cut-Clique Problem

Giovanna Fortez[(✉)], Franco Robledo[(✉)], Pablo Romero[(✉)], and Omar Viera[(✉)]

Instituto de Matemática y Estadística IMERL, Facultad de Ingeniería, Universidad de la República, Montevideo, Uruguay
{giovanna.fortez,frobledo,promero,viera}@fing.edu.uy

Abstract. In Marketing, the goal is to understand the psychology of the customer in order to maximize sales. A common approach is to combine web semantic, sniffing, historical information of the customer, and machine learning techniques.

In this paper, we exploit the historical information of sales in order to assist product placement. The rationale is simple: if two items are sold jointly, they should be close. This concept is formalized in a combinatorial optimization problem, called *Max Cut-Clique* or *MCC* for short.

The hardness of the *MCC* promotes the development of heuristics. The literature offers a GRASP/VND methodology as well as an Iterated Local Search (ILS) implementation. In this work, a novel Genetic Algorithm is proposed to deal with the *MCC*. A comparison with respect to previous heuristics reveals that our proposal is competitive with state-of-the-art solutions.

Keywords: Marketing · Combinatorial optimization problem · Max cut-clique · Metaheuristics

1 Motivation

Large-scale corporations keep massive databases from their customers. Nevertheless, decision-makers and practitioners sometimes do not have the knowledge to combine Machine Learning techniques with Optimization, in order to exploit the benefits of Big Data.

An effective dialogue between academics and practitioners is crucial [7]. A bridge between the science-practice division can be found in Market Basket Analysis, or MBA [1]. In synthesis, MBA is a Data Mining technique originated in the field of Marketing. It has recent applications to other fields, such as bioinformatics [3,5], World Wide Web [15], criminal networks [6] and financial networks [17]. The goal of MBA is to identify relationships between groups of products, items, or categories.

The information obtained from MBA is of paramount importance in the business strategy and operations. In Marketing, we can find valuable applications such as product placement, optimal product-line offering, personalized marketing

G. Nicosia et al. (Eds.): LOD 2020, LNCS 12565, pp. 528–539, 2020.
https://doi.org/10.1007/978-3-030-64583-0_47

campaigns and product promotions. The analysis is commonly supported by Machine Learning, Optimization and Logical rules for association.

This work is focused on a specific combinatorial optimization methodology to assist product placement. Incidentally, it finds nice applications in biological systems as well. The problem under study is called Max Cut-Clique (MCC), and it was introduced by P. Martins [19].

Given a simple undirected graph $\mathcal{G} = (V, E)$ (where the nodes are items and links represent correlations), we want to find the clique $\mathcal{C} \subseteq V$ such that the number of links shared between $\mathcal{C}$ and $V - \mathcal{C}$ is maximized. Basically, the goal is to identify a set of *key-products* (clique $\mathcal{C}$), that are correlated with a massive number of products ($V - \mathcal{C}$). The MCC has an evident application to product-placement. For instance, the manager of a supermarket must decide how to locate the different items in different compartments. In a first stage, it is essential to determine the correlation between the different pairs of items, for psychological/attractive reasons. Then, the priceless/basic products (bread, rice, milk and others) could be hidden on the back, in order to give the opportunity for other products in a large corridor. Chocolates should be at hand by kids. Observe that the MCC appears in the first stage, while marketing/psychological aspects play a key role in a second stage for product-placement in a supermarket.

In [19], the author states that the MCC is presumably hard, since related problems such as $MAX - CUT$ and $MAX - CLIQUE$ are both $\mathcal{NP}$-Complete. A formal proof had to wait until 2018, where a reduction from $MAX - CLIQUE$ was provided [4]. Therefore, the MCC is systematically addressed by the scientific community with metaheuristics and exact solvers that run in exponential time. The first heuristic available in the literature on the MCC develops an Iterated Local Search [20]. Integer Linear Programming methods are also available [13]. The literature in the MCC is not abundant, since the problem has been posed recently.

A formal proof of complexity [4] is here included, since it is simple and it supports the applicability of heuristics. Furthermore, the main concepts of the previous GRASP/VND heuristic is given. A fair comparison between the novel proposal and this heuristic takes place.

This paper is organized in the following manner. Section 2 formally presents the hardness of the MCC. Section 3 briefly presents the previous GRASP/VND methodology. The novel Genetic Algorithm is introduced in Sect. 4. A fair comparison between both heuristics is presented using DIMACS benchmark in Sect. 5. Section 6 contains concluding remarks and trends for future work.

The reader is invited to consult the authoritative books on Graph Theory [14], Computational Complexity [10] and Metaheuristics [11] for the terminology used throughout this work.

2   Computational Complexity

The cornerstone in computational complexity is Cook's Theorem [8] and Karp reducibility among combinatorial problems [18].

Stephen Cook formally proved that the joint satisfiability of an input set of clauses in disjunctive form is the first $\mathcal{NP}$-Complete decision problem [8]. Furthermore, he provided a systematic procedure to prove that a certain problem is $\mathcal{NP}$-Complete. Specifically, it suffices to prove that the decision problem belongs to set $\mathcal{NP}$, and that it is at least as hard as an $\mathcal{NP}$-Complete problem. Richard Karp followed this hint, and presented the first 21 combinatorial problems that belong to this class [18]. In particular, $MAX - CLIQUE$ belongs to this list. The reader is invited to consult an authoritative book in Complexity Theory, which has a larger list of $\mathcal{NP}$-Complete problems and a rich number of bibliographic references [10].

Here, we formally prove that the MCC is at least as hard as $MAX - CLIQUE$. Let us denote $|\mathcal{C}|$ the cardinality of a clique $\mathcal{C}$, and $\delta(\mathcal{C})$ denotes the corresponding cutset induced by the clique (or the set) $\mathcal{C}$.

Definition 1 (MAX-CLIQUE). *GIVEN: a simple graph $G = (V, E)$ and a real number K.*
QUESTION: is there a clique $\mathcal{C} \subseteq V$ such that $|\mathcal{C}| \geq K$?

For convenience, we describe MCC as a decision problem:

Definition 2 (MCC). *GIVEN: a simple graph $G = (V, E)$ and a real number K.*
QUESTION: is there a clique $\mathcal{C} \subseteq G$ such that $|\delta(\mathcal{C})| \geq K$?

Theorem 1 was established for the first time in [4]. For a matter of completeness, the proof is here included.

Theorem 1. *The MCC belongs to the class of $\mathcal{NP}$-Complete problems.*

Proof. We prove that the MCC is at least as hard as $MAX - CLIQUE$. Consider a simple graph $G = (V, E)$ with order $n = |V|$ and size $m = |E|$. Let us connect a large number of M hanging nodes, to every single node $v \in V$. The resulting graph is called H (see Fig. 1 for an example). If we find a polynomial-time algorithm for MCC, then we can produce the max cut-clique in H. But observe that the Max Cut-Clique $\mathcal{C}$ in H cannot include hanging nodes, thus it must belong entirely to G. If a clique $\mathcal{C}$ has cardinality c, then the clique-cut has precisely $c \times M$ hanging nodes. By construction, the cut-clique must maximize the number of hanging nodes, if we choose $M \geq m$. As a consequence, c must be the $MAX - CLIQUE$. We proved that the MCC is at least as hard as $MAX - CLIQUE$, as desired. Since MCC belongs to the set of $\mathcal{NP}$ Decision problems, it belongs to the $\mathcal{NP}$-Complete class. ∎

Theorem 1 promotes the development of heuristics in order to address the MCC.

3   GRASP/VND Heuristic

GRASP and Tabu Search are well known metaheuristics that have been successfully used to solve many hard combinatorial optimization problems. GRASP is

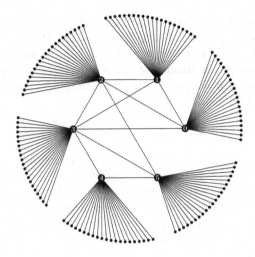

Fig. 1. Construction of H with $M = 21$ hanging nodes.

an iterative multi-start process which operates in two phases [22]. In the Construction Phase a feasible solution is built whose neighborhood is then explored in the Local Search Phase. Tabu Search [2,12] is a strategy to prevent local search algorithms getting trapped in locally optimal solutions. A penalization mechanism called Tabu List is considered to avoid returning to previously visited solutions. For a complete description of these methods the reader is referred to the works of Glover and Laguna [12] and Resende and Ribeiro [22]. The reader is invited to consult the comprehensive Handbook of Metaheuristic for further information [11].

The full GRASP/VND implementation for the MCC is proposed in [4]. It strictly follows a traditional two-phase GRASP template with a Variable Neighborhood Descent (VND) as the local search phase, followed by an UPDATE of a Tabu list.

The goal in VND to combine a rich diversity of neighborhoods in order to obtain a solution that is both feasible and locally optimum for every feasible neighborhood. In [4], the authors consider five neighborhood structures considered to build the VND:

- **Remove:** a singleton $\{i\}$ is removed from a clique $\mathcal{C}$.
- **Add:** a singleton $\{i\}$ is added from a clique $\mathcal{C}$.
- **Swap:** if we find $j \notin \mathcal{C}$ such that $\mathcal{C} - \{i\} \subseteq N(j)$, we can include j in the clique and delete i (swap i and j).
- **Cone:** generalization of Swap for multiple nodes. The clique $\mathcal{C}$ is replaced by $\mathcal{C} \cup \{i\} - \mathcal{A}$, being $\mathcal{A}$ the nodes from $\mathcal{C}$ that are non-adjacent to i.
- **Aspiration:** this movement offers the opportunity of nodes belonging to the Tabu List to be added.

The previous neighborhoods take effect whenever the resulting cut-clique is increased. It is worth to remark that **Add**, **Swap**, and **Aspiration** were already considered in the previous ILS implementation [20]. The current VND is enriched with 2 additional neighborhood structures, named **Remove** and **Cone**. The Tabu list works during the potential additions during **Add**, **Swap** and **Cone**. On the other hand, **Aspiration** provides diversification with an *opportunistic unchoking* process: it picks nodes from the Tabu List instead. The reader is invited to find further details in [4].

4   Genetic Algorithm

Genetic Algorithms GA is a well-known family of metheuristics to solve hard combinatorial optimization problems. They belong to Evolutionary Computing, a wider family of metaheuristics. The goal is to emulate principles of the natural evolution of biological species, where the most adaptive individuals survive [16]. The reader can find a generous number of applications of GA to Engineering in [23].

Algorithm 1. GA: BASIC ALGORITHM

    Input: $\mathcal{G}$
    Output: $\mathcal{C}$

1: $generation = 0$
2: $Pop(generation) \leftarrow Initialize(\mathcal{G})$
3: **while** $(generation < \text{MAX\_GENERATION})$ **do**
4:     $fitnessPop \leftarrow$ **evaluate** $(Pop(generation))$
5:     $Parents \leftarrow$ **tournamentSelection** $(Pop(generation))$
6:     $Offspring \leftarrow$ **Crossover2x** $(Parents)$
7:     $Offspring \leftarrow$ **simpleMutation** $(Offspring)$
8:     $newpop \leftarrow$ **replacement** $(Offspring, Parents)$
9:     $generation = generation + 1$
10:    $P(generation) = newpop$
11: **end while**
12: **return** $\mathcal{C}$ ▷ Best $\mathcal{C}$ Ever Found

In general, the algorithm randomly generates a set of feasible solutions, which are the *individuals* of the initial population. Then, combining crossover and mutation, this initial population evolves, and new generations are defined until a halting criterion is met. While the evolution is running, the exploration and exploitation of the solution space occurs, and an adequate operation selection must be applied [21]. Furthermore, the definition of the next generation trades between the selection of the best solutions and the preservation of the main characteristics of the population.

We strictly followed the traditional template of Genetic Algorithms, in Algorithm 1. In this case, the input is a simple undirected graph $\mathcal{G} = (V, E)$ and the result is a clique $\mathcal{C}$. The development is based on Malva Project, a collaborative and open source framework for computational intelligence in C++ [9]. The main reason of this choice is performance. The particular functions that are suitable for the *MCC* are detailed in the following subsections.

4.1 Fitness and Notation

Consider a simple graph $\mathcal{G} = (V, E)$. Let us sort the node-set $V = \{v_1, \ldots, v_n\}$. Each individual is represented by a bitmap, which is a binary word $\mathcal{X} = (x_1, \ldots, x_n)$ such that $x_i = 1$ means that node v_i belongs to the clique $\mathcal{C}$.

The *fitness* function is responsible for the survivability of the individuals in order to promote to the next generation. In this case, the fitness is precisely our objective function:

$$|\delta(\mathcal{C})| = \sum_{v \in \mathcal{C}} deg(v) - |\mathcal{C}| \times |\mathcal{C} - 1| \tag{1}$$

The fitness evaluation is performed by Algorithm 2, where Expression (1) is considered.

Algorithm 2. EVALUATE

    Input: $\mathcal{X}, \mathcal{G}$
    Output: *fitness*
1: *fitness* $= 0$
2: **for** $i = 1$ **to** $|V|$ **do**
3:     **if** $(\mathcal{X}[i]) = 1$ **then** $\forall i \in \mathcal{X}$
4:          *fitness* $=$ *fitness* $+ deg(i)$
5:     **end if**
6: **end for**
7: *fitness* $=$ *fitness* $- |\mathcal{C}| * |\mathcal{C} - 1|$
8: **return** *fitness*

Since we store the degree-sequence in a vector, the evaluation $deg(i)$ can be accessed fast, and using memory $O(1)$.

4.2 Selection: Tournament Selection

Selection determines which individuals will be parents of the next generation. In brief, the idea is to propagate good characteristics for the survival for the next generations. However, elitism selection must be avoided in order to prevent premature convergence of the method.

For that reason, the Goldberg Tournament Selection of size 2, plays a fundamental role in determining a trade-off between exploitation and exploration. Here, two individuals are randomly taken from the whole generation and their respective fitness compared, the one who maximizes this value is effectively selected as a parent of the new generation (Line 5, Algorithm 1).

4.3   Crossover

The evolutionary operators combine the genetic information of two or more individuals into new individuals with better fitness. Precisely, the recombination is executed by $Crossover2X$. Given two parents $\mathcal{X}$, $\mathcal{Y}$ and a crossover probability p_c, $Crossover2X$ returns two descendants $\mathcal{X}^*$ and $\mathcal{Y}^*$.

The operation takes place with probability p_c (Lines 1–2). In order to perform the crossover, 2 index-nodes $h < k$ are uniformly picked at random from the labels $\{1,\ldots,n\}$ (Line 3) and define two individuals, $\mathcal{X} = (x_1,\ldots,x_n)$ and $\mathcal{Y} = (y_1,\ldots,y_n)$. The children X^* and Y^* keep identical genetic information from their corresponding parents, for the elements $i < h$ (Lines 4–7) or $i \geq k$ (Lines 12–15), but a crossing occurs between the indices $i : h \leq i < k$ (Lines 8–11). Algorithm 3 returns the descendants $\mathcal{X}^*$ and $\mathcal{Y}^*$.

Algorithm 3. $Crossover2X$

    Input: $\mathcal{X}$, $\mathcal{Y}$, $p_{crossover}$
    Output: $\mathcal{X}^*$, $\mathcal{Y}^*$
1: $prob = random(0,1)$
2: **if** $prob < p_c$ **then**
3:     $(h,k) = random(0, |V| - 1)$
4:     **for** $i = 1$ **to** $h - 1$ **do**
5:         $\mathcal{X}^*[i] = \mathcal{X}[i]$
6:         $\mathcal{Y}^*[i] = \mathcal{Y}[i]$
7:     **end for**
8:     **for** $i = h$ **to** $k - 1$ **do**
9:         $\mathcal{X}^*[i] = \mathcal{Y}[i]$
10:        $\mathcal{Y}^*[i] = \mathcal{X}[i]$
11:     **end for**
12:     **for** $i = k$ **to** $|V| - 1$ **do**
13:       $\mathcal{X}^*[i] = \mathcal{X}[i]$
14:       $\mathcal{Y}^*[i] = \mathcal{Y}[i]$
15:     **end for**
16: **end if**

The traditional crossover operation considers a single index to cross the genetic information. Here, we used two indices for diversification. The parameter p_c is obtained using a statistical inference from a training set. Section 4.7 presents further details of the parametric selection.

4.4   Mutation

This technique applies an operation called mutation that works locally in a solution, changing one or more bits uniformly at random. In brief, consider a solution $\mathcal{X} = (x_1,\ldots,x_n)$. This operator performs a random walk in the neighborhood

---

Algorithm 4. SIMPLE MUTATION

    Input: $\mathcal{X}, p_m$
    Output: $\mathcal{X}^*$
1: **for** $i = 1$ **to** $|V| - 1$ **do**
2:     $\mathcal{X}^*[i] = \mathcal{X}[i]$
3: **end for**
4: $prob = random(0, 1)$
5: **if** $prob < p_m$ **then**
6:     $k = random(0, |V| - 1)$
7:     $new\_k = random(0, 1)$
8:     $\mathcal{X}^*[k] = new\_k$
9: **end if**
10: **return** $\mathcal{X}^*$

---

of $\mathcal{X}$. This is known as a *Simple Mutation*, which modifies a single bit from $\mathcal{X}$ at random, with probability p_m.

Among the Evolutionary Computing models, GA applies crossover with a greater probability than mutation ($p_c > p_m$). The rationale behind this decision is that the exploration of the solution space is maximized. The adjustment of p_m is explained in Sect. 4.7.

4.5 Replacement: New Generation

The replacement of the generation can be performed in various manners. Since our solution is diversity-driven, we selected the next generation using from the joint-set of descendants and parents (Line 8, Algorithm 1).

4.6 Feasible Solution

Observe that a bitmap can either represent a feasible or unfeasible solution. An unfeasible solution has different treatments according to the domain of the problem at hand. As the feasible solution must fulfill the adjacency-relation in the graph, every time its structure is manipulated by operations, a shaking algorithm is applied in order to preserve feasibility. Furthermore, the evaluation only can be practical over a feasible solution.

In this sense Algorithm 5 is a fundamental piece of code. It is inspired by the Construction Phase of the GRASP/VND introduced in [4].

4.7 Parameters Adjustment

A preliminary training set of instances were performed in order to tune the parameters using statistical analysis. Concretely, Rank Test was considered including a 30 independent runs set, finding optimal values in the cartesian product for p_m, p_c, *pop_size*.

Algorithm 5. FEASIBLE SOLUTION CONSTRUCTION

    Input: $\mathcal{X}$, $\mathcal{G}$

    Output: $\mathcal{X}$

1:   $\mathcal{C} \leftarrow \emptyset, \quad \mathcal{C}' \leftarrow \emptyset$

2:   $improving = \text{MAX\_ATTEMPTS}$

3: **while** $improving > 0$ **do**

4:     $i \leftarrow selectRandom(\mathcal{X})$

5:     $\mathcal{C}' \leftarrow [\mathcal{C} \cap N(i)] \cup \{i\}$

6:     **if** $|E'(\mathcal{C}')| > |E'(\mathcal{C})|$ **then**

7:        $\mathcal{C} \leftarrow \mathcal{C}'$

8:        $improving = \text{MAX\_ATTEMPTS}$

9:     **else**

10:       $improving = improving - 1$

11:    **end if**

12: **end while**

13: $\mathcal{X} \leftarrow \mathcal{C}$

The performance is measured in terms of the quality of the solution and computational efficiency. Each set of parameters was applied over DIMACS benchmark, for graphs with different link-densities. The selected instances for this preprocessing state were *p_hat300-1*, *MANN_a9* and *keller4*.

5   Computational Results

In order to test the performance of the algorithm, a fair comparison between both heuristics is carried out using DIMACS benchmark. The test was executed on an Intel Core i3, 2.2 GHz, 3GB RAM. Table 1 reports the performance of our GA for each instance[1].

All instances were tested using 60 runs and the stop criterion defined as $MAX\_GENERATION = 1000$, since the optimal value was reached from generation 250 to 850 in the numerous execution of the algorithm. Meanwhile calibration results for algorithm parameters was: $p\_mutex = 0.1$, $p\_crossover = 0.8$, $pop\_size = 200$. For the construction of a feasible solution in Algorithm 5, the value of $MAX\_ATTEMPTS = \lfloor \frac{|V|}{10} \rfloor * 2$.

Table 1 is divided into three main vertical areas for each instance. The leftmost one indicates the best solution known and reached according to [20]; then, the best solution reached and its computational time for [4] and the third the results for GA. An additional column indicates the optimal gap between GRASP/VND and GA.

The reader can appreciate that for dense instances the GA maintains its computational times under 5 min in the worst case. In the cases where the best solution was reached at least one time over 60 runs, the optimal reported is close

---

[1] All the scripts are available at the following URL: https://drive.google.com/drive/folders/1mCTaJM4SA62rFhIutam1xDU-PZlXKtJF.

to the optimum value. Furthermore, a fair comparison between both approaches is conducted when averages values are reported.

Table 1. Comparative results.

ILS		GRASP /VND		Genetic Algorithm		GAP
Instances	$\|\delta(C)\|$	$\|\delta(C)\|$	$T(s)$	$\|\delta(C)\|$	$T(s)$	(%)
		avg	avg	avg	avg	
c-fat200-1	81	81	0.37	81	6.4	0.0
c-fat200-2	306	306	0.81	306	7.5	0.0
c-fat200-5	1892	1892	4.94	1892	12.5	0.0
c-fat500-1	110	110	2.46	110	16.15	0.0
c-fat500-2	380	380	5.83	380	14.3	0.0
c-fat500-5	2304	2304	10.85	2304	20.36	0.0
c-fat500-10	8930	8930	65.74	8930	32.59	0.0
p_hat300-2	4637	4636.2	3659.39	4633.40	171.9	0.0
p_hat300-3	7740	7726.8	3992.42	7387.27	279.8	0.04
keller5	15184	15183.24	1167.64	12382	50.57	0.18
c125_9	2766	2766	253.25	2737.2	5.0	0.01
MANN_a27	31284	31244.10	548.54	30405	46.49	0.03

The results described in this section reflect that our GA methodology is competitive with state-of-the-art solutions for the MCC, as well as having a quality solution and computational time efficiency. We underscore the accuracy reached and the considerable reduction in computational effort, finding optimal or near-optimal solutions in reasonable times.

6 Conclusions and Trends for Future Work

A deep understanding of item-correlation is of paramount importance for decision makers. Product-placement is a practical example in which the dialogue between operational researchers and decision makers is essential.

In this work we study a hard optimization problem, known as the Max Cut-Clique or MCC for short. Basically, the goal is to identify a set of *key-products*, that are correlated with a massive number of products. This idea has been successfully implemented in clothing stores. We believe that it is suitable for supermarkets and even for web sales.

Given the hardness of the MCC, a full Genetic Algorithm was introduced in order to solve the problem. A fair comparison with previous proposals reveals that our heuristic is both competitive and faster than a state-of-the-art solutions. This fact provides a room for potential online applications, where the number of items and customers is dynamic.

As future work, we want to implement our solution into a real-life product-placement scenario. After the real implementation, the feedback of sales in a period is a valuable metric of success.

Acknowledgements. This work is partially supported by MATHAMSUD 19-MATH-03 Raredep, *Rare events analysis in multi-component systems with dependent components*, STICAMSUD 19-STIC-01 ACCON, *Algorithms for the capacity crunch problem in optical networks* and Fondo Clemente Estable *Teoría y Construcción de Redes de Máxima Confiabilidad*.

References

1. Aguinis, H., Forcum, L.E., Joo, H.: Using market basket analysis in management research. J. Manag. **39**(7), 1799–1824 (2013)
2. Amuthan, A., Deepa Thilak, K.: Survey on Tabu Search meta-heuristic optimization. In: 2016 International Conference on Signal Processing, Communication, Power and Embedded System (SCOPES), pp. 1539–1543, October 2016
3. Bader, G.D., Hogue, C.W.V.: An automated method for finding molecular complexes in large protein interaction networks. BMC Bioinf. **4**, 2 (2003)
4. Bourel, M., Canale, E.A., Robledo, F., Romero, P., Stábile, L.: A GRASP/VND heuristic for the max cut-clique problem. In: Nicosia, G., Pardalos, P.M., Giuffrida, G., Umeton, R., Sciacca, V. (eds.) Machine Learning, Optimization, and Data Science - 4th International Conference, LOD 2018, Volterra, Italy, 13–16 September 2018, Revised Selected Papers, LNCS, vol. 11331, pp. 357–367. Springer, Cham (2018). https://doi.org/10.1007/978-3-030-13709-0_30
5. Brohée, S., van Helden, J.: Evaluation of clustering algorithms for protein-protein interaction networks. BMC Bioinf. **7**(1), 488 (2006)
6. Bruinsma, G., Bernasco, W.: Criminal groups and transnational illegal markets. Crime Law Soc. Change **41**(1), 79–94 (2004)
7. Cascio, W.F., Aguinis, H.: Research in industrial and organizational psychology from 1963 to 2007: changes, choices, and trends. J. Appl. Psychol. **93**(5), 1062–1081 (2008)
8. Cook, S.A.: The complexity of theorem-proving procedures. In: Proceedings of the Third Annual ACM Symposium on Theory of Computing, STOC 1971, pp. 151–158. ACM, New York (1971)
9. Fagundez, G.: The Malva Project. A framework of artificial intelligence EN C++. GitHub Inc. https://themalvaproject.github.io
10. Garey, M.R., Johnson, D.S.: Computers and Intractability: A Guide to the Theory of NP-Completeness. W. H. Freeman and Company, New York (1979)
11. Gendreau, M., Potvin, J.-Y.: Handbook of Metaheuristics, 2nd edn. Springer Publishing Company Incorporated (2010)
12. Glover, F., Laguna, M.: Tabu Search. Kluwer Academic Publishers, Norwell (1997)
13. Gouveia, L., Martins, P.: Solving the maximum edge-weight clique problem in sparse graphs with compact formulations. EURO J. Comput. Optim. **3**(1), 1–30 (2015)
14. Harary, F.: Graph Theory. Addison Wesley Series in Mathematics. Addison-Wesley (1971)
15. Henzinger, M., Lawrence, S.: Extracting knowledge from the world wide web. Proc. Nat. Acad. Sci. **101**(suppl 1), 5186–5191 (2004)

16. Holland, J.H.: Adaption in natural and artificial systems (1975)
17. Hüffner, F., Komusiewicz, C., Moser, H., Niedermeier, R.: Enumerating isolated cliques in synthetic and financial networks. In: Yang, B., Du, D.-Z., Wang, C.A. (eds.) COCOA 2008. LNCS, vol. 5165, pp. 405–416. Springer, Heidelberg (2008). https://doi.org/10.1007/978-3-540-85097-7_38
18. Karp, R.M.: Reducibility among combinatorial problems. In: Miller, R.E., Thatcher, J.W. (eds.) Complexity of Computer Computations, pp. 85–103. Plenum Press (1972). https://doi.org/10.1007/978-1-4684-2001-2_9
19. Martins, P.: Cliques with maximum/minimum edge neighborhood and neighborhood density. Comput. Oper. Res. **39**, 594–608 (2012)
20. Martins, P., Ladrón, A., Ramalhinho, H.: Maximum cut-clique problem: ILS heuristics and a data analysis application. Int. Trans. Oper. Res. **22**(5), 775–809 (2014)
21. Reeves, C.R.: Genetic algorithms. In: Glover, F., Kochenberger, G.A. (eds.) Handbook of Metaheuristics, International Series in Operations Research & Management Science, vol. 57, pp. 109–139. Springer, Boston (2010). https://doi.org/10.1007/0-306-48056-5_3
22. Resende, M.G.C., Ribeiro, C.C.: Optimization by GRASP - Greedy Randomized Adaptive Search Procedures. Computational Science and Engineering. Springer, New York (2016). https://doi.org/10.1007/978-1-4939-6530-4
23. Slowik, A., Kwasnicka, H.: Evolutionary algorithms and their applications to engineering problems. Neural Comput. Appl. **32**, 12363–12379 (2020)

The Intellectual Structure of Business Analytics: 2002–2019

Hyaejung Lim$^{(\boxtimes)}$ and Chang-Kyo Suh

School of Business Administration, Kyungpook National University, Daegu, South Korea
limhyaejung@gmail.com, ck@knu.ac.kr

Abstract. This paper identifies the intellectual structure of business analytics using document co-citation analysis and author co-citation analysis. A total of 333 research documents and 17,917 references from the Web of Science were collected. A total of 15 key documents and nine clusters were extracted from the analysis to clarify the sub-areas. Furthermore, burst detection and timeline analysis were conducted to gain a better understanding of the overall changing trends in business analytics. The main implication of the research results is in its ability to provide the state of past and present standards to practitioners and to suggest further research into business analytics to researchers.

Keywords: Business analytics · Intellectual structure · CiteSpace · Document co-citation analysis · Author co-citation analysis

1 Introduction

Upon the opening of the digital era, data collection from a variety of fields became a relatively easier task. The amount of data available also became enormously large, resulting in so-called 'big data.' The uses for big data are potentially infinite and the importance of 'business analytics' has rapidly increased. Business analytics enables big data to be used most effectively and efficiently. It detects meaningful patterns in data and interprets these patterns to provide an organization better information for decision-making. This process can also provide an organization's prediction system with better insight.

As organizations applied prediction functions based on the statistics of analyzed data, the increased usage of business analytics became inevitable and research on business analytics increased steadily. However, despite the accumulation of numerous amounts of research on business analytics over the last two decades, a precise concept of its intellectual structure has not yet been established. This leads to difficulty in understanding its sub-areas and trends in the research field. 'Intellectual structure' means that there is an invisible structure that emerges from research that builds and systematizes one concept. Through the analysis of the intellectual structure of business analytics, we can identify the current state of the field and simultaneously provide detailed subject areas and directions in which the field might proceed in the future.

© Springer Nature Switzerland AG 2020
G. Nicosia et al. (Eds.): LOD 2020, LNCS 12565, pp. 540–550, 2020.
https://doi.org/10.1007/978-3-030-64583-0_48

This paper is comprised of five sections. The second section reviews previous research papers that analyzed the intellectual structures of various fields and used CiteSpace to visualize intellectual structures. The third section discusses how to analyze the intellectual structure of business analytics. In Sect. 4, the results of the analysis are provided and interpreted. Lastly, Section 5 summarizes the analysis and discusses future research directions in business analytics as the conclusion.

2 Literature Review

The most common metrics used to measure intellectual structures are document co-citation analysis (DCA) and author co-citation analysis (ACA). In 1981, White and Griffith [22] first applied ACA to their research in order to identify its intellectual structure. Since their pioneering research, various fields of research have used this method. White and Griffith [22] identified five sub-areas and 39 major authors in the information field. The five sub-areas identified are science communication, statistics, general, information search, and leadership. Acedo and Casillas [1] applied ACA to the international management field in 2005. Based on 583 documents published between 1997 and 2000, researchers concluded that there were eight sub-areas, including joint ventures, economic approaches, and process approaches. Shiau et al. [18] identified a total of 13 sub-areas by applying ACA in the management information systems field. The sub-areas are applying techniques, information techniques, company performance, competitive advantage, and company structure.

In 2004, Chen [5] introduced CiteSpace and applied this tool to the knowledge domain visualization field. Based on 'Institute for Scientific Information' data collected between 1985 and 2003, he effectively identified changes in the knowledge domain field by using CiteSpace. Cui et al. [8] analyzed the I-model based on social commerce documents and provided the intellectual structure of the field. A total of 12,089 data were collected between the years 2005 and 2017. Cui et al. [8] identified the most influential authors and visualized the network using keywords.

Since the appearance of CiteSpace, researchers have steadily utilized the program to analyze their research more objectively and clearly, including university-industry collaborations [10], e-learning and social learning [24], ecological assets and values [16], information science [23], hospitality [15], and others.

3 Analyzing the Research

3.1 Analysis Methods and Data Collection

This research identifies the intellectual structure of business analytics research. To analyze the intellectual structure of business analytics research, we performed DCA and ACA. DCA maps a study based on the number of co-citations of the documents. ACA posits that if two authors are cited in one document simultaneously, then the two authors are academically related to each other [22]. These two analysis methods can identify academic study areas and sub-areas, as well as changes in trends in the domain. As the size of the data becomes larger, the results of the analysis become more detailed and meaningful [14].

The data used in this research were collected from the Web of Science using the keyword 'business analytics.' The field is best defined by the term itself and helps to collect a broad range of data that is related to the field. Among the total of 386 articles published between 2002 and 2019, 333 documents with 17,917 references were extracted as the final data. As indicated in Table 1, the number of documents increased every year since 2002. The data is extracted as a text file in order to insert into CiteSpace for analyzing. The data has all the information of the documents itself, like the authors (AU), title (TI), journal (SO), abstract (AB), keywords (DE), references (CR), etc. Part of the data is shown in Fig. 1.

Table 1. Document counts for each year

Year	2002	2003	2004	2005	2006
Document Counts	1	1	0	1	0
Reference Counts	9	10	0	9	0
Year	2007	2008	2009	2010	2011
Document Counts	4	4	0	5	5
Reference Counts	182	158	0	212	379
Year	2012	2013	2014	2015	2016
Document Counts	8	8	23	41	43
Reference Counts	403	252	1271	1757	2323
Year	2017	2018	2019		Total
Document Counts	50	67	72		333
Reference Counts	2533	3529	4890		17917

3.2 Analyzing with CiteSpace

In this research we used CiteSpace, developed by Chen [6], to analyze intellectual structure by applying document co-citation analysis and ACA. CiteSpace is a java program that appropriately visualizes the network and analyzes the changes in trends. This analyzing program has been applied to numerous exploratory researches which has proved its reliability and validity of its results. When the data are inserted into CiteSpace, they appear as one network. Each node expresses a document or the author of the data and the node's size represents its centrality.

The structural analysis of the network contains the betweenness centrality, modularity Q, and the silhouette. The betweenness centrality is first formally defined by Freeman (1977) [11]. In the network structure, there can be more than one link that connects the nodes. The betweenness centrality expresses itself as a node and will be calculated by each shortest link that pass through each node. If a node has more links to other nodes, the node has a higher betweenness centrality. If a node has a high betweenness centrality, the node has a high level of control of the overall network. The modularity Q measures how well the network is separated into modules (clusters). This measure has a numerical value from 0 to 1; the closer it is to 1, the better the network is. A low modularity Q

FN Clarivate Analytics Web of Science
VR 1.0
PT J
AU Handfield, R
   Jeong, S
   Choi, T
AF Handfield, Robert
   Jeong, Seongkyoon
   Choi, Thomas
TI Emerging procurement technology: data analytics and cognitive analytics
SO INTERNATIONAL JOURNAL OF PHYSICAL DISTRIBUTION & LOGISTICS MANAGEMENT
LA English
DT Article
DE Big data; Data analytics; Cognitive analytics; Procurement technology;
   Technology roadmap
ID SUPPLY CHAIN VISIBILITY; BIG DATA ANALYTICS; BUSINESS ANALYTICS; RFID
   TECHNOLOGIES; MANAGEMENT; INFORMATION; PERFORMANCE; INTEGRATION; MATURITY; BENEFITS
AB Purpose The purpose of this paper is to elucidate the emerging landscape of procurement analytics.
C1 [Handfield, Robert] North Carolina State Univ, Dept Business Management, Raleigh, NC 27695 USA.
   [Jeong, Seongkyoon; Choi, Thomas] Arizona State Univ, Dept Supply Chain Management, WP Carey
RP Handfield, R (reprint author), North Carolina State Univ, Dept Business Management, Raleigh, NC
EM rbhandfi@ncsu.edu
RI Handfield, Robert B/D-3200-2015
OI Handfield, Robert B/0000-0003-3895-1955
CR Acito F, 2014, BUS HORIZONS, V57, P565, DOI 10.1016/j.bushor.2014.06.001
   Anand J, 2016, STRATEGIC MANAGE J, V37, P1395, DOI 10.1002/smj.2401

Fig. 1. Data (part)

means that the network cannot form a well-separated cluster. The silhouette metric is a method that measures the interpretation and validation of the uncertainty of the clusters. This refers of how well the clusters are classified. The silhouette value is a numerical value between −1 and 1; the closer it is to 1, the more well-performing the clustering analysis is [6].

It is also possible to detect bursts using CiteSpace. A burst is an indicator that detects a great change in the research occurring during a short time period. For example, when the terror attacks of 9/11 happened in the United States, research regarding the Oklahoma City bombing suddenly received attention. Therefore, past research that is related to the current situation becomes important for predicting the future.

4   Results of the Analysis

4.1   Clustering Analysis

One of the best ways to review the intellectual structure is to analyze the sub-research areas. To definitize the sub areas in business analytics, we performed clustering analysis using CiteSpace. Clustering analysis classifies data with similar characteristics by calculating similarities in distance and forming clusters from data close to each other [13]. Each cluster refer to one sub area of the field. A total of nine clusters of analysis appeared on the network as a result of the clustering analysis (*see* Fig. 2). By default, only the top 10% of the data are contained in the clustering analysis.

The modularity Q of this research is 0.6169, which means the structure is well enough separated. The silhouette value is 0.2476, which refers that the clusters are well divided. Numerical information about the nine clusters is summarized in Table 2. In this study,

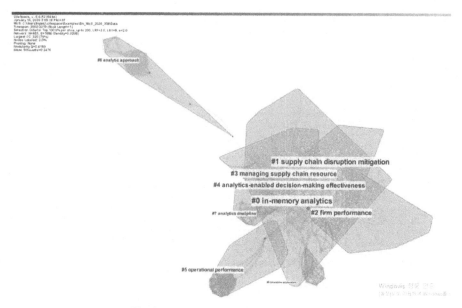

Fig. 2. Business analytics cluster network

the top two clusters (Cluster #0 and Cluster #1), as well as Cluster #6, are analyzed in detail (*see* Tables 3, 4, and 5).

Table 2. Nine clusters of business analytics research

Cluster ID	Size	Silhouette	Mean(Year)	Label(log-likelihood ratio)
0	62	0.443	2013	In-Memory Analytics
1	54	0.695	2016	Supply Chain Disruption Mitigation
2	47	0.783	2015	Firm Performance
3	44	0.843	2014	Managing Supply Chain Resource
4	41	0.769	2013	Analytics-Enabled Decision-Making Effectiveness
5	26	0.912	2011	Operational Performance
6	19	0.986	2011	Analytic Approach
7	19	0.915	2011	Analytics Discipline
9	8	0.988	2009	Interactive Exploration

Cluster #0. Cluster #0 (In-Memory Analytics) is the largest cluster with 62 references. The five selected citing articles cited 10–22% of these references. "Practitioners understanding of big data and its applications in supply chain management" [4] has the highest citation coverage and the most cited article is "Business intelligence and analytics: From big data to big impact" [7]. The 'In-Memory Analytics' research proposes the general framework of business intelligence (BI) research, distinguishes the major issues, and suggests the usage of data to apply business analytics properly to the management system (*see* Table 3).

Table 3. Cluster #0 – in-memory analytics

			Cluster #0 In-Memory Analytics						
	Cited References					Citing Articles			
Cites	Author	Year	Title	Journal	Coverage(%)	Author	Year	Title	Journal
74	Chen, HC	2012	Business intelligence and analytics: From big data to big impact.	MIS Quarterly	22	Brinch, M	2018	Practitioners understanding of big data and its applications in supply chain management.	Int J Logis Manage
42	McAfee, A	2012	Big data: the management revolution.	Harvard Review	18	Akter, S	2019	Analytics-based decision-making for service systems: a qualitative study and agenda for future research.	Int J Inf Manage
38	Lavalle, S	2011	Big data, analytics and the path from insights to value.	MIT Sloan	16	Mitri, M	2015	Toward a model undergraduate curriculum for the emerging business intelligence and analytics discipline.	Comm the Assoc Infor Syst
38	Manyika, J	2011	Big data: The next frontier for innovation, competition, and productivity.	Big Data	13	Akter, S	2016	Big data analytics in e-commerce: a systematic review and agenda for future research.	Electronic Markets
29	Sharma R	2014	Transforming decision-making processes: a research agenda for understanding the impact of business analytics on organisations.	Eur J Information Systems	10	Phillips-Wren, G	2015	Business analytics in the context of big data: a roadmap for research.	Comm the Assoc Infor Syst

Cluster #1. Cluster #1 (Supply Chain Disruption Mitigation) is composed of 54 references. The five selected citing articles cited 14–31% of these references. "The role of business analytics capabilities in bolstering firms' agility and performance" [3] has the highest citation coverage and the most cited article is "How to improve firm performance using big data analytics capability and business strategy alignment?" [2]. 'Supply Chain Disruption Mitigation' research exhibits the limitations of resource-based management and its business analytics suggests a way to lower the odds of business analytics application's failure, especially in supply chain management (*see* Table 4).

Table 4. Cluster #1 – supply chain disruption mitigation

			Cluster #1 Supply Chain Disruption Mitigation						
	Cited References					Citing Articles			
Cites	Author	Year	Title	Journal	Coverage(%)	Author	Year	Title	Journal
12	Akter, S	2016	How to improve firm performance using big data analytics capability and business strategy alignment?	Int J Prod Econ	31	Ashrafi, A	2019	The role of business analytics capabilities in bolstering firms' agility and performance.	Int J Inf Manage
10	Gupta, M	2016	Toward the development of a big data analytics capability	Infor Manage-Amster	19	Brinch, M	2018	Practitioners understanding of big data and its applications in supply chain management.	Int J Logis Manage
9	Gunasekaran, A	2017	Big data and predictive analytics for supply chain and organizational performance.	J Bus Res	15	Singh, N	2019	Building supply chain risk resilience: role of big data analytics in supply chain disruption mitigation.	Benchmarking-Int J
7	Popovic, A	2018	The impact of big data analytics on firms' high value business performance.	Infor Syst Front	14	Akter, S	2019	Analytics-based decision-making for service systems: a qualitative study and agenda for future research.	Int J Inf Manage
7	Corte-Real, N	2017	Assessing business value of Big Data Analytics in European firms.	J Bus Res	14	Kamble, S	2019	Big data-driven supply chain performance measurement system: a review and framework for implementation.	Int J Production Research

Cluster #6. Although Cluster #6 (Analytic Approach) is a relatively small cluster with only 19 references, it has more documents with a high betweenness centrality than do other clusters. The five selected citing articles cited 3–14% of the references. "A data analytic approach to forecasting daily stock returns in an emerging market" [17] has the highest citation coverage and the most cited article is "Data mining: Concepts and techniques" [12]. Cluster #6 explains how the raw data should be converted into usable and meaningful data. Moreover, it exhibits the best way to effectively deal with converted data. The research is based on machine learning and data mining and focuses mainly on prediction analysis.

Table 5. Cluster #6 – analytic approach

Cluster #6 Analytic Approach								
Cited References					Citing Articles			
Cites	Author	Year	Title	Journal	Coverage(%)	Author	Year	Title	Journal
5	Han, J	2012	Data mining: concepts and techniques.	Elsevier	14	Oztekin, A	2016	A data analytic approach to forecasting daily stock returns in an emerging market.	Eur J Oper Res
4	Turban, E	2011	Decision support and business intelligence systems.	Pearson Prentice Hall	13	Oztekin, A	2016	A hybrid data analytic approach to predict college graduation status and its determinative factors.	Ind Manag&Data Syst
4	Delen, D	2010	A machine learning-based approach to prognostic analysis of thoracic transplantations.	Artif Intell Med	4	Chongwatpol, J	2015	Integration of rfid and business analytics for trade show exhibitors.	Eur J Oper Res
3	Lu, CJ	2010	Combining independent component analysis and growing hierarchical self-organizing maps with support vector regression in product demand forecasting.	Int J Prod Econ	4	Chongwatpol, J	2015	Prognostic analysis of defects in manufacturing.	Ind Manag&Data Syst
3	Shiue, YR	2009	Data-mining-based dynamic dispatching rule selection mechanism for shop floor control systems using a support vector machine approach.	Int J Prod Res	3	Kartal, H	2016	An integrated decision analytic framework of machine learning with multi-criteria decision making for multi-attribute inventory classification.	Comp&Ind Eng

4.2 Key Documents

As shown in the results of document co-citation analysis, most of the key documents were steadily published between 2002 and 2018 (*see* Table 6). Landmark articles are usually the most cited because of the significance of their contribution.

Table 6. 15 Key documents

	Author	Year	Title	Journal	Citation Counts
1	Wang, G	2016	Big data analytics in logistics and supply chain management: Certain investigations for research and applications	IN J OF PRODUCTION ECONOMICS	186
2	Trkman, P	2010	The impact of business analytics on supply chain performance	DECISION SUPPORT SYSTEMS	133
3	Wang, YC	2018	Big data analytics: Understanding its capabilities and potential benefits for healthcare organizations	TECHN FORECASTING AND SOCIAL CHANGE	106
4	Jourdan, Z	2008	Business intelligence: An analysis of the literature	INFORMATION SYSTEMS MANAGEMENT	105
5	Gallino, S	2014	Integration of Online and Offline Channels in Retail: The Impact of Sharing Reliable Inventory Availability Information	MANAGEMENT SCIENCE	100
6	Delen, D	2013	Data, information and analytics as services	DECISION SUPPORT SYSTEMS	96
7	Kohavi, R	2002	Emerging trends in business analytics	COMMUNICATIONS OF THE ACM	90
8	Akter, S	2016	Big data analytics in E-commerce: a systematic review and agenda for future research	ELECTRONIC MARKETS	89
9	Chang, V	2014	The Business Intelligence as a Service in the Cloud	THE INT J OF E-SCIENCE	73
10	Holsapple, C	2014	A unified foundation for business analytics	DECISION SUPPORT SYSTEMS	73
11	Elbashir, MZ	2011	The Role of Organizational Absorptive Capacity in Strategic Use of Business Intelligence to Support Integrated Management Control Systems	ACCOUNTING REVIEW	67
12	Tan, TF	2014	When Does the Devil Make Work? An Empirical Study of the Impact of Workload on Worker Productivity	MANAGEMENT SCIENCE	61
13	Tambe, P	2014	Big Data Investment, Skills, and Firm Value	MANAGEMENT SCIENCE	60
14	Kache, F	2017	Challenges and opportunities of digital information at the intersection of Big Data Analytics and supply chain management	INT J OF OPER & PROD MANAG	50
15	Lacity, MC	2011	Business process outsourcing studies: a critical review and research directions	J OF INFORMATION TECHNOLOGY	49

The most cited article is "Big data analytics in logistics and supply chain management: Certain investigations for research and applications" [20] with 186 citations, followed by "The impact of business analytics on supply chain performance" [19] with 133 citations. The third-most-cited article is "Big data analytics: Understanding its capabilities and potential benefits for healthcare organizations" [21]. Wang *et al.* [20] suggested a framework on how to use the benefits of big data suitably with supply chain management and applied big data business analytics to logistics and supply chain management. Trkman *et al.* [19] conducted an experiment on the influence of business analytics on supply chain management. Wang [21] argued that the usage of big data will generate significant competitiveness because healthcare is related directly with peoples' lives.

4.3 Burst Detection

Generally, a highly influential document gets its reputation due to the steady accumulation of citation counts. Documents selected by burst detection do not have enough citation counts to be noticed. However, the article receives a large number of citations counts during a short period of time. These sudden citation counts result mostly from a newly arranged environment or an incident that occurred during a certain period. The burst detection results of this research are shown in Fig. 3.

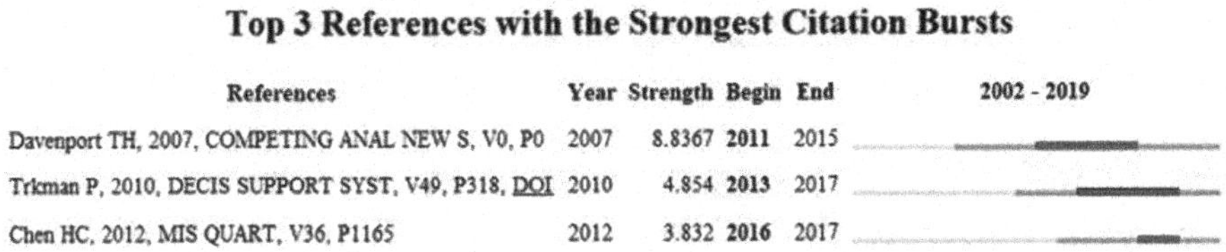

Top 3 References with the Strongest Citation Bursts

References	Year	Strength	Begin	End	2002 - 2019
Davenport TH, 2007, COMPETING ANAL NEW S, V0, P0	2007	8.8367	2011	2015	
Trkman P, 2010, DECIS SUPPORT SYST, V49, P318, DOI	2010	4.854	2013	2017	
Chen HC, 2012, MIS QUART, V36, P1165	2012	3.832	2016	2017	

Fig. 3. Burst detection

The reference with the strongest citation burst is "Competing on analytics: the new science of winning" [9] with 8.7715, followed by "The impact of business analysis on supply chain performance" [19] with 4.5279. Davenport and Harris [9] proposed that organizations should be superior in analysis in order to have a strong competitive advantage in the market. Because statistics are based on reliable facts to inform a better decision, Davenport and Harris's research seems to have a great impact as a helpful reference to practitioners. Trkman *et al.* [19] analyzed the impact of business analytics on supply chain management and business performance. As the research on the relationship between business analytics and supply chain management increased, Trkman *et al.*'s [19] research suddenly began receiving more attention in 2013.

4.4 Timeline Analysis

After the cluster analysis has been performed well, we can analyze how clusters are separated and connected chronologically with each other through timeline analysis. With the result of the timeline analysis, we can delineate which sub area has caught the most interest in a certain period. The timeline analysis gives an interpretation of the trends in the past and foresee how it can change in the near future. Figure 4 shows the result of timeline analysis based on 18 years (2002–2019) of data. The red nodes indicate the documents that are detected as bursts and the nodes are the documents that have strong or weak citation counts. Cluster #0 (In-Memory Analytics) has a majority of strong citation counted documents; however, their impact became weaker in recent times. Cluster #1 (Supply Chain Disruption Mitigation), Cluster #2 (Firm Performance), Cluster #3 (Managing Supply Chain Resource) and Cluster #4 (Analytics-enabled Decision-making Effectiveness) shows the most active connections until very recently, which can be seen as a rising trend. Cluster #5 (Operational Performance), Cluster #6 (Analytic Approach), Cluster #7 (Analytic Discipline), and Cluster #9 (Interactive Exploration) are declining clusters because the connections between these clusters hardly can be seen recently.

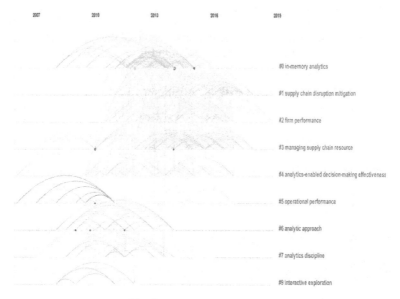

Fig. 4. Timeline (2002–2019)

5   Conclusion

Global companies like Walmart, Amazon, and Netflix have applied business analytics and produced satisfying performances. It seems to be necessary to analyze the field to give a better understanding to the researchers and practitioners. This research identifies the intellectual structure of the field business analytics of accumulated studies based on business analytics data. The intellectual structure itself present a scientific and statistical contribution to the field and propose a new insight. There are more than 1,140,000 research studies about business analytics on Google Scholar as of February 2020. This verifies the fact that an enormous number of studies about business analytics have been conducted and the accumulated research is considerable in quantity, which became a valuable data to use in the research.

The major results of this research can be summarized as follows:

- Three hundred and thirty-three documents about 'business analytics' were extracted from the 'Web of Science' database as data for this research.
- DCA and ACA were conducted using the program CiteSpace.
- A total of nine clusters were shown on the network through cluster analysis. Cluster #0 (In-Memory Analytics) has the largest and most influencing cluster and Cluster #9 (Interactive Exploration) the smallest.
- Fifteen key documents were extracted. Supply chain management appears to be the domain most related to business analytics and a rising trend was identified in healthcare.
- Three burst documents were detected.
- Two different groups of clusters were identified: rising clusters (Cluster #0 through #5) and declining clusters (Cluster #6 through #9).

Understanding the intellectual structure of a domain means that we can predict the future of research because we can visualize the flow of the research domain from the past to the present. Well-performing business analytics is required to have a positive impact on the decision-making process and performance. To do so, researchers and practitioners require a good understanding of business analytics. These are some suggestions for further research.

- We collected the data from the Web of Science only. Extracting a larger amount of data from additional databases might lead to more precise results.
- ACA only considers the first author as data. The elimination of the co-author is a known issue in ACA, an issue that needs to be addressed and improved upon in future research.
- Recent research papers may not have enough co-citation counts in order to check the influence of the factor. However, this limitation is another known issue for review articles.

References

1. Acedo, F.J., Casillas, J.C.: Current paradigms in the international management field: an author co-citation analysis. Int. Bus. Rev. **14**(5), 619–639 (2005)
2. Akter, S., Wamba, S.F., Gunasekaran, A., Dubey, R., Childe, S.J.: How to improve firm performance using big data analytics capability and business strategy alignment? Int. J. Prod. Econ. **182**, 113–131 (2016)
3. Ashrafi, A., Ravasan, A.Z., Trkman, P., Afshari, S.: The role of business analytics capabilities in bolstering firms' agility and performance. Int. J. Inf. Manag. **47**, 1–15 (2019)
4. Brinch, M., Stentoft, J., Jensen, J.K., Rajkumar, C.: Practitioners understanding of big data and its applications in supply chain management. Int. J. Logist. Manag. **29**(2), 555–574 (2018)
5. Chen, C.: Searching for intellectual turning points: Progressive knowledge domain visualization. Proc. Natl. Acad. Sci. **101**(1), 5303–5310 (2004)
6. Chen, C.: CiteSpace II: detecting and visualizing emerging trends and transient patterns in scientific literature. J. Am. Soc. Inform. Sci. Technol. **57**(3), 359–377 (2006)
7. Chen, H., Chiang, R.H., Storey, V.C.: Business intelligence and analytics: from big data to big impact. MIS Q. **36**(4), 1165–1188 (2012)
8. Cui, Y., Mou, J., Liu, Y.: Knowledge mapping of social commerce research: a visual analysis using CiteSpace. Electron. Commer. Res. **18**(4), 837–868 (2018). https://doi.org/10.1007/s10660-018-9288-9
9. Davenport Thomas, H., Harris, J.G.: Competing on Analytics: The New Science of Winning (2007)
10. Feng, F., Zhang, L., Du, Y., Wang, W.: Visualization and quantitative study in bibliographic databases: a case in the field of university–industry cooperation. J. Inform. **9**(1), 118–134 (2015)
11. Freeman, L.: A set of measures of centrality based on betweenness. Sociometry **40**(1), 35–41 (1977)
12. Han, J., Pei, J., Kamber, M.: Data Mining: Concepts and Techniques. Elsevier, Amsterdam (2011)
13. Kim, Y.J., Kim, C.H.: Mapping the intellectual structure of communication research field in Korea: an author co-citation analysis, 1989-2006. J. Korean Commun. **15**(3), 155–184 (2007)

14. Lee, M.O.: Major themes and trends in Korea women's Studies: the results of author co-citation analysis. Korea Assoc. Women's Stud. **12**(1), 180–203 (1996)
15. Li, X., Ma, E., Qu, H.: Knowledge mapping of hospitality research – a visual analysis using CiteSpace. Int. J. Hosp. Manag. **60**, 77–93 (2017)
16. Lin, Z., Wu, C., Hong, W.: Visualization analysis of ecological assets/values research by knowledge mapping. Acta Ecol. Sin. **35**(5), 142–154 (2015)
17. Oztekin, A., Kizilaslan, R., Freund, S., Iseri, A.: A data analytic approach to forecasting daily stock returns in an emerging market. Eur. J. Oper. Res. **253**(3), 697–710 (2016)
18. Shiau, W.L., Chen, S.Y., Tsai, Y.C.: Management information systems issues: co-citation analysis of journal articles. Int. J. Electron. Commer. Stud. **6**(1), 145–162 (2015)
19. Trkman, P., McCormack, K., De Oliveira, M.P.V., Ladeira, M.B.: The impact of business analytics on supply chain performance. Decis. Supp. Syst. **49**(3), 318–327 (2010)
20. Wang, G., Gunasekaran, A., Ngai, E.W., Papadopoulos, T.: Big data analytics in logistics and supply chain management: certain investigations for research and applications. Int. J. Prod. Econ. **176**, 98–110 (2016)
21. Wang, G.: Big data analytics: understanding its capabilities and potential benefits for healthcare organization. Technol. Forecast. Soc. Chang. **126**, 3–13 (2018)
22. White, H.D., Griffith, B.C.: Author co-citation: a literature measure of intellectual structure. J. Am. Soc. Inf. Sci. **32**(3), 163–171 (1981)
23. Yu, D., Xu, Z., Pedrycz, W., Wang, W.: Information sciences 1968–2016: a retrospective analysis with text mining and bibliometric. Inf. Sci. **424**, 619–634 (2017)
24. Zhang, X., Gao, Y., Yan, X., de Pablos, P.O., Sun, Y., Cao, X.: From e-learning to social-learning: mapping development of studies on social media-supported knowledge management. Comput. Hum. Behav. **51**, 803–811 (2015)

A New Approach to Early Warning Systems for Small European Banks

Michael Bräuning[1], Despo Malikkidou[2], Stefano Scalone[1(✉)],
and Giorgio Scricco[1]

[1] European Central Bank, Sonnemanstrasse 20, 60314 Frankfurt am Main, Germany
stefano.scalone@ecb.europa.eu
[2] European Banking Authority, 20 Avenue Andre Prothin, 92927 Paris, France

Abstract. This paper describes a machine learning technique to timely identify cases of individual bank financial distress. Our work represents the first attempt in the literature to develop an early warning system for small European banks.

We employ a machine learning technique, and build a decision tree model using a dataset of official supervisory reporting, complemented with qualitative banking sector and macroeconomic features. We propose a new and wider definition of financial distress, in order to capture bank distress cases at an earlier stage with respect to the existing literature on bank failures; in this way we identify bank crises at an early stage, therefore leaving a time window for supervisory intervention.

We use the Quinlan's C5.0 algorithm to estimate the model, whose final form comprises 12 features and 19 nodes, and outperforms a logit model estimation which we use to benchmark our analysis.

Keywords: Machine learning · Bank distress · Decision tree · Logit model

1 Introduction

With this work, we propose an early warning model for the timely identification of distress of financial institutions, based on a large sample of small European banks.

Models for the early identification of bank financial distress represent a useful tool both for the theoretical work of the researcher and the practical use of the supervisor. They in fact help the researcher understand what is that drives a bank into distress and tailor its investigation on bank crises, but also allow for the timely intervention of the supervisor and in most cases the triggering of policy actions before the financial situation of an institution deteriorates further.

Existing and comparable models are usually based on conventional modeling techniques such as multivariate logit models, and are calibrated using only a very small number of default (and not just distress) events. We here propose to innovate the current theoretical framework along two lines: first, we create a

© Springer Nature Switzerland AG 2020
G. Nicosia et al. (Eds.): LOD 2020, LNCS 12565, pp. 551–562, 2020.
https://doi.org/10.1007/978-3-030-64583-0_49

new and wider definition of distress event inspired by literature and European regulation, and obtain a sample of distress events in our dataset of small European banks that is significantly larger than most other works in the literature. Second, we propose a machine learning methodology to build a decision tree model, which notably improves the predictive performance with respect to the most usual techniques (we benchmark our decision tree with a logit estimation).

This theoretical framework is applied to a unique dataset of more than 3,000 small European banks, the so called Less Significant Institutions (LSIs[1]).

The remainder of this paper is organised as follows: Sect. 2 presents the literature on the topic and explains the context for our work, Sects. 3 and 4 describe our dataset, the new definition of financial distress and the methodology used, Sect. 5 presents our results and Sect. 6 concludes.

2   Literature Review

All attempts in the literature to develop a model for the early detection of distress cases are relatively recent [25], as researchers tend to focus on bank failures or on systemic bank crises[2] rather than single-bank distress: despite representing an undoubtedly useful theoretical exercise, the practical usefulness of such models is limited as once a bank is defaulting the situation is often irreversible. The reason for such a hole in the literature is also technical, as it is often the lack of data that pushes researchers to analyse the problem from the systemic point of view or to focus on bank default; we instead make extensive use of supervisory data to estimate our model, which is significantly more granular and of high frequency. The change of perspective from bank default to bank distress that we propose in this work helps to detect cases of financial difficulties early enough to allow for a time frame for supervisory intervention.

As said, our model is developed at the level of the single bank, an approach that differs with the general strand of literature: in economics, in fact, much more diffused are early warning systems (EWS) to recognise signs of potential systemic risks, both in the banking sector (with examples of works analysing the area under the ROC curve to identify early warning indicators that might represent potential source of vulnerabilities [1,7]) and at the country level (creating some composite indicators of financial distress, based on the commonly shared hypothesis that economic distress and fragility of an economy are good predictors of financial crises [15]).

The literature on single indicators of financial distress is vast, and often exceeds (or consciously diverges from) the scope of canonical early warning systems. A fundamental reference on the topic is Kaminsky and Reinhard [14], whose work focuses on currency crises and makes a first step in the design of an early warning system that would help detecting a domestic financial crisis.

---

[1] As defined by the Single Supervisory Mechanism (SSM) of the European Central Bank.

[2] On this see the extensive literature reviews by Kumar Ravi and Ravi [18] and Davis and Karim [6] for banking crises and Gramith et al. [10] for systemic banking crises.

The main signals are indeed currency and exchange rate expectations, but the authors underline how micro data on banking would be the natural complement of their analysis: they find that currency crises deepen banking crises, and analyse the deep interlinkages between these two sectors.

Regarding models on banking crises, there is a tendency in the literature to identify indicators signalling banking problems, rather than building a fully specified model to predict bank distress. This is the approach taken by Honohan [11], with the author proposing a set of features, each of which with a threshold that if exceeded can successfully predict a situation of bank distress. This apparently dogmatic approach however leaves room for the discretionary intervention of the supervisor, the *expert judgement* procedure that is extremely common in this strand of work.

One of the first milestones in the literature on bank distress is represented by the pioneering work by Sinkey [24], who adapted the work by Altman [3] to predict bank crises in a framework of multiple discriminant analysis. As outcome of the model, the features indicated as good predictors for bank distress are asset composition, loan characteristics, capital adequacy, sources of revenues, efficiency and profitability. Many of these features correspond to the widely used CAMELS indicators,[3] which represent a useful starting point for the feature selection of any banking model (including ours).

Not surprisingly, the interest in the literature on predicting bank distress events peaked after the 2007 financial crisis: Jin et al. [12,13] further developed the CAMELS approach, and complemented the six indicators with data on banks' internal controls on risk-taking and audit quality features to find an improved predictive rate. Cole and White [5] found that measures of commercial real estate investments are also relevant for predicting bank distress while Betz et al. [4] use the CAMELS indicators as a starting point for their early warning model, but with a narrower definition of bank distress.

A good recap of the features most widely used in the literature is in Oet et al. [21] who, despite focusing on explanatory features for systemic risk and financial distress, depict a list that contains many of the features also contained in our model.

3   Data and Sample of Distress Events

Our dataset is composed of supervisory reporting, micro- and macro-economic data[4] resulting in a wide panel covering the whole universe of around 3,000 European less significant institutions over a period of four years.

For each institution, the available bank-level data comes from Common Reporting (COREP, containing information on capital adequacy and risk-specific

---

[3] Initially published by the Federal Financial Institutions Examination Council in 1979 and further developed by the Federal Reserve in 1995, and commonly used in the early warning system literature [4].

[4] Mainly obtained from the ECB Statistical Data Warehouse, complemented by Eurostat and the OECD for regional data.

information), available on quarterly basis since December 2014, and the Financial Reporting (FINREP, which includes balance sheet items, the statement of profit and loss and detailed breakdowns of assets and liabilities by product and counterparty), available with different frequency since December 2014 and on quarterly basis since March 2017.

In order for an early warning system on bank distress to be useful in practice for the supervisor and the policy maker, the recognition of the distress event must be timely enough to allow a buffer of time for intervention. If we consider the failure or liquidation of a bank as triggering event [5,9,12,14], we lose the practical validity of the model, which would in turn only be helpful for ex post calibrations. Moreover, bank failures in Europe are relatively rare, this making the estimation of such an early warning system even more challenging.

We therefore relax the traditional hypothesis of considering only bank crises or defaults as positive events in the sample, and instead consider all financial distress cases. By doing so, the sample of distress events significantly grows in size, allowing us to obtain precise estimations despite the short time horizon on which we span our model.

We use a mixed approach, and base our definition of distress on the Banking Recovery and Resolution Directive (BRRD[5]) and the four conventional types of financial distress in Betz et al. [4].

We consider a bank to be in financial distress if:

- It is deemed to be failing or likely to fail within the meaning of Article 32 of the BRRD. For categorizing a bank as failing or likely to fail, indicators assessing whether a bank has breached the minimum capital requirements or capital buffers are constructed;
- It meets the conditions for early intervention pursuant to Article 27 of the BRRD. The triggers used to meet the conditions of early interventions consist of indicators for assessing if a bank is close to breaching minimum capital requirements;
- In case of the removal of the senior management and management body of the institution, in its entirely or with regard to individuals, in line with article 28 of the BRRD;
- It is placed under special administration and/or is appointed of a temporary administrator pursuant to Article 29 of the BRRD;
- There is a rapid and significant deterioration of its financial situation according to Article 96 of the Framework Regulation. This is based on expert judgement by national central banks and in-house qualitative and data;
- When there is an indication that the supervised LSI can no longer be relied upon to fulfil their obligations towards their creditors or where there is an indication of circumstances that could lead to a determination that the LSI concerned is unable to repay the deposits as referred to in Article 2 (1) (8) of Directive 2014/49/EU;

---

[5] The Bank Recovery and Resolution Directive, establishing a framework for the recovery and resolution of credit institutions and investment firms in Europe.

– One of the four types of conventional bank distress events proposed by Betz et al. (2014) (i.e. bankruptcies, liquidations, state interventions and forced mergers) is met.

Following this approach we obtain of more than 350 distress events throughout a sample of only six quarters (consecutive, and starting from 2014 Q4).

4 Methodology

4.1 Data Pre-processing

Data pre-processing steps are required to ensure that unreliable and noisy data as well as irrelevant and redundant information is eliminated prior to the modelling phase. As such, the final training dataset used for the analysis is of high quality, thus increasing the efficiency and performance of the final model. This represents a key step in our process, as we start from the extremely vast dataset of supervisory reporting consisting of more than 3,000 features, and end up with a final sample of only 12.

We start by cleaning the data, in order to eliminate incomplete or uninformative data; despite using mainly supervisory reporting data for our analysis, data quality is still an issue, the dataset contains several missing values and the data pre-processing procedure represents a fundamental step for our analysis. Moreover, many institutions may not report some data points, following the proportionality criteria that inspires the supervisory reporting framework (which allows a lighter scope of data reporting for smaller banks). We therefore eliminate features for which the majority of values is missing, or where the variance across the sample is close to zero. This first step already reduces the number of total indicators to less than 500, this giving a hint of the importance of the data pre-processing procedure.

The second step is a simple transformation of the data, with the goal of increasing consistency and comparability across institutions. More than one accounting standard coexists in Europe, this complicating the job of the supervisor; we apply a transformation technique, based on a mapping of these heterogeneous data points to make different data sources somehow coherent. In a following stage, we normalise our features through the creation of ratios in order to increase comparability.

We remove explanatory features which are too highly correlated, using a simple threshold of 0.9 and select the final set of indicators based on their ability to predict distress. Features are ranked according to their importance, captured by the individual Area under the Receiver Operating Characteristic curve (AUC) for each indicator; using this technique, we select the top 100 features in terms of predictive performance which we use as starting point for the decision tree.

4.2 Decision Trees

Our early warning system is based on a decision tree methodology. The predictive performance of this technique is very high, both in and out of sample, as demonstrated by the data on accuracy presented in Sect. 4; moreover, this methodology well handles missing values, a common issue in the early warning literature [20], and one of the main flaws in our dataset. Finally, the decision tree methodology is relatively transparent, and allows supervisors to interpret the output tree and understand which indicators affect bank distress, thus minimizing the risk of creating a *black box* model.

A decision tree is a classification technique commonly used in machine learning. The tree recursively identifies the significant indicators and their respective thresholds which best split the sample into the pre-determined classes (in our case distress and no distress).

We employ Quinlan's C5.0 algorithm to build the classification tree model [16,23]. The C5.0 algoritm is one of the most commonly used, as it is relatively fast and accurate, as well as efficient in handling missing data and removing unhelpful attributes.[6] The algoritm represents an improvement to the canonical C4.5, with the advantages of being faster and improving memory efficiency, and allowing for asymmetric costs for specific errors.

5   Results

In training model, we select a number of specification options:

- We impose a relatively short prediction horizon (1–3 months ahead of distress as starting point), given the short term (<1 year) scope of this early warning system. By considering pre-default events as target feature, we ensure that the system has a forward looking perspective;
- We impose asymmetric misclassification costs when assessing the performance of explanatory features: we consider Type I errors (missing a distress events) to be twice as costly as Type II errors (issuing a false alarm). In principle, this assumes that when faced with a tradeoff of issuing more false alarms or missing a distress event, the policymaker would take a conservative stance and choose the former;
- To increase the robustness of simple decision trees (which in general is relatively low [2]), we employ a boosting technique á la Freund et al. [8] to identify which features to include in the final version of the tree.[7]

As there is no univocal rule to choose one particular tree among the estimated ones, we use the boosting technique to simulate the creation of a large number of trees, and select the 20 features that rank highest as of importance (measured in terms of how often they appear in the trials).

---

[6] For a literature review of Data Mining Algoritms see Wu et al. [27]. The relative R environment used in this paper refers to Kuhn et al. [17].

[7] Boosting is a technique for generating and combining multiple classifiers to improve the predictive accuracy of the model. Instead of using a single tree, n separate decision trees (trials) are grown and combined to make predictions. The error rate of the boosted classifier is often substantially lower than that of single trees.

We complement the feature selection with both expert judgement and quantitative measures: in particular, for evaluating the performance of the model, we rely on the area under the Receiver-Operating-Characteristic curve (AUC) and Matthews Correlation Coefficient (MCC), both standard measures of accuracy in the early warning system literature [22].

The final tree is composed of 19 nodes, covering 11 different explanatory features, and is represented in Fig. 1.[8]

The features included in the model are:

- Adjusted profitability;
- Non-performing loans (NPL) ratio;
- Non-performing loans coverage ratio;
- Deficit-to-GDP ratio;
- GDP growth;
- Liquidity coverage ratio (LCR);
- Leverage ratio;
- Equity exposures;
- Exposures in default;
- Two proxies for market risk;[9]
- Membership of an institutional protection scheme (IPS).

The indicator of the root node is profitability, adjusted for the different accounting standards of the banks in the sample. The node splits the banks between profit (right branch) and loss (left branch) making. The remaining features are a mix of macro-economic indicators (deficit-to-GDP ratio, and real GDP growth) and banking indicators covering the most important risks: credit (non-performing loans ratio, non-performing loans coverage ratio, exposure in default), liquidity (liquidity coverage ratio), market (captured by the sum of trading financial assets and financial liabilities held for trading over total asset and net gains on financial assets and liabilities held for trading over total operating income), capital (leverage ratio and equity exposure), together with the qualitative information of whether a bank is member of an institutional protection scheme (IPS).

In the framework of supervised learning, the role of our tree is not only to find an efficient method to identify which banks are in financial distress, but also to suggest which features are significant and how they model the distress. In this perspective, it is interesting to analyse the main paths through the tree: if a bank is making profits (root node to the right), profitability is likely to not be an issue, and the model suggests to investigate credit risk (first node to the right is the NPL ratio). If credit risk is deemed as material, i.e. if the bank has a high level of non-performing loans, the model moves to analyse whether these

---

[8] Please note that the trees represented in this version of the paper are somehow anonymised, i.e. without the precise splitting thresholds of each node.

[9] The sum of trading financial assets and financial liabilities held for trading over total assets, and net gains on financial assets and liabilities held for trading over total operating income.

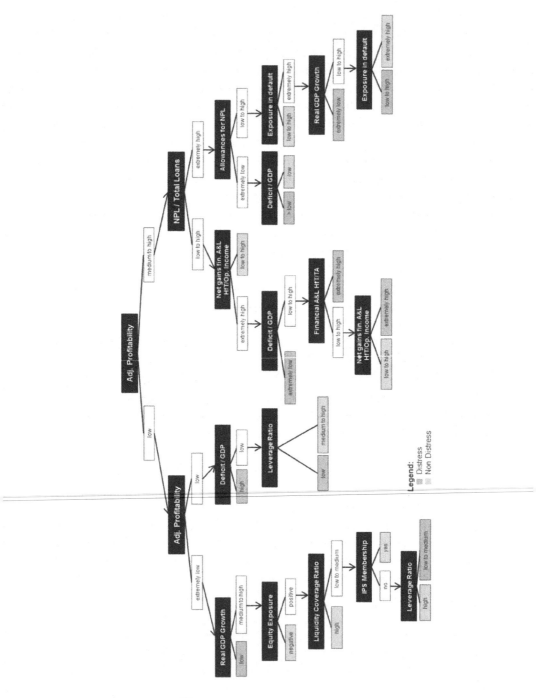

Fig. 1. Decision tree for full sample

NPLs are covered by sufficient allowances. On the other hand, for banks with low non-performing loans (NPL ratio node to the left), market risk becomes relevant in capturing distress: this is intuitive, as banks strongly relying on income from market activities are subject to higher volatility and potential distress, especially in countries with fragile economic fundamentals (high level of deficit over GDP ratio).

On the opposite side of the tree, if a bank is deeply unprofitable (root node to the left, first node to the left), if it is operating in a country with a low growth of GDP it is automatically labelled as being in distress. When instead the bank has relatively high equity exposures combined with a weak LCR ratio, then the distress will be determined by the eventual membership in an IPS; Institution Protecting Schemes in fact protect banks from financial distress, therefore making member institutions less vulnerable. Finally, for moderately unprofitable banks, the economic conditions of the country in which they operate (proxied by deficit over GDP) and the leverage ratio indicate whether a bank is in distress or not.

The predictive performance of the model is very high, both in and out of sample (which we conduct via a 10-fold cross-validation testing). As depicted in Table 1, in the training data the true positive rate is 0.89, while the false positive rate (Type II error) and false negative rate (Type I error) equal 0.01 and 0.11, respectively. The AUC is 0.95, much closer to the unity value of a perfect classifier, than to the 0.5 value of a purely random one. The MCC is also high, at a value of 0.89. The out of sample performance (25% of the initial observations), based on a random split is also satisfactory, with an AUC of 0.92 and a MCC of 0.81. Type I and II errors are therefore comparable to the in-sample error rates, and remain at adequately low levels.

Table 1. Validation results full sample

Measures	In sample (train)	Out of sample (test)
Type I error rate	0.01	0.03
Type II error rate	0.11	0.10
AUC	0.95	0.92
MCC	0.89	0.81

It is useful to remark how the model is estimated by introducing an unbalance between level-1 and level-2 errors: false positives are in fact considered to be less problematic than missed distress events, and therefore weight half.

We benchmark the results of our decision tree model with a Logit regression, a typical approach in the literature when predicting a binary categorical target features (in our case the event of a financial distress).

For the data pre-processing we instead follow the same exact steps of the decision tree: we clean the dataset by eliminating features with too many missing values, with almost zero variance and with too high correlation[10]. We select the model features with a LASSO regression (á la Tibshirani [26]), in order to prevent overfitting; this methodology minimises the sum of the squared errors just like a normal linear estimation process, but bounds on the sum of the (absolute) value of the coefficients.

Data is then randomly split into training (75%) and test (25%) sample, in order to allow for independent performance validation of the model. For the logit, we follow Lang et al. [19] applying a 10-fold cross-validation to identify the best lambda for our purposes, in order to obtain the model which optimises the out-of-sample forecasting performance. The model so obtained is then compared to the cross-validated Quinlan.

The results of the logit are still relatively accurate, but the Logit model misses many more distress events than the decision tree (see Table 2 below). The Logit is in fact significantly more sensitive to missing values, and therefore fails to detect distress events relative to institutions reporting a sufficient number of NAs, unlike the decision tree.

Table 2. Validation results logit model

Measures	In sample (train)	Out of sample (test)
Type I error rate	0.01	0.02
Type II error rate	0.18	0.19
AUC	0.95	0.90
MCC	0.32	0.46

The supervised learning estimation framework of the decision tree allows us to perform a precise *ex post* back-testing analysis. In the course of 2017 the model correctly identified 79% of distress events. The missing financial deterioration cases were mainly triggered by qualitative features (e.g. governance issues), which our purely quantitative model fails to capture.

6   Conclusions

This paper develops an innovative model to identify cases of bank financial distress, using a subsample of 3,000 small European institutions[11] for a time period of six quarters (between 2014 and 2016).

We build a sample of distress cases based on European regulation, to early detect future cases of financial deterioration rather than simply referring to banks

---

[10] We build the correlation matrix and proceed to reduce pairwise correlation by selecting the couples of too highly correlated features, and eliminating the one with the largest mean absolute correlation.

[11] Excluding the so called Significant Institutions, as defined by the SSM.

that are already in or close to default. With a broad definition of financial deterioration, our sample of distress events is significantly larger than any other work in the literature, despite a relatively short time series of data.

On this sample we construct a decision tree model, which accurately classifies banks into distressed or non-distressed; the prediction horizon of the model is one-three months, a time span that would in our view give the supervisor enough time to trigger supervisory action.

We find that the predictive power of our model is extremely high, and the decision tree steadily outperforms the Logit approach, the most widely used methodology to predict binary classifications which we use as benchmark.

As a final remark, further extensions of this work should go in the direction of increasing the prediction horizon (currently 1–3 months), and could include the development of an *ad hoc* tree for each business model. Unfortunately, the data does not allow us to do this, yet, as given the polarisation of the dataset towards retail lenders, the sample of remaining business models is simply not numerous enough to allow for a proper estimation.

A clear limitation of this work is the length of the time series, which we try to partly overcome with the size of the panel and the granularity of the data. However, we plan to re-estimate the model with a longer time series when it will be available, to continue to back test the predicting performance of the model and to analyse the changes in the business models of the banks; the latter in particular could be achieved in two ways: first, by re-estimating the model to understand how the environment changed, and what new features contribute to describing the business of an institution; second, by conducting a case-by-case analysis for institutions for which the model changes classification: in this way, it would be possible to get some insights on how the changes in monetary policy and the economic environment influence the behaviour of banks (think for example of the current prolonged period of low interest rates, which might be pushing retail banks that see their interest margins eroded into finding new sources of income).

References

1. Aldasoro, I., Borio, C., Drehmann, M.: Early warning indicators of Aldasoro: banking crises: expanding the family. BIS Q. Rev. 29–45 (2018)
2. Alessi, L., Detken, C.: Identifying excessive credit growth and leverage. ECB Working Paper 1723 (2014)
3. Altman, E.I.: Financial ratios, discrimimant analysis and the prediction of corporate bankruptcy. J. Finan. **23**, 589–609 (1968)
4. Betz, F., Oprica, S., Peltonen, T.A., Sarlin, P.: Predicting distress in European banks. J. Bank. Finan. **45**, 225–241 (2014)
5. Cole, R.A., White, L.J.: Deja Vu all over again: the causes of U.S. commercial bank failures this time around. J. Finan. Serv. Res. **42**(1–2), 5–29 (2012)
6. Davis, P.E., Karim, D.: Comparing early warning systems for banking crises. J. Finan. Stabil. **4**(2), 89–120 (2011)
7. Drehmann, M., Juselius, M.: Evaluating early warning indicators of banking crises: satisfying policy requirements. BIS Working Papers 421 (2013)

8. Freund, Y., Schapire, R., Abe, N.: A short introduction to boosting. J.-Jpn. Soc. Artif. Intell. **14**(1612), 771–780 (1999)

9. Gissel, J.L., Giacomino, D., Akers, M.D.: A review of bankruptcy prediction studies: 1930-present. J. Finan. Educ. **33**, 1–42 (2007)

10. Gramith, D., Miller, G.L., Oet, M.V., Ong, S.J.: Early warning systems for systemic banking risk. Banks Bank Syst. **5**(2), 199–211 (2010)

11. Honohan, P.: Banking system failures in developing and transition countries: diagnosis and predictions. BIS Working Papers 39 (1997)

12. Jin, J., Kanagaretnam, K., Lobo, G.: Ability of accounting and audit quality variables to predict bank failure during the financial crisis. J. Bank. Finan. **31**(11), 2811–2819 (2011)

13. Jin, J., Kanagaretnam, K., Lobo, G., Mathieu, R.: Impact of FDICIA internal controls on bank risk taking. J. Bank. Finan. **37**(2), 614–624 (2013)

14. Kaminsky, G.L., Reinhart, C.M.: The twin crises: the causes of banking and balance-of-payments problems. Am. Econ. Rev. **89**(3), 473–500 (1999)

15. Kaminsky G.L.: Currency and banking crises: the early warning of distress. IMF Working Paper WP/99/178 (1999)

16. Kuhn, M., Johnson, K.: Applied Predictive Modeling. Springer, New York (2013). https://doi.org/10.1007/978-1-4614-6849-3

17. Kuhn, M., Weston, S., Coulter, N., Culp, M.: C5.0 decision trees and rule-based models (2015)

18. Kumar Ravi, P., Ravi, V.: Bankruptcy prediction in banks and firms via statistical and intelligent techniques - a review. Eur. J. Oper. Res. **180**(1), 1–28 (2006)

19. Lang, J.H., Peltonen, T.A., Sarlin, P.: A framework for early-warning modeling with an application to banks. ECB Working Papers 2182 (2018)

20. Mitchell, T.: Machine Learning. McGraw-Hill, New York (1997)

21. Oet, M.V., Bianco, T., Gramlich, D., Ong, S.J.: SAFE: an early warning system for systemic banking risk. J. Bank. Finan. **37**(11), 4510–4533 (2006)

22. Peltonen, T.A., Piloiu, A., Sarlin, P.: Network linkages to predict bank distress. ECB Working Paper 1828 (2015)

23. Quinlan, R.: C4.5: Programs for Machine Learning. Morgan Kaufmann Publishers, Burlington (1993)

24. Sinkey, J.F.: A multivariate statistical analysis of the characteristics of problem banks. J. Finan. **30**(1), 21–36 (1975)

25. Rosa, P.S., Gartner, R.I.: Financial distress in Brazilian banks: an early warning model. Revista Contabilidade & Finanças **30**(1), 312–331 (2018)

26. Tibshirani, R.: Regression shrinkage and selection via the LASSO. J. R. Stat. Soc. B **58**(1), 267–288 (1996)

27. Wu, X., Kumar, V., Quinlan, J.R., Ghosh, J., Yang, Q., et al.: Top 10 algorithms in data mining. Knowl. Inf. Syst. **14**(1), 1–37 (2008). https://doi.org/10.1007/s10115-007-0114-2

Evaluating the Impact of Training Loss on MR to Synthetic CT Conversion

Moiz Khan Sherwani[1]([✉]) [iD], Paolo Zaffino[2] [iD], Pierangela Bruno[1] [iD],
Maria Francesca Spadea[2] [iD], and Francesco Calimeri[1] [iD]

[1] Department of Mathematics and Computer Science, University of Calabria,
Rende, Italy
sherwani@mat.unical.it
[2] Department of Experimental and Clinical Medicine, University of Catanzaro,
Catanzaro, Italy

Abstract. Radiation therapy is one of the most important strategies for treating patients with tumor. The rationale is to deliver high radiation doses to the tumor in order to damage its DNA while sparing, at the same time, healthy tissues. In order to optimize such a process, biomedical images play a fundamental role; in particular, Magnetic Resonance (MR) produces well-contrasted images for precisely contouring tumors and organs at risk. However, due to the physical information stored in it, Computed Tomography (CT) is mandatory for accurate radiation dose calculation. To overcome this limitation, several algorithms, usually based on deep learning techniques, were proposed for converting MR–to–synthetic CT (sCT). In this paper, we report about the evaluation of the impact of three different train losses, commonly used for non–medical applications. Tests were ran on a cohort of 15 brain MR/CT image pairs. An algorithm for MR-to–sCT conversion, previously developed for MR-only radiotherapy, was used as benchmark platform. Predicted sCT images were compared on the basis of intensities and edges reconstruction. Results show that potential improvements can be achieved if non–medical image loss will be adapted to this application.

Keywords: Synthetic CT · Radiotherapy · Medical imaging · Deep learning

1 Introduction

Magnetic Resonance (MR) images and Computed Tomography (CT) scans are extensively used in clinical routines, both for diagnosis and treatment. In particular, one of the most common image-guided therapy procedure makes use of radiations, thus making radiation therapy a widely used strategy for treating cancer [1]; the aim is to hit the tumor with ionizing radiations in order to damage its DNA. During radiotherapy planning treatment, cancerous tissue (to treat) and Organs At Risk (OARs) (to spare), are identified and contoured on CT

© Springer Nature Switzerland AG 2020
G. Nicosia et al. (Eds.): LOD 2020, LNCS 12565, pp. 563–573, 2020.
https://doi.org/10.1007/978-3-030-64583-0_50

images. Afterwards, radiation dose to be delivered to the patient is computed on the basis of density information extracted from CT. In this step, the contours are used as a constraint in order to maximize the dose to tumor and, at the same time, to minimize the irradiation of OARs. However, this image modality, even if mandatory for dose calculation, does not offer an optimal contrast between soft tissues, making the accurate delineation of anatomical structures a challenging task. On the contrary, MR, without taking advantage of ionizing radiation, provides an excellent contrast between tissues, although the intensities stored in it does not allow a reliable dose calculation. To overcome such a limitation, several algorithms [2–6] have been proposed to convert MR–to–synthetic CT (sCT). In this way, once contours are drawn on MR image, it will be possible to generate a sCT on which will be possible to obtain a reliable dose computation. Several of those algorithms are based on deep learning strategies, where the chosen loss functions play fundamental role in achieving optimal conversion.

In this work, we present a performance comparison of different loss functions for MR–to–sCT conversion tasks, in order to assess the effectiveness of those metrics also for medical application. In particular, to achieve our aim, we made use of an algorithm for MR–to–sCT conversion, previously developed for a different purpose, as benchmark platform. We also included in experiments two losses that are commonly used for non–medical images where intensities and feature are different.

The remainder of the paper is structured as follows. After discussing the related work in Sect. 2, we describe the material and methods in Sect. 3 and the performed experiments in Sect. 4. Results are presented in Sect. 5 and discussed in Sect. 6. Section 7 draws our conclusions by summarizing the main results of the work.

2   Related Work

Image processing in general proved to improve several clinical tasks [7–20]. In this section, we surveyed the related work on sCT generation. Several methods have been developed to convert MR–to–sCT; some approaches rely on the relation of MR intensities to CT Hounsfield Units (HU), especially for head [21] and pelvis [22]. In particular, MR-to-sCT techniques are usually classified as atlas-based, voxel-based or hybrid methods. Johnstone et al. [23] and Edmund et al. [24] provided a systematic review based on the above approaches (atlas, voxel and hybrid methods) for MR–to–sCT. The atlas-based method [2,25] relies on standard MR sequences (T1 and/or T2 weighted only)[1] This method automatically defines the organs and map realistic electron densities; however, it is not able to generate a statistically significant sCT HU better than CT, for the organs of interest. The Mean Absolute Error (MAE) showed the significant loss in terms of comparison between the ground truth and the generated sCT. With

---

[1] Two types of MR acquisitions that, by taking into account different matter properties, lead to different contrast among tissues.

the rise of deep learning applications, Xiang et al. [26] proposed a Deep Convolutional Neural Network (DCNN) approach to generate sCT for T1 MR. This method provided more accurate MR–to–CT end to end mapping, as well as it provided faster convergence of the training of the deep network. This approach outperformed previous atlas based proposals in literature in terms of quality and processing time, but the limitation of this approach is the training time. Indeed, this approach requires more than 2 days to train a model while the traditional methods take only one day.

Another DCNN-based approach is provided by Han et al. [6]. This approach outperformed all previous atlas-based approaches and DCNN approach by Xiang et al. [26] in terms of MAE and other metrics, i.e. Mean Squared Error (MSE) and Pearson correlation coefficient (CC).

Wolverink et al. [27] provided a Generative Adversarial Neural Network (GANN)-based method trained on unpaired data. The network was trained on MR and CT scans of 24 patients and it outperformed Han [6] approach in terms of MAE value, computed by comparing CT to the generated CT. Indeed, the authors showed that using unpaired data as training factor, the resulting sCT were more realistic with less blurring and more detailed w.r.t. using paired data. However, direct comparison was not possible in this research due to different datasets but based on the computation time and the training losses, this approach was effective. Emami et al. [28] provided a comparison among GANN and DCNN based methods. The authors show that GANN is proved to be slightly superior as compared to DCNN on the same testing dataset. The authors described a novel approach for generating brain sCT based on GANN, trained two competing networks simultaneously and compared this approach with DCNN approach. GANN performance was compared to CNN based on MAE, Structural SIMilarity (SSIM) and Peak signal to noise ratio metrics between ground truth CT and generated synthetic CT. GANN outperformed in all metrics evaluated and it preserved details better than CNN, and abnormal anatomy regions were very well represented by sCTs generated by GANN.

3   Materials and Methods

3.1   Dataset Description

The data-set features 15 pairs of T1-weighted MR and CT image volumes. All patients, treated with radiation treatment gave written approval, and the nearby institutional audit board affirmed the examination. MR were obtained with a 3 T scanner, with a voxel size of $1 \times 1 \times 1$ mm^3. Maximum CT voxel size was $0.67 \times 0.67 \times 2.5$ mm^3. CT scans were procured using a facial mask to immobilize the patient head position for treatment purposes. The inter-scan interim among MR and CT procurement went from 11 to 20 days. Each MR was rigidly aligned on its CT, in order to reach an optimal match between the same anatomical structures depicted in the two different scans. Plastimatch [29], an open source software for medical image registration, was used to align the volumes.

3.2   Proposed Approach

We performed an ablation study to compare the performance gains of the approach and assess the impact of three different losses on the training step. The underlying hypothesis of the present work is that the use of different losses would improve both intensity matching and, especially, edges reconstruction, which is not completely tackled by well-know approaches. To increase the algorithm performance, we rely on SSIM and perceptual loss [30] as loss metrics with MAE. In this work, we make use of a DCNN multi-plane approach proposed by Spadea et al. [22]. This workflow proved to be highly generalizable, since it is able to convert also cone beam CT in sCT [31].

3.3   Losses Description

For each testing patient, 3 different networks were trained by using different losses. The investigated losses are defined below:

Mean Absolute Error. MAE is a loss function used for the regression model, it is the sum of absolute differences between our target and predicted variables. In other words it measures the average magnitude of errors in a set of predictions, without considering their directions. Ideally, we aim to have MAE equal to 0. MAE is defined as:

$$MAE = \frac{1}{M}\Sigma_{i=1}^{M}|sHU_n - HU_n|$$

where, M is defined as total number of samples, HU is defined as Hounsfield Unit in CT and sHU is defined as Hounsfield unit in synthetic CT.

Structural Similarity. SSIM is a metric that quantifies the perceived quality of an image. It can be used as metric to assess the similarity between two images. SSIM is defined as:

$$SSIM(x,y) = \frac{(2\mu_x\mu_y + c_1)(2\sigma_{xy} + c_2)}{(\mu_x^2 + \mu_y^2 + c_1)(\sigma_x^2 + \sigma_y^2 + c_2)}$$

with, μ_x as the average of x, μ_y as the average of y, σ_x^2 as the variance of x, σ_y^2 as the variance of y, σ_{xy} as the co-variance of x and y, $c_1 = (k_1 L)^2$, $c_2 = (k_2 L)^2$, two variables to stabilize the division with weak denominator, L the dynamic range of the pixel-values (typically this is $2^{number\_of\_bits\_per\_pixel} - 1$), and $k_1 = 0.01$ and $k_2 = 0.03$ by default.

Perceptual Loss. The rationale behind this metric is that similar images will activate similar high features represented into an activation maps. In this work, the first layer of a pre-trained VGG16 was used for this purpose. The VGG network was feed first with the synthetic image and than with the GT slice. The mean value of the difference between the two activation maps of the first layer was computed and used as metric.

4   Experiments

In this section we illustrate our experimental setting; in particular, we first describe the hardware setup; then, we discuss techniques and methodologies.

4.1   Benchmark Platform

An already developed algorithm for MR–to–sCT conversion [22] was used as benchmark platform. In particular, the code was modified by adding two new losses to be used during the training step. The chosen conversion algorithm operates on 2D slices, and the final 3D volume is obtained by stacking the images along the selected axis. Network architecture details are provided in the next section. Code was implemented in Pytorch and it was ran on a GNU/Linux workstation equipped with Asus Rampage V Extreme with CPU (Intel core i7-5930K), ram memory (32GB) and GPU (Nvidia Quadro P6000 24GB).

4.2   Deep Convolutional Neural Network (DCNN) Model for MR to sCT

Our DCNN model is the improvement of the model given by Han [6]. 2D DCNN model is used to directly learn a mapping function to convert a 2D MR slice to its corresponding 2D CT. This model trains by collecting all 2D MR slices with corresponding 2D CT slices from each training subject's 3D MR/CT pair. Once this model is trained, it can be used to convert slice by slice and the results can be assembled to get the final 3D synthetic CT. This CNN model is the modification of U-net architecture proposed by Ronneberger [32]. This model contains encoding and decoding part. Encoding part in this work follows the same architecture as VGG-16 [33] for image classification. The proposed DCNN model has 27 convolutional layers interleaved with pooling and unpooling layers and 35 million free parameters, which can be trained to learn a direct end-to-end mapping from MR images to their corresponding CTs. Training such a large model on our limited data is made possible through the principle of transfer learning and by initializing model weights from a pre-trained model.

4.3   Training

Training was run using axial slices pair. More in detail, MR slices were provided as input, and corresponding CT slices were given as Ground Truth (GT). Data augmentation was used and it consisted in different combinations of mirroring and translation along each axis. A single train was performed by using 1 patient as testing, 2 as validation and 12 as training. 5 different training sessions were ran, each of them executed by shuffling the testing and train/validation sets. As a result, 5 different patients were converted as testing cases. Our training was based on the empirically identified parameters given as follow: batch size = 20, epoch = 60, weights initialization = normal, optimizer = stochastic gradient descent, learning rate = 0.0005, $\lambda = 0.0004$ and dropout = 0.1. For each training, same learning parameters were used.

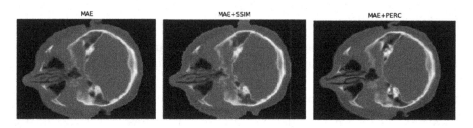

Fig. 1. Slice for the randomly selected patient extracted from the generated sCT images.

4.4   Implementation of Losses for MR–to–sCT Conversion

Since both SSIM and perceptual loss do not include information about absolute intensities, a contribution coming from MAE was also considered. As a result, SSIM and perceptual loss for MR–to–sCT conversion were implemented as described:

$$SSIM\_loss = SSIM(gt, comp)$$

$$loss = MAE + (SSIM\_loss * weight)$$

$$features = VGG(image)$$

$$Perc\_loss = features(gt) - features(synthetic)$$

$$loss = MAE + (Perc\_loss * weight)$$

For each loss, optimal weight was empirically identified. In particular, the optimal weights for MAE+SSIM and MAE+Perceptual are given as follow: SSIM weight equal to 500 and perceptual weight equal to 15000.

4.5   Metric for Result Evaluation

In order to quantify the conversion quality, three different metrics were used. MAE and Mean Error (ME) were used to assess the remapping accuracy, since voxel intensities are crucial for radiation dose calculation. ME is defined as:

$$ME = \frac{1}{M} \Sigma_{n=1}^{M}(sHU_n - HU_n)$$

To evaluate the edge reconstruction, instead, a canny edge detector algorithm [34] was applied to each slice, both for GT and for synthetic volume. Finally, Dice Similarity Index (DSC) [35] was computed between binary images obtained as output from edge detection filter.

Table 1. Mean Absolute Error ± standard deviation for each patient and for each losses used for training.

PATIENT	MAE [HU]	(MAE + SSIM) [HU]	(MAE + PERCEPTUAL) [HU]
1	69.2 ± 132.3	76.5 ± 144.4	68.0 ± 132.0
2	61.7 ± 113.3	66.5 ± 122.4	63.0 ± 117.5
3	56.2 ± 103.8	63.8 ± 118.9	56.3 ± 103.9
4	75.5 ± 144.6	81.1 ± 154.5	75.5 ± 139.4
5	70.5 ± 130.1	75.2 ± 138.9	71.5 ± 126.9

Table 2. Mean Error ± standard deviation for each patient and for each losses used for training.

PATIENT	MAE [HU]	(MAE + SSIM) [HU]	(MAE + PERCEPTUAL) [HU]
1	18.5 ± 148.1	9.4 ± 153.1	16.8 ± 147.6
2	12.2 ± 128.5	10.1 ± 139.0	18.7 ± 132.0
3	1.6 ± 118.0	−0.1 ± 134.9	2.7 ± 118.1
4	−18.7 ± 162.0	−33.7 ± 171.2	−20.8 ± 157.1
5	30.5 ± 144.9	17.3 ± 157.0	32.7 ± 142.0

5   Results

Remapping accuracy, quantified in terms of MAE and ME, are reported in Table 1 and 2 respectively.

Ideally, both MAE and ME should be as close as possible to 0, that means perfect match between intensities. Edges reconstruction accuracy, quantified in terms of DSC, is reported in Table 3. A DSC equal to 0 means completely different boundaries, DSC equal to 1 means perfectly matching edges An exemplary case of differences in reconstruction by using the three losses is shown in Fig. 1.

6   Discussion

To execute a radiotherapy treatment, which is currently one of the most important strategies in cancer treatment, CT is *de facto* mandatory for an accurate dose computation. However, it provides the patient with additional ionizing radiation, and the produced images feature poor contrast between soft tissues, thus possibly severely impacting on contouring and distinction between anatomical structures to treat and to spare. On the opposite, MR provides an higher contrast, but does not contain the information for executing a reliable dose computation. To overcome such a limitation, several approaches have been proposed for the MR–to–sCT conversion task, in order to be able to both precisely contour the tissues of interest (on MR) and compute doses (on the generated sCT).

Table 3. Edges reconstruction accuracy, quantified it terms of DSC, for each patient and for each losses used for training.

PATIENT	MAE	MAE + SSIM	MAE + PERCEPTUAL
1	0.450	0.437	0.454
2	0.499	0.484	0.495
3	0.490	0.481	0.485
4	0.380	0.386	0.384
5	0.473	0.460	0.468

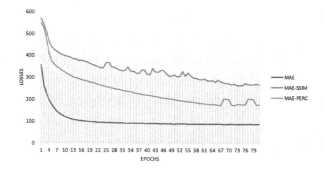

Fig. 2. Left panel: MAE plot comparison for each patient. Right panel: DSC plot comparison for each patient.

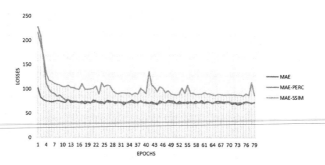

Fig. 3. Upper panel: Training loss with respect to epochs for MAE, MAE+PERC, and MAE+SSIM. Lower panel: Validation loss with respect to epochs for MAE, MAE+PERC, and MAE+SSIM.

An already developed framework for sCT generation was used as benchmark platform for the experiments.

The chosen losses include the most common one used for this type of task (MAE), and two metrics usually adopted in non–medical image processing (SSIM and perceptual loss). 5 patients were converted by stacking 2D slices along the axial direction. Each slice was obtained by means of a 2D network trained with 12 patients (2 for validation). Since a single volume is made of several slices, the amount of shown cases was in the order of some thousand. Accuracy output was quantified in terms of intensity difference (fundamental for accurate dose computation) and edge reconstruction. Since both SSIM and perceptual loss do not take in account absolute intensity, they were always used in combination with MAE. Plots of the results are reported in Fig. 2. Plots for the training and validation losses evolved with respect to each epoch is also reported in Fig. 3

From a general point of view, MAE is still the choice to prefer, both in terms of intensity and edges accuracy; however, MAE plus perceptual loss in some cases performs better than MAE alone. Interestingly, including SSIM led always to a worst result. Even if SSIM and perceptual loss proved to be effective in non–medical image processing, MR–to–sCT conversion represent a more challenging task and, for this reason, it is not an unforeseen result. On the contrary, since perceptual loss in some cases performs better than MAE, further investigation will be done in this direction. Replacing the VGG16 with another network, trained on medical images, could be the key to further improve the conversion quality. We are fully aware that GANN are specifically designed for this type of application, but our aim is to develop a loss simple to be implemented and that enables a more stable training, limiting, at the same time, the computational effort. Our future efforts will be to enroll more patients into the study, in order to have a larger statistical population. This will allow to minimize randomness in the network training and to execute statistical tests that will make our conclusion stronger.

7   Conclusion

In this work, the impact of 3 different losses for improving deep learning- based MR–to–sCT conversion was assessed. The analysis has been carried out by taking into account both intensity and edges reconstruction accuracy. Results suggest that MAE is still the best solution, even if the use of perceptual loss shown to be able to perform better in some cases. The issue to overcome could be that these losses were initially developed for tasks not related to medical image processing. On this regard, future work will focus on adapting this loss, designed for non–medical application, in order to improve sCT generation and, as a consequence, the radiotherapy process.

References

1. Yang, X., et al.: MRI-based proton radiotherapy for prostate cancer using deep convolutional neural networks. Int. J. Radiat. Oncol. Biol. Phys. **105**, S200 (2019)
2. Sjölund, J., Forsberg, D., Andersson, M., Knutsson, H.: Generating patient specific pseudo-CT of the head from MR using atlas-based regression. Phys. Med. Biol. **60**, 825–839 (2015)

3. Dowling, J.A., et al.: Automatic substitute computed tomography generation and contouring for magnetic resonance imaging (MRI)-alone external beam radiation therapy from standard MRI sequences. Int. J. Radiat. Oncol. Biol. Phys. **93**, 1144–1153 (2015)

4. Arabi, H., Koutsouvelis, N., Rouzaud, M., Miralbell, R., Zaidi, H.: Atlas-guided generation of pseudo-CT images for MRI-only and hybrid PET-MRI-guided radiotherapy treatment planning. Phys. Med. Biol. **61**, 6531–6552 (2016)

5. Burgos, N., et al.: Iterative framework for the joint segmentation and CT synthesis of MR images: application to MRI-only radiotherapy treatment planning. Phys. Med. Biol. **62**, 4237 (2017)

6. Han, X.: MR-based synthetic CT generation using a deep convolutional neural network method. Med. Phys. **44**, 02 (2017)

7. Fritscher, K., Raudaschl, P., Zaffino, P., Spadea, M.F., Sharp, G.C., Schubert, R.: Deep neural networks for fast segmentation of 3D medical images. In: Ourselin, S., Joskowicz, L., Sabuncu, M.R., Unal, G., Wells, W. (eds.) MICCAI 2016. LNCS, vol. 9901, pp. 158–165. Springer, Cham (2016). https://doi.org/10.1007/978-3-319-46723-8_19

8. Zaffino, P., et al.: Multi atlas based segmentation: should we prefer the best atlas group over the group of best atlases? Phys. Med. Biol. **63**(12), 12NT01 (2018)

9. Zaffino, P., et al.: Radiotherapy of Hodgkin and non-Hodgkin lymphoma: a nonrigid image-based registration method for automatic localization of prechemotherapy gross tumor volume. Technol. Cancer Res. Treat. **15**(2), 355–364 (2016)

10. Spadea, M.F., et al.: Contrast-enhanced proton radiography for patient set-up by using x-ray CT prior knowledge. Int. J. Radiat. Oncol. Biol. Phys. **90**(3), 628–636 (2014)

11. Zaffino, P., et al.: Fully automatic catheter segmentation in MRI with 3D convolutional neural networks: application to MRI-guided gynecologic brachytherapy. Phys. Med. Biol. **64**(16), 165008 (2019)

12. Bruno, P., et al.: Using CNNs for designing and implementing an automatic vascular segmentation method of biomedical images. In: Ghidini, C., Magnini, B., Passerini, A., Traverso, P. (eds.) AI*IA 2018. LNCS (LNAI), vol. 11298, pp. 60–70. Springer, Cham (2018). https://doi.org/10.1007/978-3-030-03840-3_5

13. Tappeiner, E., et al.: Multi-organ segmentation of the head and neck area: an efficient hierarchical neural networks approach. Int. J. Comput. Assist. Radiol. Surg. **14**(5), 745–754 (2019). https://doi.org/10.1007/s11548-019-01922-4

14. Fritscher, K.D., Peroni, M., Zaffino, P., Spadea, M.F., Schubert, R., Sharp, G.: Automatic segmentation of head and neck CT images for radiotherapy treatment planning using multiple atlases, statistical appearance models, and geodesic active contours. Med. Phys. **41**(5), 051910 (2014)

15. Raudaschl, P.F., et al.: Evaluation of segmentation methods on head and neck CT: auto-segmentation challenge 2015. Med. Phys. **44**(5), 2020–2036 (2017)

16. Ciardo, D., et al.: Atlas-based segmentation in breast cancer radiotherapy: evaluation of specific and generic-purpose atlases. Breast **32**, 44–52 (2017)

17. Milletari, F., Navab, N., Ahmadi, S.-A.: V-net: fully convolutional neural networks for volumetric medical image segmentation. In: 2016 Fourth International Conference on 3D Vision (3DV), pp. 565–571. IEEE (2016)

18. Litjens, G., et al.: A survey on deep learning in medical image analysis. Med. Image Anal. **42**, 60–88 (2017)

19. Lee, H., Chen, Y.-P.P.: Image based computer aided diagnosis system for cancer detection. Expert Syst. Appl. **42**(12), 5356–5365 (2015)

20. Moccia, S., Penza, V., Vanone, G.O., De Momi, E., Mattos, L.S.: Automatic workflow for narrow-band laryngeal video stitching. In: 2016 38th Annual International Conference of the IEEE Engineering in Medicine and Biology Society (EMBC), pp. 1188–1191. IEEE (2016)
21. Pileggi, G., et al.: Proton range shift analysis on brain pseudo-CT generated from T1 and T2 MR. Acta Oncologica **57**(11), 1521–1531 (2018)
22. Spadea, M.F., et al.: Deep convolution neural network (DCNN) multiplane approach to synthetic CT generation from MR images-application in brain proton therapy. Int. J. Radiat. Oncol. Biol. Phys. **105**(3), 495–503 (2019)
23. Johnstone, E., et al.: Systematic review of synthetic computed tomography generation methodologies for use in magnetic resonance imaging-only radiation therapy. Int. J. Radiat. Oncol. Biol. Phys. **100**(1), 199–217 (2018)
24. Edmund, J.M., Nyholm, T.: A review of substitute CT generation for MRI-only radiation therapy. Radiat. Oncol. **12**(1), 28 (2017). https://doi.org/10.1186/s13014-016-0747-y
25. Dowling, J.A., et al.: An atlas-based electron density mapping method for magnetic resonance imaging (MRI)-alone treatment planning and adaptive MRI-based prostate radiation therapy. Int. J. Radiat. Oncol. Biol. Phys. **83**(1), e5–e11 (2012)
26. Xiang, L., Wang, Q., Nie, D., Qiao, Y., Shen, D.: Deep embedding convolutional neural network for synthesizing CT image from t1-weighted MR image. CoRR, vol. abs/1709.02073 (2017)
27. Wolterink, J.M., Dinkla, A.M., Savenije, M.H.F., Seevinck, P.R., van den Berg, C.A.T., Išgum, I.: Deep MR to CT synthesis using unpaired data. In: Tsaftaris, S.A., Gooya, A., Frangi, A.F., Prince, J.L. (eds.) SASHIMI 2017. LNCS, vol. 10557, pp. 14–23. Springer, Cham (2017). https://doi.org/10.1007/978-3-319-68127-6_2
28. Emami, H., Dong, M., Nejad-Davarani, S., Glide-Hurst, C.: Generating synthetic CTs from magnetic resonance images using generative adversarial networks. Med. Phys. **45**, 06 (2018)
29. Zaffino, P., Raudaschl, P., Fritscher, K., Sharp, G.C., Spadea, M.F.: Plastimatch mabs, an open source tool for automatic image segmentation. Med. Phys. **43**(9), 5155–5160 (2016)
30. Johnson, J., Alahi, A., Fei-Fei, L.: Perceptual losses for real-time style transfer and super-resolution. In: Leibe, B., Matas, J., Sebe, N., Welling, M. (eds.) ECCV 2016. LNCS, vol. 9906, pp. 694–711. Springer, Cham (2016). https://doi.org/10.1007/978-3-319-46475-6_43
31. Thummerer, A., et al.: Comparison of CBCT based synthetic CT methods suitable for proton dose calculations in adaptive proton therapy. Phys. Med. Biol. **65**, 095002 (2020)
32. Ronneberger, O., Fischer, P., Brox, T.: U-net: convolutional networks for biomedical image segmentation. CoRR, vol. abs/1505.04597 (2015)
33. Simonyan, K., Zisserman, A.: Very deep convolutional networks for large-scale image recognition. arXiv preprint arXiv:1409.1556 (2014)
34. Canny, J.: A computational approach to edge detection. IEEE Trans. Pattern Anal. Mach. Intell. **6**, 679–698 (1986)
35. Dice, L.R.: Measures of the amount of ecologic association between species. Ecology **26**(3), 297–302 (1945)

Global Convergence of Sobolev Training for Overparameterized Neural Networks

Jorio Cocola[1(✉)] and Paul Hand[1,2]

[1] Department of Mathemathics, Northeastern University, Boston, MA, USA
[2] Khoury College of Computer Sciences, Northeastern University, Boston, MA, USA
{cocola.j,p.hand}@northeastern.edu

Abstract. Sobolev loss is used when training a network to approximate the values and derivatives of a target function at a prescribed set of input points. Recent works have demonstrated its successful applications in various tasks such as distillation or synthetic gradient prediction. In this work we prove that an overparameterized two-layer relu neural network trained on the Sobolev loss with gradient flow from random initialization can fit any given function values and any given directional derivatives, under a separation condition on the input data.

Keywords: Gradient flow · Neural networks · Sobolev training

1 Introduction

Deep neural networks are ubiquitous and have established state of the art performances in a wide variety of applications and fields. These networks often have a large number of parameters which are tuned via gradient descent (or its variants) on an empirical risk minimization task. In particular in supervised learning it is often required that the output of the network fits certain values/labels that can be thought as coming from an unknown target function. In many settings, though, additional prior information on the task or target function might be available, and enforcing them might be of interest. One such example is the case of high order derivatives of the unknown target function, which, as shown in [5], naturally arises in problems such as distillation, in which a large teacher network is used to train a more compact student network, or prediction of synthetic gradients for training deep complex models. Therefore [5] proposed the *"Sobolev training"* which given training inputs $\{x_i\}_{i=1}^n$, attempts to minimizes the following empirical risk:

$$L(W) = \sum_{i=1}^{n} \left[\ell(f(W, x_i), f^*(x_i)) + \sum_{j=1}^{K} \ell_j(D_x^j f(W, x_i), D_x^j f^*(x_i)) \right] \quad (1)$$

where $f(W, x)$ is a neural network with input x and parameters W, f^* denotes the target function, ℓ is a loss penalizing the deviation from the outputs of f,

© Springer Nature Switzerland AG 2020
G. Nicosia et al. (Eds.): LOD 2020, LNCS 12565, pp. 574–586, 2020.
https://doi.org/10.1007/978-3-030-64583-0_51

and ℓ_j are loss functions penalizing the deviations of the j-th derivative $D_x^j f$ of the network f with respect to x from the j-th derivative $D_x^j f^*$ of the target f^*.

The empirical successes of Sobolev training have been demonstrated in a number of works. In [5] it was shown that Sobolev training leads to smaller generalization errors than standard training, in tasks such as distillation and synthetic gradient prediction especially in the low data regime. Similar results were also obtained for transfer learning via Jacobian matching in [14]. Earlier Sobolev training was applied in [13] in order to enforce invariance to translations and small rotations. More recently, instead, Sobolev training has been used in the context of anisotropic hyperelasticity in order to improve the predictions on the stress tensor (derivative of the network with respect to the input deformation tensor) in [16]. Finally, the idea of Sobolev training is also tightly connected to other techniques which have been recently successfully employed, such as attention matching in student distillation [18] and [5], or convex data augmentation for generalization and robustness improvement [19].

On the theoretical side, justification for Sobolev training was given by [5], extending the classical work of Hornik [9] and giving universal approximation properties of neural networks with relu activation function in Sobolev spaces. This result was then further improved for deep networks in [7]. While these works motivated the use of the Sobolev loss (1), conditions under which it can be successfully minimized were not given. In particular even though the network used in Sobolev training are usually shallow, the resulting loss (1) is highly non-convex and therefore the success of first order methods is not a priori guaranteed.

In this paper we study a two-layer relu neural network trained with a Sobolev loss when at each input point the output values and a set of directional derivatives of the target function are given. Leveraging recent results on training with standard losses [1,2,12,20,20] we show that if the network is sufficiently overparameterized, the weights are randomly initialized, and the data satisfy certain natural non-degeneracy assumptions, Gradient Flow achieves a global minimum.

2   Main Result

We study the training of neural networks with *"Directional Sobolev Training"*. In particular we assume we are given training data

$$\left\{x_i, y_i, V_i, h_i\right\}_{i=1}^{n}, \tag{2}$$

where $x_i \in \mathbb{R}^d$, $y_i \in \mathbb{R}$, $h_i \in \mathbb{R}^k$ and $V_i \in \mathbb{R}^{d \times k}$ with orthonormal columns (unit Euclidean norm and pairwise orthogonal). This training data can be thought as being generated by a differentiable function $f^* : \mathbb{R}^d \to \mathbb{R}$ according to

$$y_i = f^*(x_i), \quad \text{and} \quad h_i = V_i^T \nabla f^*(x_i) \qquad \text{for } i = 1, \ldots, n \tag{3}$$

so that each entry of the vector h_i corresponds to a directional derivative of f^* in the direction given by the corresponding column of the matrix V_i. We will denote by $y \in \mathbb{R}^n$ and $h \in \mathbb{R}^{nk}$ the vectors with n entries y_i and n blocks h_i respectively.

In this work we study the training of a two-layer neural network with *width* m:

$$f(W, x) = \sum_{r=1}^{m} a_r \sigma(w_r^T x), \tag{4}$$

where a_r are fixed at initialization, $\sigma(z) = \max(0, z)$ is the relu activation function, and W is the weight matrix with rows $\{w_r^T\}_{r=1}^{m}$. The network weights $\{w_r\}_{r=1}^{m}$ are learned by minimizing the *Directional Sobolev Loss*

$$\min_{W \in \mathbb{R}^{m \times d}} L(W) := \frac{1}{2} \sum_{i=1}^{n} (f(W, x_i) - y_i)^2 + \frac{1}{2} \sum_{i=1}^{n} \|V_i^T \nabla_x f(W, x_i) - h_i\|_2^2.$$

via the *Gradient Flow*

$$\frac{dw_r(t)}{dt} = -\frac{\partial L(W(t))}{\partial w_r}. \tag{5}$$

Note that even though the relu activation function σ is not differentiable, we let $\sigma'(z) = \mathbb{I}\{z > 0\}$ and $\sigma''(z) = 0$. This corresponds to the choice made in most of the deep learning libraries, and the dynamical system (5) can then be seen as the one followed in practice when using Sobolev training. Explicit formulas for the partial derivatives are given in the next section.

In this work we prove that for wide enough networks, gradient flow converges to a global minimizer of $L(W)$. In particular define the vectors of residuals $e(t) \in \mathbb{R}^n$ and $S(t) \in \mathbb{R}^{k \cdot n}$ with coordinates

$$[e(t)]_i = y_i - f(W(t), x_i) \qquad [S(t)]_i = h_i - V_i^T \nabla_x f(W(t), x_i). \tag{6}$$

We show that $e(t) \to 0$ and $S(t) \to 0$ as $t \to \infty$, under the following assumption of non-degeneracy of the training data.

Assumption 1. *The data are normalized so that $\|x_i\|_2 = 1$ and there exist $0 < \delta_1$ and $0 \leq \delta_2 < k^{-1}$ such that the following hold:*

$$\min_{i \neq j}(\|x_i - x_j\|_2, \|x_i + x_j\|_2) \geq \delta_1, \tag{7}$$

and for every $i = 1, \ldots, n$:

$$\max_{1 \leq j \leq k} |v_{i,j}^T x_i| \leq \delta_2, \tag{8}$$

where $v_{i,j}$ are the columns of V_i.

Given $w \in \mathbb{R}^d$ define the following "feature maps":

$$\phi_w(x_i) := \sigma'(w^T x_i) x_i,$$

$$\psi_w(x_i) := \sigma'(w^T x_i) V_i,$$

and matrix:

$$\Omega(w) = [\phi_w(x_1), \ldots, \phi_w(x_n), \psi_w(x_1), \ldots \psi_w(x_n)] \in \mathbb{R}^{d \times (k+1)n}.$$

The next quantity plays an important role in the proof of convergence of the gradient flow (5).

Definition 1. *Define the matrix* $H^\infty \in \mathbb{R}^{n(k+1) \times n(k+1)}$ *with entries given by* $[H^\infty]_{i,j} = \mathbb{E}_{w \sim \mathcal{N}(0, I_d)}[\Omega(w)^T \Omega(w)]_{i,j}$, *and let* λ_* *be its smallest eigenvalue.*

Under the non-degeneracy of the training set we show that H^∞ is strictly positive definite.

Proposition 1. *Under the Assumptions 1 the minimum eigenvalue of* H^∞ *obeys:*

$$\lambda_* \geq \frac{(1 - k\delta_2)\delta_1}{100\, n^2}.$$

We are now ready to state the main result of this work.

Theorem 1. *Assume Assumption 1 is satisfied. Consider a one hidden layer neural network (4), let* $\gamma = \|y\|_2 + \|h\|_2$, *set the number of hidden nodes to* $m = \Omega(n^6 k^4 \gamma^2 / (\lambda_*^4 \delta^3))$ *and i.i.d. initialize the weights according to:*

$$w_r \sim \mathcal{N}(0, I_d) \quad and \quad a_r \sim \mathrm{unif}\{-\frac{1}{m^{1/2}}, \frac{1}{m^{1/2}}\} \qquad for\, r = 1, \ldots, m. \qquad (9)$$

Consider the Gradient Flow (5), then with probability $1 - \delta$ *over the random initialization of* $\{w_r\}_r$ *and* $\{a_r\}_r$, *for every* $t \geq 0$:

$$\|e(t)\|_2^2 + \|S(t)\|_2^2 \leq \exp(-\lambda_* t)\,(\|e(0)\|_2^2 + \|S(0)\|_2^2)$$

and in particular $L(W(t)) \to 0$ *as* $t \to \infty$.

The proof of this theorem is given in Sect. 3, below we will show how to extend this result to a network with bias.

2.1 Consequences for a Network with Bias

Given training data (2) generated by a target function f^* according to (3), in this section we demonstrate how the previous theory can be extended to the Sobolev training of a two-layer network with width m and bias term b:

$$g(W, b, x) = \sum_{r=1}^{m} a_r \sigma(\alpha\, w_r^T x + \beta\, b_r) \qquad (10)$$

where[1] $\alpha = 1/(2k)$ and $\beta = \sqrt{1 - \alpha^2}$.

---

[1] Notice that the introduction of the constants α and β does not change the expressivity of the network.

Similarly as before, the network weights $\{w_r\}_{r=1}^m$ and biases $\{b_r\}_{r=1}^m$ are learned by minimizing the *Directional Sobolev Loss*

$$\min_{W \in \mathbb{R}^{m \times d}, b \in \mathbb{R}^d} L(W, b) := \frac{1}{2} \sum_{i=1}^n (g(W, b, x_i) - y_i)^2 + \frac{1}{2} \sum_{i=1}^n \| \frac{1}{\alpha} V_i^T \nabla_x g(W, b, x_i) - \frac{h_i}{\alpha} \|_2^2.$$

(11)

via the *Gradient Flow*

$$\frac{dw_r(t)}{dt} = -\frac{\partial L(W(t))}{\partial w_r} \quad \text{and} \quad \frac{db_r(t)}{dt} = -\frac{\partial L(W(t))}{\partial b_r}.$$

(12)

Based on the following separation conditions on the input point $\{x_i\}_i$ we will prove convergence to zero training error of the Sobolev loss.

Assumption 2. *The data are normalized so that $\|x_i\|_2 = 1$ and there exists $\hat{\delta}_1 > 0$ such that the following holds*

$$\min_{i \neq j}(\|x_i - x_j\|_2) \geq \hat{\delta}_1.$$

(13)

Define the vectors of residuals $e(t) \in \mathbb{R}^n$ and $S(t) \in \mathbb{R}^{k \cdot n}$ with coordinates

$$[e(t)]_i = y_i - g(W(t), b(t), x_i), \quad [S(t)]_i = (h_i - V_i^T \nabla_x g(W(t), b(t), x_i))/\alpha,$$

then the next theorem follows readily from the analysis in the previous section.

Theorem 2. *Assume Assumption 2 is satisfied. Consider a two-layer neural network* (10), *let $\gamma = \|y\|_2 + \|h/\alpha\|_2$, set the number of hidden nodes to $m = \Omega(n^6 k^4 \gamma^2/(\lambda_0^4 \delta^3))$ and i.i.d. initialize the weights according to:*

$$w_r \sim \mathcal{N}(0, I_d), \quad b_r \sim \mathcal{N}(0, 1) \quad \text{and} \quad a_r \sim \text{unif}\{-\frac{1}{m^{1/2}}, \frac{1}{m^{1/2}}\}.$$

Consider the Gradient Flow (12), *then with probability $1 - \delta$ over the random initialization of $\{b_r\}_r$, $\{w_r\}_r$ and $\{a_r\}_r$, for every $t \geq 0$*

$$\|e(t)\|_2^2 + \|S(t)\|_2^2 \leq \exp(-\lambda_* t)(\|e(0)\|_2^2 + \|S(0)\|_2^2)$$

where $\lambda_* \geq \frac{\delta}{200\,n^2}$ and $\delta = \min(\alpha \hat{\delta}_1, 2\beta)$. In particular $L(W(t), b(t)) \to 0$ as $t \to \infty$.

2.2   Discussion

Theorem 2 establishes that the gradient flow (12) converges to a global minimum and therefore that a wide enough network, randomly initialized and trained with the Sobolev loss (11) can interpolate any given function values and directional derivatives. We observe that recent works in the analysis of standard training [12,21] have shown that using more refined concentration results and control on the weight dynamics, the polynomial dependence on the number of samples n can be lowered. We believe that by applying similar techniques to the Sobolev training, the dependence of m from the number of samples n and derivatives k can be further improved.

Regarding the assumptions on the input data, we note that [2,6,12] have shown convergence of gradient descent to a global minimum of the standard ℓ_2 loss, when the input points $\{x_i\}_{i=1}^n$ satisfy the separation conditions (7). These conditions ensure that no two input points x_i and x_j are parallel and reduce to (13) for a network with bias. While the separation condition (7) is also required in Sobolev training, the condition (8) is only required in case of a network without bias as a consequence of its homogeneity.

Finally, the analysis of gradient methods for training overparameterized neural networks with standard losses has been used to study their inductive bias and ability to learn certain classes of functions (see for example [2,3]). Similarly, the results of this paper could be used to shed some light on the superior generalization capabilities of networks trained with a Sobolev loss and their use for knowledge distillation.

3   Proof of Theorem 1

We follow the lines of recent works on the optimization of neural networks in the Neural Tangent Kernel regime [4,10,12] in particular the analysis of [2,6,17]. We investigate the dynamics of the residuals error $e(t)$ and $S(t)$, beginning with that of the predictions. Let $\bar{F}(W(t), x_i) = V_i^T \nabla_x f(W(t), x_i)$, then:

$$
\frac{d}{dt} f(W(t), x_i) = \sum_{r=1}^{m} \frac{\partial f(W(t), x_i)}{\partial w_r}^T \frac{dw_r(t)}{dt}
$$

$$
= \sum_{j=1}^{n} A_{ij}(t)(y_j - f(W(t), x_j)) + \sum_{j=1}^{n} B_{ij}(t)(h_j - \bar{F}(W(t), x_j)),
$$

$$
\frac{d}{dt} \bar{F}(W(t), x_i) = \sum_{r=1}^{m} \frac{\partial \bar{F}(W(t), x_i)}{\partial w_r}^T \frac{dw_r(t)}{dt}
$$

$$
= \sum_{j=1}^{n} B_{ji}(t)^T (y_j - f(W(t), x_j)) + \sum_{j=1}^{n} C_{ij}(t)(h_j - \bar{F}(W(t), x_j)),
$$

where we defined the matrices $A(t) = \sum_{r=1}^{m} A_r(t) \in \mathbb{R}^{n \times n}$, $B(t) = \sum_{r=1}^{m} B_r(t) \in \mathbb{R}^{n \times n \cdot k}$, $C(t) = \sum_{r=1}^{m} C_r(t) \in \mathbb{R}^{n \cdot k \times n \cdot k}$, with block structure:

$$
[A_r(t)]_{ij} := \frac{\partial f(W(t), x_i)}{\partial w_r}^T \frac{\partial f(W(t), x_j)}{\partial w_r} = \frac{1}{m} \sigma'(w_r^T x_i) \sigma'(w_r^T x_j) x_i^T x_j,
$$

$$
[B_r(t)]_{ij} := \frac{\partial f(W(t), x_i)}{\partial w_r}^T \frac{\partial \bar{F}(W(t), x_j)}{\partial w_r} = \frac{1}{m} \sigma'(w_r^T x_i) \sigma'(w_r^T x_j) x_i^T V_j,
$$

$$
[C_r(t)]_{ij} := \frac{\partial \bar{F}(W(t), x_i)}{\partial w_r}^T \frac{\partial \bar{F}(W(t), x_j)}{\partial w_r} = \frac{1}{m} \sigma'(w_r^T x_i) \sigma'(w_r^T x_j) V_i^T V_j.
$$

The residual errors (6) then follow the dynamical system:

$$\frac{d}{dt}\begin{pmatrix} e \\ S \end{pmatrix} = -H(t)\begin{pmatrix} e \\ S \end{pmatrix} \tag{14}$$

where $H(t) \in \mathbb{R}^{n(k+1)\times n(k+1)}$ is given by:

$$H(t) = \begin{bmatrix} A(t) & B(t) \\ B(t)^T & C(t) \end{bmatrix}.$$

We moreover observe that if we define:

$$\Omega_r(t) := [\frac{\partial f(W(t), x_1)}{\partial w_r}, \dots, \frac{\partial f(W(t), x_n)}{\partial w_r}, \frac{\partial \bar{F}(W(t), x_1)}{\partial w_r}, \dots, \frac{\partial \bar{F}(W(t), x_n)}{\partial w_r}]$$

and $H_r(t) := \Omega_r(t)^T \Omega_r(t)$, then direct calculations show that $H(t) = \sum_r H_r(t)$ and $H(t)$ is symmetric positive semidefinite for all $t \geq 0$. In the next section we will show that $H(t)$ is strictly positive definite in a neighborhood of initialization, while in Sect. 3.2 we will show that this holds for large enough time leading to global convergence to zero of the errors.

3.1   Analysis Near Initialization

In this section we analyze the behavior of the matrix $H(t)$ and the dynamics of the errors $e(t), S(t)$ near initialization. We begin by bounding the output and directional derivatives of the network for every t.

Lemma 1. *For all $t \geq 0$ and $1 \leq i \leq n$, it holds:*

$$\|\frac{\partial f(W(t), x_i)}{\partial w_r}\|_2 \leq \frac{1}{\sqrt{m}}$$

$$\|\frac{\partial \bar{F}(W(t), x_i)}{\partial w_r}\|_2 \leq \sqrt{\frac{k}{m}}$$

$$\|H_r(t)\|_2 \leq \frac{n(k+1)}{m}.$$

We now lower bound the smallest eigenvalue of $H(0)$.

Lemma 2. *Let $\delta \in (0, 1)$, and $m \geq \frac{32}{\lambda_*} n(k+1) \ln(n(k+1)/\delta)$ then with probability $1 - \delta$ over the random initialization:*

$$\lambda_{min}(H(0)) \geq \frac{3}{4}\lambda_*$$

We now provide a bound on the expected value of the residual errors at initialization.

Lemma 3. *Let $\{w_r\}_r$ and $\{a_r\}$ be randomly initialized as in (9) then the residual errors (6) at time zero satisfy with probability at least $1 - \delta$:*

$$\|e(0)\|_2 + \|S(0)\|_2 \leq \frac{2\sqrt{nk} + \gamma}{\delta^{1/2}}$$

where $\gamma = \|y\|_2 + \|h\|_2$.

Next define the neighborhood around initialization:

$$B_0 := \left\{ W : \|H(W) - H(W(0))\|_F \leq \frac{\lambda_*}{4} \right\}$$

and the escape time

$$\tau_0 := \inf\{t : W(t) \notin B_0\}. \tag{15}$$

We can now prove the main result of this section which characterizes the dynamics of $e(t)$, $S(t)$ and the weights $w_r(t)$ in the vicinity of $t = 0$.

Lemma 4. *Let $\delta \in (0, 1)$ and $m \geq \frac{32}{\lambda_*} n(k+1) \ln(n(k+1)/\delta)$ then with probability $1 - \delta$ over the random initialization, for every $t \in [0, \tau_0]$:*

$$\|e(t)\|_2^2 + \|S(t)\|_2^2 \leq \exp(-\lambda_* t)\,(\|e(0)\|_2^2 + \|S(0)\|_2^2)$$

and

$$\|w_r(t) - w_r(0)\|_2 \leq \frac{4}{\lambda_*}\sqrt{\frac{kn}{m}}\,(\|e(0)\|_2 + \|S(0)\|_2) =: R$$

Proof. Observe that if $t \in [0, \tau_0]$, by Lemma 2 with probability $1 - \delta$:

$$\lambda_{\min}(H(t)) \geq \lambda_{\min}(H(0)) - \|H(t) - H(0)\|_F \geq \frac{\lambda_*}{2}.$$

Therefore using (14) it follows that for any $t \in [0, \tau_0]$:

$$\frac{\mathrm{d}}{\mathrm{d}t}\frac{1}{2}(\|e(t)\|_2^2 + \|S(t)\|_2^2) \leq -\frac{\lambda_*}{2}(\|e(t)\|_2^2 + \|S(t)\|_2^2),$$

which implies the first claim by Gronwall's lemma. Next, using (5), the bounds in Lemma 1 and the above inequality we obtain:

$$\|\frac{\mathrm{d}}{\mathrm{d}t}w_r(t)\|_2 = \|\frac{\partial}{\partial w_r}L(W(t))\|_2$$

$$= \|\sum_{i=1}^{n}(y_i - f(W(t), x_i))\frac{\partial f(W(t), x_i)}{\partial w_r} + \sum_{i=1}^{n}\frac{\partial \bar{F}_i(t)}{\partial w_r}(h_i - \bar{F}_i(t))\|$$

$$\leq \frac{1}{\sqrt{m}}\sum_{i=1}^{n}|y_i - f(W(t), x_i)| + \sqrt{\frac{k}{m}}\sum_{i=1}^{n}\|h_i - \bar{F}_i(W(t))\|$$

$$\leq 2\sqrt{\frac{kn}{m}}e^{-\frac{\lambda_*}{2}t}(\|e(0)\| + \|S(0)\|).$$

We can therefore conclude by bounding the distance from initialization as:

$$\|w_r(t) - w_r(0)\| \leq \int_0^t \|\frac{\mathrm{d}}{\mathrm{d}t}w_r(s)\|_2 \mathrm{d}t \leq \frac{4}{\lambda_*}\sqrt{\frac{kn}{m}}(\|e(0)\| + \|S(0)\|)$$

3.2 Proof of Global Convergence

In order to conclude the proof of global convergence, according to Lemma 4, we need only to show that $\tau_0 = \infty$ where τ_0 is defined in (15). Arguing by contradiction, assume this is not the case and $\tau_0 < \infty$. Below we bound $\|H(\tau_0) - H(0)\|_F$.

Let $Q_{ijr} := |\sigma'(w_r^T(\tau_0)x_i)\sigma'(w_r^T(\tau_0)x_j) - \sigma'(w_r^T(0)x_i)\sigma'(w_r^T(0)x_j)|$, then from the formulas for $[A_r]_{ij}$, $[B_r]_{ij}$ and $[C_r]_{ij}$ in the previous sections, we have:

$$|[A(\tau_0) - A(0)]_{ij}| \leq \frac{1}{m}\sum_{r=1}^{m} Q_{ijr},$$

$$\|[B^T(\tau_0) - B^T(0)]_{ij}\|_2 \leq \frac{\sqrt{k}}{m}\sum_{r=1}^{m} Q_{ijr},$$

$$\|[C(\tau_0) - C(0)]_{ij}\|_F \leq \frac{k}{m}\sum_{r=1}^{m} Q_{ijr}.$$

Let $R > 0$ as in in Lemma 4, then with probability at least $1 - \delta$ for all $1 \leq r \leq m$ we have $\|w_r(\tau_0) - w_r(0)\| \leq R$. Moreover observe that if $\|w_r(\tau_0) - w_r(0)\| \leq R$ and $\sigma'(w_r(0)^T x_i) \neq \sigma'(w_r(\tau_0)^T x_i)$, then $|w_r(0)^T x_i| \leq R$. Therefore, for any $i \in [n]$ and $r \in [m]$ we can define the event $E_{i,r} = \{|w_r(0)^T x_i| \leq R\}$ and observe that:

$$\mathbb{1}\{\sigma'(w_r(0)^T x_i) \neq \sigma'(w_r(\tau_0)^T x_i)\} \leq \mathbb{1}\{E_{i,r}\} + \mathbb{1}\{\|w_r(\tau_0) - w_r(0)\| > R\}.$$

Next note that $w_r(0)^T x_i \sim \mathcal{N}(0,1)$, so that $\mathbb{P}(E_{i,r}) \leq \frac{2R}{\sqrt{2\pi}}$ and in particular:

$$\frac{1}{m}\sum_{r=1}^{m}\mathbb{E}[Q_{ijr}] \leq \frac{1}{m}\sum_{r=1}^{m}(\mathbb{P}[E_{i,r}] + \mathbb{P}[E_{j,r}]) + 2\mathbb{P}[\cup_r\{\|w_r(\tau_0) - w_r(0)\| > R\}] \leq \frac{4R}{\sqrt{2\pi}} + 2\delta.$$

By Markov inequality we can conclude that with probability at least $1 - \delta$:

$$\|A(\tau_0) - A(0)\|_F \leq \sum_{i,j}|[A(\tau_0)]_{ij} - [A(0)]_{ij}| \leq \frac{4n^2}{\sqrt{2\pi\delta}}R + \frac{2n^2}{m},$$

$$\|B^T(\tau_0) - B^T(0)\|_F \leq \sum_{i,j}\|[B^T(\tau_0)]_{ij} - [B^T(0)]_{ij})\|_2 \leq \frac{4n^2\sqrt{k}}{\sqrt{2\pi\delta}}R + \frac{2n^2}{m},$$

$$\|C(\tau_0) - C(0)\|_F \leq \sum_{i,j}\|[C(\tau_0)]_{ij} - [C(0)]_{ij})\|_F \leq \frac{4n^2 k}{\sqrt{2\pi\delta}}R + \frac{2n^2}{m},$$

and using Lemma 3 together with the definition of H and R :

$$\|H(\tau_0) - H(0)\|_F \leq \frac{16n^2 k}{\sqrt{2\pi}\,\delta}R + \frac{8n^2}{m} = \mathcal{O}\left(\frac{n^3 k^2}{\sqrt{m}\delta^2}\frac{\max(1,\gamma)}{\lambda_*}\right)$$

Then choosing $m = \Omega(n^6 k^4\gamma^2/(\lambda_*^4\delta^3))$ we obtain

$$\|H(\tau_0) - H(0)\|_F < \frac{\lambda_*}{4}$$

which contradicts the definition of τ_0 and therefore $\tau_0 = \infty$.

Acknowledgements. PH is supported in part by NSF CAREER Grant DMS-1848087.

A   Supplementary proofs for Sect. 3.1

In this section we provide the remaining proofs of the results in Sect. 3.1. We begin recalling the following matrix Chernoff inequality (see for example [15, Theorem 5.1.1]).

Theorem 3 (Matrix Chernoff). *Consider a finite sequence X_k of $p \times p$ independent, random, Hermitian matrices with $0 \preceq X_k \preceq LI$. Let $X = \sum_k X_k$, then for all $\epsilon \in [0,1)$*

$$\mathbb{P}\left[\lambda_{min}(X) \le \epsilon\lambda_{min}\big(\mathbb{E}[X]\big)\right] \le pe^{-(1-\epsilon)^2\lambda_{min}(\mathbb{E}[X])/2L} \tag{16}$$

In order to lower bound the smallest eigenvalue of $H(0)$ we use Lemma 1 together with the previous concentration result.

Proof (Lemma 2). We first note that $\mathbb{E}[H(0)] = \mathbb{E}[\sum_r H_r(0)] = H^\infty$, and moreover $H_r(0)$ is symmetric positive semidefinite with $\lambda_{max}(H_r) \le n(k+1)/m$ by Lemma 1. Applying then the concentration bound (16) with the assumption $m \ge \frac{32}{\lambda_*} n(k+1) \ln(n(k+1)/\delta)$ gives the thesis.

We next upper bound the errors at initialization.

Proof (Lemma 3). Note that for any x_i, due the the assumption on the independence of the weights at initialization and the normalization of the data:

$$\mathbb{E}[(f(W,x_i))^2] = \sum_{r=1}^{m} \frac{1}{m}\mathbb{E}[\sigma(w_r^T x_i)^2] \le 1$$

and similarly for the directional derivatives

$$\mathbb{E}[\|\bar{F}(W,x_i)\|_2^2] = \mathbb{E}_{g\sim\mathcal{N}(0,I)}[\|\sigma'(g^T x_i)V_i^T g\|_2^2] \le \sum_{j=1}^{k}\mathbb{E}[(v_{i,j}^T g)^2] \le k.$$

We conclude the proof by using Jensen's and Markov's inequalities.

B   Proof of Proposition 1

Consider the $d \times (k+1)$ matrices $\mathbf{X}_i = [x_i, V_i]$, and for $w \in \mathbb{R}^d$ define

$$\hat{\psi}_w(x_i) = \sigma'(w^T x_i)\mathbf{X}_i.$$

and the $d \times (k+1)n$ matrix:

$$\widehat{\Omega}(w) = [\hat{\psi}_w(x_1), \ldots, \hat{\psi}_w(x_n)] \in \mathbb{R}^{d \times n(k+1)}$$

which corresponds to a column permutation of $\Omega(w)$. Next observe that the matrix $\widehat{H}^\infty = \mathbb{E}_{w \sim \mathcal{N}(0,I_d)}[\widehat{\Omega}(w)^T \widehat{\Omega}(w)]$ is similar to H^∞ and therefore has the same eigenvalues. In this section we lower bound λ_* by analyzing $\widehat{H}^\infty$.

We begin recalling some facts about the spectral properties of the products of matrices.

Definition 2 ([8]). *Let $\mathbf{A} = [A_{\alpha\beta}]_{\alpha=1,\ldots,n}^{\beta=1,\ldots,n}$ and $\mathbf{B} = [B_{\alpha\beta}]_{\alpha=1,\ldots,n}^{\beta=1,\ldots,n}$ be $np \times np$ matrices in which each block is in $p \times p$. Then we define the block Hadamard product of $\mathbf{A}\square\mathbf{B}$ as the $np \times np$ matrix with:*

$$\mathbf{A}\square\mathbf{B} := [A_{\alpha\beta}B_{\alpha\beta}]_{\alpha=1,\ldots,n}^{\beta=1,\ldots,n}$$

where $A_{\alpha\beta}B_{\alpha\beta}$ denotes the usual matrix product between $A_{\alpha\beta}$ and $B_{\alpha\beta}$.

Generalizing Schur's Lemma one has the following regarding the eigenvalues of the block Hadamard product of two block matrices.

Proposition 2 ([8]). *Let $\mathbf{A} = [A_{\alpha\beta}]_{\alpha=1,\ldots,n}^{\beta=1,\ldots,n}$ and $\mathbf{B} = [B_{\alpha\beta}]_{\alpha=1,\ldots,n}^{\beta=1,\ldots,n}$ be $np \times np$ positive semidefinite matrices. Assume that every $p \times p$ block of $\mathbf{A}$ commutes with every $p \times p$ block of $\mathbf{B}$, then:*

$$\lambda_{min}(\mathbf{B}\square\mathbf{A}) = \lambda_{min}(\mathbf{A}\square\mathbf{B}) \geq \lambda_{min}(A) \cdot \min_\alpha \lambda_{min}(B_{\alpha\alpha})$$

We finally recall the following on the eigenvalues of Kronecker product of matrices.

Proposition 3 ([11]). *Let $A \in \mathbb{R}^{n \times n}$ with eigenvalues $\{\lambda_i\}$ and $B \in \mathbb{R}^{m \times m}$ with eigenvalues $\{\mu_i\}$, then Kronecker product $A \otimes B$ between A and B has eigenvalues $\{\lambda_i \mu_j\}$.*

We next define the following random kernel matrix.

Definition 3. *Let $w \sim \mathcal{N}(0, I)$ then define the random matrix $\mathcal{M}(w) \in \mathbb{R}^{n \times n}$ with entries $[\mathcal{M}(w)]_{ij} = \sigma'(w^T x_i)\sigma'(w^T x_j)$.*

The next result from [12] establishes positive definiteness of this matrix in expectation, under the separation condition (7).

Lemma 5 ([12]). *Let $x_1, \ldots, x_d$ in $\mathbb{R}^d$ with unit Euclidean norm and assume that (7) is satisfied for all $i = 1, \ldots d$. Then the following holds:*

$$\mathbb{E}_{w \sim \mathcal{N}(0,I)}[\mathcal{M}(w)] \succeq \frac{\delta_1}{100n^2}$$

Finally let $\mathbf{X} \in \mathbb{R}^{d \times n(k+1)}$ block matrix with $d \times (k+1)$ blocks $\mathbf{X}_i$. Thanks to the assumption (8) the following result on the Gram matrices $\mathbf{X}_i^T \mathbf{X}_i$ holds.

Lemma 6 *Assume that the condition (8) is satisfied, then for any $i = 1, \ldots, n$ we have $\lambda_{min}(\mathbf{X}_i^T \mathbf{X}_i) \geq 1 - k\delta_2 > 0$.*

Proof. The claim follows by observing that by *Gershgorin's Disk Theorem*:

$$|\lambda_{\min}(\mathbf{X}_i^T \mathbf{X}_i) - 1| \leq \sum_{1 \leq j \leq k} |x_i^T v_{i,j}| \leq k\delta_2.$$

Finally observe that we can write:

$$\widehat{H}^\infty = \mathbb{E}_{w \sim \mathcal{N}(0,I)} \left[(\mathbf{X}^T \mathbf{X}) \square (\mathcal{M}(w) \otimes I) \right].$$

so that Proposition 2, Proposition 3, Lemma 5 and Lemma 6 allow to derive the thesis of Proposition 1.

References

1. Allen-Zhu, Z., Li, Y., Song, Z.: A convergence theory for deep learning via over-parameterization. arXiv preprint arXiv:1811.03962 (2018)
2. Arora, S., Du, S.S., Hu, W., Li, Z., Wang, R.: Fine-grained analysis of optimization and generalization for overparameterized two-layer neural networks. arXiv preprint arXiv:1901.08584 (2019)
3. Bietti, A., Mairal, J.: On the inductive bias of neural tangent kernels. In: Advances in Neural Information Processing Systems. pp. 12873–12884 (2019)
4. Chizat, L., Oyallon, E., Bach, F.: On lazy training in differentiable programming. arXiv preprint arXiv:1812.07956 (2018)
5. Czarnecki, W.M., Osindero, S., Jaderberg, M., Swirszcz, G., Pascanu, R.: Sobolev training for neural networks. In: Advances in Neural Information Processing Systems, pp. 4278–4287 (2017)
6. Du, S.S., Zhai, X., Poczos, B., Singh, A.: Gradient descent provably optimizes over-parameterized neural networks. arXiv preprint arXiv:1810.02054 (2018)
7. Gühring, I., Kutyniok, G., Petersen, P.: Error bounds for approximations with deep relu neural networks in $w^{s,p}$ norms. arXiv preprint arXiv:1902.07896 (2019)
8. Günther, M., Klotz, L.: Schur's theorem for a block Hadamard product. Linear Algebra Appl. **437**(3), 948–956 (2012)
9. Hornik, K.: Approximation capabilities of multilayer feedforward networks. Neural Netw. 4(2), 251–257 (1991)
10. Jacot, A., Gabriel, F., Hongler, C.: Neural tangent Kernel: convergence and generalization in neural networks. In: Advances in Neural Information Processing Systems, pp. 8571–8580 (2018)
11. Laub, A.J.: Matrix Analysis for Scientists and Engineers, vol. 91. SIAM (2005)
12. Oymak, S., Soltanolkotabi, M.: Towards moderate overparameterization: global convergence guarantees for training shallow neural networks. arXiv preprint arXiv:1902.04674 (2019)
13. Simard, P., Victorri, B., LeCun, Y., Denker, J.: Tangent prop-a formalism for specifying selected invariances in an adaptive network. In: Advances in Neural Information Processing Systems, pp. 895–903 (1992)
14. Srinivas, S., Fleuret, F.: Knowledge transfer with Jacobian matching. arXiv preprint arXiv:1803.00443 (2018)
15. Tropp, J.A., et al.: An introduction to matrix concentration inequalities. Found. Trends® Mach. Learn. **8**(1–2), 1–230 (2015)
16. Vlassis, N., Ma, R., Sun, W.: Geometric deep learning for computational mechanics part i: anisotropic hyperelasticity. arXiv preprint arXiv:2001.04292 (2020)

17. Weinan, E., Ma, C., Wu, L.: A comparative analysis of optimization and generalization properties of two-layer neural network and random feature models under gradient descent dynamics. Sci. China Math. **63**, 1235–1258 (2020). https://doi.org/10.1007/s11425-019-1628-5

18. Zagoruyko, S., Komodakis, N.: Paying more attention to attention: improving the performance of convolutional neural networks via attention transfer. arXiv preprint arXiv:1612.03928 (2016)

19. Zhang, H., Cisse, M., Dauphin, Y.N., Lopez-Paz, D.: mixup: beyond empirical risk minimization. arXiv preprint arXiv:1710.09412 (2017)

20. Zou, D., Cao, Y., Zhou, D., Gu, Q.: Stochastic gradient descent optimizes overparameterized deep ReLUnetworks. arXiv preprint arXiv:1811.08888 (2018)

21. Zou, D., Gu, Q.: An improved analysis of training over-parameterized deep neural networks. In: Advances in Neural Information Processing Systems, pp. 2053–2062 (2019)

Injective Domain Knowledge in Neural Networks for Transprecision Computing

Andrea Borghesi$^{(\boxtimes)}$, Federico Baldo, Michele Lombardi, and Michela Milano

DISI, University of Bologna, Bologna, Italy
{andrea.borghesi3,federico.baldo2,
michele.lombardi2,michela.milano}@unibo.it

Abstract. Machine Learning (ML) models are very effective in many learning tasks, due to the capability to extract meaningful information from large data sets. Nevertheless, there are learning problems that cannot be easily solved relying on pure data, e.g. scarce data or very complex functions to be approximated. Fortunately, in many contexts domain knowledge is explicitly available and can be used to train better ML models.

This paper studies the improvements that can be obtained by integrating prior knowledge when dealing with a context-specific, non-trivial learning task, namely precision tuning of transprecision computing applications. The domain information is injected in the ML models in different ways: I) additional features, II) ad-hoc graph-based network topology, III) regularization schemes. The results clearly show that ML models exploiting problem-specific information outperform the data-driven ones, with an average improvement around 38%.

Keywords: Machine learning · Domain knowledge · Transprecision computing

1 Introduction

In recent years, ML approaches have been exhaustively proved to be successful with a wide range of learning tasks. Typically, ML models are sub-symbolic, black-box techniques capable of effectively exploiting the information contained in large amounts of data. Part of their usefulness is their adaptability, that is the fact that ML models with the same architecture and training algorithm can be applied in very different contexts with good results. This happens because most ML approaches make very few assumptions on the underlying data and the functions that they are trying to learn.

However, data-driven models can be not ideal if, for instance, the data is relatively expensive to obtain and the function to be learned is very hard. At the same time, in many areas domain-specific information is available (e.g. structured data, knowledge about the data generation process, domain experts experience, etc.) but not exploited. In such cases, it makes sense to take advantage of this

© Springer Nature Switzerland AG 2020
G. Nicosia et al. (Eds.): LOD 2020, LNCS 12565, pp. 587–600, 2020.
https://doi.org/10.1007/978-3-030-64583-0_52

information to improve the performance of the ML techniques, so they do not have to start from scratch while dealing with difficult learning tasks. In other words, *why learn again something that you already know?*.

In this paper we discuss a strategy to inject domain knowledge expressed as constraints in an ML model, namely a Neural Network (NN). We limit the experimental evaluation to a specific area, namely *transprecision computing*, a novel paradigm that allows trade-offs between the energy required to perform the computation and the accuracy of its outcome [8]. We consider different sources of prior information and adopt suitable injection approaches for each of them: I) feature extraction, II) ad-hoc NN structure, and III) data augmentation combined with a regularization strategy. The supervised learning task is very hard, due to non-linearity, non-monotonicity and relatively small data sets (a few thousands of samples). The experimental results clearly show that exploiting prior information leads to remarkable gains. On average over all benchmarks, the knowledge injection provides a 38% improvement in terms of prediction error (error decrease). The rest of the paper is structured as follows: after the discussion about related works (Sect. 2), Sect. 3 introduces the injection approaches; Sect. 4 details transprecision computing and the specific learning task, highlighting its difficulty and the domain knowledge that can be extracted; Sect. 5 summarizes the experimental results; finally, Sect. 6 concludes the paper.

2   Related Works

The combination of sub-symbolic models with domain knowledge is an area explored by previous research in many fields [12]. For instance, feature engineering [6] is a common method for improving the accuracy of data-driven ML models by selecting useful features and/or transform the original ones to facilitate the learner's task. In general, this is not a trivial problem and requires much effort, both from system expert and ML practitioners. In this paper, we use a different approach, as we employ domain knowledge to create novel features that render *explicit* the information hidden in the raw data.

Another research direction aims at training NNs while forcing constraints which can be drawn from knowledge domain. [9] presents a method for translating logical constraints in loss functions that guide the training towards the desired output. [11] proposes a different approach to incorporate domain knowledge in an NN by adopting a loss function that merges mean squared error and a penalty measuring whether the NN output respect a set of constraints derived from the domain; the method is limited to constraints enforcing monotonicity and bounds on the target variable. [14] introduces a method to integrate semantic knowledge in deep NNs, again exploiting a loss function; in this case the approach is targeted at semi-supervised learning and not well suited for supervised tasks. Acting on the loss function with a regularization term has been proposed also by [3], with their work on Semantic-Based Regularization (SBR), a method to merge high-level domain information expressed as first-order logic in ML models. We have exploited their technique in combination to a data augmentation strategy to enhance an ML model.

Graph Convolutional Neural Networks (GCNN) [2,7] are a type of neural networks specialized for learning tasks involving graphs. GCNNs have been recently used in several fields [15], owning to their capability to deal with data whose structure can be described via graphs, thanks to a generalization in the spectral domain of the convolutional layers found in many deep learning networks. GCNNs most common applications involve semi-supervised classification tasks, with the goal of predicting the class of unlabeled nodes in a graph – a case of graph learning.

3   Domain Knowledge Injection

The main goal of this paper is the exploration of how an ML model can be improved through the exploitation of domain knowledge. We claim that data-driven ML models can benefit from the injection of prior knowledge provided by domain experts; Sect. 5 will report the results of the experimental evaluation, conducted on a specific learning task where domain knowledge is available (details in Sect. 4). We consider domain knowledge that can be expressed in the form of logical constraints between variables (input and output features of the ML models) and/or encoded in a graph. Let X be the training set and y the targets, either continuous values (regression) or categorical labels (classification), f the model trained to learn the relation between X and y. A DL model is a family of parametric functions $f(x|\omega)$, i.e. $\{f(x|\omega)|\omega \in \Omega\}$ where the set Ω is called *hypothesis space*. *Training* exploits data to find the parameter values in Ω that minimize a so called *loss function*. In the most general case this is a function $\mathcal{L}(y|\theta)$, where y corresponds to the output of DNN and θ is the available empirical knowledge (e.g. the labels).

In general, domain knowledge can be expressed as a set of logical constraints between the input features X and the target y. For instance, the monotonicity property holds if $x_1 \leq x_2 \implies y_1 \leq y_2$ for every pair in X. We propose a multi-faceted domain knowledge injection strategy and we introduce three different approaches, each one addressing a specific weakness encountered by data-driven techniques (Fig. 1a portrays the three injection mechanisms):

1. *feature space* manipulation for information implicit but hidden in the raw data – if the examples available in the data set are not sufficient nor informative enough to train accurate ML models, a set of additional features can be created using the domain knowledge and reasoning about the relationships among the original features;
2. *ad-hoc network topology* for learning tasks where the relationships among the features and the data structure can be encoded with graphs (hypothesis space exploration);
3. *data augmentation* and *regularization function* for a twofold scope: I) learning with very few data (e.g. active learning), by generating artificial examples, II) enforcing desired properties in the output of the ML model by adding a regularization term to the loss function $\mathcal{L}$.

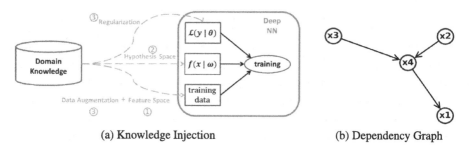

(a) Knowledge Injection   (b) Dependency Graph

Fig. 1. (a) Domain knowledge injection in DNNs - (b) Example of dependency graph

The feature extraction (1) takes into account prior knowledge that can be expressed via a set of binary constraints C among the input features X; these constraints can be used to obtain an extended training set X' by checking if every example in X satisfies them or not. The regularization method (3) assumes that the knowledge can be expressed as first-order logic constraints between input features X and the target y; data augmentation helps to cope with scarce data and amplify the effect of the regularization.

We introduce the knowledge injection strategy and present three different techniques, each tailored for a specific source of information. At the current stage we were more interested in measuring the specific contribution of each method, thus they were tested separately, but we plan to explore hybrid solutions in future works. As a case study we consider a complex supervised learning task and then we tackle it with multiple data-driven ML models, and in particular we use neural networks (NN). Subsequently, we inject the domain knowledge and then we experimentally evaluate the obtained improvements.

4   Transprecision Computing

There exist many techniques for transprecision computing and in this paper we focus on an approach targeting floating-point (FP) variables and operations, as their execution and data transfer can require a large share of the energy consumption for many applications; decreasing the number of bits used to represent FP variables can lead to energy savings, with the side-effect of reduced accuracy on the outcome of the application (also referred to as *benchmark*). Deciding the optimal number of bits for FP variables while respecting a bound on the computation accuracy is referred to as *precision tuning*. In this context, understanding the relationship between assigned precision and accuracy is a critical issue, and not an easy one, as this relationship cannot be analytically expressed for non-trivial benchmarks [10]. Therefore, we address this problem via a ML model, that is *learning* the relationship between precision and accuracy. For this scope, we use a transprecision library for precision tuning called *FlexFloat* [1] to create a suitable data set; this means running a benchmark with multiple precision configurations and store the associated error. As this is a highly time-consuming task, we work with data sets of relatively limited size (5000 samples at maximum)[1], an issue that complicates the learning task.

4.1 Problem Description

We consider numerical benchmarks where multiple FP variables partake in the computation of the result for a given input set, which includes a structured set of FP values (typically a vector or a matrix). The number of variables with controllable precision in a benchmark B is n_{var}^{B}; these variables are the union of the original variables of the program and the additional variables inserted by FlexFloat to handle intermediate results (see [13] for details). FlexFloat allows to run a benchmark with different precision (different numbers of bits assigned to the FP variables) and to measure the reduction in output quality due to the adjusted precision (reduction w.r.t. the output obtained with maximum precision) – we will refer to this reduction as *error*. If O indicates the result computed with the tuned precision and O^M the one obtained with maximum precision, the error E is given by $E = \max_i \frac{(o_i - o_i^M)^2}{(o_i^M)^2}$ – this is the metric adopted in the transprecision community [5]. As a case study we selected a representative subset of the benchmarks studied in the context of transprecision computing. At this stage, we do not focus on whole applications but rather on micro-benchmarks, in particular the following ones: 1) *FWT*, Fast Walsh Transform for real vectors, from the domain of advanced linear algebra ($n_{var}^{FWT} = 2$); 2) *saxpy*, a generalized vector addition, basic linear algebra ($n_{var}^{saxpy} = 3$); 3) *convolution*, convolution of a matrix, ML ($n_{var}^{conv} = 4$); 4) *dwt*, Discrete wavelet transform, from signal processing ($n_{var}^{dwt} = 7$); 5) *correlation*, compute correlation matrix of input, data mining ($n_{var}^{corr} = 7$). 6) *BlackScholes*, estimates the price for a set of options applying Black-Scholes equation, from computational finance ($n_{var}^{BScholes} = 15$).

Beside the precision configuration, another element that impacts a benchmark's output, and thus the error, is the input set fed to the application (e.g. the actual values of the FP variables). The vast majority of transprecision tuning approaches consider the single input set case [5]: a fixed input set is given to the benchmark and the precision of the variables is tuned for that particular input set (no guarantee that the configuration found will suit different input sets). We opted for "stochastic" approach: we consider multiple input sets, so that a distribution of errors is associated to each configuration, rather than a single value. The learning task is then not to predict the error associated to a specific input set but to learn the relation between precision configuration and *mean error* over all input sets. Learning the relationship between variable precision and error is a hard problem. First, the error metric is very susceptible to

---

[1] The learning task is only a part of a larger project aiming at solving the precision tuning problem with optimization techniques; state-of-the-art algorithms for FP precision tuning (e.g. [5]) dictate a bound on the time to solve the optimization problem – hence, the need of a low data set creation time.

differences between output at maximum precision and output at reduced precision, due to the maximization component. Secondly, the precision-error space is non-smooth, non-linear, non-monotonic, and with many peaks (local optima). In practice, increasing the precision of all variables does not guarantee an error reduction.

4.2   Data Set Creation

As a first step, we created a collection of data sets containing examples of the benchmarks run at different precision, with the corresponding error values. We call *configuration* the assignment of a precision to each FP variable. The configuration space was explored via Latin Hypercube Sampling (LHS). As described in the previous section, for each configuration the benchmarks were run with 30 different input sets[2] and the error associated to each combination of <configuration, input set> was computed. As target we then use the average over the 30 input-specific errors. The majority of configurations lead to small errors, from 10^{-1} to 10^{-30}. However, in a minority of cases lowering the precision of critical variables generates extremely large errors; in the transprecision computing context, error larger than 1 are deemed excessive.

After a preliminary analysis, we realized that for an ML model it is very hard to discern between small and relatively close errors (i.e. $e-20$ and $e-15$); we therefore opted to predict the negative of the logarithm of the error, thus magnifying the relative differences. Moreover, a careful examination revealed that overly large error values were usually due to numerical issues arising during computation (e.g. overflow, underflow, division by zero, or not-a-number exceptions). This intuitively means that the large-error configurations are likely to follow a distinct pattern w.r.t. the configurations having a more reasonable error value. We are much more interested in relatively small error (e.g. $E \leq 0.95$, not in logarithmic scale) as in transprecision computing the largest accepted error is typically 0.1 (meaning an output accuracy higher than 90%). Hence, we decide to level out all the errors in the data set above the 0.95 threshold; if the <configuration, input set> combination produced an error $E \geq 0.95$, after pre-processing its error is set to 0.95 (before the conversion to logarithmic scale).

4.3   Knowledge Injection

As the benchmarks are programs composed by a set of interdependent FP variables, the variables' interactions represent a source of valuable information for learning the relationship between precision and error. This domain-level knowledge is encoded in the *dependency graph* of the benchmark, which specifies how the program variables are related. For instance, consider the expression $V_1 = V_2 + V_3$; this corresponds to four precisions that need to be decided $x_i, i \in [1, 4]$. The first three precision-variables x_1, x_2, and x_3 represent the precision of the actual variables of the expression, respectively V_1, V_2, and V_3;

---

[2] Long vectors and matrices containing different real values.

the last variable x_4 is a *temporary* variable introduced by FlexFloat to handle the (possibly) mismatching precision of the operands V_2 and V_3 (FlexFloat performs a cast from x_2 and x_3 to the intermediate precision x_4). Each variable is a node in the dependency graph, and the relations among variables are directed edges, as depicted in Fig. 1b; an edge entering a node means that the precision of the source-variable is linked to the precision of the destination-variable.

Additional Features Extraction. As we have seen, the prior information on the benchmarks is encoded in directed graphs; for explanatory purposes, we will take as example the micro-benchmark represented by the graph in Fig. 1b. Using the encoded knowledge, a set of additional features characterizing the precision configurations can be obtained. We consider only one type of relation, that is assignments (e.g. $x_4 \rightarrow x_1$). In this kind of expression, granting a larger number of bits to the value to be assigned x_4 would be pointless since the final precision of the expression is ultimately governed by the precision of the result variable x_1. Configurations that respect this relationship have a higher probability to lead to smaller errors w.r.t. configurations that do not respect this constraint. In practice, configurations where $x_4 \leq x_1$ are associated to smaller errors.

This information can be added to the training set as a collection of additional features. For each couple of variables involved in an assignment operation $x_i \rightarrow x_j$ we compute the feature $F_{ji} = x_j - x_i$[3], which is then added to the data set. Each feature corresponds to one of the logic binary constraints used to express the domain knowledge. For instance, if we consider again the example of Fig. 1b there are three additional features, one for each assignment expression (highlighted by the three arrows in the graph): $F_{43} = x_4 - x_3$, $F_{42} = x_4 - x_2$, $F_{14} = x_1 - x_4$. Thanks to these additional features an extended data set can be obtained. If we consider two possible configurations for the micro-benchmark in Fig. 1b, $C_1 = [27, 45, 35, 40]$ and $C_2 = [42, 23, 4, 10]$, the original data set would be composed by four features (one for each FP variable) plus the associate error (the target of the regression task). Instead, the extended data set contains seven features plus the error: $C_1^{ext} = [27, 45, 35, 40, 13, -5, -5]$ and $C_2^{ext} = [42, 23, 4, 10, -32, -13, -6]$.

Graphical Convolutional Neural Networks. The transprecision learning task is a supervised regression problem whose prior information can be expressed through a graph, that is the dependency graph that links the variable in the benchmark. As mentioned in Sect. 3, GCNNs are well suited to deal with graph-structured problems. As our problem is different from those considered in the literature, we did not adopt the standard approach but we exploited the main component of GCNNs, the *graph convolution*, implemented via Graph Convolutional Layers (GCL), and applied to the transprecision task. The GCNN has the following structure: first, from the dependency graph we compute the adjacency

---

[3] If $F_{ji} \leq 0$ it means that the $x_i \leq x_j$ is not respected, hence a higher error is associated to the configuration.

matrix; then the adjacency matrix and the input feature matrix are combined to form the input of a first GCL, which is then fed to a second one. Its output becomes the input for a fully connected dense layer with 128 neurons, followed by two other fully connected layers of decreasing dimension (respectively, 32 and 8). The final layer is, again, a dense layer with a single neuron, that is the network output.

Data Augmentation and Regularization. As mentioned in Sect. 4, the learning task is made more difficult by the presence of non-mononicity: situations where the normal precision-error relationship is not respected. They arise due to numerical instability, and their presence is magnified by the use of small data sets and a limited number of different input sets; with sufficiently large data sets they would be discarded as outliers. As mentioned before, the learning task addressed in this paper is a step towards an optimization model for precision tuning; with this scope in mind, it would be preferable to have an ML model that does not reproduce non-monotonicity events in its predictions. This is a domain knowledge about an undesirable property that should be corrected. The problem with non-monotonicity would be solved if we could have more training examples, but this is not easily attainable as we should run a benchmark to compute the error associated to a configuration. However, generating new configurations without computing the error is trivial; we can exploit this advantage in conjunction with an appropriate regularization scheme in order to impose monotonicity on the ML model predictions. This process is a form of *data augmentation*. Injecting the monotonicity constraint in the training process may allow to mitigate the noise and improve generalization, even with smaller training sets. We take into account such constraints at training time by exploiting ideas from SBR [3], that advocates to the use of (differentiable) constraints as regularizers in the loss function. Let us write $x_i \prec x_j$ if configuration x_j dominates x_i, i.e. if every variable in x_j has precision at least as high as x_i; let P be the set of dominated-dominating pairs in our training set X, $P = (i,j)|x_i \prec x_j$. Then, we can formulate the following regularized loss function: $MSE(X,y) + \lambda \sum_{i,j \in P} max(0, f(x_j) - f(x_i))$, where f is the error predictor being trained, and MSE is the mean squared error. Each regularization term is associated to a pair in P and has non-zero value iff the error for the dominating pair is larger than for the dominated pair, i.e. if the monotonicity constraint is violated. New configurations in P can be generated in order to get a much more stable regularization factor without the need of a bigger train set. SBR is orthogonal to the use of additional features hence the two methods can be combined; we plan to explore the benefits of merging multiple methods in future work.

5   Experimental Evaluation

We selected 5 different data-driven models to obtain a baseline: I) a black-box optimization method (*AutoSklearn*); II) an NN composed of 4 dense layers with $10 \times n_{var}^B$ neurons each, that is, the number of variables in a benchmark multiplied by 10 (*NN-1*); III) an NN composed of 4 dense layers with $100 \times n_{var}^B$ neurons each (*NN-2*); IV) an NN composed of 10 dense layers with $10 \times n_{var}^B$ neurons each (*NN-3*); V) a NN composed of 20 dense layers with $10 \times n_{var}^B$ neurons each (*NN-4*). All NNs have a single-neuron output layer fully connected with the previous one. The black-box method used was drawn from the *AutoML* area, namely a framework called *autosklearn* [4] which uses Bayesian optimization for algorithm configuration and selection. Our problem can be cast in the AutoML mold if we treat the variables' precision as the algorithm configurations to be explored and the associated computation error as the target.

The code used to run the experiments was written in Python, using Keras and TensorFlow for the implementation of the neural networks. *Autosklearn* is distributed as a Python library and we used the version available online[4], with default parameters. The GCNN model was created using the *Spektral* library[5]. All the results presented in this section were run on 20 different instances and we report the average values. Both input feature and targets were normalized. The code used to run the experiments is available in an online repository[6]. To evaluate the impact of the additional features, the four different neural networks previously defined (NN-1, NN-2, NN-3, NN-4) were trained and tested both with and without the extended data set. At this stage we focus on the number of layers and their width and discarded other hyperparameters; their exploration will be the subject of future research works. In this paper, these are the values for the main hyperparameter used with all methods: number of epochs = 1000; batch size = 32; as training algorithm we opted for *Adam* with standard parameters; Mean Squared Error as the loss function. The data augmentation and SBR approach is used on top of a neural network with the same number of layers and neurons as NN-1. The new configurations are injected in each batch during the training, with a fixed size of 256 elements; the amount of data generated is specified by a ratio, which represents the percentage of samples introduced by the data augmentation (Table 1).

5.1 Models Accuracy

We begin by evaluating the prediction accuracy of the proposed approaches. We measure the accuracy using the MAE. In Table 2 we compare the results obtained using a training and test set size of, respectively, 5000 and 1000 examples; test and training set are randomly drawn from the samples generated through LHS. The reported accuracy measures refer to the test set. The first column of the table identifies the benchmark (the last row corresponds to the average over all of them); the second column contains the MAE obtained with the black-box approach, *AutoSklearn*; columns 3 and 4 report the MAE with the first NN (NN-1), respectively without and with the additional features; the three following couples of columns are the results with the other NNs (NN-2, NN-2,

---

[4] https://automl.github.io/auto-sklearn/master/.
[5] https://danielegrattarola.github.io/spektral/.
[6] https://github.com/AndreaBorghesi/knowInject_transComputing.

NN-3), again split between the base and the extended data set; the final two columns correspond respectively to MAE obtained with GCNN and with SBR. For this table, we consider the SBR approach with 75% of augmented examples.

Table 1. Knowledge injection approaches comparison: MAE – train set size: 5k

Benchmark	AutoSklearn	NN-1		NN-2		NN-3		NN-4		GCNN	SBR
		Base	Ext.	Base	Ext.	Base	Ext.	Base	Ext.		
FWT	0.394	0.315	0.251	0.056	0.054	0.104	0.061	0.070	0.105	0.351	0.243
saxpy	0.003	0.000	0.000	0.000	0.000	0.000	0.000	0.000	0.000	0.000	0.001
convolution	0.020	0.005	0.005	0.002	0.002	0.003	0.003	0.003	0.003	0.004	0.006
correlation	0.397	0.139	0.120	0.091	0.092	0.111	0.098	0.114	0.102	0.262	0.139
dwt	0.422	0.057	0.034	0.011	0.012	0.029	0.020	0.031	0.022	0.072	0.068
BlackScholes	0.411	0.238	0.047	0.184	0.035	0.239	0.038	0.297	0.172	0.307	0.220
Average	0.274	0.126	0.076	0.057	0.033	0.081	0.037	0.086	0.067	0.166	0.113

Table 2. Knowledge injection approaches comparison: Root Mean Squared Error – set size 5k

Benchmark	AutoSklearn	NN-1		NN-2		NN-3		NN-4		GCNN	SBR
		Base	Ext.	Base	Ext.	Base	Ext.	Base	Ext.		
FWT	1.185	0.628	0.347	0.069	0.161	0.119	0.146	0.217	0.251	0.401	285
saxpy	0.004	0.001	0.001	0.000	0.001	0.000	0.001	0.000	0.001	0.004	0.003
convolution	0.049	0.006	0.011	0.006	0.004	0.008	0.006	0.005	0.006	0.024	0.012
correlation	0.406	0.400	0.356	0.141	0.288	0.269	0.206	0.284	0.145	0.332	0.152
dwt	0.559	0.062	0.040	0.012	0.038	0.029	0.042	0.076	0.054	0.134	0.094
BlackScholes	0.656	0.394	0.096	0.508	0.046	0.488	0.071	0.593	0.187	0.552	0.308
Average	0.477	0.248	0.142	0.123	0.090	0.152	0.079	0.196	0.107	0.188	0.148

The black-box model *AutoSklearn* has clearly the worst performance, which is not entirely surprising given the complexity of the learning task. The first unexpected and disappointing result is the poor performance of the GCNN, that is outperformed by all other approaches in almost all benchmarks. We remark that this was a novel application of GCNN and this preliminary analysis merely suggests that a more careful exploration is needed. Changing the network type can produce good results: using a wider NN (from NN-1 to NN-2) greatly reduces the MAE, while deeper NNs provide smaller improvements (e.g. NN-3 and NN-4). Very interestingly, a major MAE reduction is obtained by using the additional features (column *Ext.*): for all NN types and over all benchmarks, the approach using the extended data greatly outperforms the baseline, with an average improvement of 39.7% (considering all four NN types). The results obtained with data augmentation and SBR show that this method performs better than *AutoSklearn* and the simplest NN without the additional features (NN-1), but it has a higher MAE compared to all the approaches with the extended data set. This is not an issue as SBR benefits were not expected in terms of prediction accuracy but rather on the enforcing of the monotonicity (see Sect. 5.2). We are also interested in measuring the results with smaller training sets, again using MAE as metric; we keep the test set size fixed at 1000 elements. Table 3 reports the experimental results; it has the same structure of Table 2. As expected, the prediction accuracy decreases with the training set size, but the benefits brought by the domain knowledge remain – over all training set size, the improvement brought by the engineered features is 38.7%.

Table 3. Knowledge injection approaches comparison: average on all benchmarks MAE (rows 1 = −4) and RMSE (rows 5–8) – varying train set size

Train set size	AutoSklearn	NN-1		NN-2		NN-3		NN-4		GCNN	SBR
		Base	Ext.	Base	Ext.	Base	Ext.	Base	Ext.		
500 (MAE)	0.288	0.196	0.131	0.100	0.064	0.140	0.078	0.144	0.134	0.316	0.190
1000 (MAE)	0.285	0.178	0.107	0.087	0.048	0.108	0.056	0.142	0.117	0.256	0.181
2000 (MAE)	0.278	0.155	0.085	0.077	0.041	0.094	0.047	0.119	0.060	0.210	0.162
5000 (MAE)	0.274	0.126	0.076	0.057	0.033	0.081	0.037	0.086	0.067	0.166	0.133
500 (RMSE)	0.797	0.550	0.245	0.236	0.153	0.317	0.148	0.249	0.304	0.316	0.23
1000 (RMSE)	0.769	0.435	0.139	0.152	0.121	0.206	0.095	0.375	0.235	0.25	0.21
2000 (RMSE)	0.530	0.284	0.155	0.168	0.072	0.105	0.096	0.260	0.156	0.210	0.184
5000 (RMSE)	0.477	0.248	0.142	0.123	0.090	0.152	0.079	0.196	0.107	0.166	0.148

5.2 Semantic Based Regularization Impact

This section provides additional details on the experiments on data augmentation and SBR. The model was tested on the previous benchmarks and different ratios of data injected, i.e. 25% and 75%. To have a more precise evaluation of the approach, we relied on another metric beside MAE, that is the number of violated monotonicity constraints – the goal of this approach is to *reduce* their number. We underline that not every benchmark had monotonicity issues (as they are outliers), and in these cases the regularization factor is of no use and might keep the model from a good approximation. For this reason, Table 4 and Table 4, report just the values from significant benchmarks (i.e. benchmarks that exhibit the most marked non-monotonic behavior), these are *convolution* and *correlation*. The third column reports the result obtained with NN-1 without the additional features. Columns 4–6 correspond to the results obtained with data augmentation and SBR, with different percentages of injected data (0%, 25%, 75%).

Table 4. SBR: MAE and number of violated constraints (#Viol.)

Benchmark	Size	NN-1		SBR		SBR 25%		SBR 75%	
		MAE	#Viol.	MAE	#Viol.	MAE	#Viol.	MAE	#Viol.
convolution	500	0.012	168	0.019	156	0.013	171	0.011	126
	5000	0.005	0	0.006	13	0.006	12	0.006	6
correlation	500	0.263	111	0.265	120	0.263	116	0.262	98
	5000	0.139	91	0.059	59	0.059	71	0.139	92

With larger training sets, the benefits of data augmentation and SBR are marginal: the additional constraint on the loss function is not very useful, given the abundance of training samples allowing for better generalization. Similarly, larger training sets lead to a natural decrease in the number of monotonicity constraints violated (as their proportion in the training set diminishes). Nevertheless, the more interesting results can be observed when fewer data points are available, since the models show a decrease in the number of violated constraints opposed to the network without regularization. Furthermore, the networks performed better with higher ratios of data injected, i.e. 18%, on average. Finally, the MAE seems to have values compatible to the results obtained with *NN-1*, a good result since prediction accuracy was not SBR's scope. These results encourage the idea of a hybrid model merging data augmentation plus SBR and additional features (both approaches enabled by the injection of domain knowledge), as future development of this work.

6   Conclusion and Future Works

In this paper we present a strategy for injecting domain knowledge in an ML model. As a case of study, we considered a learning task from the transprecision computing field, namely predicting the computation error associated to the precision used for handling a set of FP variables composing a benchmark. This is a difficult regression problem, hard to be addressed with pure data-driven ML methods; we have shown how critical improvements can be reached by injecting domain knowledge in the ML models.

We introduced three knowledge-injection approaches and applied them on top of NNs with varying structures: feature engineering, a GCNN, and a data augmentation scheme enabled by SBR. The GCNN approach did not improve the accuracy of the ML model w.r.t. the baseline and it should be explored more in detail. Conversely, the creation of extended data set was revealed to be extremely useful, leading to remarkable reduction in prediction error (39.7% on average and up to 47.5% in the best case). Data augmentation plus SBR showed its potential with training sets of limited size, in terms of reduced number of violated monotonicity constraints while preserving the ML models' prediction accuracy. In future works we plan to integrate the learners in an optimization model for solving the FP tuning precision problem. In this regard we will explore active learning strategies and we expect SBR to have good result, especially when combined with the additional features (the methods are orthogonal). Moreover, we will perform experiments with other domain knowledge injection approaches, for instance by building data sets in accordance with the prior information and by exploiting the knowledge to guide the training of the NN by constraining its output.

Acknowledgments. This work has been partially supported by European H2020 FET project OPRECOMP (g.a. 732631).

References

1. Borghesi, A., Tagliavini, G., et al.: Combining learning and optimization for transprecision computing. In: Proceedings of the 17th ACM Conference on Computing Frontiers (2020)
2. Defferrard, M., Bresson, X., Vandergheynst, P.: Convolutional neural networks on graphs with fast localized spectral filtering. In: NIPS, pp. 3844–3852 (2016)
3. Diligenti, M., Gori, M., Sacca, C.: Semantic-based regularization for learning and inference. Artif. Intell. **244**, 143–165 (2017)
4. Feurer, M., Klein, A., Eggensperger, K., et al.: Efficient and robust automated machine learning. In: Advances in Neural Information Processing Systems, pp. 2962–2970 (2015)
5. Ho, N.M., Manogaran, E., et al.: Efficient floating point precision tuning for approximate computing. In: Design Automation Conference (ASP-DAC), pp. 63–68. IEEE (2017)
6. Khurana, U., Samulowitz, H., Turaga, D.: Feature engineering for predictive modeling using reinforcement learning. In: 32nd AAAI Conference on Artificial Intelligence (2018)
7. Kipf, T.N., Welling, M.: Semi-supervised classification with graph convolutional networks. In: Proceedings of the 5th ICLR, ICLR 2017 (2017)
8. Malossi, A.C.I., Schaffner, M., et al.: The transprecision computing paradigm: concept, design, and applications. In: Design, Automation & Test in Europe Conference & Exhibition (DATE), pp. 1105–1110. IEEE (2018)
9. Fischer, M., Balunovic, M., Drachsler-Cohen, D., Gehr, T., Zhang, C., Vechev, M.: Dl2: training and querying neural networks with logic. In: International Conference on Machine Learning (2019)
10. Moscato, M., Titolo, L., Dutle, A., Muñoz, C.A.: Automatic estimation of verified floating-point round-off errors via static analysis. In: Tonetta, S., Schoitsch, E., Bitsch, F. (eds.) SAFECOMP 2017. LNCS, vol. 10488, pp. 213–229. Springer, Cham (2017). https://doi.org/10.1007/978-3-319-66266-4_14
11. Muralidhar, N., Islam, M.R., et al.: Incorporating prior domain knowledge into deep neural networks. In: 2018 IEEE International Conference on Big Data, pp. 36–45. IEEE (2018)
12. von Rueden, L., Mayer, S., et al.: Informed machine learning - a taxonomy and survey of integrating knowledge into learning systems. arXiv preprint arXiv:1903.12394v2 (2020)
13. Tagliavini, G., Mach, S., et al.: A transprecision floating-point platform for ultra-low power computing. In: Design, Automation & Test in Europe Conference & Exhibition (DATE), pp. 1051–1056. IEEE (2018)

14. Xu, J., Zhang, Z., Friedman, T., Liang, Y., Broeck, G.V.: A semantic loss function for deep learning with symbolic knowledge. arXiv preprint arXiv:1711.11157 (2017)
15. Zhang, S., Tong, H., Xu, J., Maciejewski, R.: Graph convolutional networks: a comprehensive review. Comput. Soc. Netw. **6**(1), 11 (2019). https://doi.org/10.1186/s40649-019-0069-y

XM_HeatForecast: Heating Load Forecasting in Smart District Heating Networks

Federico Bianchi$^{(\boxtimes)}$, Francesco Masillo, Alberto Castellini,
and Alessandro Farinelli

Department of Computer Science, Verona University, Strada Le Grazie 15,
37134 Verona, Italy
{federico.bianchi,francesco.masillo,
alberto.castellini,alessandro.farinelli}@univr.it

Abstract. Forecasting is an important task for intelligent agents involved in dynamical processes. A specific application domain concerns district heating networks, in which the future heating load generated by centralized power plants and distributed to buildings must be optimized for better plant maintenance, energy consumption and environmental impact. In this paper we present XM_HeatForecast a Python tool designed to support district heating network operators. The tool provides an integrated architecture for *i)* generating and updating in real-time predictive models of heating load, *ii)* supporting the analysis of prediction performance and errors, *iii)* inspecting model parameters and analyzing the historical dataset from which models are trained. A case study is presented in which the software is used on a synthetic dataset of heat loads and weather forecast from which a regression model is generated and updated every 24 h, while predictions of load in the next 48 h are performed every hour.

Software available at: https://github.com/XModeling

Video available at: https://youtu.be/JtInizI4e_s.

Keywords: Forecasting · Smart grids · Predictive modeling · Interpretability

1 Introduction

Forecasting future behaviours of complex dynamical systems is a key functionality of intelligent agents involved in dynamical processes [18]. It entails the ability to learn patterns and variable relationships from past data (usually in the form of a multivariate time series), to generate a model of these patterns, and to use this model to infer future values of some variables of interest [3–5]. In the context of the recently proposed paradigm of smart grids, forecasting has gained

---

F. Bianchi and F. Masillo—Contributed equally to this work.

© Springer Nature Switzerland AG 2020
G. Nicosia et al. (Eds.): LOD 2020, LNCS 12565, pp. 601–612, 2020.
https://doi.org/10.1007/978-3-030-64583-0_53

a key role [17], since it allows to predict future loads of the network in order to improve network maintenance and efficiency [15]. A specific type of smart grids, which have proven important in recent years for environmental sustainability are District Heating Networks (DHNs) [9], in which the heat is generated in one or more centralized power plants and distributed through an insulated network of pipe system to commercial and residential buildings [10]. Short-Term Load Forecasting (STLF) [8,12–14], namely, the prediction of the system load over time intervals ranging from one hour to one week, is particularly important in DHN management because it supports operators in various decision-making tasks, including supply planning, generation reserve, system security and financial planning [8]. Different kinds of methods are proposed in the literature for time-series forecasting applied to DHNs and related domains. In [1], for instance, authors investigate the application of support vector regression, regression trees, feed forwards and multiple linear regression models. Considering models based on deep learning, Recurrent Neural Networks (RNN), Long Short-Term Memory (LSTM) [11] and convolutional LSTM [19] are among the most popular techniques for time series forecasting. Computational tools for STLF in DHNs consist of statistical and machine learning methods able to *i)* learn predictive models from past data, *ii)* update predictive models online while new data are available, *iii)* make load predictions, *iv)* analyze prediction performance, *v)* analyze model parameters. Our aim is to develop of an open source prototype for research in heating load forecasting which can easily be released in production. To the best of our knowledge the literature does not provide similar tools, while there exist commercial software.

In this paper we present a novel open source Python software for STLF in smart district heating networks. Its name is *XM_HeatForecast* and it extends the open source suite *eXplainable Modeling* (XM) [6,7]. *XM_HeatForecast* is designed to support DHN operators in all phases of load forecasting and it can be interfaced with other tools for planning and scheduling. The tool provides *i)* a module for predictive model generation based on time series, *ii)* a graphical user interface for integrated and interactive analysis of predictions, models performance and data. The current (first) version of the software uses linear autoregressive models for prediction, where variables are generated from prior knowledge on the application domain, but the tool is independent of the prediction method and the data sources. The main contribution of this work to the state-of-the-art is the presentation of the software *XM_HeatForecast* and all its components. Moreover, we provide a case study in which the tool is tested on a synthetic but realistic dataset showing the differences in prediction performance depending on the length of the historical data. Model coefficients are also investigated to show how the software supports model interpretability through integration of different kinds of information. Other original elements are the online approach by which the model is updated and the open communication interface for data exchange with other tools, such as planners.

The rest of the manuscript is organized as follows. Section 2 presents an overview of *XM_HeatForecast*. In Sect. 3 a detailed description of modules for model and forecast generation is provided. In Sect. 4 modules for visualization and analysis are presented. In Sect. 5 communication interface is described. Section 6 contains a complete case study. Conclusions and future directions are discussed in Sect. 7.

2 XM_HeatForecast: System Overview

In this section we provide a high-level description of the *XM_HeatForecast* architecture, and an overview of its main modules and testing dataset.

2.1 Main Modules and Software Structure

XM_HeatForecast has four main components, namely, *Data processor*, *Model generator*, *Forecaster* and *Graphical user interface* (GUI). In Fig. 1 each part is identified by a color to highlight different functionalities (i.e., green, gray, light blue and red). In the current version the tool is designed to re-train models every 24 h and predict future heating loads every hour. When the tool is started[1] a forecasting horizon between 24 or 48 h and a starting instant (date/time) for the predictions are asked to the user. The values of the first parameter have been suggested by operators of a real DHN plant because of their usefulness in plant operations (e.g., production and maintenance planning). The second parameter allows to perform tests on previously acquired datasets. In this case also a refresh time can be provided by the user to speed-up the simulation. Namely, instead of waiting one hour for each update, simulations can be performed with a time interval of few seconds, as shown in the attached video.

Input/Output: The input and output of the tool consists of time series data read from folders *Weather forecast, Data* and *Model*, and predictions stored in folder *Forecast* (see the folder icons in Fig. 1). The folders contain comma separated (*csv*) files automatically read and written in real-time. **Data processor:** this module reads batches of data from folders *Weather forecast* and *Data* (green arrow 1.a in Fig. 1), processes and re-stores them into folder *Data* (green arrow 1.b in Fig. 1). Processed data are then loaded by *Model generator* and *Forecaster* for training and forecasting. **Model generator:** it reads past load data from folder *Data* (purple arrow 2.a), trains the model and saves it into folder *Model* (purple arrow 2.b). Past models are stored in the same folder for analysis purposes. **Forecaster:** every hour, it loads processed data and the last trained model from directories *Data* and *Model* (blue arrow 3.a), respectively, and it predicts the future heating load (blue arrow 3.b). **Performance measures:** Performance is computed at run-time when forecasting from time series of predicted load and real load. The current model is then evaluated by root mean square error on testing data (i.e., data not used for training the model). The coefficient of determination (R^2) of the models are computed on training set and model

---

[1] *XM_HeatForecast* requires Python 3.X. A script is provided to automate the installation process. The script is executed by command *./install.sh*.

parameters are extracted from the current model and saved into the *Performance* folder (orange arrows 4.a and 4.b in Fig. 1). **GUI**: it is a graphical user-friendly interface that updates in real-time showing current predictions, errors, model parameters and historical data. **Communication interface**: module synchronization is performed at run-time by file exchange (red arrows 5.a, 5.b and 5.c in Fig. 1). Figure 2 shows folder organization.

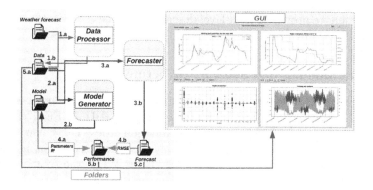

Fig. 1. Overview of the XM_HeatForecast. (Color figure online)

2.2　Testing Dataset

The current version of tool provides a dataset for testing the software, which it includes weather forecasts provided by an external forecast repository [2], calendar events (social factors) [3] and synthetic heating loads. Main weather factors that affect heating load are considered, namely temperature (T), relative humidity (RH), rainfall (R), wind speed (WS), wind direction (WD). From these signals we computed variables with strong predictive capabilities for heating load, according to the literature [2,16]. In particular, from T we compute squared of temperature T^2, moving average of temperature on last 7 days $T_{ma(7)}$, maximum of temperature T_M, squared of maximum temperature T_M^2, maximum temperature of a day ago T_{Mp} and square of maximum temperature of a day ago T_{Mp}^2. Historical load variables used by the model are heating load l_i of i days ago for $1 \leq i \leq 7$ and load peak at previous day l_p. The software is however independent of the set of variables used by the model.

2.3　Data Pre-processing

The *Data processor* module transforms raw data to data in a models-readable format. Source data are *.csv* files. *Data processor* reads from these files the historical heating load and current weather forecast, merges them and it computes

---

[2] https://rp5.ru/.
[3] https://pypi.org/project/workalendar/.

the variables described in the previous subsection. Finally it saves processed data in the *Data* folder. This procedure is displayed by green arrows 1.a and 1.b in Fig. 1.

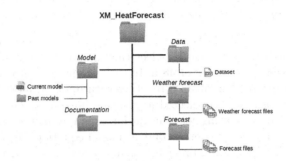

Fig. 2. Folder organization.

3 Model Generation and Forecasting

In this section we describe the modules that manage predictive models generation, parameters updating and forecasting of heating load.

3.1 Model Generator

The *XM_HeatForecast* architecture is designed to be independent from the choice of a specific model. In the current version of the tool we used autoregressive linear models generated by the approach proposed in [2]. In particular, the model consists of multiple linear regression equations, one for each pair (day, hour), therefore we have 168 equations in total. The model estimates future values of heating load l using a linear combination of weather variables, past loads and social factors described in the previous section. The equations have the following form: $M_{ij} = f(weather, dload, social)$, where $i \in \{1,7\}$ indicates the day of week, $j \in \{0, \ldots, 23\}$ the hour of the day, *weather* are weather variables, *dload* includes past heating load, and *social* consists of variables related to calendar events, such as holidays.

Given a series of pre-processed data read from the *Data* folder, the *Model generator* generates the model and updates parameters during the re-training process performed every 24 h. Each equation M_{ij} is trained using corresponding day/hour data. The current version of the software uses the Ordinary Least Squares (OLS) function of *statsmodels*[4] (version 0.11.1) since it natively provides statistics such as parameter p-values, but other libraries (such as *scikit-learn*[5])

---

[4] https://www.statsmodels.org.
[5] https://scikit-learn.org.

can be used to employ other modeling frameworks. Models are stored in the *Model* folder using the *dump* function of the *Pickle* library. The procedure is displayed by purple arrows 2.a and 2.b in Fig. 1.

3.2   Forecaster

The *Forecaster* is responsible of predictions of future heating load. The architecture of the tool allows to easily select the amount of forecasting jobs and a forecasting horizon between 24 or 48 h. In the current version we schedule a job every hour and batch of input data together with updated model are loaded by this module. According to a selected forecasting horizon h, *Forecaster* predicts the future h-hours of heating load, using for each day of the week (i) and for each hour of the day (j) the correspondent model equation M_{ij} and the data of the last week. The forecasting procedure is displayed by blue arrows 3.a and 3.b in Fig. 1.

3.3   Performance Evaluation

Two standard metrics are provided to evaluate performance measures, namely, the root-mean-squared error (RMSE) computed on the test set, and the coefficient of determination R^2 computed on the training set. Given an observed time-series with n observations $y_1, \ldots, y_n$ and predictions $\hat{y}_1, \ldots, \hat{y}_n$, the formula for the root-mean-squared error is: $\text{RMSE} = \sqrt{\dfrac{1}{n} \sum_{t=1}^{N} (\hat{y}_t - y_t)^2}$. The RMSE is used in two ways. First, to evaluate the error of the last prediction every time a new forecasting is performed and real heating load are available. Second, to show the error trend on a rolling basis using a sliding window of h-hours, where h is the forecasting horizon. Every time the model is re-trained the GUI is refreshed to show updated R^2 coefficients and model parameters. The performance evaluation procedure is displayed by orange arrows 4.a and 4.b in Fig. 1.

4   Model and Forecast Visualization and Analysis

The *XM_HeatForecast* graphical user interface is divided into four main sections, identified by numbers from 1 to 4 in Fig. 3 and described in the following. All charts can be separately visualized in independent windows that allow zooming and saving to file.

4.1   Visualization of Heating Load Predictions

The first section of the GUI (top-left) contains future heating load prediction for h-hours after the current date/time instant. The current date and time is displayed in the top-central part of the interface. The chart shows the prediction in orange and the real load in blue (if available). The RMSE of the prediction is also displayed on top of the chart. The menu on top of this chart allows to select and visualize weather variables, i.e., T, RH, WS, WD and R. If a variable is selected, then the chart is refreshed to display the values of the corresponding variable during the h hours of the last prediction. This allows operators to discover relationships between heating load predictions, prediction errors and weather factors.

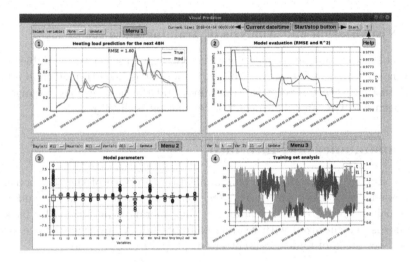

Fig. 3. Main elements of the *XM_HeatForecast* graphical user interface.

4.2   Visualization of Model and Prediction Performance

The second section of the GUI (top-right) displays the time evolution of two performance indicators, namely, the coefficient of determination R^2 (orange line) and the RMSE of h-hours prediction (blue line). This chart is updated every hour, namely whenever a new prediction is performed, but the R^2 statistics is updated only every 24 h, when the model is re-trained. Figure 3 shows, for instance, 15 days of prediction errors. These chart is useful to assess the prediction quality over time, as shown in the case study of Sect. 6.

4.3   Visualization of Model Parameters

The third section of the GUI (bottom-left) shows model parameters. These values are important for DHN operators because they provide insight about variables contribution to the prediction. This insight can strongly support decision making by DHN operators. The box plot displayed in Fig. 3 shows, for each variable, the distribution of related coefficient values across the 168 equations (e.g., temperature T has a negative median value, see the red horizontal line in the corresponding box plot). Menu 2, on top of this chart, allows the operator to select various combinations of the three factors *day, hour* and *variable*, to deepen the analysis of coefficients. By selecting a specific day (e.g., Monday) the box plot is updated by a heatmap in which rows represent variables, columns represent hours and cell colours represent coefficient values for specific day, hour and variable. Heatmaps are used to visualize coefficients when single factors are selected in menu 2. When two factors are selected a table is visualized to show related coefficient values.

4.4   Visualization and Analysis of Training Data

The last section (bottom-right) allows the operator to select two variables of the training set (i.e., the dataset on which the model has been trained) and analyze them. In Fig. 3 the temperature (blue line) and the heating load (orange line) have been selected, and their negative correlation is displayed.

5   Communication Interface

In this section we describe the communication interface of the tool consisting of back-end with *Data processor, Model generator* and *Forecaster* and front-end which is the GUI. Each reading/writing operation performed by back-end modules is synchronized with reading of front-end. An overview of the system and a description of folders content are displayed in Figs. 1 and 2.

5.1   Back-End

In the following we describe the back-end and a step-by-step process that allows the core of the tool to be synchronized with front-end. Every hour when a forecasting job is scheduled, a new file containing weather forecast for the next hour and past data are read. Past data consists of historical heating load of the last week (i.e., l_i of i days ago ($1 \leq i \leq 7$), heating load peaks at previous day l_p and calendar events such holidays H. Current weather forecast and past data are then merged and processed by *Data processor*, which returns a batch having new variables computed from current value of T (see Sect. 2.2). The new batch of data is saved into *Data* folder in order to be read by *Forecaster* in forecasting phase or GUI for graphical visualizations and it has 168 rows and 20 columns, which are a week of past observations for 20 variables in accordance with model requirements (see Sect. 3.2). Every 24 h a whole set of historical data updated with the observations of the previous day is passed to *Model generator* for the training and the update of model parameters. *Model* folder contains current model and a chronology of past models saved as *.pickle* objects. Every hour, *Forecaster* generates an output file which is saved into *Forecast* folder. Each file contains the future h-hours of predicted heating load.

The metrics for the evaluation of performance are computed at run-time in two moments, the first, every 24 h R^2 coefficients and model parameters are saved into *Performace* folder and they are updated and available as soon as a training instance is completed. Due to the type of model, a *csv* file containing R^2 coefficients has the following structure: 24 rows and 7 columns, one for each pair day of the week (j) and hour of the day (i). Each position i, j provides a coefficient of determination of model in a specific day (j) and hour (i). Model parameters are also saved in *csv* files. A file containing the parameters of a specific variable has 24 rows and 7 columns. As the previous metric, each position contains a weight for specific day of the week and hour (i, j). The second moment, every hour RMSE is computed given the last series of predicted and real load.

5.2   Front-End

GUI checks if new data are available using a polling technique on *Forecast*, *Performance* and *Data* folders. It reads last forecasting file and if available, real heating load to display the top-left graph in Fig. 3, that it consists of a comparison between real and predicted load with related RMSE. It loads R^2 coefficients and RMSE values to display the graph on the top-right, which shows the performance of model since tool has been starting. It loads model parameters to display box-plot charts, heatmaps or tables with model parameters on the bottom-left. Finally, GUI reads dataset used for training the models to support data analysis on the bottom-right.

6   Case Study

We provide a case study showing the standard use of *XM_HeatForecast* and the advantages it introduces from the data analysis point of view by integrating machine learning and visualization tools. The analysis here presented is based on a real-world dataset[6] provided by AGSM[7] a utility company that manages a DHN in Verona (Italy). The dataset contains variables described in Sect. 2.2 acquired from 01.01.2016 to 30.04.2018. In these tests we compute multi-equation linear autoregressive models with one equation for each weekday-hour, as described in Sect. 3.1. Further details about models are provided in [2]. The experiment focuses, in particular, on the analysis of model performance depending on the number of observations in the training set. As explained above, *XM_HeatForecast* performs model re-training every 24-hours. We evaluate eight training intervals displayed in Fig. 4a and test the model on twelve 48-hours forecasts starting from a fixed date, namely, 19.02.2018 at 12 am. The table in Fig. 4a shows, in each row, the training length (in months and number of observations) and related $\overline{RMSE}$. The chart in Fig. 4b displays the same data on a graphical basis. Interestingly, performance reaches a plateau after about one year and a half, showing that this time horizon is needed to have good prediction performance. We provide also the training time which is nearly constant. Another interesting analysis allowed by *XM_HeatForecast* concerns the relationship between independent variables (i.e., weather and social factors) and the

---

[6] The dataset is not available for privacy reasons.

[7] https://www.agsm.it/.

heating load prediction (and error). Operators are interested to understand, in particular, the reasons of model errors to increase the knowledge about the network and improve model performance. The red line in Fig. 4c shows, for instance, a case in which the maximum temperature strongly increases between 03.07.2018 (about 10°C) and 04.07.2018 (about 16°C) (see the red area). This change seems to be the cause of a deviation between the true and the predicted load due to the slowness of the model to adapt to temperature changes. In fact, in the temperature peak the model predicts a higher load than the true one for about 7 h, as if it expected lower temperature at that time.

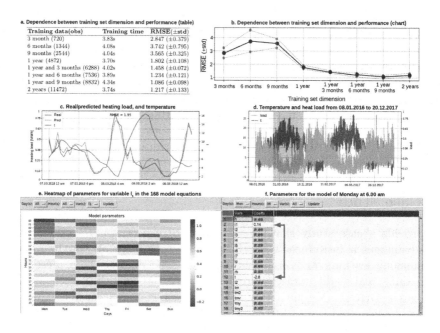

Fig. 4. Example of data analysis with *XM_HeatForecast* (Color figure online)

Finally, we show how *XM_HeatForecast* can support knowledge extraction from model parameters. The multi-equation autoregressive linear model generated by the current version of the software consists of 168 linear regression equations, one for each weekday and hour. With 20 independent features this results in 3360 coefficients, which are difficult to display efficiently. The bottom-left panel of *XM_HeatForecast* (described in Sect. 4.3) provides useful support on the analysis of these parameters. Selecting a variable from the menu of this panel a heatmap of model parameters is shown. For instance the heatmap displayed in Fig. 4e displays 168 coefficients of variable l_1, namely the heating load with one day lag. Heatmap rows correspond to the 24 h of the day, and columns correspond to the 7 days of the week. The chart allows to easily observe patterns and other properties of model parameters. For instance, red cells in the figure correspond to pairs weekday-hour in which the predicted heating load highly depends on the load of the previous day at the same hour. In fact, these cells represent high coefficients (i.e., high correlation) while green cells low coefficients. Notice that Sunday and Monday have low parameters on average. In the first case this is due to the strong change of behaviour between working days and Sunday (mainly for commercial buildings), in the second case it is due to the strong change of behaviour between Sunday (weekend) and Monday (working day), which makes the load of the previous day not informative to predict future load. *XM_HeatForecast* allows also to display numerical values of model parameters, as shown in Fig. 4f, by selecting a specific day of the week and hour of the day in the menu. The figure shows coefficients for each variable on Monday at 6.00 and the red arrow indicates opposite signs of temperature T and previous day load l_1 coefficients, which highlights that the temperature T is anti-correlated with the predicted load, and the load of the previous day is little but positively correlated with the predicted load. The anti-correlation between temperature and heating load can be clearly shown also in the bottom-right panel of *XM_HeatForecast* (see Sect. 4.4). Figure 4d displays the time evolution of the two variables over a time period of two years in the training set. It shows a strong seasonality with low temperature and high heating load in winter and high temperature and low heating load in spring. By zooming on the chart a similar pattern is visualized also on a daily basis, because of the temperature variation between day and night.

7   Conclusions

XM_HeatForecast integrates in a single tool both machine learning and data visualization capabilities for heating load forecasting in DHN. The tool allows to generate predictive models and analyze their parameters, predictions and performance, supporting interactive and integrated analysis in real-time. This integration provides new possibilities for DHN operators to improve their knowledge about the network and, consequently, to improve operational performance. The paper describes the tool and provides a novel case study on a real dataset where the software tools are used to collect knowledge about the real network and to show the prediction ability. Future developments concern the extensions to a other predictive modeling frameworks (e.g., LSTM) and the application to novel domains, such as behaviour prediction in intelligent robotic platforms.

Acknowledgments. The research has been partially supported by the projects "Dipartimenti di Eccellenza 2018-2022", funded by the Italian Ministry of Education, Universities and Research (MIUR), and "GHOTEM/CORE-WOOD, POR-FESR 2014-2020", funded by Regione del Veneto.

References

1. Bandyopadhyay, S., Hazra, J., Kalyanaraman, S.: A machine learning based heating and cooling load forecasting approach for DHC networks. In: IEEE Power Energy Society Innovation Smart Grid Technology Conference (ISGT), pp. 1–5 (2018)

2. Bianchi, F., Castellini, A., Tarocco, P., Farinelli, A.: Load forecasting in district heating networks: model comparison on a real-world case study. In: Nicosia, G., Pardalos, P., Umeton, R., Giuffrida, G., Sciacca, V. (eds.) LOD 2019. LNCS, vol. 11943, pp. 553–565. Springer, Cham (2019). https://doi.org/10.1007/978-3-030-37599-7_46

3. Castellini, A., et al.: Activity recognition for autonomous water drones based on unsupervised learning methods. In: Proceedings of 4th Italian Workshop on Artificial Intelligence and Robotics (AI*IA 2017), vol. 2054, pp. 16–21 (2018)

4. Castellini, A., Bicego, M., Masillo, F., Zuccotto, M., Farinelli, A.: Time series segmentation for state-model generation of autonomous aquatic drones: a systematic framework. Eng. Appl. Artif. Intell. **90**, 103499 (2020)

5. Castellini, A., et al.: Subspace clustering for situation assessment in aquatic drones. In: Proceedings of the Symposium on Applied Computing, SAC 2019, page To appear. ACM (2019)

6. Castellini, A., Masillo, F., Sartea, R., Farinelli, A.: eXplainable Modeling (XM): Data analysis for intelligent agents. In: Proceedings of 18th International Conference on Autonomous Agents and MultiAgent Systems, AAMAS 2019, pp. 2342–2344 (2019)

7. eXplainable Modeling code (2019). https://github.com/XModeling/XM

8. Fallah, S.N., Ganjkhani, M., Shamshirband, S., Chau, K.-W.: Computational intelligence on short-term load forecasting: a methodological overview. Energies **12**, 393 (2019)

9. Fang, T.: Modelling district heating and combined heat and power (2016)

10. Fang, T., Lahdelma, R.: Evaluation of a multiple linear regression model and SARIMA model in forecasting heat demand for district heating system. Appl. Energy **179**, 544–552 (2016)

11. Goodfellow, I., Bengio, Y., Courville, A.: Deep Learning. MIT Press, Cambridge (2016)

12. Gross, G., Galiana, F.D.: Short-term load forecasting. In: Proceedings of the IEEE, pp. 1558–1573 (1987)

13. Hagan, M.T., Behr, S.M.: The time series approach to short term load forecasting. IEEE Trans. Power Syst. **2**, 785–791 (1987)

14. Jacob, M., Neves, C., Vukadinović Greetham, D.: Short term load forecasting. Forecasting and Assessing Risk of Individual Electricity Peaks. MPE, pp. 15–37. Springer, Cham (2020). https://doi.org/10.1007/978-3-030-28669-9_2

15. Mirowski, P., Chen, S., Ho, T.K., Yu, C.N.: Demand forecasting in smart grids. Bell Labs Tech. J. **18**, 135–158 (2014)

16. Ramanathan, R., Engle, R., Granger, C.W.J., Vahid-Araghi, F., Brace, C.: Short-run forecast of electricity loads and peaks. Int. J. Forecast. **13**, 161–174 (1997)

17. Raza, M., Khosravi, A.: A review on artificial intelligence based load demand forecasting techniques for smart grid and buildings. Renew. Sustain. Energy Rev. **50**, 1352–1372 (2015)

18. Russell, S., Norvig, P.: Artificial Intelligence: A Modern Approach, 2nd edn. Pearson Education, London (2003)

19. Sainath, T., Vinyals, O., Senior, A., Sak, H.: Convolutional, long short-term memory, fully connected deep neural networks. In: 2015 IEEE - (ICASSP), pp. 4580–4584 (2015)

Robust Generative Restricted Kernel Machines Using Weighted Conjugate Feature Duality

Arun Pandey$^{(\boxtimes)}$, Joachim Schreurs, and Johan A. K. Suykens

Department of Electrical Engineering ESAT-STADIUS, KU Leuven,
Kasteelpark Arenberg 10, 3001 Leuven, Belgium
{arun.pandey,joachim.schreurs,johan.suykens}@esat.kuleuven.be

Abstract. Interest in generative models has grown tremendously in the past decade. However, their training performance can be adversely affected by contamination, where outliers are encoded in the representation of the model. This results in the generation of noisy data. In this paper, we introduce weighted conjugate feature duality in the framework of Restricted Kernel Machines (RKMs). The RKM formulation allows for an easy integration of methods from classical robust statistics. This formulation is used to fine-tune the latent space of generative RKMs using a weighting function based on the Minimum Covariance Determinant, which is a highly robust estimator of multivariate location and scatter. Experiments show that the weighted RKM is capable of generating clean images when contamination is present in the training data. We further show that the robust method also preserves uncorrelated feature learning through qualitative and quantitative experiments on standard datasets.

Keywords: Machine learning · Generative models · Robustness · Kernel methods · Restricted kernel machines

1 Introduction

A popular choice for generative models in machine learning are latent variable models such as Variational Auto-Encoders (VAE) [18], Restricted Boltzmann Machines (RBM) [24,27] and Generative Adversarial Networks (GAN) [2,11,19]. These latent spaces provide a representation of the input data by embedding into an underlying vector space. Exploring these spaces allows for deeper insights in the structure of the data distribution, as well as understanding relationships between data points. The interpretability of the latent space is enhanced when the model learns a disentangled representation [5,13]. In a disentangled representation, a single latent feature is sensitive to changes in a single generative factor, while being relatively invariant to changes in other factors [3]. For example hair color, lighting conditions or orientation of faces.

A. Pandey and J. Schreurs—Authors contributed equally to this work.

© Springer Nature Switzerland AG 2020
G. Nicosia et al. (Eds.): LOD 2020, LNCS 12565, pp. 613–624, 2020.
https://doi.org/10.1007/978-3-030-64583-0_54

In generative modelling, training data is often assumed to be ground truth, therefore outliers can severely degrade the learned representations and performance of trained models. The same issue arises in generative modelling where contamination of the training data results in encoding of the outliers. Consequently, the network generates noisy images when reconstructing out-of-sample extensions. To solve this problem, multiple robust variants of generative models were proposed in [6,10,31]. However, these generative models require clean training data or only consider the case where there is label noise. In this paper, we address the problem of *contamination on the training data itself*. This is a common problem in real-life datasets, which are often contaminated by human error, measurement errors or changes in system behaviour. To the best of our knowledge, this specific problem is not addressed in other generative methods. The Restricted Kernel Machine (RKM) formulation [28] allows for a straightforward integration of methods from classical robust statistics to the RKM framework. The RKM framework yields a representation of kernel methods with visible and hidden units establishing links between kernel methods [29] and RBMs. [26] showed how kernel PCA fits into the RKM framework. A tensor-based multiview classification model was developed in [15]. In [21], a multi-view generative model called Generative RKM (Gen-RKM) is introduced which uses explicit feature-maps for joint feature-selection and subspace learning. Gen-RKM learns the basis of the latent space, yielding uncorrelated latent variables. This allows to generate data with specific features, i.e. a disentangled representation.

Contributions: This paper introduces a weighted Gen-RKM model that detects and penalizes the outliers to regularize the latent space. Thanks to the introduction of weighted conjugate feature duality, a RKM formulation for weighted kernel PCA is derived. This formulation is used within the Gen-RKM training procedure to fine-tune the latent space using different weighting schemes. A weighting function based on Minimum Covariance Determinant (MCD) [23] is proposed. Qualitative and quantitative experiments on standard datasets show that the proposed model is unaffected by large contamination and can learn meaningful representations.

2   Weighted Restricted Kernel Machines

2.1   Weighted Conjugate Feature Duality

For a comprehensive overview of the RKM framework, the reader is encouraged to refer [21,28]. In this section, we extend the notion of conjugate feature duality by introducing a weighting matrix. Assuming $D \succ 0$ to be a positive-definite diagonal weighting matrix, the following holds for any two vectors $e, h \in \mathbb{R}^n$, $\lambda > 0$:

$$\frac{1}{2\lambda} e^\top D e + \frac{\lambda}{2} h^\top D^{-1} h \geq e^\top h. \tag{1}$$

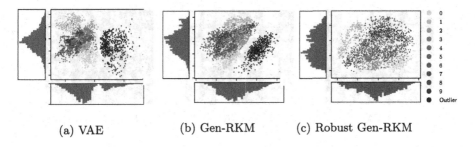

(a) VAE   (b) Gen-RKM   (c) Robust Gen-RKM

Fig. 1. Illustration of robustness against outliers on the MNIST dataset. 20% of the training data is contaminated with noise. The models are trained with a 2-dimensional latent space in the standard setup, see Sect. 4. The presence of outliers distorts the distribution of the latent variables for the Gen-RKM and VAE, where the histogram of the latent variables is skewed. By down-weighting the outliers, the histogram resembles a Gaussian distribution again.

The inequality could be verified using the Schur complement by writing the above in its quadratic form:

$$\frac{1}{2} \begin{bmatrix} e^\top & h^\top \end{bmatrix} \begin{bmatrix} \frac{1}{\lambda} DI & -I \\ -I & \lambda D^{-1}I \end{bmatrix} \begin{bmatrix} e \\ h \end{bmatrix} \geq 0. \tag{2}$$

It states that for a matrix $Q = \begin{bmatrix} A & B \\ B^\top & C \end{bmatrix}$, one has $Q \succeq 0$ if and only if $A \succ 0$ and the Schur complement $C - B^\top A^{-1} B \succeq 0$ [4], which proves the above inequality. This is also known as the Fenchel-Young inequality for quadratic functions [22].

We assume a dataset $\mathcal{D} = \{x_i\}_{i=1}^N$, with $x_i \in \mathbb{R}^d$ consisting of N data points. For a feature-map $\phi : \mathbb{R}^d \mapsto \mathbb{R}^{d_f}$ defined on input data points, the weighted kernel PCA objective [25] in the Least-Squares Support Vector Machine (LS-SVM) setting is given by [1]:

$$\min_{U,e} J(U, e) = \frac{\eta}{2} \mathrm{Tr}(U^\top U) - \frac{1}{2\lambda} e^\top De \quad \text{s.t.} \quad e_i = U^\top \phi(x_i), \forall i = 1, \ldots, N, \tag{3}$$

where $U \in \mathbb{R}^{d_f \times s}$ is the unknown interconnection matrix. By using (1), the error variables e_i are conjugated to latent variables $h_i \in \mathbb{R}^s$ and substituting the constraints into the objective function yields

$$J \leq \mathcal{J}_t^D := \sum_{i=1}^N \left\{ -\phi(x_i)^\top U h_i + \frac{\lambda}{2} D_{ii}^{-1} h_i^\top h_i \right\} + \frac{\eta}{2} \mathrm{Tr}(U^\top U). \tag{4}$$

The stationary points of $\mathcal{J}_t^D$ are given by:

$$\begin{cases} \frac{\partial \mathcal{J}_t^D}{\partial h_i} = 0 \implies \lambda D_{ii}^{-1} h_i = U^\top \phi(x_i), \ \forall i = 1, \ldots, N \\ \frac{\partial \mathcal{J}_t^D}{\partial U} = 0 \implies U = \frac{1}{\eta} \sum_{i=1}^N \phi(x_i) h_i^\top. \end{cases} \tag{5}$$

Eliminating U and denoting the kernel matrix $K := [k(x_i, x_j)]_{ij}$ with kernel function $k(x, y) = \langle \phi(x), \phi(y) \rangle$, the eigenvectors $H := [h_1, \ldots, h_N]$, $\Lambda :=$

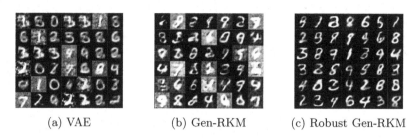

(a) VAE   (b) Gen-RKM   (c) Robust Gen-RKM

Fig. 2. Illustration of robust generation on the MNIST dataset. 20% of the training data is contaminated with noise. The images are generated by random sampling from a fitted Gaussian distribution on the learned latent variables. When using a robust training procedure, the model does not encode the noisy images. As a consequence, no noisy images are generated.

$\text{diag}\{\lambda_1, \ldots, \lambda_s\} \in \mathbb{R}^{s \times s}$ such that $\lambda_1 \geq \ldots \geq \lambda_s$ with s the dimension of the latent space, we get the weighted eigenvalue problem:

$$\frac{1}{\eta}[DK]H^\top = H^\top \Lambda. \tag{6}$$

One can verify that each eigenvalue-eigenvector pair lead to the value $\mathcal{J}_t^D = 0$. Using the weighted kernel PCA potential outliers can be penalized, which is discussed in more detail in Sect. 3.

2.2   Generation

Given the learned interconnection matrix U, and a latent variable $h^\star$, consider the following objective function

$$\mathcal{J}_g = -\phi(x^\star)^\top U h^\star + \frac{1}{2}\phi(x^\star)^\top \phi(x^\star), \tag{7}$$

with a regularization term on the input data. Here $\mathcal{J}_g$ denotes the objective function for generation. To reconstruct or denoise a training point, $h^\star$ can be one-of-the corresponding hidden units of the training point. Random generation is done by fitting a normal distribution on the learned latent variables, afterwards a random $h^\star$ is sampled from the distribution which is put through the decoder network. Note that in the training objective (ref. (4)) we are imposing a soft-Gaussian prior over latent variables through quadratic regularization on $\{h_i\}_{i=1}^N$. The stationary points of (7) yields the *generated feature vector* [21,26] $\varphi(x^\star)$, given by the corresponding $h^\star$. With slight abuse of notation, we denote the generated feature-vector by $\varphi(x^\star) = [\frac{1}{\eta}\sum_{i=1}^N \phi(x_i)h_i^\top]h^\star$, which is a point in the feature-space corresponding to an unknown $x^\star$ in data space. To obtain the generated data in the input space, the inverse image of the feature map $\phi(\cdot)$ should be computed. In kernel methods, this is known as the pre-image problem. We seek to find the function $\psi\colon \mathbb{R}^{d_f} \mapsto \mathbb{R}^d$, such that $(\psi \circ \varphi)(x^\star) \approx x^\star$, where $\varphi(\boldsymbol{x}^\star)$ is calculated from above. The pre-image problem is known to be ill-conditioned [20], and consequently various approximation techniques have been proposed [14]. Another approach is to explicitly define pre-image maps and learn the parameters in the training procedure [21]. In the experiments, we use (convolutional) neural networks as the feature maps $\phi_\theta(\cdot)$, where the notation extends to $\varphi_\theta(\cdot)$. Another (transposed convolutional) neural network is used for the pre-image map $\psi_\zeta(\cdot)$ [8]. The parameters θ and ζ correspond to the network parameters. These parameters are learned by minimizing the reconstruction error in combination with the weighted RKM objective function. The training algorithm is described in more detail in Sect. 3.2.

Remark on Out-of-Sample Extension: To reconstruct or denoise an out-of-sample test point $\boldsymbol{x}^\star$, the data is projected on the latent space using:

$$h^\star = \lambda^{-1} \boldsymbol{U}^\top \phi(\boldsymbol{x}^\star) = \frac{1}{\lambda\eta} \sum_{i=1}^{N} h_i k(\boldsymbol{x}_i, \boldsymbol{x}^\star). \tag{8}$$

The latent point is reconstructed by projecting back to the input space by first computing the generated feature vector followed by its pre-image map $\psi_\zeta(\cdot)$.

3 Robust Estimation of the Latent Variables

3.1 Robust Weighting Scheme

In this paper, we propose a weighting scheme to make the estimation of the latent variables more robust against contamination. The weighting matrix is a diagonal matrix with a weight $\boldsymbol{D}_{ii}$ corresponding to every h_i such that:

$$\boldsymbol{D}_{ii} = \begin{cases} 1 & \text{if } d_i^2 \le \chi_{s,\alpha}^2 \\ 10^{-4} & \text{otherwise,} \end{cases} \tag{9}$$

with s the dimension of the latent space, α the significance level of the Chi-squared distribution and d_i^2 the Mahalanobis distance for the corresponding h_i:

$$d_i^2 = (\boldsymbol{h}_i - \hat{\boldsymbol{\mu}})^\top \hat{\boldsymbol{S}}^{-1} (\boldsymbol{h}_i - \hat{\boldsymbol{\mu}}), \tag{10}$$

with $\hat{\boldsymbol{\mu}}$ and $\hat{\boldsymbol{S}}$ the robustly estimated mean and covariance matrix respectively. In this paper, we propose to use the Minimum Covariance Determinant (MCD) [23]. The MCD is a highly robust estimator of multivariate location and scatter which has been used in many robust multivariate statistical methods [16,17]. Given a data matrix of N rows with s columns, the objective is to find the $N_{\mathrm{MCD}} < N$ observations whose sample covariance matrix has the lowest determinant. Its influence function is bounded [7] and has the highest possible breakdown value when $N_{\mathrm{MCD}} = \lfloor (N+s+1)/2 \rfloor$. In the experiments, we typically take $N_{\mathrm{MCD}} = \lfloor N \times 0.75 \rfloor$ and $\alpha = 0.975$ for the Chi-squared distribution. The user could further tune these parameters according to the estimated contamination degree in the dataset. Eventually, the reweighting procedure can be repeated

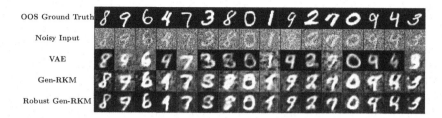

Fig. 3. Illustration of robust denoising on the MNIST dataset. 20% of the training data is contaminated with noise. The first and second row show the clean and noisy test images respectively. The third, fourth and fifth row show the denoised image using the VAE, Gen-RKM and robust Gen-RKM respectively.

iteratively, but in practice one single additional weighted step will often be sufficient. Kernel PCA can take the interpretation of a one-class modeling problem with zero target value around which one maximizes the variance [30]. The same holds in the Gen-RKM framework. This is a natural consequence of the regularization term $\frac{\lambda}{2} \sum_{i=1}^{N} h_i^\top h_i$ in the training objective (see (4)), which implicitly puts a Gaussian prior on the hidden units. When the training of feature map is done correctly, one expects the latent variables to be normally distributed around zero [29]. Gaussian distributed latent variables are essential for having a *continuous* and *smooth* latent space, allowing easy interpolation. This property is also essential for VAEs and was studied in [18], where a regularization term, in the form of the Kullback-Leibler divergence between the encoder's distribution and a unit Gaussian as a prior on the latent variables was used. When training a non-robust generative model in the presence of outliers, the contamination can severely distort the distribution of the latent variables. This effect is seen in Fig. 1, where a discontinuous and skewed distribution is visible.

3.2   Algorithm

We propose to use the above described reweighting step within the Gen-RKM framework [21]. The algorithm is flexible to incorporate both kernel-based, (deep) neural network and Convolutional based models within the same setting, and is capable of jointly learning the feature maps and latent representations. The Gen-RKM algorithm consists out of two phases: a training phase and a generation phase which occurs one after another. In the case of explicit feature maps, the training phase consists of determining the parameters of the explicit feature and pre-image map together with the hidden units $\{h_i\}_{i=1}^{N}$.

We propose an adapted algorithm of [21] with an extra re-weighting step wherein the system in (6) is solved. Furthermore, the reconstruction error is weighted to reduce the effect of potential outliers on the pre-image maps. The loss function now becomes:

$$\min_{\theta,\zeta} \ \mathcal{J}_c^D(\theta,\zeta) = \mathcal{J}_t^D + \frac{c_{\text{stab}}}{2}(\mathcal{J}_t^D)^2 + \frac{c_{\text{acc}}}{N}\sum_{i=1}^{N} D_{ii}\mathcal{L}(\boldsymbol{x}_i,\psi_\zeta(\varphi_\theta(\boldsymbol{x}_i))), \qquad (11)$$

where $c_{stab} \in \mathbb{R}^+$ is a stability constant [28] and $c_{acc} \in \mathbb{R}^+$ is a regularization constant to control the stability with reconstruction accuracy. In the experiments, the loss function is equal to the mean squared error (MSE), however other loss functions are possible. The generation algorithm is the same as in [21].

Table 1. FID Scores [12] over 10 iterations for 4000 randomly generated samples when the training data is contaminated with 20% outliers. (smaller is better).

Dataset	FID score			
	VAE	β-VAE ($\beta = 3$)	RKM	Rob Gen-RKM
MNIST	142.54 ± 0.73	187.21 ± 0.11	134.95 ± 1.61	$\mathbf{87.32 \pm 1.92}$
F-MNIST	245.84 ± 0.43	291.11 ± 1.6	163.51 ± 1.24	$\mathbf{153.32 \pm 0.05}$
SVHN	168.21 ± 0.23	234.87 ± 1.45	112.45 ± 1.4	$\mathbf{98.14 \pm 1.2}$
CIFAR-10	201.21 ± 0.71	241.23 ± 0.34	187.08 ± 0.58	$\mathbf{132.6 \pm 0.21}$
Dsprites	234.51 ± 1.10	298.21 ± 1.5	182.65 ± 0.57	$\mathbf{160.56 \pm 0.96}$
3Dshapes	233.18 ± 0.94	252.41 ± 0.38	177.29 ± 1.60	$\mathbf{131.18 \pm 1.45}$

4 Experiments

In this section, we evaluate the robustness of the weighted Gen-RKM on the MNIST, Fashion-MNIST (F-MNIST), CIFAR-10, SVHN, Dsprites and 3Dshapes dataset[1]. The last two datasets will be used in disentanglement experiments since they include the ground truth generating factors which are necessary to quantify the performance. Training of the robust Gen-RKM is done using the algorithm proposed in Sect. 3.2, where we take $N_{\text{MCD}} = \lfloor N \times 0.75 \rfloor$ and $\alpha = 0.975$ for the Chi-squared distribution (see (9)). Afterwards we compare with the standard Gen-RKM [21], VAE and β-VAE. The models have the same encoder/decoder architecture, optimization parameters and are trained until convergence. Information on the training settings and model architectures is given in the Appendix.

Generation and Denoising: Figure 2 shows the generation of random images when models were trained on the contaminated MNIST dataset. The contamination consists of adding Gaussian noise $\mathcal{N}(0.5, 0.5)$ to 20% of the data. The images are generated by random sampling from a fitted Gaussian distribution on the learned latent variables. As we can see, when using a robust training procedure, the model does not encode the noisy images. As a consequence, no noisy

---

[1] http://yann.lecun.com/exdb/mnist/, https://github.com/zalandoresearch/fashion-mnist, https://github.com/deepmind/dsprites-dataset, https://github.com/deepmind/3d-shapes, https://www.cs.toronto.edu/~kriz/cifar.html, http://ufldl.stanford.edu/housenumbers/.

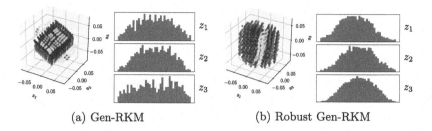

(a) Gen-RKM   (b) Robust Gen-RKM

Fig. 4. Illustration of disentanglement on the 3DShapes dataset. Clean data is depicted in purple, outliers in yellow. The training subset is contaminated with a third generating factor (20% of the data is considered as outliers). The outliers are down-weighted in the robust Gen-RKM, which moves them to the center.

images are generated and the generation quality is significantly better. This is also confirmed by the Fréchet Inception Distance (FID) scores [12] in Table 1, which quantifies the quality of generation. The robust Gen-RKM clearly outperforms the other methods when the data has contamination. Moreover VAEs are known to generate samples closer to the mean of dataset. This negatively affects the FID scores which also takes into account the diversity within the generated images. The scores for β-VAE are worst due to the inherent emphasis on imposing a Gaussian distribution on latent variables trading-off with the reconstruction quality [13]. The classical RKM performs slightly better than VAE and its variant. This is attributed to the presence of kernel PCA during training, which is often used in denoising applications and helps to some extent in dealing with contamination in the dataset.

Next, we use generative models in the denoising experiment. Image denoising is accomplished by projecting the noisy test set observations on the latent space, afterwards projecting back to the input space. Because there is a latent bottleneck, the most important features of the images are retained while insignificant features like noise are removed. Figure 3 shows an illustration of robust denoising on the MNIST dataset. The robust Gen-RKM does not encode the noisy images within the training procedure. Consequently, the model is capable of denoising the out-of-sample test images. When comparing the denoising quality on the full test set (5000 images sampled uniformly at random), the Mean Absolute Error (MAE) of Gen-RKM: MAE = 0.415 and VAE: MAE = 0.434 is much higher than the robust version: MAE = 0.206. The experiments show that basic generative models like Gen-RKM and VAE are highly affected by outliers, while the robust counterpart can cope with a significant fraction of contamination.

Effect on Disentanglement: In this experiment, contamination is an extra generating factor which is not present in the majority of the data. The goal is to train a disentangled representation, where the robust model only focuses on the most prominent generating factors. We subsample a 'clean' training subset which consists of cubes with different floor, wall and object hue. The scale and orientation are kept constant with minimum scale and 0° orientation respectively.

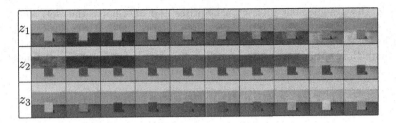

Fig. 5. Illustration of latent traversals along the 3 latent dimensions for 3DShapes dataset using the robust Gen-RKM model. The first, second and third row distinctly captures the floor-hue, wall-hue and object-hue respectively while keeping other generative factors constant.

Afterwards, the training data is contaminated by cylinders with maximum scale at 30° orientation (20% of the data is considered as outliers). The training data now consist out of 3 'true' generating factors (floor, wall and object hue) which appear in the majority of the data and 3 'noisy' generating factors (object, scale and orientation) which only occur in a small fraction. To illustrate the effect of the weighting scheme, Fig. 4 visualizes the latent space of the (robust) Gen-RKM model. The classical Gen-RKM encodes the outliers in the representation, which results in a distorted Gaussian distribution of the latent variables. This is not the case for the robust Gen-RKM, where the outliers are downweighted. An illustration of latent traversals along the 3 latent dimensions using the robust Gen-RKM model is given in Fig. 5, where the robust model is capable of disentangling the 3 'clean' generating factors.

To quantify the performance of disentanglement, we use the proposed framework[2] of [9], which consists of 3 metrics: disentanglement, completeness and informativeness. The framework could be used when the *ground-truth latent structure is available*, which is the case for 3Dshapes and DSprites dataset. The results are shown in Table 2, where the robust method outperforms the Gen-RKM. The above experiment is repeated on the DSprites dataset. The 'clean' training subset consists of ellipse shaped datapoints with minimal scale and 0° angle at different x and y positions. Afterwards, the training data is contaminated with a random sample of different objects at larger scales, different angles at different x and y positions. The training data now consist out of 2 'true' generating factors (x and y positions) which appear in the majority of the data and 3 'noisy' generating factor (orientation, scale and shape) which only occur in a small fraction. In addition to RKM, the results of β-VAE are shown in Table 2.

---

[2] Code and dataset available at https://github.com/cianeastwood/qedr.

Table 2. Disentanglement Metric on DSprites and 3D Shapes dataset. The training subset is contaminated with extra generating factors (20% of the data is considered as outliers). The framework of [9] with Lasso and Random Forest regressor [9] is used to evaluate the learned representation. For disentanglement and completeness higher score is better, for informativeness, lower is better.

Dataset	h_{dim}	Algorithm	Lasso			Random forest		
			Disent.	Comple.	Inform.	Disent.	Comple.	Inform.
DSprites	2	β-VAE ($\beta = 3$)	0.19	0.16	6.42	0.13	0.32	**1.39**
		Gen-RKM	0.07	0.07	**5.82**	0.25	0.27	5.91
		Rob Gen-RKM	**0.21**	**0.21**	9.13	**0.36**	**0.38**	5.95
3DShapes	3	β-VAE ($\beta = 3$)	0.24	0.28	**2.72**	0.12	0.13	2.15
		Gen-RKM	0.14	0.14	3.03	0.15	0.15	1.09
		Rob Gen-RKM	**0.47**	**0.49**	3.13	**0.44**	**0.45**	**1.02**

5   Conclusion

Using a weighted conjugate feature duality, a RKM formulation for weighted kernel PCA is proposed. This formulation is used within the Gen-RKM training procedure to fine-tune the latent space using a weighting function based on the MCD. Experiments show that the weighted RKM is capable of generating denoised images in spite of contamination in the training data. Furthermore, being a latent variable model, robust Gen-RKM preserves the disentangled representation. Future work consists of exploring various robust estimators and other weighting schemes to control the effect of sampling bias in the data.

Acknowledgments. EU: The research leading to these results has received funding from the European Research Council under the European Union's Horizon 2020 research and innovation program/ERC Advanced Grant E-DUALITY (787960). This paper reflects only the authors' views and the Union is not liable for any use that may be made of the contained information. Research Council KUL: Optimization frameworks for deep kernel machines C14/18/068 Flemish Government: FWO: projects: GOA4917N (Deep Restricted Kernel Machines: Methods and Foundations), PhD/Postdoc grant Impulsfonds AI: VR 2019 2203 DOC.0318/1QUATER Kenniscentrum Data en Maatschappij Ford KU Leuven Research Alliance Project KUL0076 (Stability analysis and performance improvement of deep reinforcement learning algorithms).

Appendix

Table 3 shows the details on training settings used in this paper. The PyTorch library in Python was used with a 8GB NVIDIA QUADRO P4000 GPU.

Table 3. Model architectures. All convolutions and transposed-convolutions are with stride 2 and padding 1. Unless stated otherwise, layers have Parametric-RELU ($\alpha = 0.2$) activation functions, except output layers of the pre-image maps which have sigmoid activation functions. $N_{\text{sub}} \leq N$ is the training subset size, s the latent space dimension and m the minibatch size.

Dataset	Optimizer (Adam)	Architecture			Parameters	
MNIST/F-MNIST/ CIFAR-10/SVHN	1e-4	Feature-map (fm)	Conv 32×4×4; Conv 64×4×4; FC 228 (Linear)	N_{sub}	3000
		Pre-image map	Reverse of fm	s	10
		Latent space dim	10	m	200
Dsprites/3DShapes	1e-4	Feature-map (fm)	Conv 20×4×4; Conv 40×4×4; Conv 80×4×4; FC 228 (Linear)	N_{sub}	1024/1200
		Pre-image map	Reverse of fm	s	2
		Latent space dim	2/3	m	200

References

1. Alzate, C., Suykens, J.A.K.: Multiway spectral clustering with out-of-sample extensions through weighted kernel PCA. IEEE Trans. Pattern Anal. Mach. Intell. **32**(2), 335–347 (2008)
2. Arjovsky, M., Chintala, S., Bottou, L.: Wasserstein generative adversarial networks. In: ICML, pp. 214–223 (2017)
3. Bengio, Y., Courville, A., Vincent, P.: Representation learning: a review and new perspectives. IEEE Trans. Pattern Anal. Mach. Intell. **35**(8), 1798–1828 (2013)
4. Boyd, S., Vandenberghe, L.: Convex Optimization. Cambridge University Press, New York (2004)
5. Chen, X., Duan, Y., Houthooft, R., Schulman, J., Sutskever, I., Abbeel, P.: InfoGAN: interpretable representation learning by information maximizing generative adversarial nets. In: Advances in Neural Information Processing Systems, pp. 2172–2180 (2016)
6. Chrysos, G.G., Kossaifi, J., Zafeiriou, S.: Robust conditional generative adversarial networks. In: International Conference on Learning Representations (2019)
7. Croux, C., Haesbroeck, G.: Influence function and efficiency of the minimum covariance determinant scatter matrix estimator. J. Multivar. Anal. **71**(2), 161–190 (1999)
8. Dumoulin, V., Visin, F.: A guide to convolution arithmetic for deep learning. arXiv preprint arXiv:1603.07285 (2016)
9. Eastwood, C., Williams, C.K.I.: A framework for the quantitative evaluation of disentangled representations. In: International Conference on Learning Representations (2018)
10. Futami, F., Sato, I., Sugiyama, M.: Variational inference based on robust divergences. In: 21st International Conference on Artificial Intelligence and Statistics (2018)
11. Goodfellow, I.J., et al.: Generative adversarial nets. In: Advances in Neural Information Processing Systems, vol. 27, pp. 2672–2680 (2014)
12. Heusel, M., Ramsauer, H., Unterthiner, T., Nessler, B., Hochreiter, S.: GANs trained by a two time-scale update rule converge to a local nash equilibrium. In: Proceedings of the 31st NIPS, pp. 6629–6640 (2017)

13. Higgins, I., et al.: Beta-VAE: learning basic visual concepts with a constrained variational framework. In: International Conference on Learning Representations (2017)
14. Honeine, P., Richard, C.: Preimage problem in kernel-based machine learning. IEEE Signal Process. Mag. **28**(2), 77–88 (2011)
15. Houthuys, L., Suykens, J.A.K.: Tensor learning in multi-view kernel PCA. In: Kůrková, V., Manolopoulos, Y., Hammer, B., Iliadis, L., Maglogiannis, I. (eds.) ICANN 2018. LNCS, vol. 11140, pp. 205–215. Springer, Cham (2018). https://doi.org/10.1007/978-3-030-01421-6_21
16. Hubert, M., Rousseeuw, P.J., Vanden Branden, K.: ROBPCA: a new approach to robust principal component analysis. Technometrics **47**(1), 64–79 (2005)
17. Hubert, M., Rousseeuw, P.J., Verdonck, T.: A deterministic algorithm for robust location and scatter. J. Comput. Graph. Stat. **21**(3), 618–637 (2012)
18. Kingma, D.P., Welling, M.: Auto-encoding variational bayes. In: 2nd International Conference on Learning Representations, Banff, AB, Canada, 14–16 April 2014, Conference Track Proceedings (2014)
19. Liu, M.Y., Tuzel, O.: Coupled generative adversarial networks. In: Advances in Neural Information Processing Systems, pp. 469–477 (2016)
20. Mika, S., Schölkopf, B., Smola, A., Müller, K.R., Scholz, M., Rätsch, G.: Kernel PCA and De-noising in Feature Spaces. In: Advances in Neural Information Processing Systems, pp. 536–542 (1999)
21. Pandey, A., Schreurs, J., Suykens, J.A.K.: Generative restricted kernel machines. arXiv preprint arXiv:1906.08144 (2019)
22. Rockafellar, R.T.: Conjugate Duality and Optimization. SIAM (1974)
23. Rousseeuw, P.J., Driessen, K.V.: A fast algorithm for the minimum covariance determinant estimator. Technometrics **41**(3), 212–223 (1999)
24. Salakhutdinov, R., Hinton, G.: Deep Boltzmann machines. In: Proceedings of the 12th International Conference on Artificial Intelligence and Statistics (2009)
25. Schölkopf, B., Smola, A., Müller, K.-R.: Kernel principal component analysis. In: Gerstner, W., Germond, A., Hasler, M., Nicoud, J.-D. (eds.) ICANN 1997. LNCS, vol. 1327, pp. 583–588. Springer, Heidelberg (1997). https://doi.org/10.1007/BFb0020217
26. Schreurs, J., Suykens, J.A.K.: Generative kernel PCA. In: European Symposium on Artificial Neural Networks, pp. 129–134 (2018)
27. Smolensky, P.: Information processing in dynamical systems: foundations of harmony theory. In: Parallel Distributed Processing: Explorations in the Microstructure of Cognition, vol. 1, pp. 194–281. MIT Press, Cambridge (1986)
28. Suykens, J.A.K.: Deep restricted kernel machines using conjugate feature duality. Neural Comput. **29**(8), 2123–2163 (2017)
29. Suykens, J.A.K., Van Gestel, T., De Brabanter, J., De Moor, B., Vandewalle, J.: Least Squares Support Vector Machines. River Edge, NJ (2002)
30. Suykens, J.A.K., Van Gestel, T., Vandewalle, J., De Moor, B.: A support vector machine formulation to PCA analysis and its kernel version. IEEE Trans. Neural Netw. **14**(2), 447–450 (2003)
31. Tang, Y., Salakhutdinov, R., Hinton, G.: Robust Boltzmann machines for recognition and denoising. In: 2012 IEEE Conference on Computer Vision and Pattern Recognition, pp. 2264–2271. IEEE (2012)

A Location-Routing Based Solution Approach for Reorganizing Postal Collection Operations in Rural Areas

Maurizio Boccia[1], Antonio Diglio[2], Adriano Masone[1(✉)], and Claudio Sterle[1,3]

[1] Department of Electrical Engineering and Information Technology,
University "Federico II of Naples", Naples, Italy
{maurizio.boccia,adriano.masone,claudio.sterle}@unina.it
[2] Department of Industrial Engineering,
University "Federico II of Naples", Naples, Italy
antonio.diglio@unina.it
[3] Istituto di Analisi dei Sistemi ed Informatica A. Ruberti, IASI-CNR, Rome, Italy

Abstract. Postboxes generally represent a cost for a postal service provider since they require to be maintained and daily visited by an operator to collect the letters. On the other hand, for the users, they are the main access points to the postal network and so the reduction of their number could result in a deterioration of the quality of the service for them. This work addresses a real problem involving the reduction of the number of postboxes located in rural areas to reorganize the collection operations of a postal service provider. We formulate this problem as a location-routing problem where the number of postmen employed is minimized guaranteeing the user accessibility. Exact and heuristic solution methods have been developed, tested and compared to existing solution methods on instances based on real data of rural instances showing the effectiveness of the proposed methodologies.

Keywords: Postal service · Location-routing · Equity

1 Introduction

Postal services are generally provided through a dedicated logistics network capable of performing all typical operations such as reception, collection, transportation, sorting, and delivery of postal items (i.e., letters or parcels) over a given territory. In the last few years, the amount of correspondence has drastically decreased due to the impact of the new ways of communications based on the latest technological developments [12]. This fall in mail volume has driven to a broad underutilization of the postboxes thus making the collection phase of the postal service providers highly inefficient. Indeed, an open postbox affects the total operational costs of the postal provider since an operator has to visit it to collect the letters and eventually maintain it.

© Springer Nature Switzerland AG 2020
G. Nicosia et al. (Eds.): LOD 2020, LNCS 12565, pp. 625–636, 2020.
https://doi.org/10.1007/978-3-030-64583-0_55

Currently, the main Italian postal provider is evaluating the opportunity of reducing the number of existing postboxes without significantly affecting the user accessibility to the network and fulfilling the requirements on the quality of service [2]. In this context, examples of reorganization processes can be found in many application fields: healthcare, communication and manufacturer industry, to cite only a few [5,14,17]. However, the reorganization of the postal logistic network has to take into account some peculiarities due to the universal nature of the postal service [8], and therefore specific approaches have to be devised. The majority of the literature on the reorganization of postal service logistic network is focused on transportation [4,22], delivery [1,10] or integrated collection-delivery operations [13,21]. To the best of authors knowledge only three contributions address the problem of postboxes/post offices closures arising from the reorganization of the collection operations. In particular, a problem that seeks to select postbox sites minimizing the routing costs of the collection operations and an inconvenience cost incurred by users is proposed in [15]. An ex-post spatial analysis of the impact on the post office closures on user accessibility in urban and rural areas due to the reconfiguration of the postal service in Wales is proposed in [11]. The impact of the reduction of the number of postboxes in urban areas on management costs, in terms of postmen employed, is evaluated in [6]. The resulting problem is solved by an iterative two phases solution method where: in the first phase, given p the number of postmen, the postboxes are clustered solving a large-scale p-median problem [16] with an additional constraint on the compactness of the resulting clusters (assumed as a proxy of the routing time); in the second phase, the duration of the route within each cluster is computed. If the duration of at least one route exceeds the duration of a postman shift, then, in the next iteration, the upper bound on the cluster compactness is reduced. Otherwise, if the model leads to infeasibility, the value of p is increased by one.

In this work, we investigated the impact of postbox closure in terms of number of deployed postmen in rural areas. In contrast with [6], we developed a method that is based on the solution of an original integrated location-routing model. This choice is motivated by different reasons. Indeed, rural areas are characterized by small-medium dimensions, compared to urban area dimensions, that can be effectively tackled by exact and heuristic approaches. Moreover, unlike urban areas, the routing aspect is very important in rural areas due to the issues connected to the rural road networks that are generally characterized by a low density and high travel times. Therefore, the choice to keep open or to close a postbox rather than another and subsequently the assignment postbox-postman have a huge impact on the routing component. Several variants of location-routing models exist in literature [7,18,20] but to the best of authors' knowledge, this work is the first dealing with the minimization of the number of collection tours guaranteeing a minimum threshold distance for every customer. To this aim, we developed an exact solution method based on the solution of an integer programming (IP) formulation of an original location-routing problem and two heuristic approaches based on the determination of a lower bound for the problem object of study relaxing the postman shift duration. Computational results on instances coherent with real data were carried out comparing them with the results of the solution method proposed in [6] to prove the effectiveness of the presented approaches.

The remainder of this paper is structured as follow: a formal description of the problem is given in Sect. 2; the formulation of the integrated location-routing model is provided in Sect. 3; the two heuristic approaches are described in Sect. 4; computational results are showed in Sect. 5; conclusions and future research perspectives are given in Sect. 6.

2   Problem Description

The postal provider seeks to reduce the operational costs of its logistic network by closing some postboxes but minimizing the deterioration of users' accessibility. On the other hand, postboxes represent the main users' access points to the postal network and their distribution is currently regulated in Italy by a demographic criterion. This criterion expresses the maximum number of users that can be served by a single postbox; however, due to the sharp fall in letter mail volumes, it can be now considered outdated. Therefore, we refer to a spatial criterion expressing the maximum distance that a customer should cover in order to reach the nearest postbox.

In this work, we modelled this problem using the number of postmen dedicated to the collection operations as a measure of the operational costs arising from the postbox reorganization, as in [6]. To this aim, we defined an original location-routing problem that minimizes the number of postmen employed for the postal collection and guaranteeing operational constraints as the maximum duration of a postman shift and accessibility constraints as the maximum distance customer-postbox. In particular, this distance is conceived as a threshold parameter that will splits the set of customers into two subsets: the subset of customers that have at least one postbox within the threshold and the subset of customers that do not have any postbox within the threshold. The customers that belong to the second subset are defined disadvantaged users. The nearest postbox to each of these customers will be kept open since we do not want to worse their accessibility level. On the other hand, each customer of the first subset will have at least one postbox open between those within the threshold distance. An example of two different solutions obtained for the same instance (10 postboxes and 5 customers indicated with red squares and blue triangles, respectively) considering two threshold distances is reported in Fig. 1. In Figs. 1a and 1c the blue circles represent the threshold distance of each customer. In Figs. 1b and 1d we show the corresponding solutions assuming that the maximum duration of a postman shift is equal to 100. The solution reported in Fig. 1b requires only one postman and 6 postboxes can be closed (indicated by the black squares). The solution reported in Fig. 1d requires instead two postmen since the threshold distance is lower and so only 5 postboxes can be closed.

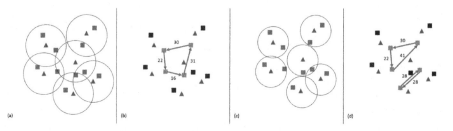

Fig. 1. An example of an instance considering two different threshold distances and the corresponding solutions

3   Problem Formulation

The problem object of study involves three kinds of decisions: 1) identification of the postboxes to be closed; 2) assignment of open postboxes to the postmen; 3) definition of collection tours for the postmen. To formally introduce the problem, let I be the set of customers and J the set of postboxes, with J_i being the subset of postboxes reachable by the client $i, \forall i \in I$. Moreover, let J^* be the subset of J representing the postboxes that cannot be closed since they are located within a post office. If k_{max} is the number of postmen available, then we can introduce three kinds of decision variables:

- s_k binary variable equal to 1 if the k-th postman is employed, 0 otherwise.
- y_j^k binary variable equal to 1 if postbox j is open and assigned to the k-th postman, 0 otherwise.
- x_{ij}^k binary variable equal to 1 if postbox i precedes postbox j in the k-th postman collection route, 0 otherwise.

On the basis of this notation, being T the working hours of each postman and t_{ij} the time required for a postman to move from postbox i to postbox j, we can introduce the following IP formulation:

$$\min \sum_{k=1}^{k_{max}} s_k \tag{1}$$

subject to:

$$s_k \leq s_{k-1}, \forall k \in \{2, ..., k_{max}\} \tag{2}$$

$$y_j^k \leq s_k, \forall k \in \{1, ..., k_{max}\}, j \in J \tag{3}$$

$$\sum_{k=1}^{k_{max}} y_j^k \leq 1, \forall j \in J \setminus J^* \tag{4}$$

$$\sum_{k=1}^{k_{max}} y_j^k = 1, \forall j \in J^* \tag{5}$$

$$\sum_{j \in J_i} \sum_{k=1}^{k_{max}} y_j^k \geq 1, \forall i \in I \tag{6}$$

$$\sum_{i \in J} x_{ij}^k = y_j^k, \quad \forall k \in \{1, ..., k_{max}\}, j \in J \tag{7}$$

$$\sum_{i \in J} x_{ji}^k = y_j^k, \quad \forall k \in \{1, ..., k_{max}\}, j \in J \tag{8}$$

$$\sum_{i \in S} \sum_{j \in J \setminus S} x_{ij}^k \geq y_u^k + y_v^k - 1, \forall k \in \{1, ..., k_{max}\}, S \subseteq J, u \in S, v \in J \setminus S \tag{9}$$

$$\sum_{i \in J} \sum_{j \in J} t_{ij} x_{ij}^k \leq T, \quad \forall k \in \{1, ..., k_{max}\} \tag{10}$$

$$\sum_{i \in J} \sum_{j \in J} t_{ij} x_{ij}^k \leq \sum_{i \in J} \sum_{j \in J} t_{ij} x_{ij}^{k-1}, \quad \forall k \in \{2, ..., k_{max}\} \tag{11}$$

$$s_k \in \{0, 1\}, \quad \forall k \in \{1, ..., k_{max}\} \tag{12}$$

$$y_j^k \in \{0, 1\}, \quad \forall k \in \{1, ..., k_{max}\}, j \in J \tag{13}$$

$$x_{ij}^k \in \{0, 1\}, \quad \forall k \in \{1, ..., k_{max}\}, i, j \in J \tag{14}$$

The objective function (1) minimizes the number of employed postmen. Constraints (2) guarantee that a postman cannot be employed if the postman associated to the previous route is not dispatched. Constraints (3) ensure that a postbox can be assigned only to a performed collection route. The assignment of a postbox to at most one of the performed collection routes is given by constraints (4). Constraints (5) ensure that the postbox offices cannot be closed and will be visited by a postman. The requirement of the presence of at least one open postbox between those reachable by the client i is provided by the constraints (6). Constraints (7, 8) guarantee that if a postbox is visited then the corresponding node has exactly one entering and exiting arc, respectively. Subtour elimination is guaranteed by constraints (9). Constraints (10) represent the work shift of the postmen. We highlight that since the time required to collect the mail from a postbox is generally significant lower than the time required to move between two postbox location we have not explicitly considered it. However, these times can be easily taken into account adding a term $\sum_{j \in J} t_j y_j^k$ to the left side of these constraints, where t_j is the time required to visit the postbox j. Constraints (11) avoid symmetries in the solution. Constraints (12–14) define the domain of the variables. Finally, we underline that it is also possible to take into account the space limitation of the postal vehicles adding a constraint for each postman ensuring that the sum of the mail volume collected is lower than the vehicle capacity. However, the aforementioned drastic decline in volumes makes it reasonable to neglect this aspect in the proposed model.

4   An IP Based Solution Method

The formulation proposed in the previous section can be hard to solve using a commercial MIP solver even for small instances. Therefore, a solution method able to compute a lower bound and an upper bound for the problem under investigation has been developed. The lower bound is computed solving a variant of the Prize Collecting TSP (PCTSP), [3]. This problem is obtained relaxing the constraints related to the shifts of the postmen. Then, the upper bound is determined on the basis of the solution of the PCTSP. In particular, two ways of computing the upper bound will be described. The first one is a greedy heuristic based on the route-first cluster-second framework. The second one exploits the formulation described in Sect. 3.

4.1   Lower Bound Determination

A valid lower bound for the problem can be computed assuming that there is only one postman employed minimizing the collection time and without constraints related to his shift. To this aim, let y_j be a binary variable equal to 1 if the postbox j is visited, 0 otherwise and let x_{ij} be a binary variable equal to 1 if the postman goes from postbox i to postbox j, 0 otherwise. On the basis of the introduced notation, it is possible to formulate the following IP model:

$$\min \sum_{i \in J} \sum_{j \in J} t_{ij} x_{ij} \tag{15}$$

subject to:

$$y_j = 1, \forall j \in J^* \tag{16}$$

$$\sum_{j \in J_i} y_j \geq 1, \quad \forall i \in I \tag{17}$$

$$\sum_{i \in J} x_{ij} = y_j, \quad \forall j \in J \tag{18}$$

$$\sum_{i \in J} x_{ji} = y_j, \quad \forall j \in J \tag{19}$$

$$\sum_{i \in S} \sum_{j \in J \setminus S} x_{ij} \geq y_u + y_v - 1, \forall S \subseteq J, u \in S, v \in J \setminus S \tag{20}$$

$$y_j \in \{0, 1\}, \quad \forall j \in J \tag{21}$$

$$x_{ij} \in \{0, 1\}, \quad \forall i, j \in J \tag{22}$$

The objective function (15) minimizes the collection time. Constraints (16) and (17) are related to the post offices and the customer quality of service, respectively. Constraints (18–20) ensure the feasibility of the postman route. Finally, constraints (21–22) define the integrality of the variables. The objective function value of the optimal solution is the minimum time necessary to serve a subset of postboxes satisfying the requirement on the post offices and the quality of service of the customers. Therefore, if we divide this value for the duration of a postman shift, we obtain the minimum number of postmen/tour needed for the collection phase. This lower bound can be used to improve the effectiveness in addressing the problem. Indeed, it can be used as valid inequalities for the *tour* variables in the general model, setting each $s_k = 1, \forall k \leq \lceil LB/T \rceil$.

4.2 Upper Bound Determination

The tour obtained solving the lower bound formulation can be exploited to compute a feasible solution for the original problem. In the following, two heuristic methods will be described. The first one splits the tour in several subtours each one with a duration that respects the maximum postman working hours. The second one solves a reduced problem obtained keeping open the postboxes visited in the lower bound solution and closing the others.

Solution Method Based on the Lower Bound Route. The first method is based on the route first and cluster second method framework. Route first and cluster second methods are very well-known algorithms for multi-vehicles problem as the Vehicle Routing Problem (VRP), [19]. These algorithms construct a giant TSP tour in a first phase and decompose it into feasible subtours in a second phase. Our method focuses only on the route splitting phase since the giant tour will be the one determined by the solution of the lower bound formulation. The splitting problem is tackled through a partitioning approach. Indeed, given a fixed route, our objective is to find the minimum number of subtours, constrained by the work shift, such that each postbox belongs exactly to a subtour. We formulated this problem in terms of a simple IP model that results to be the well-known Set Covering Problem, [9]. To this aim, we define the binary variable z_{ij} that is equal to 1 if the generated sub-route begins in the node i and ends in the node j of the giant route that are connected through the arc ij, 0 otherwise. On this basis, it is possible to formulate the problem as follows:

$$\min \sum_{i \in J} \sum_{j \in J} z_{ij} \tag{23}$$

subject to:

$$\sum_{ij \in Z_l} z_{ij} = 1, \quad \forall l \in J \tag{24}$$

$$z_{ij} \in \{0, 1\}, \quad \forall i, j \in J \tag{25}$$

The objective function (23) minimizes the number of subtours generated. The constraints (24) ensure that each customer belongs exactly to one subtour. The constraints (25) express the integrality of the variables. To better understand the correspondence between the z variables and the subtours, in Fig. 2 is reported an example with six visited postboxes. In particular, Fig. 2a shows the tour obtained solving the lower bound formulation. This tour exceeds the maximum route duration and has to be split in smaller subtours. The splitting showed in Fig. 2b is not feasible since it defines two overlapping subtours since both visit nodes 4 and 5. Figure 2c shows another unfeasible splitting. Indeed, nodes 3, 5 and 6 are visited twice since the subtours defined by variables z_{25} and z_{53} are overlapping. Finally, Fig. 2d shows a feasible splitting since each customer is served exactly once.

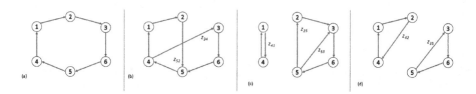

Fig. 2. An example of feasible and unfeasible splitting of the tour obtained by solving the lower bound formulation

Solution Method Based on the Lower Bound Open Postboxes. It is possible to exploit the lower bound solution also in terms of open postboxes without fixing the routing component. Indeed, being O the set of open postboxes, we can solve the formulation (1–14) replacing constraints (4–5) with the following ones:

$$\sum_{k=1}^{k_{max}} y_j^k = \begin{cases} 1, & \forall j \in O \\ 0, & \forall j \in J \setminus O \end{cases} \tag{26}$$

These constraints fix the open and closed postboxes. Therefore, this reduced problem requires two types of decisions: the assignment postbox-postman and the routing of each postman. It is straightforward that the upper bound provided by this method will always be lower or equal to the one obtained fixing also the routing component since in the worst case it will output the same solution.

5   Computational Results

The proposed approaches have been implemented in Python 3.6 and linked with a commercial MIP solver (GUROBI optimizer 8.1) to solve the IP formulations. The computational experiments were carried out with a time limit of 3600 seconds on an Intel Core I7-8700 CPU 3.20 GHz workstation with 16 Gb of RAM. Our test bed consists of four instances generated coherently with real data of rural instances and it is available online[1]. For each instance, we determine the threshold distance d on the basis of the percentage of disadvantaged users α. For example, if $\alpha = 0$ then it means that all the users have at least one postbox within the threshold distance; if $\alpha = 0.1$ then the 10% of users do not have a

---

[1] http://www.ing.unisannio.it/boccia/LRP-Postboxes.htm.

postbox within threshold distance and so on. We recall that, due to the accessibility constraints, a disadvantaged user will always have the closest postbox open. The details on our test instances are given in Table 1 where: Name is the instance name; K, P, O and C are the number of postmen, postboxes, post offices and customers, respectively; D is the threshold distance, function of the parameter α

Table 1. Instance details

Name	K	P	O	C	D(α)				
					$\alpha = 0.0$	$\alpha = 0.1$	$\alpha = 0.2$	$\alpha = 0.3$	$\alpha = 1.0$
Ins_1	9	42	7	422	27603	3771	2761	1564	0
Ins_2	9	44	13	397	5312	2305	1830	1381	0
Ins_3	9	50	16	344	4717	1493	667	488	0
Ins_4	9	61	28	483	8323	2703	1549	708	0

The results obtained by each proposed solution method and the approach proposed in [6] are reported in Table 2 for each instance and for each value of experimented α. In particular, the first and the second columns report the instance name and the value α, respectively. The third column reports the best lower bound found between the solution of the lower bound formulation and the whole problem formulation. In particular, if the reported value is obtained only by the problem formulation (i.e., the lower bound of the problem formulation is greater than the one obtained solving the lower bound formulation) then an asterisk (*) is reported. The successive eight columns show the upper bound (first four columns) and the difference between the upper bound and the best lower bound (successive four columns) obtained by solving the set partitioning (SP), the reduced formulation (RF), the whole formulation (PF), and the two phases solution method ($2P$) proposed in [6], respectively. Finally, the last five columns report the running times of each approach in seconds.

The results show that the best lower bound is determined through the solution of the lower bound formulation for all the instances except for the instance Ins_1 with $\alpha = 0$. Moreover, we can observe that the RF and PF are not able to determine a better solution than the one obtained using SP. This is due to two reasons: the solution found by SP is the optimal solution (as for (Ins_1 with $\alpha = 0$ and Ins_2 with $\alpha = 0.3$); the difficulty of solving RF and PF does not allow to obtain a better solution within the time limit. However, the methods are able to determine near optimal solutions (the difference between the upper and lower bound is equal to 1) for almost all the instances. Comparing our results with those obtained by $2P$ in terms of quality of the solution, we can observe that we are able to obtain a better solution on four instances: Ins_1 with $\alpha = \{0.3, 1\}$; Ins_4 with $\alpha = \{0.1, 1\}$. These results confirm that the proposed approaches can output better solution since, unlike the $2P$ approach, they focus also on the

Table 2. Results of the proposed solution methods

Name	α	BLB	Upper Bounds				Δ				Running Times [s]				
			SP	RF	PF	2P	SP	RF	PF	2P	LB	SP	RF	PF	2P
Ins_1	0	4*	4	4	4	4	0	0	0	0	0.46	0.02	1.74	11.66	16.9
Ins_1	0.1	6	8	8	8	8	2	2	2	2	11.98	0.09	3600	3600	59.09
Ins_1	0.2	6	8	8	8	8	2	2	2	2	28.3	0.08	3600	3600	59.08
Ins_1	0.3	6	8	8	8	9	2	2	2	3	134.05	0.1	3600	3600	62.89
Ins_1	1	6	8	8	8	9	2	2	2	3	247.31	0.25	3600	3600	119.89
Ins_2	0	3	4	4	4	4	1	1	1	1	1.18	0.03	3.79	16.85	16.04
Ins_2	0.1	4	5	5	5	5	1	1	1	1	86.14	0.06	309.34	3600	32.24
Ins_2	0.2	4	5	5	5	5	1	1	1	1	144	0.3	3600	3600	27.9
Ins_2	0.3	5	5	5	5	5	0	0	0	0	113.72	0.25	3600	3600	32.51
Ins_2	1	5	6	6	6	6	1	1	1	1	172.01	0.26	3600	3600	41.97
Ins_3	0	4	5	5	5	5	1	1	1	1	240.31	0.04	3600	3600	37.02
Ins_3	0.1	5	6	6	6	6	1	1	1	1	196.7	0.19	3600	3600	36.31
Ins_3	0.2	5	6	6	6	6	1	1	1	1	107.02	0.19	3600	3600	47.85
Ins_3	0.3	5	6	6	6	6	1	1	1	1	329.66	0.25	3600	3600	45.85
Ins_3	1	5	6	6	6	6	1	1	1	1	198.36	0.3	3600	3600	31.87
Ins_4	0	6	7	7	7	7	1	1	1	1	177.2	0.09	3600	3600	185.82
Ins_4	0.1	7	8	8	8	9	1	1	1	2	701.5	0.32	3600	3600	958.59
Ins_4	0.2	7	8	8	8	8	1	1	1	1	1534.2	0.53	3600	3600	151.52
Ins_4	0.3	7	8	8	8	8	1	1	1	1	1536.2	0.45	3600	3600	156.05
Ins_4	1	7	8	8	8	9	1	1	1	2	2818.4	1.48	3600	3600	303.84

routing component that is a key features of rural instances. In terms of number of postmen employed varying the percentage of disadvantaged customers, we can observe that for each instance there are no significative differences except for $\alpha = 0$, i.e. considering that each customer has at least one postbox within the threshold distance. This means that, not considering the case with $\alpha = 0$, a postal provider can choose equally between the two extreme situations (α equal to 0.1 or 1.0) in the reorganization of its collection operations.

In terms of running times, we can observe that the complexity of the problem increases when the parameter α rises. In particular, it is possible to solve the lower bound formulation and SP for all the instances while RF and PF are solved to optimality for few instances within the time limit. Moreover, we highlight that, comparing RF and PF in terms of instances solved within 1 h (3 and 2 instances solved, respectively), even fixing the postboxes that have to be open (i.e., solving RF rather than solving PF) the complexity of the problem is still high. Comparing the running times of our approaches and those obtained by $2P$, we can observe that our approaches are generally slower. We expected these results since we take directly into account the routing component unlike $2P$. However, when the value of α is low the complexity of the routing problem generally decreases and SP is faster than $2P$ (Ins_1 with $\alpha = \{0, 0.1, 0.2\}$; Ins_2 with $\alpha = 0$; Ins_4 with $\alpha = \{0, 0.1\}$).

6   Conclusions

In this work we presented an original location-routing problem arising in the reorganization of collection operations of a service postal provider with particular focus on rural areas. This problem consists in the definition of the minimum number of collection tours needed to collect the letters from the postboxes such as the level of accessibility for the customers defined by the regulatory authority is guaranteed. Exact and heuristics approaches based on the solution of IP formulations have been developed to tackle the problem and experimented on instances derived from real data. The computational results showed the applicability and the effectiveness of the proposed methodologies. These methodologies can be used by a postal provider company in a negotiation phase with the regular authority to define the new settlements for the definition of the user quality of service.

Future research perspectives may include an extension of the proposed approaches to the case where the workload of the operators has to be equally distributed. Moreover, it is possible to develop a framework that addresses different type of postal operations in rural areas (e.g., collection-transportation, collection-delivery, collection-transportation-delivery) integrating the proposed approaches. Finally, further studies may explore: the possibility to exploit methods proposed in literature for different location-routing problems adapting them to the tackled problem; the possibility to modify the proposed approaches to tackle large scale instances to apply them also in urban areas; the extension of the model to stochastic cases where one or more problem aspects, such as customer demands or travel times are nondeterministic.

References

1. Abbatecola, L., Fanti, M.P., Mangini, A.M., Ukovich, W.: A decision support approach for postal delivery and waste collection services. IEEE Trans. Autom. Sci. Eng. **13**(4), 1458–1470 (2016)
2. AGCOM (2014). Public consultation on the criteria for the distribution of access points to the public postal network. (Italian Authority for Communication Guarantees). http://www.agcom.int. Accessed 4 February 2020
3. Balas, E.: The prize collecting traveling salesman problem. Networks **19**(6), 621–636 (1989)
4. Boysen, N., Fedtke, S., Weidinger, F.: Truck scheduling in the postal service industry. Transp. Sci. **51**(2), 723–736 (2017)
5. Bruno, G., Diglio, A., Piccolo, C., Cannavacciuolo, L.: Territorial reorganization of regional blood management systems: evidences from an Italian case study. Omega **89**, 54–70 (2019)
6. Bruno, G., Cavola, M., Diglio, A., Laporte, G., Piccolo, C.: Reorganizing postal collection operations in urban areas as a result of declining mail volumes - a case study in Bologna. J. Oper. Res. Soc. https://doi.org/10.1080/01605682.2020.1736446. to appear (2020)
7. Drexl, M., Schneider, M.: A survey of variants and extensions of the location-routing problem. Eur. J. Oper. Res. **241**(2), 283–308 (2015)

8. European Parliament and Council of European Union. (1998). Directive 97/67/EC of the European Parliament and of the Council of 15 December 1997. http://www.eur-lex.europa.eu. Accessed 4 February 2020

9. Garfinkel, R.S., Nemhauser, G.L.: The set-partitioning problem: set covering with equality constraints. Oper. Res. **17**(5), 848–856 (1969)

10. Grunert, T., Sebastian, H.-J.: Planning models for long-haul operations of postal and express shipment companies. Eur. J. Oper. Res. **122**(2), 289–309 (2000)

11. Higgs, G., Langford, M.: Investigating the validity of rural-urban distinctions in the impacts of changing service provision: the example of postal service reconfiguration in Wales. Geoforum **47**, 53–64 (2013)

12. Hong, S.-H., Wolak, F.A.: Relative prices and electronic substitution: Changes in household-level demand for postal delivery services from 1986 to 2004. J. Econometrics **145**(1–2), 226–242 (2008)

13. Jung, H., Lee, K., Chun, W.: Integration of GIS, GPS, and optimization technologies for the effective control of parcel delivery service. Comput. Ind. Eng. **51**(1), 154–162 (2006)

14. Iellamo, S., Alekseeva, E., Chen, L., Coupechoux, M., Kochetov, Y.: Competitive location in cognitive radio networks. 4OR-Q. J. Oper. Res. **13**, 81–110 (2015)

15. Labbe, M., Laporte, G.: Maximizing user convenience and postal service efficiency in post box location. Belgian J. Oper. Res. Stat. Comput. Sci. **26**(2), 21–36 (1986)

16. Masone, A., Sforza, A., Sterle, C., Vasilyev, I.: A graph clustering based decomposition approach for large scale p-median problems. Int. J. Artif. Intell. **16**(1), 116–129 (2018)

17. Masone, A., Sterle, C., Vasilyev, I., Ushakov, A.: A three-stage p-median based exact method for the optimal diversity management problem. Networks **74**(2), 174–189 (2019)

18. Nagy, G., Salhi, S.: Location-routing: issues, models and methods. Eur. J. Oper. Res. **177**(2), 649–672 (2007)

19. Prins, C., Lacomme, P., Prodhon, C.: Order-first split-second methods for vehicle routing problems: a review. Transp. Res. Part C: Emerg. Technol. **40**, 179–200 (2014)

20. Prodhon, C., Prins, C.: A survey of recent research on location-routing problems. Eur. J. Oper. Res. **238**(1), 1–17 (2014)

21. Qu, Y., Bard, J.F.: A grasp with adaptive large neighborhood search for pickup and delivery problems with transshipment. Comput. Oper. Res. **39**(10), 2439–2456 (2012)

22. Zenker, M., Boysen, N.: Dock sharing in cross-docking facilities of the postal service industry. J. Oper. Res. Soc. **69**, 1–16 (2018)

Dynamic Selection of Classifiers Applied to High-Dimensional Small-Instance Data Sets: Problems and Challenges

Alexandre Maciel-Guerra$^{(\boxtimes)}$, Grazziela P. Figueredo ,
and Jamie Twycross

School of Computer Science, University of Nottingham,
Computer Science Building, Wollaton Road, Nottingham NG8 1BB, UK
{Alexandre.MacielGuerra,Grazziela.Figueredo,
Jamie.Twycross}@nottingham.ac.uk

Abstract. Dynamic selection (DS) of classifiers have been explored by researchers due to their overall ability to obtain higher accuracy on low-sample data sets when compared majority voting. Little literature, however, has employed DS to high-dimensional data sets with substantially more features than samples. Since, several studies have reported the benefits of applying feature selection methods to high-dimensional data sets, raised the following open research questions: 1. How DS methods perform for such data sets? 2. Do they perform better than majority voting? and 3. Does feature selection as a pre-processing step improve their performance? The performance of 21 DS methods was statistically compared against the performance of majority voting on 10 high-dimensional data sets and with a filter feature selection method. We found that majority voting is among the best ranked classifiers and none of the DS methods perform statistically better than it with and without feature selection. Moreover, we demonstrated that feature selection does improve the performance of DS methods.

Keywords: Ensemble learning · Dynamic integration of classifiers · Dynamic selection · Machine learning · Majority voting · High dimensional data sets

1 Introduction

Over the past decades, multiple classifier systems (MCS) became a very active area in pattern recognition. One of the most promising approaches involves dynamic selection (DS), in which different classifiers are selected for each unseen sample. Several authors have recently shown that dynamic selection (DS) methods obtain high performances in terms of accuracy on low dimensional datasets [7,9]. Nevertheless, many authors observed that DS techniques are still far from the upper bound performance of the oracle, which always predicts the correct label if at least one classifier in the ensemble predicts the correct

© Springer Nature Switzerland AG 2020
G. Nicosia et al. (Eds.): LOD 2020, LNCS 12565, pp. 637–649, 2020.
https://doi.org/10.1007/978-3-030-64583-0_56

label. With this in mind, some authors have reported and proposed solutions to improve the quality of the region of competence in low-dimensional datasets to increase their performance [7,20]. Over the past decade, DS techniques have been evaluated on low dimensional datasets and, to the best of our knowledge, there is no comprehensive work in the literature that verifies the performance of the state-of-art DS methods when dealing with high-dimensional small-instances datasets. Maciel-Guerra *et al.* (2019) [18] studied the performance of DS methods over a single high-dimensional protein microarray data set.

High-dimensional data sets with a small number of samples are typical in some domains, such as biology, medicine, bioinformatics and neuroimaging. Often in these areas data do not exist in abundance or is expensive to acquire [4]. In high dimensional data sets, many dimensions are irrelevant and/or redundant which can directly impact the quality of the regions of competence [16,21]. Feature selection methods have been employed to remove irrelevant features and filter methods are usually chosen due to their low computational cost [16,21].

The focus of this paper is, therefore, to evaluate how DS methods perform on high-dimensional small-instance data sets and compare it to majority voting which is the simplest MCS method. Despite the large number of papers published in DS, there is no comprehensive study available verifying the use of this methods on this specific type of data set. Following the recent study of Maciel-Guerra *et al.* (2019) [18] that studied the performance of DS methods over a single small instance high dimensional data set, we have three research questions, namely:

1. How DS methods perform in terms of accuracy?
2. Do they perform statistically better than majority voting?
3. Does feature selection as a pre-processing step improve their performance?

To answer these questions, 10 real-world benchmark data sets with a high number of features and a low number of samples are selected. Four data sets are text based while six are biomedical data sets relating to different types of cancer (lung, prostate, leukemia, colon, glioma and ovarian). Twenty-one DS methods available in the literature are compared against majority voting. The Iman-Davenport extension of the Friedman test [14] is used to statistically verify the performance of the classifiers over all data sets and the Bonferroni-Dunn test [10] is used as a post-hoc test to evaluate if any of the methods outperform statistically majority voting.

This paper is organised as follows. Section 2 provides background on the main topics of this paper. Section 3 introduces the experiments design with the data sets and statistical methods used. A discussion between the performance of DS methods and other MCS methods is conducted in Sect. 4. The conclusion and future research are given in Sect. 5.

2   Background

The quantity of data collected from multiple sources have increased greatly in the past decade, particularly in medicine and life sciences, which brings challenges and opportunities. Heterogeneity, scalability, computational time and complexity are some of the challenges that impede progress to extract meaningful information from data [2,3]. High-dimensional data sets with a small number of samples are typical in some domains, such as biology, medicine, bioinformatics and neuroimaging [12]. We believe that approaches such as DS can improve the classification and increase knowledge discovery in high-dimensional data.

2.1 Dynamic Selection

An important task regarding classification ensembles is the decision as to which classifiers are required to be included to achieve high prediction accuracy. Static Selection (SS), Dynamic Classifier Selection (DCS) and Dynamic Ensemble Selection (DES) are the techniques commonly employed to determine the set of classifiers within the ensemble. SS works by selecting a group of classifiers for all new samples, while DCS and DES select a single or a group of classifiers for each new sample, respectively. Recently, DS methods have been preferred over static methods due to their ability to create different classifier configurations, i.e. different groups of classifiers are experts in different local regions of the feature space. As for many cases, different samples are associated with different classification difficulties and the ability to choose a group of classifiers can possibly overcome static selection methods limitations [6,9,15].

Table 1. DS methods information

Name	Selection criteria	DS Method	Region of Competence	Year
Classifier Rank (CR)	Ranking	DCS	k-NN	1993
Modified Classifier Rank (MCR)	Ranking	DCS	k-NN	1997
Overall Local Accuracy (OLA)	Accuracy	DCS	k-NN	1997
Local Class Accuracy (LCA)	Accuracy	DCS	k-NN	1997
A Priori	Probabilistic	DCS	k-NN	1999
A Posteriori	Probabilistic	DCS	k-NN	1999
Multiple Classifier Behaviour (MCB)	Behaviour	DCS	k-NN	2002
Modified Local Accuracy (MLA)	Accuracy	DCS	k-NN	2002
DES - kMeans	Accuracy & Diversity	DES	k-Means	2006
DES - K-Nearest Neighbour (DES-kNN)	Accuracy & Diveristy	DES	k-NN	2006

(*continued*)

Table 1. (*continued*)

Name	Selection criteria	DS Method	Region of Competence	Year
k-Nearest ORAcles Elimimante (KNORA-E)	Oracle	DES	k-NN	2008
k-Nearest ORAcles Union (KNORA-U)	Oracle	DES	k-NN	2008
DES - Exponential (DES-EXP)	Probabilistic	DES	All samples	2009
DES - Randomised Reference Classifier (DES-RRC)	Probabilistic	DES	All samples	2011
DES - Minimal Difference (DES-MD)	Probabilistic	DES	All samples	2011
DES - Kullback-Leibler Divergence (DES-KL)	Probabilistic	DES	All samples	2012
DES - Performance (DES-P)	Probabilistic	DES	All samples	2012
k-Nearest Output Profiles Elimiante (KNOP-E)	Behaviour	DES	k-NN	2013
k-Nearest Output Profiles Union (KNOP-U)	Behaviour	DES	k-NN	2013
Meta-learning - DES (Meta-DES)	Meta-learning	DES	k-NN	2015
Dynamic Selection on Complexity (DSOC)	Accuracy & Complexity	DCS	k-NN	2016

For DS methods to achieve optimum recognition rates they need to select the most competent classifiers for any given test sample, which can be done by measuring different selection criteria depending on the technique used (accuracy, ranking, behaviour, diversity, probabilistic, complexity and meta-learning). More information about each one of this different criteria can be found on the recent review by [6,9]. A local region of the feature space surrounding the test sample (Region of Competence) is used to estimate the competence of each classifier according to any selection criteria. The majority of DS techniques relies on k-Nearest Neighbours (k-NN) algorithms (Table 1) and the quality of the neighbourhood can have a huge impact on the performance of DS methods [6,7,9].

Table 1 shows the different DS methods found in the literature which were presented in the most recent review by Cruz *et al.* [9]. More information about each one can be found on their respective reference or on the recent reviews done by [6,9]. These methods were chosen due to their differences in the selection criteria and because they present the most important breakthroughs in the area over the past three decades.

2.2   High-Dimensional Data

Financial, risk management, computational biology, health studies are some of the areas where high-dimensional data sets can be produced. However, in some of these areas, such as biology and medicine, it might not be feasible to have thousands or millions of samples due to the nature of the disease or the access to samples [29]. DNA microarray is one example of these types of data sets where data collected from tissue and cell samples are used to measure the levels of gene expression. The number of genes is usually far higher than the number of patients in cancer research for instance [4].

Data sets with a high number of features usually poses challenges that are commonly referred as the "curse of dimensionality". One of the main aspects of this curse is *distance concentration*, which can directly affect machine learning application, specially the ones that deal with distance metrics such as k-NN. Concentration of distance refers to the tendency of distance to all points to become almost equal in high-dimensional spaces [1, 24, 25].

For these reasons, for any classifier to be successful (have a high accuracy level), it is usually necessary to have sufficient data to cover the feature space during training, so it can have as much information possible on the feature space to find the correct learning function to predict the output associated with new inputs [4, 29]. If this is not the case, researchers frequently apply different feature selection techniques to remove unwanted (redundant, irrelevant, noisy) features and, consequently, improve the performance of classifiers [4].

2.3   Feature Selection

In two recent reviews Bólon-Canedo *et al.* [4, 5] reported the benefits of applying feature selection methods to high-dimensional data sets, and highlighted the fact that feature selection methods are considered a *de facto* standard in machine learning and data analysis since its introduction.

Feature selection maps $\mathbf{x} \in \mathbb{R}^d \rightarrow \mathbf{y} \in \mathbb{R}^p$ where $p < d$. The reduction criteria usually either maintains or improves the accuracy of classifiers trained with this data, while simplifying the complexity of the model [4, 5, 13]. Feature selection methods can be classified into three main groups:

1. Filter methods: perform the feature selection as a pre-processing step. It is independent from the learning stage and relies only on the attributes of the data [5]. Despite the lower time consumption, one of the main disadvantages of filters is the fact that they do not interact with the learning method; which usually leads to worse performance when compared to other methods [4].
2. Wrapper methods: use a learning algorithm as a subroutine, measuring the usefulness of each subset of features with the prediction performance of the learning algorithm over a validation set [5]. Although usually wrapper methods show a better performance when compared with filter methods, they have a much higher computation cost which increases as the number of features in the data increases [4].

3. Embedded methods: the feature selection process is built into the learning method, so it can use the core of the learning method to rank the features by their importance [4,5].

Tsymbal *et al.* [28] and Pechenizkiy *et al.* [22] demonstrated the benefits of integrating feature selection methods to the DS framework. However, the data sets used had a sample-feature ratio higher than one. In addition, the filtering method proposed by Almeida (2014) [20] achieved higher performances in terms of accuracy only on datasets with less than 20 features and 3 classes. These authors were able to show that feature selection methods incorporated to the DS framework can improve the performance of DS methods on some data sets and overcome some of the problems related to high-dimensional data sets. In addition, Maciel-Guerra *et al.* (2019) studied a protein microarray data set to evaluate the performance of DS methods. The authors demonstrated that for this single data set, DS methods do not outperform majority voting.

3   Experiments

3.1   Data Sets

The experiments are conducted on 10 real-world high-dimensional data sets (Table 2. Nine of those data sets are obtained from the *Feature Selection data sets* (Arizona State University [17]) and another from the UCI machine learning repository [11]. We considered only data sets with small sample sizes

Table 2. Data sets attributes

Data set	Sample (s)	Features (f)	Ratio (s/f)	No. of classes	Distribution	Type	Source
Leukemia/ALLAML	72	7129	0.0101	2	65.3 - 34.7%	Microarary	[17]
Arcene	200	10000	0.02	2	56 - 44%	Mass spectrometry	[17]
Basehock	1993	4862	0.4099	2	49.9 - 50.1%	Text	[17]
Colon	62	2000	0.031	2	64.5 - 35.5%	Microarary	[17]
Dexter	600	20000	0.03	2	50 - 50%	Text	[11]
Gli85	85	22283	0.0038	2	30.6 - 69.4%	Microarray	[17]
Pcmac	1943	3289	0.5907	2	50.5 - 49.5%	Text	[17]
Prostate	102	5966	0.0171	2	49 - 51%	Microarary	[17]
Relathe	1427	4322	0.3302	2	54.6 - 45.4%	Text	[17]
Smk-Can	187	19993	0.0094	2	48.1 - 51.9%	Microarary	[17]

3.2   Experimental Design

All techniques are implemented using the *scikit-learn* [23] and the *DESlib* [8] libraries in Python. The experiments are conducted using 30 replicates. For each replicate, the data sets are randomly divided in 50% for the training set, 25% for the Region of Competence set and 25% for the test set as suggested by Cruz *et al.* [9]. These divisions are performed preserving the proportion of samples for each class by using the stratified k-fold cross validation function in the *scikit-learn* [23] library.

The pool of classifiers is composed of 11 decision trees, as suggested by Woloszynski *et al.* [31], with pruning level set to 10. The pool is generated using the bagging technique, similarly to the methodology followed by Woloszynski in [30,31]. An odd number of classifiers is chosen to overcome decision ties. These classifiers are used due to their instability when trained with different sets of data, i.e., small differences on the training set can create different trees [31]. Following the recent survey on DS techniques [9], the size of the Region of Competence K is set to 7 neighbours for all the techniques based on k-NN. Moreover, as suggested by Cruz and Soares in [9,26,27], 30% of the base classifiers are selected using accuracy and diversity for the techniques DES-kNN and DES-kMeans. In addition, the number of clusters of DES-kMeans is set to 5.

3.3 Comparison of Techniques

The Friedman test F_F with Iman-Davenport correction [14] is employed for statistical comparison of multiple classifier system techniques as suggested by Cruz and Demsar in [9,10]. F_F ranks the algorithms for each data set separately, i.e. the best algorithm gets ranking 1, the second best ranking 2, and so on. In case of ties, average ranks are assigned. F_F is distributed according to the $\mathcal{X}_F^2$ distribution and the Iman-Davenport extension (Eq. 1) is distributed according to the F-distribution (Eq. 2) with $k-1$ and $(N-1) \times (k-1)$ degrees of freedom. The null-hypothesis states that all algorithms are equivalent and so their average ranks should be equal.

$$\mathcal{X}_F^2 = \frac{12N}{k(k+1)} \left[\sum_j R_j^2 - \frac{k(k+1)^2}{4} \right] \tag{1}$$

$$F_F = \frac{(N-1)\mathcal{X}_F^2}{N(k-1) - \mathcal{X}_F^2} \tag{2}$$

where R_j is the average rank of the j-th classifier, k is the number of classifiers and N is the number of data sets.

The rank of each method is calculated using the weighted ranking approach proposed by Yu in [32], which considers the differences among the average performance metric values between classifiers for each data set [32]. The best performing algorithm is the one with the lowest average rank. Next, as suggested by [10], to compare all classifiers against a control, we use the Bonferroni-Dunn test with the following test equation to compare two classifiers:

$$z = (R_i - R_j) \Big/ \sqrt{\frac{k(k+1)}{6N}} \tag{3}$$

where R_i is the rank of i-th classifier, k is the number of classifiers and N is the number of data sets. The z value is than used to find the corresponding p-value from the two-tailed normal distribution table, which is subsequently compared to an appropriate significance level α. The Bonferroni-Dunn test subsequently divides α by $k-1$ to control the family-wise error rate. The level of $\alpha = 0.05$ is considered as significance level. Hence, the level of $p < 0.0022$ was considered as statistically significant.

4   Results and Discussion

Accuracy is calculated for all experiments and averaged over the 10 replications. In addition, the rank of all classifiers for each data set is calculated according to the weighted ranking approach proposed by Yu in [32] and averaged to measure the Z-score to find its respective p-value. With 22 classifiers and 10 data sets, the Friedman test is distributed according to the F distribution with $22-1 = 21$ and $(10-1) \times (22-1) = 189$ degrees of freedom. The critical value of F $(21,189)$ for $\alpha = 0.0001$ is 2.8165.

Table 1 shows the 21 DS methods used. Nine are dynamic classifier selection methods (the first eight and the last one based on the date the paper was published) which select a single classifier from the pool of classifiers. The remaining methods are dynamic ensemble selection techniques, which select an ensemble of classifiers from the initial pool. These techniques are selected because they incorporate all the major breakthroughs in the area of dynamic selection on the past three decades as highlighted by Cruz *et al.* [9], i.e, the papers which proposed different selection techniques to be incorporated into the DS framework. We compare the average rank obtained by the majority voting method (static selection) against the 21 DS methods.

The first experiment assesses classifier performance without feature selection. Table 3 shows the average accuracy and standard deviation for each data set, the average rank, Z-score and p-value results for all the classifiers that had a rank lower than majority voting without feature selection. The F_F statistic is 4.7468, so the null-hypothesis can be rejected with 99.99% confidence. To compare all classifiers against a control, majority voting, the Bonferroni-Dunn test is used to measure the Z-score for each classifier. Even though there are 3 classifiers (KNORA-U, KNOP-U and DES-P) with a better rank than majority voting, none of them is statistically different from majority voting.

The second experiment (Table 4) employs the univariate feature selection method. Instead of selecting a specific number of features, a p-value is computed using the ANOVA F-test and a family wise error rate is used to select them with a 95% confidence level. For high-dimensional data sets it is necessary to compute a feature selection method to reduce the complexity of the problem. Nonetheless, this is not an easy task due to the "curse of dimensionality". Therefore, the feature selection method chosen must be fast to compute because of the large number of features. This is the reasoning for choosing a filter method as the feature selection approach. For this experiment, the F_F statistical value was 5.6171. Aposteriori, KNORA-U and KNOP-U had a lower rank when compared with majority voting, nevertheless, these ranks are not statistically different.

The aforementioned results show that for all the data sets we tested with more features than samples dynamic selection methods are statistically equivalent to

Table 3. Average accuracy, ranking, z-score and respective p-value for the classifiers that had a lower rank when compared with majority without feature selection and the oracle results

	knop u	knora u	des p	majority voting	oracle
Allaml	0.9333 ± 0.0563	0.9296 ± 0.0573	0.9296 ± 0.0573	0.9278 ± 0.0576	1 ± 0
Arcene	0.7067 ± 0.0646	0.7173 ± 0.0667	0.704 ± 0.0576	0.71 ± 0.0586	0.996 ± 0.0095
Basehock	0.9045 ± 0.0138	0.8922 ± 0.0132	0.8923 ± 0.0132	0.8917 ± 0.0128	0.9625 ± 0.013
Colon	0.7542 ± 0.0926	0.7604 ± 0.0983	0.7417 ± 0.1067	0.7438 ± 0.0932	0.9896 ± 0.0233
Dexter	0.8789 ± 0.0357	0.8722 ± 0.0366	0.8731 ± 0.0368	0.8729 ± 0.0367	0.992 ± 0.0111
Gli	0.8136 ± 0.0678	0.8212 ± 0.0713	0.8273 ± 0.0737	0.8152 ± 0.0713	0.9924 ± 0.0169
Pcmac	0.8648 ± 0.0162	0.8582 ± 0.0158	0.8576 ± 0.016	0.8575 ± 0.016	0.9421 ± 0.0261
Prostate	0.8782 ± 0.0724	0.8833 ± 0.064	0.8821 ± 0.062	0.8821 ± 0.0688	0.9936 ± 0.0143
Relathe	0.825 ± 0.0205	0.8085 ± 0.0226	0.8121 ± 0.0228	0.8076 ± 0.0217	0.9525 ± 0.0193
Smkcan	0.6298 ± 0.0637	0.6255 ± 0.0551	0.6262 ± 0.0661	0.6135 ± 0.053	0.9986 ± 0.0053
Rank	5,60	6,49	6,82	7,47	−
z score	0,6413	0,3374	0,2220	0	−
p-value	0,5213	0,7358	0,8243	1	−

Table 4. Average accuracy, ranking, z-score and respective p-value for the classifiers that had a lower rank when compared with majority with univariate feature selection based on the ANOVA-F test with Family-wise Error rate and the oracle results

	aposteriori	knop u	knora u	majority voting	oracle
Allaml	0.9111 ± 0.0682	0.9315 ± 0.0652	0.9333 ± 0.0664	0.9333 ± 0.0664	1 ± 0
Arcene	0.6907 ± 0.0593	0.7727 ± 0.065	0.7693 ± 0.0655	0.766 ± 0.0687	0.9913 ± 0.0123
Basehock	0.9048 ± 0.0144	0.9063 ± 0.0128	0.895 ± 0.0137	0.8929 ± 0.0136	0.9633 ± 0.0125
Colon	0.8625 ± 0.0987	0.7896 ± 0.0876	0.7896 ± 0.0876	0.7938 ± 0.0886	0.9688 ± 0.0419
Dexter	0.8949 ± 0.0262	0.9009 ± 0.0174	0.8976 ± 0.0166	0.8962 ± 0.0204	0.99 ± 0.0089
Gli	0.8864 ± 0.0721	0.8515 ± 0.0778	0.8545 ± 0.0866	0.8606 ± 0.0813	0.9924 ± 0.0169
Pcmac	0.8737 ± 0.0138	0.8684 ± 0.0186	0.8666 ± 0.0184	0.8641 ± 0.0153	0.9198 ± 0.0368
Prostate	0.8949 ± 0.0432	0.8936 ± 0.0656	0.8885 ± 0.0669	0.8897 ± 0.0657	0.9885 ± 0.0202
Relathe	0.8313 ± 0.0166	0.8274 ± 0.0204	0.8183 ± 0.0209	0.8139 ± 0.0204	0.9198 ± 0.0374
Smkcan	0.7553 ± 0.0568	0.7333 ± 0.0688	0.7369 ± 0.0685	0.7355 ± 0.074	0.9872 ± 0.0224
Rank	5,89	5,9	7,3	7,81	−
z score	0,6606	0,6585	0,1756	0	−
p-value	0,5089	0,5102	0,8606	1	−

a simple method such as majority voting. This result differs from the recent reviews in the literature [6,9] that showcased the higher performance of DS methods over majority voting on low-dimensions data sets. Nonetheless, the filter feature selection method chosen was able to reduce drastically the number of features (Table 5) and increase the performance of most classifiers over all data sets.

The type of data sets used in our work might explain the reasons of our findings. The data sets investigated have a far larger number of features compared to the number of instances. This situation poses a problem for machine learning techniques for some reasons: (1) wrapper methods require a reasonable computational time to select a subset of features in a large search space, hence the selection of a filter technique to reduce the dimensionality; (2) it is likely that there is insufficient data to cover the entire feature space, because the reduction of dimensionality increased the performance of 97% of 22 classifiers over 10 data sets; (3) Euclidean distance does not work on high-dimensional spaces since points are equally distance from one another.

Table 5. Number of features after applying the filter univariate feature selection based on the ANOVA-F test with Family-wise Error rate

Data sets	Features before filter	Features after filter	Reduction
Allaml	7129	130	98,18%
Arcene	10000	937	90,63%
Basehock	4862	286	94,12%
Colon	2000	16	99,20%
Dexter	20000	36	99,82%
Gli	22283	265	98,81%
Pcmac	3289	59	98,21%
Prostate	5966	198	96,68%
Relathe	4322	126	97,08%
Smkcan	19993	63	99,68%

We focused on demonstrating that DS methods did not have high performance levels on data sets with high-dimensionality and low sample sizes when compared with a simple MCS method such as majority voting. The results suggest that the Euclidean distance used by most of the methods is not working and therefore an alternative must be proposed for these types of data set. Moreover, feature selection could be incorporate to the DS framework to select the most important features for each sample. Although the results suggest an increase in performance, they are still far from the oracle. This indicates that the features selected might still not be the best subset.

In addition, due to the properties of high-dimensional spaces, clusters can be masked [21]; and a phenomena called *local feature relevance* happens, i.e., different subsets of features are relevant for different clusters [16]. This might explain the reason why the accuracy after feature selection was still further apart from the oracle and further investigations must be conducted to overcome this issue and improve even further the results.

5   Conclusions

In this paper, we investigated how DS methods perform on high dimensional data sets, more specifically those with a sample-feature ratio below one. We compared

21 DS methods against the majority voting method. Our approach used the Friedman test with the Iman-Davenport correction to compare the averaged weighted ranking of each classifier for all data sets. If the null-hypothesis is rejected, the Bonferroni-Dunn test is used as a post-hoc test to compare all classifiers against a control (majority voting). Experiments with and without feature selection were performed and showed that for high dimensional data sets the DS methods are statistically equivalent to the majority voting. For both studies, with and without feature selection, the null-hypothesis of the F_F statistic was reject with a confidence of 99.99%. Moreover, on both studies, the Bonferroni-Dunn test showed that none of the best ranked classifiers are statistically different from the majority voting classifier, which contradicts most of the results in the literature. This paper extends the research done by Maciel-Guerra *et al.* (2019) [18] by using a more comprehensive list of data sets. Our results indicate that modifications to the traditional DS framework could be beneficial.

The future work will extend the study of DS methods on high dimensional data sets with modifications proposed to the way the region of competence works. As suggested by Aggarwal *et al.* [1], the use of L_1 norm and the natural extension the authors provide is more preferable for high dimensional spaces when compared with Euclidean distance. Therefore, it would be important to investigate whether different distance metrics can improve the region of competence. As suggested by Maciel-Guerra *et al.* (2020) [19], we will focus on subspace clustering which localise their search not only in terms of samples but in terms of features as well to overcome the issues presented by k-NN on high-dimensional data sets.

References

1. Aggarwal, C.C., Hinneburg, A., Keim, D.A.: On the surprising behavior of distance metrics in high dimensional space. In: Van den Bussche, J., Vianu, V. (eds.) ICDT 2001. LNCS, vol. 1973, pp. 420–434. Springer, Heidelberg (2001). https://doi.org/10.1007/3-540-44503-X_27
2. Agrawal, D., et al.: Challenges and opportunities with big data: a white paper prepared for the computing community consortium committee of the computing research association. Computing Research Association (2012)
3. Ballard, C., Wang, W.: Dynamic ensemble selection methods for heterogeneous data mining. In: 2016 12th World Congress on Intelligent Control and Automation (WCICA), pp. 1021–1026, June 2016. https://doi.org/10.1109/WCICA.2016.7578244
4. Bolón-Canedo, V., Sánchez-Marono, N., Alonso-Betanzos, A., Benítez, J.M., Herrera, F.: A review of microarray datasets and applied feature selection methods. Inf. Sci. **282**, 111–135 (2014). https://doi.org/10.1016/j.ins.2014.05.042
5. Bolón-Canedo, V., Sánchez-Marono, N., Alonso-Betanzos, A.: Feature selection for high-dimensional data. Prog. Artif. Intell. **5**(2), 65–75 (2016). https://doi.org/10.1007/s13748-015-0080-y
6. Britto Jr., A.S., Sabourin, R., Oliveira, L.E.S.: Dynamic selection of classifiers - a comprehensive review. Pattern Recogn. **47**(11), 3665–3680 (2014). https://doi.org/10.1016/j.patcog.2014.05.003

7. Cruz, R.M., Zakane, H.H., Sabourin, R., Cavalcanti, G.D.: Dynamic ensemble selection vs k-NN: why and when dynamic selection obtains higher classification performance? In: The Seventh International Conference on Image Processing Theory, Tools and Applications (IPTA), Montreal, Canada (2017)

8. Cruz, R.M.O., Hafemann, L.G., Sabourin, R., Cavalcanti, G.D.C.: DESlib: A Dynamic ensemble selection library in Python. arXiv preprint arXiv:1802.04967 (2018)

9. Cruz, R.M., Sabourin, R., Cavalcanti, G.D.: Dynamic classifier selection: recent advances and perspectives. Inf. Fusion **41**(Supplement C), 195–216 (2018). https://doi.org/10.1016/j.inffus.2017.09.010

10. Demšar, J.: Statistical comparisons of classifiers over multiple data sets. J. Mach. Learn. Res. **7**, 1–30 (2006)

11. Dheeru, D., Karra Taniskidou, E.: UCI machine learning repository (2017). http://archive.ics.uci.edu/ml

12. Donoho, D.L.: High-dimensional data analysis: the curses and blessings of dimensionality. In: AMS Conference on Math Challenges of the 21st century, pp. 1–33 (2000)

13. Ghojogh, B., et al.: Feature selection and feature extraction in pattern analysis: A literature review. ArXiv abs/1905.02845 (2019)

14. Iman, R.L., Davenport, J.M.: Approximations of the critical region of the fbietkan statistic. Commun. Stat. - Theory Methods **9**(6), 571–595 (1980). https://doi.org/10.1080/03610928008827904

15. Ko, A.H.R., Sabourin, R., Britto Jr., A.S.: From dynamic classifier selection to dynamic ensemble selection. Pattern Recogn. **41**(5), 1718–1731 (2008). https://doi.org/10.1016/j.patcog.2007.10.015

16. Kriegel, H.P., Kröger, P., Zimek, A.: Clustering high-dimensional data: a survey on subspace clustering, pattern-based clustering, and correlation clustering. ACM Trans. Knowl. Disc. Data **3**(1), 1–57 (2009)

17. Li, J., et al.: Feature selection datasets at arizona state university. http://featureselection.asu.edu/datasets.php. Accessed August 2018

18. Maciel-Guerra, A., Figueredo, G.P., Zuben, F.J.V., Marti, E., Twycross, J., Alcocer, M.J.C.: Microarray feature selection and dynamic selection of classifiers for early detection of insect bite hypersensitivity in horses. In: IEEE Congress on Evolutionary Computation, CEC 2019, Wellington, New Zealand, 10–13 June 2019, pp. 1157–1164. IEEE (2019). https://doi.org/10.1109/CEC.2019.8790319

19. Maciel-Guerra, A., Figueredo, G.P., Zuben, F.J.V., Marti, E., Twycross, J., Alcocer, M.J.C.: Subspace-based dynamic selection: a proof of concept using protein microarray data. In: WCCI - World Congress on Computational Intelligence, The International Joint Conference on Neural Networks (IJCNN) 2020, Glasgow, UK, 19–24 July 2020. IEEE (2020)

20. de Menezes Sabino Almeida, H.A.: Selecao dinamica de classificadores baseada em filtragem e em distancia adaptativa. Master's thesis, Federal University of Pernambuco, Recife, Brazil (2014)

21. Parsons, L., Haque, E., Liu, H.: Subspace clustering for high dimensional data: a review. SIGKDD Explor. **6**, 90–105 (2004). https://doi.org/10.1145/1007730.1007731

22. Pechenizkiy, M., Tsymbal, A., Puuronen, S., Patterson, D.: Feature extraction for dynamic integration of classifiers. Fundamenta Informaticae **77**(3), 243–275 (2007)

23. Pedregosa, F., et al.: Scikit-learn: machine learning in python. J. Mach. Learn. Res. **12**, 2825–2830 (2011)

24. Radovanovic, M., Nanopoulos, A., Ivanovic, M.: Nearest neighbors in high-dimensional data: the emergence and influence of hubs. In: Proceedings of the 26th International Conference on Machine Learning, ICML 2009, vol. 382, p. 109, January 2009. https://doi.org/10.1145/1553374.1553485

25. Radovanovic, M., Nanopoulos, A., Ivanovic, M.: Hubs in space: popular nearest neighbors in high-dimensional data. J. Mach. Learn. Res. **11**, 2487–2531 (2010)

26. Soares, R.G.F., Santana, A., Canuto, A.M.P., de Souto, M.C.P.: Using accuracy and diversity to select classifiers to build ensembles. In: The 2006 IEEE International Joint Conference on Neural Network Proceedings, pp. 1310–1316 (2006). https://doi.org/10.1109/IJCNN.2006.246844

27. de Souto, M.C.P., Soares, R.G.F., Santana, A., Canuto, A.M.P.: Empirical comparison of dynamic classifier selection methods based on diversity and accuracy for building ensembles. In: 2008 IEEE International Joint Conference on Neural Networks (IEEE World Congress on Computational Intelligence), pp. 1480–1487, June 2008. https://doi.org/10.1109/IJCNN.2008.4633992

28. Tsymbal, A., Puuronen, S., Skrypnyk, I.: Ensemble feature selection with dynamic integration of classifiers. In: International Congress on Computational Intelligence Methods and Applications CIMA2001 (2001). https://doi.org/10.1007/3-540-39963-1_44

29. Verleysen, M., François, D.: The curse of dimensionality in data mining and time series prediction. In: Cabestany, J., Prieto, A., Sandoval, F. (eds.) IWANN 2005. LNCS, vol. 3512, pp. 758–770. Springer, Heidelberg (2005). https://doi.org/10.1007/11494669_93

30. Woloszynski, T., Kurzynski, M.: A probabilistic model of classifier competence for dynamic ensemble selection. Pattern Recogn. **44**(10–11), 2656–2668 (2011). https://doi.org/10.1016/j.patcog.2011.03.020

31. Woloszynski, T., Kurzynski, M., Podsiadlo, P., Stachowiak, G.W.: A measure of competence based on random classification for dynamic ensemble selection. Inf. Fusion **13**(3), 207–213 (2012). https://doi.org/10.1016/j.inffus.2011.03.007

32. Yu, Z., et al.: A new kind of nonparametric test for statistical comparison of multiple classifiers over multiple datasets. IEEE Trans. Cybern. **47**(12), 4418–4431 (2017). https://doi.org/10.1109/TCYB.2016.2611020

Caching Suggestions Using Reinforcement Learning

Mirco Tracolli[1,2,3]([⊠]), Marco Baioletti[1]([⊠]), Valentina Poggioni[1]([⊠]),
Daniele Spiga[3]([⊠]), and on behalf of the CMS Collaboration

[1] Università degli Studi di Perugia, Perugia, Italy
m.tracolli@gmail.com, {marco.bailotti,valentina.poggioni}@unipg.it
[2] Università degli Studi di Firenze, Florence, Italy
[3] INFN Sezione di Perugia, Perugia, Italy
daniele.spiga@pg.ionfn.it

Abstract. Big data is usually processed in a decentralized computational environment with a number of distributed storage systems and processing facilities to enable both online and offline data analysis. In such a context, data access is fundamental to enhance processing efficiency as well as the user experience inspecting the data and the caching system is a solution widely adopted in many diverse domains. In this context, the optimization of cache management plays a central role to sustain the growing demand for data. In this article, we propose an autonomous approach based on a Reinforcement Learning technique to implement an agent to manage the file storing decisions. Moreover, we test the proposed method in a real context using the information on data analysis workflows of the CMS experiment at CERN.

Keywords: Cache · Optimization · Intelligent system · Big data · Data science workflow · Reinforcement learning · Addition policy

1 Introduction

The Big Data volume continues to grow in several different domains, from business to scientific fields. There are several open challenges from several perspectives: from the infrastructural point of view due to the continuous data growth and the constantly increasing amount of computational resources needed; from the user perspective because the desired pattern is to deal with these distributed resources as a single logical entity.

In this context the data lake emerged as a possible solution and it seems an ideal way to store and analyze huge amounts of data in one location. Its definition changed during the years and became a synonym of a complex infrastructure that deals with unstructured data. The difference between a data lake and a classical data warehouse is that in a data warehouse, the data is pre-processed and categorized to a fixed structure when it's stored. Instead, a data lake stores

G. Nicosia et al. (Eds.): LOD 2020, LNCS 12565, pp. 650–662, 2020.
https://doi.org/10.1007/978-3-030-64583-0_57

information of any type and structure in their native formats and make those data available for future reporting and analysis.

Thus, a popular solution in the data lake environment is to involve several centers that preserve the original data and an intermediate layer of caches that serve the user requests. In a typical scenario, these centers are connected and all together store the whole information the clients want to access.

Interesting considerations and pro-cons evaluations of data lakes can be found in [7,12]. In literature, different data lakes architectures have been proposed, designed with different purposes in different contexts. Among the others, Terrizano et al. presented and discussed the architecture proposed by IBM in [18], Kadochnikov et al. introduced in [3] a data lake prototype for HL-LHC, Skluzacek et al. proposed in [16] the system Klimatic to retrieve, index, and store geospatial data in a virtual data lake.

In this work, we want to demonstrate that the cache layer in the data lake architecture can be optimized and improved. The well-known caching systems (like in [2,4,9,19]) are neither designed nor optimized for this paradigm. Moreover, we do not want to consider only the hit rate as a general optimization criterion due to uneven and unprocessed data.

For this reason, we propose a method to optimize the related content management of the cache layer in a data lake architecture. Despite it is not the first attempt using artificial intelligence to enhance the performance of a cache, the added value of the proposed work is to manage Big Data improving the user experience, using a generalized approach and not solving a specific caching situation.

Finally, we also choose to benchmark the model using a real Big Data use case that involves the data analysis workflow of the Compact Muon Solenoid (CMS [5]) experiment at Large Hadron Collider (LHC) at the European Organization for Nuclear Research (CERN) [6].

2 Data Lake Architecture

The data lake architecture we refer to in this work is depicted in Fig. 1, where the data servers are the core of the system and they are surrounded by a cache layer composed of several caches.

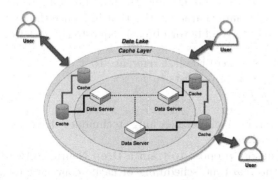

Fig. 1. Data lake model

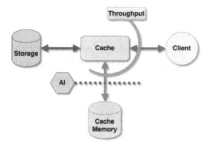

Fig. 2. Schema of the environment

The presented architecture has similarities with a Content Delivery Network (CDN) [1] and also with the web content caching (especially with video file streaming [1]) due to the intermediate cache layer, with the difference that it has to deal with unstructured data. Nevertheless, the data lake cache system has a more fine grade organization: it could be distributed in distinct regions in which each cache instance does not foresee duplicate files within its neighbor caches. However, the problem of managing a consistent amount of data persist.

With the intent to measure an improvement of the cache layer we defined a throughput measure, with the main goal to serve better the clients. Consequently, the main target is to write fewer data into the cache while maximizing the amount of data read by the clients from the cache (Fig. 2).

3   Related Works

The main policies adopted for the cache management are LRU (Least Recently Used) and LFU (Least Frequently Used). These policies are known to underperform in the big data environments because they were initially designed for evicting fixed-size pages from buffer caches [2]. Other algorithms are based on file sizes and choose to remove a file if it is the biggest (Size Big) or the smallest (Size Small) into the cache. Because these algorithms are easy to implement and do not require high computation, they are often used as a baseline to test new approaches, especially to compare new Machine Learning techniques.

Nevertheless, the main characteristics that influence the replacement process are still related only to the file (or object) requested f [4,9,19]:

- **recency**: time of (since) the last reference to f
- **frequency**: number of requests to f
- **size**: the size of f

Recently, several Machine Learning techniques have been proposed to improve the file caching.

In [11] the authors propose to train a Deep Neural Network in advance to manage better the real-time scheduling of cache content into a heterogeneous network. This article has a file set with a lower number of files compared to the problem we want to tackle, plus, it uses a cost function related to the network repeaters and it is not a generic approach that can be adapted to any type of file cache. In [14] a deep recurrent neural network is applied to predict the cache accesses and make a better caching decision. However, that solution is not feasible for our problem due to the low number of requests and unique files. Also, techniques like the Gradient Boosting Tree are used in [8] to automate cache management of distributed data. But, in this example they have a small cache size with file distributed as chunks in the storage system (Hadoop Distributed File System, HDFS) and still, the number of unique files is not high. There are also Deep Reinforcement Learning approaches like [15], where a small number of files have to be placed in an optimal hierarchical cache system. The most recent work proposing a DRL approach for smart caching is [20], where the authors propose a Wolpertinger architecture for smart caching optimizing the hit rate. This approach is different from the one proposed here because in that case the authors assume that all the files in the cache will have the same size, while we consider the file size as the key for a better data management.

Thus, the problems solved by the cited works are not comparable in size with the one we have chosen, where we have a higher number of files to manage and a huge amount of requests per day (more details in the use case). Moreover, a direct comparison is not trivial because they mostly use synthetic datasets not released online and generated with a specific distribution that match their use cases, such as network traffic or HDFS data requests, etc.

4   Proposed Approach

The proposed approach is based on Reinforcement Learning (RL), a Machine Learning paradigm that learns how to dynamically operate in any environment. The main technique used is the associative search task, known also as *contextual bandits* in literature [17]. We use the trial-and-error technique to search for the best actions in a given situation. As a control algorithm, we chose the off-policy Q-Learning to update the action values. Furthermore, we integrated a delayed reward mechanisms to improve the action decisions due to the cache environment changes. The environment is described in Fig. 3 and, in detail, we have a cache with a specific size S and a network bandwidth B.

The state s is composed using the basic information (features) taken from the file f statistics collected during the environment lifetime. The statistic history traces 14 days of the file's requests and it is deleted if the file is no more requested and it is not present into the cache memory.

To decide whether a file f should be stored in the cache we selected the following features of f: the size, the number of requests, and the delta time, i.e. the elapsed time between the current time and the time of the last request.

Since the number of states must be finite, the file features are discretized in a finite number of classes (using a simple binning technique with ranges) and hence the state S is represented as a triple (σ, ϕ, δ), where σ, ϕ, δ are the labels

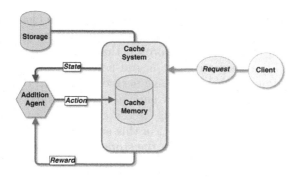

Fig. 3. Generic Reinforcement Learning technique schema

of the class to which the size, the frequency and the delta time of f belongs, respectively.

The underlying cache uses the Least Recently Used (LRU) criterion to decide which files have to be removed.

The system also takes into account of two cache watermarks, a higher W_{high} watermark and a lower W_{low} watermark. When the size of the files stored in the cache reaches W_{high}, the least recently used files are removed until W_{low} is reached. This mechanism prevents that the cache memory becomes too full. The parameters to W_{high} and W_{low} are set according to the amount of available space.

Each time the agent requests a file f, the state s is computed in terms of the statistics of f. For each possible state, there are two actions: *Store* and *DontStore*. We consider as next state s' (Formula 1) the same input state s (Contextual Bandit). As a consequence, the agent has to learn which action is the best for the current state through the delayed rewards used as stimuli of past decision traces with the intent to allow only the valuable files to be stored. To assign the delayed rewards we memorize the last action a taken for the input state s and we reward or penalize that action basing on cache constraints when the same state s occur in the next requests.

The agent updates the action values of the state using the Q-Learning method, where the quality Q (goodness) of the state is a combination of the state and the action value:

$$Q(s,a) \leftarrow Q(s,a) + \alpha \cdot (r + \gamma \cdot Q(s', best) - Q(s,a)) \tag{1}$$

where s is the state, s' is the next state and a the action. The α parameter is the learning rate and γ is the discount factor. In our experiment we fixed $\alpha = 0.9$ and $\gamma = 0.5$. The reward/punishment r is computed as follows.

The cache environment will assign a negative or positive reward on the base of several constraints that involve cache statistics. In our method, we choose to penalize those actions that store files which do not increment the value of Read on Hit, in particular, we want that the amount of hits is higher than the miss files.

The positive reward r is the size of f, while the negative reward is $-f$. These rewards are assigned to the last action decided for the state where the action was taken and it remains for the next requests until a new request of the file changes the stored action. In the Algorithm 1 a simplified description of the simulation process is described. The algorithm flow is divided in two branches: the training phase and the normal execution. During the training phase the exploration of new actions is done with probability ϵ (initialized to 1.0), that is decreased with an exponential decay rate of $5 \cdot 10^{-8}$ (with a lower limit of 0.1) each time the *Update* method of the *additionTable* is called. The choices made during the training phase are collected in a map and this information is used to provide late rewards to our agent as explained before.

The normal phase consists in performing the best action from the *additionTable* for the corresponding state.

```
Data: file requests
Result: a cache simulation
Initialize cache
Initialize historical statistic table stats ; // to collect file statistics
Initialize Q-Table additionTable ; // all action values to 0
Initialize map doneActions ; // history of taken actions
for each requested file req do
    stats.update(req.filename)
    hit ← cache.check(req.filename)
    if training = true then
        currentState ← cache.State
        if hit = false then
            if map.check(req.filename) = true then
                prevState, prevAction ← map.get(req.filename)
                reward ← makeReward(cache.Statistics, prevAction, req)
                additionTable.Update(prevState, prevAction, reward)
            end
            if randomNumber() < ε then
                currentAction ← additionTable.peekRandomAction(currentState)
            else
                currentAction ← additionTable.peekBestAction(currentState)
            end
            reward ← makeReward(cache.Statistics, currentAction, req)
            additionTable.Update(curState, curAction, reward)
            if currentAction = Store then
                cache.addFile(req.filename, req.size)
            end
            map.insert(req.filename, curState, curAction)
        else
            prevState, prevAction ← map.get(req.filename)
            reward ← makeReward(cache.Statistics, prevAction, req)
            additionTable.Update(prevState, prevAction, reward)
        end
    else
        if hit = false then
            currentState ← cache.State
            bestAction ← cache.peekBestAction(currentState)
            if bestAction = Store then
                cache.addFile(req.filename, req.size)
            end
        end
    end
end
```

Algorithm 1: Algorithm SCDL

4.1   Evaluation Measures

The main measure we focus, as mentioned in Sect. 2, is the *Throughput*, defined in the following way:

$$Throughput = \frac{readOnHitData}{writtenData} \tag{2}$$

where *readOnHitData* is the sum of the size of the files on which the cache had hit and the *writtenData* is the sum of the size of the files written into the cache.

   To get a wider picture of the effects of our method, we have chosen to monitor other characteristics of the cache. As a consequence, we defined the following metrics to verify the impact of the proposed strategies:

– Cache disk space: this measure, named *Cost*, aims at quantifying how much the cache is working, i.e. when it deletes or writes files:

$$Cost = \frac{writtenData + deletedData}{cacheSize} \tag{3}$$

As a result, if the cache maintains a good file composition in memory, it will write and delete fewer data.
– Network bandwidth: to measure the impact on this constraint we simply take into account the miss data over the bandwidth of the cache. If the network is saturated, the cache cannot retrieve files anymore:

$$BandSaturation = \frac{readOnMissData}{cacheBandwidth} \tag{4}$$

– CPU efficiency: it is also possible to access to the CPU time and Wall time of each request on the log file used in our experiments. This allow us to measure the CPU efficiency as follows:

$$CPUEfficiency = \frac{\sum CPUTime}{\sum WallTime} \tag{5}$$

These data are taken as they are from the request logs because they are strictly related to the use case. In particular, the request contains a variable denoting whether the file was served locally or remotely. We assumed in the simulation that a file served by the cache has no impact to the CPU efficiency and, as a consequence, it will have a CPU efficiency equal to the local request, otherwise, it will be lower due to the time the file is acquired (remote file). In the latter case, we penalize the CPU efficiency subtracting the difference between the local and remote access efficiency to simulate the loss in performances. This difference is calculated using the average values of the local and remote efficiencies in the whole dataset.

5   Use Case

As stated in Sect. 1, our experiments refer to a real use case in the domain of the High Energy physics, in particular we will test our method within the CMS workflow. In the current computing model of the CMS experiment at LHC recorded and simulated data are distributed and replicated via a central management system across a worldwide network of custodial storage sites. There are different typologies of datasets, created for different purposes, called "tiers" that have to be managed with different policies. For instance, the most frequently used samples might be kept on disk ready to be accessed with the lowest latency possible, while custodial data, and more in general legacy data that are rarely accessed can be stored on tape facilities. In this model the computing payloads are sent and distributed across the grid sites based on the input data location (pre-placement) in order to reduce inefficiencies from network latency. Nevertheless there are few cases where CMS payloads run at one site while reading data from a remote location. Several metadata of the CMS jobs are collected and stored on a dedicated monitoring system. For the purpose of this work, data relative to the information of the input datasets requested by each job have been analyzed for the whole 2018.

The agent was trained in 4 months of the year 2018: January, February, March, and April. The training phase includes watching these four months in several episodes. For the remaining period, the Q-table was only applied and tested. During the tests, we also compared different policies other than LRU in order to verify the effect of each algorithm in the context of data analysis.

5.1   Data Analysis

The data used in this work are requests of user analysis that contain several information such as the file name, the file size or the file type. The information contained in the log database refers to the data popularity research made in CMS [10,13]. The collected data are archived per day and, to simulate an environment with a single cache within the current data analysis workflow, we decided to filter the requests by region, choosing the Italy (IT) region.

A preliminary analysis of data gives us useful information about the dimension of the problem. In the Italian region, the number of requested files is very high with respect to the number of users and sites where the jobs are run (Fig. 4, 5). The workflow provides that the user can launch tasks that have more jobs, and those jobs can have several files. Nevertheless, the number of requests per file is not too high and the average value is small even if we do not consider the files requested only 1 time (Fig. 6). The file size we found in the logs is most concentrated in the range between $2GB$ to $4GB$ but spaced from $1GB$ to $40GB$.

Moreover, this is confirmed in the other regions. As a consequence, we expect to have comparable results regardless of the region. For example, the main difference is with the USA region is that due to the higher number of users and sites there is a higher order of magnitude in the number of requests, tasks, and jobs. Still, the number of the unique files per day is $\sim 10^5$, the same of the Italian

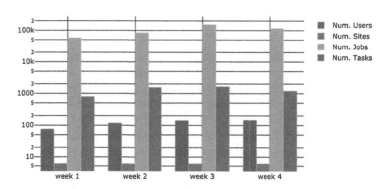

Fig. 4. Data general statistics of Italy requests in January 2018

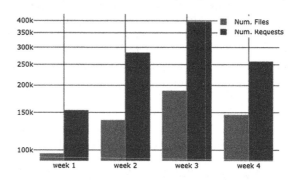

Fig. 5. Number of request statistics of Italy requests in January 2018

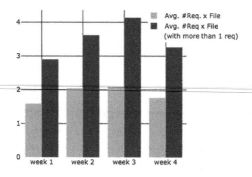

Fig. 6. Average number of request statistics of Italy requests in January 2018

region and the average number of requests per file (mean value ∼6) is comparable with the Italian region (mean value ∼4).

6   Experimental Results

SCDL algorithm has been implemented and tested over a simulation produced with the data described in Sect. 5. The test run in order to demonstrate the ability of SCDL to improve the measures described in Sect. 4.1 with respect other caching policies described in Sect. 3. Because LRU is the main algorithm adopted in a cache layer it is considered as the baseline for these results.

The results shown in Table 1 attest that the throughput and the cost of SCDL approach are higher compared to the other policies, also to the most used LRU. The measure results are shown in percentage respect to the denominator defined in 4.1.

Table 1. Test results of IT region.

Cache type	Throughput	Cost	Read on hit ratio	Band sat.	CPU Eff.
SCDL	**79.43%**	**50.68%**	21.22%	58.94%	58.75%
LFU	65.01%	104.73%	**33.29%**	**51.00%**	**60.92%**
Size Big	49.02%	111.73%	28.55%	54.40%	60.41%
LRU	47.15%	112.84%	27.64%	54.93%	59.90%
Size Small	46.71%	113.01%	27.39%	55.01%	59.73%

We scored the best result for the throughput measurement. Moreover, the cost of the Reinforcement Learning approach is an half compared to the other policies. Nevertheless, the read on hit ratio is not the best, due to our request filtering. We also made a control check by looking at the CPU efficiency and we have not to be penalized so much because we scored just ∼2% less. Hence, our decision to filter the requests is working quite well, without a heavy impact on the user. Even though the CPU efficiency is only measured to verify the side effects in this particular use case and it could not be considered a trustful measure in this simulation because we used historical information of a different environment to approximate the cache effect measurement.

During the simulation, we also monitored the other metrics values using several plots. In the following part, we will see a specific view of the results. We can see in the throughput plot (Fig. 7, Eq. 2) the different behavior of each policy used. Our higher values demonstrate that we respect the constraint to write less and maintain a high as possible output towards the clients.

With a bandwidth (Eq. 4) of $10Gbit$ we saw that it is always slightly higher (∼5%) than the other methods because of our mechanism file filtering by importance and then, the cache has to serve them remotely using the network.

The target to have less work (from the cache perspective) is fulfilled. The mainstream LRU policy can not compete also with LFU function in cost measure in these tests (Eq. 3). But, our RL approach makes the cache low active and do

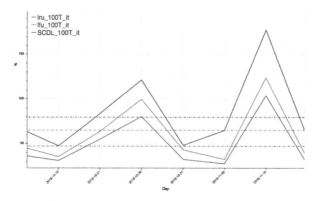

Fig. 7. Throughput detail of the period between October and November

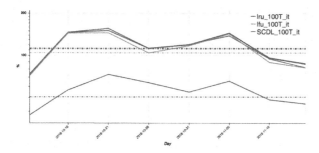

Fig. 8. Cost detail of the period between October and November

only the minimum operations. This can be seen in a lower amount of written and deleted data but there are several missed files not stored that affect the network because the cache will serve those files in proxy mode.

7 Conclusions

In this work the algorithm Smart Cache for Data Lake (SCDL) has been presented. The well-known cache policies do not fit well in the data lake architecture and in general in the Big Data environment. In particular, if we want to improve the user experience with a transparent layer of cache to reduce the infrastructure costs, the well-known cache algorithms are not the answer. SCDL has better overall performances to fulfill the client's expectations and, if we can improve these aspects without changing or upgrading the hardware, it would be the best result we can have to face future challenges. This kind of cache could adapt itself to the context and the users, choosing the best files from the requests. Further studies will focus to improve the hit and miss ratio, to optimize the network utilization and, in general, the costs to maintain the cache system. Moreover, we would experiment with an agent that can choose also an eviction policy and compare our solution with other RL approaches. The desired side effect of our optimization could be a considerable decrease in bandwidth utilization.

References

1. Adhikari, V.K., et al.: Unreeling netflix: understanding and improving multi-CDN movie delivery. In: 2012 Proceedings IEEE INFOCOM, pp. 1620–1628. IEEE (2012)
2. Ali, W., Shamsuddin, S.M., Ismail, A.S., et al.: A survey of web caching and prefetching. Int. J. Adv. Soft Comput. Appl. **3**(1), 18–44 (2011)
3. Bird, I., Campana, S., Girone, M., Espinal, X., McCance, G., Schovancová, J.: Architecture and prototype of a WLCG data lake for HL-LHC. In: EPJ Web of Conferences, vol. 214, p. 04024. EDP Sciences (2019)
4. Chen, T.: Obtaining the optimal cache document replacement policy for the caching system of an EC website. Eur. J. Oper. Res. **181**(2), 828–841 (2007)
5. Collaboration, C., et al.: The CMS experiment at the CERN LHC (2008)
6. Fanfani, A., et al.: Distributed analysis in CMS. J. Grid Comput. **8**(2), 159–179 (2010)
7. Fang, H.: Managing data lakes in big data era: what's a data lake and why has it became popular in data management ecosystem. In: 2015 IEEE International Conference on Cyber Technology in Automation, Control, and Intelligent Systems (CYBER), pp. 820–824. IEEE (2015)
8. Herodotou, H.: Autocache: employing machine learning to automate caching in distributed file systems. In: International Conference on Data Engineering Workshops (ICDEW), pp. 133–139 (2019)
9. Koskela, T., Heikkonen, J., Kaski, K.: Web cache optimization with nonlinear model using object features. Comput. Netw. **43**(6), 805–817 (2003)
10. Kuznetsov, V., Li, T., Giommi, L., Bonacorsi, D., Wildish, T.: Predicting dataset popularity for the CMS experiment. arXiv preprint arXiv:1602.07226, 2016
11. Lei, L., You, L., Dai, G., Vu, T.X., Yuan, D., Chatzinotas, S.: A deep learning approach for optimizing content delivering in cache-enabled HetNet. In: 2017 International Symposium on Wireless Communication Systems (ISWCS), pp. 449–453. IEEE (2017)
12. Madera, C., Laurent, A.: The next information architecture evolution: the data lake wave. In: Proceedings of the 8th International Conference on Management of Digital EcoSystems, pp. 174–180 (2016)
13. Meoni, M., Perego, R., Tonellotto, N.: Dataset popularity prediction for caching of CMS big data. J. Grid Comput. **16**(2), 211–228 (2018)
14. Narayanan, A., Verma, S., Ramadan, E., Babaie, P., Zhang, Z.-L.: Deepcache: a deep learning based framework for content caching. In: Proceedings of the 2018 Workshop on Network Meets AI & ML, pp. 48–53 (2018)
15. Sadeghi, A., Wang, G., Giannakis, G.B.: Deep reinforcement learning for adaptive caching in hierarchical content delivery networks. IEEE Trans. Cognit. Commun. Netw. **5**(4), 1024–1033 (2019)
16. Skluzacek, T.J., Chard, K., Foster, I.: Klimatic: a virtual data lake for harvesting and distribution of geospatial data. In: 2016 1st Joint International Workshop on Parallel Data Storage and Data Intensive Scalable Computing Systems (PDSW-DISCS), pp. 31–36. IEEE (2016)

17. Sutton, R.S., Barto, A.G.: Reinforcement Learning: An Introduction. MIT Press, Cambridge (2018)
18. Terrizzano, I.G., Schwarz, P.M., Roth, M., Colino, J.E.: Data wrangling: the challenging yourney from the wild to the lake. In: CIDR (2015)
19. Tian, G., Liebelt, M.: An effectiveness-based adaptive cache replacement policy. Microprocess. Microsyst. **38**(1), 98–111 (2014)
20. Zhong, C., Gursoy, M.C., Velipasalar, S.: A deep reinforcement learning-based framework for content caching. In: 2018 52nd Annual Conference on Information Sciences and Systems (CISS), pp. 1–6. IEEE (2018)

Image Features Anonymization for Privacy Aware Machine Learning

David Nizar Jaidan$^{(\boxtimes)}$ and Le Toan Duong$^{(\boxtimes)}$

Innovation LAB Scalian, Labège, France
`david.jaidan@gmail.com`

Abstract. Data privacy is a major public concern in the digital age, especially image data that provides a large amount of information. The wariness about the use of image data affect the sharing and publication of these data. In this context, Differential Privacy is one of the most effective approaches to preserve privacy. However, the scope of Differential Privacy applications remains very limited (e.g. structured data) and focuses on a straightforward approaches (e.g. image anonymization). In this study, we investigate and evaluate an approach that anonymize image feature using a PCA-based differentially private algorithm. We show that Differential Privacy can be more effective once it is introduced after feature extraction. Using this approach, we show that the image re-constructor and facial recognition models are fooled within specified conditions. We believe that our study will significantly motivate the use of Differential Privacy techniques to anonymize image features, which may lead to increase data sharing and privacy preservation.

Keywords: Image anonymization · Privacy · Machine learning · Transfer learning · Features anonymization · Face recognition · Differential privacy

1 Introduction

Nowadays, a huge amount of image data has been captured by personal and commercial cameras. By way of indication, there are on average 4000 photos uploaded on Facebook every second [1] including a large amount of labeled data. On one hand, sharing data can be in fact very useful for firmes, researchers and governments. Indeed, it allows researchers to have a high quality data and to share resources in case of collaboration. Face images can be shared to develop facial recognition technologies for use in identification and criminal investigations [2]. Images obtained from video surveillance or conventional cameras are used to train pedestrian detection models [3]. One the other hand, the processing of image data raises privacy concerns and sensitive data can be exposed quite easily. In fact, information about individuals such as location, habits or health condition can be leaked when these data are handled. The new European Union's General Data Protection Regulation (GDPR) [4] states that personal data controllers

© Springer Nature Switzerland AG 2020
G. Nicosia et al. (Eds.): LOD 2020, LNCS 12565, pp. 663–675, 2020.
https://doi.org/10.1007/978-3-030-64583-0_58

must set up appropriate organizational and technical measures to ensure the data protection principles. Therefore, business processes that handle personal data must be designed and built with consideration of these principles and must provide guarantees to protect data (e.g. using anonymization).

Several studies proposed solutions for privacy-preserving images sharing based on cryptography [5,6]. While encrypted images can keep the entire content and be secured by a secret key, they are not suitable for data publishing. Other privacy-preserving image methods are based on the idea that the private parts (e.g. faces and texts) are blurred. Some blurring techniques are proposed such as pixelization and blurring. However, several depixelization techniques have been developed such as cubic convolution interpolation [7] and reconstruction attacks [8]. In addition, two facial recognition algorithms by the PCA (Principal Component Analysis) and LDA (Linear Discriminant Analysis) were used to evaluate these anonymization techniques and the results show that they were ineffective [9]. The standard anonymization approaches can sometimes be effective, but without formal mathematical guarantees. To deal with this lack, Dwork [10,11] introduced a new approach named ϵ-differential privacy that provides formal privacy guarantees. A number of studies have involved Differential Privacy (DP) in order to anonymize image data. Fan [12] developed a differentially private methods for pixelization and Gaussian blur, Senekane [13] report a scheme for privacy-preserving image classification using Support Vector Machine (SVM) and DP, and Abadi [14] demonstrate how it is possible to train effectively a deep neural networks with non-convex objectives (under a modest privacy budget). However, the scope of DP applications focuses on a straightforward approaches and aim to anonymize image or model within the training process without focusing on image feature.

In this paper, we investigate and evaluate an approach that anonymize image feature using a PCA-based differentially private algorithm. The paper is organized as follows: Sect. 2 provides a brief description of data and the different studied approaches, feature extraction, feature anonymization and Machine Learning tools. Then, the evaluation and the results analysis are discussed in Sect. 3. Finally, Sect. 4 concludes the paper and gives an outlook of future works.

2   Methodology

In this study, we investigate and evaluate an approach that anonymize images using a PCA-based differentially private algorithm. Figure 1 shows a schematic view of the different steps of our study. We furthermore aimed to anonymize image data in two iterations:

Image Anonymization: We evaluated image anonymization by directly applying the anonymisation method to the image as a 2-D array of pixels. Since the anonymization algorithm input is an X_{np} matrix of n rows and p columns, each row refers to a sample (one image). For this, each image represented by a matrix of pixels is transformed to a one-dimensional vector before anonymization. Once anonymization is performed, we reshape the vector in order to obtain the same size as the original image. To evaluate the anonymization utility, we used two CNNs (Convolutional Neural Network) for MNIST and CIFAR-10 datasets (Fig. 2).

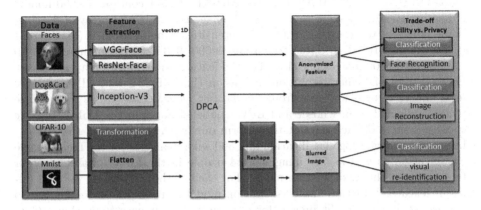

Fig. 1. Schematic view of the study methodology.

Fig. 2. Architecture of Convolutional Neural Network models used to classify MNIST (CNN1) and CIFAR-10 (CNN2).

Image Feature Anonymization: Using different Feature extractor models, we extract features from the *Dogs & Cats* and *Faces* image datasets. We use three extractor models: Inception-V3, VGG-Face and ResNet-Face. These models are described in Sect. 2.2. Then, we use a differential privacy algorithm, DPCA (detailed in Sect. 2.3) to anonymize the extracted features. Finally, we evaluate the Machine Learning algorithms performances (utility) and the privacy preserving quality trade-off. On the one hand, in order to measure the utility, we evaluate the performance of the Machine Learning models (Feed-Forward Neural Network and XGBoost; detailed in Sect. 2.4) to classify the images (cat/dog classification for the *Dogs&Cats* dataset and male/female classification for the *Faces* dataset). On the other hand, to measure the privacy preserving quality, we try to reconstruct the original image using an extractor (Encoder-Decoder) that we have trained (detailed in Sect. 2.5) for the *Dogs&Cats* dataset and we try to recognize faces using facial recognition models (detailed in Sect. 2.5).

2.1   Data Description

In this study, we propose an image anonymization framework based on DP. To that end, we tested our approach on four image datasets that are handled in order of increasing complexity and tasks (see Fig. 1). We first evaluated the direct application of anonymization on image data. For that, we chose low-resolution datasets (MNIST and CIFAR-10). The MNIST dataset contains 70,000 images of handwritten digits (zero to nine). Each image is a 28 * 28 * 1 array of floating-point numbers representing gray-scale intensities ranging from 0 (black) to 1 (white). The CIFAR-10 dataset contains 60000 of 32 * 32 colour images in 10 classes, with 6000 images per class. For the feature anonymization approach, we handled data of higher resolution dimension than MNIST and CIFAR-10, we used images of dogs and cats from the dataset *Dogs&Cats* [15] as well as face images collected from wikiset and LFW (Labeled Faces in the Wild) [16]. We chose a sample of each dataset to perform our experiments: 7500 *Dogs&Cats* images for the cat/dog classification and 500 wikiset images for the male/female classification as well as 727 images of 14 people in the LFW dataset in order to investigate the identification model. For each of these datasets, we used different feature extraction and classification models. Image pre-processing is also performed in order to fit images for our study. The images from *Dogs&Cats* are resized to 256 * 256 * 3. For wikiset and LFW images, poor quality images are removed, then the face part is captured, cropped and resized to 224 * 224 * 3 before the feature extraction.

2.2   Feature Extraction Techniques

In this section, we focus on models used to extract image features. We chose to use pre-trained models (Transfer Learning). We note that training the model on a large dataset exhibit good generalization and reduces over-fitting. Thus, in this study, we are using three pre-trained models: Inception-V3 for dogs&cats images, VGG-Face and ResNet-Face for face images.

Inception-V3: The Inception-v3 introduced by Google [17] is the 3rd version in a series of Deep Learning Convolutional Architectures model. This model was trained using a dataset of 1,000 classes obtained from the original ImageNet dataset which is composed of 1,331,167 images split into two sets each containing 1,281,167 and 50,000 images for training and evaluation. We have extracted the front part of the dense layers and introduced a GlobalAveragePooling2D layer at the end to extract features. Thus, the feature size is 2048. We selected this model as a feature extractor for *Dogs&Cats* images because among other pre-trained models on ImageNet, Inception-V3 is one of the best in terms of prediction score, model complexity and GPU inference time. Inception-V3 was trained for the ImageNet Large Visual Recognition Challenge where it was a first runner up.

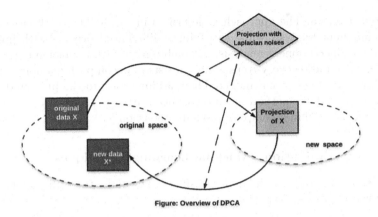

Figure: Overview of DPCA

Fig. 3. Schematic view of the DPCA anonymization approach.

VGG-Face: The VGG-Face CNN descriptors are computed using our CNN implementation based on the VGG-Very-Deep-16 CNN architecture introduced by [18] and are evaluated on the Labeled Faces in the Wild [16] and the YouTube Faces [19] dataset. We use the VGG-Face which was trained on a large dataset containing 2,622 subjects to generate features from faces. First, the model is trained for a classification task using a SoftMax activation function in the output layer to classify faces as individuals. Then, the output layer is removed in order to produce a face vector representation, called "face embedding". Then, the model is fine-tuned in order to reduce Euclidean distance between vectors that are generated based on the same person and also to extend the distance between vectors generated by different persons. Finally, all extracted features using the VGG-Face extractor have a dimension of 2622.

ResNet-Face: As part of this study, we also used another feature extractor model for faces that is integrated into the Python *face_recognition* module of dlib. The model is based on ResNet-34 from [20], but with fewer layers and about half the number of filters. It is trained on about 3 million images in the same way as for VGG-Face. This extractor provides a vector of 128 features for each image. Regarding our anonymization method using DPCA, we chose this extractor instead of the VGG-Face because the feature vector in ResNet-34 is smaller.

2.3 Feature Anonymization Techniques: Differential Privacy

Differential Privacy (DP) is a highly efficient approach to preserving privacy which is not based on an encryption approach. It is a promising mechanism of data collection, use and privacy. In this study, we chose Differential-Private Data Publishing Through Component Analysis (DPCA) algorithm [21] because this method focuses on optimizing the application of noise and maximizing the utility of the output data. The approach is described in Fig. 3. The method uses principal component analysis while satisfying ϵ-differential privacy with improved utility through component analysis, by combining it with Laplacian noise during the projection and recovery steps. This consists of decomposing a noisy variance matrix instead of the exact matrix. Then, adding noise to the projected matrix before recovery. This method has already been used to anonymize textual data and time series within a complex anomaly detection task [22].

2.4   Utility Evaluation : Machine Learning Techniques

In this study, we used two different classifiers to evaluate the utility of data after anonymization:

eXtreme Gradient Boosting (XGBoost): XGBoost [23] is considered as the "state-of-the-art" for structured data. It a scalable algorithm that has recently been dominating Machine Learning application. It is an implementation of the gradient boosted decision trees designed for speed and high performance. XGBoost creates new models that predict the residues or errors of previous models, then combine these models to make the final prediction. XGBoost uses a gradient descent algorithm to minimize the loss function when adding new models.

Multi-layer Feed-Forward Neural Network: A Feed-Forward Neural Network [24] is structured in several layers in which information flows only from the input layer to the output layer. The model has at least three layers (input layer, hidden layer and output layer). Each node is a neuron that uses a nonlinear activation function except for the input nodes. Feed-Forward Neural Network uses a supervised learning technique for training "back-propagation". Each layer of the model is composed of a flexible number of neurons; the neurons of the last layer are the outputs of the model. It is able to separate data that is not linearly separable.

2.5   Privacy Validation

Image Reconstruction. We designed a re-constructor model based on the work of [25]. The model is trained as a decoder within an auto-encoder. However, the encoder component (Inception-v3) and the decoder component are not symmetric. In addition, only the decoder was trained, the parameters of the encoder are fixed during the training. The architecture of the re-constructor is shown in Fig. 4.

Face Recognition. In this part, we focused on facial identification (recognition). For that, we used identification models to assess if they are able to retrieve the person's identity from the anonymized features. The identification model

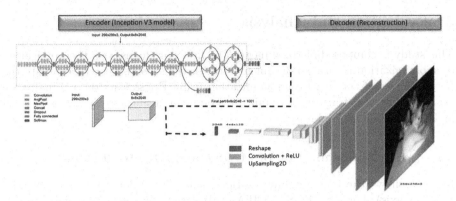

Fig. 4. Schematic view of the image reconstruction methodology. The Inception-V3 graphic was designed by Google.

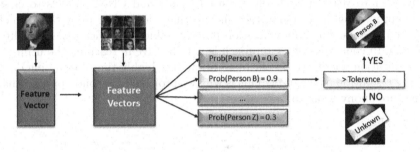

Fig. 5. Schematic view of the the face recognition methodology.

takes face features as input. We also designed a reference database for identification models with 727 images of 14 persons labeled by name. We deployed two different approaches: In the first one, the reference database is used to train classification models (Neural network and XGBoost), while in the second approach, the database is used to compare and evaluate the similarity between feature vectors in order to predict identity. We defined two identifiers based on this second approach. The first one uses the *compare_faces* function in the *face_recognition* module. The feature vector from the test is compared to the labeled feature vectors in the reference database. Then, the person with the highest similarity probability is chosen and if this probability is greater than a given threshold (0.6 by default in the *compare_faces* function), the identifier predicts this person, otherwise it predicts an unknown (see Fig. 5). The second identifier is based upon the idea that the feature vector is certainly generated by one of the individuals in the reference database. This is the case in a closed environment such as, employees in a company or passengers on a ship. To do this, we compare the feature vector we want to identify with all labeled vectors in the reference base and we chose the closest individual to the feature vector. Thus, there are no unknowns in the prediction.

3   Evaluation and Analysis

This study is composed of three parts. The first part focuses on the application of anonymization algorithm on image data. In the second part, we evaluate the performances of the feature anonymization using Dogs&Cats data. Finally, we evaluate how well feature anonymization performs using face image data from Wikiset and the LFW dataset.

3.1   Image Anonymization of MNIST and CIFAR-10 Data

Figure 6 shows the image anonymization using DPCA for different k (y-axis) and ϵ (x-axis). For the MNIST (CIFAR-10) dataset, we note that with an $\epsilon = 0,1$ and $\epsilon = 1$ ($\epsilon = 1$ and $\epsilon = 10^3$), the image is completely corrupted regardless of the value of k. The noise is more important and spread with increasing k. In order to reach a trade-off, we used CNN1 and CNN2 models to classify anonymized images. We found (not shown) that the models are not able to classify the datasets. The anonymization process reduces significantly the utility of the data. In conclusion, this approach is not effective in reaching the utility and anonymization quality trade-off. In addition, it is highly time consuming despite the low resolution of the data. This is why we proposed a second approach based on feature anonymization.

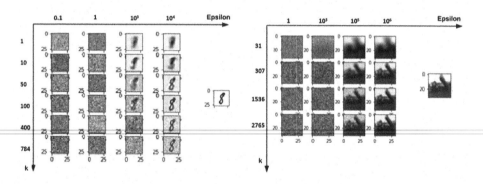

Fig. 6. Anonymized images according to epsilon and k for (left) MNIST and (right) CIFAR-10 datasets

3.2   Feature Anonymization of *Dogs&Cats* Data

We extract features from the *Dogs&Cats* dataset using Inception-V3 feature extractor, then we applied DPCA anonymization algorithm, and we evaluate the Machine Learning algorithms performances (utility) and the privacy preserving quality (reconstruction) trade-off. The Table 1 shows the classification model scores for different k-value and ϵ-value. In this case, we show results from the

Table 1. The utility quality (accuracy scores) on the *Dogs&Cats* dataset according to different $k - value$ and $\epsilon - value$

	$\epsilon = 1e3$	$\epsilon = 5e3$	$\epsilon = 3e4$	$\epsilon = 1e5$
$k = 516$	0.49	0.68	0.97	0.99
$k = 1024$	0.49	0.64	0.97	0.99
$k = 2000$	0.49	0.63	0.97	0.99

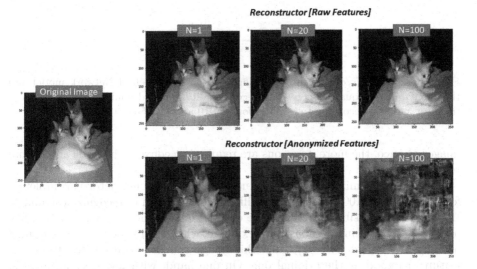

Fig. 7. The reconstructed image from the raw and anonymized features for three different scenarios based on the number of trained images ($N = 1, 20, 100$).

Feed-Forward Neural Network which gives better scores than XGBoost. We note that with an $\epsilon = 3e4$, the accuracy of the model is 97%, versus a model accuracy of 99% for the raw image data. We assume that, at this privacy level, the 2% loss of accuracy is acceptable. We note also that accuracy increases with increasing ϵ. Furthermore, we are not able to see clearly the effect of the k-parameter on the utility of anonymized features. We used the reconstruction model (presented in the Sect. 2.5) to assess the anonymization quality. The model was trained to reconstruct the image from the raw features. Then, the model is evaluated on its efficiency to reconstruct similar images from the anonymized features. Here, we chose $k = 1024$ and $\epsilon = 7e4$ in order to preserve a high accuracy score. The re-constructor was trained in three different scenarios based on the number of trained images ($N = 1, 20, 100$). The Fig. 7 shows the reconstructed images from both the raw and anonymized features for $N = 1, 20, 100$. For raw features, The constructor is able to reconstruct the original image. We note that the reconstruction quality decreases slightly with increasing number of trained images N. In fact, the more we train the re-constructor on a larger number of

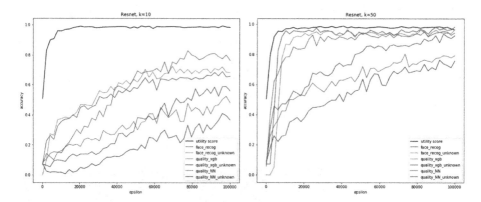

Fig. 8. The utility score (accuracy) using Feed-Forward Neural Network model and the anonymization quality score (face_recog, quality_xgb and quality_NN) using identification models, according to ϵ values for k = 10 and k = 50.

images at once, the more iterations are required to converge towards an optimal result.

For the anonymized features, we note that the re-constructor is completely fooled when N = 100, it seems very difficult for a human to recognize the image. However, it is possible to retrieve the original image when there are only a few images in the training database. For example, when the re-constructor is trained on only one image (N = 1), then this image can be reconstructed from raw features as good as the original one. On one hand, with few training images, the re-constructor is able to recognize feature vectors easily. So, it is not fooled after the feature anonymization. On the other hand, if there are several images in the training database (which is often the case), the distance between feature vectors is reduced. In addition, the feature vectors are still closer to each other after anonymization using DP.

3.3 Feature Anonymization of Wikiset and LFW Data

In this section, we focus on the feature anonymization of face images. In this case, we deal with a facial recognition and gender classification (M&F) task. We show results for features extracted using ResNet-Face, the feature vectors are 128-dimensional. In addition, we note that VGG-Face extractor results (not presented here) are similar to those of the ResNet-Face. Figure 8 shows the utility score (accuracy) and the anonymization quality score (face_recog, quality_xgb and quality_NN; see Sect. 2.5) according to ϵ values for k = 10 and k = 50. On one hand, we evaluated the effect of the k and ϵ parameters on the utility of anonymized features, which is measured by the prediction (accuracy) score using a classification model. We note that accuracy increases quickly with increasing ϵ. We note also that the k parameter has no significant effect on the utility as the privacy is poor (large epsilon). On the other hand, we evaluated the effect of the k and ϵ parameters on anonymization quality, which is measured by the

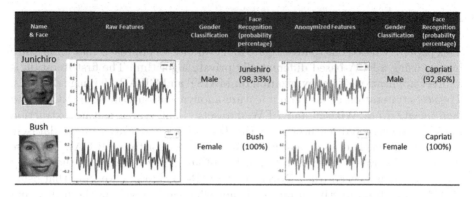

Fig. 9. Example of the face recognition model fooled using differential privacy anonymization.

accuracy score using identification models (detailed in Sect. 2.5). We note that accuracy increases slightly (compared to utility) with increasing ϵ. This finding is more pronounced when k is small (k = 10). This means that the first principal components (k = 10) provide enough features in order to classify a gender. This is an important result because it shows that reducing the k-parameter value in the DPCA leads to reducing the noise level without affecting the privacy quality. This result is consistent because the added noise depends on the ϵ value. Furthermore, we note the significant effect of k within the identification score. The higher k is, the more identifiers are able to recognize the faces. However, there is a significant gap between the classification score and the identification score (when $k = 10$). This means that the optimal ϵ value can be found in order to have a high accuracy score for the classification model (more utility) and a low accuracy score for the identification model (more privacy). But if k is large ($k = 50$), the accuracy curves for all models show a similar behavior. In this case, we are not able to get both secure and useful data.

The Fig. 9 shows an example (from the sample of results shown in Fig. 8) of how we were able to fool the facial recognition model using the feature anonymization approach. For the raw features, the face recognition and the classification models are able to identify the two figures (Junichiro and Bush) as well as their gender, respectively. For the anonymized features, the gender classification is right in both cases while the identification model is misleading with a high confidence score (92.86% and 100%). The face recognition model confused the two individuals with another one existing in the face database (Capriati). In conclusion, our results show that within the feature vector extracted from images, there are features characterizing gender and other features allowing facial recognition. PCA allows us to find the relevant combination of features for classification. Thus, we can preserve the utility of anonymized data. It is important to find an optimal pair (k; ϵ) for each anonymization task.

4   Conclusions and Future Work

This paper is carried out to investigate and evaluate an approach that anonymize images using a PCA-based differentially private algorithm. The first iteration is a direct application of anonymization on MNIST and CIFAR-10 image datasets. The second iteration is based on a feature anonymization using Dogs&Cats and face images from Wikiset and LFW datasets. Using three feature extraction models, we extracted features from the *Dogs&Cats* and *Faces* image datasets and anonymized them using a differentially private algorithm (DPCA). Then, we examined the trade-off between the utility of anonymized data and the privacy preservation quality. For the MNIST and CIFAR-10 datasets, we show that using DPCA algorithm directly on images is not effective in reaching the utility and the anonymization quality trade-off. For the *Dogs&Cats* image dataset, we show that the image re-constructor is completely fooled as long as the training image dataset number is greater than N = 100. However, it is possible to retrieve the original image when there are only a few images in the training database. For the *Faces* image dataset, we show that we can find a trade-off and an optimal pair (k; epsilon) for each classification task. This means that we can preserve the utility of anonymized data with a high accuracy score while misleading the facial recognition models.

In this study, we evaluate and compare two different anonymization approaches using Differential Privacy techniques and we have shown that it is possible to reach a trade-off on anonymized features.

As future work, we plan to apply anonymization on video-based data combining Convolutional Neural Network (CNN) with Differential Privacy methods. In order to preserve privacy and motivate data sharing, we emphasize the need to apply, explore and compare alternative anonymization methods on various images and to evaluate the anonymization approach using more complex task classification.

Acknowledgments. This work is supported by Scalian.

References

1. FacebookStats, 28th March 2020. https://www.omnicoreagency.com/facebook-statistics/
2. Agaga, S.: Facial recognition in criminal investigation, January 2018
3. Antonio, J.A., Romero, M.: Pedestrians' detection methods in video images: a literature review. In: 2018 International Conference on Computational Science and Computational Intelligence (CSCI), pp. 354–360. IEEE (2018)
4. Jan Philipp Albrecht: How the GDPR will change the world. Eur. Data Prot. L. Rev. **2**, 287 (2016)
5. Wang, Q., Hu, S, Ren, K., Wang, J., Wang, Z., Du, M.: Catch me in the dark: effective privacy-preserving outsourcing of feature extractions over image data. In: IEEE INFOCOM 2016-the 35th Annual IEEE International Conference on Computer Communications, pp. 1–9. IEEE (2016)

6. Zhang, L., Jung, T., Liu, C., Ding, X., Li, X., Liu, Y.: POP: privacy-preserving outsourced photo sharing and searching for mobile devices. In: 2015 IEEE 35th International Conference on Distributed Computing Systems, pp. 308–317 (2015)
7. Keys, R.: Cubic convolution interpolation for digital image processing. IEEE Trans. Acoust. Speech Signal Process. **29**(6), 1153–1160 (1981)
8. Cavedon, L., Foschini, L., Vigna, G.: Getting the face behind the squares: reconstructing pixelized video streams. In: WOOT, pp. 37–45 (2011)
9. Dufaux, F., Ebrahimi, T.: A framework for the validation of privacy protection solutions in video surveillance. In: 2010 IEEE International Conference on Multimedia and Expo, pp. 66–71. IEEE (2010)
10. Dwork, C.: Differential privacy. In: Encyclopedia of Cryptography and Security, pp. 338–340 (2011)
11. Dwork, C.: Differential privacy. In: Bugliesi, M., Preneel, B., Sassone, V., Wegener, I. (eds.) ICALP 2006. LNCS, vol. 4052, pp. 1–12. Springer, Heidelberg (2006). https://doi.org/10.1007/11787006_1
12. Fan, L.: Differential privacy for image publication (2019)
13. Senekane, M.: Differentially private image classification using support vector machine and differential privacy. Mach. Learn. Knowl. Extract. **1**(1), 483–491 (2019)
14. Abadi, M., et al.: Deep learning with differential privacy. In: Proceedings of the 2016 ACM SIGSAC Conference on Computer and Communications Security, pp. 308–318 (2016)
15. Dogs & Cats (2013). https://www.kaggle.com/c/dogs-vs-cats/dat. Accessed 18 May 2020
16. Huang, G., Mattar, M., Berg, T., Learned-Miller, E.: Labeled faces in the wild: a database forstudying face recognition in unconstrained environments. Technical report, October 2008
17. Szegedy, C., Vanhoucke, V., Ioffe, S., Shlens, J., Wojna, Z.: Rethinking the inception architecture for computer vision. In: Proceedings of the IEEE Conference on Computer Vision and Pattern Recognition, pp. 2818–2826 (2016)
18. Parkhi, O.M., Vedaldi, A., Zisserman, A.: Deep face recognition (2015)
19. Wolf, L., Hassner, T., Maoz, I.: Face recognition in unconstrained videos with matched background similarity. In: CVPR 2011, pp. 529–534. IEEE (2011)
20. He, K., Zhang, X., Ren, S., Sun, J.: Deep residual learning for image recognition. arxiv e-prints. arXiv preprint arXiv:1512.0338, 6(7) (2015)
21. Jiang, X., Ji, Z., Wang, S., Mohammed, N., Cheng, S., Ohno-Machado, L.: Differential-private data publishing through component analysis. Trans. Data Priv. **6**(1), 19 (2013)
22. Jaidan, D.N., Carrere, M., Chemli, Z., Poisvert, R.: Data anonymization for privacy aware machine learning. In: Nicosia, G., Pardalos, P., Umeton, R., Giuffrida, G., Sciacca, V. (eds.) LOD 2019. LNCS, vol. 11943, pp. 725–737. Springer, Cham (2019). https://doi.org/10.1007/978-3-030-37599-7_60
23. Tianqi, C., Carlos, G.: Xgboost: a scalable tree boosting system. In: Proceedings of the 22Nd ACM SIGKDD International Conference on Knowledge Discovery and Data Mining, KDD 2016, pp. 785–794. ACM, New York (2016)
24. Svozil, D., Kvasnicka, V., Pospichal, J.: Introduction to multi-layer feed-forward neural networks. Chemometr. Intell. Lab. Syst. **39**(1), 43–62 (1997)
25. Radford, A., Metz, L., Chintala, S.: Unsupervised representation learning with deep convolutional generative adversarial networks. arXiv preprint arXiv:1511.06434 (2015)

Prediction of Spot Prices in Nord Pool's Day-Ahead Market Using Machine Learning and Deep Learning

Mika Rantonen$^{(\boxtimes)}$ ⓘ and Joni Korpihalkola$^{(\boxtimes)}$ ⓘ

Institute of Information Technology, JAMK University of Applied Sciences (JAMK), Piippukatu 2, 40100 Jyväskylä, Finland
{mika.rantonen,joni.korpihalkola}@jamk.fi

Abstract. Aim of this paper is to describe and compare the machine learning and deep learning based forecasting models that predict Spot prices in Nord Pool's Day-ahead market in Finland with open-source software. The liberalization of electricity markets has launched an interest in forecasting future prices and developing models on how the prices will develop. Due to the improvements in computing capabilities, more and more complex machine learning models and neural networks can be trained faster as well as the growing amount of open data enables to collect of the large and relevant dataset. The dataset consist of multiple different features ranging from weather data to production plans was constructed. Different statistical models generated forecasts from Spot price history and machine learning models were trained on the constructed dataset. The forecasts were compared to a baseline model using three different error metrics. The result was an ensemble of statistical and machine learning models, where the models' forecasts were combined and given weights by a neural network acting as a metalearner. The results also prove that the model is able to forecast the trend and seasonality of Spot prices but unable to predict sudden price spikes.

Keywords: Machine learning · Deep learning · SPOT price prediction

1 Introduction

Nowadays, the liberalization of power markets enables to trade electricity, it can be bought and sold as any other commodity. The participants of electricity market have to optimize profits and risks for example to accurately forecast of short-term future electricity prices [4].

Finnish area spot prices are defined on Nord Pool's Elspot market. Nord Pool is a public company that operates on two different markets, Elspot market and Elbas market. Elspot market is the day-ahead market, which is the focus of this publication. The participants in the Nord Pool Elspot market are mostly electricity retailers and large production plants on the buyer side and owners of large power plants on the seller side. Elspot market is divided into bidding areas,

© Springer Nature Switzerland AG 2020
G. Nicosia et al. (Eds.): LOD 2020, LNCS 12565, pp. 676–687, 2020.
https://doi.org/10.1007/978-3-030-64583-0_59

each with a transmission system operator responsible for monitoring congestion in their grid. Physical power contracts are traded in Elspot market for the next day, which is why it is also referred to as a day-ahead market. Participants in the Elspot market have until 12:00 CET to submit their purchase and sell orders for the 24-h period on the next day. The Spot market members can submit hourly orders, block orders, exclusive group orders and flexi orders [14].

The electricity prices are affected by many variables, such as electricity consumption, outside temperature and electricity production plans. The relevant variables are defined using data analysis methods to build a relevant dataset. The dataset will be used to train machine learning models and neural networks to predict future spot prices. Statistical models will also be used to forecast spot prices based on price history data [14].

2   Related Works

Based on the market needs and the evolution of neural networks, several methods already been proposed in the literature to model the electricity price. Li et al. [9] presents that the methods can be categorized into equilibrium analysis, simulation methods, time series, econometric methods, intelligent system methods, and volatility analysis. The publication compares these methods and techniques based on the model classification, time horizon and prediction accuracy. The summarized information is helpful to the verification, comparison and improvement of a specific method or hybrid method for electricity price forecasting in the competitive environment. Tan et al. [12] proposes a price forecasting method based on wavelet transform combined with Autoregressive Integrated Moving Average (ARIMA) and generalized autoregressive conditional heteroskedasticity (GARCH) models. The results show that the proposed method is far more accurate than the other compared forecast methods. Knapik [8] uses the autoregressive ordered probit, a Markov model, and an autoregressive conditional multinomial model to analyze the drivers of the process and to forecast extreme price events. The best forecasts of the extreme price events are obtained based on the ACM (1; 1) model. Amor et al. [3] proposes the predictive performance of the proposed hybrid k -factor GARMA-LLWNN model which provides evidence of the power compared to the hybrid ARFIAM-LLWNN model, the individual LLWNN model and the k-factor GARMAFIGARCH model resulting a robust forecasting method. Karabiber et al. [7] presents the three individual models for forecasting in the Danish Day-Ahead Market. The used models were Trend and Seasonal Components (TBATS), ARIMA and Artificial Neural Networks (ANN) in which ARIMA and ANN methods are used with external regressors. All three models have surpassed the benchmark seasonal naïve model and ANN have provided the best results among the three models. Aggarwal et al. [2] presents an overview of different price-forecasting methodologies and key issues have been analyzed. They concluded there is no systematic evidence of out-performance of one model over the other models on a consistent basis. Beigait et al. [5] presents, there are many approaches which can be used for electricity price forecasting, features such as multiply seasonality, high volatility and spikes make it difficult to achieve high accuracy of prediction in Lithuania's electricity price zone. The highest average accuracy during forecasting experiments was achieved using Elman neural network, however the most accurate prediction (MAPE error equal = 2.94 %) was made by using Jordan network. Voronin [14] presents a model which aims not only to predict dayahead electricity prices and high degree of accuracy but also price spikes in Nord Pool. The proposed models are based on an iterative forecasting strategy implemented as a combination of two modules separately applied to normal price and price spike prediction in which the normal price module employed the previously applied forecasting technique that was a mixture of Wavelet transform (WT), linear SARIMA and nonlinear Neural Network (NN). The price spike module was a combination of the spike probability and the spike value forecasting models. Wang et al. [15] propose an electricity price forecasting framework which consists of two-stages feature processing and an improved Support Vector Machine (SVM) classifier has been proposed to solve this problem. A new hybrid feature selector based on Grey CorrelationAnalysis (GCA) is used to process the n-dimensional time sequence as an input. Furthermore, Kernel Principle Component Analysis (KPCA) is applied to extract new features with less redundancy to boost SVM classifier in accuracy and speed. Furthermore, the differential evolution (DE) algorithm obtains the appropriate super parameters for differential evolution DE based Support Vector Machine (DESVM) automatically and efficiently. Conejo et al. [6] presents the recommended set of variables are demand, differenced demand, and electricity prices lagged by 24 and 168 h. Furthermore, for week-ahead forecasts, the regression model with an hourly approach is recommended, while for day-ahead forecasts, the Seasonal Autoregressive Integrated Moving Average (SARIMAX)-model is recommended, which also includes electricity prices lagged by 1 and 2 h as input variables and includes electricity prices lagged by 1 and 2 h as input variables. Su et al. [11] proposed data-driven predictive models for natural gas price forecasting based on common machine learning tools such a as ANN, SVM, gradient boosting machines (GBM), and Gaussian process regression (GPR). Results show that the ANN gets better prediction performance compared with SVM, GBM, and GPR. Brusaferri et al. [1] proposed a probabilistic energy price forecast based on Bayesian deep learning techniques. The results show the capability of the proposed method to achieve robust performances in out-of-sample conditions while providing forecast uncertainty indications.

3   Dataset

3.1   Data Sources and Feature Engineering and Importance

The electricity prices are affected by many variables therefore the dataset has be to collected from many different resources. Therefore the data can be various format and the granularity of data can vary drastically. Electricity prices are highly volatile due to the fact that there is no economically viable way to store large amounts of electricity. Unexpected demands in electricity, shortages, transmission failures, generator failures are some of the usual causes for price spikes [10]. Since the spot price is dictated by the market, some price spikes may occur due to market gaming or false speculations [8].

Since the spot price is calculated for every hour of the day, and the price data is available in an hourly frequency, it is necessary for all others variables to be in an hourly frequency. The fluctuating spot price is clearly cyclical. There is a clear 24-h cycle of how the price usually behaves, a weekly cycle with the price being lower on the weekends and a yearly cycle with price being higher in winter times. Simply using a number to describe the time does not convey its cyclical nature to a machine learning model. Feature engineering principles are used to create new features. With time series data, rolling mean and standard deviation with varying window sizes was used as features. A rolling mean with a window size of 36 for example would calculate the mean of the latest 36 spot prices. A lagged version of spot price can also be used as a feature. Spot prices lagged by one day and one week were added as features, because the adequate forecast results can be achieved by just using the prices from the day before or week before. Having too many features can negatively affect the accuracy of the models. A list of features to be deleted was created from the result of random forest algorithm's permutative importances function.

Feature importances were listed from the trained random forest model. This provides a percentage for each feature meaning how much the feature affects the final prediction. To compare the random forest's Gini importances and permutative importances, both were retrieved from the same random forest model. According to Gini importance, the precipitation and cyclical month features had zero impact on the final prediction. The most significant features were the spot price one week ago, spot price 24 h ago, Prophet's daily seasonality and Prophet's additive terms. All these features had an importance score of 0.05 or higher. Permutation importance, calculated Prophet's daily seasonality to be the most important feature with a 0.15 importance score. The second-best feature was Prophet's additive terms, all other features were below 0.07 score. Interestingly, energy consumption in Sweden, temperature in Sweden and Finland and yearly seasonality of Prophet gained a negative importance score. The results of Gini and permutation importances are vastly different. Both agree that precipitation data at its current form is not useful. However, no further assumptions regarding what features to cut can be made. Testing of one-hot-encoded data, the importance scores claimed that the one-hot encoded features had no importance on the prediction. However, when the random forest was trained without one-hot encoded features, the losses were higher. This most likely means that the feature importance score is not compatible with one-hot encoded features and may not always provide informative results.

4   Predicting the Day-Ahead Hourly Prices

4.1   Autoregression

An autoregression (AR) model was trained that forecasts the next 24 h. The model is evaluated on the whole test dataset using the walk-forward validation method. As would be expected, the model reacts a day late to the sudden upward or downward spikes in spot prices as seen in Fig. 1. However, if there are no sudden spikes, the model is able to produce an acceptable forecast on normal workdays. For a simple model, it achieves quite low mean absolute error (MAE) and root-mean-square errors (RMSE).

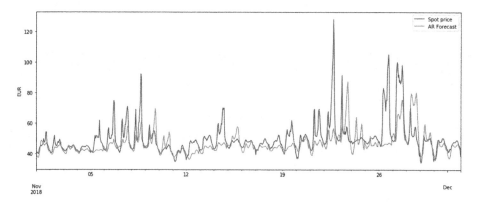

Fig. 1. Autoregression model predictions compared to real spot prices

4.2   Prophet

The prophet model can be tuned by changing the following hyperparameters: *changepoints*, *n_changepoints*, *changepoint_range* and *changepoint_prior_scale*. Changepoint is when the time series trend changes suddenly and *changepoint_prior_scale* determines how much the changepoints fit into the data. In case of the spot price, the spot price weekly trend lowers during the weekends and yearly trend lowers during the summers, at least usually. The effects of holidays can also be changed. It was seen earlier that the spot price trend lowers during winter holidays. Prophet model forecasts from spot price history data alone, since the use of exogenous variables is not supported. The prophet model is fast to train; therefore, a grid search was implemented to find the best hyperparameters. According to [13] the model can perform worse with more data, since a longer history can mean that the model is overfit to past data that is maybe not as relevant in the future. Thus, the length of the training data was also included in the grid search. After grid search, the best parameters for the model were found. The changepoint prior scale was at 0.05, n_changepoints at 48, holidays_prior_scale at 10, seasonali-ty_prior_scale at 10, seasonality mode in additive and the model was only given data from 2018. The model was made to predict on the test dataset, and the forecasts were compared to actual spot prices. The model predicts spot prices on average to be 7.2 euros lower or higher than the real value. When plotting the forecast values and actual values in a line graph in Fig. 2, it can be seen that the model has learned the daily and weekly seasonality well. The weakness of this model is that it is unable to predict sudden spikes in the spot price. The Prophet model provides a data frame of yearly, monthly, weekly and daily seasonality along with additive terms.

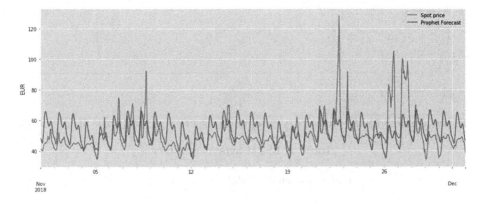

Fig. 2. Prophet's forecasts and spot prices

4.3   Onestep LSTM

A multivariate Long short-term memory (LSTM) network uses multiple features to predict the spot price. In total there are 26 features in the training dataset. The created network uses the spot price and other features at timestep $t-1$ to predict the spot price at timestep t. The network is a simple network with a single hidden layer, which is an LSTM layer that is fed the input directly and that connects to a fully-connected layer. The fully-connected layer then gives the spot price value as output. The model uses the last 48 timesteps, which means that the input shape will be 48 rows long with 26 columns. The model provides good results, as can be seen in Fig. 3, however it is not suited well for spot price forecasting, since the buy and sell orders are sent the day before, which means that predicting one hour ahead has no real-world application. This model should be modified to make predictions more than $t+1$ timesteps ahead. Hyperparameter optimization was conducted in a brute force way of simply trying different combinations of hyperparameters and calculating RMSE and MAE values to determine the best hyperparameters. The tested hyperparameters are listed in Table 1.

4.4   Encoder-Decoder LSTM

In order to forecast more than one hour ahead of the current time, the model needs to have a multistep output. In other words, the previous model was made

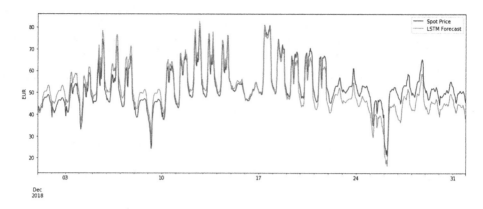

Fig. 3. Onestep LSTM forecasts (yellow) and spot prices (blue) (Color figure online)

Table 1. LSTM hyperparameters

Hyperparameters	Values
EPOCHS	20, 50
NEURONS	10, 20, 40, 60, 80
OPTIMIZERS	ADAM, RMSPROP, SGD
LEARNING RATES	0.001, 0.0005, 0.0001
BATCH SIZED	1, 4, 12, 24, 48, 168

to predict, the output was a single float number, the output needs to be modified to be a sequence of numbers that would be the forecast horizon. The encoder-decoder LSTM contains two sub-models, the encoder and decoder. The encoder creates an internal representation of the input sequence, which the decoder will then use to predict a sequence. This is called a sequence-to-sequence model which has been used to automate translations between languages. The input and output timesteps that the model is trained on can be modified *i.e.* the model can take as input a week's worth of data or only a day's worth of data and forecast either the next day or the next week spot prices. This was used to change the forecast horizon of the model. The network has two hidden layers and utilizes a repeat vector to repeat the input between two LSTM layers. The CuDNNLSTM is an optimized version of the regular LSTM layer, which can be only trained with Nvidia's graphics processing units (GPU). In the final version, two more hidden LSTM layers were added. The network training time increased slightly, but test results are better. Input weight regularization and low learning rate of 0.00001 were also found to reduce loss and improve forecast accuracy. Learning rates of 0.01 or higher made the model ignore daily seasonality and draw flat lines across the forecast horizon as a result. This problem also appeared when the forecast horizon was a week or longer. The hyperparameters that achieved the best results are listed below in Table 2.

Table 2. Best LSTM Encoder-decoder training hyperparameters

Hyperparameters	Values
OPTIMIZER	ADAM
LEARNING RATE	0.00001
CU_DNNLSTM_2 LAYER	100
CU_DNNLSTM_3 LAYER	100
DENSE_2 LAYER	100
DENSE_3 LAYER	100
EPOCHS	50
BATCH SIZE	12

4.5   CNN-LSTM

The previous LSTM Encoder-decoder was changed to include convolution layers. This was achieved by adding two one-dimensional convolution layers and using one-dimensional max pooling, as visualized in Fig. 4. The output from the convolutional layers is flattened and sent to the LSTM layer. This architecture is similar to the encoder-decoder, in this case the encoder is the convolutional neural network (CNN) and the decoder is the LSTM.

4.6   Ensemble Learning

Ensemble learning was implemented by creating a stacking model, where predictions of multiple models were combined. The best combination method was found to be the metalearner, where the neural network was a fully-connected neural network with two hidden layers. The models to combine where selected from previous test results. For a 24-h forecast, the best result was achieved by combining the forecasts of the LSTM network, ARe model and Prophet. A 36-h forecast was also created to test how the performance compared to a 24-h forecast. In the 36-h ensemble model, three LSTM models were combined with AR and prophet model forecasts. In the 48-h forecast, a combination of four LSTM models achieved the best results.

4.7   Forecast Accuracy Evaluation

The accuracy of forecasting methods and trained forecasting models will be measured on a test dataset. The test dataset will be separate from the training data and will be a continuum of the training data timewise. The test dataset will be ten weeks of data before the end of the year 2018. This means that the test data is from October 22 to December 31 as shown in Fig. 5.

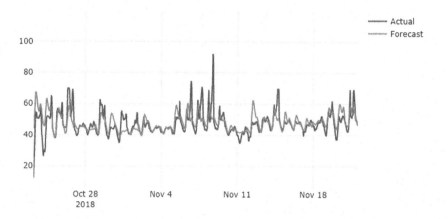

Fig. 4. CNN-LSTM results with forecast horizon of 48

Fig. 5. Training and test data split

5   Results

The best forecasts were, depending on the forecast horizon, usually achieved with neural networks or the ensemble model, where the forecasts of models' autoregression, Prophet and multiple LSTMs were combined, and the forecasts were given weights with a separate metalearner neural network. After many rounds of feature engineering, the dataset that was constructed had the following features:

- Cyclical hour, day, weekday and month.
- Electricity consumption in Finland and Sweden
- Emission allowances price
- Hydro power production in Sweden
- Hydro reservoir levels in Norway and Sweden
- Nuclear power production in Finland and Sweden
- Precipitation near Harsprånget
- Prophet's daily, weekly and yearly seasonality
- Spot price, moving average and standard deviation from X hours ago, where X is the forecast horizon
- Temperature in Finland and Sweden
- Wind power production in Sweden

The best results from all the forecasting models were documented and written on the Tables 3, 4 and 5. The lowest errors per error metric are bolded. The results from all the 24 h ahead forecast models were written down on the Table 3. The

Table 3. 24 h ahead forecast errors.

Model name	MAE	MAPE	RMSE
AR	4.90	**9.04**	8.10
Prophet	6.84	9.15	13.99
LSTM	**4.80**	9.17	7.45
CNN-LSTM	5.40	9.98	7.99
Ensemble	4.96	9.41	**7.12**

best performing models were autoregressive, LSTM, CNN-LSTM and ensemble, where their error values were quite close to each other.

Table 4 shows the results for models with forecast horizon 36. The encoder-decoder LSTM is the best model in this horizon on almost all metrics. The CNN-LSTM and ensemble were the second-best models and had almost similar error values.

Table 4. 36 h ahead forecast errors.

Model name	MAE	MAPE	RMSE
AR	6.39	11.88	9.53
Prophet	6.84	**9.15**	13.99
LSTM	**5.08**	9.57	**8.08**
CNN-LSTM	5.41	10.10	8.40
Ensemble	5.30	10.30	8.24

The results with forecast horizon 48 were recorded in Table 5. In this case the CNN-LSTM model was the best based on MAE and RMSE metrics. The autoregressive, LSTM and ensemble models share the second place with mostly similar error values.

To summarize, the best 24 h ahead forecast was made by the LSTM model, where the forecast differs ±4.80 euros from the spot price on average. The best 36 h ahead forecast was performed by LSTM again, where the forecast error is ±5.08 euros. The best 48 h ahead was achieved by the CNN-LSTM model, where forecasts were ±5.0 euros different from spot prices on average. The best one-week ahead forecasts were achieved by the ensemble model, where the forecasts were off by ±5.08 euros on average.

Table 5. 48 h ahead forecast errors.

Model name	MAE	MAPE	RMSE
AR	5.57	10.22	9.24
Prophet	6.84	**9.15**	13.99
LSTM	5.41	10.04	8.17
CNN-LSTM	**5.08**	9.76	**7.77**
Ensemble	5.44	10.25	8.14

6   Conclusion

The aim of the paper was to present the developed the machine learning and deep learning based forecasting models that predict Spot prices in Nord Pool's Day-ahead market in Finland with open-source software. The measurements show that the depending on the forecast horizon the neural networks or the ensemble model, where the forecasts of models' autoregression, Prophet and multiple LSTMs were combined, and the forecasts were given weights with a separate metalearner neural network. The developed machine learning model is able to forecast the trend and seasonality of Spot prices but unable to predict sudden price spikes. The price spikes may happen for a variety of reasons such as a breakdown of a transformer. The best course of action would be to design a separate model to predict price spikes and connect real time event observer. This price spike prediction model could then be combined with current models to improve performance. Further feature engineering and optimizing the used neural networks or switching to a new architecture could improve model accuracy and lower forecast error. When using sequence that were the length of a week or longer, the neural networks predictions would flatten to an almost even line. This happened because the sequence was too long and instead of fluctuating daily, the neural network thinks the best way to reduce loss is to just draw a flat line. This could be fixed by allowing length of input sequence and output sequence to be different. Future improvements will contains the data from machine failure due to the sudden peaks can be recognized.

References

1. Bayesian deep learning based method for probabilistic forecast of day-ahead electricity prices. Appl. Energy **250**, 1158–1175 (2019)
2. Aggarwal, S., Saini, L., Kumar, A.: Electricity price forecasting in deregulated markets: a review and evaluation. Int. J. Electr. Power Energy Syst. **31**(1), 13–22 (2009)
3. Amor, S., Boubaker, H., Belkacem, L.: Forecasting electricity spot price for nord pool market with a hybrid k-factor GARMA-LLWNN model. J. Forecast. **37**(8), 832–851 (2018)
4. Beigaite, R., krilavicius, T., Man, K.: Electricity price forecasting for nord pool data, pp. 1–6, January 2018

5. Beigaite, R.: Electricity price forecasting for nord pool data using recurrent neural networks (2018)

6. Conejo, A., Contreras, J., Espínola, R., Plazas, M.: Forecasting electricity prices for a day-ahead pool-based electric energy market. Int. J. Forecast. **21**, 435–462 (2005)

7. Karabiber, O.A., Xydis, G.: Electricity price forecasting in the danish day-ahead market using the TBATS, ANN and ARIMA Methods. Energies **12**(5), 1–29 (2019)

8. Knapik, O.: Modeling and forecasting electricity price jumps in the nord poolpower market. Technical report, Aarhus University, Department of Economics and Business Economics, Aarhus V, Denmark (2017)

9. Li, G., Lawarree, J., Liu, C.C.: State-of-the-art of electricity price forecasting in a grid environment. In: Rebennack, S., Pardalos, P., Pereira, M., Iliadis, N. (eds.) Handbook of Power Systems II. Energy Systems, pp. 161–187. Springer, Heidelberg (2010). https://doi.org/10.1007/978-3-642-12686-4_6

10. Proietti, T., Haldrup, N., Knapik, O.: Spikes and memory in (nord pool) electricity price spot prices. Technical report, Tor Vergata University, CEIS, Rome, Italy (2017)

11. Su, M., Zhang, Z., Zhu, Y., Donglan, Z., Wen, W.: Data driven natural gas spot price prediction models using machine learning methods. Energies **12**, 1680 (2019)

12. Tan, Z., Zhang, J., Wang, J., Xu, J.: Day-ahead electricity price forecasting using wavelet transform combined with ARIMA and GARCH models. Appl. Energy **87**(11), 3606–3610 (2010)

13. Taylor, S., Letham, B.: Forecasting at scale. Am. Stat. **72**, 37–45 (2017)

14. Voronin, S.: Price spike forecasting in a competitive day-ahead energy market. Ph.D. thesis, LUT School of Technology, Lappeenranta University of Technology (2013)

15. Wang, K., Xu, C., Zhang, Y., Guo, S., Zomaya, A.Y.: Robust big data analytics for electricity price forecasting in the smart grid. IEEE Trans. Big Data **5**(1), 34–45 (2019)

Unit Propagation by Means
of Coordinate-Wise Minimization

Tomáš Dlask[(✉)] [ID]

Faculty of Electrical Engineering, Czech Technical University in Prague,
Prague, Czech Republic
dlaskto2@fel.cvut.cz

Abstract. We present a novel theoretical result concerning the applicability of coordinate-wise minimization on the dual problem of linear programming (LP) relaxation of weighted partial Max-SAT that shows that every fixed point of this procedure defines a feasible primal solution. In addition, this primal solution corresponds to the result of a certain propagation rule applicable to weighted Max-SAT. Moreover, we analyze the particular case of LP relaxation of SAT and observe that coordinate-wise minimization on the dual problem resembles unit propagation and also has the same time complexity as a naive unit propagation algorithm. We compare our theoretical results with max-sum diffusion which is a coordinate-wise minimization algorithm that is used to optimize the dual of the LP relaxation of the Max-Sum problem and can in fact perform a different kind of constraint propagation, namely deciding whether a given constraint satisfaction problem (CSP) has non-empty arc consistency closure.

Keywords: LP relaxation · Unit propagation · Coordinate-wise optimization · Constraint propagation

1 Introduction

Coordinate-wise minimization is an iterative method for optimizing a given function over a given set. It usually proceeds by cyclic updates over the variables where in each update, we set each variable to be in the set of minimizers of the given function while the other variables are fixed. It was proposed in [14] that one should perform the coordinate-wise updates so that each variable is assigned a value from the relative interior of optimizers.

Formally, given a closed convex set $X \subseteq \mathbb{R}^n$, a convex function $f \colon X \to \mathbb{R}$, and an initial point $x \in X$, coordinate-wise minimization will iterate over indices of variables $i \in [n] = \{1, ..., n\}$ and update each x_i to some value from the set

$$\operatorname*{ri\ argmin}_{y \in P_i(X, x)} f(x_1, ..., x_{i-1}, y, x_{i+1}, ..., x_n) \tag{1}$$

where $P_i(X, x) = \{y \in \mathbb{R} \mid (x_1, ..., x_{i-1}, y, x_{i+1}, ..., x_n) \in X\}$ is the set of possible values for x_i such that the point x stays in X after changing its i-th coordinate

© Springer Nature Switzerland AG 2020
G. Nicosia et al. (Eds.): LOD 2020, LNCS 12565, pp. 688–699, 2020.
https://doi.org/10.1007/978-3-030-64583-0_60

and ri Y denotes the relative interior of a convex set Y. As defined in [14], a point $x \in X$ is called an interior local minimum (ILM) of f on X if for all $i \in [n]$ it holds that x_i is in the set (1). It naturally follows that interior local minima are the fixed points of coordinate-wise minimization with relative interior rule.

However, interior local minima need not be global minima of f on X in general [4,14]. Nevertheless, it is interesting to analyze the properties of these points. In the next section, we are going to recall the Max-Sum problem (resp. weighted CSP) and review the fact that interior local minima of the dual of its LP relaxation do not need to be the global optima, but they have a connection to arc consistency of an underlying CSP and coordinate-wise optimization in fact corresponds to enforcing its arc consistency. Furthermore, other studies of its LP relaxation resulted in discoveries in the field of weighted CSPs, such as tractability of new classes [8] or the notion of Optimal Soft Arc Consistency and Virtual Arc Consistency for weighted CSPs [2].

Seeing that the analysis of the LP relaxation of Max-Sum problem and its coordinate-wise optimization resulted in better understanding of the original problem motivates us to study the LP relaxation of weighted partial Max-SAT. Even though coordinate-wise optimization of its dual LP may not reach a global optimum, it provides an upper bound on its value and we prove that it can also in a precise sense perform unit propagation and a form of dominating unit-clause rule which are important constraint propagation methods for solving SAT, resp. weighted Max-SAT, similarly as arc consistency is important for CSP.

1.1 Max-Sum Problem and Arc Consistency

As reviewed in [12], the Max-Sum problem (also called MAP inference in graphical models or weighted CSP) is defined on an undirected graph (V, E) with a set of labels K as an optimization problem

$$\max \sum_{i \in V} g_i(x_i) + \sum_{\{i,j\} \in E} g_{ij}(x_i, x_j) \tag{2a}$$

$$x_i \in K \quad \forall i \in V \tag{2b}$$

where $g_i \colon K \to \mathbb{R} \cup \{-\infty\}$ and $g_{ij} \colon K^2 \to \mathbb{R} \cup \{-\infty\}$ are given and x_i are variables, where each variable corresponds to a node $i \in V$ and takes on a value from K. Finding a maximizer of the problem (2) is known to be NP-hard, which is a motivation to relax it into an LP which reads

$$\max \sum_{\substack{i \in V \\ k \in K}} g_i(k)\alpha_i(k) + \sum_{\substack{\{i,j\} \in E \\ k,l \in K}} g_{ij}(k,l)\alpha_{ij}(k,l) \tag{3a}$$

$$\sum_{l \in K} \alpha_{ij}(k,l) = \alpha_i(k) \qquad \forall \{i,j\} \in E, k \in K \tag{3b}$$

$$\sum_{k \in K} \alpha_i(k) = 1 \qquad \forall i \in V \tag{3c}$$

$$\alpha_i(k) \geq 0 \qquad \forall i \in V, k \in K \tag{3d}$$

$$\alpha_{ij}(k,l) \geq 0 \qquad \forall \{i,j\} \in E, k, l \in K. \tag{3e}$$

The fact that (3) is a relaxation of (2) is obvious since choosing $x_i = k$ in (2) corresponds to setting $\alpha_i(k) = 1$ in (3) and $\alpha_{ij}(k,l) = 1$ if $x_i = k$ and $x_j = l$ and otherwise $\alpha_{ij}(k,l) = 0$. The dual LP to (3) can be formulated as

$$\min \sum_{i \in V} \max_{k \in K} \left(g_i(k) + \sum_{j \in N_i} \varphi_{ij}(k)\right) + \sum_{\{i,j\} \in E} \max_{k,l \in K} \left(g_{ij}(k,l) - \varphi_{ij}(k) - \varphi_{ji}(l)\right) \quad (4)$$

where N_i is the set of neighbors of a node $i \in V$ in graph (V, E). The dual (4) is optimized over variables $\varphi_{ij}(k), \varphi_{ji}(k) \in \mathbb{R}$ for each $\{i, j\} \in E, k \in K$. For a more detailed description of the LP formulations, we refer to [12].

For practical applications, there exist various specialized algorithms that solve different forms of the dual (4) by (block-)coordinate optimization and the fixed points of the algorithms are not far from the global optima. These algorithms are called convex message-passing and are very successfully used in computer vision for approximately solving the Max-Sum problem [6]. Such algorithms are, e.g., MPLP [5], TRW-S [7], or max-sum diffusion [9].

The Max-Sum problem (2) can be also interpreted as a weighted CSP, where the infinite[1] node values $g_i(k) = -\infty$ indicate that variable x_i can not be assigned label k. Infinite edge values $g_{ij}(k,l) = -\infty$ denote that variable x_i can not have label k simultaneously with variable x_j having label l. The finite values of g state preferences on which assignments are more desirable.

Any ILM of the dual (4) forms an arc consistent CSP which is defined over the so-called maximal nodes and maximal edges [12,13]. We remark that this special kind of arc consistency for weighted CSPs was independently discovered in [3] as approximate optimization of (4) which is of practical interest due to its ability to provide useful upper bounds. The connection between this approach and coordinate-wise updates was given in [2].

If we considered (2) to be a classical CSP, where the range of g functions is $\{0, -\infty\}$, then (3) can be feasible with optimal value 0 or infeasible. To find out which case it is, one can use max-sum diffusion[2] [9,12] which performs coordinate-wise minimization on (4). If the dual objective does not decrease below 0 after a known finite number of iterations, then the CSP (2) has non-empty arc consistency (AC) closure and we can even obtain the nodes and edges that form the closure from the dual solution. On the other hand, if the dual objective decreases below 0, then the original problem (2) has empty AC closure and both (2) and (3) are infeasible. In other words, for a CSP given as (2), we can use coordinate-wise minimization on the dual (4) to find its AC closure or determine that it is empty, this follows from [12,13].

This is the motivation to search for other problems where coordinate-wise optimization can also be used as a form of constraint propagation combined with approximate optimization. We present such results for the dual LP relaxation of Max-SAT in the following sections.

---

[1] We define $-\infty \cdot 0$ to be 0 by convention in (3a) if infinite values are used.

[2] Max-sum diffusion is usually defined for finite values only, but a generalization based on the relative interior rule is straightforward. Alternatively, one could replace $-\infty$ with arbitrary finite negative number to find the AC closure.

2 LP Relaxation of Weighted Partial Max-SAT

Recall from, e.g., [4] the LP relaxation of weighted partial Max-SAT which is defined by a set of soft clauses $\mathcal{S}$ and a set of hard clauses $\mathcal{H}$ where each soft clause $c \in \mathcal{S}$ is assigned a positive weight $w_c > 0$, and the task is to find an assignment of variables x_i, $i \in [n]$ such that all the hard clauses are satisfied and the sum of weights of satisfied soft clauses is maximized. For brevity, we define matrix $S \in \{-1, 0, 1\}^{\mathcal{S} \times n}$ encoding the soft clauses similarly as in [4], i.e.,

$$S_{ci} = \begin{cases} 1, & \text{if literal } x_i \text{ is in soft clause } c \\ -1, & \text{if literal } \neg x_i \text{ is in soft clause } c \\ 0, & \text{otherwise} \end{cases} \tag{5}$$

and analogously with $H \in \{-1, 0, 1\}^{\mathcal{H} \times n}$ for hard clauses. Furthermore, we define $s \in \mathbb{Z}^{\mathcal{S}}$ to be a vector containing the number of negated variables in each soft clause and similarly with $h \in \mathbb{Z}^{\mathcal{H}}$. The LP relaxation can be stated as

$$\max \; w^T z \tag{6a}$$
$$S^c x + s_c \geq z_c \qquad \forall c \in \mathcal{S} \tag{6b}$$
$$H^c x + h_c \geq 1 \qquad \forall c \in \mathcal{H} \tag{6c}$$
$$x_i \in [0, 1] \qquad \forall i \in [n] \tag{6d}$$
$$z_c \in [0, 1] \qquad \forall c \in \mathcal{S} \tag{6e}$$

where S^c is the row of S corresponding to c and similarly with H^c. The dual linear program to (6) reads

$$\min \left(\sum_{i \in [n]} \max\{S_i^T y + H_i^T q, 0\} + \sum_{c \in \mathcal{S}} \max\{w_c - y_c, 0\} + s^T y + (h - 1)^T q \right) \tag{7a}$$

$$y_c \geq 0 \quad \forall c \in \mathcal{S} \tag{7b}$$
$$q_c \geq 0 \quad \forall c \in \mathcal{H} \tag{7c}$$

where S_i is the i-th column of S (similarly with H_i) and $(h - 1) \in \mathbb{Z}^{\mathcal{H}}$ denotes the vector h with subtracted 1 in each component.

2.1 Coordinate-Wise Minimization

The special form of coordinate-wise update satisfying the relative interior rule for (7) was not shown in [4] where it was only discussed in general setting. We will derive the update for the specific case of the dual of weighted partial Max-SAT here. To do so, we first analyze the restriction of the objective (7a) to a single q_c, $c \in \mathcal{H}$, which (after a simple reformulation and up to constant) reads

$$\sum_{\substack{i \in [n] \\ H_{ci} = 1}} \max\{k_{ci} + q_c, 0\} + \sum_{\substack{i \in [n] \\ H_{ci} = -1}} \max\{k_{ci}, q_c\} - q_c \tag{8}$$

where $k_{ci} = S_i^T y + \sum_{c' \in \mathcal{H} - \{c\}} H_{c'i} q_{c'}$ are constants w.r.t. q_c.

Function (8) has breakpoints $-H_{ci} k_{ci}$ for each $i \in [n]$ with $H_{ci} \neq 0$, we sort these breakpoints into a non-decreasing sequence $b_1 \leq b_2 \leq ... \leq b_m$.

Observe that (8) is a convex piecewise-affine function which is a sum of shifted[3] ReLU functions minus q_c. By this observation, the function (8) is decreasing on $(-\infty, b_1]$, constant on $[b_1, b_2]$, and increasing on $[b_2, \infty)$. The set of optima of (8) subject to $q_c \geq 0$ is then the projection of $[b_1, b_2]$ onto $[0, \infty)$, i.e.,

$$[\max\{b_1, 0\}, \max\{b_2, 0\}]. \tag{9}$$

If a hard clause $c \in \mathcal{H}$ contains only one variable, then the function (8) has only one breakpoint b_1 and is decreasing on $(-\infty, b_1]$ and constant on $[b_1, \infty)$. The set of optimizers of (8) subject to $q_c \geq 0$ is thus

$$[\max\{b_1, 0\}, \infty). \tag{10}$$

The analysis for updates of variables y_c, $c \in \mathcal{S}$, is similar. The restriction of the objective (7a) to a single variable y_c reads

$$\sum_{\substack{i \in [n] \\ S_{ci}=1}} \max\{k'_{ci} + y_c, 0\} + \sum_{\substack{i \in [n] \\ S_{ci}=-1}} \max\{k'_{ci}, y_c\} + \max\{w_c, y_c\} - y_c \tag{11}$$

where k'_{ci} are constants w.r.t. y_c given by $k'_{ci} = H_i^T q + \sum_{c' \in \mathcal{S} - \{c\}} S_{c'i} y_{c'}$. Function (11) has not only breakpoints $-S_{ci} k'_{ci}$ for each $i \in [n]$ with $S_{ci} \neq 0$, but also w_c as its breakpoint. The reasoning is analogous to the previous case: we find the two smallest[4] breakpoints b_1, b_2 and the set of optimizers is again determined by (9). Notice that due to the fact that w_c is also a breakpoint, function (11) has always at least two breakpoints in the sense above.

To sum up, the updates of the variables are the following:

- If we update q_c for hard clause $c \in \mathcal{H}$ with at least two literals, then choose a value from the relative interior of (9). If the clause has only one literal, choose a value from the relative interior of (10). The values b_1, b_2 are the two smallest among $-H_{ci} k_{ci}$ for $i \in [n]$ with $H_{ci} \neq 0$.
- If we update y_c for soft clause $c \in \mathcal{S}$, choose a value from the relative interior of (9), where values b_1, b_2 are the two smallest among w_c and all $-S_{ci} k'_{ci}$ for $i \in [n]$ with $S_{ci} \neq 0$.

We remark that if an interval $[l, u]$ contains only one value (i.e., $l = u$), then its relative interior is a set containing this value only. If $l < u$, then the relative interior of the interval is (l, u), i.e., it contains the whole interval except for its boundary.

---

[3] By shifted ReLU, we understand $f(x) = \max\{a + x, b\}$ for some constants a, b. The function is constant on $(-\infty, b - a]$ and increasing on $[b - a, \infty)$, where $b - a$ is its breakpoint (point of non-differentiability).

[4] If the values of the breakpoints with minimal value coincide, we allow $b_1 = b_2$.

2.2   Feasibility of Primal Solutions

It was proven in [4] that if all the clauses have length at most 2, then any ILM of (7) is a global optimum and the optimal solution of (6) is determined by

$$
x_i = \begin{cases} 0, & \text{if } S_i^T y + H_i^T q < 0 \\ \frac{1}{2}, & \text{if } S_i^T y + H_i^T q = 0 \\ 1, & \text{if } S_i^T y + H_i^T q > 0 \end{cases} \qquad z_c = \min\{S^c x + s_c, 1\}. \tag{12}
$$

For instances with clauses of length 3 or more, the ILM is no longer guaranteed to be globally optimal. In this paper, we extend these results by Theorem 1, which is valid for clauses of arbitrary length. Before we present the theorem, let us define the value of a variable x_i in clause $c \in \mathcal{H}$ as

$$
v(x, i, c) = \begin{cases} x_i, & \text{if } H_{ci} = 1 \\ 1 - x_i, & \text{if } H_{ci} = -1 \\ 0, & \text{otherwise} \end{cases} \tag{13}
$$

and we define $v(x, c) = H^c x + h_c = \sum_{i \in [n]} v(x, i, c)$ as the value of the clause c.

Proposition 1. *Let $c \in \mathcal{H}$ be a hard clause, $i \in [n]$ satisfy $H_{ci} \neq 0$, and let (y, q) be feasible for (7). Then for x_i defined by (12) it holds that*

$$
v(x, i, c) = \begin{cases} 1, & \text{if } b < q_c \\ \frac{1}{2}, & \text{if } b = q_c \\ 0, & \text{if } b > q_c \end{cases} \tag{14}
$$

where $b = -H_{ci} k_{ci}$ is the breakpoint of function (8) corresponding to x_i.

Proof. Follows directly from the definition of k_{ci}, definition of primal solution (12), and analysis of cases for the sign of H_{ci}.

As an example, we show it for $b < q_c$, $H_{ci} = -1$, the other cases are analogous. Since $b < q_c$, we can substitute $b = -H_{ci} k_{ci}$ to obtain $-H_{ci} k_{ci} < q_c$. Multiplying both sides by $H_{ci} < 0$ results in $-k_{ci} > H_{ci} q_c$ and thus $0 > H_{ci} q_c + k_{ci}$. Substituting k_{ci} by its definition yields $0 > H_{ci} q_c + k_{ci} = S_i^T y + H_i^T q$ which means that $x_i = 0$ by (12). Since $H_{ci} = -1$, x_i is negated in clause c and thus has value 1, i.e., $v(x, i, c) = 1$.   □

Theorem 1. *For any instance of weighted partial Max-SAT, given an ILM (y, q) of (7), the assignment (12) is feasible for (6).*

Proof. Constraints (6b), (6d) and (6e) are clearly satisfied, so we focus on (6c) for some $c \in \mathcal{H}$ and show that (12) is feasible. Since (y, q) is an ILM, q_c must be in the relative interior of optimizers that was derived in Sect. 2.1. We will denote by $x_{d(j)}$ the variable that determines breakpoint b_j. If hard clause c contains only one literal, then $q_c \in (\max\{b_1, 0\}, \infty)$ by the result in Sect. 2.1. Therefore $q_c > b_1$ and by Proposition 1, $v(x, d(1), c) = v(x, c) = 1$.

If there are at least two literals in c, then we consider the following cases:

- If $b_1 = b_2$, then there is a unique optimizer $q_c = \max\{b_1, 0\} = \max\{b_2, 0\} \geq b_1 = b_2$, which means that $v(x, c) \geq v(x, d(1), c) + v(x, d(2), c) \geq \frac{1}{2} + \frac{1}{2} = 1$ by Proposition 1.
- If $b_1 < b_2 \leq 0$, then there is a unique optimizer $q_c = 0 \geq b_2 > b_1$, hence $v(x, c) \geq v(x, d(1), c) + v(x, d(2), c) \geq 1 + \frac{1}{2} \geq 1$ by Proposition 1.
- If $b_1 < b_2$ and $b_2 > 0$, then $q_c \in (\max\{b_1, 0\}, b_2)$, which results in $b_1 < q_c < b_2$, which means that $v(x, c) \geq v(x, d(1), c) + v(x, d(2), c) = 1 + 0 = 1$ by Proposition 1.

In all cases, it was shown that the value of the clause is at least 1 because the values of the other variables in the clause are non-negative. The constraints (6c) are therefore satisfied.   □

3   LP Relaxation of SAT

As a consequence of Theorem 1, we could consider a simplified setting, i.e., deciding feasibility of a linear programming problem with constraints (6c) and (6d) only. This is the LP relaxation of SAT problem which reads

$$H^c x + h_c \geq 1 \qquad \qquad \forall c \in \mathcal{H} \qquad \qquad (15a)$$
$$x_i \in [0,1] \qquad \qquad \forall i \in [n]. \qquad \qquad (15b)$$

We will focus on the corresponding dual (7) where it holds that $\mathcal{S} = \emptyset$. The dual of (15) is always feasible and the primal (15) is always bounded (due to no objective function), therefore if the primal is infeasible, then the dual is unbounded. And if the primal is feasible, then the dual has optimal value zero.

Assume that we initialize the dual with $q = 0$ and sequentially update each q_c, $c \in \mathcal{H}$ to be in the relative interior of optimizers. If after $|\mathcal{H}| + 1$ cycles[5] of updates, the objective of the dual improves (i.e., decreases below 0), then the primal is infeasible. On the other hand, by [14, Theorem 18], if the objective stays 0, then we are in an ILM and we can use (12) to obtain a feasible primal solution. To decide this, coordinate minimization needs to perform $O(|\mathcal{H}|^2)$ updates of variables (i.e., visits to hard clauses).

3.1   Unit Propagation

The feasibility of (15) can also be determined by unit propagation modified for this specific setting. Unit propagation is a method that can simplify a set of clauses by gradually assigning values to logical variables that can take only a single specific value in order to result in a feasible solution.

Unit propagation proceeds by finding a unit clause (i.e., a clause of length 1). If the unit clause corresponds to a negated variable, i.e., $\neg x_i$, then x_i must be

---

[5] By a cycle of updates, we mean a single loop of updates in which each variable is updated once with the relative interior rule.

set to be 0 in any feasible assignment. Then, this value is propagated into the remaining clauses where it occurs. If a clause contains the literal $\neg x_i$, then it can be marked as satisfied. If a clause contains the literal x_i (i.e., not negated occurrence of x_i), then the literal can not contribute to satisfaction of such clause and can be removed from it, but the rest of the clause remains. The case when the unit clause corresponds to a non-negated variable x_i is analogous. Notice that by removing a literal from clauses when propagating, a clause may become unit and the process of unit propagation continues with a different variable. More on unit propagation can be found, e.g., in [16].

Observe that the feasibility of (15) can be determined by the following algorithm that performs unit propagation:

1. Set $x_i = \frac{1}{2}$ for all $i \in [n]$
2. Iterate over all hard clauses $c \in \mathcal{H}$:
   (a) If $v(x, c) = 0$ return infeasible
   (b) If $v(x, c) = \frac{1}{2}$ then set the only fractional x_i in the clause to 1 (if it is not negated) or 0 (if it is negated).
3. If there was a change in some component of x, go to step 2. Otherwise terminate as feasible with current assignment of values x.

The algorithm above proceeds by assigning determined integral values for the propagated variables and keeps the other values of variables as $\frac{1}{2}$, which is considered as a label for 'undecided' variables. It can be easily seen that if the procedure returns a solution x, then this solution is feasible because the values of all clauses are at least 1 (otherwise, there would be update or infeasible termination in steps 2a or 2b). It can be also seen that if a feasible solution exists, it will be found by the algorithm because all assignments caused by unit clauses are enforced and any clause of length at least 2 (after omitting the variables whose value was propagated) is satisfied by assigning the value $\frac{1}{2}$ to the remaining variables in it.

To more intuitively see how unit propagation is done in coordinate-wise minimization, consider that there is a unit clause $c \in \mathcal{H}$ that reads[6] $x_i \geq 1$. Then, by the reasoning in Sect. 2.1, we need to set q_c to some value from $(\max\{b_1^c, 0\}, \infty)$, where b_1^c is the corresponding breakpoint, as mentioned in Sect. 2.1. Assume that q_c is set to some large value, then by definition (12), x_i is forced to 1 due to

$$S_i^T y + H_i^T q = k_{ci} + H_{ci} q_c = k_{ci} + q_c = -b_1^c + q_c > 0. \tag{16}$$

Now, analyze some clause $c' \in \mathcal{H}$ that also contains literal x_i, i.e., $H_{c'i} = 1$. Then, $k_{c'i}$ contains $H_{ci} q_c = q_c$ as one of its addends and is therefore large. By $H_{c'i} = 1$, we obtain that the position of the breakpoint $b_i^{c'}$ corresponding to x_i of the function (7a) restricted to $q_{c'}$ is $b_i^{c'} = -H_{c'i} k_{c'i} = -k_{c'i}$, which should be a negative number which is large in absolute value. This therefore means that $b_i^{c'} < q_{c'}$, which results in satisfying the constraint (15a) corresponding to clause c'. The analysis for clauses c'' that contain the literal $\neg x_i$ is analogous and simply results in moving the corresponding breakpoint and $v(x, i, c'') = 0$.

---

[6] The reasoning in case that the unit clause is in form $1 - x_i \geq 1$ is analogous.

It is interesting to observe that the asymptotic time complexity of the naive unit propagation algorithm described above is $O(|\mathcal{H}|^2)$, i.e., quadratic in the number of clauses, which is the same as the number of updates required by coordinate-wise minimization to decide whether (15) is feasible.

We are aware of the fact that there exist also faster algorithms for unit propagation, such as [16], but it is an open question whether a more involved coordinate-wise minimization algorithm could imitate their behavior.

4   Propagation in Weighted Max-SAT

In the previous section, we analyzed the case of LP relaxation of SAT problem. Now, we would like to answer the question whether coordinate-wise optimization in the dual (7) also corresponds to some generalized propagation rule in weighted[7] Max-SAT.

Propagation of values can be done by the dominating unit-clause rule, which was introduced in [10] for unweighted Max-SAT and later generalized for weighted Max-SAT in [15]. The dominating unit-clause rule is the following:

- If $n_1(i) \geq \sum_{k=1}^{\infty} p_k(i)$, then set $x_i = 0$.
- If $p_1(i) \geq \sum_{k=1}^{\infty} n_k(i)$, then set $x_i = 1$.

where $i \in [n]$ and for $k \in \mathbb{N}$, we define $p_k(i) = \displaystyle\sum_{\substack{c \in \mathcal{S} \\ |c|=k \\ S_{ci}=1}} w_c$ and $n_k(i) = \displaystyle\sum_{\substack{c \in \mathcal{S} \\ |c|=k \\ S_{ci}=-1}} w_c$

and $|c|$ is the number of literals in clause c. In [15], there is also an additional remark which says that if both conditions in the rule hold for some $i \in [n]$, then x_i is present only in unit clauses and $p_1(i) = n_1(i)$ and the choice for the value of x_i can be arbitrary.

We are going to show that coordinate-wise minimization in the dual (7) corresponding to weighted Max-SAT (i.e., $\mathcal{H} = \emptyset$) also performs the dominating unit-clause rule under an additional condition that the inequalities in the rule are strict. By assuming the strict version of the rule, we eliminate the possibility of x_i being set ambiguously.

Theorem 2. *Let y be an ILM of (7) corresponding to weighted Max-SAT (i.e., without hard clauses, $\mathcal{H} = \emptyset$). The values x_i, $i \in [n]$ defined by (12) satisfy: If $n_1(i) > \sum_{k=1}^{\infty} p_k(i)$, then $x_i = 0$. If $p_1(i) > \sum_{k=1}^{\infty} n_k(i)$, then $x_i = 1$.*

The proof of Theorem 2 is lengthy and requires a number of auxiliary propositions, that is why we present it in Appendix A.

In a similar way as unit propagation was performed on (15), we could repeatedly apply the dominating unit-clause rule on a weighted Max-SAT instance by gradually setting some variables to 0 or 1 until it reaches a fixed point, where the rule can no longer be applied on the remaining unsatisfied clauses. From

---

[7] Weighted partial Max-SAT can be reduced into weighted Max-SAT by assigning large weights to the hard clauses, this is mentioned, e.g., in [11].

Theorem 2, it follows that any ILM y of coordinate-wise minimization on (7) without hard clauses defines through (12) a feasible solution x for the primal problem (6) where all components x_i decided by the dominating unit-clause rule with strict inequalities have the same value.

To further validate our theoretical results, we evaluated our approach on 500 smallest[8] instances from Max-SAT Evaluation 2019 [1]. In each case, those variables that were decided by the strict version of the dominating unit-clause rule were also decided by (12) for a given ILM of (7) and their values were equal.

5   Conclusion

We presented our result in Theorem 1 that extends knowledge from [4] in the sense that for any instance of weighted partial Max-SAT, any ILM of the dual (7) defines a feasible primal solution. It was discussed in Sect. 3 that the process of coordinate-wise minimization on the dual of LP relaxation of SAT problem (15) resembles unit propagation and can decide feasibility of the primal problem. Moreover, in Theorem 2, we have shown that the fixed points of coordinate-wise minimization on the dual define primal solutions where the components that would be determined by the strict version of dominating unit-clause rule equal to the value that would be assigned by the rule. In other words, coordinate-wise minimization can in a precise sense apply these propagation rules.

This result is of theoretical importance because, as mentioned in Sect. 1.1, it is known that coordinate-wise optimization can also decide whether a given CSP has non-empty AC closure. Since both arc consistency and unit propagation can be viewed as methods of constraint propagation, it is an open question for further research whether there are other propagation methods applicable to some combinatorial problems that could be imitated by coordinate-wise optimization.

Acknowledgement. Research described in the paper was supervised by doc. Ing. Tomáš Werner, Ph.D., FEE CTU in Prague. This work has been supported by Czech Science Foundation project 19-09967S and Grant Agency of CTU in Prague, grant SGS19/170/OHK3/3T/13.

A   Proof of Theorem 2

All of the following propositions assume that y is an ILM of (7) with $\mathcal{H} = \emptyset$, $c \in \mathcal{S}$ is an arbitrary clause, and $i \in [n]$ is also arbitrary.

Proposition 2. *If $S_{ci} = 1$ and $S_i^T y > 0$, then $w_c \geq y_c$.*

Proof. Proof by contradiction – assume $w_c < y_c$. From $S_i^T y > 0$, it follows from definition of k'_{ci} that $S_{ci} y_c + k'_{ci} > 0$, thus $y_c > -S_{ci} k'_{ci}$. Since there are two breakpoints ($-S_{ci} k'_{ci}$ and w_c) that are strictly lower than y_c and $w_c > 0$, y_c is not in the relative interior of optimizers by reasoning in Sect. 2.1.   □

---

[8] Smallest in terms of the size of the file. The considered 500 instances had up to 287 thousand variables and up to 1.4 million clauses.

Proposition 3. *If $S_{ci} \neq 0$ and $S_i^T y = 0$, then $w_c \geq y_c$.*

Proof. Following the reasoning in Sect. 2.1, y_c should be in the relative interior of (9) for two smallest breakpoints b_1, b_2. Since y_c is equal to breakpoint $b_j = -S_{ci}k'_{ci}$, it must hold that $b_j = b_2 = b_1$ (or $b_j = b_2 = 0 \geq b_1$, which trivially leads to $y_c = 0 < w_c$), otherwise y_c is not in the relative interior of optimizers. Since $w_c > 0$ is also a breakpoint, it must be greater or equal to the smallest breakpoint b_1, this yields $w_c \geq b_1 = b_j = -S_{ci}k'_{ci} = y_c$. □

Proposition 4. *If $S_{ci} = -1$, $|c| = 1$, and $S_i^T y \geq 0$, then $w_c \leq y_c$.*

Proof. Function (7a) restricted to y_c has two breakpoints, $-S_{ci}k'_{ci}$ and w_c. Since $S_i^T y = -y_c + k'_{ci} \geq 0$, it holds that $y_c \leq -S_{ci}k'_{ci}$. By the reasoning in Sect. 2.1, it can either hold that $w_c < y_c < -S_{ci}k'_{ci}$ or $w_c = y_c = -S_{ci}k'_{ci}$, hence $w_c \leq y_c$. □

Proposition 5. *If $S_{ci} = -1$ and $S_i^T y < 0$, then $w_c \geq y_c$.*

Proof. Similar to Proposition 2. It holds that $y_c > -S_{ci}k'_{ci}$, which follows from $S_{ci} = -1$ and $S_i^T y < 0$. Assuming $w_c < y_c$ leads to a contradiction. □

Proposition 6. *If $S_{ci} = 1$, $|c| = 1$, and $S_i^T y \leq 0$, then $w_c \leq y_c$.*

Proof. Similarly to Proposition 4, function (7a) restricted to y_c has two breakpoints, $-S_{ci}k'_{ci}$ and w_c. Since $S_i^T y = y_c + k'_{ci} \leq 0$, it holds that $y_c \leq -S_{ci}k'_{ci}$. By the reasoning in Sect. 2.1, it can either hold that $w_c < y_c < -S_{ci}k'_{ci}$ or $w_c = y_c = -S_{ci}k'_{ci}$, hence $w_c \leq y_c$. □

*Proof (**Theorem** 2).* We will prove the two parts of Theorem 2 separately, both by contradiction. For the first part, assume that $S_i^T y \geq 0$ (which should be contradictory because it results in $x_i = 1$ or $x_i = \frac{1}{2}$). It holds that

$$\sum_{\substack{c' \in S \\ |c'|=1 \\ S_{c'i}=-1}} y_{c'} \geq \sum_{\substack{c' \in S \\ |c'|=1 \\ S_{c'i}=-1}} w_{c'} > \sum_{\substack{c \in S \\ S_{ci}=1}} w_c \geq \sum_{\substack{c \in S \\ S_{ci}=1}} y_c \tag{17}$$

where inequality on the left follows from Proposition 4, inequality in the middle is equivalent to the assumption in Theorem 2 and inequality on the right follows from Propositions 2 and 3. This results in

$$\sum_{\substack{c' \in S \\ S_{c'i}=-1}} y_{c'} = \sum_{\substack{c' \in S \\ |c'|>1 \\ S_{c'i}=-1}} y_{c'} + \sum_{\substack{c' \in S \\ |c'|=1 \\ S_{c'i}=-1}} y_{c'} \geq \sum_{\substack{c' \in S \\ |c'|=1 \\ S_{c'i}=-1}} y_{c'} > \sum_{\substack{c \in S \\ S_{ci}=1}} y_c \tag{18}$$

where inequality on the right follows from (17) and inequality in the middle follows from non-negativity of y_c, hence $\sum_{\substack{c' \in S \\ S_{c'i}=-1}} y_{c'} > \sum_{\substack{c \in S \\ S_{ci}=1}} y_c$, i.e., $S_i^T y < 0$, which is contradictory with $S_i^T y \geq 0$.

For the second part, assume that $S_i^T y \leq 0$ (which should be contradictory, because it results in $x_i = 0$ or $x_i = \frac{1}{2}$). Similarly as in the previous part, it holds that

$$\sum_{\substack{c' \in \mathcal{S} \\ |c'|=1 \\ S_{c'i}=1}} y_{c'} \geq \sum_{\substack{c' \in \mathcal{S} \\ |c'|=1 \\ S_{c'i}=1}} w_{c'} > \sum_{\substack{c \in \mathcal{S} \\ S_{ci}=-1}} w_c \geq \sum_{\substack{c \in \mathcal{S} \\ S_{ci}=-1}} y_c \qquad (19)$$

where the left inequality is given by Proposition 6, the middle inequality follows from the condition in Theorem 2, and the inequality on the right follows from Propositions 3 and 5. We proceed analogously as in the first part and obtain that $S_i^T y > 0$, which is contradictory.   □

References

1. Bacchus, F., Järvisalo, M., Martins, R.: MaxSAT Evaluation 2019: solver and benchmark descriptions (2019). http://hdl.handle.net/10138/306989
2. Cooper, M.C., de Givry, S., Sanchez, M., Schiex, T., Zytnicki, M., Werner, T.: Soft arc consistency revisited. Artif. Intell. **174**(7–8), 449–478 (2010)
3. Cooper, M.C., de Givry, S., Sánchez, M., Schiex, T., Zytnicki, M.: Virtual arc consistency for weighted CSP. In: AAAI, pp. 253–258 (2008)
4. Dlask, T., Werner, T.: A class of linear programs solvable by coordinate-wise minimization. In: Kotsireas, I.S., Pardalos, P.M. (eds.) LION 2020. LNCS, vol. 12096, pp. 52–67. Springer, Cham (2020). https://doi.org/10.1007/978-3-030-53552-0_8
5. Globerson, A., Jaakkola, T.S.: Fixing max-product: convergent message passing algorithms for MAP LP-relaxations. In: Advances in Neural Information Processing Systems, pp. 553–560 (2008)
6. Kappes, J., Andres, B., et al.: A comparative study of modern inference techniques for discrete energy minimization problems. In: Proceedings of the IEEE Conference on Computer Vision and Pattern Recognition, pp. 1328–1335 (2013)
7. Kolmogorov, V.: Convergent tree-reweighted message passing for energy minimization. IEEE Trans. Pattern Anal. Mach. Intell. **28**(10), 1568–1583 (2006)
8. Kolmogorov, V., Thapper, J., Živný, S.: The power of linear programming for general-valued CSPs. SIAM J. Comput. **44**(1), 1–36 (2015)
9. Kovalevsky, V., Koval, V.: A diffusion algorithm for decreasing energy of max-sum labeling problem. Glushkov Institute of Cybernetics, Kiev, USSR (1975)
10. Niedermeier, R., Rossmanith, P.: New upper bounds for maximum satisfiability. J. Algorithms **36**(1), 63–88 (2000)
11. Pipatsrisawat, K., Palyan, A., Chavira, M., Choi, A., Darwiche, A.: Solving weighted max-sat problems in a reduced search space: a performance analysis. J. Satisfiab. Boolean Model. Comput. **4**(2–4), 191–217 (2008)
12. Werner, T.: A linear programming approach to max-sum problem: a review. IEEE Trans. Pattern Anal. Mach. Intell. **29**(7), 1165–1179 (2007)
13. Werner, T., Průša, D., Dlask, T.: Relative interior rule in block-coordinate descent. In: Conference on Computer Vision and Pattern Recognition (2020)
14. Werner, T., Průša, D.: Relative interior rule in block-coordinate minimization. ArXiv.org (2019)
15. Xing, Z., Zhang, W.: MaxSolver: an efficient exact algorithm for (weighted) maximum satisfiability. Artif. Intell. **164**(1–2), 47–80 (2005)
16. Zhang, H., Stickel, M.E.: An efficient algorithm for unit propagation. In: Proceedings of the 4th International Symposium on Artificial Intelligence and Mathematics, vol. 96 (1996)

A Learning-Based Mathematical Programming Formulation for the Automatic Configuration of Optimization Solvers

Gabriele Iommazzo[1,2]([⊠]), Claudia D'Ambrosio[1], Antonio Frangioni[2], and Leo Liberti[1]

[1] LIX CNRS, École Polytechnique, Institut Polytechnique de Paris, Palaiseau, France
{giommazz,dambrosio,liberti}@lix.polytechnique.fr
[2] Dipartimento di Informatica, Università di Pisa, Pisa, Italy
frangio@di.unipi.it

Abstract. We propose a methodology, based on machine learning and optimization, for selecting a solver configuration for a given instance. First, we employ a set of solved instances and configurations in order to learn a performance function of the solver. Secondly, we formulate a mixed-integer nonlinear program where the objective/constraints explicitly encode the learnt information, and which we solve, upon the arrival of an unknown instance, to find the best solver configuration for that instance, based on the performance function. The main novelty of our approach lies in the fact that the configuration set search problem is formulated as a mathematical program, which allows us to a) enforce hard dependence and compatibility constraints on the configurations, and b) solve it efficiently with off-the-shelf optimization tools.

Keywords: Automatic algorithm configuration · Mathematical programming · Machine learning · Optimization solver configuration · Hydro unit committment

1 Introduction

We address the problem of finding instance-wise optimal configurations for general Mathematical Programming (MP) solvers. We are particularly motivated by state-of-the-art general-purpose solvers, which combine a large set of diverse algorithmic components (relaxations, heuristics, cutting planes, branching, ...) and therefore have a long list of user-configurable parameters; tweaking them can have a significant impact on the quality of the obtained solution and/or on

This paper has received funding from the European Union's Horizon 2020 research and innovation programme under the Marie Sklodowska-Curie grant agreement n. 764759 "MINOA".

G. Nicosia et al. (Eds.): LOD 2020, LNCS 12565, pp. 700–712, 2020.
https://doi.org/10.1007/978-3-030-64583-0_61

the efficiency of the solution process (see, e.g., [14]). Good solvers have effective default parameter configurations, carefully selected to provide good performances in most cases. Furthermore, tuning tools may be available (e.g., [18, Ch. 10]) which run the solver, with different configurations, on one or more instances within a given time limit, and record the best parameter values encountered. Despite all this, the produced parameter configurations may still be highly suboptimal with specific instances. Hence, a manual search for the best parameter values may be required. This is a highly nontrivial and time-consuming task, due to the large amount of available parameters (see, e.g., [19]), which requires a profound knowledge of the application at hand and an extensive experience in solver usage. Therefore, it is of significant interest to develop general approaches, capable of performing it efficiently and effectively in an automatic way.

This setting is an instance of the Algorithm Configuration Problem (ACP) [27], which is defined as follows: given a target algorithm, its set of parameters, a set of instances of a problem class and a measure of the performance of the target algorithm on a pair (instance, algorithmic configuration), find the parameter configuration providing optimal algorithmic performance according to the given measure, on a specific instance or instance set. Several domains can benefit from automating this task. Some possible applications, beyond MP solvers (see, e.g., [12] and references therein), are: solver configuration for the propositional satisfiability problem, hyperparameter tuning of ML models or pipelines, algorithm selection, administering ad-hoc medical treatment, etc. Our approach for addressing the ACP on MP solvers is based on a two-fold process:

(i) in the *Performance Map Learning Phase* (PMLP), supervised Machine Learning (ML) techniques [25] are used to learn a *performance function*, which maps some features of the instance being solved, and the parameter configuration, into some measure of solver efficiency and effectiveness;

(ii) the formal model underlying the ML methodology used in the PMLP is translated into MP terms; the resulting formulation, together with constraints encoding the compatibility of the configuration parameter values, yields the *Configuration Set Search Problem* (CSSP), a Mixed-Integer Nonlinear Program (MINLP) which, for a given instance, finds the configuration providing optimal performance with respect to the performance function.

The main novelty of our approach lies in the fact that we explicitly model and optimize the CSSP using the mathematical description of the PMLP technique. This is in contrast to most of the existing ACP approaches, which instead employ heuristics such as local searches [2,16], genetic algorithms [3], evolutionary strategies [8] and other methods [24]. Basically, most approaches consider the performance function as a black box, even when it is estimated by means of some ML technique and, therefore, they cannot reasonably hope to find a global minimum when the number of parameters grows. Rather, one of the strengths of our methodology is that it exploits the mathematical structure of the CSSP, solving it with sophisticated, off-the-shelf MP solvers. Moreover, formulating the CSSP by MP is advantageous as it allows the seamless integration of the compatibility constraints on the configuration parameters, which is something that other ACP methods may struggle with. The idea of using a ML predictor to define the unknown components (constraints, objective) of a MP has been already explored in *data-driven optimization*. In general, it is possible to represent the ML model of a mapping/relation as a MP (or, equivalently, a Constraint Programming model) and optimize upon this [23]; however, while this is in principle possible, the set of successful applications in practice is limited. Indeed, using this approach in the ACP context is, to the best of our knowledge, new; it also comes with some specific twists. We tested this idea with the following components: we configured nine parameters of the IBM ILOG CPLEX solver [18], which we employed to solve instances of the Hydro Unit Commitment (HUC) problem [7], we chose Support Vector Regression (SVR) [28] as the PMLP learning methodology, and we used the off-the-shelf MINLP solver Bonmin [5] to solve the CSSP.

The paper is structured as follows: in Sect. 2 we will review existing work on algorithm configuration; in Sect. 3 we will detail our approach and provide the explicit formulation of the CSSP with SVR; in Sect. 4 we will discuss some computational results.

2   The Algorithm Configuration Problem

Most algorithms have a very high number of configurable parameters of various types (boolean, categorical, integer, continuous), which usually makes the ACP very hard to solve in practice. Notably, this issue significantly affects MP solvers: they are highly complex pieces of software, embedding several computational components that tackle the different phases of the solution process; the many available algorithmic choices are exposed to the user as a long list of tunable parameters (for example, more than 150 in CPLEX [19]).

Approaches to the ACP can be compared based on how they fit into the following two categories: Per-Set (PS) or Per-Instance (PI); offline or online. In PS approaches, the optimal configuration is defined as the one with the best overall performance over a set of instances belonging to the same problem class. Therefore, PS approaches first find the optimal configuration for a problem class and then use it for any instance pertaining to that class. The exploration of the configuration set is generally conducted by means of heuristics, such as various local search procedures [2,16], racing methods [24], genetic algorithms [3] or other evolutionary algorithms [26]. In this context, an exception is, e.g., the approached described in [15], which predicts the performance of the target algorithm by random forest regression and then uses it to guide the sampling in an iterative local search. PS approaches, however, struggle when the target algorithm performance varies considerably among instances belonging to the same problem class. In these cases, PI methodologies, which assume that the optimal algorithmic configuration depends on the instance at hand, are likely to produce better configurations. However, while PS approaches are generally problem-agnostic, PI ones require prior knowledge of the problem at hand, to efficiently encode each instance by a set of features. PI approaches typically focus on learning a good surrogate map of the performance function, generally by performing regression: this approximation is used to direct the search in the configuration set. In [17], for example, linear basis function regression is used to approximate the target algorithm runtime, which is defined as a map of both features and configurations; then, for a new instance with known features, the learnt map is evaluated at all configuration points in an exhaustive search, to find the estimated best one. However, other approaches may be used: in [8], for example, a map from instance features to optimal configuration is learnt by a neural network; in [6] the ACP is restricted to a single binary parameter of CPLEX, and a classifier is then trained to predict it. In [21], instead, CPLEX is run on a given instance for a certain amount of computational resources, then a ranking ML model is trained, on-the-fly, to learn the ordering of branch and bound variables, and it is then used to predict the best branching variable, at each node, for the rest of the execution. Another approach, presented in [20], first performs clustering on a set of instances, then uses the PS methodology described in [3] to find one good algorithmic configuration for each cluster. In [31], instead, instances are automatically clustered in the leaves of a trained decision tree, which also learns the best configuration for each leaf; at test time, a new instance is assigned to a leaf based on its features, and it receives the corresponding configuration. The purpose of an ACP approach is to provide a good algorithmic configuration upon the arrival of an unseen instance. We call a methodology offline if the learning happens before that moment, which is the case for all the approaches cited above. Otherwise, we call an ACP methodology online; these approaches normally use reinforcement learning techniques (see, e.g., [10]) or other heuristics [4].

In our approach, we define the performance of the target algorithm as a function of both features and controls, in order to account for the fact that the best configuration of a solver may vary among instances belonging to the same class of problems; this makes our approach PI. Moreover, we perform the PMLP only once, offline, which allows us to solve the resulting CSSP for any new instances in a matter of seconds. What makes our approach stand out from other methodologies is that the learning phase is treated as white-box: the prediction problem of the PMLP is formulated as a MP, which conveniently allows the explicit embedding of a mathematical encoding of the estimated performance into the CSSP, as its objective/constraints. This is opposed to treating the learned predictor as a black-box, and therefore using it as an oracle in brute-force searches or similar heuristics, that typically do not scale as well as optimization techniques.

3   The PMLP and the CSSP

Let $\mathcal{A}$ be the target algorithm, and:

- $\mathcal{C}_\mathcal{A}$ be the set of feasible configurations of $\mathcal{A}$. We assume that each configuration $c \in \mathcal{C}_\mathcal{A}$ can be encoded into a vector of binary and/or discrete values representing categorical and numerical parameters, and $\mathcal{C}_\mathcal{A}$ can be described by means of linear constraints;

- Π be the problem to be solved, consisting of an infinite set of instances, and $\Pi' \subset \Pi$ be the (finite) set of instances used for the PMLP;
- F_Π be the set of feature vectors used to describe instances, encoded by vectors of continuous or discrete/categorical values (in the latter case they are labelled by reals);
- $p_{\mathcal{A}} : F_\Pi \times \mathcal{C}_{\mathcal{A}} \longrightarrow \mathbb{R}$ be the performance function which maps a pair (f, c) (instance feature vector, configuration) to the outcome of an execution of $\mathcal{A}$ (say in terms of the integrality gap reported by the solver after a time limit, but other measures are possible).

With the above definitions, the PMLP and the CSSP are detailed as follows.

3.1   Performance Map Learning Phase

In the PMLP we use a supervised ML predictor, e.g., SVR, to learn the coefficient vector θ^* providing the parameters of a prediction model $\bar{p}_{\mathcal{A}}(\cdot, \cdot, \theta) : F_\Pi \times \mathcal{C}_{\mathcal{A}} \rightarrow \mathbb{R}$ of the performance function $p_{\mathcal{A}}(\cdot, \cdot)$. The training set for the PMLP is

$$\mathcal{S} = \left\{ (f_i, c_i, p_{\mathcal{A}}(f_i, c_i)) \mid i \in \{1 \ldots s\} \right\} \subseteq F_{\Pi'} \times \mathcal{C}_{\mathcal{A}} \times \mathbb{R}, \tag{1}$$

where $s = |\mathcal{S}|$ and the training set labels $p_{\mathcal{A}}(f_i, c_i)$ are computed on the training vectors (f_i, c_i). The vector θ^* is chosen as to hopefully provide a good estimate of $p_{\mathcal{A}}$ on points that do not belong to $\mathcal{S}$, with the details depending on the selected ML technology.

3.2   Configuration Space Search Problem

For a given instance f and parameter vector θ, $\mathsf{CSSP}(f, \theta)$ is the problem of finding the configuration with best estimated performance $\bar{p}_{\mathcal{A}}(f, c, \theta)$:

$$\mathsf{CSSP}(f, \theta) \equiv \min_{c \in \mathcal{C}_{\mathcal{A}}} \bar{p}_{\mathcal{A}}(f, c, \theta). \tag{2}$$

The actual implementation of $\mathsf{CSSP}(f, \theta)$ depends on the MP formulation selected to encode $\bar{p}_{\mathcal{A}}$, which may require auxiliary variables and constraints to define the properties of the ML predictor. If $\bar{p}_{\mathcal{A}}$ yields an accurate estimate of $p_{\mathcal{A}}$, we expect the optimum c^*_{cssp} of $\mathsf{CSSP}(f, \theta)$ to be a good approximation of the true optimal configuration c^* for solving f. However, we remark that a) $\mathsf{CSSP}(f, \theta)$ can be hard to solve, and b) it needs to be solved quickly (otherwise one might as well solve the instance f directly). Hence, incurring the additional computational overhead for solving the CSSP may be advantageous only when the instance at hand is "hard". Achieving a balance between PMLP accuracy and CSSP cost is one of the challenges of this research.

4   Experimental Results

We tested our approach on 250 instances of the HUC problem and on 9 parameters of CPLEX, version 12.7. The PMLP and CSSP experiments were conducted on an Intel Xeon CPU E5-2620 v4 @ 2.10 GHz architecture, while CPLEX was run on an Intel Xeon Gold 5118 CPU @ 2.30 GHz. The pipeline was implemented in Python 3.6.8 [29] and AMPL Version 20200110 [13]. In the following, we detail the algorithmic set-up that we employed.

4.1 Building the Dataset

1. *Features.* The HUC is the problem of finding the optimal scheduling of a pump-storage hydro power station, where the commitment and the power generation of the plant must be decided in a short term period, in which inflows and electricity prices are previously forecast. The goal is to maximize the revenue given by power selling (see, e.g., [1]). The time horizon is fixed at 24 hours and the underlying hydro system is also fixed, so that all the instances have the same size. Thus, only 54 elements which vary from day to day are features: the date, 24 hourly prices, 24 hourly inflows, initial and target water volumes, upper and lower bound admitted on the water volumes. We encode them in a vector f of 54 continuous/discrete components. All the instances have been randomly generated with an existing generator that accurately reproduces realistic settings.

2. *Configuration parameters.* Thanks to preliminary tests, we select a subset of 9 discrete CPLEX parameters (fpheur, dive, probe, heuristicfreq, startalgorithm and subalgorithm from mip.strategy; crossover from barrier; mircuts and flowcovers, from mip.cuts), for each of which we consider between 2 and 4 different values. We then combine them so as to obtain 2304 parameter configurations. A configuration is encoded by a vector $c \in \{0, 1\}^{23}$, where each categorical parameter is represented by its incidence vector.

3. *Performance measure.* We use the integrality gap to define $p_{\mathcal{A}}(f, c)$. It has been shown that MIP solvers can be affected by performance variability issues (see, e.g., [22]), due to executing the solver on different computing platforms, permuting rows/columns of a model, adding valid but redundant constraints, performing apparently neutral changes to the solution process, etc. In order to tackle this issue, first we sample three different random seeds. For each instance feature vector f and each configuration c, we then carry out the following procedure: (i) we run CPLEX (using the Python API) three times on the instance, using the different random seeds, for 60 seconds; (ii) we record the middle out of the three obtained performance values, to be assigned to the pair (f, c). At this point, our dataset contains $250 \times 2304 = 576000$ records. The performance measure thus obtained from CPLEX output, which we call $p_{cpx}(f, c)$, usually contains some extremely large floating point values (e.g., whenever the CPLEX gap has a value close to zero in the denominator), which unduly bias the learning process. We deal with this issue as follows: we compute the maximum $\bar{p}_{cpx}$, over all values of (the range of) p_{cpx}, lower than a given threshold (set to 1e+5 in our experiments), re-set all values of p_{cpx} larger than the threshold to $\bar{p}_{cpx} + 100$, then rescale p_{cpx} so that it lies within the interval $[0, 1]$. The resulting performance measure, which

in the following we call $p_{ml}(f, c)$, is also the chosen PMLP label. Moreover, we solve each HUC instance and we record the value of its optimum; then, $\forall(f, c)$, we compute the *primal gap* ϱ_{prim} and the *dual gap* ϱ_{dual}, i.e.: the distance between the optimal value of f and the value of the feasible solution found, and the distance between the value of the tightest relaxation found and the optimal value (both over the optimum). We save p_{cpx}, p_{ml}, ϱ_{prim} and ϱ_{dual} in our dataset. We remark that setting the time-limit, imposed on CPLEX runs, to 60 seconds provides the solver enough time to move past the preliminary processing and to begin working on closing the gap, even for very hard instances (i.e., the ones with long pre-processing times); this allows us to measure the actual impact that different parameter configurations have on the chosen performance measure.

4. *Feature engineering.* We process the date in order to extract the season, the week-day, the year-day, two flags called `isHoliday` and `isWeekend`, and we perform several sine/cosine encodings, that are customarily used to treat cyclical features. Moreover, we craft new features by computing statistics on the remaining 54 features. This task takes around 12 min to complete for the whole data set.

5. *Splitting the dataset.* We randomly divide the instances into 187 In-Sample (IS) and 63 Out-of-Sample (OS), and split the dataset rows accordingly (430848 IS and 145152 OS). We use the IS data to perform Feature Selection (FS) and to train the SVR predictor; then, we assess the performance of the PMLP-CSSP pipeline both on OS instances, to test its generalization capabilities to unseen input, and on IS instances, to evaluate its performance on the data that we learn from, as detailed below.

6. *Feature selection.* We use Python's `Pandas DataFrame`'s `corr` function to perform Pearson's Linear Correlation and `sklearn RandomForestRegressor`'s `feature_importances_` attribute to perform decision trees' Feature Importance, in order to get insights on which features contribute the most to yield accurate predictions. A detailed explanation of the employed FS techniques falls outside of the scope of this document. In the following, we use the shorthand "variables" to refer to the whole list of columns of the learning dataset. In order to perform FS, we use a dedicated subset of the IS dataset, composed of 19388 records and only employed for this task; performing the selected FS techniques on this dataset takes around 8 minutes, and reduces f to 22 components. For the configuration vectors we consider three FS scenarios: `noFS`, `kindFS` and `aggFS`, yielding c vectors with, respectively, 23, 14 and 10 components. We then filter the PMLP dataset according to the FS scenario at hand. However, after this filtering, the dataset may contain points with the same (f, c) but different labels $p_{ml}(f, c)$. Thus, for each instance: a) we delete the dataset columns that FS left out; b) we perform `Pandas`'s `group_by` on the 22 columns chosen by FS, then c) compute the average p_{ml} of each group and use this as the new label (at this point, rows with the same (f, c) have the same label); d) we remove the duplicate rows of the dataset by `Pandas drop_duplicates`, keeping only one row. Lastly, we select $\sim$11200 points for the PMLP.

4.2   PMLP Experimental Setup

The PMLP methodology of choice in this paper is SVR. Its advantages are: (a) the PMLP for training an SVR can be formulated as a convex Quadratic Program (QP), which can be solved efficiently; (b) even complicated and possibly nonlinear performance functions can be learned by using the "kernel trick" [28]; (c) the solution of the PMLP for SVR provides a closed-form algebraic expression of the performance map $\bar{p}_\mathcal{A}$, which yields an easier formulation of $\mathsf{CSSP}(f, \theta)$. We use a Gaussian kernel during SVR training, which is the default choice in absence of any other meaningful prior [9]. We assess the prediction error of the predictor by Nested Cross Validation (NCV) [30]; furthermore, our training includes a phase for determining and saving the hyperparameters and the model coefficients of the SVR. These two tasks take approximately 4 hours in the `aggFS` and in the `kindFS` scenarios, and 5 hours in the `noFS` one. We use Python's `sklearn.model_selection.RandomizedSearchCV` for the inner loop of the NCV and a customized implementation for the outer loop, and `sklearn.svm.SVR` as the implementation of choice for the ML model.

A common issue in data-driven optimization is that using customary ML error metrics may not lead to good solutions of the optimization problem (see, for example, [11]). We tackled this issue by comparing the following metrics for the CV-based hyperparameter tuning phase, both computed on $p_{\mathtt{ml}}$: the classical Mean Absolute Error $\mathsf{MAE} = \sum_{i \in S} |p_i - \bar{p}_i|$, where $p_i = p_{\mathtt{ml}}(f_i, c_i)$ and $\bar{p}_i = \bar{p}_\mathcal{A}(f_i, c_i)$; the custom metric $\mathsf{cMAE}_\delta = \sum_{i \leq s} L_\delta(p_i, \bar{p}_i)$, $\delta \in \{0.2, 0.3, 0.4\}$, where

$$
L_\delta(p_i, \bar{p}_i) = \begin{cases} (\bar{p}_i - p_i) \cdot (1 + \frac{1}{1+exp(p_i - \bar{p}_i)}) & \text{if } p_i \leq \delta \text{ and } \bar{p}_i > p_i \\ (p_i - \bar{p}_i) \cdot (1 + \frac{1}{1+exp(\bar{p}_i - p_i)}) & \text{if } p_i \geq 1 - \delta \text{ and } \bar{p}_i < p_i \\ (p_i - \bar{p}_i) & \text{if } \delta \leq p_i \leq 1 - \delta \\ 0 & \text{otherwise.} \end{cases}
$$

4.3   CSSP Experimental Setup

The choice of a Gaussian kernel in the SVR formulation makes the CSSP a MINLP with a nonconvex objective function $\bar{p}_\mathcal{A}$. More precisely, for a given instance with features $\bar{f}$, our CSSP is:

$$
\min_{c \in \mathcal{C}_\mathcal{A}} \sum_{i=1}^{s} \alpha_i \exp\left(-\gamma \|(f_i, c_i) - (\bar{f}, c)\|_2^2 \right) \tag{3}
$$

where, for all $i \leq s$, (f_i, c_i) belong to the training set, α_i are the dual solutions of the SVR, γ is the scaling parameter of the Gaussian kernel, and $\mathcal{C}_\mathcal{A}$ is defined by mixed-integer linear programming constraints encoding the dependences/compatibility of the configurations. We use AMPL to formulate the CSSP, and the nonlinear solver Bonmin [5], manually configured (with settings `heuristic_dive_fractional yes`, `algorithm B-Hyb`, `heuristic_feasibility _pump yes`) and with a time limit of 60 seconds, to solve it; then we retrieve, for each instance f, the Bonmin solution $c^*_{\mathtt{bm}}$. Since we have enumerated all possible configurations, we can also compute the "true" global optimum $c^*_{\mathtt{cssp}}$

$= \arg\min\{\bar{p}_A(f,c), c \in \mathcal{C}_A\}$ for sake of comparison. In Table 1, we report the percentage of cases where $c^*_{bm} = c^*_{cssp}$ ("%glob. mins") and, for all the instances where this is not true, the average distance between $\bar{p}_A(c^*_{bm})$ and $\bar{p}_A(c^*_{cssp})$, over all the instances of the considered set ("avg loc. mins"); we also report the average CSSP solution time ("CSSP time"). The kindFS and the aggFS scenarios achieve better results, in terms of "%glob mins", than the noFS one. Furthermore, the local optima found by Bonmin are quite good ones: they are never larger than 3.2% (see "avg loc. mins") of c^*_{cssp}. The time that Bonmin takes to solve the CSSP is reduced from the noFS scenario to the aggFS one. This is due to the fact that the kindFS and the aggFS CSSP formulations have less variables than the noFS one, and so they are easier to solve. Bonmin needs, on average, less than 16 seconds to solve any CSSP; however, devising more efficient techniques to solve the CSSP (say, reformulations, decomposition, ...) might be necessary if our approach is scaled to considerably more algorithmic parameters.

Table 1. Quality of Bonmin's solutions, w.r.t. $\bar{p}$

set	FS	%glob. mins	avg loc. mins	CSSP time
IS	noFS	83.69	**2.013e−02**	15.36
	kindFS	**87.70**	2.758e−02	10.73
	aggFS	84.49	3.122e−02	5.95
OS	noFS	85.32	**1.594e−02**	13.44
	kindFS	90.48	2.014e−02	12.23
	aggFS	**93.25**	1.957e−02	6.15

4.4 Results

In order to assess the performance of the approach, we retrieve $p_{cpx}(c^*_{bm})$, $p_{cpx}(c_{cpx})$ (CPLEX default configuration), the primal and dual gap of c^*_{bm} and c_{cpx} from the filtered dataset, for every IS and OS instance. Table 2 shows: the wins "%w" and the non-worsenings "%w+d", i.e., the percentage of instances such that $p_{cpx}(c^*_{bm})$ is $<$ or $\leq$ than $p_{cpx}(c_{cpx})$, by the first sixteen decimal digits of p_{cpx}, in scientific notation; the wins-over-nondraws "%w$_{nond}$", i.e., the percent wins over the instances such that $p_{cpx}(c^*_{bm}) \neq p_{cpx}(c_{cpx})$; the average $|p_{cpx}(c_{cpx}) - p_{cpx}(c^*_{bm})|$, over all the instances which score a win ("avg w") or a loss ("avg l"); the average $|p_{cpx}(c^*_{bm}) - \underline{p}_{cpx}|$ over all the other instances, where $\underline{p}_{cpx} = \min_{c \in \mathcal{C}_A} p_{cpx}(f,c)$ for a given f ("avg d"). The "%w", "%w+d" and "%w$_{nond}$" are higher on IS instances than on the OS ones. The IS instances are used as the training set, so it is not surprising that $\bar{p}_A$ is less accurate at OS instances; this results in worse CSSP solutions for OS instances. The fact that "avg d" is always 0 implies that, whenever CPLEX—configured by

Table 2. Pipeline quality w.r.t. $p_{\mathrm{cpx}}(c^*_{\mathrm{bm}})$, by FS scenario and IS/OS set

set	FS	%w	%w+d	%w$_{\mathrm{nond}}$	avg d	avg w	avg l
IS	noFS	47.06	96.12	92.39	0	3.400e+14	**6.493e+19**
	kindFS	48.91	98.02	**96.12**	0	3.415e+14	8.895e+19
	aggFS	**48.98**	97.79	95.70	0	**3.438e+14**	8.530e+19
OS	noFS	33.73	83.73	67.50	0	3.148e+14	7.042e+19
	kindFS	**36.64**	82.61	**68.06**	0	3.051e+14	5.385e+19
	aggFS	33.86	77.05	59.78	0	**3.335e+14**	**4.486e+19**

Table 3. Quality of the solutions attained by CPLEX, configured by c^*_{bm} and c_{cpx}, solving HUC instances, aggregated by IS/OS set, FS scenario and PMLP metric

set	FS	metric	%feas c^*_{bm}	%feas c_{cpx}	$\varrho^{\mathrm{bm}}_{\mathrm{prim}}$	$\varrho^{\mathrm{cpx}}_{\mathrm{prim}}$	$\varrho^{\mathrm{bm}}_{\mathrm{dual}}$	$\varrho^{\mathrm{cpx}}_{\mathrm{dual}}$
IS	noFS	cMAE$_{.2}$	96.79	100	1.504e−01	5.812e−01	9.708e−02	1.018e−01
		cMAE$_{.3}$	95.72		1.193e−01		8.104e−02	
		cMAE$_{.4}$	98.40		1.143e−01		7.253e−02	
		MAE	**98.93**		**1.058e−01**		**4.698e−02**	
	kindFS	cMAE$_{.2}$	98.75		**6.975e−02**		**3.792e−02**	
		cMAE$_{.3}$	**99.38**		8.271e−02		4.245e−02	
		cMAE$_{.4}$	98.48		7.085e−02		5.015e−02	
		MAE	96.43		7.183e−02		4.159e−02	
	aggFS	cMAE$_{.2}$	**98.04**		7.850e−02		5.179e−02	
		cMAE$_{.3}$	97.86		**7.669e−02**		5.111e−02	
		cMAE$_{.4}$	97.59		9.267e−02		6.680e−02	
		MAE	97.59		7.747e−02		**4.998e−02**	
OS	noFS	cMAE$_{.2}$	87.30	100	**6.968e−02**	4.633e−01	7.931e−02	8.121e−02
		cMAE$_{.3}$	92.06		1.231e−01		8.500e−02	
		cMAE$_{.4}$	88.89		1.305e−01		9.687e−02	
		MAE	**93.65**		1.371e−01		**6.611e−02**	
	kindFS	cMAE$_{.2}$	89.68		1.471e−01		**6.471e−02**	
		cMAE$_{.3}$	**87.83**		**1.261e−01**		1.189e−01	
		cMAE$_{.4}$	91.01		1.541e−01		1.283e−01	
		MAE	92.86		1.486e−01		1.078e−01	
	aggFS	cMAE$_{.2}$	**89.42**		2.054e−01		9.062e−02	
		cMAE$_{.3}$	88.89		2.061e−01		9.395e−02	
		cMAE$_{.4}$	88.10		2.075e−01		9.874e−02	
		MAE	88.62		**1.792e−01**		**7.733e−02**	

c_{cpx}—can close the gap, our c^*_{bm} proves to be just as efficient. In the nondraws, our approach shows consistent gains with respect to c_{cpx}, both on IS and OS instances. From this we gather that $\bar{p}_{\mathcal{A}}$ provides an accurate approximation of p_{cpx}'s global minima, even at points outside the training set. The noFS scenario presents the worst results. The aggFS scenario achieves the highest percent wins on IS instances; it also provides the largest average wins and the smallest average loss, overall. However, the kindFS scenario presents the best performance, by the highest "%w" on OS instances and by the best overall "$\%\text{w}_\text{nond}$". In Table 3, we report the percentage of IS and OS instances such that CPLEX, configured by c^*_bm and c_cpx, manages to find a feasible solution ("%feas c^*_bm" and "%feas c_cpx"); for those instances, the columns "$\varrho^\text{bm}_\text{prim}$", "$\varrho^\text{cpx}_\text{prim}$", "$\varrho^\text{bm}_\text{dual}$" and "$\varrho^\text{cpx}_\text{dual}$" report the average primal and dual gap achieved by the solver. CPLEX's default configuration always allows the solver to obtain a feasible solution within the time-limit. Our approach presents similar results on IS instances ("%feas c^*_bm" is approximately 98%), but it has slightly worse performances on the OS ones ("%feas c^*_bm" is around 90%). However, the primal/dual gaps provided by our methodology are always better than those achieved by using CPLEX's default setting; actually, they are up to an order of magnitude smaller, in IS instances. The cMAE$_{.4}$ and the MAE metrics provide the highest "%feas c^*_bm", respectively, on IS instances (above 98%) and OS instances (around 92%). On average, the MAE also provides the best primal/dual gaps for IS instances, while the cMAE$_{.4}$ is the best choice for OS instances. Lastly, while the kindFS scenario prevails on IS instances, the noFS one dominates on the OS ones.

5   Conclusions

The methodology presented in this paper conflates ML and MP techniques to solve the ACP. All in all, the results show that the approach is promising, in that the configurations that it provides typically have better primal/dual gaps than CPLEX's. The fact that, in a small fraction of the cases, no feasible solution is found depends on how infeasibility is encoded by p_ml; an easy fix to this issue would be to use a performance measure promoting feasibility. Notably, it is interesting that, in some cases, the custom cMAE error metric outperforms the classical MAE. We also observed that using FS techniques is conducive to much easier CSSP formulations, without overly affecting the quality of the solutions. Overall, since choices taken at any point in the pipeline affect its final outcome, a number of details have to be carefully considered for the approach to work.

References

1. van Ackooij, W., et al.: Shortest path problem variants for the hydro unit commitment problem. In: Proceedings of the Joint EURO/ALIO International Conference 2018 on Applied Combinatorial Optimization (EURO/ALIO 2018), Electronic Notes in Discrete Mathematics, vol. 69, pp. 309–316 (2018)
2. Adenso-Díaz, B., Laguna, M.: Fine-tuning of algorithms using fractional experimental design and local search. Oper. Res. **54**(1), 99–114 (2006)
3. Ansótegui, C., Sellmann, M., Tierney, K.: A gender-based genetic algorithm for the automatic configuration of algorithms. In: Gent, I.P. (ed.) CP 2009. LNCS, vol. 5732, pp. 142–157. Springer, Heidelberg (2009). https://doi.org/10.1007/978-3-642-04244-7_14

4. Battiti, R., Brunato, M.: Reactive search: machine learning for memory-based heuristics. University of Trento, Technical report (2005)
5. Bonami, P., et al.: An algorithmic framework for convex mixed integer nonlinear programs. Discrete Optim. **5**(2), 186–204 (2008)
6. Bonami, P., Lodi, A., Zarpellon, G.: Learning a classification of mixed-integer quadratic programming problems. In: van Hoeve, W.-J. (ed.) CPAIOR 2018. LNCS, vol. 10848, pp. 595–604. Springer, Cham (2018). https://doi.org/10.1007/978-3-319-93031-2_43
7. Borghetti, A., D'Ambrosio, C., Lodi, A., Martello, S.: An MILP approach for short-term hydro scheduling and unit commitment with head-dependent reservoir. IEEE Trans. Power Syst. **23**(3), 1115–1124 (2008)
8. Brendel, M., Schoenauer, M.: Instance-based parameter tuning for evolutionary AI planning. In: Proceedings of the 13th Annual Conference Companion on Genetic and Evolutionary Computation, GECCO 2011, pp. 591–598. ACM (2011)
9. Cristianini, N., Shawe-Taylor, J.: An Introduction to Support Vector Machines: And Other Kernel-based Learning Methods. Cambridge University Press (2000)
10. Degroote, H., Bischl, B., Kotthoff, L., De Causmaecker, P.: Reinforcement learning for automatic online algorithm selection - an empirical study. In: Proceedings of the 16th ITAT Conference Information Technologies - Applications and Theory, pp. 93–101 (2016)
11. Demirović, E., et al.: An investigation into prediction + optimisation for the Knapsack problem. In: Rousseau, L.-M., Stergiou, K. (eds.) CPAIOR 2019. LNCS, vol. 11494, pp. 241–257. Springer, Cham (2019). https://doi.org/10.1007/978-3-030-19212-9_16
12. Eggensperger, K., Lindauer, M., Hutter, F.: Pitfalls and best practices in algorithm configuration. CoRR abs/1705.06058 (2017)
13. Fourer, R., Gay, D.: The AMPL Book. Duxbury Press, Pacific Grove (2002)
14. Hutter, F., Hoos, H.H., Leyton-Brown, K.: Automated configuration of mixed integer programming solvers. In: Lodi, A., Milano, M., Toth, P. (eds.) CPAIOR 2010. LNCS, vol. 6140, pp. 186–202. Springer, Heidelberg (2010). https://doi.org/10.1007/978-3-642-13520-0_23
15. Hutter, F., Hoos, H.H., Leyton-Brown, K.: Sequential model-based optimization for general algorithm configuration. In: Coello, C.A.C. (ed.) LION 2011. LNCS, vol. 6683, pp. 507–523. Springer, Heidelberg (2011). https://doi.org/10.1007/978-3-642-25566-3_40
16. Hutter, F., Hoos, H.H., Leyton-Brown, K., Stützle, T.: ParamILS: an automatic algorithm configuration framework. J. Artif. Intell. Res. **36**(1), 267–306 (2009)
17. Hutter, F., Youssef, H.: Parameter adjustment based on performance prediction: towards an instance-aware problem solver. Technical report, Technical report: MSR-TR-2005125, Microsoft Research (2005)
18. IBM: IBM ILOG CPLEX Optimization Studio, CPLEX 12.7 User's Manual. IBM (2016)
19. IBM: IBM ILOG CPLEX Optimization Studio CPLEX Parameters Reference (2016)
20. Kadioglu, S., Malitsky, Y., Sellmann, M., Tierney, K.: ISAC: Instance specific algorithm configuration. In: Proceedings of the 2010 Conference on ECAI 2010: 19th European Conference on Artificial Intelligence, pp. 751–756. IOS Press, Amsterdam (2010)
21. Khalil, E.B., Bodic, P.L., Song, L., Nemhauser, G., Dilkina, B.: Learning to branch in mixed integer programming. In: Proceedings of the Thirtieth AAAI Conference on Artificial Intelligence, AAAI 2016, pp. 724–731. AAAI Press (2016)

22. Lodi, A., Tramontani, A.: Performance variability in mixed-integer programming. Tutor. Oper. Res. **10**, 1–12 (2013)
23. Lombardi, M., Milano, M., Bartolini, A.: Empirical decision model learning. Artif. Intell. **244**, 343–367 (2017)
24. López-Ibáñez, M., Dubois-Lacoste, J., Pérez Cáceres, L., Birattari, M., Stützle, T.: The irace package: iterated racing for automatic algorithm configuration. Oper. Res. Perspect. **3**, 43–58 (2016)
25. Mohri, M., Rostamizadeh, A., Talwalkar, A.: Foundations of Machine Learning. 2nd edn. The MIT Press (2018)
26. Nannen, V., Eiben, A.E.: Relevance estimation and value calibration of evolutionary algorithm parameters. In: Proceedings of the 20th International Joint Conference on Artifical Intelligence, IJCAI 2007, pp. 975–980. Morgan Kaufmann Publishers Inc. (2007)
27. Rice, J.R.: The algorithm selection problem. Adv. Comput. **15**, 65–118 (1976)
28. Smola, A.J., Schölkopf, B.: A tutorial on support vector regression. Stat. Comput. **14**(3), 199–222 (2004)
29. Van Rossum, G., Drake, F.L.: Python 3 Reference Manual. CreateSpace, Scotts Valley (2009)
30. Varma, S., Simon, R.: Bias in error estimation when using cross-validation for model selection. BMC Bioinf. **7**, 91 (2006)
31. Vilas Boas, M.G., Gambini Santos, H., de Campos Merschmann, L.H., Vanden Berghe, G.: Optimal decision trees for the algorithm selection problem: Integer programming based approaches. CoRR abs/1907.02211 (2019)

Reinforcement Learning for Playing WrapSlide

J. E. Knoll and T. Schmidt-Dumont$^{(\boxtimes)}$ (ID)

Data Science Research Group, Department of Industrial Engineering,
Stellenbosch University, Stellenbosch 7600, South Africa
jacques@knollnet.co.za, thorstens@sun.ac.za

Abstract. The toroidal puzzle game *WrapSlide* shares a lot of characteristics with the famous Rubik's cube in that it consists of a grid of tiles that has to be slid in different directions to get to a solved state. *WrapSlide* is a mobile game that may be played with a $4 \times 4, 6 \times 6$ or 8×8 grid with either two, three or four colours in the grid. The objective of the game is to group the tiles of the same colour into the four quadrants of the grid. In this paper, a reinforcement learning approach towards playing the game *WrapSlide* is implemented in order to ascertain whether a reinforcement learning agent is capable of solving the puzzle in a near-optimal number of moves.

Keywords: Reinforcement learning · Combinatorial optimisation · Games

1 Introduction

Since the inception of the Rubik's cube by Ernö Rubik it has become one of the most influential puzzles ever to be invented. The Rubik's cube has sparked many studies into discovering its so-called God's Number (the smallest number of moves required to solve the puzzle from any starting position). This has proven to be an arduous task, given that the state space comprises 43 252 003 274 489 856 000 possible permutations [16]. Research on this topic began in 1981 when Morgan Thistlewaite, a British mathematician, proved that through the use of an algorithm, only 52 moves were required to solve any of the Rubik's Cube's combinations [16]. It was not until 29 years later, in 2010, that the God number was proven to be 20 in a process that required 35 years worth of CPU processing power to find the solution [16].

The venture into solving for the God's number of the Rubik's cube coincided with the establishment of the field of *artificial intelligence* (AI), which has, in many ways, revolutionised the modern world. There have been countless investigations into the capabilities of AI, and *machine learning* (ML) in particular, for developing strategies for playing board games, including Arthur Samuel's Checkers player [3], Deep Blue, IBM's Chess player [6] and perhaps most recently Deepmind's AlphaGo, which defeated the world champion at the

© Springer Nature Switzerland AG 2020
G. Nicosia et al. (Eds.): LOD 2020, LNCS 12565, pp. 713–724, 2020.
https://doi.org/10.1007/978-3-030-64583-0_62

game of Go [19]. Board and combinatorial games have proved to be an excellent testing ground for measuring the progress made within AI and ML as it is thought that an intelligent strategy is required to excel.

The toroidal slide puzzle game *WrapSlide* may be seen as an attempt to modernise the Rubik's cube for the 21st Century. *WrapSlide* is a grid-based game in which the aim is to group blocks of the same colour together in each quadrant by sliding the top, bottom, left or right half of the grid left or right, or up or down, respectively. The blocks slid off the board on one side reappear on the opposite side.

As *WrapSlide* was only released in 2014, there have been a limited number of investigations performed on the puzzle. The God numbers for the 4×4 grid have been determined, while upper and lower bounds on the God numbers for the 6×6 grid have been established. Building state enumeration trees, van Zyl [22] showed that the God numbers for the two-, three- and four-colour 4×4 grids are 6, 10 and 12, respectively. Furthermore, upper bounds for the three- and four-colour 6×6 grids have been determined as 21 and 28, respectively [22]. In another study on *WrapSlide*, Geelhoud and Meester proved that any state of the 6×6 *WrapSlide* grid is reachable from the solved state [9]. The exact God numbers have, to the best of the authors knowledge, not been determined for the 6×6 and 8×8 *WrapSlide* grids.

The aim in this paper is to develop and implement a *reinforcement learning* (RL) algorithm capable of solving the *WrapSlide* puzzle in a near-optimal number of moves. The methodological approach followed in developing the RL agent, including a description of the identification of class representatives, and the development of the *WrapSlide* environment is outlined in Sect. 2. This is followed by a description of the numerical results achieved during the training and testing phases of the RL agent in Sect. 3. The paper closes in Sect. 4 with a conclusion highlighting the limitations of the adopted approach, as well as providing ideas for future work that may be pursued to address these limitations.

2 Methodological Approach

The development of the methodological approach is detailed in this section. This approach includes descriptions of a technique for complexity reduction by determination of mathematically equivalent states through the identification of class representatives, an introduction to RL, the mathematical implementation of *WrapSlide* as an RL environment and the configuration of the RL agent. Although the approach developed in this paper is specific to solving the puzzle *WrapSlide*, the application of ML algorithms for playing games has found application in the field of planning and search [4], where such games have been employed in order to assess the effectiveness of planning algorithms.

2.1 Symmetries, Equivalence Classes and Tilings on a Torus

A symmetry may be defined as a transformation of an object that appears to leave the object unchanged. Essentially, the object appears to be in the same position as it was before, although it has, in fact, moved [20]. There exist a number of transformation processes which may result in symmetries. These are *reflection, rotation, translation* and *glide symmetry*.

Reflection occurs when an object is mirrored across a pre-specified plane. Rotation typically occurs around an axis that is perpendicular to the object (in the case of a two-dimensional shape). In the case of a square object, such as a *WrapSlide* grid, each time the object is rotated by 90°, a new symmetry is produced. Translation, on the other hand, occurs when every part of the object has been moved by the same distance in the same direction. Finally, glide symmetry occurs when an object is shifted a certain distance in some direction and thereafter, when reflected about a line matches up with the original object. An object may be defined as *unique* if, once it has undergone a transformation, at least one point of the object is not in its original position.

The concept of symmetries may be employed in order to define so-called *symmetry groups*. A *group* is defined as a closed system consisting of transformations where any combination of these transformations forms another member in its group. If each of these transformations produces a symmetry, then it may be defined as a symmetry group [20].

The above-mentioned transformations all occur in the Euclidean plane, which may be characterised by ordered pairs of real numbers, (x, y), and denoted by $\mathbf{R}^2$. An *isometry* is defined as a function $h : \mathbf{R}^2 \leftarrow \mathbf{R}^2$ which preserves distance in the Euclidean plane: $||h(\mathbf{v}) - h(\mathbf{w})|| = ||\mathbf{v} - \mathbf{w}||$ for all vectors $\mathbf{v}$ and $\mathbf{w}$ in $\mathbf{R}^2$, where $|| \cdot ||$ represents a norm [8]. The same concept of symmetry groups may be applied to isometries, in that the combination of two or more isometries produces another isometry. There are, however, a limited number of unique objects that can be produced. Grouping all objects with its symmetries and isometries results in an *equivalence class* of objects.

Using a set $\mathbf{X}$, an *equivalence relation* on $\mathbf{X}$, denoted by $\sim$, satisfies three properties. These properties are $x \sim x$ for all x, $x \sim y$ if $y \sim x$ and that $x \sim y$ and $y \sim z$ implies $x \sim z$. As such, an *equivalence class* is a subset $\mathbf{S} = \{y \in \mathbf{X} | y \sim x\}$, meaning that $\mathbf{S}$ is a set of which all objects are equivalent to x [18]. As all objects in $\mathbf{S}$ are equivalent, a single object may be used as the *class representative* for the purpose of analysis. The behaviour of all objects in the class may be predicted by calculations performed on the class representative.

Schwartz [18] defined the notion of a square torus as a compromise between a unit sphere in $\mathbf{R}^3$ and the Euclidean plane in $\mathbf{R}^2$. This may be further explained in that a square torus is bounded like a sphere. Every point on the plane is, however, flat. Employing the example of a virtual, two-dimensional bug. When crawling across a square torus and the bug crawls over the top edge of the torus, it would "reappear" at the bottom edge [18]. This may be represented graphically in the unit cell in Fig. 1(a) embedded in a torus in Fig. 1(b). Naturally, larger structures may be created, as illustrated in Fig. 1(c). For a structure with m horizontal unit cells and n vertical unit cells, the $m \times n$ plane shown is embedded in a torus in Fig. 1(d).

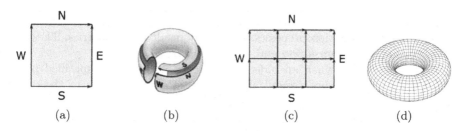

Fig. 1. The sub-units of a square torus. (a) A unit cell, (b) A unit cell embedded in a torus, (c) 2 × 3 unit cells, and (d) m × n unit cells embedded in a torus [13].

Decomposing a torus into subsets, as illustrated in Fig. 1 may also be performed by viewing the unit cells as tilings. A tiling may be defined as non-overlapping shapes that cover a surface without any gaps. Classifying the features of a torus as being covered in tiles allows for the identification of two characteristics. The first of these is that subsets, or unit cells do not have to be identical. The second characteristic is that tiling symmetries may occur [10]. Therefore the concept of class representatives may be employed in order to significantly reduce the size of the state space when performing an analysis of the puzzle game *WrapSlide*. When solving for the God number of the Rubik's cube, rotation and reflection, which may be employed when working with a cube were used in order to reduce the size of the state space [16]. In this study, methods for generating grid symmetries resulting from translation and glide symmetry were also developed. In order to illustrate the effect of employing the equivalence classes on the size of the state space, consider Table 1 below. A complete description of the process employed for generating symmetries, identifying isometries, equivalence classes and class representatives for the *WrapSlide* grid may be found in [12].

Table 1. The effect of employing class representatives in reducing the size of the state space.

4 × 4 Grid	Two colours	Three colours	Four colours
Complete state space	1.820×10^3	9.009×10^5	6.306×10^7
Reduced state space	7.800×10^1	14.523×10^3	84.174×10^3

The results presented in Table 1 are based on the state enumeration trees built while employing the concept of equivalence classes to ensure that each class representative occurs only once in the tree, while the size of the complete state space was determined mathematically. Due to the exponential increase in complexity, as illustrated by the increase in the sizes of the state spaces for both the complete and reduced state spaces, state enumeration trees for the 6 × 6 and 8 × 8 grids were not constructed. As a result, the sizes of the reduced state spaces for those grids are, to the authors best knowledge, still unknown.

2.2   Reinforcement Learning

In RL, an agent attempts to learn an optimal policy by trial and error [21]. At each time step, the agent receives information about the current state of the environment it finds itself in, based on which it takes an action, transforming the environment into a new state. The mapping according to which the agent chooses its action is called the *policy*, which is used to define the agent's behaviour. Based on the action chosen, as well as the resulting next state, the agent receives a scalar *reward* which provides an indication of the quality of the action chosen. In RL, the aim of an agent is to discover a policy which maximises the accumulated reward over time (*i.e.* finding a policy according to which the agent always chooses a best action with respect to the long-term reward achieved) [21].

   Throughout the trial and error search procedure, the agent's actions are typically guided by a pre-specified policy, with the aim of achieving a suitable trade-off between exploration of the state-action space, and exploitation of what the agent has already learnt. Perhaps the simplest action selection policy is the so-called ϵ-greedy policy [21]. According to this policy, the agent will choose the action with the largest associated value with probability $(1 - \epsilon)$, while choosing a random action with probability ϵ, where $\epsilon \in (0, 1]$ denotes a small positive number.

Q-Learning. Numerous RL algorithms which exhibit favourable learning speeds and contain easily customised parameters have been proposed in the literature. One of the most notable of these is Q-Learning, proposed by Watkins and Dayan [23]. In Q-Learning, a scalar value, denoted by $Q(s, a)$, is assigned to each state-action pair (s, a) so as to provide an indication of the quality of that combination. The aim is then to

$$\text{maximise} \sum_{t=0}^{\infty} \gamma^t r_t, \tag{1}$$

where r_t denotes the reward obtained at time step t, and γ denotes the so-called discount factor employed to determine the relative importance of future rewards. The Q-value is then updated according to the rule

$$Q(s_t, a_t) = Q(s_t, a_t) + \alpha \left(r_{t+1} + \gamma \max_{a_{t+1}} Q\left(s_{t+1}, a_{t+1}\right) - Q\left(s_t, a_t\right) \right), \tag{2}$$

where r_{t+1} denotes the reward received after performing action a_t when the system is in state s_t, resulting in the new state s_{t+1}, and where $\alpha \in (0, 1]$ is a user-defined parameter known as the *learning rate*. A more detailed description of the algorithm is provided by Watkins and Dayan [23].

Function Approximation. In the simplest scenario, the Q-values are kept and updated in a lookup table. This approach is, however, limiting when state and action spaces become large, such as due to the combinatorial nature of the

WrapSlide puzzle. *Artificial Neural Networks* (ANNs) have found wide application when performing function approximation in conjunction with RL. In such a scenario, the error term E employed in the loss function when training the ANN, typically employing a backpropagation approach, is based on the Q-Leaning update rule, and given by

$$E = r_t + \gamma \max_a Q(s_{t+1}, a) - Q(s_t, a_t). \tag{3}$$

Employing this update rule, the ANN may be employed as a general function approximator in order to learn the optimal Q-values for large combinatorial or continuous problems [2].

2.3   A Mathematical Implementation of the *WrapSlide* Puzzle

The so-called *gym* library [5], developed in the programming language Python [15], is an open source toolkit for comparing and implementing RL algorithms, originally developed by *OpenAI* in 2016 [5]. The gym library provides a number of standardised environments that are compatible with various numerical computation libraries such as *TensorFlow* [1], without assuming anything about the RL agent. In order to develop an RL agent capable of solving the puzzle *WrapSlide*, a gym-compatible environment of the puzzle *WrapSlide* was developed.

Any gym-compatible library comprises three essential components and one optional component. The three essential components are (1) an initialisation procedure, (2) a reset procedure, and (3) a step function. The optional component of the environment comprises a visualisation component.

Initialisation. In order to initialise the environment the state and action spaces are defined. The solved *WrapSlide* state is generated subject to the specific criteria, including the size of the grid and the number of colours. Each colour is simply denoted by an integer 1–4. For a 4×4 grid this initialisation step yields a 4×4 matrix. The class representative of the solved state is then found, given that there may be either two, three or four different states which may be classified as solved, depending on the number of colours selected. Furthermore the action space is defined. The action space consists of $4(n-1)$ distinct actions, which represent the possible sliding moves that may be performed, and where n denotes the size of the grid.

Reset. Once the *WrapSlide* environment has been initialised, the environment is reset to a random initial state, which represents the first observation passed to the learning agent. This starting state, like the solved state is generated subject to the selected grid size, as well as the number of colours selected. For purposes of simplicity, the $n \times n$ matrix is read lexicographically into a n^2-element vector which is passed to the learning agent.

Step. The step function presents arguably the most important component of the environment, as this function performs the action selected by the RL agent and returns the resulting observation and its associated reward. The agent selects one of the $4(n-1)$ discrete actions, which is passed to the step function. Within the step function, the current grid formation is read from the environment, the specified action is performed and the next grid configuration is determined. Thereafter, the class representative for the new state is determined. This class representative is then compared to the class representative of the solved state to determine whether the puzzle has been solved successfully. If the agent has, indeed, solved the puzzle, the agent is notified that the terminal state has been reached and awarded a reward of 1. Otherwise, the new state, in the form of the class representative is communicated to the agent and this procedure is repeated.

2.4 The Reinforcement Learning Agent Configuration

In this study, an ANN approach towards function approximation was adopted, in conjunction with a Q-learning [23] RL agent. The ANN was implemented employing the well-known *Keras* deep learning library [7] developed for the Python programming language.

A *multi-layer perceptron* (MLP) feedforward ANN was implemented as function approximator. While there exist more complex ANN architectures that may be more suited to the underlying problem at hand, the MLP ANN was chosen as a starting point in this study in order to ascertain the limitations of this approach, to develop more complex tailor-made approaches in future.

The input layer of the ANN takes the shape of the size of the n^2-element vector which represents the state of the environment. The output layer of the ANN comprises $4(n-1)$ neurons, the size of the action space, dependent on the size of the *WrapSlide* grid. The number of fully connected hidden layers, and the number of neurons contained in each of the hidden layers were varied as the complexity of the problem increased, in order to adjust and adapt to the increase in complexity. Based on empirical experimentation, the sigmoid activation function, in conjunction with the Adam [11] backpropagation technique as included in the Keras library [7] were employed throughout the study.

The Q-learning agent is defined in conjunction with the $Keras - RL$ library [14]. The agent receives as input the MLP ANN, the size of the action space and the action selection policy to be employed.

3 Numerical Results

Throughout the training procedure, ϵ, γ and α were set to 0.1, 0.99, and 0.001, respectively. Furthermore, based on empirical experimentation, the sigmoid activation function was employed in all neurons of the ANN. The weights of the ANN were updated with the aid of the well-known *Adam* backpropagation method. Batch updating was employed in the training of the ANN, with the default batch size of 32, as suggested in the *Keras-RL* library [14]. The code of the algorithmic implementations and the *WrapSlide* environment may be found at [17].

3.1   The Reinforcement Learning Training Procedure

Due to the relative simplicity presented by the 4 × 4 two-colour grid, the agent was capable of solving the puzzle by trial-and-error, and thus able to learn to solve the puzzle effectively, even with very simple ANN architectures. The agent was trained for 1 000 000 steps, and, dependent on the number of hidden neurons capable of solving the puzzle up to 12 000 times during those 1 000 000 training steps. The training progression of the agent over the first 5 000 episodes is shown in Fig. 2(a). Furthermore, experimentation as to the number of neurons required in the single hidden layer for sufficient generalisation ability was performed. The number of hidden neurons was varied from 5 to 20. The results of this experimentation are presented in Table 2.

Table 2. The average number of steps per episode during the training procedure for the 4 × 4 two-colour grid. The average over all episodes, as well as the average after convergence are shown.

Neurons	Mean	Mean at Steady State
5	87.56	38.35
10	22.51	7.50
15	9.07	6.25
20	8.16	5.97

As may be seen in the table, at steady state the agent was able to solve the puzzle in an average number of 5.97 steps. Given that the God number for the 4 × 4 two-colour grid has been confirmed as 6, it is suggested that the agent typically followed the optimal move sequence when solving the puzzle (given that the agent would naturally chose suboptimal actions 10% of the time due to the ϵ-greedy action selection policy being in place).

Training the agent to solve the 4 × 4 three- and four-colour grids proved significantly more challenging. It was observed that when following the pure trial-and-error approach the agent was often unable to solve the puzzle within the first 2 500 000 steps. This is attributed to the fact that, on the contrary to the two-player games at which RL has excelled, there is a very limited number of ways in which the solved state may be reached, and the possibility exists that the agent never reaches the solved state by trial-and-error.

Due to the difficulty in defining an effective cost function based on the *Wrap-Slide* states, reward function shaping was not considered in order to guide the search of the agent. Instead, an adaptation to the training procedure was introduced, with the aim that the agent would still be capable of learning to play the game based on its own strategy, rather than one shaped by human thinking. The aim during the development of this training procedure was to increase the probability of the agent reaching the solved state by trial-and-error due to the increased proximity of the starting state to the solved state, but not altering the reward function in the process. The strategy developed works as follows:

1. The bottom half (starting from the solved state) of a state enumeration tree is generated.
2. The agent is trained for 50 000 steps on each layer of the state enumeration tree (starting only one step away from the solved state, then two *etc.*) Furthermore, each episode is terminated if the agent has not solved the puzzle within 1 000 steps. If the agent is unsuccessful no reward is awarded.
3. Once the agent has been trained on states within the bottom half of the state enumeration tree, training on the entire state space commences.

As a result of the increase in complexity, ANNs containing 50 and 100 hidden neurons were employed when solving the 4×4 three-colour grid. The training progression, once the training on the complete state space commenced, is shown in Fig. 2(b). Note that due to the fact that the agent had already learnt to solve the puzzle when starting in one of the first five layers of the state enumeration tree the exponential decay in the number of steps per episode is not as drastic as it was for the 4×4 two-colour grid. These results are presented in Table 3.

Based on the fact that the God number of the 4×4 three-colour grid has been confirmed as 10 [22], it was concluded that the agent achieved sufficient effectiveness in finding a near-optimal solution path with an average length of 15.40 moves, while employing the ϵ-greedy method.

The same training procedure as followed for the 4×4 three-colour grid was employed for the 4×4 four-colour grid. Due to the increase in complexity, the number of neurons in the hidden layer, as well as the number of hidden layers required to be able to solve the puzzle were once again varied. Based on extensive numerical experimentation, the combinations as shown in Table 3 were employed.

Table 3. The average number of steps per episode during training for the 4×4 three- and four-colour grids. The average over all episodes, and the average after convergence are shown.

Colours	Hidden layers	Neurons per layer	Mean	Mean at Steady state
3	1	50	20.28	19.59
3	1	100	15.66	15.40
4	2	50	276.41	160.99
4	2	100	67.04	53.29
4	3	50	71.42	62.12
4	3	100	33.94	29.71

The training progression once training commenced on the entire state space for the configuration with two hidden layers, each containing 100 neurons is shown in Fig. 2(c) below. Further increases in the number of neurons per layer, as well as the number of hidden layers employed were investigated, but these did not lead to significantly improved training results.

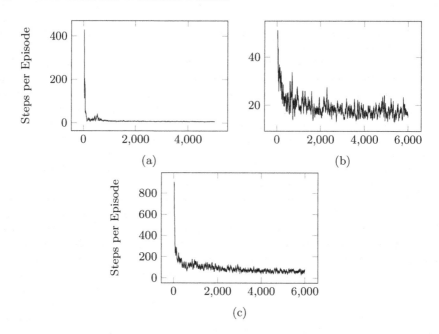

Fig. 2. The training progression for the 4 × 4 two-colour grid with a single hidden layer of 20 neurons in (a), the three-colour 4 × 4 grid with a single hidden layer of 100 neurons in (b), and the four-colour 4 × 4 grid with two hidden layers of 100 neurons each in (c). In order to filter out some noise, a moving average over 30 episodes is shown.

Given that the bottom half of the state enumeration tree for the 4 × 4 three-colour grid contains 591 states, and the bottom half of the state enumeration tree for the 4 × 4 four-colour grid contains 697 states, the agent was initially trained on 4.06% and 0.83% of the state space for the three- and four-colour grids, respectively.

3.2   The Reinforcement Learning Testing Procedure

The steady state results recorded when training the RL agents are influenced by the ϵ-greedy approach employed during training. Therefore, in order to fairly assess the performance of the RL agent when solving the *WrapSlide* puzzle, a greedy testing procedure was designed. In order to assess the performance of the RL agent on each of the 4 × 4 *WrapSlide* grids, the RL agent was required to solve the puzzle 100 times for randomly generated *WrapSlide* states.

Initially, a purely greedy approach was employed. It was found, however, that when the greedy approach was employed, the agent may "get stuck" infinitely cycling between a select few states. As a result, the action selection policy was amended, to allow the agent to escape the loop. In the resulting policy the agent greedily selects the action with the highest Q-value, except if the resulting state s_{t+1} has already been visited in that episode (indicating that a cycle occurs). In that case the agent will select the action with the next largest Q-value which does not result in a cycle. The results achieved for each of the ANN configurations for the *WrapSlide* incarnations are shown in Table 4.

Table 4. The average number of steps per episode during testing.

Colours	Hidden layers	Neurons per layer	Mean	Median	God number
2	1	5	64.26	19	6
2	1	10	5.62	5	6
2	1	15	5.22	5	6
2	1	20	5.36	5	6
3	1	20	323.26	243	10
3	1	50	17.04	14	10
3	1	100	11.64	11	10
4	2	50	105.40	105	12
4	2	100	36.87	36	12
4	3	50	37.29	36	12
4	3	100	17.37	17	12

As is evident from the results, the mean number of steps per episode when employing the amended greedy action selection policy is, in fact, smaller than when employing only the ϵ-greedy method. As indicated when considering the mean and the median, the agent was typically able to follow a near-optimal path when solving the puzzle, based on the God numbers of 6, 10 and 12 that have been found for the 4×4 two-, three- and four-colour grids respectively.

4   Conclusion

In this paper an approach towards solving the puzzle game *WrapSlide* with a *Reinforcement Learning* (RL) approach was proposed. The RL agent was capable of successfully solving all of the 4×4 incarnations of the *WrapSlide* puzzle in a near-optimal number of moves. It was found that the approach of employing the relatively simple feedforward neural network for value function approximation will reach its limitation soon. Increasing the number of hidden layers and hidden neurons in order to achieve better generalisation for more complex cases may prove naïve and ineffective. Therefore, the authors suggest the implementation of neural networks with more complex structures which may be tailored to the underlying combinatorial problem. Two such approaches include the use of Convolutional Neural Networks, where the input is treated as a two-dimensional image (as seen by a human player), and the use of tree search protocols, such as Monte Carlo Tree Search which have been employed successfully when solving two-player games such as Go.

References

1. Abadi, M., et al.: TensorFlow: a system for large-scale machine learning. In: Proceedings 12th USENIX Symposium on Operating Systems Design and Implementation, pp. 265–283. USENIX Association, Savannah (2016)
2. Aggarwal, C.C.: Neural Networks and Deep Learning. Springer, Cham (2018). https://doi.org/10.1007/978-3-319-94463-0_9
3. Berliner, H.: Backgammon program beats world champ. ACM SIGART Bull. **69**, 6–9 (1980)
4. Bonet, B., Geffner, H.: Planning as heuristic search. Artif. Intell. **129**, 5–33 (2001)
5. Brockman, G., et al.: OpenAI Gym. arXiv 1606:01540 (2016)
6. Campbell, M.A., Hoane, A.J., Hsu, F.H.: Deep blue. Artif. Intell. **134**(1–2), 57–83 (2002)
7. Chollet, F.: Keras. https://github.com/fchollet/keras. Accessed 24 Apr 2020
8. Conrad, K.: Plane Isometries and the Complex Numbers, Technical report, Department of Mathematics, University of Connecticut, Connecticut (CT)
9. Geelhoud, D., Meester, L.: De WrapSlide-puzzle algebraïsch bekeken, Technical report, Leiden University, Leiden
10. Kell, B.: Symmetries and Tilings. http://math.cmu.edu/~bkell/21110-2010s/symmetry-tilings.html. Accessed 24 Apr 2020
11. Kingma, D.P., Ba, J.: Adam: a method for stochastic optimisation. arXiv 1412:6980 (2014)
12. Knoll, J.E.: Reinforcement Learning for Playing WrapSlide, Honour's thesis, Industrial Engineering, Stellenbosch University (2019)
13. Nelson, A., Newman, H., Shipley, M.: 17 Plane Symmetry Groups. https://caicedoteaching.files.wordpress.com/2012/05/nelson-newman-shipley.pdf. Accessed 24 Apr 2020
14. Plappert, M.: Keras-RL. https://github.com/keras-rl/keras-rl. Accessed 24 Apr 2020
15. Python Software Foundation. https://www.python.org/psf/. Version 3.7. Accessed 6 Apr 2020
16. Rokicki, T., Kociemba, H., Davidson, M., Dethride, J.: God's Number is 20. http://cube20.org/. Accessed 24 Apr 2020
17. Schmidt-Dumont, T.: All the code pertaining to the ongoing project on WrapSlide at Stellenbosch University. https://github.com/thorstenschmidt-dumont/WrapSlide. Accessed 22 May 2020
18. Schwartz, R.E.: Mostly Surfaces, 1st edn. American Mathematical Society, Providence, RI (2011)
19. Silver, D., et al.: Mastering the game of go with deep reinforcement learning and tree search. Nature **529**(7587), 484–504 (2016)
20. Stewart, I., Golubitsky, M.: Fearful Symmetry: Is God a Geometer?, 1st edn. Courier Corporation, Cambridge (2010)
21. Sutton, R.S., Barto, A.G.: Reinforcement Learning: An Introduction, 1st edn. MIT Press, Cambridge (1998)
22. van Zyl, G.. In Pursuit of God Numbers for the Puzzle WrapSlide, Honour's thesis, Industrial Engineering, Stellenbosch University (2016)
23. Watkins, C.J., Dayan, P.: Q-learning. Mach. Learn. **8**(3–4), 279–292 (1992)

Discovering Travelers' Purchasing Behavior from Public Transport Data

Francesco Branda(✉) , Fabrizio Marozzo , and Domenico Talia

DIMES, University of Calabria, Rende, Italy
{fbranda,fmarozzo,talia}@dimes.unical.it

Abstract. In recent years, the demand for collective mobility services is characterized by a significant growth. The long-distance coach market has undergone an important change in Europe since FlixBus adopted a dynamic pricing strategy, providing low-cost transport services and an efficient and fast information system. This paper presents a methodology, called *DA4PT (Data Analytics for Public Transport)*, aimed at discovering the factors that influence travelers in booking and purchasing a bus ticket. Starting from a set of 3.23 million user-generated event logs of a bus ticketing platform, the methodology shows the correlation rules between travel features and the purchase of a ticket. Such rules are then used to train a machine learning model for predicting whether a user will buy or not a ticket. The results obtained by this study reveal that factors such as occupancy rate, fare of a ticket, and number of days passed from booking to departure, have significant influence on traveler's buying decisions. The methodology reaches an accuracy of 93% in forecasting the purchase of a ticket, showing the effectiveness of the proposed approach and the reliability of results.

Keywords: Public transport · Bus · Travelers' buying behaviour · Ticketing platform · Machine learning · Dynamic pricing

1 Introduction

The long-distance bus industry has traditionally been slow to evolve and it is quite resistant to change. While countries like UK, Sweden and Norway liberalized their coach transport market beyond high-speed rail a long time ago, other important markets like France, Italy, and Germany opened up recently [7]. A turning point has occurred in 2015 when FlixBus entered the European market, significantly increasing the supply of interregional buses and practicing aggressive pricing policies, to which many other local operators decided to adapt [6]. Therefore, thanks to relatively low cost, rapidly increasing convenience and routing flexibility, the bus transportation offers an added value to passengers over airlines and trains in the last years. In particular, the far-away locations of airports and the strict security procedures have made flying slower and tedious, are pushing more travelers to avoid airlines on short-to-medium distances. The

© Springer Nature Switzerland AG 2020
G. Nicosia et al. (Eds.): LOD 2020, LNCS 12565, pp. 725–736, 2020.
https://doi.org/10.1007/978-3-030-64583-0_63

train, on the other hand, appears to be rather expensive in many countries and though faster than the bus, it is also potentially more vulnerable to delays and missed connections. The rise in competition, the greater attention to customer experience and the possibility of offering long-distance journeys, lead transport companies to use intelligent tools to plan and manage their mobility offer in a dynamic and adaptive manner.

This paper presents a methodology, called *DA4PT* (*Data Analytics for Public Transport*), aimed at discovering the factors that influence travelers' behaviour in ticket purchasing. In particular, DA4PT uses Web scraping techniques and process mining algorithms to understand behaviours of users while searching and booking bus tickets. Starting from a set of user-generated event logs of a bus ticketing platform, the methodology shows the correlation rules between travel attributes and the purchase of a ticket. Then, such rules are used to train a machine learning model for predicting whether a user will buy or not a ticket.

The proposed methodology has been applied on a dataset composed by 3.23 million event logs of an Italian bus ticketing platform, collected from August 1st, 2018 to October 20st, 2019. The results obtained by this study reveal that factors such as occupancy rate, fares of tickets, and number of days passed from booking to departure, have significant influence on traveler's buying decisions. We experimentally evaluated the accuracy of our methodology comparing some of the most relevant machine learning algorithms used in the literature [15]. Among them, Random Forest proved to be the best classification algorithm with an accuracy of 93% and low variance in results than other algorithms in the demand forecasting domain.

Compared to the state of the art, the presented methodology analyzes the user behavior on a bus ticketing platform for understanding if a user will buy a ticket after visiting the website and the main factors that influence her/his decision. The model obtained could be used to allow bus platforms to switch to (or improve) dynamic pricing strategies that maximize the percentage of occupation of a bus, the number of tickets sold, and the total revenue.

The rest of the paper is organized as follows. Section 2 discusses related work. Section 3 outlines the main concepts and goals of our analysis. Section 4 presents the proposed methodology. Section 5 illustrates the case study. Section 6 concludes the paper.

2   Related Work

Several approaches concerning demand forecasting have been proposed in the literature. In this section we briefly review some of the most representative related work in the area of demand forecasting, discussing differences and similarities with the methodology we designed.

Liu et al. [9] proposed a multi-factor travel prediction framework, which fuses complex factors of the market situation and individual characteristics of customers, to predict airline customers' personalized travel demands. With respect to our work, this is not focused on bus travels, however it could be applied to predict travel demands on the basis of the factors that influence travelers' buying decisions.

Szopiński and Nowacki [14] discovered that flight duration affects the price dispersion of airline tickets and the price dispersion increases as the date of departure approaches. Similarly, our work discovered that the low cost of a ticket can result in more sales. Therefore, it is advisable for a bus company to adopt a dynamic price strategy.

Other studies have attempted to predict the demand for transport services on the basis of price elasticity, i.e. a measure used to understand how demand can be affected by changes in price. Mumbower et al. [11] estimated the change in flight prices by using factors such as departure day of the week, time of departure, and date of booking. In particular, a linear regression method has been used to predict the number of bookings for a specific flight by date of departure, route and number of days before the departure date. Escobari [5] studied that consumers become more price sensitive as time to departure approaches and the number of active consumers increases closer to departure. These two papers cover only the Step 3 of our methodology and may be considered as alternative method for defining the correlation between the factors that influence travelers' buying decisions.

Abdelghany and Guzhva [1] used a time-series modeling approach for airport short-term demand forecasting. The model assesses how various external factors such as seasonality, fuel price, airline strategies, incidents and financial conditions, affect airport activity levels. In [17] Yeboah et al. developed an explanatory model of pre-travel information-seeking behaviours in a British urban environment, using binomial logistic regression. The considered factors include socio-demographics, trip context, frequency of public transport use, used information sources, and smartphone ownership and use. The two models proposed can be integrated in the methodology we designed.

3   Problem Definition

Let $D = \{d_1, d_2, \ldots\}$ be a dataset collecting trip instances of a bus company, where each d_i is a tuple described by the following features: *trip itinerary* identifier; *origin* and *destination* cities; *booking* and *departure* date; *fare* of a ticket; number of *bus seats*.

Let $EL = \{e_1, e_2, \ldots\}$ be a set of event logs generated by users of the bus ticketing platform, where an event e_i is a tuple defined by the following fields: *cookie* of the user; description of the user *action*; *timestamp*; *trip itinerary* identifier; number of *bus seats* required by the user. For instance, a single event e_i may be: (*i*) find a trip, or (*ii*) calculate fare of a ticket for a given trip, or (*iii*) select a seat on the bus, or (*iv*) pay the booked trip. Some users finalize their search by purchasing the ticket (*purchased*) while others abandon the platform without purchasing the ticket (*abandoned*).

The main goal of this work is to infer patterns and trends about users behaviour for training a machine learning model that can predict whether a user will buy a ticket or not. More specifically, the data analysis we carried out aims at achieving the following goals:

1. Discover a set F of *factors*, $F = \{F_1, ..., F_J\}$, where a *factor* F_j influences travelers' purchasing behaviour;
2. Train a machine learning model on the basis of F, to predict whether or not a user will buy a ticket.

4   Proposed Methodology

As shown in Fig. 1, the proposed methodology consists of four main steps:

1) Data collection through *Web scraping* techniques;
2) Pre-processing of event data and execution of *process mining* algorithms on event data;
3) Identification of *the main factors* influencing traveler's purchasing behaviour;
4) *Data analysis* and *machine learning* for purchase prediction.

For each step, a formal description and a use case are illustrated in the following sections.

4.1   Steps 1–2: Web Scraping and Process Mining

The first two steps aim at defining the event logs EL used for understanding the behaviour of users while searching and booking bus trips. Specifically, during step 1, data collection is carried out by using Web scraping techniques (i.e., a set of techniques used to automatically extract information from a website), to know all the interactions of a user with the bus ticketing portal, for instance whether a user buys or not a ticket, or in which step of the buying decision process s/he leaves the platform.

Step 2 is aimed at exploiting process mining algorithms for learning and support in the (re)design of purchasing processes by automatically discovering models that explain the events registered in log traces provided as input [4]. The set of event logs EL is pre-processed in order to clean, select and transform data, making it suitable for analysis. In particular, we first clean the collected data for identifying the most compliant model with the event logs. Then, we proceed

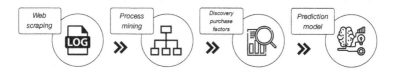

Fig. 1. The main steps of the DA4PT methodology.

by selecting only the events that end successfully and unsuccessfully (e.g., the purchase of a ticket and the abandonment of the platform by users, respectively). Finally, we transform data by keeping one event per user in a day.

The output of step 2 is the dataset $\hat{D} = \{\hat{d}_1, \hat{d}_2, ... \}$, where $\hat{d}_i$ is a tuple $\langle u_i, \{e_{i1}, e_{i2}, ..., e_{ik}\}\rangle$ in which e_{ij} is the jth event generated by user u_i.

4.2   Step 3: Discovery of Purchase Factors

The goal of step 3 is to identify the key factors that push a user to buy a ticket. Specifically, we perform exploratory factor analysis for reducing a large set of attributes to a more coherent number which can explain travelers' purchasing behaviour. Then, we apply the correlation analysis to define the conditions that tend to occur simultaneously, or the patterns that recur in certain conditions.

In particular, the goal of our analysis is to generate correlation rules like $f \rightarrow e_{purchased}$ (if factor $f \in F$ occurs, then it is likely that also event $e_{purchased}$ occurs). The correlations between an attribute and the class attribute (*purchased*) or (*abandoned*) is evaluated using the Pearson's correlation coefficient [12]. The values of the Pearson's correlation can be in the range [-1,1], where the value of 1 represents a strong linear relationship, 0 no linear correlation, while -1 corresponds to a negative linear correlation.

Below we report the meaning of the attributes that we have added into the dataset $\hat{D}$ before performing exploratory factor analysis. The value of these attributes has been calculated from other attributes in D: *i*) *Days before departure* (*DBD*), by calculating the difference between booking and departure date; *ii*) *Booking day of the week* (*BDOW*), by extracting the day from a booking date; *iii*) *Occupancy rate for a bus* (*OCCR*), by evaluating the number of required bus seats per passenger; *iv*) *Fare of a ticket* (*HML*), by dividing the price of each trip itinerary into three bands (high, medium, and low).

4.3   Step 4: Prediction Model

After defining the purchase factors, we start the process of learning by analyzing the dataset $\hat{D}$ in order to train a model capable of automatically learning whether or not a user will finalize a purchase. In particular, the model has been trained on information which depends on the route, departure date and date of booking (e.g., ticket fare, occupancy rate of a bus).

The accuracy of our approach is evaluated comparing the most relevant machine learning algorithms used in the literature (Naïve Bayes [10], Logistic Regression [16], Decision Tree [13], Random Forest [3]). The performance of the machine learning models has been evaluated through a confusion matrix. Specifically, tickets that are correctly predicted as *purchased* are counted as True Positive (*TP*), whereas tickets that are predicted as *purchased* but are actually *abandoned* are counted as False Positive (*FP*). Similarly, tickets that are correctly predicted as *abandoned* are counted as True Negative (*TN*), whereas tickets that are predicted as *abandoned* but are actually *purchased* are counted as False Negative (*FN*). Starting from the confusion matrix we can compute metrics such as *accuracy, precision, recall* and *F1-score*.

It is worth noticing that our dataset is unbalanced because the two classes, *purchased* and *abandoned*, are not equally represented. In particular, there is a high percentage of users who visit the bus website without buying any tickets, and a low percentage who instead purchases tickets. In order to get accurate prediction models and correctly evaluate them, we need to use balanced *training sets* and *test sets* in which half the logs lead to the purchase of a ticket and half to abandonment. We used to this purpose the random under-sampling algorithm [8], which balances class distribution through random discarding of major class tuples as described in [2].

The machine learning algorithms have been implemented in Python using the library *sklearn* for producing the confusion matrix and the resulting measures, and its library *imblearn* that has been used to deal with the class-imbalance problem.

5   A Case Study

This section reports the results obtained by the analysis of event logs of an Italian bus ticketing platform. Specifically, we extracted more than 3 million of event logs about trip itineraries and price tickets of an Italian bus company, collected from August 1st, 2018 to October 20st, 2019. The total size of the final dataset $\hat{D}$ is about 700 MB. Data have been analyzed using the DA4PT methodology to discover the main factors influencing customers in purchasing bus tickets and, based on those factors, to train a machine learning model for predicting whether or not a customer will buy a ticket.

5.1   Steps 1-2: Web Scraping and Process Mining

At *Step 1*, the set of event logs *EL* is composed by all interactions of the users with the bus ticketing platform. In particular, the buying decision process of a user is described by four types of event:

- *list_trips*, to find the routes between the origin and destination locations;
- *estimate_ticket*, to determine the itinerary cost on the basis of the route select by user;
- *choice_seat*, to find available seats on the bus chosen;
- *purchased_ticket*, to confirm the payment of the booked trip.

Specifically, each event $e_i \in EL$ is defined as shown in Fig. 2. For example, the user with ID *1JYASX* queried the system asking for the list of trips of the route Soverato-Rome (*line 1*). Then, he estimated the cost of the trip (*line 2*), and selected a seat on the bus (*line 3*). Finally, he paid for the ticket (*line 4*). The user with ID *28UAKS* logged on to the bus ticketing platform to estimate the cost of the route Milan-Lamezia Terme (*line 5-6*), but hasn't finalized the purchase of the ticket (*line 7*).

COOKIE	ACTION	TIMESTAMP	TRIP ID	DEPARTURE DATE	BOOKING DATE	ORIGIN CITY	DESTINATION CITY	No. SEAT	BUS SEAT	FARE	BOUGHT
1JYASX	list_trips	2018-10-16 11:31:19		2018-10-22		Soverato	Rome				
1JYASX	estimate_ticket	2018-10-16 11:31:37	141772	2018-10-22		Soverato	Rome	1	45	35 €	
1JYASX	choice_seat	2018-10-16 11:36:28	141772	2018-10-22		Soverato	Rome	1	45	35 €	
1JYASX	purchased_ticket	2018-10-16 11:42:20	141772	2018-10-22	2018-10-16	Soverato	Rome	1	45	35 €	YES
28UAKS	list_trips	2019-02-24 18:15:07		2019-02-26		Milan	Lamezia Terme				
28UAKS	estimate_ticket	2019-02-24 18:15:40	408003	2019-02-26		Milan	Lamezia Terme	2	52	64 €	
28UAKS	choice_seat	2019-02-24 18:20:05	408003	2019-02-26		Milan	Lamezia Terme	2	52	64 €	NO

Fig. 2. Example of the log traces extracted from the bus ticketing Web portal.

At *Step 2*, after having defined the set of event logs *EL*, we applied process mining algorithms with the aim of identifying trends and human patterns, and understanding behaviours of users while searching and booking bus trips.

Figure 3 shows the navigation paths corresponding to those produced by human navigation on the bus ticketing portal. There are three type of paths: the green and red paths related to events that end with the purchase of a ticket (*purchased*) and the abandonment of the platform (*abandoned*) respectively, whereas the blue paths are related to redundant events that generate loops. The percentage present on the edges describes the users who leave a state to reach a previous/next state, or a terminal state (*abandoned* or *purchased*). For example, 100% of the users of *Start* state are distributed as follows: 70% look for a trip, then 14% continue browsing estimating the cost of the chosen trip, and 16% leave the platform without buying a ticket. We cleaned collected data by removing for each user all blue paths. Then, we selected only paths related to *purchased* and *abandoned* events. Finally, we transformed data by keeping the last event of the booking life cycle per user. At the end of this step, we built the final dataset $\hat{D}$ that results composed of about 300,000 tuples, each one containing the *purchased* or *abandoned* event by a single user.

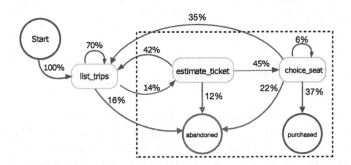

Fig. 3. Process mining algorithms applied to user event logs.

To analyze the behaviour of a user who searched the possible routes between two localities in a given date (*list_trips*), we focused on all the events he/she generates when chooses one of the routes. As shown in the dashed rectangle in Fig. 3, we focused on the events *estimate_ticket*, *choice_seat*, *purchased* and *abandoned*. In this range, only 17% of users purchase a ticket (45% choose a seat, of which only 37% purchase) while 83% abandon the platform without buying. In the next section we study what are the main factors that lead a user to the purchase of a ticket.

5.2   Step 3: Discovery of Purchase Factors

In the following we present the main results of the exploratory factor and correlation analysis discussed above.

First, we describe some statistical indications of travelers' purchasing behaviour obtained from several attributes included in D. In particular, Fig. 4 reports the number of purchased tickets considering *departure months*, *departure weekdays*, and *routes* attributes.

Specifically, Fig. 4(A) suggests that most people travel in spring and summer, with a significant drop in passengers in autumn and winter, except for the Christmas holidays. Figure 4(B) indicates that the number of trips in the weekend is higher than on working days, whereas Fig. 4(C) shows how some routes are more popular than others. For example, the number of purchased tickets for the route <Rome - Lamezia Terme> is about 3,400, because the Calabria region is a tourist destination and many workers/students live outside the region.

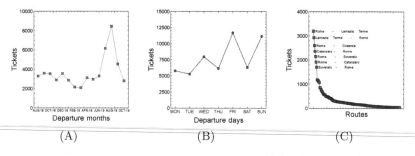

Fig. 4. No. of purchased tickets considering (A) *departure month*, (B) *departure day of the week*, and (C) *route* attributes.

Another result of our analysis was discovering the correlation between the four derived attributes and the class attribute (*purchased* or *abandoned*) as described in Sect. 4.2. Specifically, for each derived attribute, we measure the numbers and the percentage of purchased tickets and the correlation.

Starting from the *DBD* attribute, Fig. 5(A) shows the number of purchased tickets versus the number of days before departure. It clearly shows that a few days before departure, users buy more frequently. For example, 30 K tickets have been purchased on the platform between 0 and 9 days before the trip, 20K between 10 and 20 days before, and so on. By observing the trend line over histogram in Fig. 5(B), it can be noted that the percentage of purchasing a ticket is pretty high a few days passed from booking to departure, then there is a decreasing when the date of departure is far away. For example, 22% of users complete the purchase of the ticket between 0 and 19 days before the trip, 10% between 20 and 29 days before, and so on.

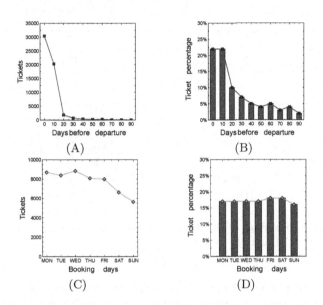

Fig. 5. No. and percentage of purchased tickets considering the *days before departure* (*DBD*) and the *booking day of the week* (*BDOW*).

Considering the *BDOW* attribute, Fig. 5(C) shows the days of the week when users prefer to book a ticket. In the first three days of the week (MON-TUE-WED) most tickets are sold, while in the other days the number of tickets sold drops drastically. Histogram in Fig. 5(D) shows that the probability of purchasing a ticket is slightly higher on Fridays and Saturdays compared to other days of the week.

We also evaluated how the occupancy rate (*OCCR*) attribute influences the buying behaviour of the users. As shown in Fig. 6(A), the tickets are mostly bought when the percentage of available seats is between 10% and 30%, whereas the trend line shown in Fig. 6(B), describes that the probability of purchasing a ticket lightly increases when the bus seats are running out. Note that several buses do not reach the full occupancy because many tickets are not bought on the platform, and then are not registered in the event logs.

Finally, we show the impact that the *HML* attribute had on users' purchasing choices. In particular, for each trip itinerary we have divided the price into high, medium and low, discovering that most users are pushed to buy a ticket when the price is low (Fig. 6(C)). In fact, the probability of buying a ticket in low range (about 20%) is much higher than buying it in medium and high ranges (about 15% and 13% respectively), as shown in Fig. 6(D).

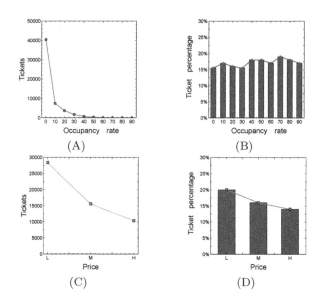

Fig. 6. No. and percentage of purchased tickets considering the *occupancy rate for a bus* (*OCCR*) and the *fare of a ticket* (*HML*).

To define the potential purchase factors, a correlation analysis was executed. The *DBD* (days before departure) attribute have the highest correlation coefficient (r) with a value of 0.86. The other attributes also have a high correlation with the class attribute: $r = 0.74$ for *BDOW* (booking day of the week) and *OCCR* (occupancy rate) and, $r = 0.68$ for *HML* (fare of a ticket).

5.3   Step 4: Prediction Model

Before running the learning algorithms, we used the random under-sampling algorithm to balance class distribution in $\hat{D}$. In our case, we have a total of 247,525 samples: 42,995 *purchased*, and 204,530 *abandoned*.

The following parameters have been used for the evaluation tests: *i*) target dataset $\hat{D}$, *ii*) purchase factors, as described in Sect. 5.2, and *iii*) number of routes considered. As performance indicators we used the accuracy and *weighted-average* F1-score. The goal is to maximize accuracy with balanced values of F1-score. Moreover, to measure the quality of a classifier with respect to a given class, for each algorithm we evaluated the *purchased* recall (R_p) and *abandoned* recall (R_a).

Table 1 summarizes the results obtained by the four machine learning algorithms we used. Specifically, Random Forest proved to be the best classification model with $R_p = 0.95$ and $R_a = 0.85$, showing better accuracy and low variance in results than other algorithms. A high value of accuracy is also obtained by Decision Tree with $R_p = 0.87$ and $R_a = 0.84$, showing good robustness and stability.

Table 1. Performance evaluation.

Algorithms	Accuracy	Precision	Recall	F1-score
Naïve Bayes	0.615	0.644	0.615	0.595
Logistic Regression	0.615	0.616	0.615	0.615
Decision Tree	0.864	0.865	0.864	0.864
Random Forest	0.930	0.928	0.930	0.928

A similar value of accuracy is observed for Naïve Bayes and Logistic Regression ($R_p = 0.56$ and $R_a = 0.65$), but Naïve Bayes is less accurate on the *purchased* class than *abandoned* class ($R_p = 0.38$ and $R_a = 0.84$).

Figure 7(A) shows a time plot of the collected tickets data, in which the accuracy performance of the four machine learning algorithms is plotted versus the number of routes. The trend is quite evident: the accuracy of Random Forest stably ranging from 0.91 to 0.96, followed by Decision Tree (0.81-0.88), Logistic Regression (0.50-0.63), and Naïve Bayes (0.52-0.59).

Figure 7(B) shows the number of tickets correctly predicted related to the *purchased* class. Also in this case, the accuracy of Random Forest is the highest in all routes considered, confirming its very good prediction performance with respect to the other algorithms in the demand forecasting domain. Please notice that for both examples, we considered the first thirty routes based on the number of purchased tickets.

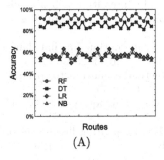
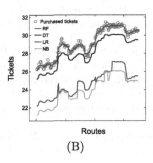

(A)   (B)

Fig. 7. Comparison among Naïve Bayes (NB), Logistic Regression (LR), Decision Tree (DT) and Random Forest (RF).

6 Conclusions

This paper proposes a methodology (DA4PT) that through Web scraping techniques and process mining algorithms allows to discover the factors that influence the behaviour of bus travelers in ticket booking and to learn a model for predicting ticket purchasing.

DA4PT has been validated through a real case study based on 3.23 million event logs of an Italian bus ticketing platform, collected from August 1st, 2018 to October 20st, 2019. The results obtained by this study reveals that factors such as occupancy rate, fare of a ticket, and number of days passed from booking to departure, have significant influence on traveler's buying decisions. We experimentally evaluated the accuracy of our methodology and Random Forest proved to be the best classification algorithm, showing an accuracy of 93% and a low variance.

Using the methodology discussed in this work, the buying behaviour of large communities of people can be analyzed for providing valuable information and high-quality knowledge that are fundamental for the growth of business and organization systems.

References

1. Abdelghany, A., Guzhva, V.: A time-series modelling approach for airport short-term demand forecasting. J. Airport Manag. **5**(1), 72–87 (2010)
2. Belcastro, L., Marozzo, F., Talia, D., Trunfio, P.: Using scalable data mining for predicting flight delays. ACM Trans. Intell. Syst. Technol. **8**(1), 5:1–5:20 (2016)
3. Breiman, L.: Random forests. Mach. Learn. **45**(1), 5–32 (2001)
4. Diamantini, C., Genga, L., Marozzo, F., Potena, D., Trunfio, P.: Discovering mobility patterns of Instagram users through process mining techniques. In: IEEE International Conference on Information Reuse and Integration, pp. 485–492 (2017)
5. Escobari, D.: Estimating dynamic demand for airlines. Econ. Lett. **124**(1), 26–29 (2014)
6. Gremm, C.: Impacts of the German interurban bus market deregulation on regional railway services (2017)
7. Grimaldi, R., Augustin, K., Beria, P., et al.: Intercity coach liberalisation. The cases of Germany and Italy. In: World Conference on Transport Research-WCTR 2016, pp. 474–490. Elsevier BV (2017)
8. Kotsiantis, S., Kanellopoulos, D., Pintelas, P., et al.: Handling imbalanced datasets: a review. GESTS Int. Trans. Comput. Sci. Eng. **30**(1), 25–36 (2006)
9. Liu, J., et al.: Personalized air travel prediction: a multi-factor perspective. ACM Trans. Intell. Syst. Technol. (TIST) **9**(3), 1–26 (2017)
10. Maron, M.E.: Automatic indexing: an experimental inquiry. J. ACM (JACM) **8**(3), 404–417 (1961)
11. Mumbower, S., Garrow, L.A., Higgins, M.J.: Estimating flight-level price elasticities using online airline data. Transp. Res. Part A: Policy Pract. **66**, 196–212 (2014)
12. Pearson, K.: Determination of the coefficient of correlation. Science **30**(757), 23–25 (1909)
13. Safavian, S.R., Landgrebe, D.: A survey of decision tree classifier methodology. IEEE Trans. Syst. Man Cybern. **21**(3), 660–674 (1991)
14. Szopiński, T., Nowacki, R.: The influence of purchase date and flight duration over the dispersion of airline ticket prices. Contemp. Econ. **9**(3), 253–366 (2015)
15. Talia, D., Trunfio, P., Marozzo, F.: Data Analysis in the Cloud: Models. Techniques and Applications (2015). https://doi.org/10.1016/C2014-0-02172-7
16. Walker, S.H., Duncan, D.B.: Estimation of the probability of an event as a function of several independent variables. Biometrika **54**(1–2), 167–179 (1967)
17. Yeboah, G., Cottrill, C.D., Nelson, J.D., Corsar, D., Markovic, M., Edwards, P.: Understanding factors influencing public transport passengers' pre-travel information-seeking behaviour. Pub. Transp. **11**(1), 135–158 (2019)

Correction to: Machine Learning, Optimization, and Data Science

Giuseppe Nicosia⑩, Varun Ojha⑩, Emanuele La Malfa⑩,
Giorgio Jansen, Vincenzo Sciacca, Panos Pardalos⑩,
Giovanni Giuffrida⑩, and Renato Umeton⑩

Correction to:
G. Nicosia et al. (Eds.): *Machine Learning, Optimization, and Data Science*, LNCS 12565, https://doi.org/10.1007/978-3-030-64583-0

The original version of chapter 2 was inadvertently published with wrong RTS values in Table 3: "Results comparison with RTS, S, and SVM[light] with standard linear loss with a 10-fold cross validation procedure."
The RTS values were corrected by replacing the wrong values with the appropriate ones.
The original version of the book was inadvertently published with a typo in the author's e-mail address. In the contribution it read "fportera2@gail.com" but correctly it should have read "filippo.portera@unive.it".
The footnote reads "[1]Code is available at: https://osf.io/fbzsc/" and has been added to the last sentence in the abstract.

The original version of chapter 34 was inadvertently published with incorrect allocations between authors and affiliations resp. one affiliation was entirely missing.
The affiliations have been corrected and read as follows: [1]University of Primorska, Koper, Slovenia; [2]Jožef Stefan Institute, Ljubljana, Slovenia; and [3]Urgench State University, Urgench, Uzbekistan. The authors' affiliations are: Jamolbek Mattiev[1,3] and Branko Kavšek[1,2].

The updated version of these chapters can be found at
https://doi.org/10.1007/978-3-030-64583-0_2
https://doi.org/10.1007/978-3-030-64583-0_34

Author Index

Printed in the United States
by Baker & Taylor Publisher Services